Tips for Customizing Your RSMeans Cost Data By Location

The cost data found throughout the book reflects national averages.

Customize your cost data for over 930 locations throughout the U.S. and Canada.

☑ To adjust costs to your region, use the Location Factors found in the back of the book.

☑ To compare costs from region-to-region, use the City Cost Indexes found in the back of the book.

☑ For additional guidance, refer to the How RSMeans Data Works section towards the front of the book.

RSMeans
Reed Construction Data

Metric Conversion Formulas

Length

in. = 2.5400 cm.	cm = 0.3937 in.
ft. = 0.3048 m.	meter = 3.2808 ft.
yd. = 0.9144 m.	meter = 1.0936 yd.
mile = 1.6093 km.	Km. = 0.6214 mile

Area

sq. cm. = 0.1550 sq. in.	sq.in. = 6.4516 sq. cm.
sq. m. = 10.7639 sq. ft.	sq. ft. = 0.0929 sq. m.
sq. m. = 1.1960 sq. yd.	sq. yd. = 0.8361 sq. m.
hectare = 2.4710 acres	acre = 0.4047 hectare
sq. km. = 0.3861 sq. mile	sq. mile = 2.5900 sq. km.

Weight

grain = 0.0648 g.	gram = 15.4324 grains
oz. = 28.3495 g.	gram = 0.0353 oz.
lb. = 0.4536 kg.	kg. = 2.2046 lbs.

kg. = 0.0011 ton (short)	
ton (met.) = 1.1023 ton (short)	
ton (met.) = 0.9842 ton (large)	

Weight, cont.

ton (short) = 907.1848 kg.	
ton (short) = 0.9072 ton (met.)	
ton (large) = 1.0160 ton (met.)	

Common Units of Measure

Abbreviation	Measurement	Formula
SF	Square feet	Length (in feet) x Width (in feet)
SY	Square yards	Square feet / 9
CF	Cubic feet	Length (in feet) x Width (in feet) x Depth (in feet)
CY	Cubic yards	Cubic feet / 27
BF	Board foot	Length (in inches) x width (in inches) x thickness (in inches)/144
MBF	Thousand board foot	Board foot / 1,000
LF	Linear feet	12 inches of the item
SFCA	Square foot of contact area	A square foot of concrete formwork
SQ	Square	100 square feet
Ton	Ton	2,000 pounds

$197.95 per copy (in United States)
Price is subject to change without prior notice.

0053

RSMeans®
Square Foot Costs

34TH ANNUAL EDITION

2013

RSMeans®
A division of Reed Construction Data, LLC
Construction Publishers & Consultants
700 Longwater Drive
Norwell, MA 02061
USA

1-800-334-3509
www.rsmeans.com

Copyright 2012 by RSMeans
A division of Reed Construction Data, LLC
All rights reserved.

Printed in the United States of America
ISSN 1540-6326
ISBN 978-1-936335-74-9

Senior Editor
Marilyn Phelan, AIA

Contributing Editors
Christopher Babbitt
Ted Baker
Adrian C. Charest, PE
Gary W. Christensen
David G. Drain, PE
Cheryl Elsmore
Robert J. Kuchta
Robert C. McNichols
Genevieve Medeiros
Melville J. Mossman, PE
Jeannene D. Murphy
Stephen C. Plotner
Eugene R. Spencer
Phillip R. Waier, PE

Cover Design
Wendy McBay
Paul Robertson
Wayne Engebretson

President & CEO
Iain Melville

Chief Operating Officer
Richard Remington

Chief Customer Officer
Lisa Fiondella

Vice President, Marketing & Business Development
Steven Ritchie

Vice President, Finance & Operations
Renée Pinczes

Engineering Director
Bob Mewis, CCC

Director, Product Management
Jennifer Johnson

Product Manager
Andrea Sillah

Product Marketing Manager
Ryan Cummings

Senior Director, Project & Process Management
Rodney L. Sharples

Production Manager
Michael Kokernak

Editing Supervisor
Jill Goodman

Production Supervisor
Paula Reale-Camelio

Production
Jonathan Forgit
Mary Lou Geary
Debbie Panarelli
Caroline Sullivan

Technical Support
Christopher Anderson
Gary L. Hoitt
Kathryn S. Rodriguez

Related RSMeans Products and Services

The residential section of this book is aimed at new residential construction projects valued at up to $1,000,000. The labor rate used in the calculations is open shop (non-union). The commercial/industrial/institutional section is aimed primarily at new construction projects valued at $1,500,00 and up. The labor rate in the calculations is union. The engineers at RSMeans suggest the following products and services as companion information resources to *RSMeans Square Foot Costs:*

Construction Cost Data Books

Assemblies Cost Data 2013
Building Construction Cost Data 2013
Residential Cost Data 2013
Light Commercial Cost Data 2013

Reference Books

Estimating Building Costs
RSMeans Estimating Handbook
Green Building: Project Planning & Estimating
How to Estimate with RSMeans Data
Plan Reading & Material Takeoff
Project Scheduling & Management for Construction
Universal Design for Style, Comfort & Safety

Seminars and In-House Training

Conceptual Estimating Using RSMeans CostWorks® CD
Means CostWorks® Training
RSMeans Data for Job Order Contracting (JOC)
Plan Reading & Material Take-Off

RSMeans on the Internet

Visit RSMeans at **www.rsmeans.com** for a one-stop portal for the most reliable and current resources available on the market. Here you can learn about our more than 20 annual cost data books, also available on CD and online versions, as well as our libary of reference books and RSMeans seminars.

RSMeans Electronic Data

Get the information found in RSMeans cost books electronically at **MeansCostWorks.com**.

RSMeans Business Solutions

Cost engineers and analysts conduct facility life cycle and benchmark research studies and predictive cost modeling, as well as offer consultation services for conceptual estimating, real property management, and job order contracting. Research studies are designed with the application of proprietary cost and project data from Reed Construction Data and RSMeans' extensive North American databases. Analysts offer building product manufacturers qualitative and quantitative market research, as well as time/motion studies for new products, and Web-based dashboards for market opportunity analysis. Clients are from both the public and private sectors, including federal, state, and municipal agencies; corporations; institutions; construction management firms; hospitals; and associations.

RSMeans for Job Order Contracting (JOC)

Best practice job order contracting (JOC) services support the owner in the development of their program. RSMeans develops all required contracting documents, works with stakeholders, and develops qualified and experienced contractor lists. RSMeans JOCWorks® software is the best of class in the functional tools that meet JOC specific requirements. RSMeans data is organized to meet JOC cost estimating requirements.

Construction Costs for Software Applications

More than 25 unit price and assemblies cost databases are available through a number of leading estimating and facilities management software providers (listed below). For more information see the "Other RSMeans Products" pages at the back of this publication.
RSMeansData™ is also available to federal, state, and local government agencies as multi-year, multi-seat licenses.

- 4Clicks-Solutions, LLC
- ArenaSoft Estimating
- Assetworks
- Beck Technology
- BSD – Building Systems Design, Inc.
- CMS – Construction Management Software
- Corecon Technologies, Inc.
- CorVet Systems
- Hard Dollar Corporation
- Maxwell Systems
- MC²
- Parsons Corporation
- ProEst
- Sage Timberline Office
- UDA Technologies, Inc.
- US Cost, Inc.
- VFA – Vanderweil Facility Advisers
- WinEstimator, Inc.
- R & K Solutions
- Tririga

Table of Contents

Foreword

Our Mission

Since 1942, RSMeans has been actively engaged in construction cost publishing and consulting throughout North America.

Today, more than 70 years after RSMeans began, our primary objective remains the same: to provide you, the construction and facilities professional, with the most current and comprehensive construction cost data possible.

Whether you are a contractor, owner, architect, engineer, facilities manager, or anyone else who needs a reliable construction cost estimate, you'll find this publication to be a highly useful and necessary tool.

With the constant flow of new construction methods and materials today, it's difficult to find the time to look at and evaluate all the different construction cost possibilities. In addition, because labor and material costs keep changing, last year's cost information is not a reliable basis for today's estimate or budget.

That's why so many construction professionals turn to RSMeans. We keep track of the costs for you, along with a wide range of other key information, from city cost indexes . . . to productivity rates . . . to crew composition . . . to contractor's overhead and profit rates.

RSMeans performs these functions by collecting data from all facets of the industry and organizing it in a format that is instantly accessible to you. From the preliminary budget to the detailed unit price estimate, you'll find the data in this book useful for all phases of construction cost determination.

The Staff, the Organization, and Our Services

When you purchase one of RSMeans' publications, you are, in effect, hiring the services of a full-time staff of construction and engineering professionals.

Our thoroughly experienced and highly qualified staff works daily at collecting, analyzing, and disseminating comprehensive cost information for your needs. These staff members have years of practical construction experience and engineering training prior to joining the firm. As a result, you can count on them not only for accurate cost figures, but also for additional background reference information that will help you create a realistic estimate.

The RSMeans organization is always prepared to help you solve construction problems through its variety of data solutions, including online, CD, and print book formats, as well as cost estimating expertise available via our business solutions, training, and seminars.

Besides a full array of construction cost estimating books, RSMeans also publishes a number of other reference works for the construction industry. Subjects include construction estimating and project and business management, special topics such as green building and job order contracting, and a library of facility management references.

In addition, you can access all of our construction cost data electronically in convenient CD format or on the web. Visit **RSMeansonline.com** for more information on our 24/7 online cost data.

What's more, you can increase your knowledge and improve your construction estimating and management performance with an RSMeans construction seminar or in-house training program. These two-day seminar programs offer unparalleled opportunities for everyone in your organization to become updated on a wide variety of construction-related issues.

RSMeans is also a worldwide provider of construction cost management and analysis services for commercial and government owners.

In short, RSMeans can provide you with the tools and expertise for constructing accurate and dependable construction estimates and budgets in a variety of ways.

Robert Snow Means Established a Tradition of Quality That Continues Today

Robert Snow Means spent years building RSMeans, making certain he always delivered a quality product.

Today, at RSMeans, we do more than talk about the quality of our data and the usefulness of our books. We stand behind all of our data, from historical cost indexes to construction materials and techniques to current costs.

If you have any questions about our products or services, please call us toll-free at 1-800-334-3509. Our customer service representatives will be happy to assist you. You can also visit our website at **www.rsmeans.com**

How the Book Is Built: An Overview

A Powerful Construction Tool

You have in your hands one of the most powerful construction tools available today. A successful project is built on the foundation of an accurate and dependable estimate. This book will enable you to construct just such an estimate.

For the casual user the book is designed to be:

- quickly and easily understood so you can get right to your estimate.
- filled with valuable information so you can understand the necessary factors that go into the cost estimate.

For the regular user, the book is designed to be:

- a handy desk reference that can be quickly referred to for key costs.
- a comprehensive, fully reliable source of current construction costs for typical building structures when only minimum dimensions, specifications, and technical data are available. It is specifically useful in the conceptual or planning stage for preparing preliminary estimates which can be supported and developed into complete engineering estimates for construction.
- a source book for rapid estimating of replacement costs and preparing capital expenditure budgets.

To meet all of these requirements, we have organized the book into the following clearly defined sections.

How to Use the Book: The Details

This section contains an in-depth explanation of how the book is arranged and how you can use it to determine a reliable construction cost estimate. It includes information about how we develop our cost figures and how to completely prepare your estimate.

Residential Section

Model buildings for four classes of construction—economy, average, custom, and luxury—are developed and shown with complete costs per square foot.

Commercial, Industrial, Institutional Section

This section contains complete costs for 77 typical model buildings expressed as costs per square foot.

Assemblies Section

The cost data in this section has been organized in an "Assemblies" format. These assemblies are the functional elements of a building and are arranged according to the 7 divisions of the UNIFORMAT II classification system. The costs in this section include standard markups for the overhead and profit of the installing contractor. The assemblies costs are those used to calculate the total building costs in the Commercial/Industrial/Institutional section.

Reference Section

This section includes information on General Conditions, Location Factors, Historical Cost Indexes, a Glossary, and a listing of Abbreviations used in this book.

Location Factors: You can adjust the total project costs to over 900 locations throughout the U.S. and Canada by using the data in this section.

Historical Cost Indexes: These indexes provide you with data to adjust construction costs over time.

Glossary: A listing of common terms related to construction and their definitions is included.

Abbreviations: A listing of the abbreviations used throughout this book, along with the terms they represent, is included.

The Scope of This Book

This book is designed to be as comprehensive and as easy to use as possible. To that end we have made certain assumptions and limited its scope in two key ways:

1. We have established material prices based on a national average.
2. We have computed labor costs based on a 30-city national average of union wage rates.

All costs and adjustments in this manual were developed from the *RSMeans Building Construction Cost Data, Assemblies Cost Data, Residential Cost Data,* and *Light Commercial Cost Data.* Costs include labor, material, overhead and profit, and general conditions and architects' fees where specified. No allowance is made for sales tax, financing, premiums for material and labor, or unusual business and weather conditions.

What's Behind the Numbers? The Development of Cost Data

The staff at RSMeans continuously monitors developments in the construction industry in order to ensure reliable, thorough, and up-to-date cost information.

While overall construction costs may vary relative to general economic conditions, price fluctuations within the industry are dependent upon many factors. Individual price variations may, in fact, be opposite to overall economic trends. Therefore, costs are continually monitored and complete updates are published yearly. Also, new items are frequently added in response to changes in materials and methods.

Costs—$ (U.S.)

All costs represent U.S. national averages and are given in U.S. dollars. The RSMeans Location Factors can be used to adjust costs to a particular location. The Location Factors for Canada can be used to adjust U.S. national averages to local costs in Canadian dollars. No exchange rate conversion is necessary.

Factors Affecting Costs

Due to the almost limitless variation of building designs and combinations of construction methods and materials, costs in this book must be used with discretion. The editors have made every effort to develop costs for "typical" building types. By definition, a square foot estimate, due to the short time and limited amount of detail required, involves a relatively lower degree of accuracy than a detailed unit cost estimate, for example. The accuracy of an estimate will increase as the project is defined in more detailed components. The user should examine closely and compare the specific requirements of the project being estimated with the components included in the models in this book. Any differences should be accounted for in the estimate.

In many building projects, there may be factors that increase or decrease the cost beyond the range shown in this manual. Some of these factors are:

- Substitution of building materials or systems for those used in the model
- Custom-designed finishes, fixtures, and/or equipment
- Special structural qualities (allowance for earthquake, future expansion, high winds, long spans, and unusual shape)
- Special foundations and substructures to compensate for unusual soil conditions
- Inclusion of typical "owner-supplied" equipment in the building cost (special testing and monitoring equipment for hospitals, kitchen equipment for hotels, and offices that have cafeteria accommodations for employees, etc.)
- An abnormal shortage of labor or materials

- Isolated building site or rough terrain that would affect the transportation of personnel, material, or equipment
- Unusual climatic conditions during the construction process

We strongly urge that, for maximum accuracy, these factors be considered each time a structure is being evaluated.

If users require greater accuracy than this manual can provide, the editors recommend that the *RSMeans Building Construction Cost Data* or the *RSMeans Assemblies Cost Data* be consulted.

Final Checklist

Estimating can be a straightforward process provided you remember the basics. Here's a checklist of some of the steps you should remember to complete before finalizing your estimate.

Did you remember to . . .

- include the Location Factor for your city?
- mark up the entire estimate sufficiently for your purposes?
- include all components of your project in the final estimate?
- double check your figures for accuracy?
- call RSMeans if you have any questions about your estimate or the data you've found in our publications?

Remember, RSMeans stands behind its publications. If you have any questions about your estimate, about the costs you've used from our books, or even about the technical aspects of the job that may affect your estimate, feel free to call the RSMeans editors at 1-800-334-3509.

Residential Section

Table of Contents

Introduction to Residential Section

The residential section of this manual contains costs per square foot for four classes of construction in seven building types. Costs are listed for various exterior wall materials which are typical of the class and building type. There are cost tables for Wings and Ells with modification tables to adjust the base cost of each class of building. Non-standard items can easily be added to the standard structures.

Cost estimating for a residence is a three-step process:
(1) Identification
(2) Listing dimensions
(3) Calculations

Guidelines and a sample cost estimating procedure are shown on the following pages.

Identification

To properly identify a residential building, the class of construction, type, and exterior wall material must be determined. Page 6 has drawings and guidelines for determining the class of construction. There are also detailed specifications and additional drawings at the beginning of each set of tables to further aid in proper building class and type identification.

Sketches for eight types of residential buildings and their configurations are shown on pages 8 and 9. Definitions of living area are next to each sketch. Sketches and definitions of garage types are on page 10.

Living Area

Base cost tables are prepared as costs per square foot of living area. The living area of a residence is that area which is suitable and normally designed for full time living. It does not include basement recreation rooms or finished attics, although these areas are often considered full time living areas by the owners.

Living area is calculated from the exterior dimensions without the need to adjust for exterior wall thickness. When calculating the living area of a 1-1/2 story, two story, three story or tri-level residence, overhangs and other differences in size and shape between floors must be considered.

Only the floor area with a ceiling height of seven feet or more in a 1-1/2 story residence is considered living area. In bi-levels and tri-levels, the areas that are below grade are considered living area, even when these areas may not be completely finished.

Base Tables and Modifications

Base cost tables show the base cost per square foot without a basement, with one full bath and one full kitchen for economy and average homes, and an additional half bath for custom and luxury models. Adjustments for finished and unfinished basements are part of the base cost tables. Adjustments for multi family residences, additional bathrooms, townhouses, alternate roofs, and air conditioning and heating systems are listed in modification tables below the base cost tables.

Costs for other modifications, including garages, breezeways and site improvements, are in the modification tables on pages 58 to 61.

Listing of Dimensions

To use this section of the manual, only the dimensions used to calculate the horizontal area of the building and additions and modifications are needed. The dimensions, normally the length and width, can come from drawings or field measurements. For ease in calculation, consider measuring in tenths of feet, i.e., 9 ft. 6 in. = 9.5 ft., 9 ft. 4 in. = 9.3 ft.

In all cases, make a sketch of the building. Any protrusions or other variations in shape should be noted on the sketch with dimensions.

Calculations

The calculations portion of the estimate is a two-step activity:
(1) The selection of appropriate costs from the tables
(2) Computations

Selection of Appropriate Costs

To select the appropriate cost from the base tables, the following information is needed:
(1) Class of construction
    (a) Economy
    (b) Average
    (c) Custom
    (d) Luxury
(2) Type of residence
    (a) 1 story
    (b) 1-1/2 story
    (c) 2 story
    (d) 3 story
    (e) Bi-level
    (f) Tri-level
(3) Occupancy
    (a) One family
    (b) Two family
    (c) Three family
(4) Building configuration
    (a) Detached
    (b) Town/Row house
    (c) Semi-detached
(5) Exterior wall construction
    (a) Wood frame
    (b) Brick veneer
    (c) Solid masonry
(6) Living areas

Modifications are classified by class, type, and size.

Computations

The computation process should take the following sequence:
(1) Multiply the base cost by the area.
(2) Add or subtract the modifications.
(3) Apply the location modifier.

When selecting costs, interpolate or use the cost that most nearly matches the structure under study. This applies to size, exterior wall construction, and class.

How to Use the Residential Square Foot Cost Pages

The following is a detailed explanation of a sample entry in the Residential Square Foot Cost Section. Each bold number below corresponds to the item being described on the facing page with the appropriate component of the sample entry following in parenthesis.

Prices listed are costs that include overhead and profit of the installing contractor. Total model costs include an additional mark-up for General Contractors' overhead and profit, and fees specific to class of construction.

RESIDENTIAL **Average 1** **2 Story 2**

- Simple design from standard plans
- Single family — 1 full bath, 1 kitchen
- No basement
- Asphalt shingles on roof
- Hot air heat
- Gypsum wallboard interior finishes
- Materials and workmanship are average

3

Note: The illustration shown may contain some optional components (for example: garages and/or fireplaces) whose costs are shown in the modifications, adjustments, & alternatives below or at the end of the square foot section.

Base cost per square foot of living area

5

Exterior Wall **4**	1000	1200	1400	1600	1800	Living Area 2000	2200	2600	3000	3400	3800
Wood Siding - Wood Frame	131.75	119.35	113.50	109.45	105.20	100.80	97.90	92.25	86.85	84.45	82.10
Brick Veneer - Wood Frame	137.90	125.05	118.80	114.55	109.95	105.40	102.30	96.20	90.50	87.90	85.35
Stucco on Wood Frame	126.55	114.50	108.95	105.10	101.15	96.90	**6**	88.80	83.75	81.45	79.25
Solid Masonry	154.75	140.60	133.35	128.40	123.10	118.00	114.25	107.10	100.55	97.35	94.45
Finished Basement, Add **7**	21.30	21.05	20.35	19.85	19.35	19.00		17.95	17.50	17.15	16.85
Unfinished Basement, Add	8.35	7.75	7.30	7.00	6.70	6.45	**8**	5.80	5.55	5.35	5.15

Modifications

Add to the total cost

Upgrade Kitchen Cabinets	$ + 3955
Solid Surface Countertops (Included) **9**	
Full Bath - including plumbing, wall and floor finishes	+ 7074
Half Bath - including plumbing, wall and floor finishes	+ 4047
One Car Attached Garage	+ 13,481
One Car Detached Garage	+ 17,772
Fireplace & Chimney	+ 6550

Adjustments

For multi family - add to total cost

Additional Kitchen	$ + 7917
Additional Bath	+ 7074
Additional Entry & Exit	+ 1576
Separate Heating	+ 1514
Separate Electric	+ 1958

For Townhouse/Rowhouse - Multiply cost per square foot by

Inner Unit **10**	.90
End Unit	.95

Alternatives

Add to or deduct from the cost per square foot of living area

Cedar Shake Roof	+ 1.45
Clay Tile Roof	+ 2.40
Slate Roof	+ 3.30
Upgrade Walls to Skim Coat Plaster	+ .55
Upgrade Ceilings to Textured Finish	+ .49
Air Conditioning, in Heating Ductwork **11**	+ 2.67
In Separate Ductwork	+ 5.22
Heating Systems, Hot Water	+ 2.12
Heat Pump	+ 2.61
Electric Heat	– .56
Not Heated	– 2.87

Additional upgrades or components

Kitchen Cabinets & Countertops	Page 93
Bathroom Vanities	94
Fireplaces & Chimneys	94
Windows, Skylights & Dormers	94
Appliances	95
Breezeways & Porches	95
Finished Attic **12**	95
Garages	96
Site Improvements	96
Wings & Ells	37

1 Class of Construction (Average)

The class of construction depends upon the design and specifications of the plan. The four classes are economy, average, custom, and luxury.

2 Type of Residence (2 Story)

The building type describes the number of stories or levels in the model. The seven building types are 1 story, 1-1/2 story, 2 story, 2-1/2 story, 3 story, bi-level, and tri-level.

3 Specification Highlights (Hot Air Heat)

These specifications include information concerning the components of the model, including the number of baths, roofing types, HVAC systems, and materials and workmanship. If the components listed are not appropriate, modifications can be made by consulting the information shown lower on the page or the Assemblies Section.

4 Exterior Wall System (Wood Siding - Wood Frame)

This section includes the types of exterior wall systems and the structural frame used. The exterior wall systems shown are typical of the class of construction and the building type shown.

5 Living Areas (2000 SF)

The living area is that area of the residence which is suitable and normally designed for full time living. It does not include basement recreation rooms or finished attics. Living area is calculated from the exterior dimensions without the need to adjust for exterior wall thickness. When calculating the living area of a 1-1/2 story, 2 story, 3 story or tri-level residence, overhangs and other differences in size and shape between floors must be considered. Only the floor area with a ceiling height of seven feet or more in a 1-1/2 story residence is considered living area. In bi-levels and tri-levels, the areas that are below grade are considered living area, even when these areas may not be completely finished. A range of various living areas for the residential model are shown to aid in selection of values from the matrix.

6 Base Costs per Square Foot of Living Area ($100.80)

Base cost tables show the cost per square foot of living area without a basement, with one full bath and one full kitchen for economy and average homes, and an additional half bath for custom and luxury models. When selecting costs, interpolate or use the cost that most nearly matches the residence under consideration for size, exterior wall system, and class of construction. Prices listed are costs that include overhead and profit of the installing contractor, a general contractor markup, and an allowance for plans that vary by class of construction. For additional information on contractor overhead and architectural fees, see the Reference Section.

7 Basement Types (Finished)

The two types of basements are finished and unfinished. The specifications and components for both are shown on the Building Classes page in the Introduction to this section.

8 Additional Costs for Basements ($19.00 or $6.45)

These values indicate the additional cost per square foot of living area for either a finished or an unfinished basement.

9 Modifications and Adjustments (Upgrade Kitchen Cabinets $3955)

Modifications and Adjustments are costs added to or subtracted from the total cost of the residence. The total cost of the residence is equal to the cost per square foot of living area times the living area. Typical modifications and adjustments include kitchens, baths, garages and fireplaces.

10 Multiplier for Townhouse/ Rowhouse (Inner Unit .90)

The multipliers shown adjust the base costs per square foot of living area for the common wall condition encountered in townhouses or rowhouses.

11 Alternatives (Skim Coat Plaster $.55)

Alternatives are costs added to or subtracted from the base cost per square foot of living area. Typical alternatives include variations in kitchens, baths, roofing, and air conditioning and heating systems.

12 Additional Upgrades or Components (Wings & Ells page 37)

Costs for additional upgrades or components, including wings or ells, breezeways, porches, finished attics, and site improvements, are shown in other locations in the Residential Square Foot Cost Section.

5

Building Classes

Economy Class

An economy class residence is usually built from stock plans. The materials and workmanship are sufficient to satisfy building codes. Low construction cost is more important than distinctive features. The overall shape of the foundation and structure is seldom other than square or rectangular.

An unfinished basement includes 7' high 8" thick foundation wall composed of either concrete block or cast-in-place concrete.

Included in the finished basement cost are inexpensive paneling or drywall as the interior finish on the foundation walls, a low cost sponge backed carpeting adhered to the concrete floor, a drywall ceiling, and overhead lighting.

Average Class

An average class residence is a simple design and built from standard plans. Materials and workmanship are average, but often exceed minimum building codes. There are frequently special features that give the residence some distinctive characteristics.

An unfinished basement includes 7'-6" high 8" thick foundation wall composed of either cast-in-place concrete or concrete block.

Included in the finished basement are plywood paneling or drywall on furring that is fastened to the foundation walls, sponge backed carpeting adhered to the concrete floor, a suspended ceiling, overhead lighting and heating.

Custom Class

A custom class residence is usually built from plans and specifications with enough features to give the building a distinction of design. Materials and workmanship are generally above average with obvious attention given to construction details. Construction normally exceeds building code requirements.

An unfinished basement includes a 7'-6" high 10" thick cast-in-place concrete foundation wall or a 7'-6" high 12" thick concrete block foundation wall.

A finished basement includes painted drywall on insulated 2" x 4" wood furring as the interior finish to the concrete walls, a suspended ceiling, carpeting adhered to the concrete floor, overhead lighting and heating.

Luxury Class

A luxury class residence is built from an architect's plan for a specific owner. It is unique both in design and workmanship. There are many special features, and construction usually exceeds all building codes. It is obvious that primary attention is placed on the owner's comfort and pleasure. Construction is supervised by an architect.

An unfinished basement includes 8' high 12" thick foundation wall that is composed of cast-in-place concrete or concrete block.

A finished basement includes painted drywall on 2" x 4" wood furring as the interior finish, suspended ceiling, tackless carpet on wood subfloor with sleepers, overhead lighting and heating.

Configurations

Detached House

This category of residence is a free-standing separate building with or without an attached garage. It has four complete walls.

Semi-Detached House

This category of residence has two living units side-by-side. The common wall is a fireproof wall. Semi-detached residences can be treated as a row house with two end units. Semi-detached residences can be any of the building types.

Town/Row House

This category of residence has a number of attached units made up of inner units and end units. The units are joined by common walls. The inner units have only two exterior walls. The common walls are fireproof. The end units have three walls and a common wall. Town houses/row houses can be any of the building types.

Building Types

One Story

This is an example of a one-story dwelling. The living area of this type of residence is confined to the ground floor. The headroom in the attic is usually too low for use as a living area.

One-and-a-half Story

The living area in the upper level of this type of residence is 50% to 90% of the ground floor. This is made possible by a combination of this design's high-peaked roof and/or dormers. Only the upper level area with a ceiling height of 7 feet or more is considered living area. The living area of this residence is the sum of the ground floor area plus the area on the second level with a ceiling height of 7 feet or more.

One Story with Finished Attic

The main living area in this type of residence is the ground floor. The upper level or attic area has sufficient headroom for use as a living area. This is made possible by a high peaked roof. The living area in the attic is less than 50% of the ground floor. The living area of this type of residence is the ground floor area only. The finished attic is considered an adjustment.

Two Story

This type of residence has a second floor or upper level area which is equal or nearly equal to the ground floor area. The upper level of this type of residence can range from 90% to 110% of the ground floor area, depending on setbacks or overhangs. The living area is the sum of the ground floor area and the upper level floor area.

Two-and-one-half Story

This type of residence has two levels of equal or nearly equal area and a third level which has a living area that is 50% to 90% of the ground floor. This is made possible by a high peaked roof, extended wall heights and/or dormers. Only the upper level area with a ceiling height of 7 feet or more is considered living area. The living area of this residence is the sum of the ground floor area, the second floor area and the area on the third level with a ceiling height of 7 feet or more.

Bi-level

This type of residence has two living areas, one above the other. One area is about 4 feet below grade and the second is about 4 feet above grade. Both areas are equal in size. The lower level in this type of residence is originally designed and built to serve as a living area and not as a basement. Both levels have full ceiling heights. The living area is the sum of the lower level area and the upper level area.

Three Story

This type of residence has three levels which are equal or nearly equal. As in the 2 story residence, the second and third floor areas may vary slightly depending on setbacks or overhangs. The living area is the sum of the ground floor area and the two upper level floor areas.

Tri-level

This type of residence has three levels of living area: one at grade level, one about 4 feet below grade, and one about 4 feet above grade. All levels are originally designed to serve as living areas. All levels have full ceiling heights. The living area is a sum of the areas of each of the three levels.

Garage Types

Attached Garage

Shares a common wall with the dwelling. Access is typically through a door between dwelling and garage.

Basement Garage

Constructed under the roof of the dwelling but below the living area.

Built-In Garage

Constructed under the second floor living space and above basement level of dwelling. Reduces gross square feet of living area.

Detached Garage

Constructed apart from the main dwelling. Shares no common area or wall with the dwelling.

Building Components

1. Excavation	11. Collar Ties	21. Fascia	31. Backfill	41. Sub-floor
2. Sill Plate	12. Ridge Rafter	22. Downspout	32. Drainage Stone	42. Finish Floor
3. Basement Window	13. Roof Sheathing	23. Shutter	33. Drainage Tile	43. Attic Insulation
4. Floor Joist	14. Roof Felt	24. Window	34. Wall Footing	44. Soffit
5. Shoe Plate	15. Roof Shingles	25. Wall Shingles	35. Gravel	45. Ceiling Strapping
6. Studs	16. Flashing	26. Building Paper	36. Concrete Slab	46. Wall Insulation
7. Drywall	17. Flue Lining	27. Wall Sheathing	37. Column Footing	47. Cross Bridging
8. Plate	18. Chimney	28. Fire Stop	38. Pipe Column	48. Bulkhead Stairs
9. Ceiling Joists	19. Roof Shingles	29. Dampproofing	39. Expansion Joint	
10. Rafters	20. Gutter	30. Foundation Wall	40. Girder	

Exterior Wall Construction

Typical Frame Construction

Typical wood frame construction consists of wood studs with insulation between them. A typical exterior surface is made up of sheathing, building paper and exterior siding consisting of wood, vinyl, aluminum or stucco over the wood sheathing.

Brick Veneer

Typical brick veneer construction consists of wood studs with insulation between them. A typical exterior surface is sheathing, building paper and an exterior of brick tied to the sheathing, with metal strips.

Stone

Typical solid masonry construction consists of a stone or block wall covered on the exterior with brick, stone or other masonry.

Residential Cost Estimate Worksheet

Worksheet Instructions

The residential cost estimate worksheet can be used as an outline for developing a residential construction or replacement cost. It is also useful for insurance appraisals. The design of the worksheet helps eliminate errors and omissions. To use the worksheet, follow the example below.

1. Fill out the owner's name, residence address, the estimator or appraiser's name, some type of project identifying number or code, and the date.
2. Determine from the plans, specifications, owner's description, photographs or any other means possible the class of construction. The models in this book use economy, average, custom and luxury as classes. Fill in the appropriate box.
3. Fill in the appropriate box for the residence type, configuration, occupancy, and exterior wall. If you require clarification, the pages preceding this worksheet describe each of these.
4. Next, the living area of the residence must be established. The heated or air conditioned space of the residence, not including the basement, should be measured. It is easiest to break the structure up into separate components as shown in the example. The main house (A), a one-and-one-half story wing (B), and a one story wing (C). The breezeway (D), garage (E), and open covered porch (F) will be treated differently. Data entry blocks for the living area are included on the worksheet for your use. Keep each level of each component separate, and fill out the blocks as shown.
5. By using the information on the worksheet, find the model, wing or ell in the following square foot cost pages that best matches the class, type, exterior finish and size of the residence being estimated. Use the *Modifications, Adjustments, and Alternatives* to determine the adjusted cost per square foot of living area for each component.
6. For each component, multiply the cost per square foot by the living area square footage. If the residence is a townhouse/rowhouse, a multiplier should be applied based upon the configuration.
7. The second page of the residential cost estimate worksheet has space for the additional components of a house. The cost for additional bathrooms, finished attic space, breezeways, porches, fireplaces, appliance or cabinet upgrades, and garages should be added on this page. The information for each of these components is found with the model being used, or in the *Modifications, Adjustments, and Alternatives* pages.
8. Add the total from page one of the estimate worksheet and the items listed on page two. The sum is the adjusted total building cost.
9. Depending on the use of the final estimated cost, one of the remaining two boxes should be filled out. Any additional items or exclusions should be added or subtracted at this time. The data contained in this book is a national average. Construction costs are different throughout the country. To allow for this difference, a location factor based upon the first three digits of the residence's zip code must be applied. The location factor is a multiplier that increases or decreases the adjusted total building cost. Find the appropriate location factor and calculate the local cost. If depreciation is a concern, a dollar figure should be subtracted at this point.
10. No residence will match a model exactly. Many differences will be found. At this level of estimating, a variation of plus or minus 10% should be expected.

Adjustments Instructions

No residence matches a model exactly in shape, material, or specifications.

The common differences are:

1. Two or more exterior wall systems:
   a. Partial basement
   b. Partly finished basement
2. Specifications or features that are between two classes
3. Crawl space instead of a basement

EXAMPLES

Below are quick examples. See pages 15–17 for complete examples of cost adjustments for these differences:

1. Residence "A" is an average one-story structure with 1,600 S.F. of living area and no basement. Three walls are wood siding on wood frame, and the fourth wall is brick veneer on wood frame. The brick veneer wall is 35% of the exterior wall area.

   Use page 28 to calculate the Base Cost per S.F. of Living Area.
   Wood Siding for 1,600 S.F.   = $103.15 per S.F.
   Brick Veneer for 1,600 S.F.   = $118.10 per S.F.
   .65 ($103.15) + .35 ($118.10) = $108.39 per S.F. of Living Area.

2. a. Residence "B" is the same as Residence "A"; However, it has an unfinished basement under 50% of the building. To adjust the $108.39 per S.F. of living area for this partial basement, use page 28.

   $108.39 + .50 ($10.45) = $113.62 per S.F. of Living Area.

   b. Residence "C" is the same as Residence "A"; However, it has a full basement under the entire building. 640 S.F. or 40% of the basement area is finished.

   Using Page 28:

   $108.39 + .40 ($30.55) + .60 ($10.45) = $126.88 per S.F. of Living Area.

3. When specifications or features of a building are between classes, estimate the percent deviation, and use two tables to calculate the cost per S.F.

   A two-story residence with wood siding and 1,800 S.F. of living area has features 30% better than Average, but 70% less than Custom.

   From pages 30 and 42:

   Custom 1,800 S.F. Base Cost   = $127.65 per S.F.
   Average 1,800 S.F. Base Cost   = $105.20 per S.F.
   DIFFERENCE   = $22.45 per S.F.

   Cost is $105.20 + .30 ($22.45) = $127.65 per S.F. of Living Area.

4. To add the cost of a crawl space, use the cost of an unfinished basement as a maximum. For specific costs of components to be added or deducted, such as vapor barrier, underdrain, and floor, see the "Assemblies" section, pages 233 to 462.

Model Residence Example

First Floor Plan

E
D
C
A
B
F

24'
8'
12'
16'
12'
28'
20'
18'
12'
24'
12'
12'
46'
18'

Second Floor Plan

A
B

46'
28'
20'
5'
5'
5'
5'
13'

A = Main House
B = 1-1/2 Story Wing
C = 1 Story Wing
D = Breezeway
E = Garage
F = Open Covered Porch

**RESIDENTIAL
COST ESTIMATE**

OWNERS NAME:	**Albert Westenberg**	APPRAISER:	**Nicole Wojtowicz**
RESIDENCE ADDRESS:	**300 Sygiel Road**	PROJECT:	**# 55**
CITY, STATE, ZIP CODE:	**Three Rivers, MA 01080**	DATE:	**Jan. 1, 2013**

CLASS OF CONSTRUCTION	RESIDENCE TYPE	CONFIGURATION	EXTERIOR WALL SYSTEM
☐ ECONOMY	☐ 1 STORY	☑ DETACHED	☑ WOOD SIDING - WOOD FRAME
☑ AVERAGE	☐ 1 1/2 STORY	☐ TOWN/ROW HOUSE	☐ BRICK VENEER - WOOD FRAME
☐ CUSTOM	☑ 2 STORY	☐ SEMI-DETACHED	☐ STUCCO ON WOOD FRAME
☐ LUXURY	☐ 2 1/2 STORY		☐ PAINTED CONCRETE BLOCK
	☐ 3 STORY	OCCUPANCY	☐ SOLID MASONRY (AVERAGE & CUSTOM)
	☐ BI-LEVEL	☑ ONE FAMILY	☐ STONE VENEER - WOOD FRAME
	☐ TRI-LEVEL	☐ TWO FAMILY	☐ SOLID BRICK (LUXURY)
		☐ THREE FAMILY	☐ SOLID STONE (LUXURY)
		☐ OTHER	

* LIVING AREA (Main Building)			* LIVING AREA (Wing or Ell)	(B)		* LIVING AREA (Wing or Ell)	(C)	
First Level	**1288**	S.F.	First Level	**360**	S.F.	First Level	**192**	S.F.
Second level	**1288**	S.F.	Second level	**310**	S.F.	Second level		S.F.
Third Level		S.F.	Third Level		S.F.	Third Level		S.F.
Total	**2576**	S.F.	Total	**670**	S.F.	Total	**192**	S.F.

* Basement Area is not part of living area.

MAIN BUILDING			COSTS PER S.F. LIVING AREA	
Cost per Square Foot of Living Area, from Page	**30**		$	92.25
Basement Addition: _____ % Finished, **100** % Unfinished			+	5.80
Roof Cover Adjustment: **Cedar Shake** Type, Page **30** (Add or Deduct)			( )	1.45
Central Air Conditioning: ☐ Separate Ducts ☑ Heating Ducts, Page **30**			+	2.67
Heating System Adjustment: _____ Type, Page _____ (Add or Deduct)			( )	—
Main Building: Adjusted Cost per S.F. of Living Area			$ #	102.17

MAIN BUILDING TOTAL COST	$ 102.17 /S.F.	x	2,576 S.F.	x		=	$ 263,190
	Cost per S.F. Living Area		Living Area		Town/Row House Multiplier (Use 1 for Detached)		TOTAL COST

WING OR ELL (B)	**1 - 1/2** STORY		COSTS PER S.F. LIVING AREA	
Cost per Square Foot of Living Area, from Page **37**			$	89.65
Basement Addition: **100** % Finished, _____ % Unfinished			+	27.60
Roof Cover Adjustment: _____ Type, Page _____ (Add or Deduct)			( )	—
Central Air Conditioning: ☐ Separate Ducts ☑ Heating Ducts, Page **30**			+	2.67
Heating System Adjustment: _____ Type, Page _____ (Add or Deduct)			( )	—
Wing or Ell (**B**): Adjusted Cost per S.F. of Living Area			$ #	119.92

WING OR ELL (B) TOTAL COST	$ 119.92 /S.F.	x	670 S.F.		=	$ 80,346
	Cost per S.F. Living Area		Living Area			TOTAL COST

WING OR ELL (C)	**1** STORY (WOOD SIDING)		COSTS PER S.F. LIVING AREA	
Cost per Square Foot of Living Area, from Page **# 37**			$	151.15
Basement Addition: _____ % Finished, _____ % Unfinished			+	—
Roof Cover Adjustment: _____ Type, Page _____ (Add or Deduct)			( )	—
Central Air Conditioning: ☐ Separate Ducts ☐ Heating Ducts, Page _____			+	—
Heating System Adjustment: _____ Type, Page _____ (Add or Deduct)			( )	—
Wing or Ell (C) Adjusted Cost per S.F. of Living Area			$	151.15

WING OR ELL (C) TOTAL COST	$ 151.15 /S.F.	x	192 S.F.		=	$ 29,021
	Cost per S.F. Living Area		Living Area			TOTAL COST

TOTAL THIS PAGE	372,557

Page 1 of 2

RESIDENTIAL
COST ESTIMATE

						QUANTITY	UNIT COST			
Total Page 1								$		372,557
						QUANTITY	UNIT COST			
Additional Bathrooms:	**2**	Full,	**1**	Half	2 @ 7074 1 @ 4047			+		18,195
Finished Attic:	**N/A**	Ft. x		Ft.			S.F.			
Breezeway: ☑ Open ☐ ıclosed		**12**	Ft. x	**12**	Ft.	144	S.F.	33.32	+	4,798
Covered Porch: ☑ Open ☐		Enclosed	**18**	Ft. x	**12**	Ft. 216	S.F.	33.20	+	7,172
Fireplace: ☑ Interior Chimney		☐ Exterior Chimney								
☑ No. of Flues (**2**) ☑ Additional Fireplaces					**1 - 2nd Story**			+		10,979
Appliances:								+		—
Kitchen Cabinets Adjustments:					(±)					—
☑ Garage ☐ Carport:	**2**	Car(s)	Description	**Wood, Attached**	(±)					22,856
Miscellaneous:								+		
					ADJUSTED TOTAL BUILDING COST			$		436,557

REPLACEMENT COST		
ADJUSTED TOTAL BUILDING COST	$	**436,557**
Site Improvements		
(A) Paving & Sidewalks	$	
(B) Landscaping	$	
(C) Fences	$	
(D) Swimming Pools	$	
(E) Miscellaneous	$	
TOTAL	$	**436,557**
Location Factor	x	**1.050**
Location Replacement Cost	$	**458,385**
Depreciation	-$	**45,838**
LOCAL DEPRECIATED COST	$	**412,547**

INSURANCE COST		
ADJUSTED TOTAL BUILDING COST	$	
Insurance Exclusions		
(A) Footings, sitework, Underground Piping	-$	
(B) Architects Fees	-$	
Total Building Cost Less Exclusion	$	
Location Factor	x	
LOCAL INSURABLE REPLACEMENT COST	$	

SKETCH AND ADDITIONAL CALCULATIONS

1 Story

© Home Planners, Inc.

1-1/2 Story

2 Story

Bi-Level

Tri-Level

©Design Basics, Inc.

		Components
1	**Site Work**	Site preparation for slab or excavation for lower level; 4' deep trench excavation for foundation wall.
2	**Foundations**	Continuous reinforced concrete footing, 8" deep x 18" wide; dampproofed and insulated 8" thick reinforced concrete block foundation wall, 4' deep; 4" concrete slab on 4" crushed stone base and polyethylene vapor barrier, trowel finish.
3	**Framing**	Exterior walls—2 x 4 wood studs, 16" O.C.; 1/2" insulation board sheathing; wood truss roof 24" O.C. or 2 x 6 rafters 16" O.C. with 1/2" plywood sheathing, 4 in 12 or 8 in 12 roof pitch.
4	**Exterior Walls**	Beveled wood siding and #15 felt building paper on insulated wood frame walls; sliding sash or double hung wood windows; 2 flush solid core wood exterior doors. **Alternates:** • Brick veneer on wood frame, has 4" veneer of common brick. • Stucco on wood frame has 1" thick colored stucco finish. • Painted concrete block has 8" concrete block sealed and painted on exterior and furring on the interior for the drywall.
5	**Roofing**	20 year asphalt shingles; #15 felt building paper; aluminum gutters, downspouts, drip edge and flashings.
6	**Interiors**	Walls and ceilings—1/2" taped and finished drywall, primed and painted with 2 coats; painted baseboard and trim; rubber backed carpeting 80%, asphalt tile 20%; hollow core wood interior doors.
7	**Specialties**	Economy grade kitchen cabinets—6 L.F. wall and base with plastic laminate counter top and kitchen sink; 30 gallon electric water heater.
8	**Mechanical**	1 lavatory, white, wall hung; 1 water closet, white; 1 bathtub, enameled steel, white; gas fired warm air heat.
9	**Electrical**	100 Amp. service; romex wiring; incandescent lighting fixtures, switches, receptacles.
10	**Overhead and Profit**	General Contractor overhead and profit.

Adjustments

Unfinished Basement:
7' high 8" concrete block or cast-in-place concrete walls.

Finished Basement:
Includes inexpensive paneling or drywall on foundation walls. Inexpensive sponge backed carpeting on concrete floor, drywall ceiling, and lighting.

- **Mass produced from stock plans**
- **Single family — 1 full bath, 1 kitchen**
- **No basement**
- **Asphalt shingles on roof**
- **Hot air heat**
- **Gypsum wallboard interior finishes**
- **Materials and workmanship are sufficient to meet codes**
- **Detail specifications on page 19**

Note: The illustration shown may contain some optional components (for example: garages and/or fireplaces) whose costs are shown in the modifications, adjustments, & alternatives below or at the end of the square foot section.

©Home Planners, Inc.

Base cost per square foot of living area

Exterior Wall	Living Area										
	600	800	1000	1200	1400	1600	1800	2000	2400	2800	3200
Wood Siding - Wood Frame	123.95	112.00	102.80	95.45	89.10	85.00	82.90	80.20	74.65	70.60	67.85
Brick Veneer - Wood Frame	129.55	116.95	107.25	99.50	92.75	88.40	86.25	83.30	77.50	73.25	70.35
Stucco on Wood Frame	116.20	105.05	96.45	89.75	83.95	80.10	78.20	75.80	70.65	66.90	64.45
Painted Concrete Block	123.80	111.85	102.65	95.35	89.00	84.90	82.80	80.10	74.55	70.55	67.80
Finished Basement, Add	30.55	28.80	27.50	26.40	25.45	24.85	24.55	24.00	23.30	22.80	22.35
Unfinished Basement, Add	13.35	11.95	10.90	10.05	9.30	8.85	8.55	8.20	7.65	7.20	6.85

Modifications

Add to the total cost

Upgrade Kitchen Cabinets	$ + 902
Solid Surface Countertops	+ 655
Full Bath - including plumbing, wall and floor finishes	+ 5659
Half Bath - including plumbing, wall and floor finishes	+ 3238
One Car Attached Garage	+ 12,126
One Car Detached Garage	+ 15,607
Fireplace & Chimney	+ 5620

Adjustments

For multi family - add to total cost

Additional Kitchen	$ + 4505
Additional Bath	+ 5659
Additional Entry & Exit	+ 1576
Separate Heating	+ 1514
Separate Electric	+ 1095

For Townhouse/Rowhouse - Multiply cost per square foot by

Inner Unit	+ .95
End Unit	+ .97

Alternatives

Add to or deduct from the cost per square foot of living area

Composition Roll Roofing	– 1.05
Cedar Shake Roof	+ 3.35
Upgrade Walls and Ceilings to Skim Coat Plaster	+ .75
Upgrade Ceilings to Textured Finish	+ .49
Air Conditioning, in Heating Ductwork	+ 4.26
In Separate Ductwork	+ 6.33
Heating Systems, Hot Water	+ 2.22
Heat Pump	+ 2.15
Electric Heat	– 1.28
Not Heated	– 3.69

Additional upgrades or components

Kitchen Cabinets & Countertops	Page 58
Bathroom Vanities	59
Fireplaces & Chimneys	59
Windows, Skylights & Dormers	59
Appliances	60
Breezeways & Porches	60
Finished Attic	60
Garages	61
Site Improvements	61
Wings & Ells	25

- **Mass produced from stock plans**
- **Single family — 1 full bath, 1 kitchen**
- **No basement**
- **Asphalt shingles on roof**
- **Hot air heat**
- **Gypsum wallboard interior finishes**
- **Materials and workmanship are sufficient to meet codes**
- **Detail specifications on page 19**

Note: The illustration shown may contain some optional components (for example: garages and/or fireplaces) whose costs are shown in the modifications, adjustments, & alternatives below or at the end of the square foot section.

Base cost per square foot of living area

Exterior Wall	Living Area										
	600	800	1000	1200	1400	1600	1800	2000	2400	2800	3200
Wood Siding - Wood Frame	142.00	117.70	105.05	99.10	94.95	88.30	85.15	81.90	75.05	72.35	69.55
Brick Veneer - Wood Frame	149.75	123.30	110.25	104.00	99.60	92.50	89.15	85.65	78.40	75.50	72.50
Stucco on Wood Frame	131.15	109.80	97.75	92.30	88.45	82.35	79.55	76.60	70.35	67.90	65.45
Painted Concrete Block	141.80	117.50	104.90	99.00	94.85	88.15	85.05	81.75	74.90	72.25	69.50
Finished Basement, Add	23.45	19.95	19.05	18.40	17.95	17.20	16.85	16.50	15.75	15.40	15.05
Unfinished Basement, Add	11.80	9.00	8.30	7.75	7.40	6.80	6.50	6.25	5.65	5.40	5.10

Modifications

Add to the total cost

Upgrade Kitchen Cabinets	$ + 902
Solid Surface Countertops	+ 655
Full Bath - including plumbing, wall and floor finishes	+ 5659
Half Bath - including plumbing, wall and floor finishes	+ 3238
One Car Attached Garage	+ 12,126
One Car Detached Garage	+ 15,607
Fireplace & Chimney	+ 5620

Adjustments

For multi family - add to total cost

Additional Kitchen	$ + 4505
Additional Bath	+ 5659
Additional Entry & Exit	+ 1576
Separate Heating	+ 1514
Separate Electric	+ 1095

For Townhouse/Rowhouse -
Multiply cost per square foot by

Inner Unit	+ .95
End Unit	+ .97

Alternatives

Add to or deduct from the cost per square foot of living area

Composition Roll Roofing	– .75
Cedar Shake Roof	+ 2.45
Upgrade Walls and Ceilings to Skim Coat Plaster	+ .75
Upgrade Ceilings to Textured Finish	+ .49
Air Conditioning, in Heating Ductwork	+ 3.18
In Separate Ductwork	+ 5.56
Heating Systems, Hot Water	+ 2.12
Heat Pump	+ 2.37
Electric Heat	– 1.02
Not Heated	– 3.39

Additional upgrades or components

Kitchen Cabinets & Countertops	Page 58
Bathroom Vanities	59
Fireplaces & Chimneys	59
Windows, Skylights & Dormers	59
Appliances	60
Breezeways & Porches	60
Finished Attic	60
Garages	61
Site Improvements	61
Wings & Ells	25

Important: See the Reference Section for Location Factors (to adjust for your city) and Estimating Forms.

- **Mass produced from stock plans**
- **Single family — 1 full bath, 1 kitchen**
- **No basement**
- **Asphalt shingles on roof**
- **Hot air heat**
- **Gypsum wallboard interior finishes**
- **Materials and workmanship are sufficient to meet codes**
- **Detail specifications on page 19**

Note: The illustration shown may contain some optional components (for example: garages and/or fireplaces) whose costs are shown in the modifications, adjustments, & alternatives below or at the end of the square foot section.

Base cost per square foot of living area

Exterior Wall	Living Area										
	1000	1200	1400	1600	1800	2000	2200	2600	3000	3400	3800
Wood Siding - Wood Frame	111.90	101.05	96.10	92.60	89.00	85.00	82.40	77.45	72.60	70.45	68.50
Brick Veneer - Wood Frame	117.90	106.65	101.25	97.55	93.65	89.50	86.70	81.30	76.20	73.85	71.70
Stucco on Wood Frame	103.50	93.30	88.80	85.65	82.50	78.70	76.45	72.05	67.60	65.70	63.95
Painted Concrete Block	111.75	100.95	95.95	92.50	88.90	84.90	82.35	77.35	72.55	70.35	68.35
Finished Basement, Add	16.00	15.35	14.80	14.45	14.10	13.80	13.50	13.00	12.65	12.40	12.15
Unfinished Basement, Add	7.25	6.70	6.25	6.00	5.65	5.45	5.25	4.80	4.55	4.35	4.15

Modifications

Add to the total cost

Upgrade Kitchen Cabinets	$ + 902
Solid Surface Countertops	+ 655
Full Bath - including plumbing, wall and floor finishes	+ 5659
Half Bath - including plumbing, wall and floor finishes	+ 3238
One Car Attached Garage	+ 12,126
One Car Detached Garage	+ 15,607
Fireplace & Chimney	+ 6210

Adjustments

For multi family - add to total cost

Additional Kitchen	$ + 4505
Additional Bath	+ 5659
Additional Entry & Exit	+ 1576
Separate Heating	+ 1514
Separate Electric	+ 1095

For Townhouse/Rowhouse - Multiply cost per square foot by

Inner Unit	+ .93
End Unit	+ .96

Alternatives

Add to or deduct from the cost per square foot of living area

Composition Roll Roofing	– .55
Cedar Shake Roof	+ 1.70
Upgrade Walls and Ceilings to Skim Coat Plaster	+ .76
Upgrade Ceilings to Textured Finish	+ .49
Air Conditioning, in Heating Ductwork	+ 2.58
In Separate Ductwork	+ 5.10
Heating Systems, Hot Water	+ 2.07
Heat Pump	+ 2.52
Electric Heat	– .89
Not Heated	– 3.21

Additional upgrades or components

Kitchen Cabinets & Countertops	Page 58
Bathroom Vanities	59
Fireplaces & Chimneys	59
Windows, Skylights & Dormers	59
Appliances	60
Breezeways & Porches	60
Finished Attic	60
Garages	61
Site Improvements	61
Wings & Ells	25

- **Mass produced from stock plans**
- **Single family — 1 full bath, 1 kitchen**
- **No basement**
- **Asphalt shingles on roof**
- **Hot air heat**
- **Gypsum wallboard interior finishes**
- **Materials and workmanship are sufficient to meet codes**
- **Detail specifications on page 19**

Note: The illustration shown may contain some optional components (for example: garages and/or fireplaces) whose costs are shown in the modifications, adjustments, & alternatives below or at the end of the square foot section.

Base cost per square foot of living area

Exterior Wall	Living Area										
	1000	1200	1400	1600	1800	2000	2200	2600	3000	3400	3800
Wood Siding - Wood Frame	103.45	93.25	88.70	85.60	82.40	78.65	76.35	72.00	67.55	65.65	63.90
Brick Veneer - Wood Frame	107.90	97.35	92.60	89.30	85.90	82.00	79.60	74.90	70.25	68.20	66.35
Stucco on Wood Frame	97.10	87.40	83.30	80.40	77.50	73.95	71.90	67.95	63.80	62.05	60.55
Painted Concrete Block	103.30	93.10	88.60	85.50	82.30	78.55	76.30	71.90	67.50	65.55	63.85
Finished Basement, Add	16.00	15.35	14.80	14.45	14.10	13.80	13.50	13.00	12.65	12.40	12.15
Unfinished Basement, Add	7.25	6.70	6.25	6.00	5.65	5.45	5.25	4.80	4.55	4.35	4.15

Modifications

Add to the total cost

Upgrade Kitchen Cabinets	$ + 902
Solid Surface Countertops	+ 655
Full Bath - including plumbing, wall and floor finishes	+ 5659
Half Bath - including plumbing, wall and floor finishes	+ 3238
One Car Attached Garage	+ 12,126
One Car Detached Garage	+ 15,607
Fireplace & Chimney	+ 5620

Adjustments

For multi family - add to total cost

Additional Kitchen	$ + 4505
Additional Bath	+ 5659
Additional Entry & Exit	+ 1576
Separate Heating	+ 1514
Separate Electric	+ 1095

For Townhouse/Rowhouse -
Multiply cost per square foot by

Inner Unit	+ .94
End Unit	+ .97

Alternatives

Add to or deduct from the cost per square foot of living area

Composition Roll Roofing	– .55
Cedar Shake Roof	+ 1.70
Upgrade Walls and Ceilings to Skim Coat Plaster	+ .72
Upgrade Ceilings to Textured Finish	+ .49
Air Conditioning, in Heating Ductwork	+ 2.58
In Separate Ductwork	+ 5.10
Heating Systems, Hot Water	+ 2.07
Heat Pump	+ 2.52
Electric Heat	– .89
Not Heated	– 3.21

Additional upgrades or components

Kitchen Cabinets & Countertops	Page 58
Bathroom Vanities	59
Fireplaces & Chimneys	59
Windows, Skylights & Dormers	59
Appliances	60
Breezeways & Porches	60
Finished Attic	60
Garages	61
Site Improvements	61
Wings & Ells	25

Important: See the Reference Section for Location Factors (to adjust for your city) and Estimating Forms.

- **Mass produced from stock plans**
- **Single family — 1 full bath, 1 kitchen**
- **No basement**
- **Asphalt shingles on roof**
- **Hot air heat**
- **Gypsum wallboard interior finishes**
- **Materials and workmanship are sufficient to meet codes**
- **Detail specifications on page 19**

Note: The illustration shown may contain some optional components (for example: garages and/or fireplaces) whose costs are shown in the modifications, adjustments, & alternatives below or at the end of the square foot section.

©Design Basics, Inc.

Base cost per square foot of living area

Exterior Wall	Living Area										
	1200	1500	1800	2000	2200	2400	2800	3200	3600	4000	4400
Wood Siding - Wood Frame	95.30	87.30	81.40	79.00	75.55	72.80	70.50	67.55	64.05	62.90	60.20
Brick Veneer - Wood Frame	99.45	91.10	84.80	82.20	78.65	75.70	73.25	70.15	66.45	65.20	62.35
Stucco on Wood Frame	89.55	82.10	76.65	74.45	71.25	68.75	66.65	63.95	60.70	59.65	57.15
Solid Masonry	95.20	87.25	81.30	78.90	75.45	72.70	70.45	67.50	64.00	62.80	60.15
Finished Basement, Add*	19.25	18.35	17.60	17.25	16.95	16.60	16.30	15.90	15.55	15.40	15.10
Unfinished Basement, Add*	8.00	7.30	6.65	6.40	6.20	5.85	5.65	5.35	5.10	4.95	4.75

*Basement under middle level only.

Modifications

Add to the total cost

Upgrade Kitchen Cabinets	$ + 902
Solid Surface Countertops	+ 655
Full Bath - including plumbing, wall and floor finishes	+ 5659
Half Bath - including plumbing, wall and floor finishes	+ 3238
One Car Attached Garage	+ 12,126
One Car Detached Garage	+ 15,607
Fireplace & Chimney	+ 5620

Adjustments

For multi family - add to total cost

Additional Kitchen	$ + 4505
Additional Bath	+ 5659
Additional Entry & Exit	+ 1576
Separate Heating	+ 1514
Separate Electric	+ 1095

For Townhouse/Rowhouse -
Multiply cost per square foot by

Inner Unit	+ .93
End Unit	+ .96

Alternatives

Add to or deduct from the cost per square foot of living area

Composition Roll Roofing	– .75
Cedar Shake Roof	+ 2.45
Upgrade Walls and Ceilings to Skim Coat Plaster	+ .66
Upgrade Ceilings to Textured Finish	+ .49
Air Conditioning, in Heating Ductwork	+ 2.21
In Separate Ductwork	+ 4.74
Heating Systems, Hot Water	+ 1.98
Heat Pump	+ 2.61
Electric Heat	– .77
Not Heated	– 3.11

Additional upgrades or components

Kitchen Cabinets & Countertops	Page 58
Bathroom Vanities	59
Fireplaces & Chimneys	59
Windows, Skylights & Dormers	59
Appliances	60
Breezeways & Porches	60
Finished Attic	60
Garages	61
Site Improvements	61
Wings & Ells	25

1 Story — Base cost per square foot of living area

Exterior Wall	Living Area							
	50	100	200	300	400	500	600	700
Wood Siding - Wood Frame	173.20	131.35	113.25	93.50	87.55	83.95	81.60	82.20
Brick Veneer - Wood Frame	187.20	141.30	121.55	99.05	92.60	88.60	86.00	86.45
Stucco on Wood Frame	153.60	117.35	101.55	85.70	80.55	77.45	75.40	76.20
Painted Concrete Block	172.85	131.10	113.00	93.35	87.45	83.90	81.50	82.05
Finished Basement, Add	46.35	37.95	34.40	28.55	27.35	26.65	26.20	25.85
Unfinished Basement, Add	25.45	18.85	16.10	11.60	10.65	10.15	9.75	9.50

1-1/2 Story — Base cost per square foot of living area

Exterior Wall	Living Area							
	100	200	300	400	500	600	700	800
Wood Siding - Wood Frame	137.50	110.25	93.55	83.25	78.15	75.70	72.60	71.75
Brick Veneer - Wood Frame	150.00	120.25	101.85	89.70	84.20	81.35	77.95	77.05
Stucco on Wood Frame	120.00	96.20	81.90	74.15	69.80	67.75	65.10	64.30
Painted Concrete Block	137.15	109.95	93.30	83.05	78.05	75.55	72.50	71.60
Finished Basement, Add	31.05	27.55	25.20	22.60	21.90	21.40	21.00	20.95
Unfinished Basement, Add	15.65	12.90	11.10	9.10	8.55	8.15	7.85	7.80

2 Story — Base cost per square foot of living area

Exterior Wall	Living Area							
	100	200	400	600	800	1000	1200	1400
Wood Siding - Wood Frame	140.85	104.20	88.25	72.10	66.90	63.80	61.70	62.50
Brick Veneer - Wood Frame	154.80	114.20	96.60	77.70	71.90	68.45	66.10	66.80
Stucco on Wood Frame	121.20	90.20	76.55	64.35	59.90	57.25	55.50	56.50
Painted Concrete Block	140.45	103.90	88.05	71.95	66.80	63.70	61.60	62.40
Finished Basement, Add	23.20	19.00	17.25	14.30	13.70	13.40	13.15	13.00
Unfinished Basement, Add	12.75	9.45	8.05	5.80	5.35	5.10	4.90	4.75

Base costs do not include bathroom or kitchen facilities. Use Modifications/Adjustments/Alternatives on pages 58-61 where appropriate.

Important: See the Reference Section for Location Factors (to adjust for your city) and Estimating Forms.

1 Story

1-1/2 Story

2 Story

2-1/2 Story

Bi-Level

Tri-Level

		Components
1	**Site Work**	Site preparation for slab or excavation for lower level; 4' deep trench excavation for foundation wall.
2	**Foundations**	Continuous reinforced concrete footing, 8" deep x 18" wide; dampproofed and insulated 8" thick reinforced concrete foundation wall, 4' deep; 4" concrete slab on 4" crushed stone base and polyethylene vapor barrier, trowel finish.
3	**Framing**	Exterior walls—2 x 4 wood studs, 16" O.C.; 1/2" plywood sheathing; 2 x 6 rafters 16" O.C. with 1/2" plywood sheathing, 4 in 12 or 8 in 12 roof pitch, 2 x 6 ceiling joists; 1/2" plywood subfloor on 1 x 2 wood sleepers 16" O.C.; 2 x 8 floor joists with 5/8" plywood subfloor on models with more than one level.
4	**Exterior Walls**	Beveled wood siding and #15 felt building paper on insulated wood frame walls; double hung windows; 3 flush solid core wood exterior doors with storms. **Alternates:** • Brick veneer on wood frame, has 4" veneer of common brick. • Stucco on wood frame has 1" thick colored stucco finish. • Solid masonry has a 6" concrete block load bearing wall with insulation and a brick or stone exterior. Also solid brick or stone walls.
5	**Roofing**	25 year asphalt shingles; #15 felt building paper; aluminum gutters, downspouts, drip edge and flashings.
6	**Interiors**	Walls and ceilings—1/2" taped and finished drywall, primed and painted with 2 coats; painted baseboard and trim, finished hardwood floor 40%, carpet with 1/2" underlayment 40%, vinyl tile with 1/2" underlayment 15%, ceramic tile with 1/2" underlayment 5%; hollow core and louvered interior doors.
7	**Specialties**	Average grade kitchen cabinets—14 L.F. wall and base with plastic laminate counter top and kitchen sink; 40 gallon electric water heater.
8	**Mechanical**	1 lavatory, white, wall hung; 1 water closet, white; 1 bathtub, enameled steel, white; gas fired warm air heat.
9	**Electrical**	100 Amp. service; romex wiring; incandescent lighting fixtures, switches, receptacles.
10	**Overhead and Profit**	General Contractor overhead and profit.

Adjustments

Unfinished Basement:
7'-6" high cast-in-place concrete walls 8" thick or 8" concrete block.

Finished Basement:
Includes plywood paneling or drywall on furring on foundation walls, sponge backed carpeting on concrete floor, suspended ceiling, lighting and heating.

- **Simple design from standard plans**
- **Single family — 1 full bath, 1 kitchen**
- **No basement**
- **Asphalt shingles on roof**
- **Hot air heat**
- **Gypsum wallboard interior finishes**
- **Materials and workmanship are average**
- **Detail specifications on p. 27**

Note: The illustration shown may contain some optional components (for example: garages and/or fireplaces) whose costs are shown in the modifications, adjustments, & alternatives below or at the end of the square foot section.

©Home Planners, Inc.

Base cost per square foot of living area

Exterior Wall	Living Area										
	600	800	1000	1200	1400	1600	1800	2000	2400	2800	3200
Wood Siding - Wood Frame	149.10	134.70	123.75	115.15	108.00	103.15	100.60	97.50	91.25	86.70	83.60
Brick Veneer - Wood Frame	166.10	151.10	139.70	130.70	123.15	118.10	115.35	112.10	105.55	100.80	97.45
Stucco on Wood Frame	155.60	141.70	131.20	123.00	116.10	111.50	109.10	106.15	100.10	95.80	92.85
Solid Masonry	181.70	165.00	152.35	142.10	133.50	127.75	124.70	120.90	113.55	108.20	104.35
Finished Basement, Add	37.00	35.75	34.10	32.60	31.35	30.55	30.10	29.50	28.55	27.85	27.20
Unfinished Basement, Add	15.15	13.70	12.65	11.70	11.00	10.45	10.20	9.80	9.25	8.80	8.45

Modifications

Add to the total cost

Upgrade Kitchen Cabinets	$ + 3955
Solid Surface Countertops (Included)	
Full Bath - including plumbing, wall and floor finishes	+ 7074
Half Bath - including plumbing, wall and floor finishes	+ 4047
One Car Attached Garage	+ 13,481
One Car Detached Garage	+ 17,772
Fireplace & Chimney	+ 5930

Adjustments

For multi family - add to total cost

Additional Kitchen	$ + 7917
Additional Bath	+ 7074
Additional Entry & Exit	+ 1576
Separate Heating	+ 1514
Separate Electric	+ 1958

For Townhouse/Rowhouse -
Multiply cost per square foot by

Inner Unit	+ .92
End Unit	+ .96

Alternatives

Add to or deduct from the cost per square foot of living area

Cedar Shake Roof	+ 2.90
Clay Tile Roof	+ 4.85
Slate Roof	+ 6.60
Upgrade Walls to Skim Coat Plaster	+ .46
Upgrade Ceilings to Textured Finish	+ .49
Air Conditioning, in Heating Ductwork	+ 4.41
In Separate Ductwork	+ 6.58
Heating Systems, Hot Water	+ 2.27
Heat Pump	+ 2.24
Electric Heat	- .71
Not Heated	- 3.08

Additional upgrades or components

Kitchen Cabinets & Countertops	Page 58
Bathroom Vanities	59
Fireplaces & Chimneys	59
Windows, Skylights & Dormers	59
Appliances	60
Breezeways & Porches	60
Finished Attic	60
Garages	61
Site Improvements	61
Wings & Ells	37

- **Simple design from standard plans**
- **Single family — 1 full bath, 1 kitchen**
- **No basement**
- **Asphalt shingles on roof**
- **Hot air heat**
- **Gypsum wallboard interior finishes**
- **Materials and workmanship are average**
- **Detail specifications on p. 27**

Note: The illustration shown may contain some optional components (for example: garages and/or fireplaces) whose costs are shown in the modifications, adjustments, & alternatives below or at the end of the square foot section.

©By Designer

Base cost per square foot of living area

Exterior Wall	Living Area										
	600	800	1000	1200	1400	1600	1800	2000	2400	2800	3200
Wood Siding - Wood Frame	166.90	139.30	124.70	117.75	112.75	105.20	101.65	97.90	90.20	87.15	83.90
Brick Veneer - Wood Frame	174.85	145.05	130.00	122.75	117.55	109.50	105.70	101.75	93.60	90.40	86.95
Stucco on Wood Frame	160.15	134.45	120.10	113.50	108.75	101.55	98.20	94.60	87.30	84.40	81.40
Solid Masonry	196.60	160.80	144.55	136.40	130.55	121.40	116.95	112.45	103.05	99.30	95.20
Finished Basement, Add	30.15	26.35	25.20	24.30	23.75	22.80	22.30	21.80	20.80	20.40	19.85
Unfinished Basement, Add	13.10	10.25	9.55	9.00	8.60	8.00	7.70	7.40	6.85	6.55	6.20

Modifications

Add to the total cost

Upgrade Kitchen Cabinets	$ + 3955
Solid Surface Countertops (Included)	
Full Bath - including plumbing, wall and floor finishes	+ 7074
Half Bath - including plumbing, wall and floor finishes	+ 4047
One Car Attached Garage	+ 13,481
One Car Detached Garage	+ 17,772
Fireplace & Chimney	+ 6550

Adjustments

For multi family - add to total cost

Additional Kitchen	$ + 7917
Additional Bath	+ 7074
Additional Entry & Exit	+ 1576
Separate Heating	+ 1514
Separate Electric	+ 1958

For Townhouse/Rowhouse -
Multiply cost per square foot by

Inner Unit	+ .92
End Unit	+ .96

Alternatives

Add to or deduct from the cost per square foot of living area

Cedar Shake Roof	+ 2.10
Clay Tile Roof	+ 3.50
Slate Roof	+ 4.75
Upgrade Walls to Skim Coat Plaster	+ .53
Upgrade Ceilings to Textured Finish	+ .49
Air Conditioning, in Heating Ductwork	+ 3.35
In Separate Ductwork	+ 5.75
Heating Systems, Hot Water	+ 2.16
Heat Pump	+ 2.48
Electric Heat	- .64
Not Heated	- 2.96

Additional upgrades or components

Kitchen Cabinets & Countertops	Page 58
Bathroom Vanities	59
Fireplaces & Chimneys	59
Windows, Skylights & Dormers	59
Appliances	60
Breezeways & Porches	60
Finished Attic	60
Garages	61
Site Improvements	61
Wings & Ells	37

- **Simple design from standard plans**
- **Single family — 1 full bath, 1 kitchen**
- **No basement**
- **Asphalt shingles on roof**
- **Hot air heat**
- **Gypsum wallboard interior finishes**
- **Materials and workmanship are average**
- **Detail specifications on p. 27**

Note: The illustration shown may contain some optional components (for example: garages and/or fireplaces) whose costs are shown in the modifications, adjustments, & alternatives below or at the end of the square foot section.

Base cost per square foot of living area

Exterior Wall	Living Area										
	1000	1200	1400	1600	1800	2000	2200	2600	3000	3400	3800
Wood Siding - Wood Frame	131.75	119.35	113.50	109.45	105.20	100.80	97.90	92.25	86.85	84.45	82.10
Brick Veneer - Wood Frame	137.90	125.05	118.80	114.55	109.95	105.40	102.30	96.20	90.50	87.90	85.35
Stucco on Wood Frame	126.55	114.50	108.95	105.10	101.15	96.90	94.15	88.80	83.75	81.45	79.25
Solid Masonry	154.75	140.60	133.35	128.40	123.10	118.00	114.25	107.10	100.55	97.35	94.45
Finished Basement, Add	21.30	21.05	20.35	19.85	19.35	19.00	18.65	17.95	17.50	17.15	16.85
Unfinished Basement, Add	8.35	7.75	7.30	7.00	6.70	6.45	6.25	5.80	5.55	5.35	5.15

Modifications

Add to the total cost

Upgrade Kitchen Cabinets	$ + 3955
Solid Surface Countertops (Included)	
Full Bath - including plumbing, wall and floor finishes	+ 7074
Half Bath - including plumbing, wall and floor finishes	+ 4047
One Car Attached Garage	+ 13,481
One Car Detached Garage	+ 17,772
Fireplace & Chimney	+ 6550

Adjustments

For multi family - add to total cost

Additional Kitchen	$ + 7917
Additional Bath	+ 7074
Additional Entry & Exit	+ 1576
Separate Heating	+ 1514
Separate Electric	+ 1958

For Townhouse/Rowhouse -
Multiply cost per square foot by

Inner Unit	+ .90
End Unit	+ .95

Alternatives

Add to or deduct from the cost per square foot of living area

Cedar Shake Roof	+ 1.45
Clay Tile Roof	+ 2.40
Slate Roof	+ 3.30
Upgrade Walls to Skim Coat Plaster	+ .55
Upgrade Ceilings to Textured Finish	+ .49
Air Conditioning, in Heating Ductwork	+ 2.67
In Separate Ductwork	+ 5.22
Heating Systems, Hot Water	+ 2.12
Heat Pump	+ 2.61
Electric Heat	– .56
Not Heated	– 2.87

Additional upgrades or components

Kitchen Cabinets & Countertops	Page 58
Bathroom Vanities	59
Fireplaces & Chimneys	59
Windows, Skylights & Dormers	59
Appliances	60
Breezeways & Porches	60
Finished Attic	60
Garages	61
Site Improvements	61
Wings & Ells	37

- **Simple design from standard plans**
- **Single family — 1 full bath, 1 kitchen**
- **No basement**
- **Asphalt shingles on roof**
- **Hot air heat**
- **Gypsum wallboard interior finishes**
- **Materials and workmanship are average**
- **Detail specifications on p. 27**

Note: The illustration shown may contain some optional components (for example: garages and/or fireplaces) whose costs are shown in the modifications, adjustments, & alternatives below or at the end of the square foot section.

Base cost per square foot of living area

Exterior Wall	Living Area										
	1200	1400	1600	1800	2000	2400	2800	3200	3600	4000	4400
Wood Siding - Wood Frame	130.55	122.65	111.80	109.80	105.90	99.50	94.50	89.55	86.90	82.25	80.85
Brick Veneer - Wood Frame	137.15	128.55	117.25	115.25	111.00	104.10	98.95	93.55	90.70	85.80	84.25
Stucco on Wood Frame	125.05	117.65	107.20	105.15	101.50	95.55	90.75	86.10	83.65	79.25	77.95
Solid Masonry	155.05	144.75	132.15	130.20	125.05	116.75	111.00	104.50	101.15	95.40	93.60
Finished Basement, Add	18.00	17.65	16.95	16.90	16.45	15.75	15.45	14.95	14.70	14.35	14.15
Unfinished Basement, Add	6.95	6.40	6.00	5.95	5.65	5.20	5.00	4.75	4.55	4.30	4.25

Modifications

Add to the total cost

Upgrade Kitchen Cabinets	$ + 3955
Solid Surface Countertops (Included)	
Full Bath - including plumbing, wall and floor finishes	+ 7074
Half Bath - including plumbing, wall and floor finishes	+ 4047
One Car Attached Garage	+ 13,481
One Car Detached Garage	+ 17,772
Fireplace & Chimney	+ 7185

Adjustments

For multi family - add to total cost

Additional Kitchen	$ + 7917
Additional Bath	+ 7074
Additional Entry & Exit	+ 1576
Separate Heating	+ 1514
Separate Electric	+ 1958

For Townhouse/Rowhouse -
Multiply cost per square foot by

Inner Unit	+ .90
End Unit	+ .95

Alternatives

Add to or deduct from the cost per square foot of living area

Cedar Shake Roof	+ 1.25
Clay Tile Roof	+ 2.10
Slate Roof	+ 2.85
Upgrade Walls to Skim Coat Plaster	+ .53
Upgrade Ceilings to Textured Finish	+ .49
Air Conditioning, in Heating Ductwork	+ 2.42
In Separate Ductwork	+ 5.04
Heating Systems, Hot Water	+ 1.91
Heat Pump	+ 2.69
Electric Heat	- 1
Not Heated	- 3.35

Additional upgrades or components

Important: See the Reference Section for Location Factors (to adjust for your city) and Estimating Forms.

- **Simple design from standard plans**
- **Single family — 1 full bath, 1 kitchen**
- **No basement**
- **Asphalt shingles on roof**
- **Hot air heat**
- **Gypsum wallboard interior finishes**
- **Materials and workmanship are average**
- **Detail specifications on p. 27**

Note: The illustration shown may contain some optional components (for example: garages and/or fireplaces) whose costs are shown in the modifications, adjustments, & alternatives below or at the end of the square foot section.

Base cost per square foot of living area

Exterior Wall	Living Area										
	1500	1800	2100	2500	3000	3500	4000	4500	5000	5500	6000
Wood Siding - Wood Frame	121.10	109.90	104.95	101.10	93.70	90.45	86.00	81.20	79.65	77.80	76.00
Brick Veneer - Wood Frame	127.25	115.60	110.30	106.20	98.30	94.80	89.95	84.85	83.20	81.25	79.20
Stucco on Wood Frame	115.90	105.10	100.45	96.80	89.85	86.75	82.65	78.05	76.65	74.95	73.25
Solid Masonry	144.10	131.15	124.85	120.05	110.95	106.75	100.85	94.85	92.90	90.55	88.00
Finished Basement, Add	15.45	15.25	14.80	14.50	13.95	13.70	13.25	12.95	12.75	12.65	12.40
Unfinished Basement, Add	5.65	5.30	5.00	4.75	4.40	4.25	4.00	3.80	3.65	3.65	3.50

Modifications

Add to the total cost

Upgrade Kitchen Cabinets	$ + 3955
Solid Surface Countertops (Included)	
Full Bath - including plumbing, wall and floor finishes	+ 7074
Half Bath - including plumbing, wall and floor finishes	+ 4047
One Car Attached Garage	+ 13,481
One Car Detached Garage	+ 17,772
Fireplace & Chimney	+ 7185

Adjustments

For multi family - add to total cost

Additional Kitchen	$ + 7917
Additional Bath	+ 7074
Additional Entry & Exit	+ 1576
Separate Heating	+ 1514
Separate Electric	+ 1958

For Townhouse/Rowhouse -
Multiply cost per square foot by

Inner Unit	+ .88
End Unit	+ .94

Alternatives

Add to or deduct from the cost per square foot of living area

Cedar Shake Roof	+ .95
Clay Tile Roof	+ 1.60
Slate Roof	+ 2.20
Upgrade Walls to Skim Coat Plaster	+ .55
Upgrade Ceilings to Textured Finish	+ .49
Air Conditioning, in Heating Ductwork	+ 2.42
In Separate Ductwork	+ 5.04
Heating Systems, Hot Water	+ 1.91
Heat Pump	+ 2.69
Electric Heat	– .77
Not Heated	– 3.12

Additional upgrades or components

Kitchen Cabinets & Countertops	Page 58
Bathroom Vanities	59
Fireplaces & Chimneys	59
Windows, Skylights & Dormers	59
Appliances	60
Breezeways & Porches	60
Finished Attic	60
Garages	61
Site Improvements	61
Wings & Ells	37

- Simple design from standard plans
- Single family — 1 full bath, 1 kitchen
- No basement
- Asphalt shingles on roof
- Hot air heat
- Gypsum wallboard interior finishes
- Materials and workmanship are average
- Detail specifications on p. 27

Note: The illustration shown may contain some optional components (for example: garages and/or fireplaces) whose costs are shown in the modifications, adjustments, & alternatives below or at the end of the square foot section.

Base cost per square foot of living area

Exterior Wall	Living Area										
	1000	1200	1400	1600	1800	2000	2200	2600	3000	3400	3800
Wood Siding - Wood Frame	122.65	110.95	105.65	101.90	98.15	94.00	91.45	86.35	81.45	79.25	77.15
Brick Veneer - Wood Frame	127.25	115.20	109.65	105.75	101.75	97.45	94.70	89.30	84.20	81.80	79.65
Stucco on Wood Frame	118.80	107.30	102.25	98.60	95.10	91.05	88.60	83.85	79.15	77.05	75.05
Solid Masonry	139.90	126.85	120.50	116.20	111.55	106.90	103.70	97.45	91.75	88.95	86.45
Finished Basement, Add	21.30	21.05	20.35	19.85	19.35	19.00	18.65	17.95	17.50	17.15	16.85
Unfinished Basement, Add	8.35	7.75	7.30	7.00	6.70	6.45	6.25	5.80	5.55	5.35	5.15

Modifications

Add to the total cost

Upgrade Kitchen Cabinets	$ + 3955
Solid Surface Countertops (Included)	
Full Bath - including plumbing, wall and floor finishes	+ 7074
Half Bath - including plumbing, wall and floor finishes	+ 4047
One Car Attached Garage	+ 13,481
One Car Detached Garage	+ 17,772
Fireplace & Chimney	+ 5930

Adjustments

For multi family - add to total cost

Additional Kitchen	$ + 7917
Additional Bath	+ 7074
Additional Entry & Exit	+ 1576
Separate Heating	+ 1514
Separate Electric	+ 1958

For Townhouse/Rowhouse - Multiply cost per square foot by

Inner Unit	+ .91
End Unit	+ .96

Alternatives

Add to or deduct from the cost per square foot of living area

Cedar Shake Roof	+ 1.45
Clay Tile Roof	+ 2.40
Slate Roof	+ 3.30
Upgrade Walls to Skim Coat Plaster	+ .51
Upgrade Ceilings to Textured Finish	+ .49
Air Conditioning, in Heating Ductwork	+ 2.67
In Separate Ductwork	+ 5.22
Heating Systems, Hot Water	+ 2.12
Heat Pump	+ 2.61
Electric Heat	– .56
Not Heated	– 2.87

Additional upgrades or components

Kitchen Cabinets & Countertops	Page 58
Bathroom Vanities	59
Fireplaces & Chimneys	59
Windows, Skylights & Dormers	59
Appliances	60
Breezeways & Porches	60
Finished Attic	60
Garages	61
Site Improvements	61
Wings & Ells	37

- Simple design from standard plans
- Single family — 1 full bath, 1 kitchen
- No basement
- Asphalt shingles on roof
- Hot air heat
- Gypsum wallboard interior finishes
- Materials and workmanship are average
- Detail specifications on p. 27

Note: The illustration shown may contain some optional components (for example: garages and/or fireplaces) whose costs are shown in the modifications, adjustments, & alternatives below or at the end of the square foot section.

©Design Basics, Inc.

Base cost per square foot of living area

Exterior Wall	Living Area										
	1200	1500	1800	2100	2400	2700	3000	3400	3800	4200	4600
Wood Siding - Wood Frame	116.55	107.30	100.50	94.50	90.75	88.50	86.15	83.95	80.10	77.05	75.50
Brick Veneer - Wood Frame	120.80	111.10	103.95	97.70	93.70	91.35	88.80	86.45	82.55	79.35	77.70
Stucco on Wood Frame	112.95	104.00	97.55	91.85	88.25	86.10	83.85	81.75	78.10	75.15	73.65
Solid Masonry	132.35	121.60	113.45	106.35	101.75	99.15	96.15	93.60	89.10	85.50	83.65
Finished Basement, Add*	24.45	24.00	23.00	22.20	21.70	21.35	20.90	20.70	20.20	19.85	19.65
Unfinished Basement, Add*	9.25	8.50	7.90	7.40	7.10	6.90	6.65	6.45	6.20	5.95	5.85

*Basement under middle level only.

Modifications

Add to the total cost

Upgrade Kitchen Cabinets	$ + 3955
Solid Surface Countertops (Included)	
Full Bath - including plumbing, wall and floor finishes	+ 7074
Half Bath - including plumbing, wall and floor finishes	+ 4047
One Car Attached Garage	+ 13,481
One Car Detached Garage	+ 17,772
Fireplace & Chimney	+ 5930

Adjustments

For multi family - add to total cost

Additional Kitchen	$ + 7917
Additional Bath	+ 7074
Additional Entry & Exit	+ 1576
Separate Heating	+ 1514
Separate Electric	+ 1958

For Townhouse/Rowhouse - Multiply cost per square foot by

Inner Unit	+ .90
End Unit	+ .95

Alternatives

Add to or deduct from the cost per square foot of living area

Cedar Shake Roof	+ 2.10
Clay Tile Roof	+ 3.50
Slate Roof	+ 4.75
Upgrade Walls to Skim Coat Plaster	+ .44
Upgrade Ceilings to Textured Finish	+ .49
Air Conditioning, in Heating Ductwork	+ 2.25
In Separate Ductwork	+ 4.92
Heating Systems, Hot Water	+ 2.04
Heat Pump	+ 2.72
Electric Heat	– .46
Not Heated	– 2.77

Additional upgrades or components

- Post and beam frame
- Log exterior walls
- Simple design from standard plans
- Single family — 1 full bath, 1 kitchen
- No basement
- Asphalt shingles on roof
- Hot air heat
- Gypsum wallborad interior finishes
- Materials and workmanship are average
- Detail specification on page 27

Note: The illustration shown may contain some optional components (for example: garages and/or fireplaces) whose costs are shown in the modifications, adjustments, & alternatives below or at the end of the square foot section.

Base cost per square foot of living area

Exterior Wall	Living Area										
	600	800	1000	1200	1400	1600	1800	2000	2400	2800	3200
6" Log - Solid Wall	165.30	149.95	138.30	129.05	121.30	116.10	113.35	109.95	103.30	98.40	95.00
8" Log - Solid Wall	155.25	140.95	130.10	121.70	114.65	109.80	107.30	104.25	98.10	93.60	90.55
Finished Basement, Add	37.00	35.75	34.10	32.60	31.35	30.55	30.10	29.50	28.55	27.85	27.20
Unfinished Basement, Add	15.15	13.70	12.65	11.70	11.00	10.45	10.20	9.80	9.25	8.80	8.45

Modifications

Add to the total cost

Upgrade Kitchen Cabinets	$ + 3955
Solid Surface Countertops (Included)	
Full Bath - including plumbing, wall and floor finishes	+ 7074
Half Bath - including plumbing, wall and floor finishes	+ 4047
One Car Attached Garage	+ 13,481
One Car Detached Garage	+ 17,772
Fireplace & Chimney	+ 5930

Adjustments

For multi family - add to total cost

Additional Kitchen	$ + 7917
Additional Bath	+ 7074
Additional Entry & Exit	+ 1576
Separate Heating	+ 1514
Separate Electric	+ 1958

*For Townhouse/Rowhouse -
Multiply cost per square foot by*

Inner Unit	+ .92
End Unit	+ .96

Alternatives

Add to or deduct from the cost per square foot of living area

Cedar Shake Roof	+ 2.90
Air Conditioning, in Heating Ductwork	+ 4.41
In Separate Ductwork	+ 6.57
Heating Systems, Hot Water	+ 2.27
Heat Pump	+ 2.23
Electric Heat	– .73
Not Heated	– 3.08

Additional upgrades or components

Kitchen Cabinets & Countertops	Page 58
Bathroom Vanities	59
Fireplaces & Chimneys	59
Windows, Skylights & Dormers	59
Appliances	60
Breezeways & Porches	60
Finished Attic	60
Garages	61
Site Improvements	61
Wings & Ells	37

- **Post and beam frame**
- **Log exterior walls**
- **Simple design from standard plans**
- **Single family — 1 full bath, 1 kitchen**
- **No basement**
- **Asphalt shingles on roof**
- **Hot air heat**
- **Gypsum wallboard interior finishes**
- **Materials and workmanship are average**
- **Detail specification on page 27**

Note: The illustration shown may contain some optional components (for example: garages and/or fireplaces) whose costs are shown in the modifications, adjustments, & alternatives below or at the end of the square foot section.

Base cost per square foot of living area

Exterior Wall	Living Area										
	1000	1200	1400	1600	1800	2000	2200	2600	3000	3400	3800
6" Log - Solid Wall	146.50	133.40	127.00	122.55	117.90	113.15	109.90	103.65	97.80	95.00	92.45
8" Log - Solid Wall	135.60	123.35	117.55	113.55	109.40	105.00	102.15	96.60	91.30	88.85	86.60
Finished Basement, Add	21.30	21.05	20.35	19.85	19.35	19.00	18.65	17.95	17.50	17.15	16.85
Unfinished Basement, Add	8.35	7.75	7.30	7.00	6.70	6.45	6.25	5.80	5.55	5.35	5.15

Modifications

Add to the total cost

Upgrade Kitchen Cabinets	$ + 3955
Solid Surface Countertops (Included)	
Full Bath - including plumbing, wall and floor finishes	+ 7074
Half Bath - including plumbing, wall and floor finishes	+ 4047
One Car Attached Garage	+ 13,481
One Car Detached Garage	+ 17,772
Fireplace & Chimney	+ 6550

Adjustments

For multi family - add to total cost

Additional Kitchen	$ + 7917
Additional Bath	+ 7074
Additional Entry & Exit	+ 1576
Separate Heating	+ 1514
Separate Electric	+ 1958

For Townhouse/Rowhouse - Multiply cost per square foot by

Inner Unit	+ .92
End Unit	+ .96

Alternatives

Add to or deduct from the cost per square foot of living area

Cedar Shake Roof	+ 1.45
Air Conditioning, in Heating Ductwork	+ 2.67
In Separate Ductwork	+ 5.22
Heating Systems, Hot Water	+ 2.12
Heat Pump	+ 2.61
Electric Heat	– .56
Not Heated	– 2.87

Additional upgrades or components

Kitchen Cabinets & Countertops	Page 58
Bathroom Vanities	59
Fireplaces & Chimneys	59
Windows, Skylights & Dormers	59
Appliances	60
Breezeways & Porches	60
Finished Attic	60
Garages	61
Site Improvements	61
Wings & Ells	37

1 Story — Base cost per square foot of living area

Exterior Wall	Living Area							
	50	100	200	300	400	500	600	700
Wood Siding - Wood Frame	196.75	151.15	131.30	110.50	104.00	100.05	97.50	98.35
Brick Veneer - Wood Frame	199.75	150.10	128.50	104.80	97.75	93.40	90.70	91.40
Stucco on Wood Frame	184.30	142.25	123.85	105.40	99.35	95.70	93.40	94.40
Solid Masonry	250.45	189.50	163.25	131.75	123.10	117.90	116.10	116.10
Finished Basement, Add	59.05	48.95	44.00	35.95	34.25	33.30	32.65	32.20
Unfinished Basement, Add	27.55	20.85	18.00	13.35	12.40	11.85	11.50	11.20

1-1/2 Story — Base cost per square foot of living area

Exterior Wall	Living Area							
	100	200	300	400	500	600	700	800
Wood Siding - Wood Frame	158.80	127.70	109.00	98.05	92.35	89.65	86.25	85.20
Brick Veneer - Wood Frame	209.60	157.00	130.20	114.15	106.10	101.75	97.10	95.35
Stucco on Wood Frame	185.95	138.00	114.40	101.85	94.75	91.05	87.00	85.30
Solid Masonry	244.70	185.05	153.60	132.45	122.95	117.70	112.15	110.30
Finished Basement, Add	39.85	36.05	32.80	29.20	28.25	27.60	26.95	26.90
Unfinished Basement, Add	17.30	14.50	12.60	10.55	10.00	9.65	9.25	9.25

2 Story — Base cost per square foot of living area

Exterior Wall	Living Area							
	100	200	400	600	800	1000	1200	1400
Wood Siding - Wood Frame	159.20	119.00	101.40	84.45	78.60	75.20	72.90	74.05
Brick Veneer - Wood Frame	213.50	149.30	119.95	96.80	88.75	84.00	80.80	81.30
Stucco on Wood Frame	187.00	130.35	104.15	86.25	79.30	75.10	72.40	73.20
Solid Masonry	252.80	177.35	143.35	112.40	102.75	97.10	93.25	93.30
Finished Basement, Add	31.40	26.40	24.00	19.95	19.15	18.65	18.30	18.05
Unfinished Basement, Add	13.95	10.60	9.15	6.80	6.40	6.10	5.90	5.75

Base costs do not include bathroom or kitchen facilities. Use Modifications/Adjustments/Alternatives on pages 58-61 where appropriate.

Important: See the Reference Section for Location Factors (to adjust for your city) and Estimating Forms.

1 Story

1-1/2 Story

2 Story

2-1/2 Story

Bi-Level

Tri-Level

	Components
1 Site Work	Site preparation for slab or excavation for lower level; 4' deep trench excavation for foundation wall.
2 Foundations	Continuous reinforced concrete footing, 8" deep x 18" wide; dampproofed and insulated 8" thick reinforced concrete foundation wall, 4' deep; 4" concrete slab on 4" crushed stone base and polyethylene vapor barrier, trowel finish.
3 Framing	Exterior walls—2 x 6 wood studs, 16" O.C.; 1/2" plywood sheathing; 2 x 8 rafters 16" O.C. with 1/2" plywood sheathing, 4 in 12, 6 in 12 or 8 in 12 roof pitch, 2 x 6 or 2 x 8 ceiling joists; 1/2" plywood subfloor on 1 x 3 wood sleepers 16" O.C.; 2 x 10 floor joists with 5/8" plywood subfloor on models with more than one level.
4 Exterior Walls	Horizontal beveled wood siding and #15 felt building paper on insulated wood frame walls; double hung windows; 3 flush solid core wood exterior doors with storms. **Alternates:** • Brick veneer on wood frame, has 4" veneer high quality face brick or select common brick. • Stone veneer on wood frame has exterior veneer of field stone or 2" thick limestone. • Solid masonry has an 8" concrete block wall with insulation and a brick or stone facing. It may be a solid brick or stone structure.
5 Roofing	30 year asphalt shingles; #15 felt building paper; aluminum gutters, downspouts and drip edge; copper flashings.
6 Interiors	Walls and ceilings—5/8" drywall, skim coat plaster, primed and painted with 2 coats; hardwood baseboard and trim; sanded and finished, hardwood floor 70%, ceramic tile with 1/2" underlayment 20%, vinyl tile with 1/2" underlayment 10%; wood panel interior doors, primed and painted with 2 coats.
7 Specialties	Custom grade kitchen cabinets—20 L.F. wall and base with plastic laminate counter top and kitchen sink; 4 L.F. bathroom vanity; 75 gallon electric water heater; medicine cabinet.
8 Mechanical	Gas fired warm air heat/air conditioning; one full bath including bathtub, corner shower, built in lavatory and water closet; one 1/2 bath including built in lavatory and water closet.
9 Electrical	100 Amp. service; romex wiring; incandescent lighting fixtures, switches, receptacles.
10 Overhead and Profit	General Contractor overhead and profit.

Adjustments

Unfinished Basement:
7'-6" high cast-in-place concrete walls 10" thick or 12" concrete block.

Finished Basement:
Includes painted drywall on 2 x 4 wood furring with insulation, suspended ceiling, carpeting on concrete floor, heating and lighting.

- A distinct residence from designer's plans
- Single family — 1 full bath, 1 half bath, 1 kitchen
- No basement
- Asphalt shingles on roof
- Forced hot air heat/air conditioning
- Gypsum wallboard interior finishes
- Materials and workmanship are above average
- Detail specifications on page 39

Note: The illustration shown may contain some optional components (for example: garages and/or fireplaces) whose costs are shown in the modifications, adjustments, & alternatives below or at the end of the square foot section.

©Design Basics, Inc.

Base cost per square foot of living area

Exterior Wall	Living Area										
	800	1000	1200	1400	1600	1800	2000	2400	2800	3200	3600
Wood Siding - Wood Frame	176.90	159.75	146.40	135.60	128.25	124.00	119.30	110.35	103.80	99.25	94.50
Brick Veneer - Wood Frame	197.35	179.35	165.30	153.80	146.05	141.65	136.55	127.10	120.25	115.35	110.30
Stone Veneer - Wood Frame	203.90	185.30	170.65	158.70	150.65	146.05	140.70	130.85	123.75	118.55	113.30
Solid Masonry	209.30	190.25	175.05	162.70	154.45	149.65	144.10	133.95	126.60	121.30	115.80
Finished Basement, Add	55.35	55.25	52.90	51.05	49.80	49.15	48.10	46.75	45.65	44.65	43.85
Unfinished Basement, Add	23.30	21.95	20.75	19.75	19.10	18.80	18.25	17.50	16.95	16.45	16.05

Modifications

Add to the total cost

Upgrade Kitchen Cabinets	$ + 1240
Solid Surface Countertops (Included)	
Full Bath - including plumbing, wall and floor finishes	+ 8489
Half Bath - including plumbing, wall and floor finishes	+ 4857
Two Car Attached Garage	+ 25,983
Two Car Detached Garage	+ 29,594
Fireplace & Chimney	+ 6200

Adjustments

For multi family - add to total cost

Additional Kitchen	$ + 15,816
Additional Full Bath & Half Bath	+ 13,346
Additional Entry & Exit	+ 1576
Separate Heating & Air Conditioning	+ 6833
Separate Electric	+ 1958

For Townhouse/Rowhouse -
Multiply cost per square foot by

Inner Unit	+ .90
End Unit	+ .95

Alternatives

Add to or deduct from the cost per square foot of living area

Cedar Shake Roof	+ 1.95
Clay Tile Roof	+ 3.85
Slate Roof	+ 5.60
Upgrade Ceilings to Textured Finish	+ .49
Air Conditioning, in Heating Ductwork	Base System
Heating Systems, Hot Water	+ 2.33
Heat Pump	+ 2.22
Electric Heat	– 2.08
Not Heated	– 3.86

Additional upgrades or components

Kitchen Cabinets & Countertops	Page 58
Bathroom Vanities	59
Fireplaces & Chimneys	59
Windows, Skylights & Dormers	59
Appliances	60
Breezeways & Porches	60
Finished Attic	60
Garages	61
Site Improvements	61
Wings & Ells	47

- **A distinct residence from designer's plans**
- **Single family — 1 full bath, 1 half bath, 1 kitchen**
- **No basement**
- **Asphalt shingles on roof**
- **Forced hot air heat/air conditioning**
- **Gypsum wallboard interior finishes**
- **Materials and workmanship are above average**
- **Detail specifications on page 39**

Note: The illustration shown may contain some optional components (for example: garages and/or fireplaces) whose costs are shown in the modifications, adjustments, & alternatives below or at the end of the square foot section.

●Donald A. Gardner Architects, Inc.

Base cost per square foot of living area

Exterior Wall	Living Area										
	1000	1200	1400	1600	1800	2000	2400	2800	3200	3600	4000
Wood Siding - Wood Frame	159.05	147.75	139.75	130.05	124.45	119.15	108.85	104.15	99.80	96.55	91.80
Brick Veneer - Wood Frame	168.10	156.20	147.75	137.35	131.40	125.65	114.65	109.70	104.85	101.45	96.40
Stone Veneer - Wood Frame	175.00	162.65	153.95	142.95	136.70	130.75	119.10	113.90	108.75	105.20	99.85
Solid Masonry	180.65	168.00	159.00	147.55	141.10	134.80	122.80	117.40	112.00	108.25	102.75
Finished Basement, Add	36.70	36.95	35.95	34.55	33.75	33.00	31.50	30.85	30.00	29.65	29.05
Unfinished Basement, Add	15.65	14.95	14.45	13.75	13.30	12.95	12.15	11.85	11.40	11.25	10.90

Modifications

Add to the total cost

Upgrade Kitchen Cabinets	$ + 1240
Solid Surface Countertops (Included)	
Full Bath - including plumbing, wall and floor finishes	+ 8489
Half Bath - including plumbing, wall and floor finishes	+ 4857
Two Car Attached Garage	+ 25,983
Two Car Detached Garage	+ 29,594
Fireplace & Chimney	+ 6200

Adjustments

For multi family - add to total cost

Additional Kitchen	$ + 15,816
Additional Full Bath & Half Bath	+ 13,346
Additional Entry & Exit	+ 1576
Separate Heating & Air Conditioning	+ 6833
Separate Electric	+ 1958

*For Townhouse/Rowhouse -
Multiply cost per square foot by*

Inner Unit	+ .90
End Unit	+ .95

Alternatives

Add to or deduct from the cost per square foot of living area

Cedar Shake Roof	+ 1.40
Clay Tile Roof	+ 2.75
Slate Roof	+ 4.05
Upgrade Ceilings to Textured Finish	+ .49
Air Conditioning, in Heating Ductwork	Base System
Heating Systems, Hot Water	+ 2.22
Heat Pump	+ 2.32
Electric Heat	– 1.84
Not Heated	– 3.56

Additional upgrades or components

Kitchen Cabinets & Countertops	Page 58
Bathroom Vanities	59
Fireplaces & Chimneys	59
Windows, Skylights & Dormers	59
Appliances	60
Breezeways & Porches	60
Finished Attic	60
Garages	61
Site Improvements	61
Wings & Ells	47

- A distinct residence from designer's plans
- Single family — 1 full bath, 1 half bath, 1 kitchen
- No basement
- Asphalt shingles on roof
- Forced hot air heat/air conditioning
- Gypsum wallboard interior finishes
- Materials and workmanship are above average
- Detail specifications on page 39

Note: The illustration shown may contain some optional components (for example: garages and/or fireplaces) whose costs are shown in the modifications, adjustments, & alternatives below or at the end of the square foot section.

Base cost per square foot of living area

Exterior Wall	Living Area										
	1200	1400	1600	1800	2000	2400	2800	3200	3600	4000	4400
Wood Siding - Wood Frame	148.80	139.90	133.40	127.65	121.65	112.80	105.35	100.50	97.60	94.50	91.90
Brick Veneer - Wood Frame	158.45	148.85	142.00	135.80	129.40	119.85	111.80	106.50	103.40	99.90	97.20
Stone Veneer - Wood Frame	165.85	155.75	148.55	141.95	135.40	125.25	116.65	111.10	107.80	104.10	101.20
Solid Masonry	171.90	161.40	153.95	147.05	140.35	129.65	120.70	114.90	111.45	107.50	104.50
Finished Basement, Add	29.45	29.70	28.95	28.15	27.65	26.50	25.50	24.90	24.55	24.10	23.80
Unfinished Basement, Add	12.60	12.00	11.60	11.25	11.00	10.40	9.90	9.60	9.40	9.15	9.00

Modifications

Add to the total cost

Upgrade Kitchen Cabinets	$ + 1240
Solid Surface Countertops (Included)	
Full Bath - including plumbing, wall and floor finishes	+ 8489
Half Bath - including plumbing, wall and floor finishes	+ 4857
Two Car Attached Garage	+ 25,983
Two Car Detached Garage	+ 29,594
Fireplace & Chimney	+ 6962

Adjustments

For multi family - add to total cost

Additional Kitchen	$ + 15,816
Additional Full Bath & Half Bath	+ 13,346
Additional Entry & Exit	+ 1576
Separate Heating & Air Conditioning	+ 6833
Separate Electric	+ 1958

For Townhouse/Rowhouse -
Multiply cost per square foot by

Inner Unit	+ .87
End Unit	+ .93

Alternatives

Add to or deduct from the cost per square foot of living area

Cedar Shake Roof	+ .95
Clay Tile Roof	+ 1.95
Slate Roof	+ 2.80
Upgrade Ceilings to Textured Finish	+ .49
Air Conditioning, in Heating Ductwork	Base System
Heating Systems, Hot Water	+ 2.16
Heat Pump	+ 2.59
Electric Heat	– 1.84
Not Heated	– 3.36

Additional upgrades or components

Kitchen Cabinets & Countertops	Page 58
Bathroom Vanities	59
Fireplaces & Chimneys	59
Windows, Skylights & Dormers	59
Appliances	60
Breezeways & Porches	60
Finished Attic	60
Garages	61
Site Improvements	61
Wings & Ells	47

- **A distinct residence from designer's plans**
- **Single family — 1 full bath, 1 half bath, 1 kitchen**
- **No basement**
- **Asphalt shingles on roof**
- **Forced hot air heat/air conditioning**
- **Gypsum wallboard interior finishes**
- **Materials and workmanship are above average**
- **Detail specifications on page 39**

Note: The illustration shown may contain some optional components (for example: garages and/or fireplaces) whose costs are shown in the modifications, adjustments, & alternatives below or at the end of the square foot section.

Base cost per square foot of living area

Exterior Wall	Living Area										
	1500	1800	2100	2400	2800	3200	3600	4000	4500	5000	5500
Wood Siding - Wood Frame	146.00	131.35	122.65	117.40	110.60	104.30	100.70	95.15	92.05	89.45	86.80
Brick Veneer - Wood Frame	156.05	140.60	130.95	125.25	118.10	111.05	107.15	101.15	97.75	94.85	91.95
Stone Veneer - Wood Frame	163.65	147.65	137.25	131.25	123.85	116.20	112.05	105.70	102.10	98.95	95.80
Solid Masonry	170.00	153.45	142.50	136.20	128.50	120.40	116.05	109.45	105.65	102.35	99.10
Finished Basement, Add	23.40	23.40	22.20	21.60	21.15	20.30	19.90	19.35	18.90	18.65	18.30
Unfinished Basement, Add	10.10	9.60	8.95	8.65	8.40	7.95	7.80	7.50	7.25	7.10	6.95

Modifications

Add to the total cost

Upgrade Kitchen Cabinets	$ + 1240
Solid Surface Countertops (Included)	
Full Bath - including plumbing, wall and floor finishes	+ 8489
Half Bath - including plumbing, wall and floor finishes	+ 4857
Two Car Attached Garage	+ 25,983
Two Car Detached Garage	+ 29,594
Fireplace & Chimney	+ 7863

Adjustments

For multi family - add to total cost

Additional Kitchen	$ + 15,816
Additional Full Bath & Half Bath	+ 13,346
Additional Entry & Exit	+ 1576
Separate Heating & Air Conditioning	+ 6833
Separate Electric	+ 1958

For Townhouse/Rowhouse -
Multiply cost per square foot by

Inner Unit	+ .87
End Unit	+ .94

Alternatives

Add to or deduct from the cost per square foot of living area

Cedar Shake Roof	+ .85
Clay Tile Roof	+ 1.65
Slate Roof	+ 2.45
Upgrade Ceilings to Textured Finish	+ .49
Air Conditioning, in Heating Ductwork	Base System
Heating Systems, Hot Water	+ 1.95
Heat Pump	+ 2.69
Electric Heat	– 3.22
Not Heated	– 3.36

Additional upgrades or components

Kitchen Cabinets & Countertops	Page 58
Bathroom Vanities	59
Fireplaces & Chimneys	59
Windows, Skylights & Dormers	59
Appliances	60
Breezeways & Porches	60
Finished Attic	60
Garages	61
Site Improvements	61
Wings & Ells	47

- **A distinct residence from designer's plans**
- **Single family — 1 full bath, 1 half bath, 1 kitchen**
- **No basement**
- **Asphalt shingles on roof**
- **Forced hot air heat/air conditioning**
- **Gypsum wallboard interior finishes**
- **Materials and workmanship are above average**
- **Detail specifications on page 39**

Note: The illustration shown may contain some optional components (for example: garages and/or fireplaces) whose costs are shown in the modifications, adjustments, & alternatives below or at the end of the square foot section.

Base cost per square foot of living area

Exterior Wall	Living Area										
	1500	1800	2100	2500	3000	3500	4000	4500	5000	5500	6000
Wood Siding - Wood Frame	145.50	130.95	123.90	118.00	108.80	104.25	98.75	92.90	90.85	88.55	86.25
Brick Veneer - Wood Frame	155.90	140.60	132.90	126.60	116.60	111.70	105.45	99.10	96.80	94.30	91.65
Stone Veneer - Wood Frame	163.90	147.90	139.75	133.20	122.60	117.35	110.55	103.85	101.40	98.70	95.85
Solid Masonry	170.40	154.00	145.45	138.55	127.50	122.00	114.85	107.75	105.20	102.35	99.25
Finished Basement, Add	20.55	20.50	19.80	19.25	18.40	17.95	17.25	16.75	16.55	16.25	16.00
Unfinished Basement, Add	8.85	8.35	7.95	7.75	7.30	7.05	6.75	6.50	6.35	6.20	6.10

Modifications

Add to the total cost

Upgrade Kitchen Cabinets	$ + 1240
Solid Surface Countertops (Included)	
Full Bath - including plumbing, wall and floor finishes	+ 8489
Half Bath - including plumbing, wall and floor finishes	+ 4857
Two Car Attached Garage	+ 25,983
Two Car Detached Garage	+ 29,594
Fireplace & Chimney	+ 7863

Adjustments

For multi family - add to total cost

Additional Kitchen	$ + 15,816
Additional Full Bath & Half Bath	+ 13,346
Additional Entry & Exit	+ 1576
Separate Heating & Air Conditioning	+ 6833
Separate Electric	+ 1958

For Townhouse/Rowhouse -
Multiply cost per square foot by

Inner Unit	+ .85
End Unit	+ .93

Alternatives

Add to or deduct from the cost per square foot of living area

Cedar Shake Roof	+ .65
Clay Tile Roof	+ 1.30
Slate Roof	+ 1.85
Upgrade Ceilings to Textured Finish	+ .49
Air Conditioning, in Heating Ductwork	Base System
Heating Systems, Hot Water	+ 1.95
Heat Pump	+ 2.69
Electric Heat	− 3.22
Not Heated	− 3.26

Additional upgrades or components

Kitchen Cabinets & Countertops	Page 58
Bathroom Vanities	59
Fireplaces & Chimneys	59
Windows, Skylights & Dormers	59
Appliances	60
Breezeways & Porches	60
Finished Attic	60
Garages	61
Site Improvements	61
Wings & Ells	47

- **A distinct residence from designer's plans**
- **Single family — 1 full bath, 1 half bath, 1 kitchen**
- **No basement**
- **Asphalt shingles on roof**
- **Forced hot air heat/air conditioning**
- **Gypsum wallboard interior finishes**
- **Materials and workmanship are above average**
- **Detail specifications on page 39**

Note: The illustration shown may contain some optional components (for example: garages and/or fireplaces) whose costs are shown in the modifications, adjustments, & alternatives below or at the end of the square foot section.

Base cost per square foot of living area

Exterior Wall	Living Area										
	1200	1400	1600	1800	2000	2400	2800	3200	3600	4000	4400
Wood Siding - Wood Frame	140.65	132.15	126.00	120.80	114.95	106.80	99.85	95.35	92.65	89.85	87.30
Brick Veneer - Wood Frame	147.85	138.85	132.50	126.85	120.85	112.10	104.65	99.90	97.05	93.95	91.30
Stone Veneer - Wood Frame	153.35	144.00	137.45	131.50	125.30	116.10	108.35	103.25	100.30	97.05	94.30
Solid Masonry	157.90	148.30	141.45	135.35	129.00	119.45	111.30	106.10	103.05	99.55	96.75
Finished Basement, Add	29.45	29.70	28.95	28.15	27.65	26.50	25.50	24.90	24.55	24.10	23.80
Unfinished Basement, Add	12.60	12.00	11.60	11.25	11.00	10.40	9.90	9.60	9.40	9.15	9.00

Modifications

Add to the total cost

Upgrade Kitchen Cabinets	$ + 1240
Solid Surface Countertops (Included)	
Full Bath - including plumbing, wall and floor finishes	+ 8489
Half Bath - including plumbing, wall and floor finishes	+ 4857
Two Car Attached Garage	+ 25,983
Two Car Detached Garage	+ 29,594
Fireplace & Chimney	+ 6200

Adjustments

For multi family - add to total cost

Additional Kitchen	$ + 15,816
Additional Full Bath & Half Bath	+ 13,346
Additional Entry & Exit	+ 1576
Separate Heating & Air Conditioning	+ 6833
Separate Electric	+ 1958

For Townhouse/Rowhouse -
Multiply cost per square foot by

Inner Unit	+ .89
End Unit	+ .95

Alternatives

Add to or deduct from the cost per square foot of living area

Cedar Shake Roof	+ .95
Clay Tile Roof	+ 1.95
Slate Roof	+ 2.80
Upgrade Ceilings to Textured Finish	+ .49
Air Conditioning, in Heating Ductwork	Base System
Heating Systems, Hot Water	+ 2.16
Heat Pump	+ 2.59
Electric Heat	– 1.84
Not Heated	– 3.26

Additional upgrades or components

Kitchen Cabinets & Countertops	Page 58
Bathroom Vanities	59
Fireplaces & Chimneys	59
Windows, Skylights & Dormers	59
Appliances	60
Breezeways & Porches	60
Finished Attic	60
Garages	61
Site Improvements	61
Wings & Ells	47

Important: See the Reference Section for Location Factors (to adjust for your city) and Estimating Forms.

- **A distinct residence from designer's plans**
- **Single family — 1 full bath, 1 half bath, 1 kitchen**
- **No basement**
- **Asphalt shingles on roof**
- **Forced hot air heat/air conditioning**
- **Gypsum wallboard interior finishes**
- **Materials and workmanship are above average**
- **Detail specifications on page 39**

Note: The illustration shown may contain some optional components (for example: garages and/or fireplaces) whose costs are shown in the modifications, adjustments, & alternatives below or at the end of the square foot section.

©Design Basics, Inc.

Base cost per square foot of living area

Exterior Wall	Living Area										
	1200	1500	1800	2100	2400	2800	3200	3600	4000	4500	5000
Wood Siding - Wood Frame	144.65	131.10	121.05	112.70	107.10	103.00	98.25	93.15	90.95	86.05	83.50
Brick Veneer - Wood Frame	151.85	137.65	126.95	118.05	112.05	107.80	102.70	97.30	94.95	89.75	86.95
Stone Veneer - Wood Frame	157.35	142.60	131.45	122.10	115.90	111.45	106.10	100.45	98.05	92.60	89.65
Solid Masonry	161.80	146.65	135.10	125.45	119.05	114.50	108.85	103.05	100.55	94.90	91.80
Finished Basement, Add*	36.95	36.85	35.30	34.05	33.25	32.65	31.90	31.15	30.85	30.15	29.70
Unfinished Basement, Add*	15.55	14.65	13.85	13.15	12.80	12.50	12.10	11.70	11.55	11.20	10.95

*Basement under middle level only.

Modifications

Add to the total cost

Upgrade Kitchen Cabinets	$ + 1240
Solid Surface Countertops (Included)	
Full Bath - including plumbing, wall and floor finishes	+ 8489
Half Bath - including plumbing, wall and floor finishes	+ 4857
Two Car Attached Garage	+ 25,983
Two Car Detached Garage	+ 29,594
Fireplace & Chimney	+ 6200

Adjustments

For multi family - add to total cost

Additional Kitchen	$ + 15,816
Additional Full Bath & Half Bath	+ 13,346
Additional Entry & Exit	+ 1576
Separate Heating & Air Conditioning	+ 6833
Separate Electric	+ 1958

For Townhouse/Rowhouse -
Multiply cost per square foot by

Inner Unit	+ .87
End Unit	+ .94

Alternatives

Add to or deduct from the cost per square foot of living area

Cedar Shake Roof	+ 1.40
Clay Tile Roof	+ 2.75
Slate Roof	+ 4.05
Upgrade Ceilings to Textured Finish	+ .49
Air Conditioning, in Heating Ductwork	Base System
Heating Systems, Hot Water	+ 2.09
Heat Pump	+ 2.71
Electric Heat	– 1.63
Not Heated	– 3.26

Additional upgrades or components

Kitchen Cabinets & Countertops	Page 58
Bathroom Vanities	59
Fireplaces & Chimneys	59
Windows, Skylights & Dormers	59
Appliances	60
Breezeways & Porches	60
Finished Attic	60
Garages	61
Site Improvements	61
Wings & Ells	47

1 Story — Base cost per square foot of living area

Exterior Wall	Living Area							
	50	100	200	300	400	500	600	700
Wood Siding - Wood Frame	232.00	181.15	159.00	135.65	128.40	124.05	121.15	122.15
Brick Veneer - Wood Frame	256.30	198.50	173.50	145.25	137.10	132.15	128.85	129.60
Stone Veneer - Wood Frame	274.85	211.75	184.55	152.60	143.70	138.30	134.70	135.25
Solid Masonry	290.10	222.70	193.65	158.70	149.15	143.40	139.60	139.95
Finished Basement, Add	88.25	75.50	68.45	56.80	54.50	53.10	52.15	51.45
Unfinished Basement, Add	64.90	44.55	35.20	26.95	24.80	23.45	22.55	21.90

1-1/2 Story — Base cost per square foot of living area

Exterior Wall	Living Area							
	100	200	300	400	500	600	700	800
Wood Siding - Wood Frame	186.35	153.50	133.60	121.85	115.80	112.95	109.25	108.15
Brick Veneer - Wood Frame	208.05	170.85	148.10	133.10	126.20	122.75	118.55	117.40
Stone Veneer - Wood Frame	224.65	184.15	159.10	141.70	134.20	130.30	125.65	124.45
Solid Masonry	238.25	195.00	168.20	148.85	140.70	136.50	131.50	130.20
Finished Basement, Add	59.10	54.75	50.15	44.95	43.60	42.60	41.80	41.65
Unfinished Basement, Add	38.65	29.30	24.85	21.15	19.80	18.90	18.15	17.90

2 Story — Base cost per square foot of living area

Exterior Wall	Living Area							
	100	200	400	600	800	1000	1200	1400
Wood Siding - Wood Frame	185.95	142.90	124.05	106.00	99.90	96.20	93.70	95.00
Brick Veneer - Wood Frame	210.25	160.20	138.55	115.65	108.55	104.30	101.45	102.45
Stone Veneer - Wood Frame	228.75	173.50	149.55	123.00	115.20	110.45	107.35	108.15
Solid Masonry	244.05	184.40	158.65	129.10	120.65	115.60	112.20	112.75
Finished Basement, Add	44.15	37.80	34.30	28.45	27.25	26.60	26.10	25.80
Unfinished Basement, Add	32.45	22.30	17.65	13.50	12.40	11.75	11.30	10.95

Base costs do not include bathroom or kitchen facilities. Use Modifications/Adjustments/Alternatives on pages 58-61 where appropriate.

1 Story

1-1/2 Story

2 Story

2-1/2 Story

Bi-Level

Tri-Level

	Components
1 Site Work	Site preparation for slab or excavation for lower level; 4' deep trench excavation for foundation wall.
2 Foundations	Continuous reinforced concrete footing, 8" deep x 18" wide; dampproofed and insulated 12" thick reinforced concrete foundation wall, 4' deep; 4" concrete slab on 4" crushed stone base and polyethylene vapor barrier, trowel finish.
3 Framing	Exterior walls—2 x 6 wood studs, 16" O.C.; 1/2" plywood sheathing; 2 x 8 rafters 16" O.C. with 1/2" plywood sheathing, 4 in 12, 6 in 12 or 8 in 12 roof pitch, 2 x 6 or 2 x 8 ceiling joists; 1/2" plywood subfloor on 1 x 3 wood sleepers 16" O.C.; 2 x 10 floor joists with 5/8" plywood subfloor on models with more than one level.
4 Exterior Walls	Face brick veneer and #15 felt building paper on insulated wood frame walls; double hung windows; 3 flush solid core wood exterior doors with storms. **Alternates:** • Wood siding on wood frame, has top quality cedar or redwood siding or hand split cedar shingles or shakes. • Solid brick may have solid brick exterior wall or brick on concrete block. • Solid stone concrete block with selected fieldstone or limestone exterior.
5 Roofing	Red cedar shingles; #15 felt building paper; aluminum gutters, downspouts and drip edge; copper flashings.
6 Interiors	Walls and ceilings—5/8" drywall, skim coat plaster, primed and painted with 2 coats; hardwood baseboard and trim; sanded and finished, hardwood floor 70%, ceramic tile with 1/2" underlayment 20%, vinyl tile with 1/2" underlayment 10%; wood panel interior doors, primed and painted with 2 coats.
7 Specialties	Luxury grade kitchen cabinets—25 L.F. wall and base with plastic laminate counter top and kitchen sink; 6 L.F. bathroom vanity; 75 gallon electric water heater; medicine cabinet.
8 Mechanical	Gas fired warm air heat/air conditioning; one full bath including bathtub, corner shower, built in lavatory and water closet; one 1/2 bath including built in lavatory and water closet.
9 Electrical	100 Amp. service; romex wiring; incandescent lighting fixtures, switches, receptacles.
10 Overhead and Profit	General Contractor overhead and profit.

Adjustments

Unfinished Basement:
8" high cast-in-place concrete walls 12" thick or 12" concrete block.

Finished Basement:
Includes painted drywall on 2 x 4 wood furring with insulation, suspended ceiling, carpeting on subfloor with sleepers, heating and lighting.

- Unique residence built from an architect's plan
- Single family — 1 full bath, 1 half bath, 1 kitchen
- No basement
- Cedar shakes on roof
- Forced hot air heat/air conditioning
- Gypsum wallboard interior finishes
- Many special features
- Extraordinary materials and workmanship
- Detail specifications on page 49

©Home Planners, Inc.

Note: The illustration shown may contain some optional components (for example: garages and/or fireplaces) whose costs are shown in the modifications, adjustments, & alternatives below or at the end of the square foot section.

Base cost per square foot of living area

Exterior Wall	Living Area										
	1000	1200	1400	1600	1800	2000	2400	2800	3200	3600	4000
Wood Siding - Wood Frame	202.00	186.15	173.45	164.75	159.70	154.10	143.65	136.10	130.75	125.25	120.85
Brick Veneer - Wood Frame	210.95	194.20	180.85	171.65	166.30	160.35	149.35	141.35	135.65	129.85	125.10
Solid Brick	226.00	207.80	193.15	183.20	177.45	170.85	158.85	150.15	143.85	137.40	132.25
Solid Stone	229.00	210.50	195.60	185.60	179.75	172.95	160.80	151.95	145.45	138.90	133.65
Finished Basement, Add	53.15	57.20	54.85	53.30	52.40	51.25	49.45	48.10	46.95	45.90	45.05
Unfinished Basement, Add	23.55	22.05	20.90	20.10	19.65	19.05	18.20	17.50	16.90	16.35	15.90

Modifications

Add to the total cost

Upgrade Kitchen Cabinets	$ + 1637
Solid Surface Countertops (Included)	
Full Bath - including plumbing, wall and floor finishes	+ 10,186
Half Bath - including plumbing, wall and floor finishes	+ 5828
Two Car Attached Garage	+ 30,226
Two Car Detached Garage	+ 34,182
Fireplace & Chimney	+ 8760

Adjustments

For multi family - add to total cost

Additional Kitchen	$ + 20,200
Additional Full Bath & Half Bath	+ 16,014
Additional Entry & Exit	+ 2306
Separate Heating & Air Conditioning	+ 6833
Separate Electric	+ 1958

**For Townhouse/Rowhouse -
Multiply cost per square foot by**

Inner Unit	+ .90
End Unit	+ .95

Alternatives

Add to or deduct from the cost per square foot of living area

Heavyweight Asphalt Shingles	– 1.95
Clay Tile Roof	+ 1.95
Slate Roof	+ 3.65
Upgrade Ceilings to Textured Finish	+ .49
Air Conditioning, in Heating Ductwork	Base System
Heating Systems, Hot Water	+ 2.51
Heat Pump	+ 2.40
Electric Heat	– 1.84
Not Heated	– 4.21

Additional upgrades or components

Kitchen Cabinets & Countertops	Page 58
Bathroom Vanities	59
Fireplaces & Chimneys	59
Windows, Skylights & Dormers	59
Appliances	60
Breezeways & Porches	60
Finished Attic	60
Garages	61
Site Improvements	61
Wings & Ells	57

- Unique residence built from an architect's plan
- Single family — 1 full bath, 1 half bath, 1 kitchen
- No basement
- Cedar shakes on roof
- Forced hot air heat/air conditioning
- Gypsum wallboard interior finishes
- Many special features
- Extraordinary materials and workmanship
- Detail specifications on page 49

Note: The illustration shown may contain some optional components (for example: garages and/or fireplaces) whose costs are shown in the modifications, adjustments, & alternatives below or at the end of the square foot section.

©Larry E. Belk Designs

Base cost per square foot of living area

Exterior Wall	Living Area										
	1000	1200	1400	1600	1800	2000	2400	2800	3200	3600	4000
Wood Siding - Wood Frame	187.55	173.95	164.20	152.85	146.20	139.85	127.85	122.40	117.15	113.25	107.95
Brick Veneer - Wood Frame	197.90	183.65	173.50	161.20	154.15	147.45	134.60	128.65	122.95	118.90	113.15
Solid Brick	215.30	199.90	188.95	175.30	167.55	160.10	145.80	139.30	132.75	128.35	121.95
Solid Stone	218.75	203.20	192.10	178.15	170.25	162.65	148.05	141.45	134.80	130.25	123.75
Finished Basement, Add	37.40	40.55	39.40	37.55	36.55	35.75	33.85	33.05	32.05	31.55	30.75
Unfinished Basement, Add	17.05	16.20	15.60	14.70	14.20	13.75	12.85	12.40	11.90	11.70	11.25

Modifications

Add to the total cost

Upgrade Kitchen Cabinets	$ + 1637
Solid Surface Countertops (Included)	
Full Bath - including plumbing, wall and floor finishes	+ 10,186
Half Bath - including plumbing, wall and floor finishes	+ 5828
Two Car Attached Garage	+ 30,226
Two Car Detached Garage	+ 34,182
Fireplace & Chimney	+ 8760

Adjustments

For multi family - add to total cost

Additional Kitchen	$ + 20,200
Additional Full Bath & Half Bath	+ 16,014
Additional Entry & Exit	+ 2306
Separate Heating & Air Conditioning	+ 6833
Separate Electric	+ 1958

**For Townhouse/Rowhouse -
Multiply cost per square foot by**

Inner Unit	+ .90
End Unit	+ .95

Alternatives

Add to or deduct from the cost per square foot of living area

Heavyweight Asphalt Shingles	– 1.40
Clay Tile Roof	+ 1.40
Slate Roof	+ 2.65
Upgrade Ceilings to Textured Finish	+ .49
Air Conditioning, in Heating Ductwork	Base System
Heating Systems, Hot Water	+ 2.40
Heat Pump	+ 2.67
Electric Heat	– 1.84
Not Heated	– 3.88

Additional upgrades or components

Kitchen Cabinets & Countertops	Page 58
Bathroom Vanities	59
Fireplaces & Chimneys	59
Windows, Skylights & Dormers	59
Appliances	60
Breezeways & Porches	60
Finished Attic	60
Garages	61
Site Improvements	61
Wings & Ells	57

Important: See the Reference Section for Location Factors (to adjust for your city) and Estimating Forms.

51

- **Unique residence built from an architect's plan**
- **Single family — 1 full bath, 1 half bath, 1 kitchen**
- **No basement**
- **Cedar shakes on roof**
- **Forced hot air heat/air conditioning**
- **Gypsum wallboard interior finishes**
- **Many special features**
- **Extraordinary materials and workmanship**
- **Detail specifications on page 49**

Note: The illustration shown may contain some optional components (for example: garages and/or fireplaces) whose costs are shown in the modifications, adjustments, & alternatives below or at the end of the square foot section.

Base cost per square foot of living area

Exterior Wall	Living Area										
	1200	1400	1600	1800	2000	2400	2800	3200	3600	4000	4400
Wood Siding - Wood Frame	173.35	162.70	155.00	148.20	141.15	130.85	122.20	116.55	113.20	109.60	106.50
Brick Veneer - Wood Frame	184.40	173.05	164.85	157.50	150.10	138.95	129.60	123.40	119.80	115.85	112.55
Solid Brick	202.90	190.45	181.40	173.05	165.15	152.50	141.90	134.95	130.95	126.30	122.70
Solid Stone	206.65	193.90	184.75	176.25	168.15	155.20	144.40	137.35	133.20	128.40	124.75
Finished Basement, Add	30.00	32.60	31.70	30.60	30.05	28.55	27.40	26.65	26.20	25.55	25.25
Unfinished Basement, Add	13.70	13.00	12.55	12.10	11.80	11.00	10.45	10.10	9.85	9.50	9.40

Modifications

Add to the total cost
Upgrade Kitchen Cabinets	$ + 1637
Solid Surface Countertops (Included)	
Full Bath - including plumbing, wall and floor finishes	+ 10,186
Half Bath - including plumbing, wall and floor finishes	+ 5828
Two Car Attached Garage	+ 30,226
Two Car Detached Garage	+ 34,182
Fireplace & Chimney	+ 9605

Adjustments

For multi family - add to total cost
Additional Kitchen	$ + 20,200
Additional Full Bath & Half Bath	+ 16,014
Additional Entry & Exit	+ 2306
Separate Heating & Air Conditioning	+ 6833
Separate Electric	+ 1958

For Townhouse/Rowhouse -
Multiply cost per square foot by
Inner Unit	+ .86
End Unit	+ .93

Alternatives

Add to or deduct from the cost per square foot of living area
Heavyweight Asphalt Shingles	– .95
Clay Tile Roof	+ .95
Slate Roof	+ 1.85
Upgrade Ceilings to Textured Finish	+ .49
Air Conditioning, in Heating Ductwork	Base System
Heating Systems, Hot Water	+ 2.33
Heat Pump	+ 2.81
Electric Heat	– 1.65
Not Heated	– 3.67

Additional upgrades or components

Kitchen Cabinets & Countertops	Page 58
Bathroom Vanities	59
Fireplaces & Chimneys	59
Windows, Skylights & Dormers	59
Appliances	60
Breezeways & Porches	60
Finished Attic	60
Garages	61
Site Improvements	61
Wings & Ells	57

- **Unique residence built from an architect's plan**
- **Single family — 1 full bath, 1 half bath, 1 kitchen**
- **No basement**
- **Cedar shakes on roof**
- **Forced hot air heat/air conditioning**
- **Gypsum wallboard interior finishes**
- **Many special features**
- **Extraordinary materials and workmanship**
- **Detail specifications on page 49**

Note: The illustration shown may contain some optional components (for example: garages and/or fireplaces) whose costs are shown in the modifications, adjustments, & alternatives below or at the end of the square foot section.

©Larry W. Garnett & Associates, Inc

Base cost per square foot of living area

Exterior Wall	Living Area										
	1500	1800	2100	2500	3000	3500	4000	4500	5000	5500	6000
Wood Siding - Wood Frame	168.30	151.20	141.05	133.60	123.50	116.10	109.10	105.60	102.60	99.50	96.10
Brick Veneer - Wood Frame	179.75	161.80	150.55	142.60	131.60	123.55	115.95	112.10	108.75	105.35	101.75
Solid Brick	199.10	179.55	166.45	157.60	145.25	135.90	127.40	123.00	119.15	115.20	111.20
Solid Stone	202.95	183.15	169.65	160.60	147.95	138.40	129.80	125.15	121.25	117.20	113.15
Finished Basement, Add	23.95	25.80	24.35	23.55	22.40	21.50	20.80	20.25	19.90	19.50	19.25
Unfinished Basement, Add	11.05	10.40	9.65	9.30	8.70	8.25	7.90	7.60	7.45	7.25	7.10

Modifications

Add to the total cost

Upgrade Kitchen Cabinets	$ + 1637
Solid Surface Countertops (Included)	
Full Bath - including plumbing, wall and floor finishes	+ 10,186
Half Bath - including plumbing, wall and floor finishes	+ 5828
Two Car Attached Garage	+ 30,226
Two Car Detached Garage	+ 34,182
Fireplace & Chimney	+ 10,510

Adjustments

For multi family - add to total cost

Additional Kitchen	$ + 20,200
Additional Full Bath & Half Bath	+ 16,014
Additional Entry & Exit	+ 2306
Separate Heating & Air Conditioning	+ 6833
Separate Electric	+ 1958

For Townhouse/Rowhouse -
Multiply cost per square foot by

Inner Unit	+ .86
End Unit	+ .93

Alternatives

Add to or deduct from the cost per square foot of living area

Heavyweight Asphalt Shingles	– .85
Clay Tile Roof	+ .85
Slate Roof	+ 1.60
Upgrade Ceilings to Textured Finish	+ .49
Air Conditioning, in Heating Ductwork	Base System
Heating Systems, Hot Water	+ 2.11
Heat Pump	+ 2.89
Electric Heat	– 3.25
Not Heated	– 3.67

Additional upgrades or components

- **Unique residence built from an architect's plan**
- **Single family — 1 full bath, 1 half bath, 1 kitchen**
- **No basement**
- **Cedar shakes on roof**
- **Forced hot air heat/air conditioning**
- **Gypsum wallboard interior finishes**
- **Many special features**
- **Extraordinary materials and workmanship**
- **Detail specifications on page 49**

Note: The illustration shown may contain some optional components (for example: garages and/or fireplaces) whose costs are shown in the modifications, adjustments, & alternatives below or at the end of the square foot section.

Base cost per square foot of living area

Exterior Wall	Living Area										
	1500	1800	2100	2500	3000	3500	4000	4500	5000	5500	6000
Wood Siding - Wood Frame	167.20	150.30	142.00	135.00	124.40	119.15	112.70	106.05	103.60	100.95	98.25
Brick Veneer - Wood Frame	179.20	161.35	152.35	144.90	133.40	127.60	120.40	113.20	110.50	107.55	104.50
Solid Brick	199.25	179.90	169.70	161.50	148.40	141.80	133.35	125.15	122.10	118.70	115.00
Solid Stone	203.20	183.60	173.15	164.80	151.45	144.65	135.90	127.50	124.40	120.90	117.10
Finished Basement, Add	21.00	22.60	21.70	21.05	20.00	19.40	18.60	17.95	17.70	17.40	17.05
Unfinished Basement, Add	9.70	9.10	8.65	8.35	7.85	7.55	7.15	6.85	6.70	6.55	6.35

Modifications

Add to the total cost

Upgrade Kitchen Cabinets	$ + 1637
Solid Surface Countertops (Included)	
Full Bath - including plumbing, wall and floor finishes	+ 10,186
Half Bath - including plumbing, wall and floor finishes	+ 5828
Two Car Attached Garage	+ 30,226
Two Car Detached Garage	+ 34,182
Fireplace & Chimney	+ 10,510

Adjustments

For multi family - add to total cost

Additional Kitchen	$ + 20,200
Additional Full Bath & Half Bath	+ 16,014
Additional Entry & Exit	+ 2306
Separate Heating & Air Conditioning	+ 6833
Separate Electric	+ 1958

For Townhouse/Rowhouse -
Multiply cost per square foot by

Inner Unit	+ .84
End Unit	+ .92

Alternatives

Add to or deduct from the cost per square foot of living area

Heavyweight Asphalt Shingles	– .65
Clay Tile Roof	+ .65
Slate Roof	+ 1.20
Upgrade Ceilings to Textured Finish	+ .49
Air Conditioning, in Heating Ductwork	Base System
Heating Systems, Hot Water	+ 2.11
Heat Pump	+ 2.89
Electric Heat	– 3.25
Not Heated	– 3.56

Additional upgrades or components

Kitchen Cabinets & Countertops	Page 58
Bathroom Vanities	59
Fireplaces & Chimneys	59
Windows, Skylights & Dormers	59
Appliances	60
Breezeways & Porches	60
Finished Attic	60
Garages	61
Site Improvements	61
Wings & Ells	57

Important: See the Reference Section for Location Factors (to adjust for your city) and Estimating Forms.

- **Unique residence built from an architect's plan**
- **Single family — 1 full bath, 1 half bath, 1 kitchen**
- **No basement**
- **Cedar shakes on roof**
- **Forced hot air heat/air conditioning**
- **Gypsum wallboard interior finishes**
- **Many special features**
- **Extraordinary materials and workmanship**
- **Detail specifications on page 49**

Note: The illustration shown may contain some optional components (for example: garages and/or fireplaces) whose costs are shown in the modifications, adjustments, & alternatives below or at the end of the square foot section.

Base cost per square foot of living area

Exterior Wall	Living Area										
	1200	1400	1600	1800	2000	2400	2800	3200	3600	4000	4400
Wood Siding - Wood Frame	164.20	154.05	146.75	140.50	133.70	124.20	116.15	110.80	107.70	104.40	101.45
Brick Veneer - Wood Frame	172.45	161.80	154.20	147.50	140.45	130.25	121.65	116.00	112.65	109.05	106.00
Solid Brick	186.40	174.80	166.60	159.15	151.70	140.35	130.90	124.65	121.00	116.95	113.60
Solid Stone	189.20	177.40	169.10	161.55	154.00	142.40	132.80	126.40	122.70	118.50	115.10
Finished Basement, Add	30.00	32.60	31.70	30.60	30.05	28.55	27.40	26.65	26.20	25.55	25.25
Unfinished Basement, Add	13.70	13.00	12.55	12.10	11.80	11.00	10.45	10.10	9.85	9.50	9.40

Modifications

Add to the total cost

Upgrade Kitchen Cabinets	$ + 1637
Solid Surface Countertops (Included)	
Full Bath - including plumbing, wall and floor finishes	+ 10,186
Half Bath - including plumbing, wall and floor finishes	+ 5828
Two Car Attached Garage	+ 30,226
Two Car Detached Garage	+ 34,182
Fireplace & Chimney	+ 8760

Adjustments

For multi family - add to total cost

Additional Kitchen	$ + 20,200
Additional Full Bath & Half Bath	+ 16,014
Additional Entry & Exit	+ 2306
Separate Heating & Air Conditioning	+ 6833
Separate Electric	+ 1958

**For Townhouse/Rowhouse -
Multiply cost per square foot by**

Inner Unit	+ .89
End Unit	+ .94

Alternatives

Add to or deduct from the cost per square foot of living area

Heavyweight Asphalt Shingles	– .95
Clay Tile Roof	+ .95
Slate Roof	+ 1.85
Upgrade Ceilings to Textured Finish	+ .49
Air Conditioning, in Heating Ductwork	Base System
Heating Systems, Hot Water	+ 2.33
Heat Pump	+ 2.81
Electric Heat	– 1.65
Not Heated	– 3.67

Additional upgrades or components

Kitchen Cabinets & Countertops	Page 58
Bathroom Vanities	59
Fireplaces & Chimneys	59
Windows, Skylights & Dormers	59
Appliances	60
Breezeways & Porches	60
Finished Attic	60
Garages	61
Site Improvements	61
Wings & Ells	57

Important: See the Reference Section for Location Factors (to adjust for your city) and Estimating Forms.

- **Unique residence built from an architect's plan**
- **Single family — 1 full bath, 1 half bath, 1 kitchen**
- **No basement**
- **Cedar shakes on roof**
- **Forced hot air heat/air conditioning**
- **Gypsum wallboard interior finishes**
- **Many special features**
- **Extraordinary materials and workmanship**
- **Detail specifications on page 49**

Note: The illustration shown may contain some optional components (for example: garages and/or fireplaces) whose costs are shown in the modifications, adjustments, & alternatives below or at the end of the square foot section.

©Home Planners, Inc.

Base cost per square foot of living area

Exterior Wall	Living Area										
	1500	1800	2100	2400	2800	3200	3600	4000	4500	5000	5500
Wood Siding - Wood Frame	154.30	142.45	132.70	126.05	121.20	115.55	109.75	107.20	101.65	98.65	95.20
Brick Veneer - Wood Frame	161.75	149.15	138.80	131.80	126.75	120.65	114.50	111.80	105.90	102.65	98.95
Solid Brick	174.25	160.50	149.10	141.40	136.00	129.25	122.45	119.55	113.00	109.35	105.30
Solid Stone	176.80	162.70	151.10	143.40	137.80	130.95	124.05	121.05	114.40	110.75	106.60
Finished Basement, Add*	35.40	38.05	36.50	35.50	34.85	33.85	32.95	32.55	31.70	31.10	30.55
Unfinished Basement, Add*	15.70	14.70	13.90	13.40	13.10	12.60	12.15	11.95	11.45	11.20	10.90

*Basement under middle level only.

Modifications

Add to the total cost

Upgrade Kitchen Cabinets	$ + 1637
Solid Surface Countertops (Included)	
Full Bath - including plumbing, wall and floor finishes	+ 10,186
Half Bath - including plumbing, wall and floor finishes	+ 5828
Two Car Attached Garage	+ 30,226
Two Car Detached Garage	+ 34,182
Fireplace & Chimney	+ 8760

Adjustments

For multi family - add to total cost

Additional Kitchen	$ + 20,200
Additional Full Bath & Half Bath	+ 16,014
Additional Entry & Exit	+ 2306
Separate Heating & Air Conditioning	+ 6833
Separate Electric	+ 1958

For Townhouse/Rowhouse -
Multiply cost per square foot by

Inner Unit	+ .86
End Unit	+ .93

Alternatives

Add to or deduct from the cost per square foot of living area

Heavyweight Asphalt Shingles	– 1.40
Clay Tile Roof	+ 1.40
Slate Roof	+ 2.65
Upgrade Ceilings to Textured Finish	+ .49
Air Conditioning, in Heating Ductwork	Base System
Heating Systems, Hot Water	+ 2.26
Heat Pump	+ 2.92
Electric Heat	– 1.46
Not Heated	– 3.56

Additional upgrades or components

Kitchen Cabinets & Countertops	Page 58
Bathroom Vanities	59
Fireplaces & Chimneys	59
Windows, Skylights & Dormers	59
Appliances	60
Breezeways & Porches	60
Finished Attic	60
Garages	61
Site Improvements	61
Wings & Ells	57

1 Story — Base cost per square foot of living area

Exterior Wall	Living Area							
	50	100	200	300	400	500	600	700
Wood Siding - Wood Frame	247.40	190.65	166.00	139.85	131.80	126.85	123.65	124.80
Brick Veneer - Wood Frame	275.25	210.60	182.60	150.95	141.75	136.20	132.55	133.30
Solid Brick	322.00	243.95	210.45	169.50	158.50	151.75	147.35	147.60
Solid Stone	331.45	250.65	216.00	173.20	161.80	154.90	150.35	150.45
Finished Basement, Add	95.20	86.25	77.40	62.50	59.60	57.75	56.60	55.75
Unfinished Basement, Add	49.35	37.95	33.25	25.35	23.75	22.80	22.20	21.70

1-1/2 Story — Base cost per square foot of living area

Exterior Wall	Living Area							
	100	200	300	400	500	600	700	800
Wood Siding - Wood Frame	196.90	160.25	138.00	124.85	118.20	115.05	110.90	109.65
Brick Veneer - Wood Frame	221.80	180.15	154.65	137.80	130.10	126.30	121.50	120.25
Solid Brick	263.50	213.55	182.40	159.50	150.15	145.25	139.35	137.95
Solid Stone	271.95	220.25	188.05	163.90	154.15	149.00	143.00	141.50
Finished Basement, Add	62.60	61.90	55.95	49.40	47.70	46.45	45.30	45.25
Unfinished Basement, Add	31.30	26.60	23.40	19.90	18.95	18.35	17.75	17.65

2 Story — Base cost per square foot of living area

Exterior Wall	Living Area							
	100	200	400	600	800	1000	1200	1400
Wood Siding - Wood Frame	193.55	145.55	124.55	104.45	97.60	93.45	90.75	92.15
Brick Veneer - Wood Frame	221.45	165.50	141.10	115.50	107.55	102.75	99.60	100.65
Solid Brick	268.15	198.90	168.95	134.00	124.30	118.30	114.45	114.95
Solid Stone	277.60	205.55	174.55	137.75	127.60	121.45	117.40	117.80
Finished Basement, Add	47.60	43.10	38.65	31.25	29.80	28.85	28.35	27.85
Unfinished Basement, Add	24.70	18.95	16.65	12.70	11.90	11.40	11.10	10.90

Base costs do not include bathroom or kitchen facilities. Use Modifications/Adjustments/Alternatives on pages 58-61 where appropriate.

Important: See the Reference Section for Location Factors (to adjust for your city) and Estimating Forms.

Kitchen cabinets - Base units, hardwood *(Cost per Unit)*

	Economy	Average	Custom	Luxury
24" deep, 35" high,				
One top drawer,				
One door below				
12" wide	$182	$242	$320	$425
15" wide	191	254	340	445
18" wide	206	275	365	480
21" wide	224	298	395	520
24" wide	244	325	430	570
Four drawers				
12" wide	311	415	550	725
15" wide	259	345	460	605
18" wide	285	380	505	665
24" wide	308	410	545	720
Two top drawers,				
Two doors below				
27" wide	266	355	470	620
30" wide	289	385	510	675
33" wide	319	425	565	745
36" wide	330	440	585	770
42" wide	349	465	620	815
48" wide	364	485	645	850
Range or sink base				
(Cost per unit)				
Two doors below				
30" wide	266	355	470	620
33" wide	281	375	500	655
36" wide	289	385	510	675
42" wide	308	410	545	720
48" wide	319	425	565	745
Corner Base Cabinet				
(Cost per unit)				
36" wide	375	500	665	875
Lazy Susan *(Cost per unit)*				
With revolving door	375	500	665	875

Kitchen cabinets - Wall cabinets, hardwood *(Cost per Unit)*

	Economy	Average	Custom	Luxury
12" deep, 2 doors				
12" high				
30" wide	$150	$200	$ 265	$ 350
36" wide	173	230	305	405
15" high				
30" wide	160	213	285	375
33" wide	171	228	305	400
36" wide	179	239	320	420
24" high				
30" wide	194	258	345	450
36" wide	212	283	375	495
42" wide	229	305	405	535
30" high, 1 door				
12" wide	140	186	245	325
15" wide	155	207	275	360
18" wide	169	225	300	395
24" wide	188	250	335	440
30" high, 2 doors				
27" wide	219	292	390	510
30" wide	220	293	390	515
36" wide	248	330	440	580
42" wide	270	360	480	630
48" wide	300	400	530	700
Corner wall, 30" high				
24" wide	161	214	285	375
30" wide	190	253	335	445
36" wide	205	273	365	480
Broom closet				
84" high, 24" deep				
18" wide	405	540	720	945
Oven Cabinet				
84" high, 24" deep				
27" wide	585	780	1035	1365

Kitchen countertops *(Cost per L.F.)*

	Economy	Average	Custom	Luxury
Solid Surface				
24" wide, no backsplash	89	119	160	210
with backsplash	97	129	170	225
Stock plastic laminate, 24" wide				
with backsplash	17	22	30	40
Custom plastic laminate, no splash				
7/8" thick, alum. molding	24	33	45	55
1-1/4" thick, no splash	33	44	60	75
Marble				
1/2" - 3/4" thick w/splash	45	60	80	105
Maple, laminated				
1-1/2" thick w/splash	65	87	115	150
Stainless steel				
(per S.F.)	119	159	210	280
Cutting blocks, recessed				
16" x 20" x 1" (each)	88	117	155	205

Vanity bases *(Cost per Unit)*

2 door, 30" high, 21" deep	Economy	Average	Custom	Luxury
24" wide	188	251	335	440
30" wide	220	293	390	515
36" wide	289	385	510	675
48" wide	341	455	605	795

Solid surface vanity tops *(Cost Each)*

Center bowl	Economy	Average	Custom	Luxury
22" x 25"	$251	$271	$293	$316
22" x 31"	287	310	335	362
22" x 37"	330	356	385	416
22" x 49"	405	437	472	510

Fireplaces & Chimneys *(Cost per Unit)*

	1-1/2 Story	2 Story	3 Story
Economy (prefab metal)			
Exterior chimney & 1 fireplace	$4650	$5140	$5635
Interior chimney & 1 fireplace	4455	4955	5185
Average (masonry)			
Exterior chimney & 1 fireplace	4635	5170	5875
Interior chimney & 1 fireplace	4345	4875	5310
For more than 1 flue, add	335	570	950
For more than 1 fireplace, add	3285	3285	3285
Custom (masonry)			
Exterior chimney & 1 fireplace	4825	5445	6150
Interior chimney & 1 fireplace	4525	5120	5520
For more than 1 flue, add	380	655	890
For more than 1 fireplace, add	3465	3465	3465
Luxury (masonry)			
Exterior chimney & 1 fireplace	6760	7415	8110
Interior chimney & 1 fireplace	6450	7055	7465
For more than 1 flue, add	560	930	1300
For more than 1 fireplace, add	5330	5330	5330

Windows and Skylights *(Cost Each)*

	Economy	Average	Custom	Luxury
Fixed Picture Windows				
3'-6" x 4'-0"	$ 514	$ 556	$ 600	$ 648
4'-0" x 6'-0"	879	949	1025	1107
5'-0" x 6'-0"	986	1065	1150	1242
6'-0" x 6'-0"	986	1065	1150	1242
Bay/Bow Windows				
8'-0" x 5'-0"	1265	1366	1475	1593
10'-0" x 5'-0"	1436	1551	1675	1809
10'-0" x 6'-0"	2315	2500	2700	2916
12'-0" x 6'-0"	2936	3171	3425	3699
Palladian Windows				
3'-2" x 6'-4"		1782	1925	2079
4'-0" x 6'-0"		2106	2275	2457
5'-5" x 6'-10"		2523	2725	2943
8'-0" x 6'-0"		3032	3275	3537
Skylights				
46" x 21-1/2"	401	433	573	619
46" x 28"	433	468	603	651
57" x 44"	530	572	712	769

Dormers *(Cost/S.F. of plan area)*

Framing and Roofing Only	Economy	Average	Custom	Luxury
Gable dormer, 2" x 6" roof frame	$24	$27	$30	$47
2" x 8" roof frame	25	28	31	50
Shed dormer, 2" x 6" roof frame	15	17	19	31
2" x 8" roof frame	17	18	20	32
2" x 10" roof frame	18	20	22	33

Appliances (Cost per Unit)

	Economy	Average	Custom	Luxury
Range				
30" free standing, 1 oven	$ 350	$1088	$1456	$1825
2 oven	1800	1875	1913	1950
30" built-in, 1 oven	580	1265	1608	1950
2 oven	1600	1875	2013	2150
21" free standing				
1 oven	390	438	461	485
Counter Top Ranges				
4 burner standard	300	538	656	775
As above with griddle	635	855	965	1075
Microwave Oven	190	420	535	650
Combination Range,				
Refrigerator, Sink				
30" wide	1175	2300	2863	3425
60" wide	2775	3192	3400	3608
72" wide	3825	4399	4686	4973
Comb. Range, Refrig., Sink,				
Microwave Oven & Ice Maker	6153	7076	7538	7999
Compactor				
4 to 1 compaction	555	623	656	690
Deep Freeze				
15 to 23 C.F.	575	668	714	760
30 C.F.	950	1063	1119	1175
Dehumidifier, portable, auto.				
15 pint	175	202	215	228
30 pint	213	245	261	277
Washing Machine, automatic	470	1048	1336	1625
Water Heater				
Electric, glass lined				
30 gal.	475	583	636	690
80 gal.	930	1190	1320	1450
Water Heater, Gas, glass lined				
30 gal.	775	950	1038	1125
50 gal.	965	1195	1310	1425
Water Softener, automatic				
30 grains/gal.	665	765	815	865
100 grains/gal.	905	1041	1109	1177
Dishwasher, built-in				
2 cycles	415	498	539	580
4 or more cycles	435	630	728	825
Dryer, automatic	495	760	893	1025
Garage Door Opener	375	453	491	530
Garbage Disposal	104	164	194	224
Heater, Electric, built-in				
1250 watt ceiling type	178	220	241	262
1250 watt wall type	212	247	265	282
Wall type w/blower				
1500 watt	240	330	375	420
3000 watt	420	462	483	504
Hood For Range, 2 speed				
30" wide	134	467	634	800
42" wide	355	590	708	825
Humidifier, portable				
7 gal. per day	176	203	216	229
15 gal. per day	212	244	260	276
Ice Maker, automatic				
13 lb. per day	990	1139	1213	1287
51 lb. per day	1375	1582	1685	1788
Refrigerator, no frost				
10-12 C.F.	500	568	601	635
14-16 C.F.	545	563	571	580
18-20 C.F.	645	823	911	1000
21-29 C.F.	850	1750	2200	2650
Sump Pump, 1/3 H.P.	236	326	370	415

Breezeway (Cost per S.F.)

Class	Type	Area (S.F.)			
		50	100	150	200
Economy	Open	$ 21.15	$ 18.00	$15.10	$15.10
	Enclosed	101.50	78.40	65.10	65.10
Average	Open	26.15	23.05	20.15	20.15
	Enclosed	111.20	82.85	67.80	67.80
Custom	Open	37.35	32.85	28.60	28.60
	Enclosed	153.35	114.10	93.25	93.25
Luxury	Open	38.55	33.80	30.45	30.45
	Enclosed	155.20	115.10	93.10	83.35

Porches (Cost per S.F.)

Class	Type	Area (S.F.)				
		25	50	100	200	300
Economy	Open	$ 60.65	$ 40.60	$ 31.70	$26.85	$22.90
	Enclosed	121.20	84.45	63.85	49.80	42.65
Average	Open	73.60	46.85	35.95	29.90	29.90
	Enclosed	145.45	98.55	74.55	57.70	48.90
Custom	Open	95.35	63.25	47.60	41.60	37.30
	Enclosed	189.55	129.50	98.55	76.75	66.10
Luxury	Open	102.05	66.70	49.25	44.25	39.45
	Enclosed	199.75	140.30	104.10	80.85	69.60

Finished attic (Cost per S.F.)

Class	Area (S.F.)				
	400	500	600	800	1000
Economy	$16.20	$15.65	$15.00	$14.75	$14.20
Average	24.90	24.35	23.75	23.45	22.80
Custom	30.40	29.70	29.05	28.60	28.00
Luxury	38.30	37.40	36.50	35.65	35.05

Alarm system (Cost per System)

	Burglar Alarm	Smoke Detector
Economy	$ 375	$ 65
Average	430	89
Custom	731	149
Luxury	1075	178

Sauna, prefabricated
(Cost per unit, including heater and controls—7' high)

Size	Cost
6' x 4'	$ 4675
6' x 5'	5225
6' x 6'	5600
6' x 9'	6950
8' x 10'	9125
8' x 12'	10,700
10' x 12'	11,400

Garages *

(Costs include exterior wall systems comparable with the quality of the residence. Included in the cost is an allowance for one personnel door, manual overhead door(s) and electrical fixture.)

Class	Type									
	Detached			Attached			Built-in		Basement	
	One Car	Two Car	Three Car	One Car	Two Car	Three Car	One Car	Two Car	One Car	Two Car
Economy										
Wood	$12,594	$19,230	$25,867	$ 9770	$16,838	$23,474	$-1733	$-3465	$1354	$1768
Masonry	17,811	25,759	33,708	13,034	21,414	29,363	-2373	-4747		
Average										
Wood	14,035	21,034	28,033	10,672	18,102	25,101	-1910	-3819	1534	2127
Masonry	17,811	25,759	33,708	13,034	21,414	29,363	-2373	-4067		
Custom										
Wood	15,764	24,140	32,516	12,313	21,296	29,673	-3255	-3402	2322	3703
Masonry	19,899	29,315	38,731	14,901	24,923	34,339	-3763	-4418		
Luxury										
Wood	17,633	27,407	37,180	13,947	24,327	34,100	-3332	-3557	3061	4842
Masonry	22,871	33,961	45,052	17,224	28,921	40,012	-3976	-4844		

*See the Introduction to this section for definitions of garage types.

Swimming pools (Cost per S.F.)

Residential	(includes equipment)
In-ground	$ 22.00 - 56.50
Deck equipment	1.30
Paint pool, preparation & 3 coats (epoxy)	3.28
Rubber base paint	3.03
Pool Cover	.78
Swimming Pool Heaters	(Cost per unit)
(not including wiring, external piping, base or pad)	
Gas	
155 MBH	$ 2325
190 MBH	2775
500 MBH	9600
Electric	
15 KW 7200 gallon pool	2675
24 KW 9600 gallon pool	3500
54 KW 24,000 gallon pool	5250

Wood and coal stoves

Wood Only	
Free Standing (minimum)	$ 1725
Fireplace Insert (minimum)	1739
Coal Only	
Free Standing	$ 1965
Fireplace Insert	2151
Wood and Coal	
Free Standing	$ 4041
Fireplace Insert	4141

Sidewalks (Cost per S.F.)

Concrete, 3000 psi with wire mesh	4" thick	$ 3.45
	5" thick	4.23
	6" thick	4.78
Precast concrete patio blocks (natural)	2" thick	9.30
Precast concrete patio blocks (colors)	2" thick	9.85
Flagstone, bluestone	1" thick	13.90
Flagstone, bluestone	1-1/2" thick	18.65
Slate (natural, irregular)	3/4" thick	13.70
Slate (random rectangular)	1/2" thick	20.95
Seeding		
Fine grading & seeding includes lime, fertilizer & seed per S.Y.		2.09
Lawn Sprinkler System	per S.F.	.83

Fencing (Cost per L.F.)

Chain Link, 4' high, galvanized	$ 15.75
Gate, 4' high (each)	162.00
Cedar Picket, 3' high, 2 rail	11.20
Gate (each)	159.00
3 Rail, 4' high	13.50
Gate (each)	170.00
Cedar Stockade, 3 Rail, 6' high	13.70
Gate (each)	169.00
Board & Battens, 2 sides 6' high, pine	19.60
6' high, cedar	27.50
No. 1 Cedar, basketweave, 6' high	15.65
Gate, 6' high (each)	195.00

Carport (Cost per S.F.)

Economy	$ 7.63
Average	11.56
Custom	17.25
Luxury	19.59

Insurance Exclusions

Insurance exclusions are a matter of policy coverage and are not standard or universal. When making an appraisal for insurance purposes, it is recommended that some time be taken in studying the insurance policy to determine the specific items to be excluded. Most homeowners insurance policies have, as part of their provisions, statements that read "the policy permits the insured to exclude from value, in determining whether the amount of the insurance equals or exceeds 80% of its replacement cost, such items as building excavation, basement footings, underground piping and wiring, piers and other supports which are below the undersurface of the lowest floor or where there is no basement below the ground." Loss to any of these items, however, is covered.

Costs for Excavation, Spread and Strip Footings and Underground Piping

This chart shows excluded items expressed as a percentage of total building cost.

Class	Number of Stories			
	1	1-1/2	2	3
Luxury	3.4%	2.7%	2.4%	1.9%
Custom	3.5%	2.7%	2.4%	1.9%
Average	5.0%	4.3%	3.8%	3.3%
Economy	5.0%	4.4%	4.1%	3.5%

Architect/Designer Fees

Architect/designer fees as presented in the following chart are typical ranges for 4 classes of residences. Factors affecting these ranges include economic conditions, size and scope of project, site selection, and standardization. Where superior quality and detail is required or where closer supervision is required, fees may run higher than listed. Lower quality or simplicity may dictate lower fees.

The editors have included architect/designer fees from the "mean" column in calculating costs.

Listed below are average costs expressed as a range and mean of total building cost for 4 classes of residences.

Class		Range	Mean
Luxury	— Architecturally designed and supervised	5%–15%	10%
Custom	— Architecturally modified designer plans	$2000–$3500	$2500
Average	— Designer plans	$800–$1500	$1100
Economy	— Stock plans	$100–$500	$300

Depreciation as generally defined is "the loss of value due to any cause." Specifically, depreciation can be broken down into three categories: **Physical Depreciation, Functional Obsolescence,** and **Economic Obsolescence.**

Physical Depreciation is the loss of value from the wearing out of the building's components. Such causes may be decay, dry rot, cracks, or structural defects.

Functional Obsolescence is the loss of value from features which render the residence less useful or desirable. Such features may be higher than normal ceilings, design and/or style changes, or outdated mechanical systems.

Economic Obsolescence is the loss of value from external features which render the residence less useful or desirable. These features may be changes in neighborhood socio-economic grouping, zoning changes, legislation, etc.

Depreciation as it applies to the residential portion of this manual deals with the observed physical condition of the residence being appraised. It is in essence *"the cost to cure."*

The ultimate method to arrive at *"the cost to cure"* is to analyze each component of the residence.

For example:

    Component, Roof Covering – Cost = $2400
    Average Life = 15 years
    Actual Life = 5 years
    Depreciation = 5 ÷ 15 = .33 or 33%

    $2400 × 33% = $792 = Amount of Depreciation

The following table, however, can be used as a guide in estimating the % depreciation using the actual age and general condition of the residence.

Depreciation Table — Residential

Age in Years	Good	Average	Poor
2	2%	3%	10%
5	4	6	20
10	7	10	25
15	10	15	30
20	15	20	35
25	18	25	40
30	24	30	45
35	28	35	50
40	32	40	55
45	36	45	60
50	40	50	65

Commercial/Industrial/ Institutional Section

Table of Contents

Introduction to the Commercial/ Industrial/Institutional Section

General

The Commercial/Industrial/Institutional section of this manual contains base building costs per square foot of floor area for 75 model buildings. Each model has a table of square foot costs for combinations of exterior wall and framing systems. This table is supplemented by a list of common additives and their unit costs. A breakdown of the component costs used to develop the base cost for the model is on the opposite page. The total cost derived from the base cost and additives must be modified using the appropriate location factor from the Reference Section.

This section may be used directly to estimate the construction cost of most types of buildings knowing only the floor area, exterior wall construction and framing systems. To adjust the base cost for components which are different than the model, use the tables from the Assemblies Section.

Building Identification & Model Selection

The building models in this section represent structures by use. Occupancy, however, does not necessarily identify the building, i.e., a restaurant could be a converted warehouse. In all instances, the building should be described and identified by its own physical characteristics. The model selection should also be guided by comparing specifications with the model. In the case of converted use, data from one model may be used to supplement data from another.

Adjustments

The base cost tables represent the base cost per square foot of floor area for buildings without a basement and without unusual special features. Basement costs and other common additives are listed below the base cost table. Cost adjustments can also be made to the model by using the tables from the Assemblies Section. This table is for example only.

Dimensions

All base cost tables are developed so that measurement can be readily made during the inspection process. Areas are calculated from exterior dimensions and story heights are measured from the top surface of one floor to the top surface of the floor above. Roof areas are measured by horizontal area covered and costs related to inclines are converted with appropriate factors. The precision of measurement is a matter of the user's choice and discretion. For ease in calculation, consideration should be given to measuring in tenths of a foot, i.e., 9 ft. 6 in. = 9.5 ft., 9 ft. 4 in. = 9.3 ft.

Floor Area

The term "Floor Area" as used in this section includes the sum of floor plate at grade level and above. This dimension is measured from the outside face of the foundation wall. Basement costs are calculated separately. The user must exercise his/her own judgment, where the lowest level floor is slightly below grade, whether to consider it at grade level or make the basement adjustment.

How to Use the Commercial/Industrial/Institutional Section

The following is a detailed explanation of a sample entry in the Commercial/Industrial/Institutional Square Foot Cost Section. Each bold number below corresponds to the item being described on the following page with the appropriate component or cost of the sample entry following in parenthesis.

Prices listed are costs that include overhead and profit of the installing contractor and additional mark-ups for General Conditions and Architects' Fees.

COMMERCIAL/INDUSTRIAL/INSTITUTIONAL	M.010 **1**	Apartment, 1-3 Story **2**

Costs per square foot of floor area

Exterior Wall	S.F. Area	8000	12000	15000	19000	22500	25000	29000	32000	36000
	L.F. Perimeter	213	280	330	350	400	433 **4**	442	480	520
Face Brick with Concrete Block Back-up	Wood Joists	188.10	178.85	175.25	167.80	166.00	164.80	160.75	160.10	158.85
3	Steel Joists	189.20	181.40	178.15	172.20	170.65	5 168.70	166.35	165.85	164.80
Stucco on Concrete Block	Wood Joists	164.25	157.25	154.40	149.40	148.05	147.20	144.45	144.00	143.10
	Steel Joists	179.35	172.30	169.45	164.25	162.90	162.00	159.25	158.75	157.80
Wood Siding	Wood Frame	163.85	156.95	154.25	149.30	147.95	147.10	144.50	144.00	143.10
Brick Veneer	Wood Frame	172.90	164.85	161.60	155.45	153.85	152.85	149.50	148.90	147.80
Perimeter Adj., Add or Deduct	Per 100 L.F.	21.80 **8**	14.60	11.60	9.15	7.75	7 7.05	6.00	5.45	4.80
Story Hgt. Adj., Add or Deduct **6**	Per 1 Ft.	4.15	3.70	3.35	2.85	2.80	7 2.75	2.35	2.35	2.20
	For Basement, add $35.05 per square foot of basement area									

The above costs were calculated using the basic specifications shown on the facing page. These costs should be adjusted where necessary for design alternatives and owner's requirements. Reported completed project costs, for this type of structure, range from $59.60 to $221.65 per S.F. **9**

Common additives

Description	Unit	$ Cost
Appliances		
Cooking range, 30" free standing		
1 oven	Each	570 - 2100
2 oven	Each	1375 - 2875
30" built-in		
1 oven	Each	920 - 2000
2 oven	Each	1825 - 2425
Counter top cook tops, 4 burner	Each	410 - 1475
Microwave oven	Each	278 - 795
Combination range, refrig. & sink, 30" wide **10**	Each	1875 - 5550
72" wide	Each	5925
Combination range, refrigerator, sink, microwave oven & icemaker	Each	6750
Compactor, residential, 4-1 compaction	Each	815 - 1200
Dishwasher, built-in, 2 cycles	Each	575 - 1125
4 cycles	Each	700 - 1825
Garbage disposer, sink type	Each	225 - 370
Hood for range, 2 speed, vented, 30" wide	Each	310 - 1225
42" wide	Each	420 - 2725
Refrigerator, no frost 10-12 C.F.	Each	520 - 660
18-20 C.F.	Each	805 - 1675

Description	Unit	$ Cost
Closed Circuit Surveillance, One station		
Camera and monitor	Each	1950
For additional camera stations, add	Each	1050
Elevators, Hydraulic passenger, 2 stops		
2000# capacity	Each	67,900
2500# capacity	Each	68,900
3500# capacity	Each	72,900
Additional stop, add	Each	8375
Emergency Lighting, 25 watt, battery operated		
Lead battery	Each	305
Nickel cadmium	Each	850
Laundry Equipment		
Dryer, gas, 16 lb. capacity	Each	945
30 lb. capacity	Each	3750
Washer, 4 cycle	Each	1175
Commercial	Each	1600
Smoke Detectors		
Ceiling type	Each	220
Duct type	Each	550

1 Model Number (M.010)

"M" distinguishes this section of the book and stands for model. The number designation is a sequential number.

2 Type of Building (Apartment, 1-3 Story)

There are 45 different types of commercial/industrial/institutional buildings highlighted in this section.

3 Exterior Wall Construction and Building Framing Options (Face Brick with Concrete Block Back-up and Open Web Steel Bar Joists)

Three or more commonly used exterior walls, and in most cases, two typical building framing systems are presented for each type of building. The model selected should be based on the actual characteristics of the building being estimated.

4 Total Square Foot of Floor Area and Base Perimeter Used to Compute Base Costs (22,500 Square Feet and 400 Linear Feet)

Square foot of floor area is the total gross area of all floors at grade, and above, and does not include a basement. The perimeter in linear feet used for the base cost is for a generally rectangular economical building shape.

5 Cost per Square Foot of Floor Area ($170.65)

The highlighted cost is for a building of the selected exterior wall and framing system and floor area. Costs for buildings with floor areas other than those calculated may be interpolated between the costs shown.

6 Building Perimeter and Story Height Adjustments

Square foot costs for a building with a perimeter or floor to floor story height significantly different from the model used to calculate the base cost may be adjusted, add or deduct, to reflect the actual building geometry.

7 Cost per Square Foot of Floor Area for the Perimeter and/or Height Adjustment ($7.75 for Perimeter Difference and $2.80 for Story Height Difference)

Add (or deduct) $7.75 to the base square foot cost for each 100 feet of perimeter difference between the model and the actual building. Add (or deduct) $2.80 to the base square foot cost for each 1 foot of story height difference between the model and the actual building.

8 Optional Cost per Square Foot of Basement Floor Area ($35.05)

The cost of an unfinished basement for the building being estimated is $35.05 times the gross floor area of the basement.

9 Range of Cost per Square Foot of Floor Area for Similar Buildings ($59.60 to $221.65)

Many different buildings of the same type have been built using similar materials and systems. RSMeans historical cost data of actual construction projects indicates a range of $59.60 to $221.65 for this type of building.

10 Common Additives

Common components and/or systems used in this type of building are listed. These costs should be added to the total building cost. Additional selections may be found in the Assemblies Section.

The following is a detailed explanation of a specification and costs for a model building in the Commercial/Industrial/Institutional Square Foot Cost Section. Each bold number below corresponds to the item being described on the following page with the appropriate component of the sample entry following in parenthesis.

Prices listed are costs that include overhead and profit of the installing contractor.

Model costs calculated for a 3 story building with 10' story height and 22,500 square feet of floor area   **1**

2   **Apartment, 1-3 Story**

			Unit	Unit Cost	Cost Per S.F.	% Of Sub-Total
A. SUBSTRUCTURE						
1010	Standard Foundations	Poured concrete; strip and spread footings	S.F. Ground	6.18	2.06	
1020	Special Foundations	N/A	—	—	—	
1030	Slab on Grade	4" reinforced concrete with vapor barrier and granular base	S.F. Slab	5.18	1.72	4.2%
2010	Basement Excavation	Site preparation for slab and trench for foundation wall and footing	S.F. Ground	.18	.06	
2020	Basement Walls	4' foundation wall	L.F. Wall	75	1.52	
B. SHELL						
B10 Superstructure						
1010	Floor Construction	Open web steel joists, slab form, concrete, interior steel columns	S.F. Floor	19.53	13.02	13.0%
1020	Roof Construction	Open web steel joists with rib metal deck, interior steel columns	S.F. Roof	10.14	3.38	
B20 Exterior Enclosure						
2010	Exterior Walls	Face brick with concrete block backup 88% of wall	S.F. Wall	23.29	10.93	
2020	Exterior Windows	Aluminum horizontal sliding 12% of wall	Each	524	2.24	10.7%
2030	Exterior Doors	Aluminum and glass	Each	1975	.35	
B30 Roofing						
3010	Roof Coverings	Built-up tar and gravel with flashing; perlite/EPS composite insulation	S.F. Roof	6.33	2.11	1.7%
3020	Roof Openings	N/A	—	—	—	
C. INTERIORS						
1010	Partitions	Gypsum board and sound deadening board on metal studs 10 S.F. of Floor/L.F. Partition	S.F. Partition	6.57	5.84	
1020	Interior Doors	15% solid core wood, 85% hollow core wood 80 S.F. Floor/Door	Each	555	6.94	
1030	Fittings	Kitchen Cabinets	S.F. Floor	2.99	2.99	
2010	Stair Construction	Concrete filled metal pan	Flight	14,150	3.14	24.0%
3010	Wall Finishes	70% paint, 25% vinyl wall covering, 5% ceramic tile	S.F. Surface			
3020	Floor Finishes	60% carpet, 30% vinyl composition tile, 10% ceramic tile	S.F. Floor			
3030	Ceiling Finishes	Painted gypsum board on resilient channels	S.F. Ceiling			
D. SERVICES						
D10 Conveying						
1010	Elevators & Lifts	One hydraulic passenger elevator	Each	108,450	4.82	3.8%
1020	Escalators & Moving Walks	N/A	—	—	—	
D20 Plumbing						
2010	Plumbing Fixtures	Kitchen, bathroom and service fixtures, supply and drainage 1 Fixture/200 S.F. Floor	Each	2972	14.86	
2020	Domestic Water Distribution	Gas fired water heater	S.F. Floor	4.67	4.67	15.8%
2040	Rain Water Drainage	Roof drains	S.F. Roof	1.38	.46	
D30 HVAC						
3010	Energy Supply	Oil fired hot water, baseboard radiation	S.F. Floor	9.22	9.22	
3020	Heat Generating Systems	N/A	—	—	—	
3030	Cooling Generating Systems	Chilled water, air cooled condenser system	S.F. Floor	9.92	9.92	15.1%
3050	Terminal & Package Units	N/A	—	—	—	
3090	Other HVAC Sys. & Equipment	N/A	—	—	—	
D40 Fire Protection						
4010	Sprinklers	Wet pipe sprinkler system	S.F. Floor	3.41	3.41	2.7%
4020	Standpipes	N/A	—	—	—	
D50 Electrical						
5010	Electrical Service/Distribution	600 ampere service, panel board and feeders	S.F. Floor	2.27	2.27	
5020	Lighting & Branch Wiring	Incandescent fixtures, receptacles, switches, A.C. and misc. power	S.F. Floor	7.12	7.12	9.0%
5030	Communications & Security	Addressable alarm systems, internet wiring, and emergency lighting	S.F. Floor	1.74	1.74	
5090	Other Electrical Systems	Emergency generator, 11.5 kW	S.F. Floor	.22	.22	
E. EQUIPMENT & FURNISHINGS						
1010	Commercial Equipment	N/A	—	—	—	
1020	Institutional Equipment	N/A	—	—	—	
1030	Vehicular Equipment	N/A	—	—	—	0.0%
1090	Other Equipment	N/A	—	—	—	
F. SPECIAL CONSTRUCTION						
1020	Integrated Construction	N/A	—	—	—	0.0%
1040	Special Facilities	N/A	—	—	—	
G. BUILDING SITEWORK	**N/A**					
			Sub-Total		126.40	100%
CONTRACTOR FEES (General Requirements: 10%, Overhead: 5%, Profit: 10%)				25%	31.61	
ARCHITECT FEES				8%	12.64	
			Total Building Cost		**170.65**	

3   **4**   **5**   **6**   **7**   **8**   **9**   **10**   **11**   **12**

1 *Building Description
(Model costs are calculated for a
three-story apartment building with
10' story height and 22,500 square
feet of floor area)*

The model highlighted is described in terms of building type,
number of stories, typical story height, and square footage.

2 *Type of Building
(Apartment, 1-3 Story)*

3 *Division C Interiors
(C1020 Interior Doors)*

System costs are presented in divisions according to the
7-division UNIFORMAT II classifications. Each of the
component systems are listed.

4 *Specification Highlights
(15% solid core wood;
85% hollow core wood)*

All systems in each subdivision are described with the
material and proportions used.

5 *Quantity Criteria
(80 S.F. Floor/Door)*

The criteria used in determining quantities for the
calculations are shown.

6 *Unit (Each)*

The unit of measure shown in this column is the unit of
measure of the particular system shown that corresponds
to the unit cost.

7 *Unit Cost ($555)*

The cost per unit of measure of each system
subdivision.

8 *Cost per Square Foot ($6.94)*

The cost per square foot for each system is the unit cost of
the system times the total number of units divided by the
total square feet of building area.

9 *% of Sub-Total (24.0%)*

The percent of sub-total is the total cost per square foot of
all systems in the division divided by the sub-total cost per
square foot of the building.

10 *Sub-Total ($126.40)*

The sub-total is the total of all the system costs
per square foot.

11 *Project Fees
(Contractor Fees) (25%)
(Architect Fees) (8%)*

Contractor Fees to cover the general requirements,
overhead and profit of the General Contractor are added
as a percentage of the sub-total. Architect Fees, also as a
percentage of the sub-total, are also added. These values
vary with the building type.

12 *Total Building Cost ($170.65)*

The total building cost per square foot of building area is
the sum of the square foot costs of all the systems plus
the General Contractor's general requirements, overhead
and profit, and the Architect fees. The total building cost
is the amount which appears shaded in the Cost per
Square Foot of Floor Area table shown previously.

Examples

Example 1

This example illustrates the use of the base cost tables. The base cost is adjusted for different exterior wall systems, different story height and a partial basement.

Example 1 (continued)

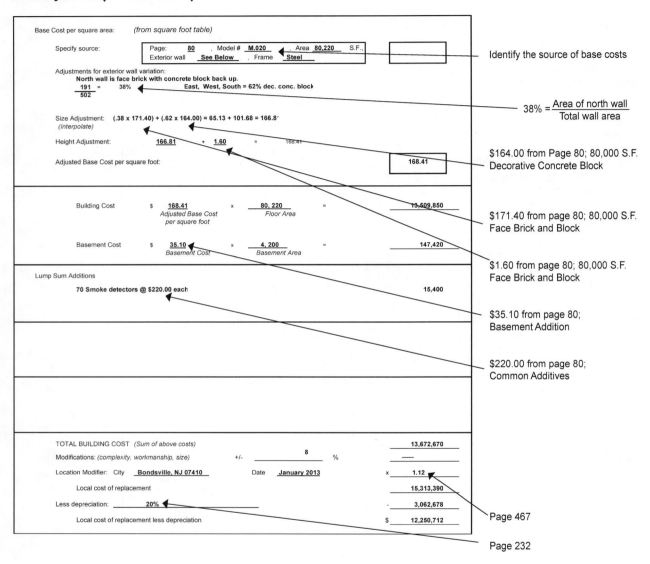

Base Cost per square area: *(from square foot table)*

Specify source:
Page: **80** , Model # **M.020** , Area **80,220** S.F.,
Exterior wall **See Below** , Frame **Steel**

→ Identify the source of base costs

Adjustments for exterior wall variation:
North wall is face brick with concrete block back up.
$\frac{191}{502}$ = **38%**   **East, West, South = 62% dec. conc. block**

→ 38% = $\frac{\text{Area of north wall}}{\text{Total wall area}}$

Size Adjustment: (.38 x 171.40) + (.62 x 164.00) = 65.13 + 101.68 = 166.8′
(Interpolate)

Height Adjustment: **166.81** + **1.60** = 168.41

→ $164.00 from Page 80; 80,000 S.F. Decorative Concrete Block

Adjusted Base Cost per square foot: **168.41**

Building Cost $ **168.41** x **80, 220** = **13,509,850**
Adjusted Base Cost per square foot   *Floor Area*

→ $171.40 from page 80; 80,000 S.F. Face Brick and Block

Basement Cost $ **35.10** x **4, 200** = **147,420**
Basement Cost   *Basement Area*

→ $1.60 from page 80; 80,000 S.F. Face Brick and Block

Lump Sum Additions
70 Smoke detectors @ $220.00 each   **15,400**

→ $35.10 from page 80; Basement Addition

→ $220.00 from page 80; Common Additives

TOTAL BUILDING COST *(Sum of above costs)*   **13,672,670**

Modifications: *(complexity, workmanship, size)* +/- _____ **8** % ___

Location Modifier: City **Bondsville, NJ 07410**   Date **January 2013**   x **1.12**

Local cost of replacement   **15,313,390**

Less depreciation: **20%** _____   - **3,062,678**

Local cost of replacement less depreciation $ **12,250,712**

→ Page 467

→ Page 232

74

Example 2

This example shows how to modify a model building.

Model #M.020, 4–7 Story Apartment, page 80, matches the example quite closely. The model specifies a 6-story building with a 10′ story height.

The following adjustments must be made:
- add stone ashlar wall
- adjust model to a 5-story building
- change partitions
- change floor finish

Two exterior wall types

CII APPRAISAL

1. SUBJECT PROPERTY:	**Westernberg Holdings, Inc.**
2. BUILDING:	**# 623**
3. ADDRESS:	**55 Pinewood Rd., Bondsville, N.J. 07410**
4. BUILDING USE:	**Apartments**

5. DATE: **Jan. 2, 2013**
6. APPRAISER: **SJS**
7. YEAR BUILT: **1992**

North wall - 4" stone Ashlar w/8" block back up (split Face)
East, West & South walls - 12" Decorative concrete block

8. EXTERIOR WALL CONSTRUCTION:
9. FRAME: **Steel**
10. GROUND FLOOR AREA: **11,000**
11. GROSS FLOOR AREA (EXCL. BASEMENT): **55,000** S.F.
12. NUMBER OF STORIES: **5**
13. STORY HEIGHT: **11 Ft.**
14. PERIMETER: **469.4**
15. BASEMENT AREA: None S.F.
16. GENERAL COMMENTS:

Example 2 (continued)

Square foot area (excluding basement) from item 11 ___55,000___ S.F.
Perimeter from item 14 ___469.4___ Lin. Ft.
Item 17-Model square foot costs (from model sub-total) ___$132.39___

From page 81, Model #M.020 Sub-total

NO.	DESCRIPTION			UNIT	UNIT COST	NEW SF COST	MODEL SF COST	+/- CHANGE
A SUBSTRUCTURE								
A 1010	Standard Foundations		Bay size: Same as model	S.F. Gnd.	7.08	1.42	1.18	+ 0.24
A 1030	Slab on Grade	Material:	Thickness: Same as model	S.F. Slab	5.18	1.04	0.86	+ 0.18
A 2010	Basement Excavation	Depth:	Area:11,000 S.F.	S.F. Gnd.	0.30	0.06	0.05	+ 0.01
A 2020	Basement Walls			L.F. Walls	79	0.67	0.66	+ 0.01
B SHELL								
B10 Superstructure								
B 1010	Floor Construction	Elevated floors: Same as model		S.F. Floor	25.63	20.50	21.36	- 0.86
B 1020	Roof Construction			S.F. Roof	9.42	1.88	1.57	+ 0.31
B20 Exterior Closure								
B 2010	Exterior Walls	Material: North wall- 4" stone ashlar on 8" CMU	Thickness: 36% of wall	S.F. Walls	28.78	11.62	10.35	+ 1.27
		Material: East, West, & South walls- 12" split rib block	Thickness: 64% of wall	S.F. Walls				
B 2020	Exterior Windows	Type:	% of wall	S.F. Wind.				
		Type:	14% of wall	Each S.F. Wind.	524	2.30	2.53	- 0.23
B 2030	Exterior Doors	Type: Same as model	Number: 4	Each	4750	0.35	0.32	+ 0.03
		Type:	Number:	Each				
B30 Roofing								
B 3010	Roof Coverings	Material: Same as model		S.F. Roof	6.24	1.25	1.04	+ 0.21
B 3020	Roof Openings	N/A		S.F. Opng.				
C INTERIORS								
C 1010	Partitions: High	Material: Gyp board on metal studs	Density: 8 S.F. Floor/L.F. Part	S.F. Part.	6.92	7.79	6.92	+ 0.87
		Material:	Density:					
C 1020	Interior Doors	Type:	Number: 80 SF	Each	676.8	8.46	8.46	---
C 1030	Fittings			S.F. Floor	3.58	3.58	3.58	---
C 2010	Stair Construction	14 Flights		Flight	11300	2.88	3.02	- 0.14
C 3010	Wall Finishes	Material: Same as model	% of Wall	9'-0" High	S.F. Walls	1.55	3.49	3.09	+ 0.40
		Material:						
C 3020	Floor Finishes	Material: Hardwood - 60% of floor		S.F. Floor	5.23	9.34	5.23	+ 4.11
		Material: Carpet - 30% of floor						
		Material: Tile -10% of floor						
C 3030	Ceiling Finishes	Material: Same as model		S.F. Ceil.	3.98	3.98	3.98	---

$\dfrac{\text{\$.30/S.F. ground area}}{\text{5 stories}}$

$\dfrac{\text{\$.30/S.F. ground area}}{\text{6 stories}}$

$\dfrac{\text{Area of Elevated Floors}}{\text{Building Floor Area}}$
$\dfrac{\$25.63/\text{S.F.} \times 4 \times 11{,}000 \text{ S.F.}}{55{,}000}$

$\dfrac{9.42/\text{S.F. roof}}{\text{5 stories}}$

Windows are 14% of wall area,
15 S.F. S.F./window

$\dfrac{524 \times .14 \times 469.4 \times 11 \times 5}{15} \div 55{,}000 = 2.30$

From Table C1010 124 page 345, C1010 126 on page 346 and C1010 128 page 347

$\dfrac{\$6.92/\text{S.F. partition} \times 9 \text{ ft. high}}{8 \text{ S.F. floor/L.F. partition}}$

From Table C3020 410 page 359 & 360

Oak Strip, (max price) @$10.03/S.F.
Carpet (26 oz.) @$7.12/S.F.
Ceramic Tile @$11.75/S.F.

.60 x 10.03 = 6.02
.30 x 7.12 = 2.14
.10 x 11.75 = 1.18
9.34

North wall is 36% of wall area, e.g.
From Table B2010 128 2350 page 306
Stone ashiar @ $42.45/S.F. wall
East, west and south walls are 64% of wall area
From Table B2010 113 1530 page 300
Split ribbed block @$21.10/S.F. wall

$\dfrac{170 \times 11}{469.4 \times 11}$

New wall cost
(.36 x $42.45/S.F.) + (.64 x $21.10/S.F.) = $28.78/S.F. wall

Convert Cost per S.F. wall to cost per S.F. of floor
Windows are 14% of wall area

$11.62 S.F. floor = $\dfrac{\$28.78/\text{S.F. wall} \times .86 \times 469.4 \text{ L.F.} \times 11 \text{ ft. high} \times 5 \text{ stories}}{55{,}000 \text{ S.F. of floor}}$

Example 2 (continued)

	NO.	SYSTEM/COMPONENT	DESCRIPTION	UNIT	UNIT COST	NEW S.F. COST	MODEL S.F. COST	+/- CHANGE
D SERVICES								
		D10 Conveying						
D	1010	Elevators & Lifts　2	Type: *Same as model*　Capacity:　Stops:	Each	230,100	8.37	7.67	+ 0.70
D	1020	Escalators & Moving Walks	Type:	Each				
		D20 Plumbing						
D	2010	Plumbing　*Kitchen, bath, + service fixtures*	*1 Fixture/215 S.F. Floor*	Each	2915.4	13.56	13.56	---
D	2020	Domestic Water Distribution		S.F. Floor	4.50	4.50	4.50	---
D	2040	Rain Water Drainage		S.F. Roof	1.32	0.26	0.22	+ 0.04
		D30 HVAC						
D	3010	Energy Supply		S.F. Floor	8.07	8.07	8.07	---
D	3020	Heat Generating Systems	Type: *Same as model*	S.F. Floor				
D	3030	Cooling Generating Systems	Type: *Same as model*	S.F. Floor	9.54	9.54	9.54	---
D	3090	Other HVAC Sys. & Equipment		Each				
		D40 Fire Protection						
D	4010	Sprinklers	Same as model	S.F. Floor	2.76	2.76	2.76	---
D	4020	Standpipes	Same as model	S.F. Floor	0.87	0.87	0.87	---
		D50 Electrical						
D	5010	Electrical Service/Distribution	Same as model	S.F. Floor	2.06	2.06	2.06	---
D	5020	Lighting & Branch Wiring		S.F. Floor	7.08	7.08	7.08	---
D	5030	Communications & Security		S.F. Floor	1.66	1.66	1.66	---
D	5090	Other Electrical Systems		S.F. Floor	0.20	0.20	0.20	---
E EQUIPMENT & FURNISHINGS								
E	1010	Commercial Equipment		Each				
E	1020	Institutional Equipment		Each				
E	1030	Vehicular Equipment		Each				
E	1090	Other Equipment		Each				
F SPECIAL CONSTRUCTION								
F	1020	Integrated Construction	N/A	S.F.				
F	1040	Special Facilities	N/A	S.F.				
G BUILDING SITEWORK								

ITEM					Total Change + 7.15
17	Model 020 total ____ $ **132.39**				
18	Adjusted S.F. cost ____ s **139.54** ____ item 17 +/- changes				
19	Building area - from item 11 ____ **55,000** ____ S.F. x adjusted S.F. cost....................		$		7,674,700
20	Basement area - from item 15 ____ S.F. x S.F. cost $		$		
21	Base building sub-total - item 20 + item 19		$		7,674,700
22	Miscellaneous addition (quality, etc.)		$		
23	Sub-total - item 22 + 21		$		7,674,700
24	General conditions -25 % of item 23		$		1,918,675
25	Sub-total - item 24 + item 23		$		9,593,375
26	Architects fees ____ **5.5** ____ % of item 25		$		527,636
27	Sub-total - item 26 + item 27		$		10,121,011
28	Location modifier .... **zip code 07410**		x		1.12
29	Local replacement cost - item 28 x item 27		$		11,335,532
30	Depreciation ____ **20** ____ % of item 29		$		2,267,106
31	Depreciated local replacement cost - item 29 less item 30		$		9,068,426
32	Exclusions		$		
33	Net depreciated replacement cost - item 31 less item 32		$		9,068,426

From page 467

Use table on page 232

Costs per square foot of floor area

Exterior Wall	S.F. Area	8000	12000	15000	19000	22500	25000	29000	32000	36000
	L.F. Perimeter	213	280	330	350	400	433	442	480	520
Face Brick with Concrete Block Back-up	Wood Joists	188.10	178.85	175.25	167.80	166.00	164.80	160.75	160.10	158.85
	Steel Joists	189.20	181.40	178.15	172.20	170.65	169.70	166.35	165.85	164.80
Stucco on Concrete Block	Wood Joists	164.25	157.25	154.40	149.40	148.05	147.20	144.45	144.00	143.10
	Steel Joists	179.35	172.30	169.45	164.25	162.90	162.00	159.25	158.75	157.80
Wood Siding	Wood Frame	163.85	156.95	154.25	149.30	147.95	147.10	144.50	144.00	143.10
Brick Veneer	Wood Frame	172.90	164.85	161.60	155.45	153.85	152.85	149.50	148.90	147.80
Perimeter Adj., Add or Deduct	Per 100 L.F.	21.80	14.60	11.60	9.15	7.75	7.05	6.00	5.45	4.80
Story Hgt. Adj., Add or Deduct	Per 1 Ft.	4.15	3.70	3.35	2.85	2.80	2.75	2.35	2.35	2.20

For Basement, add $35.05 per square foot of basement area

The above costs were calculated using the basic specifications shown on the facing page. These costs should be adjusted where necessary for design alternatives and owner's requirements. Reported completed project costs, for this type of structure, range from $59.60 to $221.65 per S.F.

Common additives

Description	Unit	$ Cost
Appliances		
Cooking range, 30" free standing		
1 oven	Each	570 - 2100
2 oven	Each	1375 - 2875
30" built-in		
1 oven	Each	920 - 2000
2 oven	Each	1825 - 2425
Counter top cook tops, 4 burner	Each	410 - 1475
Microwave oven	Each	278 - 795
Combination range, refrig. & sink, 30" wide	Each	1875 - 5550
72" wide	Each	5925
Combination range, refrigerator, sink, microwave oven & icemaker	Each	6750
Compactor, residential, 4-1 compaction	Each	815 - 1200
Dishwasher, built-in, 2 cycles	Each	575 - 1125
4 cycles	Each	700 - 1825
Garbage disposer, sink type	Each	225 - 370
Hood for range, 2 speed, vented, 30" wide	Each	310 - 1225
42" wide	Each	420 - 2725
Refrigerator, no frost 10-12 C.F.	Each	520 - 660
18-20 C.F.	Each	805 - 1675

Description	Unit	$ Cost
Closed Circuit Surveillance, One station		
Camera and monitor	Each	1950
For additional camera stations, add	Each	1050
Elevators, Hydraulic passenger, 2 stops		
2000# capacity	Each	67,900
2500# capacity	Each	68,900
3500# capacity	Each	72,900
Additional stop, add	Each	8375
Emergency Lighting, 25 watt, battery operated		
Lead battery	Each	305
Nickel cadmium	Each	850
Laundry Equipment		
Dryer, gas, 16 lb. capacity	Each	945
30 lb. capacity	Each	3750
Washer, 4 cycle	Each	1175
Commercial	Each	1600
Smoke Detectors		
Ceiling type	Each	220
Duct type	Each	550

Important: See the Reference Section for Location Factors.

Model costs calculated for a 3 story building with 10' story height and 22,500 square feet of floor area

				Unit	Unit Cost	Cost Per S.F.	% Of Sub-Total
A. SUBSTRUCTURE							
1010	Standard Foundations	Poured concrete; strip and spread footings		S.F. Ground	6.18	2.06	
1020	Special Foundations	N/A		—	—	—	
1030	Slab on Grade	4" reinforced concrete with vapor barrier and granular base		S.F. Slab	5.18	1.72	4.2%
2010	Basement Excavation	Site preparation for slab and trench for foundation wall and footing		S.F. Ground	.18	.06	
2020	Basement Walls	4' foundation wall		L.F. Wall	75	1.52	
B. SHELL							
B10 Superstructure							
1010	Floor Construction	Open web steel joists, slab form, concrete, interior steel columns		S.F. Floor	19.53	13.02	13.0%
1020	Roof Construction	Open web steel joists with rib metal deck, interior steel columns		S.F. Roof	10.14	3.38	
B20 Exterior Enclosure							
2010	Exterior Walls	Face brick with concrete block backup	88% of wall	S.F. Wall	23.29	10.93	
2020	Exterior Windows	Aluminum horizontal sliding	12% of wall	Each	524	2.24	10.7%
2030	Exterior Doors	Aluminum and glass		Each	1975	.35	
B30 Roofing							
3010	Roof Coverings	Built-up tar and gravel with flashing; perlite/EPS composite insulation		S.F. Roof	6.33	2.11	1.7%
3020	Roof Openings	N/A		—	—	—	
C. INTERIORS							
1010	Partitions	Gypsum board and sound deadening board on metal studs	10 S.F. of Floor/L.F. Partition	S.F. Partition	6.57	5.84	
1020	Interior Doors	15% solid core wood, 85% hollow core wood	80 S.F. Floor/Door	Each	555	6.94	
1030	Fittings	Kitchen Cabinets		S.F. Floor	2.99	2.99	
2010	Stair Construction	Concrete filled metal pan		Flight	14,150	3.14	24.0%
3010	Wall Finishes	70% paint, 25% vinyl wall covering, 5% ceramic tile		S.F. Surface	1.40	2.48	
3020	Floor Finishes	60% carpet, 30% vinyl composition tile, 10% ceramic tile		S.F. Floor	5.23	5.23	
3030	Ceiling Finishes	Painted gypsum board on resilient channels		S.F. Ceiling	3.68	3.68	
D. SERVICES							
D10 Conveying							
1010	Elevators & Lifts	One hydraulic passenger elevator		Each	108,450	4.82	3.8%
1020	Escalators & Moving Walks	N/A		—	—	—	
D20 Plumbing							
2010	Plumbing Fixtures	Kitchen, bathroom and service fixtures, supply and drainage	1 Fixture/200 S.F. Floor	Each	2972	14.86	
2020	Domestic Water Distribution	Gas fired water heater		S.F. Floor	4.67	4.67	15.8%
2040	Rain Water Drainage	Roof drains		S.F. Roof	1.38	.46	
D30 HVAC							
3010	Energy Supply	Oil fired hot water, baseboard radiation		S.F. Floor	9.22	9.22	
3020	Heat Generating Systems	N/A		—	—	—	
3030	Cooling Generating Systems	Chilled water, air cooled condenser system		S.F. Floor	9.92	9.92	15.1%
3050	Terminal & Package Units	N/A		—	—	—	
3090	Other HVAC Sys. & Equipment	N/A		—	—	—	
D40 Fire Protection							
4010	Sprinklers	Wet pipe sprinkler system		S.F. Floor	3.41	3.41	2.7%
4020	Standpipes	N/A		—	—	—	
D50 Electrical							
5010	Electrical Service/Distribution	600 ampere service, panel board and feeders		S.F. Floor	2.27	2.27	
5020	Lighting & Branch Wiring	Incandescent fixtures, receptacles, switches, A.C. and misc. power		S.F. Floor	7.12	7.12	9.0%
5030	Communications & Security	Addressable alarm systems, internet wiring, and emergency lighting		S.F. Floor	1.74	1.74	
5090	Other Electrical Systems	Emergency generator, 11.5 kW		S.F. Floor	.22	.22	
E. EQUIPMENT & FURNISHINGS							
1010	Commercial Equipment	N/A		—	—	—	
1020	Institutional Equipment	N/A		—	—	—	
1030	Vehicular Equipment	N/A		—	—	—	0.0%
1090	Other Equipment	N/A		—	—	—	
F. SPECIAL CONSTRUCTION							
1020	Integrated Construction	N/A		—	—	—	0.0%
1040	Special Facilities	N/A		—	—	—	
G. BUILDING SITEWORK	**N/A**						

	Sub-Total	126.40	100%
CONTRACTOR FEES (General Requirements: 10%, Overhead: 5%, Profit: 10%)	25%	31.61	
ARCHITECT FEES	8%	12.64	
Total Building Cost		**170.65**	

Costs per square foot of floor area

Exterior Wall	S.F. Area	40000	45000	50000	55000	60000	70000	80000	90000	100000
	L.F. Perimeter	366	400	433	466	500	510	530	550	594
Face Brick with Concrete Block Back-up	Steel Frame	181.45	180.00	178.85	177.85	177.05	173.60	171.40	169.60	168.90
	R/Conc. Frame	187.75	186.10	184.80	183.80	182.85	178.75	176.10	174.00	173.15
Decorative Concrete Block	Steel Frame	172.65	171.35	170.30	169.45	168.75	165.80	164.00	162.45	161.85
	R/Conc. Frame	172.55	171.20	170.10	169.30	168.55	165.55	163.70	162.20	161.55
Precast Concrete Panels	Steel Frame	187.30	185.60	184.15	183.05	182.10	177.50	174.55	172.20	171.35
	R/Conc. Frame	186.90	185.15	183.70	182.55	181.65	177.05	174.10	171.75	170.85
Perimeter Adj., Add or Deduct	Per 100 L.F.	6.85	6.00	5.35	4.95	4.55	3.90	3.45	3.00	2.65
Story Hgt. Adj., Add or Deduct	Per 1 Ft.	2.25	2.20	2.10	2.10	2.05	1.80	1.60	1.50	1.45
For Basement, add $35.10 per square foot of basement area										

The above costs were calculated using the basic specifications shown on the facing page. These costs should be adjusted where necessary for design alternatives and owner's requirements. Reported completed project costs, for this type of structure, range from $71.40 to $202.95 per S.F.

Common additives

Description	Unit	$ Cost	Description	Unit	$ Cost
Appliances			Closed Circuit Surveillance, One station		
Cooking range, 30" free standing			Camera and monitor	Each	1950
1 oven	Each	570 - 2100	For additional camera stations, add	Each	1050
2 oven	Each	1375 - 2875	Elevators, Electric passenger, 5 stops		
30" built-in			2000# capacity	Each	165,300
1 oven	Each	920 - 2000	3500# capacity	Each	171,300
2 oven	Each	1825 - 2425	5000# capacity	Each	177,300
Counter top cook tops, 4 burner	Each	410 - 1475	Additional stop, add	Each	12,700
Microwave oven	Each	278 - 795	Emergency Lighting, 25 watt, battery operated		
Combination range, refrig. & sink, 30" wide	Each	1875 - 5550	Lead battery	Each	305
72" wide	Each	5925	Nickel cadmium	Each	850
Combination range, refrigerator, sink,			Laundry Equipment		
microwave oven & icemaker	Each	6750	Dryer, gas, 16 lb. capacity	Each	945
Compactor, residential, 4-1 compaction	Each	815 - 1200	30 lb. capacity	Each	3750
Dishwasher, built-in, 2 cycles	Each	575 - 1125	Washer, 4 cycle	Each	1175
4 cycles	Each	700 - 1825	Commercial	Each	1600
Garbage disposer, sink type	Each	225 - 370	Smoke Detectors		
Hood for range, 2 speed, vented, 30" wide	Each	310 - 1225	Ceiling type	Each	220
42" wide	Each	420 - 2725	Duct type	Each	550
Refrigerator, no frost 10-12 C.F.	Each	520 - 660			
18-20 C.F.	Each	805 - 1675			

Important: See the Reference Section for Location Factors.

Model costs calculated for a 6 story building with 10'-4" story height and 60,000 square feet of floor area

				Unit	Unit Cost	Cost Per S.F.	% Of Sub-Total
A.	**SUBSTRUCTURE**						
1010	Standard Foundations	Poured concrete; strip and spread footings and 4' foundation wall		S.F. Ground	7.08	1.18	
1020	Special Foundations	N/A		—	—	—	
1030	Slab on Grade	4" reinforced concrete with vapor barrier and granular base		S.F. Slab	5.18	.86	2.1%
2010	Basement Excavation	Site preparation for slab and trench for foundation wall and footing		S.F. Ground	.30	.05	
2020	Basement Walls	4' foundation wall		L.F. Wall	79	.66	
B.	**SHELL**						
	B10 Superstructure						
1010	Floor Construction	Open web steel joists, slab form, concrete, steel columns		S.F. Floor	25.63	21.36	17.3%
1020	Roof Construction	Open web steel joists with rib metal deck, steel columns		S.F. Roof	9.42	1.57	
	B20 Exterior Enclosure						
2010	Exterior Walls	Face brick with concrete block backup	86% of wall	S.F. Wall	23.29	10.35	
2020	Exterior Windows	Aluminum horizontal sliding	14% of wall	Each	524	2.53	10.0%
2030	Exterior Doors	Aluminum and glass		Each	4750	.32	
	B30 Roofing						
3010	Roof Coverings	Built-up tar and gravel with flashing; perlite/EPS composite insulation		S.F. Roof	6.24	1.04	0.8%
3020	Roof Openings	N/A		—	—	—	
C.	**INTERIORS**						
1010	Partitions	Gypsum board and sound deadening board on metal studs	8 S.F. Floor/L.F. Partitions	S.F. Partition	6.92	6.92	
1020	Interior Doors	15% solid core wood, 85% hollow core wood	80 S.F Floor/Door	Each	677	8.46	
1030	Fittings	Kitchen cabinets		S.F. Floor	3.58	3.58	
2010	Stair Construction	Concrete filled metal pan		Flight	11,300	3.02	25.9%
3010	Wall Finishes	70% paint, 25% vinyl wall covering, 5% ceramic tile		S.F. Surface	1.55	3.09	
3020	Floor Finishes	60% carpet, 30% vinyl composition tile, 10% ceramic tile		S.F. Floor	5.23	5.23	
3030	Ceiling Finishes	Painted gypsum board on resilient channels		S.F. Ceiling	3.98	3.98	
D.	**SERVICES**						
	D10 Conveying						
1010	Elevators & Lifts	Two geared passenger elevators		Each	230,100	7.67	5.8%
1020	Escalators & Moving Walks	N/A		—	—	—	
	D20 Plumbing						
2010	Plumbing Fixtures	Kitchen, bathroom and service fixtures, supply and drainage	1 Fixture/215 S.F. Floor	Each	2915	13.56	
2020	Domestic Water Distribution	Gas fired water heater		S.F. Floor	4.50	4.50	13.8%
2040	Rain Water Drainage	Roof drains		S.F. Roof	1.32	.22	
	D30 HVAC						
3010	Energy Supply	Oil fired hot water, baseboard radiation		S.F. Floor	8.07	8.07	
3020	Heat Generating Systems	N/A		—	—	—	
3030	Cooling Generating Systems	Chilled water, air cooled condenser system		S.F. Floor	9.54	9.54	13.3%
3050	Terminal & Package Units	N/A		—	—	—	
3090	Other HVAC Sys. & Equipment	N/A		—	—	—	
	D40 Fire Protection						
4010	Sprinklers	Wet pipe sprinkler system		S.F. Floor	2.76	2.76	2.7%
4020	Standpipes	Standpipe, fire pump		S.F. Floor	.87	.87	
	D50 Electrical						
5010	Electrical Service/Distribution	1600 ampere service, panel board and feeders		S.F. Floor	2.06	2.06	
5020	Lighting & Branch Wiring	Incandescent fixtures, receptacles, switches, A.C. and misc. power		S.F. Floor	7.08	7.08	8.3%
5030	Communications & Security	Addressable alarm system, internet wiring, emergency lighting, and intercom		S.F. Floor	1.66	1.66	
5090	Other Electrical Systems	Emergency generator, 30 kW		S.F. Floor	.20	.20	
E.	**EQUIPMENT & FURNISHINGS**						
1010	Commercial Equipment	N/A		—	—	—	
1020	Institutional Equipment	N/A		—	—	—	0.0 %
1030	Vehicular Equipment	N/A		—	—	—	
1090	Other Equipment	N/A		—	—	—	
F.	**SPECIAL CONSTRUCTION**						
1020	Integrated Construction	N/A		—	—	—	0.0 %
1040	Special Facilities	N/A		—	—	—	
G.	**BUILDING SITEWORK**	**N/A**					

	Sub-Total	132.39	**100%**
CONTRACTOR FEES (General Requirements: 10%, Overhead: 5%, Profit: 10%)	25%	33.08	
ARCHITECT FEES	7%	11.58	
Total Building Cost		**177.05**	

Costs per square foot of floor area

Exterior Wall	S.F. Area	95000	112000	129000	145000	170000	200000	275000	400000	600000
	L.F. Perimeter	345	386	406	442	480	510	530	570	630
Ribbed Precast Concrete Panel	Steel Frame	238.35	235.20	230.95	229.20	225.90	222.15	214.10	207.60	202.70
	R/Conc. Frame	230.35	227.30	223.30	221.65	218.55	215.10	207.55	201.50	196.90
Face Brick with Concrete Block Back-up	Steel Frame	211.10	208.50	205.30	203.85	201.35	198.60	192.65	187.85	184.30
	R/Conc. Frame	221.10	218.50	215.25	213.85	211.35	208.55	202.65	197.85	194.25
Stucco on Concrete Block	Steel Frame	200.10	198.05	195.75	194.65	192.85	190.90	186.80	183.50	181.10
	R/Conc. Frame	210.05	208.00	205.70	204.65	202.80	200.85	196.80	193.50	191.10
Perimeter Adj., Add or Deduct	Per 100 L.F.	12.10	10.20	8.85	7.85	6.75	5.75	4.15	2.80	1.90
Story Hgt. Adj., Add or Deduct	Per 1 Ft.	3.90	3.55	3.30	3.20	3.00	2.75	2.00	1.45	1.10
For Basement, add $35.95 per square foot of basement area										

The above costs were calculated using the basic specifications shown on the facing page. These costs should be adjusted where necessary for design alternatives and owner's requirements. Reported completed project costs, for this type of structure, range from $91.40 to $214.45 per S.F.

Common additives

Description	Unit	$ Cost
Appliances		
Cooking range, 30" free standing		
1 oven	Each	570 - 2100
2 oven	Each	1375 - 2875
30" built-in		
1 oven	Each	920 - 2000
2 oven	Each	1825 - 2425
Counter top cook tops, 4 burner	Each	410 - 1475
Microwave oven	Each	278 - 795
Combination range, refrig. & sink, 30" wide	Each	1875 - 5550
72" wide	Each	5925
Combination range, refrigerator, sink,		
microwave oven & icemaker	Each	6750
Compactor, residential, 4-1 compaction	Each	815 - 1200
Dishwasher, built-in, 2 cycles	Each	575 - 1125
4 cycles	Each	700 - 1825
Garbage disposer, sink type	Each	225 - 370
Hood for range, 2 speed, vented, 30" wide	Each	310 - 1225
42" wide	Each	420 - 2725
Refrigerator, no frost 10-12 C.F.	Each	520 - 660
18-20 C.F.	Each	805 - 1675

Description	Unit	$ Cost
Closed Circuit Surveillance, One station		
Camera and monitor	Each	1950
For additional camera stations, add	Each	1050
Elevators, Electric passenger, 10 stops		
3000# capacity	Each	418,000
4000# capacity	Each	420,500
5000# capacity	Each	425,500
Additional stop, add	Each	12,700
Emergency Lighting, 25 watt, battery operated		
Lead battery	Each	305
Nickel cadmium	Each	850
Laundry Equipment		
Dryer, gas, 16 lb. capacity	Each	945
30 lb. capacity	Each	3750
Washer, 4 cycle	Each	1175
Commercial	Each	1600
Smoke Detectors		
Ceiling type	Each	220
Duct type	Each	550

Important: See the Reference Section for Location Factors.

Model costs calculated for a 15 story building with 10'-6" story height and 145,000 square feet of floor area

Apartment, 8-24 Story

				Unit	Unit Cost	Cost Per S.F.	% Of Sub-Total
A. SUBSTRUCTURE							
1010	Standard Foundations	CIP concrete pile caps		S.F. Ground	9.15	.61	
1020	Special Foundations	Steel H-piles, concrete grade beams		S.F. Ground	215	14.31	
1030	Slab on Grade	4" reinforced concrete with vapor barrier and granular base		S.F. Slab	5.18	.34	9.0%
2010	Basement Excavation	Site preparation for slab, piles and grade beam		S.F. Ground	.30	.02	
2020	Basement Walls	4' Foundation wall		L.F. Wall	75	.27	
B. SHELL							
	B10 Superstructure						
1010	Floor Construction	Open web steel joists, slab form, concrete, interior steel columns		S.F. Floor	21.96	20.50	12.1%
1020	Roof Construction	Open web steel joists with rib metal deck, interior steel columns		S.F. Roof	7.05	.47	
	B20 Exterior Enclosure						
2010	Exterior Walls	Ribbed precast concrete panel	87% of wall	S.F. Wall	47.74	19.94	
2020	Exterior Windows	Aluminum horizontal sliding	13% of wall	Each	524	2.18	14.4%
2030	Exterior Doors	Aluminum and glass		Each	3216	2.75	
	B30 Roofing						
3010	Roof Coverings	Built-up tar and gravel with flashing; perlite/EPS composite insulation		S.F. Roof	6	.40	0.2%
3020	Roof Openings	N/A		—	—	—	
C. INTERIORS							
1010	Partitions	Gypsum board on concrete block and metal studs	10 S.F. of Floor/L.F. Partition	S.F. Partition	13.91	13.91	
1020	Interior Doors	15% solid core wood, 85% hollow core wood	80 S.F. Floor/Door	Each	693	8.66	
1030	Fittings	Kitchen cabinets		S.F. Floor	3.71	3.71	
2010	Stair Construction	Concrete filled metal pan		Flight	11,300	3.35	24.2%
3010	Wall Finishes	70% paint, 25% vinyl wall covering, 5% ceramic tile		S.F. Surface	1.53	3.06	
3020	Floor Finishes	60% carpet, 30% vinyl composition tile, 10% ceramic tile		S.F. Floor	5.23	5.23	
3030	Ceiling Finishes	Painted gypsum board on resilient channels		S.F. Ceiling	3.98	3.98	
D. SERVICES							
	D10 Conveying						
1010	Elevators & Lifts	Four geared passenger elevators		Each	494,450	13.64	7.9%
1020	Escalators & Moving Walks	N/A		—	—	—	
	D20 Plumbing						
2010	Plumbing Fixtures	Kitchen, bathroom and service fixtures, supply and drainage	1 Fixture/210 S.F. Floor	Each	3259	15.52	
2020	Domestic Water Distribution	Gas fired water heater		S.F. Floor	5.18	5.18	12.1%
2040	Rain Water Drainage	Roof drains		S.F Roof	2.70	.18	
	D30 HVAC						
3010	Energy Supply	Oil fired hot water, baseboard radiation		S.F.Floor	8.07	8.07	
3020	Heat Generating Systems	N/A		—	—	—	
3030	Cooling Generating Systems	Chilled water, air cooled condenser system		S.F. Floor	9.54	9.54	10.2%
3050	Terminal & Package Units	N/A		—	—	—	
3090	Other HVAC Sys. & Equipment	N/A		—	—	—	
	D40 Fire Protection						
4010	Sprinklers	Wet pipe sprinkler system		S.F. Floor	2.80	2.80	2.6%
4020	Standpipes	Standpipe, fire pumps		S.F. Floor	1.63	1.63	
	D50 Electrical						
5010	Electrical Service/Distribution	4000 ampere service, panel board and feeders		S.F. Floor	1.79	1.79	
5020	Lighting & Branch Wiring	Incandescent fixtures, receptacles, switches, A.C. and misc. power		S.F. Floor	7.62	7.62	7.3%
5030	Communications & Security	Addressable alarm systems, internet wiring, em. lighting, intercom & security TV		S.F. Floor	3.08	3.08	
5090	Other Electrical Systems	Emergency generator, 30 kW		S.F. Floor	.22	.22	
E. EQUIPMENT & FURNISHINGS							
1010	Commercial Equipment	N/A		—	—	—	
1020	Institutional Equipment	N/A		—	—	—	0.0 %
1030	Vehicular Equipment	N/A		—	—	—	
1090	Other Equipment	N/A		—	—	—	
F. SPECIAL CONSTRUCTION							
1020	Integrated Construction	N/A		—	—	—	0.0%
1040	Special Facilities	N/A		—	—	—	
G. BUILDING SITEWORK	**N/A**						
				Sub-Total		172.96	**100%**
	CONTRACTOR FEES (General Requirements: 10%, Overhead: 5%, Profit: 10%)				25%	43.27	
	ARCHITECT FEES				6%	12.97	
				Total Building Cost		**229.20**	

Costs per square foot of floor area

Exterior Wall	S.F. Area	2000	3000	4000	5000	7500	10000	15000	25000	50000
	L.F. Perimeter	188	224	260	300	356	400	504	700	1200
Brick Veneer	Wood Frame	243.00	216.25	202.85	195.30	181.70	174.15	167.25	161.45	157.30
	Steel Frame	271.40	244.95	231.75	224.30	210.85	203.50	196.70	191.00	186.90
Brick Veneer on Block	Wood Truss	246.40	218.30	204.25	196.35	181.90	173.90	166.60	160.35	155.95
	Bearing Wall	282.15	254.05	240.00	232.10	217.60	209.55	202.20	196.05	191.55
Wood Siding	Wood Frame	228.55	204.45	192.40	185.55	173.55	167.00	160.95	155.85	152.15
Vinyl Siding	Wood Frame	227.75	204.10	192.30	185.55	173.95	167.60	161.80	156.90	153.40
Perimeter Adj., Add or Deduct	Per 100 L.F.	30.15	20.10	15.15	12.05	8.00	6.00	4.00	2.45	1.15
Story Hgt. Adj., Add or Deduct	Per 1 Ft.	3.75	2.95	2.60	2.40	1.85	1.55	1.35	1.10	0.95
For Basement, add $31.60 per square foot of basement area										

The above costs were calculated using the basic specifications shown on the facing page. These costs should be adjusted where necessary for design alternatives and owner's requirements. Reported completed project costs, for this type of structure, range from $89.75 to $225.10 per S.F.

Common additives

Description	Unit	$ Cost
Appliances		
Cooking range, 30" free standing		
1 oven	Each	570 - 2100
2 oven	Each	1375 - 2875
30" built-in		
1 oven	Each	920 - 2000
2 oven	Each	1825 - 2425
Counter top cook tops, 4 burner	Each	410 - 1475
Microwave oven	Each	278 - 795
Combination range, refrig. & sink, 30" wide	Each	1875 - 5550
60" wide	Each	4000
72" wide	Each	5925
Combination range, refrigerator, sink,		
microwave oven & icemaker	Each	6750
Compactor, residential, 4-1 compaction	Each	815 - 1200
Dishwasher, built-in, 2 cycles	Each	575 - 1125
4 cycles	Each	700 - 1825
Garbage disposer, sink type	Each	225 - 370
Hood for range, 2 speed, vented, 30" wide	Each	310 - 1225
42" wide	Each	420 - 2725

Description	Unit	$ Cost
Appliances, cont.		
Refrigerator, no frost 10-12 C.F.	Each	520 - 660
14-16 C.F.	Each	645 - 940
18-20 C.F.	Each	805 - 1675
Laundry Equipment		
Dryer, gas, 16 lb. capacity	Each	945
30 lb. capacity	Each	3750
Washer, 4 cycle	Each	1175
Commercial	Each	1600
Sound System		
Amplifier, 250 watts	Each	1975
Speaker, ceiling or wall	Each	206
Trumpet	Each	395

Model costs calculated for a 1 story building with 10' story height and 10,000 square feet of floor area

Assisted - Senior Living

				Unit	Unit Cost	Cost Per S.F.	% Of Sub-Total
A.	**SUBSTRUCTURE**						
1010	Standard Foundations	Poured concrete strip footings		S.F. Ground	2	2	
1020	Special Foundations	N/A		—	—	—	
1030	Slab on Grade	4" reinforced concrete with vapor barrier and granular base		S.F. Slab	5.18	5.18	8.7%
2010	Basement Excavation	Site preparation for slab and trench for foundation wall and footing		S.F. Ground	.55	.55	
2020	Basement Walls	4' foundation wall		L.F. Wall	79	3.16	
B.	**SHELL**						
	B10 Superstructure						
1010	Floor Construction	N/A		—	—	—	5.7 %
1020	Roof Construction	Plywood on wood trusses (pitched); fiberglass batt insulation		S.F. Roof	7.13	7.13	
	B20 Exterior Enclosure						
2010	Exterior Walls	Brick veneer on wood studs, insulated	80% of wall	S.F. Wall	23.91	7.65	
2020	Exterior Windows	Double hung wood	20% of wall	Each	630	3.16	9.8%
2030	Exterior Doors	Aluminum and glass; 18 ga. steel flush		Each	2471	1.49	
	B30 Roofing						
3010	Roof Coverings	Asphalt shingles with flashing (pitched); al. gutters & downspouts		S.F. Roof	2.85	2.85	2.3%
3020	Roof Openings	N/A		—	—	—	
C.	**INTERIORS**						
1010	Partitions	Gypsum board on wood studs	6 S.F. Floor/L.F. Partition	S.F. Partition	5.02	6.69	
1020	Interior Doors	Single leaf wood	80 S.F. Floor/Door	Each	622	7.32	
1030	Fittings	N/A		—	—	—	
2010	Stair Construction	N/A		—	—	—	28.4%
3010	Wall Finishes	Paint on drywall		S.F. Surface	.82	2.19	
3020	Floor Finishes	80% carpet, 20% ceramic tile		S.F. Floor	11	11	
3030	Ceiling Finishes	Gypsum board on wood furring, painted		S.F. Ceiling	8.46	8.46	
D.	**SERVICES**						
	D10 Conveying						
1010	Elevators & Lifts	N/A		—	—	—	0.0 %
1020	Escalators & Moving Walks	N/A		—	—	—	
	D20 Plumbing						
2010	Plumbing Fixtures	Kitchen, toilet and service fixtures, supply and drainage	1 Fixture/200 S.F. Floor	Each	3394	16.97	
2020	Domestic Water Distribution	Gas water heater		S.F. Floor	2.21	2.21	15.3%
2040	Rain Water Drainage	N/A		—	—	—	
	D30 HVAC						
3010	Energy Supply	N/A		—	—	—	
3020	Heat Generating Systems	N/A		—	—	—	
3030	Cooling Generating Systems	N/A		—	—	—	2.7 %
3050	Terminal & Package Units	Gas fired hot water boiler; terminal package heat/A/C units		S.F. Floor	3.45	3.45	
3090	Other HVAC Sys. & Equipment	N/A		—	—	—	
	D40 Fire Protection						
4010	Sprinklers	Wet pipe sprinkler system		S.F. Floor	3.51	3.51	3.4%
4020	Standpipes	Standpipes		S.F. Floor	.80	.80	
	D50 Electrical						
5010	Electrical Service/Distribution	600 ampere service, panel board, feeders and branch wiring		S.F. Floor	8.67	8.67	
5020	Lighting & Branch Wiring	Fuorescent & incandescent fixtures, receptacles, switches, A.C. and misc. power		S.F. Floor	11.28	11.28	22.6%
5030	Communications & Security	Addressable alarm, fire detection and command center, communication system		S.F. Floor	8.14	8.14	
5090	Other Electrical Systems	Emergency generator, 7.5 kW		S.F. Floor	.22	.22	
E.	**EQUIPMENT & FURNISHINGS**						
1010	Commercial Equipment	Residential washers and dryers		S.F. Floor	.41	.41	
1020	Institutional Equipment	N/A		—	—	—	
1030	Vehicular Equipment	N/A		—	—	—	1.1%
1090	Other Equipment	Residential ranges and refrigerators		S.F. Floor	1.03	1.03	
F.	**SPECIAL CONSTRUCTION**						
1020	Integrated Construction	N/A		—	—	—	0.0 %
1040	Special Facilities	N/A		—	—	—	
G.	**BUILDING SITEWORK**	**N/A**					

			Sub-Total	125.52	**100%**
CONTRACTOR FEES (General Requirements: 10%, Overhead: 5%, Profit: 10%)		25%	31.37	
ARCHITECT FEES		11%	17.26	

Total Building Cost	**174.15**

Costs per square foot of floor area

Exterior Wall	S.F. Area	12000	15000	18000	21000	24000	27000	30000	33000	36000
	L.F. Perimeter	440	500	540	590	640	665	700	732	770
Face Brick with Concrete Block Back-up	Steel Frame	197.50	191.95	186.45	183.30	180.90	177.60	175.45	173.60	172.25
	Bearing Wall	191.00	186.05	181.15	178.30	176.15	173.20	171.35	169.70	168.50
Precast Concrete	Steel Frame	204.10	198.00	192.00	188.50	185.85	182.25	179.90	177.85	176.40
Decorative Concrete Block	Bearing Wall	172.55	169.25	166.00	164.05	162.55	160.65	159.40	158.25	157.45
Concrete Block	Steel Frame	181.20	178.50	175.95	174.45	173.25	171.75	170.75	169.90	169.25
	Bearing Wall	168.35	165.35	162.50	160.80	159.50	157.85	156.65	155.70	154.95
Perimeter Adj., Add or Deduct	Per 100 L.F.	13.25	10.65	8.85	7.55	6.70	5.90	5.30	4.80	4.50
Story Hgt. Adj., Add or Deduct	Per 1 Ft.	2.05	1.90	1.70	1.60	1.55	1.40	1.35	1.25	1.25

For Basement, add $28.70 per square foot of basement area

The above costs were calculated using the basic specifications shown on the facing page. These costs should be adjusted where necessary for design alternatives and owner's requirements. Reported completed project costs, for this type of structure, range from $90.45 to $229.00 per S.F.

Common additives

Description	Unit	$ Cost
Closed Circuit Surveillance, One station		
Camera and monitor	Each	1950
For additional camera stations, add	Each	1050
Emergency Lighting, 25 watt, battery operated		
Lead battery	Each	305
Nickel cadmium	Each	850
Seating		
Auditorium chair, all veneer	Each	265
Veneer back, padded seat	Each	335
Upholstered, spring seat	Each	320
Classroom, movable chair & desk	Set	81 - 171
Lecture hall, pedestal type	Each	257 - 750
Smoke Detectors		
Ceiling type	Each	220
Duct type	Each	550
Sound System		
Amplifier, 250 watts	Each	1975
Speaker, ceiling or wall	Each	206
Trumpet	Each	395

Important: See the Reference Section for Location Factors.

Model costs calculated for a 1 story building with 24' story height and 24,000 square feet of floor area

Auditorium

			Unit	Unit Cost	Cost Per S.F.	% Of Sub-Total
A. SUBSTRUCTURE						
1010	Standard Foundations	Poured concrete; strip and spread footings	S.F. Ground	1.41	1.41	
1020	Special Foundations	N/A	—	—	—	
1030	Slab on Grade	6" reinforced concrete with vapor barrier and granular base	S.F. Slab	6.16	6.16	7.3%
2010	Basement Excavation	Site preparation for slab and trench for foundation wall and footing	S.F. Ground	.18	.18	
2020	Basement Walls	4' foundation wall	L.F. Wall	88	2.33	
B. SHELL						
	B10 Superstructure					
1010	Floor Construction	Open web steel joists, slab form, concrete (balcony)	S.F. Floor	24.72	3.09	9.0%
1020	Roof Construction	Metal deck on steel truss	S.F. Roof	9.47	9.47	
	B20 Exterior Enclosure					
2010	Exterior Walls	Precast concrete panel 80% of wall (adjusted for end walls)	S.F. Wall	45.70	23.40	
2020	Exterior Windows	Glass curtain wall 20% of wall	Each	42.05	5.38	22.2%
2030	Exterior Doors	Double aluminum and glass and hollow metal	Each	4200	2.09	
	B30 Roofing					
3010	Roof Coverings	Built-up tar and gravel with flashing; perlite/EPS composite insulation	S.F. Roof	5.84	5.84	4.4%
3020	Roof Openings	Roof hatches	S.F. Roof	.24	.24	
C. INTERIORS						
1010	Partitions	Concrete Block and toilet partitions 40 S.F. Floor/L.F. Partition	S.F. Partition	9.10	3.64	
1020	Interior Doors	Single leaf hollow metal 400 S.F. Floor/Door	Each	1212	3.03	
1030	Fittings	N/A	—	—	—	
2010	Stair Construction	Concrete filled metal pan	Flight	16,950	2.12	19.1%
3010	Wall Finishes	70% paint, 30% epoxy coating	S.F. Surface	5.53	4.42	
3020	Floor Finishes	70% vinyl tile, 30% carpet	S.F. Floor	8.97	8.97	
3030	Ceiling Finishes	Fiberglass board, suspended	S.F. Ceiling	4.35	4.35	
D. SERVICES						
	D10 Conveying					
1010	Elevators & Lifts	One hydraulic passenger elevator	Each	91,440	3.81	2.7%
1020	Escalators & Moving Walks	N/A	—	—	—	
	D20 Plumbing					
2010	Plumbing Fixtures	Toilet and service fixtures, supply and drainage 1 Fixture/800 S.F. Floor	Each	6408	8.01	
2020	Domestic Water Distribution	Gas fired water heater	S.F. Floor	3.78	3.78	9.2%
2040	Rain Water Drainage	Roof drains	S.F. Roof	1.04	1.04	
	D30 HVAC					
3010	Energy Supply	N/A	—	—	—	
3020	Heat Generating Systems	Included in D3050	—	—	—	
3030	Cooling Generating Systems	N/A	—	—	—	9.4 %
3050	Terminal & Package Units	Single zone rooftop unit, gas heating, electric cooling	S.F. Floor	13	13	
3090	Other HVAC Sys. & Equipment	N/A	—	—	—	
	D40 Fire Protection					
4010	Sprinklers	Wet pipe sprinkler system	S.F. Floor	3.51	3.51	2.8%
4020	Standpipes	Standpipe	S.F. Floor	.33	.33	
	D50 Electrical					
5010	Electrical Service/Distribution	800 ampere service, panel board and feeders	S.F. Floor	1.99	1.99	
5020	Lighting & Branch Wiring	Fluorescent fixtures, receptacles, switches, A.C. and misc. power	S.F. Floor	11.99	11.99	13.9%
5030	Communications & Security	Addressable alarm systems, emergency lighting and public address system	S.F. Floor	4.01	4.01	
5090	Other Electrical Systems	Emergency generator, 100 kW	S.F. Floor	1.37	1.37	
E. EQUIPMENT & FURNISHINGS						
1010	Commercial Equipment	N/A	—	—	—	
1020	Institutional Equipment	N/A	—	—	—	
1030	Vehicular Equipment	N/A	—	—	—	0.0 %
1090	Other Equipment	N/A	—	—	—	
F. SPECIAL CONSTRUCTION						
1020	Integrated Construction	N/A	—	—	—	0.0 %
1040	Special Facilities	N/A	—	—	—	
G. BUILDING SITEWORK	**N/A**					

					Sub-Total	138.96	**100%**

CONTRACTOR FEES (General Requirements: 10%, Overhead: 5%, Profit: 10%)		25%	34.73
ARCHITECT FEES		7%	12.16

Total Building Cost 185.85

Costs per square foot of floor area

Exterior Wall	S.F. Area	2000	2700	3400	4100	4800	5500	6200	6900	7600
	L.F. Perimeter	180	208	236	256	280	303	317	337	357
Face Brick with Concrete Block Back-up	Steel Frame	265.10	249.35	240.05	232.10	227.30	223.45	219.10	216.45	214.35
	R/Conc. Frame	280.55	264.85	255.55	247.55	242.70	238.90	234.60	231.95	229.80
Precast Concrete Panel	Steel Frame	271.95	255.15	245.20	236.70	231.50	227.40	222.80	219.90	217.65
	R/Conc. Frame	292.00	273.15	262.00	252.65	246.95	242.40	237.45	234.30	231.80
Limestone with Concrete Block Back-up	Steel Frame	303.65	282.40	269.80	258.85	252.25	247.00	241.05	237.35	234.50
	R/Conc. Frame	319.10	297.85	285.30	274.25	267.70	262.50	256.45	252.85	249.90
Perimeter Adj., Add or Deduct	Per 100 L.F.	48.85	36.10	28.70	23.75	20.25	17.75	15.75	14.10	12.80
Story Hgt. Adj., Add or Deduct	Per 1 Ft.	4.70	4.05	3.55	3.20	3.00	2.85	2.65	2.55	2.40
For Basement, add $33.20 per square foot of basement area										

The above costs were calculated using the basic specifications shown on the facing page. These costs should be adjusted where necessary for design alternatives and owner's requirements. Reported completed project costs, for this type of structure, range from $137.95 to $339.50 per S.F.

Common additives

Description	Unit	$ Cost	Description	Unit	$ Cost
Bulletproof Teller Window, 44" x 60"	Each	4925	Service Windows, Pass thru, steel		
60" x 48"	Each	6725	24" x 36"	Each	3675
Closed Circuit Surveillance, One station			48" x 48"	Each	4000
Camera and monitor	Each	1950	72" x 40"	Each	6200
For additional camera stations, add	Each	1050	Smoke Detectors		
Counters, Complete	Station	5975	Ceiling type	Each	220
Door & Frame, 3' x 6'-8", bullet resistant steel			Duct type	Each	550
with vision panel	Each	7350 - 9500	Twenty-four Hour Teller		
Drive-up Window, Drawer & micr., not incl. glass	Each	8650 - 12,600	Automatic deposit cash & memo	Each	52,500
Emergency Lighting, 25 watt, battery operated			Vault Front, Door & frame		
Lead battery	Each	305	1 hour test, 32"x 78"	Opening	5275
Nickel cadmium	Each	850	2 hour test, 32" door	Opening	5650
Night Depository	Each	9400 - 14,200	40" door	Opening	6600
Package Receiver, painted	Each	1875	4 hour test, 32" door	Opening	6000
stainless steel	Each	2875	40" door	Opening	7175
Partitions, Bullet resistant to 8' high	L.F.	320 - 505	Time lock, two movement, add	Each	2800
Pneumatic Tube Systems, 2 station	Each	31,900			
With TV viewer	Each	58,500			

Important: See the Reference Section for Location Factors.

Model costs calculated for a 1 story building with 14' story height and 4,100 square feet of floor area

Bank

				Unit	Unit Cost	Cost Per S.F.	% Of Sub-Total
A. SUBSTRUCTURE							
1010	Standard Foundations	Poured concrete; strip and spread footings		S.F. Ground	4.13	4.13	
1020	Special Foundations	N/A		—	—	—	
1030	Slab on Grade	4" reinforced concrete with vapor barrier and granular base		S.F. Slab	5.18	5.18	8.7%
2010	Basement Excavation	Site preparation for slab and trench for foundation wall and footing		S.F. Ground	.32	.32	
2020	Basement Walls	4' Foundation wall		L.F. Wall	79	5.92	
B. SHELL							
	B10 Superstructure						
1010	Floor Construction	Cast-in-place columns		L.F. Column	121	5.90	11.9%
1020	Roof Construction	Cast-in-place concrete flat plate		S.F. Roof	15.40	15.40	
	B20 Exterior Enclosure						
2010	Exterior Walls	Face brick with concrete block backup	80% of wall	S.F. Wall	32.45	22.69	
2020	Exterior Windows	Horizontal aluminum sliding	20% of wall	Each	584	6.80	17.4%
2030	Exterior Doors	Double aluminum and glass and hollow metal		Each	3305	1.62	
	B30 Roofing						
3010	Roof Coverings	Built-up tar and gravel with flashing; perlite/EPS composite insulation		S.F. Roof	7.28	7.28	4.1%
3020	Roof Openings	N/A		—	—	—	
C. INTERIORS							
1010	Partitions	Gypsum board on metal studs	20 S.F. of Floor/L.F. Partition	S.F. Partition	10.12	5.06	
1020	Interior Doors	Single leaf hollow metal	200 S.F. Floor/Door	Each	1212	6.07	
1030	Fittings	N/A		—	—	—	
2010	Stair Construction	N/A		—	—	—	13.7%
3010	Wall Finishes	50% vinyl wall covering, 50% paint		S.F. Surface	1.37	1.37	
3020	Floor Finishes	50% carpet tile, 40% vinyl composition tile, 10% quarry tile		S.F. Floor	5.06	5.06	
3030	Ceiling Finishes	Mineral fiber tile on concealed zee bars		S.F. Ceiling	6.86	6.86	
D. SERVICES							
	D10 Conveying						
1010	Elevators & Lifts	N/A		—	—	—	0.0 %
1020	Escalators & Moving Walks	N/A		—	—	—	
	D20 Plumbing						
2010	Plumbing Fixtures	Toilet and service fixtures, supply and drainage	1 Fixture/580 S.F. Floor	Each	5348	9.22	
2020	Domestic Water Distribution	Gas fired water heater		S.F. Floor	1.30	1.30	6.7%
2040	Rain Water Drainage	Roof drains		S.F. Roof	1.50	1.50	
	D30 HVAC						
3010	Energy Supply	N/A		—	—	—	
3020	Heat Generating Systems	Included in D3050		—	—	—	
3030	Cooling Generating Systems	N/A		—	—	—	6.9 %
3050	Terminal & Package Units	Single zone rooftop unit, gas heating, electric cooling		S.F. Floor	12.35	12.35	
3090	Other HVAC Sys. & Equipment	N/A		—	—	—	
	D40 Fire Protection						
4010	Sprinklers	Wet pipe sprinkler system		S.F. Floor	4.66	4.66	4.1%
4020	Standpipes	Standpipe		S.F. Floor	2.74	2.74	
	D50 Electrical						
5010	Electrical Service/Distribution	200 ampere service, panel board and feeders		S.F. Floor	2.76	2.76	
5020	Lighting & Branch Wiring	Fluorescent fixtures, receptacles, switches, A.C. and misc. power		S.F. Floor	9.34	9.34	13.2%
5030	Communications & Security	Alarm systems, internet and phone wiring, emergency lighting, and security television		S.F. Floor	10.74	10.74	
5090	Other Electrical Systems	Emergency generator, 15 kW, Uninterruptible power supply		S.F. Floor	.64	.64	
E. EQUIPMENT & FURNISHINGS							
1010	Commercial Equipment	Automatic teller, drive up window, night depository		S.F. Floor	7.74	7.74	
1020	Institutional Equipment	Closed circuit TV monitoring system		S.F. Floor	2.96	2.96	6.0%
1030	Vehicular Equipment	N/A		—	—	—	
1090	Other Equipment	N/A		—	—	—	
F. SPECIAL CONSTRUCTION							
1020	Integrated Construction	N/A		—	—	—	7.2 %
1040	Special Facilities	Security vault door		S.F. Floor	12.81	12.81	
G. BUILDING SITEWORK	**N/A**						

		Sub-Total	178.42	100%
CONTRACTOR FEES (General Requirements: 10%, Overhead: 5%, Profit: 10%)		25%	44.60	
ARCHITECT FEES		11%	24.53	
	Total Building Cost		**247.55**	

Costs per square foot of floor area

Exterior Wall	S.F. Area	12000	14000	16000	18000	20000	22000	24000	26000	28000
	L.F. Perimeter	460	491	520	540	566	593	620	645	670
Concrete Block	Steel Roof Deck	169.55	166.40	164.00	161.80	160.30	159.00	157.90	157.05	156.15
Decorative Concrete Block	Steel Roof Deck	174.10	170.60	167.80	165.40	163.65	162.25	161.00	160.00	159.05
Face Brick on Concrete Block	Steel Roof Deck	182.50	178.25	175.00	172.00	169.90	168.15	166.65	165.45	164.30
Jumbo Brick on Concrete Block	Steel Roof Deck	178.70	174.80	171.80	169.05	167.10	165.50	164.15	162.95	161.95
Stucco on Concrete Block	Steel Roof Deck	172.90	169.50	166.85	164.45	162.85	161.40	160.20	159.25	158.35
Precast Concrete Panel	Steel Roof Deck	178.55	174.65	171.60	168.90	166.95	165.40	164.00	162.90	161.85
Perimeter Adj., Add or Deduct	Per 100 L.F.	5.10	4.40	3.80	3.40	3.05	2.80	2.60	2.35	2.25
Story Hgt. Adj., Add or Deduct	Per 1 Ft.	1.15	1.05	0.95	0.90	0.85	0.80	0.80	0.70	0.80
For Basement, add $32.75 per square foot of basement area										

The above costs were calculated using the basic specifications shown on the facing page. These costs should be adjusted where necessary for design alternatives and owner's requirements. Reported completed project costs, for this type of structure, range from $61.00 to $160.65 per S.F.

Common additives

Description	Unit	$ Cost
Bowling Alleys, Incl. alley, pinsetter,		
scorer, counter & misc. supplies, average	Lane	61,500
For automatic scorer, add (maximum)	Lane	10,900
Emergency Lighting, 25 watt, battery operated		
Lead battery	Each	305
Nickel cadmium	Each	850
Lockers, Steel, Single tier, 60" or 72"	Opening	218 - 355
2 tier, 60" or 72" total	Opening	112 - 160
5 tier, box lockers	Opening	70 - 86
Locker bench, lam., maple top only	L.F.	25.50
Pedestals, steel pipe	Each	72.50
Seating		
Auditorium chair, all veneer	Each	265
Veneer back, padded seat	Each	335
Upholstered, spring seat	Each	320
Sound System		
Amplifier, 250 watts	Each	1975
Speaker, ceiling or wall	Each	206
Trumpet	Each	395

Important: See the Reference Section for Location Factors.

Model costs calculated for a 1 story building with 14' story height and 20,000 square feet of floor area

				Unit	Unit Cost	Cost Per S.F.	% Of Sub-Total
A. SUBSTRUCTURE							
1010	Standard Foundations	Poured concrete; strip and spread footings		S.F. Ground	1.21	1.21	
1020	Special Foundations	N/A		—	—	—	
1030	Slab on Grade	4" reinforced concrete with vapor barrier and granular base		S.F. Slab	5.18	5.18	7.7%
2010	Basement Excavation	Site preparation for slab and trench for foundation wall and footing		S.F. Ground	.55	.55	
2020	Basement Walls	4' Foundation wall		L.F. Wall	79	2.24	
B. SHELL							
B10 Superstructure							
1010	Floor Construction	N/A		—	—	—	
1020	Roof Construction	Metal deck on open web steel joists with columns and beams		S.F. Roof	9.54	9.54	8.0 %
B20 Exterior Enclosure							
2010	Exterior Walls	Concrete block	90% of wall	S.F. Wall	12.37	4.41	
2020	Exterior Windows	Horizontal pivoted	10% of wall	Each	2005	3.31	7.0%
2030	Exterior Doors	Double aluminum and glass and hollow metal		Each	2468	.62	
B30 Roofing							
3010	Roof Coverings	Built-up tar and gravel with flashing; perlite/EPS composite insulation		S.F. Roof	5.60	5.60	4.8%
3020	Roof Openings	Roof hatches		S.F. Roof	.17	.17	
C. INTERIORS							
1010	Partitions	Concrete block	50 S.F. Floor/L.F. Partition	S.F. Partition	10.15	2.03	
1020	Interior Doors	Hollow metal single leaf	1000 S.F. Floor/Door	Each	1212	1.22	
1030	Fittings	N/A		—	—	—	
2010	Stair Construction	N/A		—	—	—	6.1%
3010	Wall Finishes	Paint and block filler		S.F. Surface	6.08	2.43	
3020	Floor Finishes	Vinyl tile	25% of floor	S.F. Floor	1.96	.49	
3030	Ceiling Finishes	Suspended fiberglass board	25% of area	S.F. Ceiling	4.35	1.09	
D. SERVICES							
D10 Conveying							
1010	Elevators & Lifts	N/A		—	—	—	
1020	Escalators & Moving Walks	N/A		—	—	—	0.0 %
D20 Plumbing							
2010	Plumbing Fixtures	Toilet and service fixtures, supply and drainage	1 Fixture/2200 S.F. Floor	Each	3344	1.52	
2020	Domestic Water Distribution	Gas fired water heater		S.F. Floor	.24	.24	2.0%
2040	Rain Water Drainage	Roof drains		S.F. Roof	.67	.67	
D30 HVAC							
3010	Energy Supply	N/A		—	—	—	
3020	Heat Generating Systems	Included in D3050		—	—	—	
3030	Cooling Generating Systems	N/A		—	—	—	12.3 %
3050	Terminal & Package Units	Single zone rooftop unit, gas heating, electric cooling		S.F. Floor	14.75	14.75	
3090	Other HVAC Sys. & Equipment	N/A		—	—	—	
D40 Fire Protection							
4010	Sprinklers	Sprinklers, light hazard		S.F. Floor	3.51	3.51	3.4%
4020	Standpipes	Standpipe, wet, Class III		S.F. Floor	.56	.56	
D50 Electrical							
5010	Electrical Service/Distribution	800 ampere service, panel board and feeders		S.F. Floor	2.42	2.42	
5020	Lighting & Branch Wiring	High efficiency fluorescent fixtures, receptacles, switches, A.C. and misc. power		S.F. Floor	10.87	10.87	12.3%
5030	Communications & Security	Addressable alarm systems and emergency lighting		S.F. Floor	1.31	1.31	
5090	Other Electrical Systems	Emergency generator, 7.5 kW		S.F. Floor	.09	.09	
E. EQUIPMENT & FURNISHINGS							
1010	Commercial Equipment	N/A		—	—	—	
1020	Institutional Equipment	N/A		—	—	—	
1030	Vehicular Equipment	N/A		—	—	—	36.6 %
1090	Other Equipment	Bowling Alley, ballrack, automatic scorers		S.F. Floor	43.82	43.82	
F. SPECIAL CONSTRUCTION							
1020	Integrated Construction	N/A		—	—	—	0.0 %
1040	Special Facilities	N/A		—	—	—	
G. BUILDING SITEWORK	**N/A**						

			Sub-Total	119.85	100%
CONTRACTOR FEES (General Requirements: 10%, Overhead: 5%, Profit: 10%)		25%	29.96	
ARCHITECT FEES		7%	10.49	
Total Building Cost			**160.30**	

Costs per square foot of floor area

Exterior Wall	S.F. Area	6000	8000	10000	12000	14000	16000	18000	20000	22000
	L.F. Perimeter	320	386	453	520	540	597	610	660	710
Face Brick with Concrete Block Back-up	Bearing Walls	175.95	169.80	166.20	163.70	159.25	157.75	154.65	153.55	152.75
	Steel Frame	179.05	172.95	169.35	166.90	162.50	161.00	158.00	156.90	156.10
Decorative Concrete Block	Bearing Walls	166.90	161.55	158.40	156.35	152.65	151.40	148.80	147.95	147.20
	Steel Frame	169.95	164.70	161.60	159.50	155.90	154.70	152.10	151.30	150.55
Precast Concrete Panels	Bearing Walls	175.75	169.60	165.95	163.50	159.05	157.55	154.45	153.40	152.60
	Steel Frame	178.75	172.70	169.15	166.70	162.30	160.85	157.80	156.80	155.95
Perimeter Adj., Add or Deduct	Per 100 L.F.	13.95	10.45	8.35	6.95	5.95	5.25	4.65	4.20	3.80
Story Hgt. Adj., Add or Deduct	Per 1 Ft.	2.35	2.10	1.95	1.90	1.65	1.65	1.45	1.45	1.40
Basement—Not Applicable										

The above costs were calculated using the basic specifications shown on the facing page. These costs should be adjusted where necessary for design alternatives and owner's requirements. Reported completed project costs, for this type of structure, range from $73.70 to $177.70 per S.F.

Common additives

Description	Unit	$ Cost
Directory Boards, Plastic, glass covered		
30" x 20"	Each	660
36" x 48"	Each	1475
Aluminum, 24" x 18"	Each	620
36" x 24"	Each	725
48" x 32"	Each	1050
48" x 60"	Each	2175
Emergency Lighting, 25 watt, battery operated		
Lead battery	Each	305
Nickel cadmium	Each	850
Benches, Hardwood	L.F.	126 - 231
Ticket Printer	Each	7650
Turnstiles, One way		
4 arm, 46" dia., manual	Each	2025
Electric	Each	3050
High security, 3 arm		
65" dia., manual	Each	7375
Electric	Each	10,300
Gate with horizontal bars		
65" dia., 7' high transit type	Each	7200

Important: See the Reference Section for Location Factors.

Model costs calculated for a 1 story building with 14' story height and 12,000 square feet of floor area

Bus Terminal

					Unit	Unit Cost	Cost Per S.F.	% Of Sub-Total
A.	**SUBSTRUCTURE**							
1010	Standard Foundations	Poured concrete; strip and spread footings			S.F. Ground	2.53	2.53	
1020	Special Foundations	N/A			—	—	—	
1030	Slab on Grade	6" reinforced concrete with vapor barrier and granular base			S.F. Slab	7.67	7.67	11.5%
2010	Basement Excavation	Site preparation for slab and trench for foundation wall and footing			S.F. Ground	.32	.32	
2020	Basement Walls	4' Foundation wall			L.F. Wall	81	3.51	
B.	**SHELL**							
	B10 Superstructure							
1010	Floor Construction	N/A			—	—	—	
1020	Roof Construction	Metal deck on open web steel joists with columns and beams			S.F. Roof	6.58	6.58	5.4%
	B20 Exterior Enclosure							
2010	Exterior Walls	Face brick with concrete block backup	70% of wall		S.F. Wall	32.45	13.78	
2020	Exterior Windows	Store front	30% of wall		Each	31	5.64	17.4%
2030	Exterior Doors	Double aluminum and glass			Each	5650	1.88	
	B30 Roofing							
3010	Roof Coverings	Built-up tar and gravel with flashing; perlite/EPS composite insulation			S.F. Roof	6.51	6.51	5.3%
3020	Roof Openings	N/A			—	—	—	
C.	**INTERIORS**							
1010	Partitions	Lightweight concrete block	15 S.F. Floor/L.F. Partition		S.F. Partition	10.13	6.75	
1020	Interior Doors	Hollow metal	150 S.F. Floor/Door		Each	1212	8.08	
1030	Fittings	N/A			—	—	—	
2010	Stair Construction	N/A			—	—	—	27.0%
3010	Wall Finishes	Glazed coating			S.F. Surface	1.93	2.57	
3020	Floor Finishes	Quarry tile and vinyl composition tile			S.F. Floor	8.83	8.83	
3030	Ceiling Finishes	Mineral fiber tile on concealed zee bars			S.F. Ceiling	6.86	6.86	
D.	**SERVICES**							
	D10 Conveying							
1010	Elevators & Lifts	N/A			—	—	—	
1020	Escalators & Moving Walks	N/A			—	—	—	0.0%
	D20 Plumbing							
2010	Plumbing Fixtures	Toilet and service fixtures, supply and drainage	1 Fixture/850 S.F. Floor		Each	6877	8.09	
2020	Domestic Water Distribution	Electric water heater			S.F. Floor	1.53	1.53	8.8%
2040	Rain Water Drainage	Roof drains			S.F. Roof	1.15	1.15	
	D30 HVAC							
3010	Energy Supply	N/A			—	—	—	
3020	Heat Generating Systems	Included in D3050			—	—	—	
3030	Cooling Generating Systems	N/A			—	—	—	9.1%
3050	Terminal & Package Units	Single zone rooftop unit, gas heating, electric cooling			S.F. Floor	11.15	11.15	
3090	Other HVAC Sys. & Equipment	N/A			—	—	—	
	D40 Fire Protection							
4010	Sprinklers	Wet pipe sprinkler system			S.F. Floor	4.53	4.53	
4020	Standpipes	Standpipe			S.F. Floor	.94	.94	4.5%
	D50 Electrical							
5010	Electrical Service/Distribution	400 ampere service, panel board and feeders			S.F. Floor	1.88	1.88	
5020	Lighting & Branch Wiring	Fluorescent fixtures, receptacles, switches, A.C. and misc. power			S.F. Floor	9.19	9.19	11.0%
5030	Communications & Security	Alarm systems and emergency lighting			S.F. Floor	2	2	
5090	Other Electrical Systems	Emergency generator, 7.5 kW			S.F. Floor	.44	.44	
E.	**EQUIPMENT & FURNISHINGS**							
1010	Commercial Equipment	N/A			—	—	—	
1020	Institutional Equipment	N/A			—	—	—	
1030	Vehicular Equipment	N/A			—	—	—	0.0%
1090	Other Equipment	N/A			—	—	—	
F.	**SPECIAL CONSTRUCTION**							
1020	Integrated Construction	N/A			—	—	—	
1040	Special Facilities	N/A			—	—	—	0.0%
G.	**BUILDING SITEWORK**	**N/A**						

	Sub-Total	122.41	**100%**
CONTRACTOR FEES (General Requirements: 10%, Overhead: 5%, Profit: 10%)	25%	30.58	
ARCHITECT FEES	7%	10.71	
Total Building Cost		**163.70**	

Costs per square foot of floor area

Exterior Wall	S.F. Area	600	800	1000	1200	1600	2000	2400	3000	4000
	L.F. Perimeter	100	114	128	139	164	189	214	250	314
Brick with Concrete Block Back-up	Steel Frame	357.60	343.30	334.75	327.60	319.70	314.90	311.80	308.35	305.55
	Bearing Walls	351.55	337.30	328.70	321.55	313.65	308.90	305.75	302.35	299.50
Concrete Block	Steel Frame	314.85	306.80	301.95	297.90	293.40	290.70	288.90	287.00	285.40
	Bearing Walls	308.85	300.75	295.90	291.85	287.35	284.65	282.90	280.95	279.30
Galvanized Steel Siding	Steel Frame	316.60	309.50	305.20	301.65	297.70	295.35	293.80	292.10	290.70
Metal Sandwich Panel	Steel Frame	315.25	306.20	300.75	296.20	291.15	288.15	286.15	284.00	282.20
Perimeter Adj., Add or Deduct	Per 100 L.F.	98.40	73.80	59.00	49.20	36.85	29.50	24.60	19.70	14.70
Story Hgt. Adj., Add or Deduct	Per 1 Ft.	6.05	5.20	4.65	4.15	3.65	3.50	3.25	3.00	2.80
Basement—Not Applicable										

The above costs were calculated using the basic specifications shown on the facing page. These costs should be adjusted where necessary for design alternatives and owner's requirements. Reported completed project costs, for this type of structure, range from $132.30 to $375.70 per S.F.

Common additives

Description	Unit	$ Cost
Air Compressors, Electric		
1-1/2 H.P., standard controls	Each	1525
Dual controls	Each	1925
5 H.P., 115/230 volt, standard control	Each	3625
Dual controls	Each	4500
Emergency Lighting, 25 watt, battery operated		
Lead battery	Each	305
Nickel cadmium	Each	850
Fence, Chain link, 6' high		
9 ga. wire, galvanized	L.F.	28
6 ga. wire	L.F.	30
Gate	Each	370
Product Dispenser with vapor recovery		
for 6 nozzles	Each	26,900
Laundry Equipment		
Dryer, gas, 16 lb. capacity	Each	945
30 lb. capacity	Each	3750
Washer, 4 cycle	Each	1175
Commercial	Each	1600

Description	Unit	$ Cost
Lockers, Steel, single tier, 60" or 72"	Opening	218 - 355
2 tier, 60" or 72" total	Opening	112 - 160
5 tier, box lockers	Opening	70 - 86
Locker bench, lam. maple top only	L.F.	25.50
Pedestals, steel pipe	Each	72.50
Paving, Bituminous		
Wearing course plus base course	S.Y.	8.05
Safe, Office type, 1 hour rating		
30" x 18" x 18"	Each	2500
60" x 36" x 18", double door	Each	8075
Sidewalks, Concrete 4" thick	S.F.	4.28
Yard Lighting,		
20' aluminum pole		
with 400 watt		
high pressure sodium		
fixture	Each	3075

Important: See the Reference Section for Location Factors.

Model costs calculated for a 1 story building with 12' story height and 800 square feet of floor area

				Unit	Unit Cost	Cost Per S.F.	% Of Sub-Total
A. SUBSTRUCTURE							
1010	Standard Foundations	Poured concrete; strip and spread footings		S.F. Ground	3.12	3.12	
1020	Special Foundations	N/A		—	—	—	
1030	Slab on Grade	5" reinforced concrete with vapor barrier and granular base		S.F. Slab	6.78	6.78	8.2%
2010	Basement Excavation	Site preparation for slab and trench for foundation wall and footing		S.F. Ground	1.02	1.02	
2020	Basement Walls	4' Foundation wall		L.F. Wall	71	10.07	
B. SHELL							
B10 Superstructure							
1010	Floor Construction	N/A		—	—	—	
1020	Roof Construction	Metal deck, open web steel joists, beams, columns		S.F. Roof	8.88	8.88	3.5 %
B20 Exterior Enclosure							
2010	Exterior Walls	Face brick with concrete block backup	70% of wall	S.F. Wall	32.45	38.84	
2020	Exterior Windows	Horizontal pivoted steel	5% of wall	Each	779	7.40	24.8%
2030	Exterior Doors	Steel overhead hollow metal		Each	3500	17.50	
B30 Roofing							
3010	Roof Coverings	Built-up tar and gravel with flashing; perlite/EPS composite insulation		S.F. Roof	8.27	8.27	3.2%
3020	Roof Openings	N/A		—	—	—	
C. INTERIORS							
1010	Partitions	Concrete block	20 S.F. Floor/S.F. Partition	S.F. Partition	9.64	4.82	
1020	Interior Doors	Hollow metal	600 S.F. Floor/Door	Each	1212	2.02	
1030	Fittings	N/A		—	—	—	
2010	Stair Construction	N/A		—	—	—	3.2%
3010	Wall Finishes	Paint		S.F. Surface	1.28	1.28	
3020	Floor Finishes	N/A		—	—	—	
3030	Ceiling Finishes	N/A		—	—	—	
D. SERVICES							
D10 Conveying							
1010	Elevators & Lifts	N/A		—	—	—	
1020	Escalators & Moving Walks	N/A		—	—	—	0.0 %
D20 Plumbing							
2010	Plumbing Fixtures	Toilet and service fixtures, supply and drainage	1 Fixture/160 S.F. Floor	Each	2824	17.65	
2020	Domestic Water Distribution	Gas fired water heater		S.F. Floor	35.65	35.65	22.3%
2040	Rain Water Drainage	Roof drains		S.F. Roof	4.03	4.03	
D30 HVAC							
3010	Energy Supply	N/A		—	—	—	
3020	Heat Generating Systems	Unit heaters		Each	20.80	20.80	
3030	Cooling Generating Systems	N/A		—	—	—	
3050	Terminal & Package Units	N/A		—	—	—	8.1 %
3090	Other HVAC Sys. & Equipment	N/A		—	—	—	
D40 Fire Protection							
4010	Sprinklers	N/A		—	—	—	
4020	Standpipes	N/A		—	—	—	0.0 %
D50 Electrical							
5010	Electrical Service/Distribution	400 ampere service, panel board and feeders		S.F. Floor	25.19	25.19	
5020	Lighting & Branch Wiring	Fluorescent fixtures, receptacles, switches and misc. power		S.F. Floor	41.46	41.46	
5030	Communications & Security	N/A		—	—	—	26.7%
5090	Other Electrical Systems	Emergency generator, 7.5 kW		S.F. Floor	1.90	1.90	
E. EQUIPMENT & FURNISHINGS							
1010	Commercial Equipment	N/A		—	—	—	
1020	Institutional Equipment	N/A		—	—	—	
1030	Vehicular Equipment	N/A		—	—	—	0.0 %
1090	Other Equipment	N/A		—	—	—	
F. SPECIAL CONSTRUCTION							
1020	Integrated Construction	N/A		—	—	—	
1040	Special Facilities	N/A		—	—	—	0.0 %
G. BUILDING SITEWORK	**N/A**						

		Sub-Total	256.68	100%
CONTRACTOR FEES (General Requirements: 10%, Overhead: 5%, Profit: 10%)		25%	64.16	
ARCHITECT FEES		7%	22.46	
	Total Building Cost		**343.30**	

Costs per square foot of floor area

Exterior Wall	S.F. Area	2000	7000	12000	17000	22000	27000	32000	37000	42000
	L.F. Perimeter	180	340	470	540	640	740	762	793	860
Decorative Concrete Block	Wood Arch	265.05	198.70	184.35	173.85	169.95	167.50	162.65	159.35	158.00
	Steel Truss	266.10	199.70	185.30	174.85	170.95	168.45	163.65	160.40	159.05
Stone with Concrete Block Back-up	Wood Arch	325.15	231.10	210.45	195.10	189.35	185.80	178.55	173.75	171.70
	Steel Truss	319.75	225.70	205.05	189.65	183.95	180.35	173.10	168.30	166.25
Face Brick with Concrete Block Back-up	Wood Arch	296.65	215.75	198.05	185.00	180.10	177.10	170.95	166.90	165.20
	Steel Truss	291.25	210.35	192.70	179.65	174.75	171.70	165.60	161.55	159.80
Perimeter Adj., Add or Deduct	Per 100 L.F.	64.85	18.50	10.80	7.65	5.90	4.80	4.00	3.50	3.15
Story Hgt. Adj., Add or Deduct	Per 1 Ft.	3.80	2.05	1.60	1.40	1.20	1.15	0.95	0.95	0.90

For Basement, add $34.20 per square foot of basement area

The above costs were calculated using the basic specifications shown on the facing page. These costs should be adjusted where necessary for design alternatives and owner's requirements. Reported completed project costs, for this type of structure, range from $84.45 to $345.25 per S.F.

Common additives

Description	Unit	$ Cost
Altar, Wood, custom design, plain	Each	2925
Deluxe	Each	15,000
Granite or marble, average	Each	16,500
Deluxe	Each	46,300
Ark, Prefabricated, plain	Each	12,000
Deluxe	Each	156,000
Baptistry, Fiberglass, incl. plumbing	Each	7100 - 11,300
Bells & Carillons, 48 bells	Each	1,031,000
24 bells	Each	485,000
Confessional, Prefabricated wood		
Single, plain	Each	4375
Deluxe	Each	10,900
Double, plain	Each	8300
Deluxe	Each	23,400
Emergency Lighting, 25 watt, battery operated		
Lead battery	Each	305
Nickel cadmium	Each	850
Lecterns, Wood, plain	Each	1025
Deluxe	Each	6925

Description	Unit	$ Cost
Pews/Benches, Hardwood	L.F.	126 - 231
Pulpits, Prefabricated, hardwood	Each	1900 - 11,300
Railing, Hardwood	L.F.	248
Steeples, translucent fiberglass		
30" square, 15' high	Each	10,000
25' high	Each	11,600
Painted fiberglass, 24" square, 14' high	Each	8625
28' high	Each	8425
Aluminum		
20' high, 3'- 6" base	Each	9675
35' high, 8'- 0" base	Each	37,300
60' high, 14'- 0" base	Each	79,000

Important: See the Reference Section for Location Factors.

Model costs calculated for a 1 story building with 24' story height and 17,000 square feet of floor area

Church

				Unit	Unit Cost	Cost Per S.F.	% Of Sub-Total
A. SUBSTRUCTURE							
1010	Standard Foundations	Poured concrete; strip and spread footings		S.F. Ground	3.91	3.91	
1020	Special Foundations	N/A		—	—	—	
1030	Slab on Grade	4" reinforced concrete with vapor barrier and granular base		S.F. Slab	5.18	5.18	9.8%
2010	Basement Excavation	Site preparation for slab and trench for foundation wall and footing		S.F. Ground	.18	.18	
2020	Basement Walls	4' Foundation wall		L.F. Wall	79	3.76	
B. SHELL							
	B10 Superstructure						
1010	Floor Construction	N/A		—	—	—	
1020	Roof Construction	Wood deck on laminated wood arches		S.F. Roof	18.43	19.67	14.8 %
	B20 Exterior Enclosure						
2010	Exterior Walls	Face brick with concrete block backup	80% of wall (adjusted for end walls)	S.F. Wall	38.53	23.50	
2020	Exterior Windows	Aluminum, top hinged, in-swinging and curtain wall panels	20% of wall	S.F. Window	29.78	4.54	21.4%
2030	Exterior Doors	Double hollow metal swinging, single hollow metal		Each	1425	.51	
	B30 Roofing						
3010	Roof Coverings	Asphalt shingles with flashing; polystyrene insulation		S.F. Roof	5.91	5.91	4.4%
3020	Roof Openings	N/A		—	—	—	
C. INTERIORS							
1010	Partitions	Plaster on metal studs	40 S.F. Floor/L.F. Partitions	S.F. Partition	14.80	8.88	
1020	Interior Doors	Hollow metal	400 S.F. Floor/Door	Each	1212	3.03	
1030	Fittings	N/A		—	—	—	
2010	Stair Construction	N/A		—	—	—	17.6%
3010	Wall Finishes	Paint		S.F. Surface	1.16	1.39	
3020	Floor Finishes	Carpet		S.F. Floor	10.22	10.22	
3030	Ceiling Finishes	N/A		—	—	—	
D. SERVICES							
	D10 Conveying						
1010	Elevators & Lifts	N/A		—	—	—	0.0 %
1020	Escalators & Moving Walks	N/A		—	—	—	
	D20 Plumbing						
2010	Plumbing Fixtures	Kitchen, toilet and service fixtures, supply and drainage	1 Fixture/2430 S.F. Floor	Each	8286	3.41	
2020	Domestic Water Distribution	Gas fired hot water heater		S.F. Floor	.31	.31	2.8%
2040	Rain Water Drainage	N/A		—	—	—	
	D30 HVAC						
3010	Energy Supply	Oil fired hot water, wall fin radiation		S.F. Floor	10.44	10.44	
3020	Heat Generating Systems	N/A		—	—	—	
3030	Cooling Generating Systems	N/A		—	—	—	16.6%
3050	Terminal & Package Units	Split systems with air cooled condensing units		S.F. Floor	11.75	11.75	
3090	Other HVAC Sys. & Equipment	N/A		—	—	—	
	D40 Fire Protection						
4010	Sprinklers	Wet pipe sprinkler system		S.F. Floor	3.51	3.51	3.5%
4020	Standpipes	Standpipe		S.F. Floor	1.13	1.13	
	D50 Electrical						
5010	Electrical Service/Distribution	400 ampere service, panel board and feeders		S.F. Floor	.85	.85	
5020	Lighting & Branch Wiring	Fluorescent fixtures, receptacles, switches, A.C. and misc. power		S.F. Floor	8.22	8.22	9.1%
5030	Communications & Security	Alarm systems, sound system and emergency lighting		S.F. Floor	2.93	2.93	
5090	Other Electrical Systems	Emergency generator, 7.5 kW		S.F. Floor	.12	.12	
E. EQUIPMENT & FURNISHINGS							
1010	Commercial Equipment	N/A		—	—	—	
1020	Institutional Equipment	N/A		—	—	—	0.0 %
1030	Vehicular Equipment	N/A		—	—	—	
1090	Other Equipment	N/A		—	—	—	
F. SPECIAL CONSTRUCTION							
1020	Integrated Construction	N/A		—	—	—	0.0 %
1040	Special Facilities	N/A		—	—	—	
G. BUILDING SITEWORK	**N/A**						

		Sub-Total	133.35	100%
CONTRACTOR FEES (General Requirements: 10%, Overhead: 5%, Profit: 10%)		25%	33.31	
ARCHITECT FEES		11%	18.34	
Total Building Cost			**185**	

Costs per square foot of floor area

Exterior Wall	S.F. Area	2000	4000	6000	8000	12000	15000	18000	20000	22000
	L.F. Perimeter	180	280	340	386	460	535	560	600	640
Stone Ashlar Veneer On Concrete Block	Wood Truss	260.80	232.05	216.15	206.65	195.55	192.40	187.70	186.45	185.40
	Steel Joists	257.80	229.10	213.15	203.65	192.65	189.45	184.70	183.45	182.40
Stucco on Concrete Block	Wood Truss	233.40	210.75	198.90	191.90	183.90	181.55	178.20	177.30	176.50
	Steel Joists	230.85	208.20	196.35	189.30	181.35	179.00	175.65	174.75	173.95
Brick Veneer	Wood Frame	265.05	223.50	204.55	193.80	182.00	178.20	173.55	172.20	171.00
Wood Shingles	Wood Frame	253.35	214.40	197.20	187.55	177.00	173.55	169.55	168.25	167.20
Perimeter Adj., Add or Deduct	Per 100 L.F.	47.15	23.60	15.75	11.70	7.90	6.35	5.20	4.70	4.25
Story Hgt. Adj., Add or Deduct	Per 1 Ft.	5.30	4.20	3.35	2.80	2.30	2.10	1.80	1.75	1.70

For Basement, add $30.00 per square foot of basement area

The above costs were calculated using the basic specifications shown on the facing page. These costs should be adjusted where necessary for design alternatives and owner's requirements. Reported completed project costs, for this type of structure, range from $83.50 to $277.85 per S.F.

Common additives

Description	Unit	$ Cost
Bar, Front bar	L.F.	410
Back bar	L.F.	325
Booth, Upholstered, custom, straight	L.F.	237 - 435
"L" or "U" shaped	L.F.	245 - 415
Fireplaces, Brick not incl. chimney or foundation, 30" x 24" opening	Each	3050
Chimney, standard brick Single flue 16" x 20"	V.L.F.	103
20" x 20"	V.L.F.	125
2 flue, 20" x 32"	V.L.F.	191
Lockers, Steel, single tier, 60" or 72"	Opening	218 - 355
2 tier, 60" or 72" total	Opening	112 - 160
5 tier, box lockers	Opening	70 - 86
Locker bench, lam. maple top only	L.F.	25.50
Pedestals, steel pipe	Each	72.50
Refrigerators, Prefabricated, walk-in 7'-6" High, 6' x 6'	S.F.	190
10' x 10'	S.F.	150
12' x 14'	S.F.	133
12' x 20'	S.F.	117

Description	Unit	$ Cost
Sauna, Prefabricated, complete, 6' x 4'	Each	6225
6' x 9'	Each	8975
8' x 8'	Each	10,600
10' x 12'	Each	15,100
Smoke Detectors Ceiling type	Each	220
Duct type	Each	550
Sound System Amplifier, 250 watts	Each	1975
Speaker, ceiling or wall	Each	206
Trumpet	Each	395
Steam Bath, Complete, to 140 C.F.	Each	2975
To 300 C.F.	Each	3350
To 800 C.F.	Each	10,600
To 2500 C.F.	Each	13,300
Swimming Pool Complete, gunite	S.F.	85 - 105
Tennis Court, Complete with fence Bituminous	Each	38,400 - 63,000
Clay	Each	54,000 - 71,000

Model costs calculated for a 1 story building with 12' story height and 6,000 square feet of floor area

				Unit	Unit Cost	Cost Per S.F.	% Of Sub-Total
A.	**SUBSTRUCTURE**						
1010	Standard Foundations	Poured concrete; strip and spread footings		S.F. Ground	2.84	2.84	
1020	Special Foundations	N/A		—	—	—	
1030	Slab on Grade	4" reinforced concrete with vapor barrier and granular base		S.F. Slab	5.18	5.18	9.1%
2010	Basement Excavation	Site preparation for slab and trench for foundation wall and footing		S.F. Ground	.32	.32	
2020	Basement Walls	4' Foundation wall		L.F. Wall	79	5.82	
B.	**SHELL**						
	B10 Superstructure						
1010	Floor Construction	N/A		—	—	—	
1020	Roof Construction	Wood truss with plywood sheathing		S.F. Roof	6.57	6.57	4.2 %
	B20 Exterior Enclosure						
2010	Exterior Walls	Stone ashlar veneer on concrete block	65% of wall	S.F. Wall	42.44	18.76	
2020	Exterior Windows	Aluminum horizontal sliding	35% of wall	Each	524	8.31	19.2%
2030	Exterior Doors	Double aluminum and glass, hollow metal		Each	2840	2.85	
	B30 Roofing						
3010	Roof Coverings	Asphalt shingles		S.F. Roof	18.10	3.42	2.2%
3020	Roof Openings	N/A		—	—	—	
C.	**INTERIORS**						
1010	Partitions	Gypsum board on metal studs, load bearing	14 S.F. Floor/L.F. Partition	S.F. Partition	7.73	5.52	
1020	Interior Doors	Single leaf wood	140 S.F. Floor/Door	Each	622	4.44	
1030	Fittings	N/A		—	—	—	
2010	Stair Construction	N/A		—	—	—	16.5%
3010	Wall Finishes	40% vinyl wall covering, 40% paint, 20% ceramic tile		S.F. Surface	2.48	3.54	
3020	Floor Finishes	50% carpet, 30% hardwood tile, 20% ceramic tile		S.F. Floor	7.95	7.95	
3030	Ceiling Finishes	Gypsum plaster on wood furring		S.F. Ceiling	4.21	4.21	
D.	**SERVICES**						
	D10 Conveying						
1010	Elevators & Lifts	N/A		—	—	—	0.0 %
1020	Escalators & Moving Walks	N/A		—	—	—	
	D20 Plumbing						
2010	Plumbing Fixtures	Kitchen, toilet and service fixtures, supply and drainage	1 Fixture/125 S.F. Floor	Each	2200	17.60	
2020	Domestic Water Distribution	Gas fired hot water		S.F. Floor	8.03	8.03	16.5%
2040	Rain Water Drainage	N/A		—	—	—	
	D30 HVAC						
3010	Energy Supply	N/A		—	—	—	
3020	Heat Generating Systems	Included in D3050		—	—	—	
3030	Cooling Generating Systems	N/A		—	—	—	18.6 %
3050	Terminal & Package Units	Multizone rooftop unit, gas heating, electric cooling		S.F. Floor	29.03	29.03	
3090	Other HVAC Sys. & Equipment	N/A		—	—	—	
	D40 Fire Protection						
4010	Sprinklers	Wet pipe sprinkler system		S.F. Floor	5.53	5.53	4.6%
4020	Standpipes	Standpipe		S.F. Floor	1.61	1.61	
	D50 Electrical						
5010	Electrical Service/Distribution	400 ampere service, panel board and feeders		S.F. Floor	3.73	3.73	
5020	Lighting & Branch Wiring	High efficiency fluorescent fixtures, receptacles, switches, A.C. and misc. power		S.F. Floor	8.36	8.36	9.1%
5030	Communications & Security	Addressable alarm systems and emergency lighting		S.F. Floor	1.92	1.92	
5090	Other Electrical Systems	Emergency generator, 11.5 kW		S.F. Floor	.24	.24	
E.	**EQUIPMENT & FURNISHINGS**						
1010	Commercial Equipment	N/A		—	—	—	
1020	Institutional Equipment	N/A		—	—	—	0.0 %
1030	Vehicular Equipment	N/A		—	—	—	
1090	Other Equipment	N/A		—	—	—	
F.	**SPECIAL CONSTRUCTION**						
1020	Integrated Construction	N/A		—	—	—	0.0 %
1040	Special Facilities	N/A		—	—	—	
G.	**BUILDING SITEWORK**	**N/A**					

			Sub-Total	155.78	100%
CONTRACTOR FEES (General Requirements: 10%, Overhead: 5%, Profit: 10%)		25%	38.95		
ARCHITECT FEES		11%	21.42		

Total Building Cost	**216.15**

Costs per square foot of floor area

Exterior Wall	S.F. Area	4000	8000	12000	17000	22000	27000	32000	37000	42000
	L.F. Perimeter	280	386	520	585	640	740	840	940	940
Stone Ashlar on Concrete Block	Steel Joists	215.95	182.35	173.60	162.10	155.25	152.80	151.05	149.85	146.35
	Wood Joists	205.70	176.70	169.35	158.90	152.65	150.55	149.00	147.90	144.65
Face Brick on Concrete Block	Steel Joists	202.35	174.85	167.90	158.05	152.25	150.20	148.80	147.75	144.65
	Wood Joists	198.40	171.65	164.80	155.25	149.60	147.65	146.25	145.25	142.30
Decorative Concrete Block	Steel Joists	192.80	168.30	162.00	153.40	148.30	146.45	145.20	144.30	141.60
	Wood Joists	187.45	163.65	157.50	149.20	144.25	142.55	141.25	140.40	137.80
Perimeter Adj., Add or Deduct	Per 100 L.F.	26.60	13.35	8.85	6.25	4.90	3.90	3.35	2.90	2.45
Story Hgt. Adj., Add or Deduct	Per 1 Ft.	4.40	3.10	2.75	2.15	1.85	1.70	1.70	1.60	1.40

For Basement, add $30.55 per square foot of basement area

The above costs were calculated using the basic specifications shown on the facing page. These costs should be adjusted where necessary for design alternatives and owner's requirements. Reported completed project costs, for this type of structure, range from $85.40 to $245.40 per S.F.

Common additives

Description	Unit	$ Cost		Description	Unit	$ Cost
Bar, Front bar	L.F.	410		Kitchen Equipment, cont.		
Back bar	L.F.	325		Freezer, 44 C.F., reach-in	Each	4150
Booth, Upholstered, custom, straight	L.F.	237 - 435		Ice cube maker, 50 lb. per day	Each	1925
"L" or "U" shaped	L.F.	245 - 415		Lockers, Steel, single tier, 60" or 72"	Opening	218 - 355
Emergency Lighting, 25 watt, battery operated				2 tier, 60" or 72" total	Opening	112 - 160
Lead battery	Each	305		5 tier, box lockers	Opening	70 - 86
Nickel cadmium	Each	850		Locker bench, lam. maple top only	L.F.	25.50
Flagpoles, Complete				Pedestals, steel pipe	Each	72.50
Aluminum, 20' high	Each	1750		Refrigerators, Prefabricated, walk-in		
40' High	Each	3875		7'-6" High, 6' x 6'	S.F.	190
70' High	Each	10,600		10' x 10'	S.F.	150
Fiberglass, 23' High	Each	1325		12' x 14'	S.F.	133
39'-5" High	Each	3525		12' x 20'	S.F.	117
59' High	Each	7200		Smoke Detectors		
Kitchen Equipment				Ceiling type	Each	220
Broiler	Each	4175		Duct type	Each	550
Coffee urn, twin 6 gallon	Each	2625		Sound System		
Cooler, 6 ft. long, reach-in	Each	4000		Amplifier, 250 watts	Each	1975
Dishwasher, 10-12 racks per hr.	Each	5300		Speaker, ceiling or wall	Each	206
Food warmer, counter 1.2 kw	Each	560		Trumpet	Each	395

Important: See the Reference Section for Location Factors.

Model costs calculated for a 1 story building with 12' story height and 22,000 square feet of floor area

				Unit	Unit Cost	Cost Per S.F.	% Of Sub-Total
A.	**SUBSTRUCTURE**						
1010	Standard Foundations	Poured concrete; strip and spread footings		S.F. Ground	1.79	1.79	
1020	Special Foundations	N/A		—	—	—	
1030	Slab on Grade	4" reinforced concrete with vapor barrier and granular base		S.F. Slab	5.18	5.18	9.3%
2010	Basement Excavation	Site preparation for slab and trench for foundation wall and footing		S.F. Ground	.18	.18	
2020	Basement Walls	4' Foundation wall		L.F. Wall	79	3.67	
B.	**SHELL**						
	B10 Superstructure						
1010	Floor Construction	N/A		—	—	—	5.8 %
1020	Roof Construction	Metal deck on open web steel joists		S.F. Roof	6.77	6.77	
	B20 Exterior Enclosure						
2010	Exterior Walls	Stone, ashlar veneer on concrete block	65% of wall	S.F. Wall	42.44	9.63	
2020	Exterior Windows	Window wall	35% of wall	Each	46.90	5.73	13.9%
2030	Exterior Doors	Double aluminum and glass doors		Each	2840	.78	
	B30 Roofing						
3010	Roof Coverings	Built-up tar and gravel with flashing; perlite/EPS composite insulation		S.F. Roof	5.94	5.94	5.1%
3020	Roof Openings	N/A		—	—	—	
C.	**INTERIORS**						
1010	Partitions	Lightweight concrete block	14 S.F. Floor/L.F. Partition	S.F. Partition	11.20	8	
1020	Interior Doors	Single leaf wood	140 S.F. Floor/Door	Each	622	4.44	
1030	Fittings	N/A		—	—	—	
2010	Stair Construction	N/A		—	—	—	25.1%
3010	Wall Finishes	65% paint, 25% vinyl wall covering, 10% ceramic tile		S.F. Surface	1.72	2.45	
3020	Floor Finishes	60% carpet, 35% hardwood, 15% ceramic tile		S.F. Floor	7.36	7.36	
3030	Ceiling Finishes	Mineral fiber tile on concealed zee bars		S.F. Ceiling	6.86	6.86	
D.	**SERVICES**						
	D10 Conveying						
1010	Elevators & Lifts	N/A		—	—	—	0.0 %
1020	Escalators & Moving Walks	N/A		—	—	—	
	D20 Plumbing						
2010	Plumbing Fixtures	Kitchen, toilet and service fixtures, supply and drainage	1 Fixture/1050 S.F. Floor	Each	8516	8.11	
2020	Domestic Water Distribution	Gas fired hot water heater		S.F. Floor	.28	.28	7.9%
2040	Rain Water Drainage	Roof drains		S.F. Roof	.77	.77	
	D30 HVAC						
3010	Energy Supply	N/A		—	—	—	
3020	Heat Generating Systems	Included in D3050		—	—	—	
3030	Cooling Generating Systems	N/A		—	—	—	18.8 %
3050	Terminal & Package Units	Multizone rooftop unit, gas heating, electric cooling		S.F. Floor	21.84	21.84	
3090	Other HVAC Sys. & Equipment	N/A		—	—	—	
	D40 Fire Protection						
4010	Sprinklers	Wet pipe sprinkler system		S.F. Floor	4.82	4.82	4.5%
4020	Standpipes	Standpipe		S.F. Floor	.44	.44	
	D50 Electrical						
5010	Electrical Service/Distribution	400 ampere service, panel board and feeders		S.F. Floor	1.26	1.26	
5020	Lighting & Branch Wiring	High efficiency fluorescent fixtures, receptacles, switches, A.C. and misc. power		S.F. Floor	8.19	8.19	9.5%
5030	Communications & Security	Addressable alarm systems and emergency lighting		S.F. Floor	1.46	1.46	
5090	Other Electrical Systems	Emergency generator, 11.5 kW		S.F. Floor	.13	.13	
E.	**EQUIPMENT & FURNISHINGS**						
1010	Commercial Equipment	N/A		—	—	—	
1020	Institutional Equipment	N/A		—	—	—	0.0 %
1030	Vehicular Equipment	N/A		—	—	—	
1090	Other Equipment	N/A		—	—	—	
F.	**SPECIAL CONSTRUCTION**						
1020	Integrated Construction	N/A		—	—	—	0.0 %
1040	Special Facilities	N/A		—	—	—	
G.	**BUILDING SITEWORK**	**N/A**					

		Sub-Total	116.08	100%
CONTRACTOR FEES (General Requirements: 10%, Overhead: 5%, Profit: 10%)		25%	29.01	
ARCHITECT FEES		7%	10.16	
	Total Building Cost		**155.25**	

Costs per square foot of floor area

Exterior Wall	S.F. Area	15000	20000	28000	38000	50000	65000	85000	100000	150000
	L.F. Perimeter	350	400	480	550	630	660	750	825	1035
Face Brick with Concrete Block Back-up	Steel Frame	226.60	217.20	209.15	202.70	198.35	193.60	190.75	189.40	186.55
	Bearing Walls	225.25	214.55	205.45	198.00	192.95	187.35	183.90	182.40	179.10
Decorative Concrete Block	Steel Frame	218.50	210.25	203.15	197.70	193.95	190.05	187.65	186.55	184.15
	Bearing Walls	217.15	207.60	199.45	193.00	188.50	183.80	180.80	179.55	176.65
Stucco on Concrete Block	Steel Frame	217.70	209.60	202.65	197.25	193.55	189.75	187.40	186.25	183.90
	Bearing Walls	216.40	206.95	198.95	192.55	188.15	183.50	180.55	179.25	176.50
Perimeter Adj., Add or Deduct	Per 100 L.F.	9.35	7.10	5.00	3.70	2.80	2.15	1.60	1.35	0.95
Story Hgt. Adj., Add or Deduct	Per 1 Ft.	2.35	2.00	1.70	1.45	1.20	1.00	0.80	0.80	0.70

For Basement, add $35.65 per square foot of basement area

The above costs were calculated using the basic specifications shown on the facing page. These costs should be adjusted where necessary for design alternatives and owner's requirements. Reported completed project costs, for this type of structure, range from $123.55 to $290.35 per S.F.

Common additives

Description	Unit	$ Cost	Description	Unit	$ Cost
Carrels Hardwood	Each	855 - 1625	Lockers, Steel, single tier, 60" or 72"	Opening	218 - 355
Clock System			2 tier, 60" or 72" total	Opening	112 - 160
20 Room	Each	19,200	5 tier, box lockers	Opening	70 - 86
50 Room	Each	45,000	Locker bench, lam. maple top only	L.F.	25.50
Elevators, Hydraulic passenger, 2 stops			Pedestals, steel pipe	Each	72.50
1500# capacity	Each	66,000	Seating		
2500# capacity	Each	68,900	Auditorium chair, all veneer	Each	265
3500# capacity	Each	72,900	Veneer back, padded seat	Each	335
Additional stop, add	Each	8375	Upholstered, spring seat	Each	320
Emergency Lighting, 25 watt, battery operated			Classroom, movable chair & desk	Set	81 - 171
Lead battery	Each	305	Lecture hall, pedestal type	Each	257 - 750
Nickel cadmium	Each	850	Smoke Detectors		
Flagpoles, Complete			Ceiling type	Each	220
Aluminum, 20' high	Each	1750	Duct type	Each	550
40' High	Each	3875	Sound System		
70' High	Each	10,600	Amplifier, 250 watts	Each	1975
Fiberglass, 23' High	Each	1325	Speaker, ceiling or wall	Each	206
39'-5" High	Each	3525	Trumpet	Each	395
59' High	Each	7200	TV Antenna, Master system, 12 outlet	Outlet	277
			30 outlet	Outlet	216
			100 outlet	Outlet	205

Important: See the Reference Section for Location Factors.

Model costs calculated for a 2 story building with 12' story height and 50,000 square feet of floor area

College, Classroom, 2-3 Story

				Unit	Unit Cost	Cost Per S.F.	% Of Sub-Total
A. SUBSTRUCTURE							
1010	Standard Foundations	Poured concrete; strip and spread footings		S.F. Ground	1.28	.64	
1020	Special Foundations	N/A		—	—	—	
1030	Slab on Grade	4" reinforced concrete with vapor barrier and granular base		S.F. Slab	5.18	2.60	3.5%
2010	Basement Excavation	Site preparation for slab and trench for foundation wall and footing		S.F. Ground	.30	.15	
2020	Basement Walls	4' Foundation wall		L.F. Wall	75	1.53	
B. SHELL							
B10 Superstructure							
1010	Floor Construction	Open web steel joists, slab form, concrete		S.F. Floor	16.76	8.38	9.4%
1020	Roof Construction	Metal deck on open web steel joists, columns		S.F. Roof	9.64	4.82	
B20 Exterior Enclosure							
2010	Exterior Walls	Decorative concrete block	65% of wall	S.F. Wall	15.77	3.10	
2020	Exterior Windows	Window wall	35% of wall	Each	39.70	4.21	5.7%
2030	Exterior Doors	Double glass and aluminum with transom		Each	5650	.68	
B30 Roofing							
3010	Roof Coverings	Built-up tar and gravel with flashing; perlite/EPS composite insulation		S.F. Roof	5.78	2.89	2.1%
3020	Roof Openings	N/A		—	—	—	
C. INTERIORS							
1010	Partitions	Concrete block	20 S.F. Floor/L.F. Partition	S.F. Partition	14.84	7.42	
1020	Interior Doors	Single leaf hollow metal	200 S.F. Floor/Door	Each	1212	6.07	
1030	Fittings	Chalkboards, counters, cabinets		S.F. Floor	5.99	5.99	
2010	Stair Construction	Concrete filled metal pan		Flight	16,950	3.39	27.4%
3010	Wall Finishes	95% paint, 5% ceramic tile		S.F. Surface	4.03	4.03	
3020	Floor Finishes	70% vinyl composition tile, 25% carpet, 5% ceramic tile		S.F. Floor	4.86	4.86	
3030	Ceiling Finishes	Mineral fiber tile on concealed zee bars		S.F. Ceiling	6.86	6.86	
D. SERVICES							
D10 Conveying							
1010	Elevators & Lifts	Two hydraulic passenger elevators		Each	83,500	3.34	2.4%
1020	Escalators & Moving Walks	N/A		—	—	—	
D20 Plumbing							
2010	Plumbing Fixtures	Toilet and service fixtures, supply and drainage	1 Fixture/455 S.F. Floor	Each	8317	18.28	
2020	Domestic Water Distribution	Oil fired hot water heater		S.F. Floor	2.78	2.78	15.4%
2040	Rain Water Drainage	Roof drains		S.F. Roof	1.30	.65	
D30 HVAC							
3010	Energy Supply	N/A		—	—	—	
3020	Heat Generating Systems	Included in D3050		—	—	—	
3030	Cooling Generating Systems	N/A		—	—	—	14.3 %
3050	Terminal & Package Units	Multizone unit, gas heating, electric cooling		S.F. Floor	20.20	20.20	
3090	Other HVAC Sys. & Equipment	N/A		—	—	—	
D40 Fire Protection							
4010	Sprinklers	Sprinklers, light hazard		S.F. Floor	3.07	3.07	2.4%
4020	Standpipes	Standpipe, dry, Class III		S.F. Floor	.32	.32	
D50 Electrical							
5010	Electrical Service/Distribution	2000 ampere service, panel board and feeders		S.F. Floor	4.05	4.05	
5020	Lighting & Branch Wiring	High efficiency fluorescent fixtures, receptacles, switches, A.C. and misc. power		S.F. Floor	12.92	12.92	17.5%
5030	Communications & Security	Addressable alarm systems, internet wiring, communications systems and emergency lighting		S.F. Floor	7	7	
5090	Other Electrical Systems	Emergency generator, 100 kW		S.F. Floor	.74	.74	
E. EQUIPMENT & FURNISHINGS							
1010	Commercial Equipment	N/A		—	—	—	
1020	Institutional Equipment	N/A		—	—	—	
1030	Vehicular Equipment	N/A		—	—	—	0.0 %
1090	Other Equipment	N/A		—	—	—	
F. SPECIAL CONSTRUCTION							
1020	Integrated Construction	N/A		—	—	—	0.0 %
1040	Special Facilities	N/A		—	—	—	
G. BUILDING SITEWORK	**N/A**						

			Sub-Total	140.97	100%
CONTRACTOR FEES (General Requirements: 10%, Overhead: 5%, Profit: 10%)		25%	35.20		
ARCHITECT FEES		7%	12.33		
		Total Building Cost	**188.50**		

Costs per square foot of floor area

Exterior Wall	S.F. Area	10000	15000	25000	40000	55000	70000	80000	90000	100000
	L.F. Perimeter	260	320	400	476	575	628	684	721	772
Face Brick with Concrete Block Back-up	R/Conc. Frame	235.80	218.55	201.70	190.10	185.70	181.85	180.65	179.20	178.35
	Steel Frame	248.20	231.00	214.10	202.55	198.10	194.25	193.05	191.60	190.80
Decorative Concrete Block	R/Conc. Frame	219.05	204.80	191.35	182.40	178.95	176.05	175.10	174.05	173.40
	Steel Frame	231.50	217.25	203.75	194.90	191.40	188.45	187.55	186.45	185.85
Precast Concrete Panels	R/Conc. Frame	242.60	224.20	205.90	193.25	188.40	184.20	182.90	181.35	180.45
	Steel Frame	255.05	236.60	218.35	205.65	200.85	196.60	195.30	193.75	192.85
Perimeter Adj., Add or Deduct	Per 100 L.F.	18.90	12.55	7.55	4.75	3.40	2.75	2.30	2.15	1.95
Story Hgt. Adj., Add or Deduct	Per 1 Ft.	3.65	2.95	2.20	1.70	1.45	1.20	1.15	1.15	1.15

For Basement, add $34.70 per square foot of basement area

The above costs were calculated using the basic specifications shown on the facing page. These costs should be adjusted where necessary for design alternatives and owner's requirements. Reported completed project costs, for this type of structure, range from $79.80 to $235.20 per S.F.

Common additives

Description	Unit	$ Cost
Carrels Hardwood	Each	855 - 1625
Closed Circuit Surveillance, One station		
Camera and monitor	Each	1950
For additional camera stations, add	Each	1050
Elevators, Hydraulic passenger, 2 stops		
2000# capacity	Each	67,900
2500# capacity	Each	68,900
3500# capacity	Each	72,900
Additional stop, add	Each	8375
Emergency Lighting, 25 watt, battery operated		
Lead battery	Each	305
Nickel cadmium	Each	850
Furniture	Student	2800 - 5350
Intercom System, 25 station capacity		
Master station	Each	3125
Intercom outlets	Each	183
Handset	Each	500

Description	Unit	$ Cost
Kitchen Equipment		
Broiler	Each	4175
Coffee urn, twin 6 gallon	Each	2625
Cooler, 6 ft. long	Each	4000
Dishwasher, 10-12 racks per hr.	Each	5300
Food warmer	Each	560
Freezer, 44 C.F., reach-in	Each	4150
Ice cube maker, 50 lb. per day	Each	1925
Range with 1 oven	Each	3325
Laundry Equipment		
Dryer, gas, 16 lb. capacity	Each	945
30 lb. capacity	Each	3750
Washer, 4 cycle	Each	1175
Commercial	Each	1600
Smoke Detectors		
Ceiling type	Each	220
Duct type	Each	550
TV Antenna, Master system, 12 outlet	Outlet	277
30 outlet	Outlet	216
100 outlet	Outlet	205

Important: See the Reference Section for Location Factors.

Model costs calculated for a 3 story building with 12' story height and 25,000 square feet of floor area

College, Dormitory, 2-3 Story

				Unit	Unit Cost	Cost Per S.F.	% Of Sub-Total
A. SUBSTRUCTURE							
1010	Standard Foundations	Poured concrete; strip and spread footings		S.F. Ground	6.09	2.03	
1020	Special Foundations	N/A		—	—	—	
1030	Slab on Grade	4" reinforced concrete with vapor barrier and granular base		S.F. Slab	5.18	1.72	3.4%
2010	Basement Excavation	Site preparation for slab and trench for foundation wall and footing		S.F. Ground	.18	.06	
2020	Basement Walls	4' Foundation wall		L.F. Wall	79	1.26	
B. SHELL							
B10 Superstructure							
1010	Floor Construction	Concrete flat plate		S.F. Floor	24.65	16.43	13.8%
1020	Roof Construction	Concrete flat plate		S.F. Roof	13.32	4.44	
B20 Exterior Enclosure							
2010	Exterior Walls	Face brick with concrete block backup	80% of wall	S.F. Wall	32.47	14.96	
2020	Exterior Windows	Aluminum horizontal sliding	20% of wall	Each	735	3.68	13.3%
2030	Exterior Doors	Double glass & aluminum doors		Each	5900	1.42	
B30 Roofing							
3010	Roof Coverings	Single ply membrane, loose laid and ballasted; perlite/EPS composite insulation		S.F. Roof	5.43	1.81	1.2%
3020	Roof Openings	N/A		—	—	—	
C. INTERIORS							
1010	Partitions	Gypsum board on metal studs, concrete block	9 S.F. Floor/L.F. Partition	S.F. Partition	6.03	6.70	
1020	Interior Doors	Single leaf wood	90 S.F. Floor/Door	Each	622	6.91	
1030	Fittings	Closet shelving, mirrors, bathroom accessories		S.F. Floor	1.94	1.94	
2010	Stair Construction	Cast in place concrete		Flight	5050	2.63	22.2%
3010	Wall Finishes	95% paint, 5% ceramic tile		S.F. Surface	1.84	4.09	
3020	Floor Finishes	80% carpet, 10% vinyl composition tile, 10% ceramic tile		S.F. Floor	10.27	10.27	
3030	Ceiling Finishes	90% paint, 10% suspended fiberglass board		S.F. Ceiling	.99	.99	
D. SERVICES							
D10 Conveying							
1010	Elevators & Lifts	One hydraulic passenger elevator		Each	114,000	4.56	3.0%
1020	Escalators & Moving Walks	N/A		—	—	—	
D20 Plumbing							
2010	Plumbing Fixtures	Toilet and service fixtures, supply and drainage	1 Fixture/455 S.F. Floor	Each	9910	21.78	
2020	Domestic Water Distribution	Electric water heater		S.F. Floor	3.99	3.99	17.4%
2040	Rain Water Drainage	Roof drains		S.F. Roof	1.32	.44	
D30 HVAC							
3010	Energy Supply	N/A		—	—	—	
3020	Heat Generating Systems	Included in D3050		—	—	—	
3030	Cooling Generating Systems	N/A		—	—	—	8.4%
3050	Terminal & Package Units	Rooftop multizone unit system		S.F. Floor	12.65	12.65	
3090	Other HVAC Sys. & Equipment	N/A		—	—	—	
D40 Fire Protection							
4010	Sprinklers	Wet pipe sprinkler system		S.F. Floor	2.90	2.90	2.5%
4020	Standpipes	Standpipe, dry, Class III		S.F. Floor	.81	.81	
D50 Electrical							
5010	Electrical Service/Distribution	800 ampere service, panel board and feeders		S.F. Floor	2.16	2.16	
5020	Lighting & Branch Wiring	High efficiency fluorescent fixtures, receptacles, switches, A.C. and misc. power		S.F. Floor	9.66	9.66	12.7%
5030	Communications & Security	Alarm systems, internet and phone wiring, communications systems and emergency lighting		S.F. Floor	7.24	7.24	
5090	Other Electrical Systems	Emergency generator, 7.5 kW		S.F. Floor	.09	.09	
E. EQUIPMENT & FURNISHINGS							
1010	Commercial Equipment	N/A		—	—	—	
1020	Institutional Equipment	N/A		—	—	—	
1030	Vehicular Equipment	N/A		—	—	—	2.1%
2020	Other Equipment	Dormitory furniture		S.F. Floor	3.16	3.16	
F. SPECIAL CONSTRUCTION							
1020	Integrated Construction	N/A		—	—	—	
1040	Special Facilities	N/A		—	—	—	0.0%
G. BUILDING SITEWORK	**N/A**						

		Sub-Total	150.78	**100%**
CONTRACTOR FEES (General Requirements: 10%, Overhead: 5%, Profit: 10%)		25%	37.73	
ARCHITECT FEES		7%	13.19	
	Total Building Cost		**201.70**	

Costs per square foot of floor area

Exterior Wall	S.F. Area	20000	35000	45000	65000	85000	110000	135000	160000	200000
	L.F. Perimeter	260	340	400	440	500	540	560	590	640
Face Brick with Concrete Block Back-up	R/Conc. Frame	230.70	214.90	210.85	201.90	197.95	193.95	190.85	188.95	186.95
	Steel Frame	244.40	228.60	224.50	215.60	211.65	207.60	204.55	202.60	200.60
Decorative Concrete Block	R/Conc. Frame	215.65	203.65	200.55	194.05	191.15	188.20	186.00	184.65	183.25
	Steel Frame	229.35	217.35	214.20	207.75	204.85	201.90	199.70	198.35	196.85
Precast Concrete Panels With Exposed Aggregate	R/Conc. Frame	236.70	219.40	214.90	204.95	200.70	196.15	192.75	190.65	188.40
	Steel Frame	250.25	232.95	228.50	218.60	214.30	209.80	206.35	204.30	202.05
Perimeter Adj., Add or Deduct	Per 100 L.F.	18.40	10.60	8.20	5.65	4.40	3.35	2.75	2.35	1.85
Story Hgt. Adj., Add or Deduct	Per 1 Ft.	3.75	2.85	2.50	1.90	1.70	1.35	1.15	1.05	0.90

For Basement, add $34.70 per square foot of basement area

The above costs were calculated using the basic specifications shown on the facing page. These costs should be adjusted where necessary for design alternatives and owner's requirements. Reported completed project costs, for this type of structure, range from $113.40 to $273.55 per S.F.

Common additives

Description	Unit	$ Cost
Carrels Hardwood	Each	855 - 1625
Closed Circuit Surveillance, One station		
Camera and monitor	Each	1950
For additional camera stations, add	Each	1050
Elevators, Electric passenger, 5 stops		
2000# capacity	Each	165,300
2500# capacity	Each	169,300
3500# capacity	Each	169,800
Additional stop, add	Each	12,700
Emergency Lighting, 25 watt, battery operated		
Lead battery	Each	305
Nickel cadmium	Each	850
Furniture	Student	2800 - 5350
Intercom System, 25 station capacity		
Master station	Each	3125
Intercom outlets	Each	183
Handset	Each	500

Description	Unit	$ Cost
Kitchen Equipment		
Broiler	Each	4175
Coffee urn, twin, 6 gallon	Each	2625
Cooler, 6 ft. long	Each	4000
Dishwasher, 10-12 racks per hr.	Each	5300
Food warmer	Each	560
Freezer, 44 C.F., reach-in	Each	4150
Ice cube maker, 50 lb. per day	Each	1925
Range with 1 oven	Each	3325
Laundry Equipment		
Dryer, gas, 16 lb. capacity	Each	945
30 lb. capacity	Each	3750
Washer, 4 cycle	Each	1175
Commercial	Each	1600
Smoke Detectors		
Ceiling type	Each	220
Duct type	Each	550
TV Antenna, Master system, 12 outlet	Outlet	277
30 outlet	Outlet	216
100 outlet	Outlet	205

Important: See the Reference Section for Location Factors.

Model costs calculated for a 6 story building with 12' story height and 85,000 square feet of floor area

				Unit	Unit Cost	Cost Per S.F.	% Of Sub-Total
A.	**SUBSTRUCTURE**						
1010	Standard Foundations	Poured concrete; strip and spread footings		S.F. Ground	8.94	1.49	
1020	Special Foundations	N/A		—	—	—	
1030	Slab on Grade	4" reinforced concrete with vapor barrier and granular base		S.F. Slab	5.18	.86	2.0%
2010	Basement Excavation	Site preparation for slab and trench for foundation wall and footing		S.F. Ground	.30	.05	
2020	Basement Walls	4' foundation wall		L.F. Wall	75	.64	
B.	**SHELL**						
	B10 Superstructure						
1010	Floor Construction	Concrete slab with metal deck and beams		S.F. Floor	28.80	24	18.0%
1020	Roof Construction	Concrete slab with metal deck and beams		S.F. Roof	21.48	3.58	
	B20 Exterior Enclosure						
2010	Exterior Walls	Decorative concrete block	80% of wall	S.F. Wall	17.41	5.90	
2020	Exterior Windows	Aluminum horizontal sliding	20% of wall	Each	524	2.96	6.0%
2030	Exterior Doors	Double glass & aluminum doors		Each	3594	.34	
	B30 Roofing						
3010	Roof Coverings	Single ply membrane, loose laid and ballasted; perlite/EPS composite insulation		S.F. Roof	4.92	.82	0.5%
3020	Roof Openings	N/A		—	—	—	
C.	**INTERIORS**						
1010	Partitions	Concrete block	9 S.F. Floor/L.F. Partition	S.F. Partition	10.13	11.26	
1020	Interior Doors	Single leaf wood	90 S.F. Floor/Door	Each	622	6.91	
1030	Fittings	Closet shelving, mirrors, bathroom accessories		S.F. Floor	1.59	1.59	
2010	Stair Construction	Concrete filled metal pan		Flight	16,950	3.59	26.3%
3010	Wall Finishes	95% paint, 5% ceramic tile		S.F. Surface	2.53	5.63	
3020	Floor Finishes	80% carpet, 10% vinyl composition tile, 10% ceramic tile		S.F. Floor	10.27	10.27	
3030	Ceiling Finishes	Mineral fiber tile on concealed zee bars, paint		S.F. Ceiling	.99	.99	
D.	**SERVICES**						
	D10 Conveying						
1010	Elevators & Lifts	Four geared passenger elevators		Each	245,013	11.53	7.5%
1020	Escalators & Moving Walks	N/A		—	—	—	
	D20 Plumbing						
2010	Plumbing Fixtures	Toilet and service fixtures, supply and drainage	1 Fixture/390 S.F. Floor	Each	8455	21.68	
2020	Domestic Water Distribution	Oil fired water heater		S.F. Floor	1.77	1.77	15.5%
2040	Rain Water Drainage	Roof drains		S.F. Roof	2.10	.35	
	D30 HVAC						
3010	Energy Supply	Oil fired hot water, wall fin radiation		S.F. Floor	4.37	4.37	
3020	Heat Generating Systems	N/A		—	—	—	
3030	Cooling Generating Systems	Chilled water, air cooled condenser system		S.F. Floor	9.54	9.54	9.1%
3050	Terminal & Package Units	N/A		—	—	—	
3090	Other HVAC Sys. & Equipment	N/A		—	—	—	
	D40 Fire Protection						
4010	Sprinklers	Sprinklers, light hazard		S.F. Floor	2.86	2.86	2.3%
4020	Standpipes	Standpipe, dry, Class III		S.F. Floor	.63	.63	
	D50 Electrical						
5010	Electrical Service/Distribution	1200 ampere service, panel board and feeders		S.F. Floor	1.05	1.05	
5020	Lighting & Branch Wiring	High efficiency fluorescent fixtures, receptacles, switches, A.C. and misc. power		S.F. Floor	9.44	9.44	11.1%
5030	Communications & Security	Alarm systems, internet and phone wiring, communications systems and emergency lighting		S.F. Floor	6.23	6.23	
5090	Other Electrical Systems	Emergency generator, 30 kW		S.F. Floor	.23	.23	
E.	**EQUIPMENT & FURNISHINGS**						
1010	Commercial Equipment	N/A		—	—	—	
1020	Institutional Equipment	N/A		—	—	—	
1030	Vehicular Equipment	N/A		—	—	—	1.7%
2020	Other Equipment	Built-in dormitory furnishings		S.F. Floor	2.59	2.59	
F.	**SPECIAL CONSTRUCTION**						
1020	Integrated Construction	N/A		—	—	—	0.0%
1040	Special Facilities	N/A		—	—	—	
G.	**BUILDING SITEWORK**	**N/A**					

	Sub-Total	153.15	**100%**
CONTRACTOR FEES (General Requirements: 10%, Overhead: 5%, Profit: 10%)	25%	38.30	
ARCHITECT FEES	7%	13.40	
	Total Building Cost	204.85	

Costs per square foot of floor area

Exterior Wall	S.F. Area	12000	20000	28000	37000	45000	57000	68000	80000	92000
	L.F. Perimeter	470	600	698	793	900	1060	1127	1200	1320
Face Brick with Concrete Brick Back-up	Steel Frame	332.55	266.55	236.75	218.15	208.50	199.10	191.85	186.25	182.80
	Bearing Walls	327.10	261.10	231.30	212.70	203.05	193.65	186.40	180.80	177.35
Decorative Concrete Block	Steel Frame	325.20	260.95	232.10	214.10	204.80	195.60	188.80	183.45	180.05
	Bearing Walls	320.00	255.70	226.90	208.90	199.55	190.40	183.60	178.25	174.85
Stucco on Concrete Block	Steel Frame	323.75	259.80	231.15	213.35	204.05	194.95	188.15	182.95	179.60
	Bearing Walls	318.55	254.60	225.95	208.10	198.80	189.70	182.90	177.70	174.35
Perimeter Adj., Add or Deduct	Per 100 L.F.	10.80	6.45	4.65	3.50	2.90	2.25	1.90	1.60	1.40
Story Hgt. Adj., Add or Deduct	Per 1 Ft.	1.95	1.50	1.30	1.05	1.05	0.90	0.85	0.80	0.70
For Basement, add $22.40 per square foot of basement area										

The above costs were calculated using the basic specifications shown on the facing page. These costs should be adjusted where necessary for design alternatives and owner's requirements. Reported completed project costs, for this type of structure, range from $163.35 to $305.05 per S.F.

Common additives

Description	Unit	$ Cost
Cabinets, Base, door units, metal	L.F.	305
Drawer units	L.F.	600
Tall storage cabinets, open	L.F.	570
With doors	L.F.	845
Wall, metal 12-1/2" deep, open	L.F.	228
With doors	L.F.	415
Carrels Hardwood	Each	855 - 1625
Countertops, not incl. base cabinets, acid proof	S.F.	55.50 - 69
Stainless steel	S.F.	165
Fume Hood, Not incl. ductwork	L.F.	745 - 1575
Ductwork	Hood	5650 - 9275
Glassware Washer, Distilled water rinse	Each	7525 - 15,300
Seating		
Auditorium chair, all veneer	Each	265
Veneer back, padded seat	Each	335
Upholstered, spring seat	Each	320
Classroom, movable chair & desk	Set	81 - 171
Lecture hall, pedestal type	Each	257 - 750

Description	Unit	$ Cost
Safety Equipment, Eye wash, hand held	Each	445
Deluge shower	Each	840
Sink, One piece plastic		
Flask wash, freestanding	Each	2550
Tables, acid resist. top, drawers	L.F.	215
Titration Unit, Four 2000 ml reservoirs	Each	6050

Important: See the Reference Section for Location Factors.

Model costs calculated for a 1 story building with 12′ story height and 45,000 square feet of floor area

College, Laboratory

					Unit	Unit Cost	Cost Per S.F.	% Of Sub-Total
A. SUBSTRUCTURE								
1010	Standard Foundations	Poured concrete; strip and spread footings			S.F. Ground	3.08	3.08	
1020	Special Foundations	N/A			—	—	—	
1030	Slab on Grade	4″ reinforced concrete with vapor barrier and granular base			S.F. Slab	5.18	5.18	9.9%
2010	Basement Excavation	Site preparation for slab and trench for foundation wall and footing			S.F. Ground	.18	.18	
2020	Basement Walls	4′ foundation wall			L.F. Wall	75	6.15	
B. SHELL								
B10 Superstructure								
1010	Floor Construction	Metal deck on open web steel joists		(5680 S.F.)	S.F. Floor	16.77	2.18	
1020	Roof Construction	Metal deck on open web steel joists			S.F. Roof	4.96	4.96	4.8%
B20 Exterior Enclosure								
2010	Exterior Walls	Face brick with concrete block backup		75% of wall	S.F. Wall	32.44	5.84	
2020	Exterior Windows	Window wall		25% of wall	Each	39.70	2.39	7.0%
2030	Exterior Doors	Glass and metal doors and entrances with transom			Each	4854	2.15	
B30 Roofing								
3010	Roof Coverings	Built-up tar and gravel with flashing; perlite/EPS composite insulation			S.F. Roof	5.57	5.57	4.0%
3020	Roof Openings	Skylight			S.F. Roof	.34	.34	
C. INTERIORS								
1010	Partitions	Concrete block	10 S.F. Floor/L.F. Partition		S.F. Partition	11.73	11.73	
1020	Interior Doors	Single leaf-kalamein fire doors	820 S.F. Floor/Door		Each	1225	1.50	
1030	Fittings	Lockers			S.F. Floor	.05	.05	
2010	Stair Construction	N/A			—	—	—	21.5%
3010	Wall Finishes	60% paint, 40% epoxy coating			S.F. Surface	2.98	5.96	
3020	Floor Finishes	60% epoxy, 20% carpet, 20% vinyl composition tile			S.F. Floor	5.71	5.71	
3030	Ceiling Finishes	Mineral fiber tile on concealed zee runners			S.F. Ceiling	6.86	6.86	
D. SERVICES								
D10 Conveying								
1010	Elevators & Lifts	N/A			—	—	—	
1020	Escalators & Moving Walks	N/A			—	—	—	0.0 %
D20 Plumbing								
2010	Plumbing Fixtures	Toilet and service fixtures, supply and drainage	1 Fixture/260 S.F. Floor		Each	8549	32.88	
2020	Domestic Water Distribution	Gas fired water heater			S.F. Floor	3.09	3.09	24.8%
2040	Rain Water Drainage	Roof drains			S.F. Roof	.72	.72	
D30 HVAC								
3010	Energy Supply	N/A			—	—	—	
3020	Heat Generating Systems	Included in D3050			—	—	—	
3030	Cooling Generating Systems	N/A			—	—	—	13.7 %
3050	Terminal & Package Units	Multizone unit, gas heating, electric cooling			S.F. Floor	20.20	20.20	
3090	Other HVAC Sys. & Equipment	N/A			—	—	—	
D40 Fire Protection								
4010	Sprinklers	Sprinklers, light hazard			S.F. Floor	2.86	2.86	
4020	Standpipes	Standpipe, wet, Class III			S.F. Floor	.30	.30	2.1%
D50 Electrical								
5010	Electrical Service/Distribution	1000 ampere service, panel board and feeders			S.F. Floor	1.71	1.71	
5020	Lighting & Branch Wiring	High efficiency fluorescent fixtures, receptacles, switches, A.C. and misc. power			S.F. Floor	10.88	10.88	
5030	Communications & Security	Addressable alarm systems, internet wiring, and emergency lighting			S.F. Floor	3.47	3.47	11.0%
5090	Other Electrical Systems	Emergency generator, 11.5 kW, Uninterruptible power supply			S.F. Floor	.23	.23	
E. EQUIPMENT & FURNISHINGS								
1010	Commercial Equipment	N/A			—	—	—	
1020	Institutional Equipment	Cabinets, fume hoods, lockers, glassware washer			S.F. Floor	1.50	1.50	1.0 %
1030	Vehicular Equipment	N/A			—	—	—	
1090	Other Equipment	N/A			—	—	—	
F. SPECIAL CONSTRUCTION								
1020	Integrated Construction	N/A			—	—	—	0.0 %
1040	Special Facilities	N/A			—	—	—	
G. BUILDING SITEWORK	**N/A**							

			Sub-Total	147.67	100%
CONTRACTOR FEES (General Requirements: 10%, Overhead: 5%, Profit: 10%)			25%	36.92	
ARCHITECT FEES			10%	18.46	
		Total Building Cost		**203.05**	

Costs per square foot of floor area

Exterior Wall	S.F. Area	15000	20000	25000	30000	35000	40000	45000	50000	55000
	L.F. Perimeter	354	425	457	513	568	583	629	644	683
Brick Face with Concrete Block Back-up	Steel Frame	188.40	182.80	177.20	174.60	172.75	169.95	168.65	166.90	165.90
	R/Conc. Frame	177.70	172.15	166.50	164.00	162.10	159.30	158.00	156.20	155.25
Precast Concrete Panel	Steel Frame	195.00	188.85	182.45	179.60	177.45	174.25	172.80	170.65	169.55
	R/Conc. Frame	185.05	178.70	172.25	169.25	167.10	163.85	162.35	160.20	159.10
Limestone Face Concrete Block Back-up	Steel Frame	204.05	196.90	189.35	186.00	183.50	179.65	177.95	175.45	174.15
	R/Conc. Frame	193.40	186.25	178.65	175.30	172.85	168.95	167.30	164.75	163.45
Perimeter Adj., Add or Deduct	Per 100 L.F.	9.45	7.05	5.70	4.70	4.05	3.55	3.15	2.75	2.60
Story Hgt. Adj., Add or Deduct	Per 1 Ft.	2.35	2.15	1.85	1.75	1.65	1.50	1.40	1.25	1.20

For Basement, add $37.20 per square foot of basement area

The above costs were calculated using the basic specifications shown on the facing page. These costs should be adjusted where necessary for design alternatives and owner's requirements. Reported completed project costs, for this type of structure, range from $133.85 to $273.05 per S.F.

Common additives

Description	Unit	$ Cost	Description	Unit	$ Cost
Carrels Hardwood	Each	855 - 1625	Lockers, Steel, Single tier, 60" or 72"	Opening	218 - 355
Elevators, Hydraulic passenger, 2 stops			2 tier, 60" or 72" total	Opening	112 - 160
2000# capacity	Each	67,900	5 tier, box lockers	Opening	70 - 86
2500# capacity	Each	68,900	Locker bench, lam. maple top only	L.F.	25.50
3500# capacity	Each	72,900	Pedestals, steel pipe	Each	72.50
Emergency Lighting, 25 watt, battery operated			Sound System		
Lead battery	Each	305	Amplifier, 250 watts	Each	1975
Nickel cadmium	Each	850	Speaker, ceiling or wall	Each	206
Escalators, Metal			Trumpet	Each	395
32" wide, 10' story height	Each	143,400			
20' story height	Each	174,500			
48" wide, 10' Story height	Each	150,900			
20' story height	Each	184,000			
Glass					
32" wide, 10' story height	Each	137,400			
20' story height	Each	167,000			
48" wide, 10' story height	Each	143,400			
20' story height	Each	175,500			

Important: See the Reference Section for Location Factors.

Model costs calculated for a 2 story building with 12' story height and 25,000 square feet of floor area

College, Student Union

					Unit	Unit Cost	Cost Per S.F.	% Of Sub-Total
A. SUBSTRUCTURE								
1010	Standard Foundations	Poured concrete; strip and spread footings			S.F. Ground	4.52	2.26	
1020	Special Foundations	N/A			—	—	—	
1030	Slab on Grade	4" reinforced concrete with vapor barrier and granular base			S.F. Slab	5.18	2.60	5.1%
2010	Basement Excavation	Site preparation for slab and trench for foundation wall and footing			S.F. Ground	.18	.09	
2020	Basement Walls	4' foundation wall			L.F. Wall	79	1.45	
B. SHELL								
	B10 Superstructure							
1010	Floor Construction	Concrete flat plate			S.F. Floor	26.32	13.16	16.5%
1020	Roof Construction	Concrete flat plate			S.F. Roof	14.82	7.41	
	B20 Exterior Enclosure							
2010	Exterior Walls	Face brick with concrete block backup	75% of wall		S.F. Wall	32.46	10.68	
2020	Exterior Windows	Window wall	25% of wall		Each	39.70	4.35	12.4%
2030	Exterior Doors	Double aluminum and glass			Each	2625	.42	
	B30 Roofing							
3010	Roof Coverings	Built-up tar and gravel with flashing; perlite/EPS composite insulation			S.F. Roof	6.24	3.12	2.6%
3020	Roof Openings	Roof hatches			S.F. Roof	.12	.06	
C. INTERIORS								
1010	Partitions	Gypsum board on metal studs	14 S.F. Floor/L.F. Partition		S.F. Partition	5.99	4.28	
1020	Interior Doors	Single leaf hollow metal	140 S.F. Floor/Door		Each	1212	8.66	
1030	Fittings	N/A			—	—	—	
2010	Stair Construction	Cast in place concrete			Flight	7725	1.23	23.1%
3010	Wall Finishes	50% paint, 50% vinyl wall covering			S.F. Surface	2.34	3.34	
3020	Floor Finishes	50% carpet, 50% vinyl composition tile			S.F. Floor	6.87	6.87	
3030	Ceiling Finishes	Suspended fiberglass board			S.F. Ceiling	4.35	4.35	
D. SERVICES								
	D10 Conveying							
1010	Elevators & Lifts	One hydraulic passenger elevator			Each	83,500	3.34	2.7%
1020	Escalators & Moving Walks	N/A			—	—	—	
	D20 Plumbing							
2010	Plumbing Fixtures	Toilet and service fixtures, supply and drainage	1 Fixture/1040 S.F. Floor		Each	3359	3.23	
2020	Domestic Water Distribution	Gas fired water heater			S.F. Floor	.87	.87	3.7%
2040	Rain Water Drainage	Roof drains			S.F. Roof	1.12	.56	
	D30 HVAC							
3010	Energy Supply	N/A			—	—	—	
3020	Heat Generating Systems	Included in D3050			—	—	—	
3030	Cooling Generating Systems	N/A			—	—	—	16.2%
3050	Terminal & Package Units	Multizone unit, gas heating, electric cooling			S.F. Floor	20.20	20.20	
3090	Other HVAC Sys. & Equipment	N/A			—	—	—	
	D40 Fire Protection							
4010	Sprinklers	Wet pipe sprinkler system			S.F. Floor	3.07	3.07	3.1%
4020	Standpipes	Standpipe, wet, Class III			S.F. Floor	.82	.82	
	D50 Electrical							
5010	Electrical Service/Distribution	600 ampere service, panel board and feeders			S.F. Floor	1.90	1.90	
5020	Lighting & Branch Wiring	High efficiency fluorescent fixtures, receptacles, switches, A.C. and misc. power			S.F. Floor	12.58	12.58	14.5%
5030	Communications & Security	Addressable alarm systems, internet wiring, communications systems and emergency lighting			S.F. Floor	3.44	3.44	
5090	Other Electrical Systems	Emergency generator, 11.5 kW			S.F. Floor	.17	.17	
E. EQUIPMENT & FURNISHINGS								
1010	Commercial Equipment	N/A			—	—	—	
1020	Institutional Equipment	N/A			—	—	—	
1030	Vehicular Equipment	N/A			—	—	—	0.0%
1090	Other Equipment	N/A			—	—	—	
F. SPECIAL CONSTRUCTION								
1020	Integrated Construction	N/A			—	—	—	
1040	Special Facilities	N/A			—	—	—	0.0%
G. BUILDING SITEWORK	**N/A**							

	Sub-Total		124.51	**100%**
CONTRACTOR FEES (General Requirements: 10%, Overhead: 5%, Profit: 10%)			25%	31.10	
ARCHITECT FEES			7%	10.89	
Total Building Cost				**166.50**

Costs per square foot of floor area

Exterior Wall	S.F. Area	4000	6000	8000	10000	12000	14000	16000	18000	20000
	L.F. Perimeter	260	340	420	453	460	510	560	610	600
Face Brick with Concrete Block Back-up	Bearing Walls	176.20	167.25	162.90	155.80	149.25	147.30	145.85	144.70	141.20
	Steel Frame	171.75	163.95	160.05	153.95	148.35	146.65	145.40	144.45	141.35
Decorative Concrete Block	Bearing Walls	155.00	148.05	144.55	139.20	134.15	132.70	131.60	130.65	127.95
	Steel Frame	157.60	151.55	148.55	144.10	139.95	138.70	137.80	137.05	134.80
Tilt Up Concrete Wall Panels	Bearing Walls	157.70	151.15	147.85	142.90	138.35	136.90	135.90	135.05	132.60
	Steel Frame	153.25	147.75	145.05	141.05	137.45	136.30	135.40	134.75	132.80
Perimeter Adj., Add or Deduct	Per 100 L.F.	22.70	15.20	11.20	9.15	7.60	6.55	5.75	5.10	4.45
Story Hgt. Adj., Add or Deduct	Per 1 Ft.	2.90	2.55	2.30	2.00	1.70	1.60	1.60	1.55	1.25

For Basement, add *$32.50 per square foot of basement area*

The above costs were calculated using the basic specifications shown on the facing page. These costs should be adjusted where necessary for design alternatives and owner's requirements. Reported completed project costs, for this type of structure, range from $81.70 to $258.05 per S.F.

Common additives

Description	Unit	$ Cost
Bar, Front bar	L.F.	410
Back bar	L.F.	325
Booth, Upholstered, custom straight	L.F.	237 - 435
"L" or "U" shaped	L.F.	245 - 415
Bowling Alleys, incl. alley, pinsetter		
Scorer, counter & misc. supplies, average	Lane	61,500
For automatic scorer, add	Lane	10,900
Emergency Lighting, 25 watt, battery operated		
Lead battery	Each	305
Nickel cadmium	Each	850
Kitchen Equipment		
Broiler	Each	4175
Coffee urn, twin 6 gallon	Each	2625
Cooler, 6 ft. long	Each	4000
Dishwasher, 10-12 racks per hr.	Each	5300
Food warmer	Each	560
Freezer, 44 C.F., reach-in	Each	4150
Ice cube maker, 50 lb. per day	Each	1925
Range with 1 oven	Each	3325

Description	Unit	$ Cost
Movie Equipment		
Projector, 35mm	Each	12,200 - 16,700
Screen, wall or ceiling hung	S.F.	8.65 - 13.10
Partitions, Folding leaf, wood		
Acoustic type	S.F.	78 - 129
Seating		
Auditorium chair, all veneer	Each	265
Veneer back, padded seat	Each	335
Upholstered, spring seat	Each	320
Classroom, movable chair & desk	Set	81 - 171
Lecture hall, pedestal type	Each	257 - 750
Sound System		
Amplifier, 250 watts	Each	1975
Speaker, ceiling or wall	Each	206
Trumpet	Each	395
Stage Curtains, Medium weight	S.F.	10.55 - 38.50
Curtain Track, Light duty	L.F.	85.50
Swimming Pools, Complete, gunite	S.F.	85 - 105

Important: See the Reference Section for Location Factors.

Model costs calculated for a 1 story building with 12' story height and 10,000 square feet of floor area

Community Center

				Unit	Unit Cost	Cost Per S.F.	% Of Sub-Total
A. SUBSTRUCTURE							
1010	Standard Foundations	Poured concrete; strip and spread footings		S.F. Ground	2.27	2.27	
1020	Special Foundations	N/A		—	—	—	
1030	Slab on Grade	4" reinforced concrete with vapor barrier and granular base		S.F. Slab	5.18	5.18	13.7%
2010	Basement Excavation	Site preparation for slab and trench for foundation wall and footing		S.F. Ground	.32	.32	
2020	Basement Walls	4' foundation wall		L.F. Wall	75	7.85	
B. SHELL							
B10 Superstructure							
1010	Floor Construction	N/A		—	—	—	
1020	Roof Construction	Metal deck on open web steel joists		S.F. Roof	6.71	6.71	5.9 %
B20 Exterior Enclosure							
2010	Exterior Walls	Face brick with concrete block backup	80% of wall	S.F. Wall	32.45	14.11	
2020	Exterior Windows	Aluminum sliding	20% of wall	Each	623	2.12	15.1%
2030	Exterior Doors	Double aluminum and glass and hollow metal		Each	2608	1.04	
B30 Roofing							
3010	Roof Coverings	Built-up tar and gravel with flashing; perlite/EPS composite insulation		S.F. Roof	7.16	7.16	6.3%
3020	Roof Openings	Roof hatches		S.F. Roof	.09	.09	
C. INTERIORS							
1010	Partitions	Gypsum board on metal studs	14 S.F. Floor/L.F. Partition	S.F. Partition	8.16	5.83	
1020	Interior Doors	Single leaf hollow metal	140 S.F. Floor/Door	Each	1212	8.66	
1030	Fittings	Toilet partitions, directory board, mailboxes		S.F. Floor	1.61	1.61	
2010	Stair Construction	N/A		—	—	—	28.1%
3010	Wall Finishes	Paint		S.F. Surface	1.79	2.56	
3020	Floor Finishes	50% carpet, 50% vinyl tile		S.F. Floor	6.62	6.62	
3030	Ceiling Finishes	Mineral fiber tile on concealed zee bars		S.F. Ceiling	6.86	6.86	
D. SERVICES							
D10 Conveying							
1010	Elevators & Lifts	N/A		—	—	—	0.0 %
1020	Escalators & Moving Walks	N/A		—	—	—	
D20 Plumbing							
2010	Plumbing Fixtures	Kitchen, toilet and service fixtures, supply and drainage	1 Fixture/910 S.F. Floor	Each	3294	3.62	
2020	Domestic Water Distribution	Electric water heater		S.F. Floor	6.32	6.32	9.2%
2040	Rain Water Drainage	Roof drains		S.F. Roof	.53	.53	
D30 HVAC							
3010	Energy Supply	N/A		—	—	—	
3020	Heat Generating Systems	Included in D3050		—	—	—	
3030	Cooling Generating Systems	N/A		—	—	—	9.0 %
3050	Terminal & Package Units	Single zone rooftop unit, gas heating, electric cooling		S.F. Floor	10.27	10.27	
3090	Other HVAC Sys. & Equipment	N/A		—	—	—	
D40 Fire Protection							
4010	Sprinklers	Wet pipe sprinkler system		S.F. Floor	3.51	3.51	3.1%
4020	Standpipes	N/A		—	—	—	
D50 Electrical							
5010	Electrical Service/Distribution	200 ampere service, panel board and feeders		S.F. Floor	1.13	1.13	
5020	Lighting & Branch Wiring	High efficiency fluorescent fixtures, receptacles, switches, A.C. and misc. power		S.F. Floor	5.75	5.75	7.5%
5030	Communications & Security	Addressable alarm systems and emergency lighting		S.F. Floor	1.47	1.47	
5090	Other Electrical Systems	Emergency generator, 15 kW		S.F. Floor	.19	.19	
E. EQUIPMENT & FURNISHINGS							
1010	Commercial Equipment	Freezer, chest type		S.F. Floor	.92	.92	
1020	Institutional Equipment	N/A		—	—	—	
1030	Vehicular Equipment	N/A		—	—	—	2.3%
1090	Other Equipment	Kitchen equipment, directory board, mailboxes, built-in coat racks		S.F. Floor	1.66	1.66	
F. SPECIAL CONSTRUCTION							
1020	Integrated Construction	N/A		—	—	—	0.0 %
1040	Special Facilities	N/A		—	—	—	
G. BUILDING SITEWORK	**N/A**						

		Sub-Total	114.36	100%
CONTRACTOR FEES (General Requirements: 10%, Overhead: 5%, Profit: 10%)		25%	28.57	
ARCHITECT FEES		9%	12.87	

Total Building Cost 155.80

Costs per square foot of floor area

Exterior Wall	S.F. Area	10000	12500	15000	17500	20000	22500	25000	30000	40000
	L.F. Perimeter	400	450	500	550	600	625	650	700	800
Brick Veneer	Steel Frame	311.90	308.60	306.35	304.70	303.55	301.65	300.20	298.05	295.25
	Bearing Wall	325.10	320.40	317.25	315.05	313.40	310.85	308.80	305.75	301.85
Tilt Up Concrete	Steel Frame	303.65	300.30	298.10	296.45	295.30	293.40	291.95	289.75	286.95
Concrete Panels	Bearing Wall	335.25	329.65	325.80	323.15	321.15	317.95	315.45	311.80	307.05
E.I.F.S. System on Metal Studs	Steel Frame	308.20	305.45	303.65	302.35	301.40	299.85	298.70	296.90	294.65
	Bearing Wall	308.25	305.30	303.35	301.90	300.85	299.20	297.95	296.05	293.60
Perimeter Adj., Add or Deduct	Per 100 L.F.	8.20	6.50	5.45	4.75	4.15	3.70	3.30	2.70	2.05
Story Hgt. Adj., Add or Deduct	Per 1 Ft.	1.45	1.30	1.15	1.20	1.05	1.00	0.95	0.80	0.75
For Basement, add $35.75 per square foot of basement area										

The above costs were calculated using the basic specifications shown on the facing page. These costs should be adjusted where necessary for design alternatives and owner's requirements. Reported completed project costs, for this type of structure, range from $168.65 to $342.05 per S.F.

Common additives

Description	Unit	$ Cost
Clock System		
20 room	Ea.	19,200
50 room	Ea.	45,000
Closed Circuit Surveillance, one station		
Camera and monitor	Total	1950
For additional camera stations, add	Ea.	1050
For zoom lens - remote control, add	Ea.	3075 - 10,300
For automatic iris for low light, add	Ea.	2725
Directory Boards, plastic, glass covered		
30" x 20"	Ea.	660
36" x 48"	Ea.	1475
Aluminum, 24" x 18"	Ea.	620
36" x 24"	Ea.	725
48" x 32"	Ea.	1050
48" x 60"	Ea.	2175
Emergency Lighting, 25 watt, battery operated		
Lead battery	Ea.	305
Nickel cadmium	Ea.	850

Description	Unit	$ Cost
Smoke Detectors		
Ceiling type	Ea.	220
Duct type	Ea.	550
Sound System		
Amplifier, 250 watts	Ea.	1975
Speakers, ceiling or wall	Ea.	206
Trumpets	Ea.	395

Important: See the Reference Section for Location Factors.

Model costs calculated for a 1 story building with 16'-6" story height and 22,500 square feet of floor area

Computer Data Center

					Unit	Unit Cost	Cost Per S.F.	% Of Sub-Total
A. SUBSTRUCTURE								
1010	Standard Foundations	Poured concrete; strip and spread footings			S.F. Ground	1.63	1.63	
1020	Special Foundations	N/A			—	—	—	
1030	Slab on Grade	4" concrete with vapor barrier and granular base			S.F. Slab	5.18	5.18	4.1%
2010	Basement Excavation	Site preparation for slab and trench for foundation wall and footing			S.F. Ground	.18	.20	
2020	Basement Walls	4' foundation wall			L.F. Wall	79	2.19	
B. SHELL								
	B10 Superstructure							
1010	Floor Construction	N/A			—	—	—	4.2 %
1020	Roof Construction	Steel joist, girders & deck on columns			S.F. Roof	9.58	9.58	
	B20 Exterior Enclosure							
2010	Exterior Walls	Brick veneer on metal studs			S.F. Wall	23.95	10.43	
2020	Exterior Windows	Aluminum sliding		5% of Wall	Each	584	.89	5.5%
2030	Exterior Doors	Aluminum and glass			Each	3733	1	
	B30 Roofing							
3010	Roof Coverings	Single ply membrane, loose laid and ballasted; polyisocyanurate insulation			S.F. Roof	5.76	5.76	2.6%
3020	Roof Openings	Roof hatch			S.F. Roof	.10	.10	
C. INTERIORS								
1010	Partitions	Gypsum board on metal studs		13 S.F. Floor/L.F. Partition	S.F. Partition	4.66	4.30	
1020	Interior Doors	Solid core wood, metal frame		370 S.F. Floor/Door	Each	760	2.06	
1030	Fittings	Toilet partitions			S.F. Floor	.48	.48	
2010	Stair Construction	N/A			—	—	—	10.4%
3010	Wall Finishes	90% paint, 10% ceramic tile			S.F. Surface	2.02	3.73	
3020	Floor Finishes	65% carpet tile, 10% ceramic tile, 10% quarry tile			S.F. Floor	6.04	6.04	
3030	Ceiling Finishes	Mineral fiber tile on concealed zee bars			S.F. Ceiling	6.86	6.86	
D. SERVICES								
	D10 Conveying							
1010	Elevators & Lifts	N/A			—	—	—	0.0 %
1020	Escalators & Moving Walks	N/A			—	—	—	
	D20 Plumbing							
2010	Plumbing Fixtures	Toilet and service fixtures, supply and drainage		1 fixture/1180 S.F. Floor	Each	4850	4.11	
2020	Domestic Water Distribution	Gas fired water heater			S.F. Floor	.69	.69	2.6%
2040	Rain Water Drainage	Roof drains			S.F. Floor	1.08	1.08	
	D30 HVAC							
3010	Energy Supply	Hot water reheat			S.F. Floor	6.83	6.83	
3020	Heat Generating Systems	Boiler, oil fired hot water			Each	44,475	8.52	
3030	Cooling Generating Systems	Cooling tower			S.F. Floor	9.22	9.22	38.7%
3050	Terminal & Package Units	N/A			—	—	—	
3090	Other HVAC Sys. & Equipment	Heat exchanger, ductwork, air handling units, VAV terminals			S.F. Floor	63	62.64	
	D40 Fire Protection							
4010	Sprinklers	85% wet pipe sprinkler system, 15% preaction sprinkler system			S.F. Floor	4.15	4.15	2.1%
4020	Standpipes	Standpipes and hose systems			S.F. Floor	.60	.60	
	D50 Electrical							
5010	Electrical Service/Distribution	1200 ampere service, panel board and feeders			S.F. Floor	3.97	3.97	
5020	Lighting & Branch Wiring	High efficiency fluorescent fixtures, receptacles, A.C. and misc. power			S.F. Floor	16.18	16.18	28.8%
5030	Communications & Security	Addressable alarm & telephone systems, internet wiring and emergency lighting			S.F. Floor	31.15	31.15	
5090	Other Electrical Systems	Emergency generator, 350 kW; UPS system			S.F. Floor	13.69	13.69	
E. EQUIPMENT & FURNISHINGS								
1010	Commercial Equipment	N/A			—	—	—	
1020	Institutional Equipment	N/A			—	—	—	0.0 %
1030	Vehicular Equipment	N/A			—	—	—	
1090	Other Equipment	N/A			—	—	—	
F. SPECIAL CONSTRUCTION & DEMOLITION								
1020	Integrated Construction	Pedestal access floor		15% of Floor	S.F. Floor	2.27	2.27	1.0%
1040	Special Facilities	N/A			—	—	—	
G. BUILDING SITEWORK	**N/A**							

	Sub-Total	225.53	100%
CONTRACTOR FEES (General Requirements: 10%, Overhead: 5%, Profit: 10%)	25%	56.39	
ARCHITECT FEES	7%	19.73	
Total Building Cost		**301.65**	

Costs per square foot of floor area

Exterior Wall	S.F. Area	16000	23000	30000	37000	44000	51000	58000	65000	72000
	L.F. Perimeter	597	763	821	968	954	1066	1090	1132	1220
Limestone with Concrete Block Back-up	R/Conc. Frame	222.05	213.05	203.35	200.55	193.60	192.00	188.80	186.55	185.60
	Steel Frame	223.90	214.90	205.10	202.40	195.45	193.85	190.55	188.35	187.40
Face Brick with Concrete Block Back-up	R/Conc. Frame	204.30	197.20	190.20	188.05	183.30	182.00	179.80	178.20	177.50
	Steel Frame	206.05	199.05	192.05	189.90	185.05	183.85	181.55	180.05	179.30
Stone with Concrete Block Back-up	R/Conc. Frame	209.40	201.75	193.95	191.65	186.20	184.85	182.35	180.55	179.80
	Steel Frame	211.30	203.70	195.90	193.55	188.15	186.75	184.20	182.45	181.70
Perimeter Adj., Add or Deduct	Per 100 L.F.	8.60	5.85	4.50	3.65	3.10	2.70	2.30	2.10	1.95
Story Hgt. Adj., Add or Deduct	Per 1 Ft.	3.10	2.70	2.25	2.15	1.80	1.75	1.50	1.40	1.40
For Basement, add $30.70 per square foot of basement area										

The above costs were calculated using the basic specifications shown on the facing page. These costs should be adjusted where necessary for design alternatives and owner's requirements. Reported completed project costs, for this type of structure, range from $143.80 to $267.90 per S.F.

Common additives

Description	Unit	$ Cost	Description	Unit	$ Cost
Benches, Hardwood	L.F.	126 - 231	Flagpoles, Complete		
Clock System			Aluminum, 20' high	Each	1750
20 room	Each	19,200	40' high	Each	3875
50 room	Each	45,000	70' high	Each	10,600
Closed Circuit Surveillance, One station			Fiberglass, 23' high	Each	1325
Camera and monitor	Each	1950	39'-5" high	Each	3525
For additional camera stations, add	Each	1050	59' high	Each	7200
Directory Boards, Plastic, glass covered			Intercom System, 25 station capacity		
30" x 20"	Each	660	Master station	Each	3125
36" x 48"	Each	1475	Intercom outlets	Each	183
Aluminum, 24" x 18"	Each	620	Handset	Each	500
36" x 24"	Each	725	Safe, Office type, 1 hour rating		
48" x 32"	Each	1050	30" x 18" x 18"	Each	2500
48" x 60"	Each	2175	60" x 36" x 18", double door	Each	8075
Emergency Lighting, 25 watt, battery operated			Smoke Detectors		
Lead battery	Each	305	Ceiling type	Each	220
Nickel cadmium	Each	850	Duct type	Each	550

Important: See the Reference Section for Location Factors.

Model costs calculated for a 1 story building with 14' story height and 30,000 square feet of floor area

				Unit	Unit Cost	Cost Per S.F.	% Of Sub-Total
A.	**SUBSTRUCTURE**						
1010	Standard Foundations	Poured concrete; strip and spread footings		S.F. Ground	1.71	1.71	
1020	Special Foundations	N/A		—	—	—	
1030	Slab on Grade	4" reinforced concrete with vapor barrier and granular base		S.F. Slab	5.18	5.18	6.2%
2010	Basement Excavation	Site preparation for slab and trench for foundation wall and footing		S.F. Ground	.18	.18	
2020	Basement Walls	4' foundation wall		L.F. Wall	79	2.31	
B.	**SHELL**						
	B10 Superstructure						
1010	Floor Construction	Cast-in-place columns		L.F. Column	72	1.81	13.8%
1020	Roof Construction	Cast-in-place concrete waffle slab		S.F. Roof	19.15	19.15	
	B20 Exterior Enclosure						
2010	Exterior Walls	Limestone panels with concrete block backup	75% of wall	S.F. Wall	67	19.11	
2020	Exterior Windows	Aluminum with insulated glass	25% of wall	Each	735	3.07	14.9%
2030	Exterior Doors	Double wood		Each	1856	.44	
	B30 Roofing						
3010	Roof Coverings	Built-up tar and gravel with flashing; perlite/EPS composite insulation		S.F. Roof	5.58	5.58	3.7%
3020	Roof Openings	Roof hatches		S.F. Roof	.09	.09	
C.	**INTERIORS**						
1010	Partitions	Plaster on metal studs	10 S.F. Floor/L.F. Partition	S.F. Partition	13.03	15.64	
1020	Interior Doors	Single leaf wood	100 S.F. Floor/Door	Each	622	6.22	
1030	Fittings	Toilet partitions		S.F. Floor	.46	.46	
2010	Stair Construction	N/A		—	—	—	32.7%
3010	Wall Finishes	70% paint, 20% wood paneling, 10% vinyl wall covering		S.F. Surface	2.48	5.95	
3020	Floor Finishes	60% hardwood, 20% carpet, 20% terrazzo		S.F. Floor	11.68	11.68	
3030	Ceiling Finishes	Gypsum plaster on metal lath, suspended		S.F. Ceiling	9.81	9.81	
D.	**SERVICES**						
	D10 Conveying						
1010	Elevators & Lifts	N/A		—	—	—	0.0 %
1020	Escalators & Moving Walks	N/A		—	—	—	
	D20 Plumbing						
2010	Plumbing Fixtures	Toilet and service fixtures, supply and drainage	1 Fixture/1110 S.F. Floor	Each	4618	4.16	
2020	Domestic Water Distribution	Electric hot water heater		S.F. Floor	2.27	2.27	4.9%
2040	Rain Water Drainage	Roof drains		S.F. Roof	1.08	1.08	
	D30 HVAC						
3010	Energy Supply	N/A		—	—	—	
3020	Heat Generating Systems	Included in D3050		—	—	—	
3030	Cooling Generating Systems	N/A		—	—	—	13.3 %
3050	Terminal & Package Units	Multizone unit, gas heating, electric cooling		S.F. Floor	20.20	20.20	
3090	Other HVAC Sys. & Equipment	N/A		—	—	—	
	D40 Fire Protection						
4010	Sprinklers	Wet pipe sprinkler system		S.F. Floor	2.86	2.86	2.2%
4020	Standpipes	Standpipe, wet, Class III		S.F. Floor	.46	.46	
	D50 Electrical						
5010	Electrical Service/Distribution	400 ampere service, panel board and feeders		S.F. Floor	1.06	1.06	
5020	Lighting & Branch Wiring	High efficiency fluorescent fixtures, receptacles, switches, A.C. and misc. power		S.F. Floor	9.73	9.73	8.3%
5030	Communications & Security	Addressable alarm systems, internet wiring, and emergency lighting		S.F. Floor	1.66	1.66	
5090	Other Electrical Systems	Emergency generator, 11.5 kW		S.F. Floor	.14	.14	
E.	**EQUIPMENT & FURNISHINGS**						
1010	Commercial Equipment	N/A		—	—	—	
1020	Institutional Equipment	N/A		—	—	—	
1030	Vehicular Equipment	N/A		—	—	—	0.0 %
1090	Other Equipment	N/A		—	—	—	
F.	**SPECIAL CONSTRUCTION**						
1020	Integrated Construction	N/A		—	—	—	0.0 %
1040	Special Facilities	N/A		—	—	—	
G.	**BUILDING SITEWORK**	**N/A**					

		Sub-Total	152.01	100%
	CONTRACTOR FEES (General Requirements: 10%, Overhead: 5%, Profit: 10%)	25%	38.04	
	ARCHITECT FEES	7%	13.30	

Total Building Cost	**203.35**

Costs per square foot of floor area

Exterior Wall	S.F. Area	30000	40000	45000	50000	60000	70000	80000	90000	100000
	L.F. Perimeter	410	493	535	533	600	666	733	800	795
Limestone with Concrete Block Back-up	R/Conc. Frame	241.65	231.40	227.95	221.90	217.35	214.00	211.55	209.65	205.40
	Steel Frame	245.80	235.50	232.05	226.00	221.45	218.10	215.65	213.75	209.50
Face Brick with Concrete Block Back-up	R/Conc. Frame	224.80	216.20	213.30	208.75	205.00	202.25	200.20	198.65	195.55
	Steel Frame	228.95	220.35	217.45	212.90	209.15	206.40	204.40	202.85	199.70
Stone with Concrete Block Back-up	R/Conc. Frame	229.80	220.70	217.65	212.65	208.65	205.75	203.60	201.95	198.50
	Steel Frame	233.90	224.75	221.75	216.70	212.75	209.85	207.70	206.05	202.60
Perimeter Adj., Add or Deduct	Per 100 L.F.	12.65	9.40	8.45	7.55	6.35	5.40	4.75	4.25	3.75
Story Hgt. Adj., Add or Deduct	Per 1 Ft.	4.10	3.65	3.55	3.20	2.95	2.80	2.75	2.70	2.40

For Basement, add $31.35 per square foot of basement area

The above costs were calculated using the basic specifications shown on the facing page. These costs should be adjusted where necessary for design alternatives and owner's requirements. Reported completed project costs, for this type of structure, range from $144.45 to $269.45 per S.F.

Common additives

Description	Unit	$ Cost	Description	Unit	$ Cost
Benches, Hardwood	L.F.	126 - 231	Emergency Lighting, 25 watt, battery operated		
Clock System			Lead battery	Each	305
20 room	Each	19,200	Nickel cadmium	Each	850
50 room	Each	45,000	Flagpoles, Complete		
Closed Circuit Surveillance, One station			Aluminum, 20' high	Each	1750
Camera and monitor	Each	1950	40' high	Each	3875
For additional camera stations, add	Each	1050	70' high	Each	10,600
Directory Boards, Plastic, glass covered			Fiberglass, 23' high	Each	1325
30" x 20"	Each	660	39'-5" high	Each	3525
36" x 48"	Each	1475	59' high	Each	7200
Aluminum, 24" x 18"	Each	620	Intercom System, 25 station capacity		
36" x 24"	Each	725	Master station	Each	3125
48" x 32"	Each	1050	Intercom outlets	Each	183
48" x 60"	Each	2175	Handset	Each	500
Elevators, Hydraulic passenger, 2 stops			Safe, Office type, 1 hour rating		
1500# capacity	Each	66,000	30" x 18" x 18"	Each	2500
2500# capacity	Each	68,900	60" x 36" x 18", double door	Each	8075
3500# capacity	Each	72,900	Smoke Detectors		
Additional stop, add	Each	8375	Ceiling type	Each	220
			Duct type	Each	550

Important: See the Reference Section for Location Factors.

Model costs calculated for a 3 story building with 12' story height and 60,000 square feet of floor area

				Unit	Unit Cost	Cost Per S.F.	% Of Sub-Total
A. SUBSTRUCTURE							
1010	Standard Foundations	Poured concrete; strip and spread footings		S.F. Ground	3.90	1.30	
1020	Special Foundations	N/A		—	—	—	
1030	Slab on Grade	4" reinforced concrete with vapor barrier and granular base		S.F. Slab	5.18	1.72	2.6%
2010	Basement Excavation	Site preparation for slab and trench for foundation wall and footing		S.F. Ground	.18	.06	
2020	Basement Walls	4' foundation wall		L.F. Wall	75	.91	
B. SHELL							
B10 Superstructure							
1010	Floor Construction	Concrete slab with metal deck and beams		S.F. Floor	29.85	19.90	16.9%
1020	Roof Construction	Concrete slab with metal deck and beams		S.F. Roof	19.56	6.52	
B20 Exterior Enclosure							
2010	Exterior Walls	Face brick with concrete block backup	75% of wall	S.F. Wall	35.07	9.47	
2020	Exterior Windows	Horizontal pivoted steel	25% of wall	Each	748	7.48	11.1%
2030	Exterior Doors	Double aluminum and glass and hollow metal		Each	3523	.35	
B30 Roofing							
3010	Roof Coverings	Built-up tar and gravel with flashing; perlite/EPS composite insulation		S.F. Roof	5.97	1.99	1.3%
3020	Roof Openings	N/A		—	—	—	
C. INTERIORS							
1010	Partitions	Plaster on metal studs	10 S.F. Floor/L.F. Partition	S.F. Partition	13.19	13.19	
1020	Interior Doors	Single leaf wood	100 S.F. Floor/Door	Each	622	6.22	
1030	Fittings	Toilet partitions		S.F. Floor	.23	.23	
2010	Stair Construction	Concrete filled metal pan		Flight	19,725	3.28	31.6%
3010	Wall Finishes	70% paint, 20% wood paneling, 10% vinyl wall covering		S.F. Surface	2.48	4.95	
3020	Floor Finishes	60% hardwood, 20% terrazzo, 20% carpet		S.F. Floor	11.68	11.68	
3030	Ceiling Finishes	Gypsum plaster on metal lath, suspended		S.F. Ceiling	9.81	9.81	
D. SERVICES							
D10 Conveying							
1010	Elevators & Lifts	Five hydraulic passenger elevators		Each	158,160	13.18	8.4%
1020	Escalators & Moving Walks	N/A		—	—	—	
D20 Plumbing							
2010	Plumbing Fixtures	Toilet and service fixtures, supply and drainage	1 Fixture/665 S.F. Floor	Each	3099	4.66	
2020	Domestic Water Distribution	Electric water heater		S.F. Floor	2.59	2.59	5.1%
2040	Rain Water Drainage	Roof drains		S.F. Roof	2.01	.67	
D30 HVAC							
3010	Energy Supply	N/A		—	—	—	
3020	Heat Generating Systems	Included in D3050		—	—	—	
3030	Cooling Generating Systems	N/A		—	—	—	12.9 %
3050	Terminal & Package Units	Multizone unit, gas heating, electric cooling		S.F. Floor	20.20	20.20	
3090	Other HVAC Sys. & Equipment	N/A		—	—	—	
D40 Fire Protection							
4010	Sprinklers	Wet pipe sprinkler system		S.F. Floor	2.97	2.97	2.2%
4020	Standpipes	Standpipe, wet, Class III		S.F. Floor	.42	.42	
D50 Electrical							
5010	Electrical Service/Distribution	800 ampere service, panel board and feeders		S.F. Floor	1.03	1.03	
5020	Lighting & Branch Wiring	High efficiency fluorescent fixtures, receptacles, switches, A.C. and misc. power		S.F. Floor	9.86	9.86	8.1%
5030	Communications & Security	Addressable alarm systems, internet wiring, and emergency lighting		S.F. Floor	1.51	1.51	
5090	Other Electrical Systems	Emergency generator, 15 kW		S.F. Floor	.23	.23	
E. EQUIPMENT & FURNISHINGS							
1010	Commercial Equipment	N/A		—	—	—	
1020	Institutional Equipment	N/A		—	—	—	
1030	Vehicular Equipment	N/A		—	—	—	0.0 %
1090	Other Equipment	N/A		—	—	—	
F. SPECIAL CONSTRUCTION							
1020	Integrated Construction	N/A		—	—	—	0.0 %
1040	Special Facilities	N/A		—	—	—	
G. BUILDING SITEWORK	**N/A**						

			Sub-Total	156.38	**100%**
CONTRACTOR FEES (General Requirements: 10%, Overhead: 5%, Profit: 10%)			25%	39.09	
ARCHITECT FEES			7%	13.68	

Total Building Cost | **209.15**

Costs per square foot of floor area

Exterior Wall	S.F. Area	2000	5000	7000	10000	12000	15000	18000	21000	25000
	L.F. Perimeter	200	310	360	440	480	520	560	600	660
Tiltup Concrete Panel	Steel Joists	205.30	175.90	168.60	163.35	160.70	157.45	155.20	153.65	152.25
Decorative Concrete Block	Bearing Walls	203.75	173.45	165.75	160.35	157.60	154.20	151.90	150.20	148.75
Brick on Block	Steel Joists	248.80	211.30	201.55	194.60	190.95	186.30	183.25	181.00	179.05
Stucco on Concrete Block	Wood Truss	205.40	175.55	168.10	162.70	160.05	156.70	154.45	152.75	151.35
Brick Veneer	Wood Frame	215.45	183.95	175.85	170.10	167.10	163.35	160.75	158.95	157.40
Wood Siding	Wood Frame	196.55	172.40	166.25	161.90	159.70	156.90	155.00	153.65	152.50
Perimeter Adj., Add or Deduct	Per 100 L.F.	25.95	10.40	7.40	5.20	4.35	3.40	2.85	2.45	2.10
Story Hgt. Adj., Add or Deduct	Per 1 Ft.	2.55	1.60	1.35	1.10	1.05	0.85	0.80	0.75	0.65

For Basement, add $30.50 per square foot of basement area

The above costs were calculated using the basic specifications shown on the facing page. These costs should be adjusted where necessary for design alternatives and owner's requirements. Reported completed project costs, for this type of structure, range from $62.30 to $227.00 per S.F.

Common additives

Description	Unit	$ Cost
Emergency Lighting, 25 watt, battery operated		
Lead battery	Each	305
Nickel cadmium	Each	850
Flagpoles, Complete		
Aluminum, 20' high	Each	1750
40' high	Each	3875
70' high	Each	10,600
Fiberglass, 23' high	Each	1325
39'-5" high	Each	3525
59' high	Each	7200
Gym Floor, Incl. sleepers and finish, maple	S.F.	14.15
Intercom System, 25 Station capacity		
Master station	Each	3125
Intercom outlets	Each	183
Handset	Each	500

Description	Unit	$ Cost
Lockers, Steel, single tier, 60" to 72"	Opening	218 - 355
2 tier, 60" to 72" total	Opening	112 - 160
5 tier, box lockers	Opening	70 - 86
Locker bench, lam. maple top only	L.F.	25.50
Pedestals, steel pipe	Each	72.50
Smoke Detectors		
Ceiling type	Each	220
Duct type	Each	550
Sound System		
Amplifier, 250 watts	Each	1975
Speaker, ceiling or wall	Each	206
Trumpet	Each	395

Important: See the Reference Section for Location Factors.

Model costs calculated for a 1 story building with 12' story height and 10,000 square feet of floor area

Day Care Center

				Unit	Unit Cost	Cost Per S.F.	% Of Sub-Total
A. SUBSTRUCTURE							
1010	Standard Foundations	Poured concrete; strip and spread footings		S.F. Ground	2.07	2.07	
1020	Special Foundations	N/A		—	—	—	
1030	Slab on Grade	4" concrete with vapor barrier and granular base		S.F. Slab	4.66	4.66	8.4%
2010	Basement Excavation	Site preparation for slab and trench for foundation wall and footing		S.F. Ground	.32	.32	
2020	Basement Walls	4' foundation wall		L.F. Wall	79	3.48	
B. SHELL							
	B10 Superstructure						
1010	Floor Construction	Wood beams on columns		S.F. Floor	.14	.14	6.5%
1020	Roof Construction	Wood trusses		S.F. Roof	7.96	7.96	
	B20 Exterior Enclosure						
2010	Exterior Walls	Brick veneer on wood studs	85% of wall	S.F. Wall	24.26	10.89	
2020	Exterior Windows	Window wall	15% of wall	Each	46.90	3.71	15.5%
2030	Exterior Doors	Aluminum and glass; steel		Each	2494	4.74	
	B30 Roofing						
3010	Roof Coverings	Asphalt shingles, 9" fiberglass batt insulation, gutters and downspouts		S.F. Roof	4.41	4.41	3.5%
3020	Roof Openings	N/A		—	—	—	
C. INTERIORS							
1010	Partitions	Gypsum board on wood studs	8 S.F. Floor/S.F. Partition	S.F. Partition	7.52	5.01	
1020	Interior Doors	Single leaf hollow metal	700 S.F. Floor/Door	Each	1212	3.23	
1030	Fittings	Toilet partitions		S.F. Floor	.41	.41	
2010	Stair Construction	N/A		—	—	—	14.1%
3010	Wall Finishes	Paint		S.F. Surface	1.55	2.07	
3020	Floor Finishes	5% quarry tile, 95% vinyl composition tile		S.F. Floor	2.53	2.53	
3030	Ceiling Finishes	Fiberglass tile on tee grid		S.F. Ceiling	4.35	4.35	
D. SERVICES							
	D10 Conveying						
1010	Elevators & Lifts	N/A		—	—	—	0.0 %
1020	Escalators & Moving Walks	N/A		—	—	—	
	D20 Plumbing						
2010	Plumbing Fixtures	Toilet and service fixtures, supply and drainage	1 Fixture/455 S.F. Floor	Each	3035	17.05	
2020	Domestic Water Distribution	Electric water heater		S.F. Floor	1.96	1.96	15.2%
2040	Rain Water Drainage	N/A		—	—	—	
	D30 HVAC						
3010	Energy Supply	Oil fired hot water, wall fin radiation		S.F. Floor	11.49	11.49	
3020	Heat Generating Systems	N/A		—	—	—	
3030	Cooling Generating Systems	N/A		—	—	—	21.0%
3050	Terminal & Package Units	Split systems with air cooled condensing units		S.F. Floor	14.74	14.74	
3090	Other HVAC Sys. & Equipment	N/A		—	—	—	
	D40 Fire Protection						
4010	Sprinklers	Sprinkler, light hazard		S.F. Floor	3.51	3.51	3.5%
4020	Standpipes	Standpipe		S.F. Floor	.80	.80	
	D50 Electrical						
5010	Electrical Service/Distribution	200 ampere service, panel board and feeders		S.F. Floor	1.13	1.13	
5020	Lighting & Branch Wiring	High efficiency fluorescent fixtures, receptacles, switches, A.C. and misc. power		S.F. Floor	8.08	8.08	8.6%
5030	Communications & Security	Addressable alarm systems and emergency lighting		S.F. Floor	1.49	1.49	
5090	Other Electrical Systems	Emergency generator, 15 kW		S.F. Floor	.09	.09	
E. EQUIPMENT & FURNISHINGS							
1010	Commercial Equipment	N/A		—	—	—	
1020	Institutional Equipment	Cabinets and countertop		S.F. Floor	4.51	4.51	3.6 %
1030	Vehicular Equipment	N/A		—	—	—	
1090	Other Equipment	N/A		—	—	—	
F. SPECIAL CONSTRUCTION							
1020	Integrated Construction	N/A		—	—	—	0.0 %
1040	Special Facilities	N/A		—	—	—	
G. BUILDING SITEWORK	**N/A**						

		Sub-Total	124.83	100%
CONTRACTOR FEES (General Requirements: 10%, Overhead: 5%, Profit: 10%)		25%	31.23	
ARCHITECT FEES		9%	14.04	

Total Building Cost 170.10

Costs per square foot of floor area

Exterior Wall	S.F. Area	12000	18000	24000	30000	36000	42000	48000	54000	60000
	L.F. Perimeter	**460**	**580**	**713**	**730**	**826**	**880**	**965**	**1006**	**1045**
Concrete Block	Steel Frame	140.30	133.35	130.20	125.85	124.25	122.55	121.65	120.50	119.50
	Bearing Walls	137.45	130.45	127.35	122.95	121.45	119.75	118.85	117.65	116.65
Precast Concrete Panels	Steel Frame	148.40	140.15	136.50	130.95	129.15	127.00	125.95	124.45	123.15
Insulated Metal Panels	Steel Frame	143.15	135.70	132.40	127.60	125.95	124.05	123.15	121.85	120.80
Face Brick on Common Brick	Steel Frame	159.05	149.05	144.70	137.70	135.50	132.85	131.55	129.60	128.00
Tilt-up Concrete Panel	Steel Frame	142.15	134.90	131.65	127.00	125.35	123.60	122.65	121.40	120.30
Perimeter Adj., Add or Deduct	Per 100 L.F.	5.45	3.60	2.70	2.15	1.85	1.55	1.45	1.20	1.15
Story Hgt. Adj., Add or Deduct	Per 1 Ft.	0.80	0.65	0.60	0.45	0.50	0.45	0.45	0.40	0.35
For Basement, add $31.10 per square foot of basement area										

The above costs were calculated using the basic specifications shown on the facing page. These costs should be adjusted where necessary for design alternatives and owner's requirements. Reported completed project costs, for this type of structure, range from $46.25 to $178.95 per S.F.

Common additives

Description	Unit	$ Cost
Clock System		
20 room	Each	19,200
50 room	Each	45,000
Dock Bumpers, Rubber blocks		
4-1/2" thick, 10" high, 14" long	Each	85.50
24" long	Each	129
36" long	Each	120
12" high, 14" long	Each	132
24" long	Each	150
36" long	Each	173
6" thick, 10" high, 14" long	Each	140
24" long	Each	164
36" long	Each	237
20" high, 11" long	Each	204
Dock Boards, Heavy		
60" x 60" Aluminum, 5,000# cap.	Each	1400
9000# cap.	Each	1675
15,000# cap.	Each	1775

Description	Unit	$ Cost
Dock Levelers, Hinged 10 ton cap.		
6' x 8'	Each	6975
7' x 8'	Each	7850
Partitions, Woven wire, 10 ga., 1-1/2" mesh		
4' wide x 7' high	Each	196
8' high	Each	224
10' High	Each	265
Platform Lifter, Portable, 6'x 6'		
3000# cap.	Each	10,100
4000# cap.	Each	12,400
Fixed, 6' x 8', 5000# cap.	Each	13,100

Important: See the Reference Section for Location Factors.

Model costs calculated for a 1 story building with 20' story height and 30,000 square feet of floor area

Factory, 1 Story

				Unit	Unit Cost	Cost Per S.F.	% Of Sub-Total
A. SUBSTRUCTURE							
1010	Standard Foundations	Poured concrete; strip and spread footings		S.F. Ground	1.42	1.42	
1020	Special Foundations	N/A		—	—	—	
1030	Slab on Grade	4" reinforced concrete with vapor barrier and granular base		S.F. Slab	6.78	6.78	11.4%
2010	Basement Excavation	Site preparation for slab and trench for foundation wall and footing		S.F. Ground	.18	.18	
2020	Basement Walls	4' foundation wall		L.F. Wall	79	2.31	
B. SHELL							
	B10 Superstructure						
1010	Floor Construction	N/A		—	—	—	10.0 %
1020	Roof Construction	Metal deck, open web steel joists, beams and columns		S.F. Roof	9.45	9.45	
	B20 Exterior Enclosure						
2010	Exterior Walls	Concrete block	75% of wall	S.F. Wall	8.25	3.01	
2020	Exterior Windows	Industrial horizontal pivoted steel	25% of wall	Each	858	3.26	8.0%
2030	Exterior Doors	Double aluminum and glass, hollow metal, steel overhead		Each	2481	1.24	
	B30 Roofing						
3010	Roof Coverings	Built-up tar and gravel with flashing; perlite/EPS composite insulation		S.F. Roof	5.76	5.76	6.5%
3020	Roof Openings	Roof hatches		S.F. Roof	.36	.36	
C. INTERIORS							
1010	Partitions	Concrete block	60 S.F. Floor/L.F. Partition	S.F. Partition	8.60	1.72	
1020	Interior Doors	Single leaf hollow metal and fire doors	600 S.F. Floor/Door	Each	1212	2.02	
1030	Fittings	Toilet partitions		S.F. Floor	.91	.91	
2010	Stair Construction	N/A		—	—	—	7.9%
3010	Wall Finishes	Paint		S.F. Surface	4.73	1.89	
3020	Floor Finishes	Vinyl composition tile	10% of floor	S.F. Floor	2.10	.21	
3030	Ceiling Finishes	Fiberglass board on exposed grid system	10% of area	S.F. Ceiling	6.86	.69	
D. SERVICES							
	D10 Conveying						
1010	Elevators & Lifts	N/A		—	—	—	0.0 %
1020	Escalators & Moving Walks	N/A		—	—	—	
	D20 Plumbing						
2010	Plumbing Fixtures	Toilet and service fixtures, supply and drainage	1 Fixture/1000 S.F. Floor	Each	9090	9.09	
2020	Domestic Water Distribution	Gas fired water heater		S.F. Floor	.59	.59	11.3%
2040	Rain Water Drainage	Roof drains		S.F. Roof	.93	.93	
	D30 HVAC						
3010	Energy Supply	Oil fired hot water, unit heaters		S.F. Floor	9.94	9.94	
3020	Heat Generating Systems	N/A		—	—	—	
3030	Cooling Generating Systems	Chilled water, air cooled condenser system		S.F. Floor	12.18	12.18	23.5%
3050	Terminal & Package Units	N/A		—	—	—	
3090	Other HVAC Sys. & Equipment	N/A		—	—	—	
	D40 Fire Protection						
4010	Sprinklers	Sprinklers, ordinary hazard		S.F. Floor	4.01	4.01	4.9%
4020	Standpipes	Standpipe, wet, Class III		S.F. Floor	.58	.58	
	D50 Electrical						
5010	Electrical Service/Distribution	600 ampere service, panel board and feeders		S.F. Floor	1.11	1.11	
5020	Lighting & Branch Wiring	High intensity discharge fixtures, receptacles, switches, A.C. and misc. power		S.F. Floor	12.99	12.99	16.5%
5030	Communications & Security	Addressable alarm systems and emergency lighting		S.F. Floor	1.45	1.45	
5090	Other Electrical Systems	N/A		—	—	—	
E. EQUIPMENT & FURNISHINGS							
1010	Commercial Equipment	N/A		—	—	—	
1020	Institutional Equipment	N/A		—	—	—	
1030	Vehicular Equipment	N/A		—	—	—	0.0 %
1090	Other Equipment	N/A		—	—	—	
F. SPECIAL CONSTRUCTION							
1020	Integrated Construction	N/A		—	—	—	0.0 %
1040	Special Facilities	N/A		—	—	—	
G. BUILDING SITEWORK	**N/A**						

		Sub-Total	94.08	100%
CONTRACTOR FEES (General Requirements: 10%, Overhead: 5%, Profit: 10%)		25%	23.54	
ARCHITECT FEES		7%	8.23	

Total Building Cost	**125.85**

Costs per square foot of floor area

Exterior Wall	S.F. Area	20000	30000	40000	50000	60000	70000	80000	90000	100000
	L.F. Perimeter	362	410	493	576	600	628	660	700	744
Face Brick Common Brick Back-up	Steel Frame	173.30	159.00	154.20	151.20	146.70	143.65	141.55	140.05	139.00
	Concrete Frame	165.80	151.50	146.65	143.65	139.15	136.10	134.00	132.50	131.45
Face Brick Concrete Block Back-up	Steel Frame	172.25	158.20	153.40	150.45	146.10	143.10	141.00	139.55	138.50
	Concrete Frame	164.70	150.70	145.90	143.00	138.55	135.55	133.50	132.05	131.05
Stucco on Concrete Block	Steel Frame	161.25	149.90	145.95	143.50	140.05	137.65	136.05	134.85	134.05
	Concrete Frame	153.75	142.40	138.40	136.00	132.55	130.15	128.50	127.35	126.50
Perimeter Adj., Add or Deduct	Per 100 L.F.	12.70	8.45	6.30	5.05	4.25	3.70	3.15	2.80	2.55
Story Hgt. Adj., Add or Deduct	Per 1 Ft.	3.50	2.70	2.35	2.20	1.95	1.75	1.55	1.50	1.40
For Basement, add $33.90 per square foot of basement area										

The above costs were calculated using the basic specifications shown on the facing page. These costs should be adjusted where necessary for design alternatives and owner's requirements. Reported completed project costs, for this type of structure, range from $46.25 to $179.00 per S.F.

Common additives

Description	Unit	$ Cost	Description	Unit	$ Cost
Clock System			Dock Levelers, Hinged 10 ton cap.		
20 room	Each	19,200	6'x 8'	Each	6975
50 room	Each	45,000	7'x 8'	Each	7850
Dock Bumpers, Rubber blocks			Elevator, Hydraulic freight, 2 stops		
4-1/2" thick, 10" high, 14" long	Each	85.50	3500# capacity	Each	140,700
24" long	Each	129	4000# capacity	Each	147,700
36" long	Each	120	Additional stop, add	Each	10,200
12" high, 14" long	Each	132	Partitions, Woven wire, 10 ga., 1-1/2" mesh		
24" long	Each	150	4' Wide x 7" high	Each	196
36" long	Each	173	8' High	Each	224
6" thick, 10" high, 14" long	Each	140	10' High	Each	265
24" long	Each	164	Platform Lifter, Portable, 6'x 6'		
36" long	Each	237	3000# cap.	Each	10,100
20" high, 11" long	Each	204	4000# cap.	Each	12,400
Dock Boards, Heavy			Fixed, 6'x 8', 5000# cap.	Each	13,100
60" x 60" Aluminum, 5,000# cap.	Each	1400			
9000# cap.	Each	1675			
15,000# cap.	Each	1775			

Important: See the Reference Section for Location Factors.

Model costs calculated for a 3 story building with 12' story height and 90,000 square feet of floor area

				Unit	Unit Cost	Cost Per S.F.	% Of Sub-Total
A.	**SUBSTRUCTURE**						
1010	Standard Foundations	Poured concrete; strip and spread footings		S.F. Ground	6.42	2.14	
1020	Special Foundations	N/A		—	—	—	
1030	Slab on Grade	4" reinforced concrete with vapor barrier and granular base		S.F. Slab	10.56	3.52	6.3%
2010	Basement Excavation	Site preparation for slab and trench for foundation wall and footing		S.F. Ground	.18	.06	
2020	Basement Walls	4' foundation wall		L.F. Wall	79	.61	
B.	**SHELL**						
	B10 Superstructure						
1010	Floor Construction	Concrete flat slab		S.F. Floor	23.42	15.61	20.7%
1020	Roof Construction	Concrete flat slab		S.F. Roof	15.30	5.10	
	B20 Exterior Enclosure						
2010	Exterior Walls	Face brick with common brick backup	70% of wall	S.F. Wall	34.29	6.72	
2020	Exterior Windows	Industrial, horizontal pivoted steel	30% of wall	Each	1329	6.97	14.1%
2030	Exterior Doors	Double aluminum & glass, hollow metal, overhead doors		Each	2358	.44	
	B30 Roofing						
3010	Roof Coverings	Built-up tar and gravel with flashing; perlite/EPS composite insulation		S.F. Roof	5.58	1.86	2.0%
3020	Roof Openings	Roof hatches		S.F. Roof	.36	.12	
C.	**INTERIORS**						
1010	Partitions	Gypsum board on metal studs	50 S.F. Floor/L.F. Partition	S.F. Partition	3.95	.79	
1020	Interior Doors	Single leaf fire doors	500 S.F. Floor/Door	Each	1225	2.45	
1030	Fittings	Toilet partitions		S.F. Floor	.46	.46	
2010	Stair Construction	Concrete		Flight	7725	.68	8.7%
3010	Wall Finishes	Paint		S.F. Surface	.80	.32	
3020	Floor Finishes	90% metallic hardener, 10% vinyl composition tile		S.F. Floor	3.53	3.53	
3030	Ceiling Finishes	Fiberglass board on exposed grid systems	10% of area	S.F. Ceiling	4.35	.44	
D.	**SERVICES**						
	D10 Conveying						
1010	Elevators & Lifts	Two hydraulic freight elevators		Each	209,250	4.65	4.7%
1020	Escalators & Moving Walks	N/A		—	—	—	
	D20 Plumbing						
2010	Plumbing Fixtures	Toilet and service fixtures, supply and drainage	1 Fixture/1345 S.F. Floor	Each	13,652	10.15	
2020	Domestic Water Distribution	Gas fired water heater		S.F. Floor	.35	.35	10.9%
2040	Rain Water Drainage	Roof drains		S.F. Roof	1.23	.41	
	D30 HVAC						
3010	Energy Supply	Oil fired hot water, unit heaters		S.F. Floor	4.62	4.62	
3020	Heat Generating Systems	N/A		—	—	—	
3030	Cooling Generating Systems	Chilled water, air cooled condenser system		S.F. Floor	12.18	12.18	16.8%
3050	Terminal & Package Units	N/A		—	—	—	
3090	Other HVAC Sys. & Equipment	N/A		—	—	—	
	D40 Fire Protection						
4010	Sprinklers	Wet pipe sprinkler system		S.F. Floor	3.53	3.53	4.1%
4020	Standpipes	Standpipe, wet, Class III		S.F. Floor	.56	.56	
	D50 Electrical						
5010	Electrical Service/Distribution	800 ampere service, panel board and feeders		S.F. Floor	.68	.68	
5020	Lighting & Branch Wiring	High intensity discharge fixtures, switches, A.C. and misc. power		S.F. Floor	9.66	9.66	11.7%
5030	Communications & Security	Addressable alarm systems and emergency lighting		S.F. Floor	1.18	1.18	
5090	Other Electrical Systems	Emergency generator, 30 kW		S.F. Floor	.21	.21	
E.	**EQUIPMENT & FURNISHINGS**						
1010	Commercial Equipment	N/A		—	—	—	
1020	Institutional Equipment	N/A		—	—	—	0.0%
1030	Vehicular Equipment	N/A		—	—	—	
1090	Other Equipment	N/A		—	—	—	
F.	**SPECIAL CONSTRUCTION**						
1020	Integrated Construction	N/A		—	—	—	0.0%
1040	Special Facilities	N/A		—	—	—	
G.	**BUILDING SITEWORK**	**N/A**					

			Sub-Total	100	**100%**
CONTRACTOR FEES (General Requirements: 10%, Overhead: 5%, Profit: 10%)			25%	25	
ARCHITECT FEES			6%	7.50	

		Total Building Cost	**132.50**

Costs per square foot of floor area

Exterior Wall	S.F. Area	4000	4500	5000	5500	6000	6500	7000	7500	8000
	L.F. Perimeter	260	280	300	310	320	336	353	370	386
Face Brick Concrete Block Back-up	Steel Joists	185.50	181.85	178.95	175.00	171.85	169.80	168.25	166.85	165.60
	Bearing Walls	177.90	174.30	171.40	167.45	164.30	162.25	160.70	159.30	158.05
Decorative Concrete Block	Steel Joists	170.30	167.35	165.00	161.90	159.45	157.80	156.55	155.40	154.40
	Bearing Walls	162.80	159.80	157.45	154.35	151.90	150.25	149.00	147.90	146.90
Limestone with Concrete Block Back-up	Steel Joists	211.10	206.35	202.60	197.25	192.90	190.25	188.20	186.35	184.70
	Bearing Walls	203.60	198.85	195.05	189.70	185.35	182.70	180.60	178.85	177.15
Perimeter Adj., Add or Deduct	Per 100 L.F.	20.80	18.50	16.60	15.15	13.85	12.85	11.90	11.10	10.40
Story Hgt. Adj., Add or Deduct	Per 1 Ft.	2.70	2.65	2.45	2.40	2.15	2.15	2.10	2.10	2.05

For Basement, add $38.00 per square foot of basement area

The above costs were calculated using the basic specifications shown on the facing page. These costs should be adjusted where necessary for design alternatives and owner's requirements. Reported completed project costs, for this type of structure, range from $74.85 to $221.40 per S.F.

Common additives

Description	Unit	$ Cost
Appliances		
Cooking range, 30" free standing		
1 oven	Each	570 - 2100
2 oven	Each	1375 - 2875
30" built-in		
1 oven	Each	920 - 2000
2 oven	Each	1825 - 2425
Counter top cook tops, 4 burner	Each	410 - 1475
Microwave oven	Each	278 - 795
Combination range, refrig. & sink, 30" wide	Each	1875 - 5550
60" wide	Each	4000
72" wide	Each	5925
Combination range refrigerator, sink		
microwave oven & icemaker	Each	6750
Compactor, residential, 4-1 compaction	Each	815 - 1200
Dishwasher, built-in, 2 cycles	Each	575 - 1125
4 cycles	Each	700 - 1825
Garbage disposer, sink type	Each	225 - 370
Hood for range, 2 speed, vented, 30" wide	Each	310 - 1225
42" wide	Each	420 - 2725

Description	Unit	$ Cost
Appliances, cont.		
Refrigerator, no frost 10-12 C.F.	Each	520 - 660
14-16 C.F.	Each	645 - 940
18-20 C.F.	Each	805 - 1675
Lockers, Steel, single tier, 60" or 72"	Opening	218 - 355
2 tier, 60" or 72" total	Opening	112 - 160
5 tier, box lockers	Opening	70 - 86
Locker bench, lam. maple top only	L.F.	25.50
Pedestals, steel pipe	Each	72.50
Sound System		
Amplifier, 250 watts	Each	1975
Speaker, ceiling or wall	Each	206
Trumpet	Each	395

Important: See the Reference Section for Location Factors.

Model costs calculated for a 1 story building with 14' story height and 6,000 square feet of floor area

				Unit	Unit Cost	Cost Per S.F.	% Of Sub-Total
A. SUBSTRUCTURE							
1010	Standard Foundations	Poured concrete; strip and spread footings		S.F. Ground	3.29	3.29	
1020	Special Foundations	N/A		—	—	—	
1030	Slab on Grade	6" reinforced concrete with vapor barrier and granular base		S.F. Slab	6.78	6.78	11.7%
2010	Basement Excavation	Site preparation for slab and trench for foundation wall and footing		S.F. Ground	.55	.55	
2020	Basement Walls	4' foundation wall		L.F. Wall	79	4.21	
B. SHELL							
B10 Superstructure							
1010	Floor Construction	N/A		—	—	—	8.0%
1020	Roof Construction	Metal deck, open web steel joists, beams on columns		S.F. Roof	10.12	10.12	
B20 Exterior Enclosure							
2010	Exterior Walls	Face brick with concrete block backup	75% of wall	S.F. Wall	32.45	18.17	
2020	Exterior Windows	Aluminum insulated glass	10% of wall	Each	858	2.01	19.4%
2030	Exterior Doors	Single aluminum and glass, overhead, hollow metal	15% of wall	S.F. Door	40	4.48	
B30 Roofing							
3010	Roof Coverings	Built-up tar and gravel with flashing; perlite/EPS composite insulation		S.F. Roof	6.92	6.92	5.6%
3020	Roof Openings	Skylights, roof hatches		S.F. Roof	.16	.16	
C. INTERIORS							
1010	Partitions	Concrete block	17 S.F. Floor/L.F. Partition	S.F. Partition	9.11	5.36	
1020	Interior Doors	Single leaf hollow metal	500 S.F. Floor/Door	Each	1212	2.42	
1030	Fittings	Toilet partitions		S.F. Floor	.46	.46	
2010	Stair Construction	N/A		—	—	—	12.6%
3010	Wall Finishes	Paint		S.F. Surface	2.09	2.46	
3020	Floor Finishes	50% vinyl tile, 50% paint		S.F. Floor	1.92	1.92	
3030	Ceiling Finishes	Fiberglass board on exposed grid, suspended	50% of area	S.F. Ceiling	6.86	3.43	
D. SERVICES							
D10 Conveying							
1010	Elevators & Lifts	N/A		—	—	—	0.0%
1020	Escalators & Moving Walks	N/A		—	—	—	
D20 Plumbing							
2010	Plumbing Fixtures	Kitchen, toilet and service fixtures, supply and drainage	1 Fixture/375 S.F. Floor	Each	4020	10.72	
2020	Domestic Water Distribution	Gast fired water heater		S.F. Floor	2.95	2.95	11.7%
2040	Rain Water Drainage	Roof drains		S.F. Roof	1.27	1.27	
D30 HVAC							
3010	Energy Supply	N/A		—	—	—	
3020	Heat Generating Systems	Included in D3050		—	—	—	
3030	Cooling Generating Systems	N/A		—	—	—	18.4%
3050	Terminal & Package Units	Rooftop multizone unit system		S.F. Floor	23.38	23.38	
3090	Other HVAC Sys. & Equipment	N/A		—	—	—	
D40 Fire Protection							
4010	Sprinklers	Wet pipe sprinkler system		S.F. Floor	4.66	4.66	4.7%
4020	Standpipes	Standpipe, wet, Class III		S.F. Floor	1.33	1.33	
D50 Electrical							
5010	Electrical Service/Distribution	200 ampere service, panel board and feeders		S.F. Floor	1.64	1.64	
5020	Lighting & Branch Wiring	High efficiency fluorescent fixtures, receptacles, switches, A.C. and misc. power		S.F. Floor	6.67	6.67	8.0%
5030	Communications & Security	Addressable alarm systems		S.F. Floor	1.92	1.92	
5090	Other Electrical Systems	N/A		—	—	—	
E. EQUIPMENT & FURNISHINGS							
1010	Commercial Equipment	N/A		—	—	—	
1020	Institutional Equipment	N/A		—	—	—	
1030	Vehicular Equipment	N/A		—	—	—	0.0%
1090	Other Equipment	N/A		—	—	—	
F. SPECIAL CONSTRUCTION							
1020	Integrated Construction	N/A		—	—	—	0.0%
1040	Special Facilities	N/A		—	—	—	
G. BUILDING SITEWORK	**N/A**						

	Sub-Total	127.28	100%
CONTRACTOR FEES (General Requirements: 10%, Overhead: 5%, Profit: 10%)	25%	31.84	
ARCHITECT FEES	8%	12.73	
Total Building Cost		**171.85**	

Costs per square foot of floor area

Exterior Wall	S.F. Area	6000	7000	8000	9000	10000	11000	12000	13000	14000
	L.F. Perimeter	220	240	260	280	286	303	320	336	353
Face Brick with Concrete Block Back-up	Steel Joists	191.95	185.65	180.95	177.30	172.30	169.70	167.55	165.55	164.05
	Precast Conc.	200.90	194.60	190.00	186.35	181.45	178.80	176.70	174.80	173.20
Decorative Concrete Block	Steel Joists	178.35	172.95	168.95	165.75	161.75	159.50	157.70	156.00	154.70
	Precast Conc.	187.30	181.90	177.95	174.85	170.80	168.60	166.85	165.20	163.90
Limestone with Concrete Block Back-up	Steel Joists	220.55	212.40	206.35	201.60	194.65	191.25	188.35	185.80	183.70
	Precast Conc.	229.30	221.15	215.20	210.45	203.55	200.15	197.35	194.70	192.65
Perimeter Adj., Add or Deduct	Per 100 L.F.	24.00	20.55	18.05	16.00	14.45	13.10	11.95	11.15	10.30
Story Hgt. Adj., Add or Deduct	Per 1 Ft.	3.10	2.85	2.75	2.65	2.45	2.35	2.20	2.25	2.15
For Basement, add $36.90 per square foot of basement area										

The above costs were calculated using the basic specifications shown on the facing page. These costs should be adjusted where necessary for design alternatives and owner's requirements. Reported completed project costs, for this type of structure, range from $74.60 to $221.40 per S.F.

Common additives

Description	Unit	$ Cost
Appliances		
Cooking range, 30" free standing		
1 oven	Each	570 - 2100
2 oven	Each	1375 - 2875
30" built-in		
1 oven	Each	920 - 2000
2 oven	Each	1825 - 2425
Counter top cook tops, 4 burner	Each	410 - 1475
Microwave oven	Each	278 - 795
Combination range, refrig. & sink, 30" wide	Each	1875 - 5550
60" wide	Each	4000
72" wide	Each	5925
Combination range, refrigerator, sink,		
microwave oven & icemaker	Each	6750
Compactor, residential, 4-1 compaction	Each	815 - 1200
Dishwasher, built-in, 2 cycles	Each	575 - 1125
4 cycles	Each	700 - 1825
Garbage diposer, sink type	Each	225 - 370
Hood for range, 2 speed, vented, 30" wide	Each	310 - 1225
42" wide	Each	420 - 2725

Description	Unit	$ Cost
Appliances, cont.		
Refrigerator, no frost 10-12 C.F.	Each	520 - 660
14-16 C.F.	Each	645 - 940
18-20 C.F.	Each	805 - 1675
Elevators, Hydraulic passenger, 2 stops		
1500# capacity	Each	66,000
2500# capacity	Each	68,900
3500# capacity	Each	72,900
Lockers, Steel, single tier, 60" or 72"	Opening	218 - 355
2 tier, 60" or 72" total	Opening	112 - 160
5 tier, box lockers	Opening	70 - 86
Locker bench, lam. maple top only	L.F.	25.50
Pedestals, steel pipe	Each	72.50
Sound System		
Amplifier, 250 watts	Each	1975
Speaker, ceiling or wall	Each	206
Trumpet	Each	395

Important: See the Reference Section for Location Factors.

Model costs calculated for a 2 story building with 14' story height and 10,000 square feet of floor area

Fire Station, 2 Story

				Unit	Unit Cost	Cost Per S.F.	% Of Sub-Total
A. SUBSTRUCTURE							
1010	Standard Foundations	Poured concrete; strip and spread footings		S.F. Ground	3.08	1.54	
1020	Special Foundations	N/A		—	—	—	
1030	Slab on Grade	6" reinforced concrete with vapor barrier and granular base		S.F. Slab	6.78	3.40	6.4%
2010	Basement Excavation	Site preparation for slab and trench for foundation wall and footing		S.F. Ground	.55	.28	
2020	Basement Walls	4' foundation wall		L.F. Wall	88	2.50	
B. SHELL							
	B10 Superstructure						
1010	Floor Construction	Open web steel joists, slab form, concrete		S.F. Floor	14.28	7.14	8.0%
1020	Roof Construction	Metal deck on open web steel joists		S.F. Roof	4.98	2.49	
	B20 Exterior Enclosure						
2010	Exterior Walls	Decorative concrete block	75% of wall	S.F. Wall	19.40	11.65	
2020	Exterior Windows	Aluminum insulated glass	10% of wall	Each	735	2.56	14.9%
2030	Exterior Doors	Single aluminum and glass, steel overhead, hollow metal	15% of wall	S.F. Door	30.64	3.68	
	B30 Roofing						
3010	Roof Coverings	Built-up tar and gravel with flashing; perlite/EPS composite insulation		S.F. Roof	7.08	3.54	3.0%
3020	Roof Openings	N/A		—	—	—	
C. INTERIORS							
1010	Partitions	Concrete block	10 S.F. Floor/L.F. Partition	S.F. Partition	9.11	5.36	
1020	Interior Doors	Single leaf hollow metal	500 S.F. Floor/Door	Each	1225	2.45	
1030	Fittings	Toilet partitions		S.F. Floor	.41	.41	
2010	Stair Construction	Concrete filled metal pan		Flight	19,725	3.95	17.0%
3010	Wall Finishes	Paint		S.F. Surface	3.52	4.14	
3020	Floor Finishes	50% vinyl tile, 50% paint		S.F. Floor	1.92	1.92	
3030	Ceiling Finishes	Fiberglass board on exposed grid, suspended	50% of area	S.F. Ceiling	4.35	2.18	
D. SERVICES							
	D10 Conveying						
1010	Elevators & Lifts	One hydraulic passenger elevator		Each	83,400	8.34	7.0%
1020	Escalators & Moving Walks	N/A		—	—	—	
	D20 Plumbing						
2010	Plumbing Fixtures	Kitchen toilet and service fixtures, supply and drainage	1 Fixture/400 S.F. Floor	Each	4092	10.23	
2020	Domestic Water Distribution	Gas fired water heater		S.F. Floor	2.35	2.35	11.2%
2040	Rain Water Drainage	Roof drains		S.F. Roof	1.64	.82	
	D30 HVAC						
3010	Energy Supply	N/A		—	—	—	
3020	Heat Generating Systems	Included in D3050		—	—	—	
3030	Cooling Generating Systems	N/A		—	—	—	19.5 %
3050	Terminal & Package Units	Rooftop multizone unit system		S.F. Floor	23.38	23.38	
3090	Other HVAC Sys. & Equipment	N/A		—	—	—	
	D40 Fire Protection						
4010	Sprinklers	Wet pipe sprinkler system		S.F. Floor	3.74	3.74	4.3%
4020	Standpipes	Standpipe, wet, Class III		S.F. Floor	1.37	1.37	
	D50 Electrical						
5010	Electrical Service/Distribution	400 ampere service, panel board and feeders		S.F. Floor	1.73	1.73	
5020	Lighting & Branch Wiring	High efficiency fluorescent fixtures, receptacles, switches, A.C. and misc. power		S.F. Floor	6.89	6.89	8.7%
5030	Communications & Security	Addressable alarm systems and emergency lighting		S.F. Floor	1.49	1.49	
5090	Other Electrical Systems	Emergency generator, 15 kW		S.F. Floor	.26	.26	
E. EQUIPMENT & FURNISHINGS							
1010	Commercial Equipment	N/A		—	—	—	
1020	Institutional Equipment	N/A		—	—	—	
1030	Vehicular Equipment	N/A		—	—	—	0.0 %
1090	Other Equipment	N/A		—	—	—	
F. SPECIAL CONSTRUCTION							
1020	Integrated Construction	N/A		—	—	—	
1040	Special Facilities	N/A		—	—	—	0.0 %
G. BUILDING SITEWORK	**N/A**						

		Sub-Total	119.79	100%
CONTRACTOR FEES (General Requirements: 10%, Overhead: 5%, Profit: 10%)		25%	29.98	
ARCHITECT FEES		8%	11.98	
	Total Building Cost		**161.75**	

Costs per square foot of floor area

Exterior Wall	S.F. Area	4000	5000	6000	8000	10000	12000	14000	16000	18000
	L.F. Perimeter	180	205	230	260	300	340	353	386	420
Cedar Beveled Siding	Wood Frame	194.80	186.15	180.35	171.60	166.85	163.80	160.30	158.55	157.15
Aluminum Siding	Wood Frame	193.75	185.05	179.30	170.55	165.80	162.70	159.25	157.45	156.10
Board and Batten	Wood Frame	190.75	183.45	178.60	170.90	166.85	164.25	161.10	159.55	158.35
Face Brick on Block	Bearing Wall	225.25	214.95	208.10	197.05	191.35	187.60	182.85	180.60	178.95
Stucco on Block	Bearing Wall	208.10	199.30	193.55	184.60	179.85	176.70	173.15	171.35	169.95
Decorative Block	Wood Joists	198.90	189.90	183.85	174.55	169.65	166.40	162.65	160.75	159.25
Perimeter Adj., Add or Deduct	Per 100 L.F.	16.60	13.25	11.05	8.30	6.65	5.50	4.75	4.15	3.65
Story Hgt. Adj., Add or Deduct	Per 1 Ft.	2.15	1.90	1.85	1.55	1.50	1.35	1.20	1.15	1.10
For Basement, add $23.50 per square foot of basement area										

The above costs were calculated using the basic specifications shown on the facing page. These costs should be adjusted where necessary for design alternatives and owner's requirements. Reported completed project costs, for this type of structure, range from $108.50 to $237.35 per S.F.

Common additives

Description	Unit	$ Cost
Appliances		
Cooking range, 30" free standing		
1 oven	Each	570 - 2100
2 oven	Each	1375 - 2875
30" built-in		
1 oven	Each	920 - 2000
2 oven	Each	1825 - 2425
Counter top cook tops, 4 burner	Each	410 - 1475
Microwave oven	Each	278 - 795
Combination range, refrig. & sink, 30" wide	Each	1875 - 5550
60" wide	Each	4000
72" wide	Each	5925
Combination range, refrigerator, sink,		
microwave oven & icemaker	Each	6750
Compactor, residential, 4-1 compaction	Each	815 - 1200
Dishwasher, built-in, 2 cycles	Each	575 - 1125
4 cycles	Each	700 - 1825
Garbage disposer, sink type	Each	225 - 370
Hood for range, 2 speed, vented, 30" wide	Each	310 - 1225
42" wide	Each	420 - 2725

Description	Unit	$ Cost
Appliances, cont.		
Refrigerator, no frost 10-12 C.F.	Each	520 - 660
14-16 C.F.	Each	645 - 940
18-20 C.F.	Each	805 - 1675
Elevators, Hydraulic passenger, 2 stops		
1500# capacity	Each	66,000
2500# capacity	Each	68,900
3500# capacity	Each	72,900
Laundry Equipment		
Dryer, gas, 16 lb. capacity	Each	945
30 lb. capacity	Each	3750
Washer, 4 cycle	Each	1175
Commercial	Each	1600
Sound System		
Amplifier, 250 watts	Each	1975
Speaker, ceiling or wall	Each	206
Trumpet	Each	395

Important: See the Reference Section for Location Factors.

Model costs calculated for a 2 story building with 10' story height and 10,000 square feet of floor area

Fraternity/Sorority House

				Unit	Unit Cost	Cost Per S.F.	% Of Sub-Total
A. SUBSTRUCTURE							
1010	Standard Foundations	Poured concrete; strip and spread footings		S.F. Ground	3	1.50	
1020	Special Foundations	N/A		—	—	—	
1030	Slab on Grade	4" reinforced concrete with vapor barrier and granular base		S.F. Slab	5.18	2.60	5.3%
2010	Basement Excavation	Site preparation for slab and trench for foundation wall and footing		S.F. Ground	.55	.28	
2020	Basement Walls	4' foundation wall		L.F. Wall	71	2.12	
B. SHELL							
B10 Superstructure							
1010	Floor Construction	Plywood on wood joists		S.F. Floor	3.70	1.85	3.1%
1020	Roof Construction	Plywood on wood rafters (pitched)		S.F. Roof	3.38	1.89	
B20 Exterior Enclosure							
2010	Exterior Walls	Cedar bevel siding on wood studs, insulated	80% of wall	S.F. Wall	13.19	6.33	
2020	Exterior Windows	Double hung wood	20% of wall	Each	630	3.02	9.0%
2030	Exterior Doors	Solid core wood		Each	2327	1.63	
B30 Roofing							
3010	Roof Coverings	Asphalt shingles with flashing (pitched); rigid fiberglass insulation		S.F. Roof	4.92	2.46	2.0%
3020	Roof Openings	N/A		—	—	—	
C. INTERIORS							
1010	Partitions	Gypsum board on wood studs	25 S.F. Floor/L.F. Partition	S.F. Partition	7.88	2.52	
1020	Interior Doors	Single leaf wood	200 S.F. Floor/Door	Each	622	3.12	
1030	Fittings	N/A		—	—	—	
2010	Stair Construction	Wood		Flight	2207	.89	18.1%
3010	Wall Finishes	Paint		S.F. Surface	.81	.52	
3020	Floor Finishes	10% hardwood, 70% carpet, 20% ceramic tile		S.F. Floor	10.94	10.94	
3030	Ceiling Finishes	Gypsum board on wood furring		S.F. Ceiling	4.21	4.21	
D. SERVICES							
D10 Conveying							
1010	Elevators & Lifts	One hydraulic passenger elevator		Each	83,400	8.34	6.8%
1020	Escalators & Moving Walks	N/A		—	—	—	
D20 Plumbing							
2010	Plumbing Fixtures	Kitchen toilet and service fixtures, supply and drainage	1 Fixture/150 S.F. Floor	Each	2283	15.22	
2020	Domestic Water Distribution	Gas fired water heater		S.F. Floor	3.22	3.22	15.1%
2040	Rain Water Drainage	N/A		—	—	—	
D30 HVAC							
3010	Energy Supply	Oil fired hot water, baseboard radiation		S.F. Floor	8.15	8.15	
3020	Heat Generating Systems	N/A		—	—	—	
3030	Cooling Generating Systems	N/A		—	—	—	13.6%
3050	Terminal & Package Units	Split system with air cooled condensing unit		S.F. Floor	8.45	8.45	
3090	Other HVAC Sys. & Equipment	N/A		—	—	—	
D40 Fire Protection							
4010	Sprinklers	Wet pipe sprinkler system		S.F. Floor	3.74	3.74	3.9%
4020	Standpipes	Standpipe		S.F. Floor	.98	.98	
D50 Electrical							
5010	Electrical Service/Distribution	600 ampere service, panel board and feeders		S.F. Floor	4.21	4.21	
5020	Lighting & Branch Wiring	High efficiency fluorescent fixtures, receptacles, switches, A.C. and misc. power		S.F. Floor	9.96	9.96	23.3%
5030	Communications & Security	Addressable alarm, communication system, internet and phone wiring, and generator set		S.F. Floor	14.11	14.11	
5090	Other Electrical Systems	Emergency generator, 7.5 kW		S.F. Floor	.22	.22	
E. EQUIPMENT & FURNISHINGS							
1010	Commercial Equipment	N/A		—	—	—	
1020	Institutional Equipment	N/A		—	—	—	
1030	Vehicular Equipment	N/A		—	—	—	0.0%
1090	Other Equipment	N/A		—	—	—	
F. SPECIAL CONSTRUCTION							
1020	Integrated Construction	N/A		—	—	—	
1040	Special Facilities	N/A		—	—	—	0.0%
G. BUILDING SITEWORK	**N/A**						

			Sub-Total	122.48	100%
CONTRACTOR FEES (General Requirements: 10%, Overhead: 5%, Profit: 10%)			25%	30.59	
ARCHITECT FEES			9%	13.78	
		Total Building Cost		**166.85**	

131

Costs per square foot of floor area

Exterior Wall	S.F. Area	4000	6000	8000	10000	12000	14000	16000	18000	20000
	L.F. Perimeter	260	320	384	424	460	484	510	540	576
Vertical Redwood Siding	Wood Frame	186.65	165.45	155.10	147.70	142.55	138.55	135.55	133.30	131.70
Brick Veneer	Wood Frame	199.60	176.35	165.05	156.65	150.80	146.15	142.70	140.15	138.30
Aluminum Siding	Wood Frame	185.25	164.55	154.40	147.30	142.35	138.55	135.70	133.55	131.95
Brick on Block	Wood Truss	207.95	183.50	171.60	162.65	156.35	151.30	147.55	144.80	142.85
Limestone on Block	Wood Truss	239.25	209.05	194.60	182.90	174.60	167.75	162.65	159.00	156.50
Stucco on Block	Wood Truss	191.50	169.95	159.45	151.90	146.65	142.55	139.45	137.20	135.55
Perimeter Adj., Add or Deduct	Per 100 L.F.	12.15	8.10	6.05	4.80	4.05	3.50	3.05	2.70	2.45
Story Hgt. Adj., Add or Deduct	Per 1 Ft.	1.45	1.20	1.05	0.95	0.90	0.85	0.75	0.70	0.65
For Basement, add $31.30 per square foot of basement area										

The above costs were calculated using the basic specifications shown on the facing page. These costs should be adjusted where necessary for design alternatives and owner's requirements. Reported completed project costs, for this type of structure, range from $106.25 to $326.70 per S.F.

Common additives

Description	Unit	$ Cost	Description	Unit	$ Cost
Autopsy Table, Standard	Each	10,800	Planters, Precast concrete		
Deluxe	Each	18,200	48" diam., 24" high	Each	905
Directory Boards, Plastic, glass covered			7" diam., 36" high	Each	1675
30" x 20"	Each	660	Fiberglass, 36" diam., 24" high	Each	870
36" x 48"	Each	1475	60" diam., 24" high	Each	1300
Aluminum, 24" x 18"	Each	620	Smoke Detectors		
36" x 24"	Each	725	Ceiling type	Each	220
48" x 32"	Each	1050	Duct type	Each	550
48" x 60"	Each	2175			
Emergency Lighting, 25 watt, battery operated					
Lead battery	Each	305			
Nickel cadmium	Each	850			
Mortuary Refrigerator, End operated					
Two capacity	Each	13,400			
Six capacity	Each	24,100			

Important: See the Reference Section for Location Factors.

Model costs calculated for a 1 story building with 12' story height and 10,000 square feet of floor area

Funeral Home

				Unit	Unit Cost	Cost Per S.F.	% Of Sub-Total
A. SUBSTRUCTURE							
1010	Standard Foundations	Poured concrete; strip and spread footings		S.F. Ground	1.80	1.80	
1020	Special Foundations	N/A		—	—	—	
1030	Slab on Grade	4" reinforced concrete with vapor barrier and granular base		S.F. Slab	5.18	5.18	9.8%
2010	Basement Excavation	Site preparation for slab and trench for foundation wall and footing		S.F. Ground	.32	.32	
2020	Basement Walls	4' foundation wall		L.F. Wall	71	3.30	
B. SHELL							
	B10 Superstructure						
1010	Floor Construction	N/A		—	—	—	
1020	Roof Construction	Plywood on wood truss		S.F. Roof	4.76	4.76	4.4 %
	B20 Exterior Enclosure						
2010	Exterior Walls	1" x 4" vertical T & G redwood siding on wood studs	90% of wall	S.F. Wall	11.99	5.49	
2020	Exterior Windows	Double hung wood	10% of wall	Each	504	2.14	8.1%
2030	Exterior Doors	Wood swinging double doors, single leaf hollow metal		Each	1928	1.16	
	B30 Roofing						
3010	Roof Coverings	Single ply membrane, fully adhered; polyisocyanurate sheets		S.F. Roof	6.58	6.58	6.1%
3020	Roof Openings	N/A		—	—	—	
C. INTERIORS							
1010	Partitions	Gypsum board on wood studs with sound deadening board	15 S.F. Floor/L.F. Partition	S.F. Partition	9.04	4.82	
1020	Interior Doors	Single leaf wood	150 S.F. Floor/Door	Each	622	4.15	
1030	Fittings	N/A		—	—	—	
2010	Stair Construction	N/A		—	—	—	26.2%
3010	Wall Finishes	50% wallpaper, 25% wood paneling, 25% paint		S.F. Surface	3.21	3.42	
3020	Floor Finishes	70% carpet, 30% ceramic tile		S.F. Floor	11.71	11.71	
3030	Ceiling Finishes	Fiberglass board on exposed grid, suspended		S.F. Ceiling	4.35	4.35	
D. SERVICES							
	D10 Conveying						
1010	Elevators & Lifts	N/A		—	—	—	0.0 %
1020	Escalators & Moving Walks	N/A		—	—	—	
	D20 Plumbing						
2010	Plumbing Fixtures	Toilet and service fixtures, supply and drainage	1 Fixture/770 S.F. Floor	Each	3142	4.08	
2020	Domestic Water Distribution	Electric water heater		S.F. Floor	10.78	10.78	14.6%
2040	Rain Water Drainage	Roof drain		S.F. Roof	.98	.98	
	D30 HVAC						
3010	Energy Supply	N/A		—	—	—	
3020	Heat Generating Systems	Included in D3030		—	—	—	
3030	Cooling Generating Systems	Multizone rooftop unit, gas heating, electric cooling		S.F. Floor	18.70	18.70	17.3 %
3050	Terminal & Package Units	N/A		—	—	—	
3090	Other HVAC Sys. & Equipment	N/A		—	—	—	
	D40 Fire Protection						
4010	Sprinklers	Wet pipe sprinkler system		S.F. Floor	3.51	3.51	4.0%
4020	Standpipes	Standpipe, wet, Class III		S.F. Floor	.80	.80	
	D50 Electrical						
5010	Electrical Service/Distribution	400 ampere service, panel board and feeders		S.F. Floor	1.97	1.97	
5020	Lighting & Branch Wiring	High efficiency fluorescent fixtures, receptacles, switches, A.C. and misc. power		S.F. Floor	6.82	6.82	9.6%
5030	Communications & Security	Addressable alarm systems and emergency lighting		S.F. Floor	1.49	1.49	
5090	Other Electrical Systems	Emergency generator, 15 kW		S.F. Floor	.09	.09	
E. EQUIPMENT & FURNISHINGS							
1010	Commercial Equipment	N/A		—	—	—	
1020	Institutional Equipment	N/A		—	—	—	0.0 %
1030	Vehicular Equipment	N/A		—	—	—	
1090	Other Equipment	N/A		—	—	—	
F. SPECIAL CONSTRUCTION							
1020	Integrated Construction	N/A		—	—	—	0.0 %
1040	Special Facilities	N/A		—	—	—	
G. BUILDING SITEWORK	**N/A**						

		Sub-Total	108.40	**100%**
CONTRACTOR FEES (General Requirements: 10%, Overhead: 5%, Profit: 10%)		25%	27.10	
ARCHITECT FEES		9%	12.20	

Total Building Cost	**147.70**

Costs per square foot of floor area

Exterior Wall	S.F. Area	12000	14000	16000	19000	21000	23000	26000	28000	30000
	L.F. Perimeter	440	474	510	556	583	607	648	670	695
E.I.F.S. on Concrete Block	Steel Frame	127.30	124.40	122.30	119.70	118.20	116.90	115.35	114.40	113.65
Tilt-up Concrete Wall	Steel Frame	124.05	121.35	119.50	117.05	115.65	114.50	113.15	112.25	111.55
Face Brick with Concrete Block Back-up	Bearing Walls	128.25	124.85	122.40	119.30	117.50	115.90	114.15	113.05	112.10
	Steel Frame	133.90	130.45	128.00	124.85	123.10	121.55	119.75	118.60	117.65
Stucco on Concrete Block	Bearing Walls	119.80	117.00	114.95	112.45	111.05	109.75	108.30	107.40	106.65
	Steel Frame	126.00	123.15	121.15	118.60	117.15	115.90	114.50	113.55	112.75
Perimeter Adj., Add or Deduct	Per 100 L.F.	7.10	6.05	5.35	4.45	4.00	3.70	3.30	3.10	2.85
Story Hgt. Adj., Add or Deduct	Per 1 Ft.	1.70	1.50	1.50	1.30	1.25	1.20	1.15	1.15	1.05
For Basement, add $35.60 per square foot of basement area										

The above costs were calculated using the basic specifications shown on the facing page. These costs should be adjusted where necessary for design alternatives and owner's requirements. Reported completed project costs, for this type of structure, range from $56.50 to $164.45 per S.F.

Common additives

Description	Unit	$ Cost
Emergency Lighting, 25 watt, battery operated		
Lead battery	Each	305
Nickel cadmium	Each	850
Smoke Detectors		
Ceiling type	Each	220
Duct type	Each	550
Sound System		
Amplifier, 250 watts	Each	1975
Speaker, ceiling or wall	Each	206
Trumpet	Each	395

Model costs calculated for a 1 story building with 14' story height and 21,000 square feet of floor area

					Unit	Unit Cost	Cost Per S.F.	% Of Sub-Total
A.	**SUBSTRUCTURE**							
1010	Standard Foundations	Poured concrete; strip and spread footings			S.F. Ground	1.44	1.44	
1020	Special Foundations	N/A			—	—	—	
1030	Slab on Grade	4" reinforced concrete with vapor barrier and granular base			S.F. Slab	7.67	7.67	13.1%
2010	Basement Excavation	Site preparation for slab and trench for foundation wall and footing			S.F. Ground	.32	.32	
2020	Basement Walls	4' foundation wall			L.F. Wall	79	2.19	
B.	**SHELL**							
	B10 Superstructure							
1010	Floor Construction	N/A			—	—	—	11.7 %
1020	Roof Construction	Metal deck, open web steel joists, beams, columns			S.F. Roof	10.34	10.34	
	B20 Exterior Enclosure							
2010	Exterior Walls	E.I.F.S.	70% of wall		S.F. Wall	25.73	7	
2020	Exterior Windows	Window wall	30% of wall		Each	54	6.29	19.0%
2030	Exterior Doors	Double aluminum and glass, hollow metal, steel overhead			Each	4061	3.48	
	B30 Roofing							
3010	Roof Coverings	Built-up tar and gravel with flashing; perlite/EPS composite insulation			S.F. Roof	5.75	5.75	6.5%
3020	Roof Openings	N/A			—	—	—	
C.	**INTERIORS**							
1010	Partitions	Gypsum board on metal studs	28 S.F. Floor/L.F. Partition		S.F. Partition	4.25	1.82	
1020	Interior Doors	Hollow metal	280 S.F. Floor/Door		Each	1212	4.33	
1030	Fittings	N/A			—	—	—	
2010	Stair Construction	N/A			—	—	—	12.4%
3010	Wall Finishes	Paint			S.F. Surface	.82	.70	
3020	Floor Finishes	50% vinyl tile, 50% paint			S.F. Floor	1.92	1.92	
3030	Ceiling Finishes	Fiberglass board on exposed grid, suspended	50% of area		S.F. Ceiling	4.35	2.18	
D.	**SERVICES**							
	D10 Conveying							
1010	Elevators & Lifts	N/A			—	—	—	0.0 %
1020	Escalators & Moving Walks	N/A			—	—	—	
	D20 Plumbing							
2010	Plumbing Fixtures	Toilet and service fixtures, supply and drainage	1 Fixture/1500 S.F. Floor		Each	3105	2.07	
2020	Domestic Water Distribution	Gas fired water heater			S.F. Floor	1.97	1.97	7.6%
2040	Rain Water Drainage	Roof drains			S.F. Roof	2.65	2.65	
	D30 HVAC							
3010	Energy Supply	N/A			—	—	—	
3020	Heat Generating Systems	N/A			—	—	—	
3030	Cooling Generating Systems	N/A			—	—	—	10.7 %
3050	Terminal & Package Units	Single zone rooftop unit, gas heating, electric cooling			S.F. Floor	8.83	8.83	
3090	Other HVAC Sys. & Equipment	Underfloor garage exhaust system			S.F. Floor	.65	.65	
	D40 Fire Protection							
4010	Sprinklers	Wet pipe sprinkler system			S.F. Floor	4.53	4.53	5.6%
4020	Standpipes	Standpipe			S.F. Floor	.42	.42	
	D50 Electrical							
5010	Electrical Service/Distribution	200 ampere service, panel board and feeders			S.F. Floor	.49	.49	
5020	Lighting & Branch Wiring	T-8 fluorescent fixtures, receptacles, switches, A.C. and misc. power			S.F. Floor	6.60	6.60	11.8%
5030	Communications & Security	Addressable alarm systems, partial internet wiring and emergency lighting			S.F. Floor	3.29	3.29	
5090	Other Electrical Systems	Emergency generator, 7.5 kW			S.F. Floor	.07	.07	
E.	**EQUIPMENT & FURNISHINGS**							
1010	Commercial Equipment	N/A			—	—	—	
1020	Institutional Equipment	N/A			—	—	—	
1030	Vehicular Equipment	Hoists, compressor, fuel pump			S.F. Floor	1.37	1.37	1.6 %
1090	Other Equipment	N/A			—	—	—	
F.	**SPECIAL CONSTRUCTION**							
1020	Integrated Construction	N/A			—	—	—	0.0 %
1040	Special Facilities	N/A			—	—	—	
G.	**BUILDING SITEWORK**	**N/A**						

			Sub-Total	88.37	100%
	CONTRACTOR FEES (General Requirements: 10%, Overhead: 5%, Profit: 10%)		25%	22.10	
	ARCHITECT FEES		7%	7.73	

Total Building Cost 118.20

Costs per square foot of floor area

Exterior Wall	S.F. Area	85000	115000	145000	175000	205000	235000	265000	295000	325000
	L.F. Perimeter	529	638	723	823	923	951	1037	1057	1132
Face Brick with Concrete Block Back-up	Steel Frame	84.30	82.80	81.70	81.10	80.60	79.85	79.55	79.10	78.85
	Precast Concrete	66.90	65.40	64.25	63.70	63.20	62.50	62.20	61.65	61.40
Precast Concrete	Steel Frame	89.40	87.40	85.90	85.10	84.50	83.45	83.05	82.30	82.05
	Precast Concrete	71.25	69.25	67.80	66.95	66.35	65.30	64.90	64.20	63.90
Reinforced Concrete	Steel Frame	83.60	82.35	81.35	80.90	80.40	79.80	79.55	79.15	79.00
	R/Conc. Frame	68.90	67.60	66.70	66.15	65.75	65.15	64.85	64.45	64.30
Perimeter Adj., Add or Deduct	Per 100 L.F.	1.60	1.15	0.90	0.75	0.65	0.60	0.50	0.40	0.40
Story Hgt. Adj., Add or Deduct	Per 1 Ft.	0.50	0.50	0.40	0.40	0.40	0.35	0.35	0.30	0.35
Basement—Not Applicable										

The above costs were calculated using the basic specifications shown on the facing page. These costs should be adjusted where necessary for design alternatives and owner's requirements. Reported completed project costs, for this type of structure, range from $ 34.70 to $ 134.20 per S.F.

Common additives

Description	Unit	$ Cost
Elevators, Electric passenger, 5 stops		
2000# capacity	Each	165,300
3500# capacity	Each	171,300
5000# capacity	Each	177,300
Barrier gate w/programmable controller	Each	4075
Booth for attendant, average	Each	13,300
Fee computer	Each	15,700
Ticket spitter with time/date stamp	Each	7650
Mag strip encoding	Each	21,000
Collection station, pay on foot	Each	127,000
Parking control software	Each	27,200 - 116,500
Painting, Parking stalls	Stall	8.55
Parking Barriers		
Timber with saddles, 4" x 4"	L.F.	6.90
Precast concrete, 6" x 10" x 6'	Each	72.50
Traffic Signs, directional, 12" x 18", high density	Each	95.50

Model costs calculated for a 5 story building with 10' story height and 145,000 square feet of floor area

				Unit	Unit Cost	Cost Per S.F.	% Of Sub-Total
A.	**SUBSTRUCTURE**						
1010	Standard Foundations	Poured concrete; strip and spread footings		S.F. Ground	6.80	1.36	
1020	Special Foundations	N/A		—	—	—	
1030	Slab on Grade	6" reinforced concrete with vapor barrier and granular base		S.F. Slab	6.65	1.33	8.5%
2010	Basement Excavation	Site preparation for slab and trench for foundation wall and footing		S.F. Ground	.18	.04	
2020	Basement Walls	4' foundation wall		L.F. Wall	71	1.41	
B.	**SHELL**						
	B10 Superstructure						
1010	Floor Construction	Double tee precast concrete slab, precast concrete columns		S.F. Floor	27.83	22.26	45.9%
1020	Roof Construction	N/A		—	—	—	
	B20 Exterior Enclosure						
2010	Exterior Walls	Face brick with concrete block backup	40% of story height	S.F. Wall	32.39	3.23	
2020	Exterior Windows	N/A		—	—	—	6.7%
2030	Exterior Doors	N/A		—	—	—	
	B30 Roofing						
3010	Roof Coverings	N/A		—	—	—	0.0%
3020	Roof Openings	N/A		—	—	—	
C.	**INTERIORS**						
1010	Partitions	Concrete block		S.F. Partition	31.72	1.22	
1020	Interior Doors	Hollow metal		Each	24,240	.17	
1030	Fittings	N/A		—	—	—	
2010	Stair Construction	Concrete		Flight	4125	.29	12.4%
3010	Wall Finishes	Paint		S.F. Surface	1.82	.14	
3020	Floor Finishes	Parking deck surface coating		S.F. Floor	4.21	4.21	
3030	Ceiling Finishes	N/A		—	—	—	
D.	**SERVICES**						
	D10 Conveying						
1010	Elevators & Lifts	Two hydraulic passenger elevators		Each	165,300	2.28	4.7%
1020	Escalators & Moving Walks	N/A		—	—	—	
	D20 Plumbing						
2010	Plumbing Fixtures	Toilet and service fixtures, supply and drainage	1 Fixture/18,125 S.F. Floor	Each	725	.04	
2020	Domestic Water Distribution	Electric water heater		S.F. Floor	.07	.07	3.5%
2040	Rain Water Drainage	Roof drains		S.F. Roof	7.95	1.59	
	D30 HVAC						
3010	Energy Supply	N/A		—	—	—	
3020	Heat Generating Systems	N/A		—	—	—	
3030	Cooling Generating Systems	N/A		—	—	—	0.0%
3050	Terminal & Package Units	N/A		—	—	—	
3090	Other HVAC Sys. & Equipment	N/A		—	—	—	
	D40 Fire Protection						
4010	Sprinklers	Dry pipe sprinkler system		S.F. Floor	4.08	4.08	8.6%
4020	Standpipes	Standpipes and hose systems		S.F. Floor	.08	.08	
	D50 Electrical						
5010	Electrical Service/Distribution	400 ampere service, panel board and feeders		S.F. Floor	.27	.27	
5020	Lighting & Branch Wiring	T-8 fluorescent fixtures, receptacles, switches and misc. power		S.F. Floor	2.97	2.97	7.1%
5030	Communications & Security	Addressable alarm systems and emergency lighting		S.F. Floor	.12	.12	
5090	Other Electrical Systems	Emergency generator, 7.5 kW		S.F. Floor	.06	.06	
E.	**EQUIPMENT & FURNISHINGS**						
1010	Commercial Equipment	N/A		—	—	—	
1020	Institutional Equipment	N/A		—	—	—	2.7%
1030	Vehicular Equipment	Ticket dispensers, booths, automatic gates		S.F. Floor	1.29	1.29	
1090	Other Equipment	N/A		—	—	—	
F.	**SPECIAL CONSTRUCTION**						
1020	Integrated Construction	N/A		—	—	—	0.0%
1040	Special Facilities	N/A		—	—	—	
G.	**BUILDING SITEWORK**	N/A					

		Sub-Total	48.51	100%
	CONTRACTOR FEES (General Requirements: 10%, Overhead: 5%, Profit: 10%)	25%	12.10	
	ARCHITECT FEES	6%	3.64	

Total Building Cost	**64.25**

Costs per square foot of floor area

Exterior Wall	S.F. Area	20000	30000	40000	50000	75000	100000	125000	150000	175000
	L.F. Perimeter	400	500	600	650	775	900	1000	1100	1185
Reinforced Concrete	R/Conc. Frame	97.20	90.65	87.40	84.30	80.20	78.15	76.65	75.70	74.90
Perimeter Adj., Add or Deduct	Per 100 L.F.	5.60	3.75	2.85	2.25	1.50	1.10	0.90	0.80	0.70
Story Hgt. Adj., Add or Deduct	Per 1 Ft.	2.10	1.75	1.60	1.35	1.05	0.95	0.90	0.80	0.75
Basement—Not Applicable										

The above costs were calculated using the basic specifications shown on the facing page. These costs should be adjusted where necessary for design alternatives and owner's requirements. Reported completed project costs, for this type of structure, range from $50.45 to $120.15 per S.F.

Common additives

Description	Unit	$ Cost
Elevators, Hydraulic passenger, 2 stops		
1500# capacity	Each	66,000
2500# capacity	Each	68,900
3500# capacity	Each	72,900
Barrier gate w/programmable controller	Each	4075
Booth for attendant, average	Each	13,300
Fee computer	Each	15,700
Ticket spitter with time/date stamp	Each	7650
Mag strip encoding	Each	21,000
Collection station, pay on foot	Each	127,000
Parking control software	Each	27,200 - 116,500
Painting, Parking stalls	Stall	8.55
Parking Barriers		
Timber with saddles, 4" x 4"	L.F.	6.90
Precast concrete, 6" x 10" x 6'	Each	72.50
Traffic Signs, directional, 12" x 18"	Each	95.50

Important: See the Reference Section for Location Factors.

Model costs calculated for a 2 story building with 10' story height and 100,000 square feet of floor area

			Unit	Unit Cost	Cost Per S.F.	% Of Sub-Total
A. SUBSTRUCTURE						
1010	Standard Foundations	Poured concrete; strip and spread footings and waterproofing	S.F. Ground	7.94	3.97	
1020	Special Foundations	N/A	—	—	—	
1030	Slab on Grade	5" reinforced concrete with vapor barrier and granular base	S.F. Slab	6.78	3.40	22.0%
2010	Basement Excavation	Excavation 24' deep	S.F. Ground	10.75	5.38	
2020	Basement Walls	N/A	—	—	—	
B. SHELL						
	B10 Superstructure					
1010	Floor Construction	Cast-in-place concrete beam and slab, concrete columns	S.F. Floor	27.08	13.54	45.0%
1020	Roof Construction	Cast-in-place concrete beam and slab, concrete columns	S.F. Roof	25	12.50	
	B20 Exterior Enclosure					
2010	Exterior Walls	Cast-in place concrete	S.F. Wall	21.78	3.92	
2020	Exterior Windows	N/A	—	—	—	7.1%
2030	Exterior Doors	Steel overhead, hollow metal	Each	4455	.18	
	B30 Roofing					
3010	Roof Coverings	Neoprene membrane traffic deck	S.F. Roof	4.40	2.20	3.8%
3020	Roof Openings	N/A	—	—	—	
C. INTERIORS						
1010	Partitions	Concrete block	S.F. Partition	41.08	.79	
1020	Interior Doors	Hollow metal	Each	9696	.10	
1030	Fittings	N/A	—	—	—	
2010	Stair Construction	Concrete	Flight	6400	.32	2.3%
3010	Wall Finishes	Paint	S.F. Surface	2.60	.10	
3020	Floor Finishes	N/A	—	—	—	
3030	Ceiling Finishes	N/A	—	—	—	
D. SERVICES						
	D10 Conveying					
1010	Elevators & Lifts	Two hydraulic passenger elevators	Each	83,500	1.67	2.9%
1020	Escalators & Moving Walks	N/A	—	—	—	
	D20 Plumbing					
2010	Plumbing Fixtures	Drainage in parking areas, toilets, & service fixtures 1 Fixture/5000 S.F. Floor	Each	.05	.05	
2020	Domestic Water Distribution	Electric water heater	S.F. Floor	.11	.11	2.4%
2040	Rain Water Drainage	Roof drains	S.F. Roof	2.48	1.24	
	D30 HVAC					
3010	Energy Supply	N/A	—	—	—	
3020	Heat Generating Systems	N/A	—	—	—	
3030	Cooling Generating Systems	N/A	—	—	—	0.3 %
3050	Terminal & Package Units	Exhaust fans	S.F. Floor	.16	.16	
3090	Other HVAC Sys. & Equipment	N/A	—	—	—	
	D40 Fire Protection					
4010	Sprinklers	Dry pipe sprinkler system	S.F. Floor	4.09	4.09	7.3%
4020	Standpipes	Dry standpipe system, class 3	S.F. Floor	.15	.15	
	D50 Electrical					
5010	Electrical Service/Distribution	200 ampere service, panel board and feeders	S.F. Floor	.14	.14	
5020	Lighting & Branch Wiring	T-8 fluorescent fixtures, receptacles, switches and misc. power	S.F. Floor	3.21	3.21	6.2%
5030	Communications & Security	Addressable alarm systems and emergency lighting	S.F. Floor	.18	.18	
5090	Other Electrical Systems	Emergency generator, 11.5 kW	S.F. Floor	.07	.07	
E. EQUIPMENT & FURNISHINGS						
1010	Commercial Equipment	N/A	—	—	—	
1020	Institutional Equipment	N/A	—	—	—	
1030	Vehicular Equipment	Ticket dispensers, booths, automatic gates	S.F. Floor	.41	.41	0.7 %
1090	Other Equipment	N/A	—	—	—	
F. SPECIAL CONSTRUCTION						
1020	Integrated Construction	N/A	—	—	—	0.0 %
1040	Special Facilities	N/A	—	—	—	
G. BUILDING SITEWORK	**N/A**					

		Sub-Total	57.88	100%
CONTRACTOR FEES (General Requirements: 10%, Overhead: 5%, Profit: 10%)		25%	14.48	
ARCHITECT FEES		8%	5.79	
		Total Building Cost	**78.15**	

Costs per square foot of floor area

Exterior Wall	S.F. Area	2000	4000	6000	8000	10000	12000	14000	16000	18000
	L.F. Perimeter	180	260	340	420	500	580	586	600	610
Concrete Block	Wood Joists	156.90	134.85	127.50	123.80	121.60	120.15	116.30	113.70	111.55
	Steel Joists	158.15	136.90	129.85	126.35	124.20	122.75	118.90	116.40	114.30
Cast in Place Concrete	Wood Joists	167.90	143.15	134.95	130.85	128.30	126.65	121.90	118.80	116.15
	Steel Joists	170.60	145.85	137.65	133.55	131.00	129.35	124.65	121.45	118.85
Stucco	Wood Frame	162.25	137.15	128.75	124.60	122.05	120.35	115.50	112.30	109.60
Insulated Metal Panels	Steel Frame	179.75	153.65	145.00	140.70	138.05	136.30	131.15	127.80	124.95
Perimeter Adj., Add or Deduct	Per 100 L.F.	26.30	13.10	8.75	6.60	5.30	4.35	3.75	3.25	2.95
Story Hgt. Adj., Add or Deduct	Per 1 Ft.	2.00	1.45	1.25	1.20	1.15	1.05	0.95	0.85	0.80

For Basement, add $33.50 per square foot of basement area

The above costs were calculated using the basic specifications shown on the facing page. These costs should be adjusted where necessary for design alternatives and owner's requirements. Reported completed project costs, for this type of structure, range from $72.30 to $217.25 per S.F.

Common additives

Description	Unit	$ Cost
Air Compressors		
Electric 1-1/2 H.P., standard controls	Each	1525
Dual controls	Each	1925
5 H.P. 115/230 Volt, standard controls	Each	3625
Dual controls	Each	4500
Product Dispenser		
with vapor recovery for 6 nozzles	Each	26,900
Lifts, Single post		
8000# cap., swivel arm	Each	9950
Two post, adjustable frames, 12,000# cap.	Each	7650
24,000# cap.	Each	11,200
30,000# cap.	Each	51,500
Four post, roll on ramp, 25,000# cap.	Each	20,600
Lockers, Steel, single tier, 60" or 72"	Opening	218 - 355
2 tier, 60" or 72" total	Opening	112 - 160
5 tier, box lockers	Opening	70 - 86
Locker bench, lam. maple top only	L.F.	25.50
Pedestals, steel pipe	Each	72.50
Lube Equipment		
3 reel type, with pumps, no piping	Each	12,700
Spray Painting Booth, 26' long, complete	Each	21,700

Important: See the Reference Section for Location Factors.

Model costs calculated for a 1 story building with 14' story height and 10,000 square feet of floor area

					Unit	Unit Cost	Cost Per S.F.	% Of Sub-Total
A. SUBSTRUCTURE								
1010	Standard Foundations	Poured concrete; strip and spread footings			S.F. Ground	2.21	2.21	
1020	Special Foundations	N/A			—	—	—	
1030	Slab on Grade	6" reinforced concrete with vapor barrier and granular base			S.F. Slab	7.67	7.67	15.4%
2010	Basement Excavation	Site preparation for slab and trench for foundation wall and footing			S.F. Ground	.32	.32	
2020	Basement Walls	4' foundation wall			L.F. Wall	79	3.95	
B. SHELL								
B10 Superstructure								
1010	Floor Construction	N/A			—	—	—	
1020	Roof Construction	Metal deck on open web steel joists			S.F. Roof	5.51	5.51	6.0%
B20 Exterior Enclosure								
2010	Exterior Walls	Concrete block		80% of wall	S.F. Wall	12.34	6.91	
2020	Exterior Windows	Hopper type commercial steel		5% of wall	Each	524	1.22	11.2%
2030	Exterior Doors	Steel overhead and hollow metal		15% of wall	S.F. Door	20.67	2.17	
B30 Roofing								
3010	Roof Coverings	Built-up tar and gravel; perlite/EPS composite insulation			S.F. Roof	6.55	6.55	7.1%
3020	Roof Openings	Skylight			S.F. Roof	.02	.02	
C. INTERIORS								
1010	Partitions	Concrete block		50 S.F. Floor/L.F. Partition	S.F. Partition	7.50	1.50	
1020	Interior Doors	Single leaf hollow metal		3000 S.F. Floor/Door	Each	1212	.40	
1030	Fittings	Toilet partitions			S.F. Floor	.14	.14	
2010	Stair Construction	N/A			—	—	—	7.3%
3010	Wall Finishes	Paint			S.F. Surface	7.70	3.08	
3020	Floor Finishes	90% metallic floor hardener, 10% vinyl composition tile			S.F. Floor	1.24	1.24	
3030	Ceiling Finishes	Gypsum board on wood joists in office and washrooms		10% of area	S.F. Ceiling	3.69	.37	
D. SERVICES								
D10 Conveying								
1010	Elevators & Lifts	N/A			—	—	—	0.0%
1020	Escalators & Moving Walks	N/A			—	—	—	
D20 Plumbing								
2010	Plumbing Fixtures	Toilet and service fixtures, supply and drainage		1 Fixture/500 S.F. Floor	Each	1620	3.24	
2020	Domestic Water Distribution	Gas fired water heater			S.F. Floor	.53	.53	6.9%
2040	Rain Water Drainage	Roof drains			S.F. Roof	2.53	2.53	
D30 HVAC								
3010	Energy Supply	N/A			—	—	—	
3020	Heat Generating Systems	N/A			—	—	—	
3030	Cooling Generating Systems	N/A			—	—	—	10.8%
3050	Terminal & Package Units	Single zone AC unit			S.F. Floor	9.04	9.04	
3090	Other HVAC Sys. & Equipment	Garage exhaust system			S.F. Floor	.91	.91	
D40 Fire Protection								
4010	Sprinklers	Sprinklers, ordinary hazard			S.F. Floor	4.53	4.53	5.9%
4020	Standpipes	Standpipe			S.F. Floor	.87	.87	
D50 Electrical								
5010	Electrical Service/Distribution	200 ampere service, panel board and feeders			S.F. Floor	.53	.53	
5020	Lighting & Branch Wiring	T-8 fluorescent fixtures, receptacles, switches, A.C. and misc. power			S.F. Floor	8.05	8.05	13.2%
5030	Communications & Security	Addressable alarm systems, partial internet wiring and emergency lighting			S.F. Floor	3.46	3.46	
5090	Other Electrical Systems	Emergency generator, 15 kW			S.F. Floor	.09	.09	
E. EQUIPMENT & FURNISHINGS								
1010	Commercial Equipment	N/A			—	—	—	
1020	Institutional Equipment	N/A			—	—	—	
1030	Vehicular Equipment	Hoists			S.F. Floor	14.93	14.93	16.2%
1090	Other Equipment	N/A			—	—	—	
F. SPECIAL CONSTRUCTION								
1020	Integrated Construction	N/A			—	—	—	0.0%
1040	Special Facilities	N/A			—	—	—	
G. BUILDING SITEWORK	N/A							

	Sub-Total	91.97	100%
CONTRACTOR FEES (General Requirements: 10%, Overhead: 5%, Profit: 10%)	25%	23.03	
ARCHITECT FEES	8%	9.20	
Total Building Cost		**124.20**	

Costs per square foot of floor area

Exterior Wall	S.F. Area	600	800	1000	1200	1400	1600	1800	2000	2200
	L.F. Perimeter	100	120	126	140	153	160	170	180	190
Face Brick with Concrete Block Back-up	Wood Truss	276.25	259.00	238.85	230.10	223.30	215.60	210.75	206.95	203.70
	Steel Joists	278.30	261.05	240.85	232.10	225.35	217.60	212.75	208.95	205.75
Enameled Sandwich Panel	Steel Frame	260.05	244.75	227.35	219.70	213.80	207.20	202.95	199.70	196.85
Tile on Concrete Block	Steel Joists	294.55	275.70	253.15	243.55	236.00	227.40	222.05	217.75	214.15
Aluminum Siding	Wood Frame	238.70	225.20	210.40	203.80	198.70	193.10	189.45	186.65	184.20
Wood Siding	Wood Frame	239.75	226.15	211.20	204.50	199.35	193.70	190.05	187.20	184.80
Perimeter Adj., Add or Deduct	Per 100 L.F.	116.85	87.70	70.15	58.40	50.15	43.85	39.00	35.05	31.90
Story Hgt. Adj., Add or Deduct	Per 1 Ft.	7.70	6.95	5.80	5.35	5.05	4.60	4.40	4.15	4.00
Basement—Not Applicable										

The above costs were calculated using the basic specifications shown on the facing page. These costs should be adjusted where necessary for design alternatives and owner's requirements. Reported completed project costs, for this type of structure, range from $52.10 to $263.15 per S.F.

Common additives

Description	Unit	$ Cost
Air Compressors		
Electric 1-1/2 H.P., standard controls	Each	1525
Dual controls	Each	1925
5 H.P. 115/230 volt, standard controls	Each	3625
Dual controls	Each	4500
Product Dispenser		
with vapor recovery for 6 nozzles	Each	26,900
Lifts, Single post		
8000# cap. swivel arm	Each	9950
Two post, adjustable frames, 12,000# cap.	Each	7650
24,000# cap.	Each	11,200
30,000# cap.	Each	51,500
Four post, roll on ramp, 25,000# cap.	Each	20,600
Lockers, Steel, single tier, 60" or 72"	Opening	218 - 355
2 tier, 60" or 72" total	Opening	112 - 160
5 tier, box lockers	Each	70 - 86
Locker bench, lam. maple top only	L.F.	25.50
Pedestals, steel pipe	Each	72.50
Lube Equipment		
3 reel type, with pumps, no piping	Each	12,700

Important: See the Reference Section for Location Factors.

Model costs calculated for a 1 story building with 12' story height and 1,400 square feet of floor area

					Unit	Unit Cost	Cost Per S.F.	% Of Sub-Total
A. SUBSTRUCTURE								
1010	Standard Foundations	Poured concrete; strip and spread footings			S.F. Ground	4.21	4.21	
1020	Special Foundations	N/A			—	—	—	
1030	Slab on Grade	6" reinforced concrete with vapor barrier and granular base			S.F. Slab	7.67	7.67	12.5%
2010	Basement Excavation	Site preparation for slab and trench for foundation wall and footing			S.F. Ground	1.02	1.02	
2020	Basement Walls	4' foundation wall			L.F. Wall	71	7.72	
B. SHELL								
B10 Superstructure								
1010	Floor Construction	N/A			—	—	—	
1020	Roof Construction	Plywood on wood trusses			S.F. Roof	6.57	6.57	4.0%
B20 Exterior Enclosure								
2010	Exterior Walls	Face brick with concrete block backup	60% of wall		S.F. Wall	32.45	25.53	
2020	Exterior Windows	Store front and metal top hinged outswinging	20% of wall		Each	39.70	10.42	28.7%
2030	Exterior Doors	Steel overhead, aluminum & glass and hollow metal	20% of wall		Each	44.15	11.58	
B30 Roofing								
3010	Roof Coverings	Asphalt shingles with flashing; perlite/EPS composite insulation			S.F. Roof	3.79	3.79	2.3%
3020	Roof Openings	N/A			—	—	—	
C. INTERIORS								
1010	Partitions	Concrete block	25 S.F. Floor/L.F. Partition		S.F. Partition	7.50	2.40	
1020	Interior Doors	Single leaf hollow metal	700 S.F. Floor/Door		Each	1212	1.73	
1030	Fittings	Toilet partitions			S.F. Floor	1.97	1.97	
2010	Stair Construction	N/A			—	—	—	7.2%
3010	Wall Finishes	Paint			S.F. Surface	5.75	3.68	
3020	Floor Finishes	Vinyl composition tile	35% of floor area		S.F. Floor	2.03	.71	
3030	Ceiling Finishes	Painted gypsum board on furring in sales area & washrooms	35% of floor area		S.F. Ceiling	4.21	1.47	
D. SERVICES								
D10 Conveying								
1010	Elevators & Lifts	N/A			—	—	—	0.0%
1020	Escalators & Moving Walks	N/A			—	—	—	
D20 Plumbing								
2010	Plumbing Fixtures	Toilet and service fixtures, supply and drainage	1 Fixture/235 S.F. Floor		Each	2811	11.96	
2020	Domestic Water Distribution	Gas fired water heater			S.F. Floor	3.79	3.79	9.5%
2040	Rain Water Drainage	N/A			—	—	—	
D30 HVAC								
3010	Energy Supply	N/A			—	—	—	
3020	Heat Generating Systems	N/A			—	—	—	
3030	Cooling Generating Systems	N/A			—	—	—	15.9%
3050	Terminal & Package Units	Single zone AC unit			S.F. Floor	20.87	20.87	
3090	Other HVAC Sys. & Equipment	Underfloor exhaust system			S.F. Floor	5.41	5.41	
D40 Fire Protection								
4010	Sprinklers	Wet pipe sprinkler system			S.F. Floor	8.82	8.82	7.7%
4020	Standpipes	Standpipe			S.F. Floor	3.91	3.91	
D50 Electrical								
5010	Electrical Service/Distribution	200 ampere service, panel board and feeders			S.F. Floor	5.10	5.10	
5020	Lighting & Branch Wiring	High efficiency fluorescent fixtures, receptacles, switches, A.C. and misc. power			S.F. Floor	6.04	6.04	12.2%
5030	Communications & Security	Addressable alarm systems and emergency lighting			S.F. Floor	8.38	8.38	
5090	Other Electrical Systems	Emergency generator			S.F. Floor	.66	.66	
E. EQUIPMENT & FURNISHINGS								
1010	Commercial Equipment	N/A			—	—	—	
1020	Institutional Equipment	N/A			—	—	—	
1030	Vehicular Equipment	N/A			—	—	—	0.0%
1090	Other Equipment	N/A			—	—	—	
F. SPECIAL CONSTRUCTION								
1020	Integrated Construction	N/A			—	—	—	0.0%
1040	Special Facilities	N/A			—	—	—	
G. BUILDING SITEWORK	N/A							

	Sub-Total	165.41	**100%**
CONTRACTOR FEES (General Requirements: 10%, Overhead: 5%, Profit: 10%)	25%	41.35	
ARCHITECT FEES	8%	16.54	

Total Building Cost 223.30

Costs per square foot of floor area

Exterior Wall	S.F. Area	12000	16000	20000	25000	30000	35000	40000	45000	50000
	L.F. Perimeter	440	520	600	700	708	780	841	910	979
Reinforced Concrete Block	Lam. Wood Arches	167.70	163.10	160.35	158.10	154.10	152.80	151.45	150.65	150.00
	Rigid Steel Frame	171.55	166.95	164.20	161.95	157.85	156.60	155.30	154.45	153.80
Face Brick with Concrete Block Back-up	Lam. Wood Arches	194.70	187.10	182.40	178.75	171.45	169.20	166.95	165.55	164.45
	Rigid Steel Frame	198.70	191.10	186.45	182.80	175.45	173.20	171.00	169.55	168.45
Metal Sandwich Panels	Lam. Wood Arches	181.00	174.95	171.20	168.30	162.65	160.85	159.10	158.00	157.10
	Rigid Steel Frame	185.00	178.95	175.25	172.35	166.70	164.90	163.15	162.05	161.10
Perimeter Adj., Add or Deduct	Per 100 L.F.	6.95	5.25	4.15	3.40	2.80	2.30	2.10	1.85	1.65
Story Hgt. Adj., Add or Deduct	Per 1 Ft.	0.95	0.90	0.80	0.75	0.65	0.55	0.55	0.50	0.55

Basement—Not Applicable

The above costs were calculated using the basic specifications shown on the facing page. These costs should be adjusted where necessary for design alternatives and owner's requirements. Reported completed project costs, for this type of structure, range from $79.05 to $235.85 per S.F.

Common additives

Description	Unit	$ Cost
Bleachers, Telescoping, manual		
To 15 tier	Seat	134 - 187
16-20 tier	Seat	277 - 340
21-30 tier	Seat	294 - 355
For power operation, add	Seat	54 - 84.50
Gym Divider Curtain, Mesh top		
Manual roll-up	S.F.	12.90
Gym Mats		
2" naugahyde covered	S.F.	3.89
2" nylon	S.F.	8.30
1-1/2" wall pads	S.F.	7.90
1" wrestling mats	S.F.	5.90
Scoreboard		
Basketball, one side	Each	3450 - 9500
Basketball Backstop		
Wall mtd., 6' extended, fixed	Each	2400 - 2950
Swing up, wall mtd.	Each	2550 - 4000

Description	Unit	$ Cost
Lockers, Steel, single tier, 60" or 72"	Opening	218 - 355
2 tier, 60" or 72" total	Opening	112 - 160
5 tier, box lockers	Opening	70 - 86
Locker bench, lam. maple top only	L.F.	25.50
Pedestals, steel pipe	Each	72.50
Sound System		
Amplifier, 250 watts	Each	1975
Speaker, ceiling or wall	Each	206
Trumpet	Each	395
Emergency Lighting, 25 watt, battery operated		
Lead battery	Each	305
Nickel cadmium	Each	850

Model costs calculated for a 1 story building with 25' story height and 20,000 square feet of floor area

				Unit	Unit Cost	Cost Per S.F.	% Of Sub-Total
A. SUBSTRUCTURE							
1010	Standard Foundations	Poured concrete; strip and spread footings		S.F. Ground	1.72	1.72	
1020	Special Foundations	N/A		—	—	—	
1030	Slab on Grade	4" reinforced concrete with vapor barrier and granular base		S.F. Slab	5.18	5.18	8.0%
2010	Basement Excavation	Site preparation for slab and trench for foundation wall and footing		S.F. Ground	.18	.18	
2020	Basement Walls	4' foundation wall		L.F. Wall	71	2.48	
B. SHELL							
B10 Superstructure							
1010	Floor Construction	N/A		—	—	—	
1020	Roof Construction	Wood deck on laminated wood arches		S.F. Roof	18.20	18.20	15.2 %
B20 Exterior Enclosure							
2010	Exterior Walls	Reinforced concrete block (end walls included)	90% of wall	S.F. Wall	13.04	8.80	
2020	Exterior Windows	Metal horizontal pivoted	10% of wall	Each	528	3.96	11.2%
2030	Exterior Doors	Aluminum and glass, hollow metal, steel overhead		Each	2100	.62	
B30 Roofing							
3010	Roof Coverings	EPDM, 60 mils, fully adhered; polyisocyanurate insulation		S.F. Roof	5.35	5.35	4.5%
3020	Roof Openings	N/A		—	—	—	
C. INTERIORS							
1010	Partitions	Concrete block	50 S.F. Floor/L.F. Partition	S.F. Partition	9.10	1.82	
1020	Interior Doors	Single leaf hollow metal	500 S.F. Floor/Door	Each	1212	2.42	
1030	Fittings	Toilet partitions		S.F. Floor	.28	.28	
2010	Stair Construction	N/A		—	—	—	20.3%
3010	Wall Finishes	50% paint, 50% ceramic tile		S.F. Surface	9.85	3.94	
3020	Floor Finishes	90% hardwood, 10% ceramic tile		S.F. Floor	14.80	14.80	
3030	Ceiling Finishes	Mineral fiber tile on concealed zee bars	15% of area	S.F. Ceiling	6.86	1.03	
D. SERVICES							
D10 Conveying							
1010	Elevators & Lifts	N/A		—	—	—	0.0 %
1020	Escalators & Moving Walks	N/A		—	—	—	
D20 Plumbing							
2010	Plumbing Fixtures	Toilet and service fixtures, supply and drainage	1 Fixture/515 S.F. Floor	Each	4511	8.76	
2020	Domestic Water Distribution	Electric water heater		S.F. Floor	4.82	4.82	11.3%
2040	Rain Water Drainage	N/A		—	—	—	
D30 HVAC							
3010	Energy Supply	N/A		—	—	—	
3020	Heat Generating Systems	Included in D3050		—	—	—	
3030	Cooling Generating Systems	N/A		—	—	—	9.3%
3050	Terminal & Package Units	Single zone rooftop unit, gas heating, electric cooling		S.F. Floor	11.15	11.15	
3090	Other HVAC Sys. & Equipment	N/A		—	—	—	
D40 Fire Protection							
4010	Sprinklers	Wet pipe sprinkler system		S.F. Floor	3.51	3.51	3.7%
4020	Standpipes	Standpipe		S.F. Floor	.94	.94	
D50 Electrical							
5010	Electrical Service/Distribution	400 ampere service, panel board and feeders		S.F. Floor	1.13	1.13	
5020	Lighting & Branch Wiring	High efficiency fluorescent fixtures, receptacles, switches, A.C. and misc. power		S.F. Floor	9.15	9.15	11.1%
5030	Communications & Security	Addressable alarm systems, sound system and emergency lighting		S.F. Floor	2.85	2.85	
5090	Other Electrical Systems	Emergency generator, 7.5 kW		S.F. Floor	.22	.22	
E. EQUIPMENT & FURNISHINGS							
1010	Commercial Equipment	N/A		—	—	—	
1020	Institutional Equipment	N/A		—	—	—	
1030	Vehicular Equipment	N/A		—	—	—	5.5 %
1090	Other Equipment	Bleachers, sauna, weight room		S.F. Floor	6.57	6.57	
F. SPECIAL CONSTRUCTION							
1020	Integrated Construction	N/A		—	—	—	0.0 %
1040	Special Facilities	N/A		—	—	—	
G. BUILDING SITEWORK	**N/A**						

		Sub-Total	119.88	100%
CONTRACTOR FEES (General Requirements: 10%, Overhead: 5%, Profit: 10%)		25%	29.98	
ARCHITECT FEES		7%	10.49	
	Total Building Cost		**160.35**	

Costs per square foot of floor area

Exterior Wall	S.F. Area	5000	10000	15000	20000	30000	40000	50000	75000	100000
	L.F. Perimeter	300	410	500	580	710	830	930	1150	1300
Concrete Block Reinforced	Steel Frame	168.90	137.95	126.40	120.25	113.15	109.40	106.75	102.90	100.35
	Bearing Walls	172.35	141.55	130.10	123.95	116.85	113.15	110.55	106.65	104.15
Precast Concrete	Steel Frame	195.35	156.05	141.10	133.05	123.60	118.50	114.90	109.60	106.05
	Bearing Walls	199.40	160.10	145.15	137.10	127.60	122.55	118.95	113.65	110.10
Galv. Steel Siding	Steel Frame	168.15	137.45	126.00	119.85	112.85	109.10	106.50	102.65	100.15
Metal Sandwich Panel	Steel Frame	179.30	144.35	131.35	124.45	116.40	112.15	109.20	104.90	101.95
Perimeter Adj., Add or Deduct	Per 100 L.F.	17.85	8.95	6.00	4.50	3.05	2.30	1.80	1.20	0.85
Story Hgt. Adj., Add or Deduct	Per 1 Ft.	1.80	1.25	1.00	0.95	0.70	0.60	0.60	0.45	0.35
Basement—Not Applicable										

The above costs were calculated using the basic specifications shown on the facing page. These costs should be adjusted where necessary for design alternatives and owner's requirements. Reported completed project costs, for this type of structure, range from $45.35 to $206.90 per S.F.

Common additives

Description	Unit	$ Cost
Closed Circuit Surveillance, One station		
Camera and monitor	Each	1950
For additional camera stations, add	Each	1050
Emergency Lighting, 25 watt, battery operated		
Lead battery	Each	305
Nickel cadmium	Each	850
Lockers, Steel, single tier, 60" or 72"	Opening	218 - 355
2 tier, 60" or 72" total	Opening	112 - 160
5 tier, box lockers	Opening	70 - 86
Locker bench, lam. maple top only	L.F.	25.50
Pedestals, steel pipe	Each	72.50
Safe, Office type, 1 hour rating		
30" x 18" x 18"	Each	2500
60" x 36" x 18", double door	Each	8075
Sound System		
Amplifier, 250 watts	Each	1975
Speaker, ceiling or wall	Each	206
Trumpet	Each	395

Model costs calculated for a 1 story building with 24' story height and 20,000 square feet of floor area

				Unit	Unit Cost	Cost Per S.F.	% Of Sub-Total
A. SUBSTRUCTURE							
1010	Standard Foundations	Poured concrete; strip and spread footings and		S.F. Ground	1.90	1.90	
1020	Special Foundations	N/A		—	—	—	
1030	Slab on Grade	6" reinforced concrete with vapor barrier and granular base		S.F. Slab	7.67	7.67	13.6%
2010	Basement Excavation	Site preparation for slab and trench for foundation wall and footing		S.F. Ground	.32	.32	
2020	Basement Walls	4' foundation wall		L.F. Wall	79	2.30	
B. SHELL							
B10 Superstructure							
1010	Floor Construction	N/A		—	—	—	
1020	Roof Construction	Metal deck, open web steel joists, beams, columns		S.F. Roof	9.58	9.58	10.7 %
B20 Exterior Enclosure							
2010	Exterior Walls	Galvanized steel siding		S.F. Wall	11.06	3.85	
2020	Exterior Windows	Industrial horizontal pivoted steel	20% of wall	Each	2005	11.63	27.1%
2030	Exterior Doors	Steel overhead and sliding	30% of wall	Each	42	8.77	
B30 Roofing							
3010	Roof Coverings	Elastomeric membrane; fiberboard insulation		S.F. Roof	4.33	4.33	5.0%
3020	Roof Openings	Roof hatches		S.F. Roof	.14	.14	
C. INTERIORS							
1010	Partitions	Concrete block	200 S.F. Floor/L.F. Partition	S.F. Partition	9.80	.49	
1020	Interior Doors	Single leaf hollow metal	5000 S.F. Floor/Door	Each	1212	.24	
1030	Fittings	Toilet partitions		S.F. Floor	.26	.26	
2010	Stair Construction	N/A		—	—	—	1.2%
3010	Wall Finishes	Paint		S.F. Surface	1.30	.13	
3020	Floor Finishes	N/A		—	—	—	
3030	Ceiling Finishes	N/A		—	—	—	
D. SERVICES							
D10 Conveying							
1010	Elevators & Lifts	N/A		—	—	—	0.0 %
1020	Escalators & Moving Walks	N/A		—	—	—	
D20 Plumbing							
2010	Plumbing Fixtures	Toilet and service fixtures, supply and drainage	1 Fixture/1000 S.F. Floor	Each	2800	2.80	
2020	Domestic Water Distribution	Gas fired water heater		S.F. Floor	2.41	2.41	5.8%
2040	Rain Water Drainage	N/A		—	—	—	
D30 HVAC							
3010	Energy Supply	N/A		—	—	—	
3020	Heat Generating Systems	Unit heaters		Each	7.37	7.37	
3030	Cooling Generating Systems	N/A		—	—	—	8.5 %
3050	Terminal & Package Units	Exhaust fan		S.F. Floor	.27	.27	
3090	Other HVAC Sys. & Equipment	N/A		—	—	—	
D40 Fire Protection							
4010	Sprinklers	Sprinklers, extra hazard, deluge system		S.F. Floor	7.55	7.55	13.8%
4020	Standpipes	Standpipe and fire pump		S.F. Floor	4.84	4.84	
D50 Electrical							
5010	Electrical Service/Distribution	200 ampere service, panel board and feeders		S.F. Floor	1	1	
5020	Lighting & Branch Wiring	High intensity discharge fixtures, receptacles, switches and misc. power		S.F. Floor	10.33	10.33	14.3%
5030	Communications & Security	Addressable alarm systems and emergency lighting		S.F. Floor	1.31	1.31	
5090	Other Electrical Systems	Emergency generator, 7.5 kW		S.F. Floor	.15	.15	
E. EQUIPMENT & FURNISHINGS							
1010	Commercial Equipment	N/A		—	—	—	
1020	Institutional Equipment	N/A		—	—	—	0.0 %
1030	Vehicular Equipment	N/A		—	—	—	
1090	Other Equipment	N/A		—	—	—	
F. SPECIAL CONSTRUCTION							
1020	Integrated Construction	N/A		—	—	—	0.0 %
1040	Special Facilities	N/A		—	—	—	
G. BUILDING SITEWORK	**N/A**						

		Sub-Total	89.64	100%
CONTRACTOR FEES (General Requirements: 10%, Overhead: 5%, Profit: 10%)		25%	22.37	
ARCHITECT FEES		7%	7.84	
	Total Building Cost		119.85	

Costs per square foot of floor area

Exterior Wall	S.F. Area	25000	40000	55000	70000	85000	100000	115000	130000	145000
	L.F. Perimeter	388	520	566	666	766	866	878	962	1045
Face Brick with Structural Facing Tile	Steel Frame	341.35	332.90	325.25	322.80	321.15	320.05	317.40	316.65	316.05
	R/Conc. Frame	348.00	339.55	331.90	329.45	327.75	326.65	324.05	323.30	322.65
Face Brick with Concrete Block Back-up	Steel Frame	332.85	325.75	319.65	317.55	316.20	315.25	313.20	312.55	312.05
	R/Conc. Frame	332.20	325.15	319.00	316.95	315.55	314.70	312.60	311.95	311.45
Precast Concrete Panels	Steel Frame	348.25	338.65	329.85	327.00	325.15	323.90	320.80	319.95	319.20
	R/Conc. Frame	354.85	345.30	336.45	333.65	331.80	330.50	327.40	326.55	325.85
Perimeter Adj., Add or Deduct	Per 100 L.F.	9.55	5.90	4.30	3.35	2.80	2.35	2.05	1.80	1.60
Story Hgt. Adj., Add or Deduct	Per 1 Ft.	2.80	2.35	1.80	1.70	1.65	1.55	1.35	1.30	1.25

For Basement, add $34.95 per square foot of basement area

The above costs were calculated using the basic specifications shown on the facing page. These costs should be adjusted where necessary for design alternatives and owner's requirements. Reported completed project costs, for this type of structure, range from $174.40 to $440.50 per S.F.

Common additives

Description	Unit	$ Cost
Cabinet, Base, door units, metal	L.F.	305
Drawer units	L.F.	600
Tall storage cabinets, 7' high, open	L.F.	570
With doors	L.F.	845
Wall, metal 12-1/2" deep, open	L.F.	228
With doors	L.F.	415
Closed Circuit TV (Patient monitoring)		
One station camera & monitor	Each	1950
For additional camera, add	Each	1050
For automatic iris for low light, add	Each	2725
Hubbard Tank, with accessories		
Stainless steel, 125 GPM 45 psi	Each	29,500
For electric hoist, add	Each	3225
Mortuary Refrigerator, End operated		
2 capacity	Each	13,400
6 capacity	Each	24,100

Description	Unit	$ Cost
Nurses Call Station		
Single bedside call station	Each	330
Ceiling speaker station	Each	152
Emergency call station	Each	157
Pillow speaker	Each	272
Double bedside call station	Each	310
Duty station	Each	320
Standard call button	Each	173
Master control station for 20 stations	Each	6250
Sound System		
Amplifier, 250 watts	Each	1975
Speaker, ceiling or wall	Each	206
Trumpet	Each	395
Station, Dietary with ice	Each	18,200
Sterilizers		
Single door, steam	Each	133,000
Double door, steam	Each	224,000
Portable, countertop, steam	Each	3625 - 6425
Gas	Each	43,200
Automatic washer/sterilizer	Each	59,500

Model costs calculated for a 3 story building with 12' story height and 55,000 square feet of floor area

Hospital, 2-3 Story

				Unit	Unit Cost	Cost Per S.F.	% Of Sub-Total
A. SUBSTRUCTURE							
1010	Standard Foundations	Poured concrete; strip and spread footings		S.F. Ground	5.40	1.80	
1020	Special Foundations	N/A		—	—	—	
1030	Slab on Grade	4" reinforced concrete with vapor barrier and granular base		S.F. Slab	5.18	1.72	1.8%
2010	Basement Excavation	Site preparation for slab and trench for foundation wall and footing		S.F. Ground	.18	.06	
2020	Basement Walls	4' foundation wall		L.F. Wall	79	.82	
B. SHELL							
	B10 Superstructure						
1010	Floor Construction	Cast-in-place concrete beam and slab		S.F. Floor	23.31	15.54	9.0%
1020	Roof Construction	Cast-in-place concrete beam and slab		S.F. Roof	18.81	6.27	
	B20 Exterior Enclosure						
2010	Exterior Walls	Face brick and structural facing tile	85% of wall	S.F. Wall	45.63	14.37	
2020	Exterior Windows	Aluminum sliding	15% of wall	Each	735	1.78	6.8%
2030	Exterior Doors	Double aluminum and glass and sliding doors		Each	3200	.35	
	B30 Roofing						
3010	Roof Coverings	Built-up tar and gravel with flashing; perlite/EPS composite insulation		S.F. Roof	6.03	2.01	0.8%
3020	Roof Openings	Roof hatches		S.F. Roof	.03	.01	
C. INTERIORS							
1010	Partitions	Concrete block, gypsum board on metal studs	9 S.F. Floor/L.F. Partition	S.F. Partition	8.55	9.50	
1020	Interior Doors	Single leaf hollow metal	90 S.F. Floor/Door	Each	1245	13.83	
1030	Fittings	Hospital curtains		S.F. Floor	1.04	1.04	
2010	Stair Construction	Concrete filled metal pan		Flight	6125	1.45	19.6%
3010	Wall Finishes	40% vinyl wall covering, 35% ceramic tile, 25% epoxy coating		S.F. Surface	3.48	7.73	
3020	Floor Finishes	60% vinyl tile, 20% ceramic, 20% terrazzo		S.F. Floor	7.30	7.30	
3030	Ceiling Finishes	Plaster on suspended metal lath		S.F. Ceiling	6.86	6.86	
D. SERVICES							
	D10 Conveying						
1010	Elevators & Lifts	Two hydraulic hospital elevators		Each	118,250	4.30	1.8%
1020	Escalators & Moving Walks	N/A		—	—	—	
	D20 Plumbing						
2010	Plumbing Fixtures	Medical, patient & specialty fixtures, supply and drainage	1 Fixture/224 S.F. Floor	Each	6991	31.21	
2020	Domestic Water Distribution	Electric water heater		S.F. Floor	11.17	11.17	17.6%
2040	Rain Water Drainage	Roof drains		S.F. Roof	1.65	.55	
	D30 HVAC						
3010	Energy Supply	Conditioned air with hot water reheat system		S.F. Floor	5.23	5.23	
3020	Heat Generating Systems	Boiler		Each	116,750	3.24	
3030	Cooling Generating Systems	Conditioned air, chillers, cooling towers		S.F. Floor	6.63	6.63	22.4%
3050	Terminal & Package Units	N/A		—	—	—	
3090	Other HVAC Sys. & Equipment	Conditioned air, ductwork, surgery air curtain		S.F. Floor	39.54	39.54	
	D40 Fire Protection						
4010	Sprinklers	Wet pipe sprinkler system		S.F. Floor	2.98	2.98	1.5%
4020	Standpipes	Standpipe		S.F. Floor	.73	.73	
	D50 Electrical						
5010	Electrical Service/Distribution	1200 ampere service, panel board and feeders		S.F. Floor	1.63	1.63	
5020	Lighting & Branch Wiring	High efficiency hospital grade light fixtures, receptacles, switches, A.C. and misc. power		S.F. Floor	16.51	16.51	11.5%
5030	Communications & Security	Addressable alarm systems, internet wiring, communications systems, emergency lighting		S.F. Floor	5.30	5.30	
5090	Other Electrical Systems	Emergency generator, 200 kW with fuel tank, Uninterruptible power supply		S.F. Floor	4.53	4.53	
E. EQUIPMENT & FURNISHINGS							
1010	Commercial Equipment	N/A		—	—	—	
1020	Institutional Equipment	Medical gas system, kitchen equip, lab equip., curtain partitions		S.F. Floor	11.87	11.87	7.2%
1030	Vehicular Equipment	N/A		—	—	—	
2020	Other Equipment	Patient wall systems		S.F. Floor	5.73	5.73	
F. SPECIAL CONSTRUCTION							
1020	Integrated Construction	N/A		—	—	—	0.0%
1040	Special Facilities	N/A		—	—	—	
G. BUILDING SITEWORK	**N/A**						

		Sub-Total	243.59	100%
	CONTRACTOR FEES (General Requirements: 10%, Overhead: 5%, Profit: 10%)	25%	60.91	
	ARCHITECT FEES	9%	27.40	
		Total Building Cost	**331.90**	

Costs per square foot of floor area

Exterior Wall	S.F. Area	100000	125000	150000	175000	200000	225000	250000	275000	300000
	L.F. Perimeter	594	705	816	783	866	950	1033	1116	1200
Face Brick with Structural Facing Tile	Steel Frame	300.70	297.25	294.95	289.35	287.90	286.80	285.90	285.15	284.65
	R/Conc. Frame	305.35	301.75	299.40	293.70	292.25	291.10	290.15	289.45	288.85
Face Brick with Concrete Block Back-up	Steel Frame	295.30	291.95	289.75	285.00	283.65	282.60	281.75	281.05	280.55
	R/Conc. Frame	300.00	296.60	294.45	289.65	288.30	287.30	286.40	285.75	285.30
Precast Concrete Panels With Exposed Aggregate	Steel Frame	299.45	295.90	293.55	288.10	286.65	285.55	284.60	283.85	283.35
	R/Conc. Frame	304.10	300.60	298.20	292.75	291.35	290.25	289.30	288.55	288.05
Perimeter Adj., Add or Deduct	Per 100 L.F.	4.80	3.85	3.20	2.70	2.40	2.15	1.95	1.80	1.50
Story Hgt. Adj., Add or Deduct	Per 1 Ft.	2.15	2.05	1.95	1.60	1.55	1.55	1.45	1.50	1.40

For Basement, add $36.05 per square foot of basement area

The above costs were calculated using the basic specifications shown on the facing page. These costs should be adjusted where necessary for design alternatives and owner's requirements. Reported completed project costs, for this type of structure, range from $178.40 to $435.05 per S.F.

Common additives

Description	Unit	$ Cost
Cabinets, Base, door units, metal	L.F.	305
Drawer units	L.F.	600
Tall storage cabinets, 7' high, open	L.F.	570
With doors	L.F.	845
Wall, metal 12-1/2" deep, open	L.F.	228
With doors	L.F.	415
Closed Circuit TV (Patient monitoring)		
One station camera & monitor	Each	1950
For additional camera add	Each	1050
For automatic iris for low light add	Each	2725
Hubbard Tank, with accessories		
Stainless steel, 125 GPM 45 psi	Each	29,500
For electric hoist, add	Each	3225
Mortuary Refrigerator, End operated		
2 capacity	Each	13,400
6 capacity	Each	24,100

Description	Unit	$ Cost
Nurses Call Station		
Single bedside call station	Each	330
Ceiling speaker station	Each	152
Emergency call station	Each	157
Pillow speaker	Each	272
Double bedside call station	Each	310
Duty station	Each	320
Standard call button	Each	173
Master control station for 20 stations	Each	6250
Sound System		
Amplifier, 250 watts	Each	1975
Speaker, ceiling or wall	Each	206
Trumpet	Each	395
Station, Dietary with ice	Each	18,200
Sterilizers		
Single door, steam	Each	133,000
Double door, steam	Each	224,000
Portable, counter top, steam	Each	3625 - 6425
Gas	Each	43,200
Automatic washer/sterilizer	Each	59,500

Important: See the Reference Section for Location Factors.

Model costs calculated for a 6 story building with 12' story height and 200,000 square feet of floor area

				Unit	Unit Cost	Cost Per S.F.	% Of Sub-Total
A. SUBSTRUCTURE							
1010	Standard Foundations	Poured concrete; strip and spread footings		S.F. Ground	14.04	2.34	
1020	Special Foundations	N/A		—	—	—	
1030	Slab on Grade	4" reinforced concrete with vapor barrier and granular base		S.F. Slab	7.67	1.28	1.9%
2010	Basement Excavation	Site preparation for slab and trench for foundation wall and footing		S.F. Ground	.18	.03	
2020	Basement Walls	4' foundation wall		L.F. Wall	79	.34	
B. SHELL							
B10 Superstructure							
1010	Floor Construction	Concrete slab with metal deck and beams, steel columns		S.F. Floor	21.64	18.03	9.3%
1020	Roof Construction	Metal deck, open web steel joists, beams, interior columns		S.F. Roof	9.12	1.52	
B20 Exterior Enclosure							
2010	Exterior Walls	Face brick and structural facing tile	70% of wall	S.F. Wall	45.64	9.96	
2020	Exterior Windows	Aluminum sliding	30% of wall	Each	584	3.64	6.8%
2030	Exterior Doors	Double aluminum and glass and sliding doors		Each	5742	.80	
B30 Roofing							
3010	Roof Coverings	Built-up tar and gravel with flashing; perlite/EPS composite insulation		S.F. Roof	7.26	1.21	0.6%
3020	Roof Openings	Roof hatches		S.F. Roof	.18	.03	
C. INTERIORS							
1010	Partitions	Gypsum board on metal studs with sound deadening board	9 S.F. Floor/L.F. Partition	S.F. Partition	7.66	8.51	
1020	Interior Doors	Single leaf hollow metal	90 S.F. Floor/Door	Each	1245	13.83	
1030	Fittings	Hospital curtains		S.F. Floor	1.14	1.14	
2010	Stair Construction	Concrete filled metal pan		Flight	11,300	1.47	23.1%
3010	Wall Finishes	40% vinyl wall covering, 35% ceramic tile, 25% epoxy coating		S.F. Surface	3.47	7.72	
3020	Floor Finishes	60% vinyl tile, 20% ceramic, 20% terrazzo		S.F. Floor	8.19	8.19	
3030	Ceiling Finishes	Plaster on suspended metal lath		S.F. Ceiling	8.02	8.02	
D. SERVICES							
D10 Conveying							
1010	Elevators & Lifts	Six geared hospital elevators		Each	215,667	6.47	3.1%
1020	Escalators & Moving Walks	N/A		—	—	—	
D20 Plumbing							
2010	Plumbing Fixtures	Kitchen, toilet and service fixtures, supply and drainage	1 Fixture/416 S.F. Floor	Each	5965	14.34	
2020	Domestic Water Distribution	Electric water heater		S.F. Floor	4.75	4.75	9.3%
2040	Rain Water Drainage	Roof drains		S.F. Floor	3.66	.61	
D30 HVAC							
3010	Energy Supply	Oil fired hot water, wall fin radiation		S.F. Floor	4.53	4.53	
3020	Heat Generating Systems	Hot water boilers, steam boiler for services		Each	41,850	.53	
3030	Cooling Generating Systems	Chilled water units		S.F. Floor	3.01	3.01	20.9%
3050	Terminal & Package Units	N/A		—	—	—	
3090	Other HVAC Sys. & Equipment	Conditioned air with reheat, operating room air curtains		S.F. Floor	36.02	36.02	
D40 Fire Protection							
4010	Sprinklers	Wet pipe sprinkler system		S.F. Floor	2.83	2.83	1.6%
4020	Standpipes	Standpipe		S.F. Floor	.49	.49	
D50 Electrical							
5010	Electrical Service/Distribution	4000 ampere service, panel board and feeders		S.F. Floor	3.53	3.53	
5020	Lighting & Branch Wiring	High efficiency hospital grade light fixtures, receptacles, switches, A.C. and misc. power		S.F. Floor	19.30	19.30	14.1%
5030	Communications & Security	Addressable alarm systems, internet wiring, communications system, emergency lighting		S.F. Floor	2.42	2.42	
5090	Other Electrical Systems	Emergency generator, 800 kW with fuel tank, uninterruptible power supply		S.F. Floor	4.52	4.52	
E. EQUIPMENT & FURNISHINGS							
1010	Commercial Equipment	N/A		—	—	—	
1020	Institutional Equipment	Medical gases, curtain partitions		S.F. Floor	16.11	16.11	9.4%
1030	Vehicular Equipment	N/A		—	—	—	
2020	Other Equipment	Patient wall systems		S.F. Floor	3.78	3.78	
F. SPECIAL CONSTRUCTION							
1020	Integrated Construction	N/A		—	—	—	0.0%
1040	Special Facilities	N/A		—	—	—	
G. BUILDING SITEWORK	**N/A**						

		Sub-Total			211.30	**100%**
	CONTRACTOR FEES (General Requirements: 10%, Overhead: 5%, Profit: 10%)			25%	52.83	
	ARCHITECT FEES			9%	23.77	
		Total Building Cost			**287.90**	

Costs per square foot of floor area

Exterior Wall	S.F. Area	35000	55000	75000	95000	115000	135000	155000	175000	195000
	L.F. Perimeter	314	401	497	555	639	722	754	783	850
Face Brick with Concrete Block Back-up	Steel Frame	200.60	191.65	188.00	184.40	182.95	181.75	179.85	178.25	177.65
	R/Conc. Frame	200.55	191.65	188.00	184.40	182.90	181.70	179.85	178.25	177.65
Glass and Metal Curtain Walls	Steel Frame	214.70	202.90	198.05	193.15	191.15	189.65	186.95	184.75	184.00
	R/Conc. Frame	214.95	203.15	198.20	193.40	191.40	189.90	187.20	185.00	184.20
Precast Concrete Panels	Steel Frame	212.25	200.90	196.20	191.60	189.60	188.15	185.65	183.55	182.85
	R/Conc. Frame	213.45	201.90	197.15	192.40	190.45	188.95	186.40	184.25	183.50
Perimeter Adj., Add or Deduct	Per 100 L.F.	9.95	6.40	4.70	3.70	3.00	2.60	2.30	2.00	1.85
Story Hgt. Adj., Add or Deduct	Per 1 Ft.	2.75	2.30	2.10	1.80	1.75	1.75	1.50	1.40	1.40

For Basement, add $34.30 per square foot of basement area

The above costs were calculated using the basic specifications shown on the facing page. These costs should be adjusted where necessary for design alternatives and owner's requirements. Reported completed project costs, for this type of structure, range from $117.65 to $225.85 per S.F.

Common additives

Description	Unit	$ Cost	Description	Unit	$ Cost
Bar, Front bar	L.F.	410	Laundry Equipment		
Back bar	L.F.	325	Folders, blankets & sheets, king size	Each	70,500
Booth, Upholstered, custom, straight	L.F.	237 - 435	Ironers, 110" single roll	Each	37,400
"L" or "U" shaped	L.F.	245 - 415	Combination washer extractor 50#	Each	13,200
Closed Circuit Surveillance, One station			125#	Each	36,300
Camera and monitor	Each	1950	Sauna, Prefabricated, complete		
For additional camera stations, add	Each	1050	6' x 4'	Each	6225
Directory Boards, Plastic, glass covered			6' x 6'	Each	7325
30" x 20"	Each	660	6' x 9'	Each	8975
36" x 48"	Each	1475	8' x 8'	Each	10,600
Aluminum, 24" x 18"	Each	620	10' x 12'	Each	15,100
48" x 32"	Each	1050	Smoke Detectors		
48" x 60"	Each	2175	Ceiling type	Each	220
Elevators, Electric passenger, 5 stops			Duct type	Each	550
3500# capacity	Each	171,300	Sound System		
5000# capacity	Each	177,300	Amplifier, 250 watts	Each	1975
Additional stop, add	Each	12,700	Speaker, ceiling or wall	Each	206
Emergency Lighting, 25 watt, battery operated			Trumpet	Each	395
Lead battery	Each	305	TV Antenna, Master system, 12 outlet	Outlet	277
Nickel cadmium	Each	850	30 outlet	Outlet	216
			100 outlet	Outlet	205

Important: See the Reference Section for Location Factors.

Model costs calculated for a 6 story building with 10' story height and 135,000 square feet of floor area

					Unit	Unit Cost	Cost Per S.F.	% Of Sub-Total
A. SUBSTRUCTURE								
1010	Standard Foundations	Poured concrete; strip and spread footings			S.F. Ground	6.12	1.02	
1020	Special Foundations	N/A			—	—	—	
1030	Slab on Grade	4" reinforced concrete with vapor barrier and granular base			S.F. Slab	5.18	.86	2.0%
2010	Basement Excavation	Site preparation for slab and trench for foundation wall and footing			S.F. Ground	.18	.03	
2020	Basement Walls	4' foundation wall			L.F. Wall	79	.85	
B. SHELL								
	B10 Superstructure							
1010	Floor Construction	Concrete slab with metal deck, beams, columns			S.F. Floor	21.01	17.51	14.1%
1020	Roof Construction	Metal deck, open web steel joists, beams, columns			S.F. Roof	10.62	1.77	
	B20 Exterior Enclosure							
2010	Exterior Walls	Face brick with concrete block backup		80% of wall	S.F. Wall	32.45	8.33	
2020	Exterior Windows	Window wall		20% of wall	Each	49.50	3.17	8.6%
2030	Exterior Doors	Glass amd metal doors and entrances			Each	3584	.27	
	B30 Roofing							
3010	Roof Coverings	Built-up tar and gravel with flashing; perlite/EPS composite insulation			S.F. Roof	6.12	1.02	0.8%
3020	Roof Openings	Roof hatches			S.F. Roof	.06	.01	
C. INTERIORS								
1010	Partitions	Gypsum board and sound deadening board, steel studs		9 S.F. Floor/L.F. Partition	S.F. Partition	6.93	6.16	
1020	Interior Doors	Single leaf hollow metal		90 S.F. Floor/Door	Each	1212	13.47	
1030	Fittings	N/A			—	—	—	
2010	Stair Construction	Concrete filled metal pan			Flight	14,150	1.46	25.0%
3010	Wall Finishes	20% paint, 75% vinyl cover, 5% ceramic tile			S.F. Surface	1.94	3.45	
3020	Floor Finishes	80% carpet, 10% vinyl composition tile, 10% ceramic tile			S.F. Floor	5.37	5.37	
3030	Ceiling Finishes	Mineral fiber tile applied with adhesive			S.F. Ceiling	4.44	4.44	
D. SERVICES								
	D10 Conveying							
1010	Elevators & Lifts	Four geared passenger elevators			Each	230,175	6.82	5.0%
1020	Escalators & Moving Walks	N/A			—	—	—	
	D20 Plumbing							
2010	Plumbing Fixtures	Kitchen, toilet and service fixtures, supply and drainage		1 Fixture/155 S.F. Floor	Each	3856	24.88	
2020	Domestic Water Distribution	Oil fired water heater			S.F. Floor	.61	.61	18.8%
2040	Rain Water Drainage	Roof drains			S.F. Roof	1.50	.25	
	D30 HVAC							
3010	Energy Supply	Gas fired hot water, wall fin radiation			S.F. Floor	4.81	4.81	
3020	Heat Generating Systems	N/A			—	—	—	
3030	Cooling Generating Systems	Chilled water, fan coil units			S.F. Floor	13.15	13.15	13.1%
3050	Terminal & Package Units	N/A			—	—	—	
3090	Other HVAC Sys. & Equipment	N/A			—	—	—	
	D40 Fire Protection							
4010	Sprinklers	Sprinkler system, light hazard			S.F. Floor	2.83	2.83	2.3%
4020	Standpipes	Standpipes and hose systems			S.F. Floor	.35	.35	
	D50 Electrical							
5010	Electrical Service/Distribution	2000 ampere service, panel board and feeders			S.F. Floor	1.14	1.14	
5020	Lighting & Branch Wiring	High efficiency fluorescent fixtures, receptacles, switches, A.C. and misc. power			S.F. Floor	8.52	8.52	10.4%
5030	Communications & Security	Addressable alarm systems, internet wiring, communications systems and emergency lighting			S.F. Floor	4.15	4.15	
5090	Other Electrical Systems	Emergency generator, 250 kW			S.F. Floor	.48	.48	
E. EQUIPMENT & FURNISHINGS								
1010	Commercial Equipment	N/A			—	—	—	
1020	Institutional Equipment	N/A			—	—	—	
1030	Vehicular Equipment	N/A			—	—	—	0.0%
1090	Other Equipment	N/A			—	—	—	
F. SPECIAL CONSTRUCTION								
1020	Integrated Construction	N/A			—	—	—	0.0%
1040	Special Facilities	N/A			—	—	—	
G. BUILDING SITEWORK	**N/A**							

	Sub-Total	137.18	**100%**
CONTRACTOR FEES (General Requirements: 10%, Overhead: 5%, Profit: 10%)		25%	34.28
ARCHITECT FEES		6%	10.29
	Total Building Cost	**181.75**	

Costs per square foot of floor area

Exterior Wall	S.F. Area	140000	243000	346000	450000	552000	655000	760000	860000	965000
	L.F. Perimeter	403	587	672	800	936	1073	1213	1195	1312
Face Brick with Concrete Block Back-up	Steel Frame	183.00	176.55	171.55	169.65	168.65	167.85	167.35	165.50	165.15
	R/Conc. Frame	186.80	180.40	175.40	173.55	172.40	171.75	171.15	169.30	168.95
Face Brick Veneer On Steel Studs	Steel Frame	179.55	173.70	169.30	167.60	166.60	165.95	165.45	163.85	163.50
	R/Conc. Frame	186.35	180.50	176.10	174.40	173.40	172.75	172.25	170.70	170.35
Glass and Metal Curtain Walls	Steel Frame	228.95	215.10	206.05	202.35	200.20	198.75	197.65	194.75	194.05
	R/Conc. Frame	224.25	210.60	201.65	198.00	195.85	194.40	193.35	190.45	189.75
Perimeter Adj., Add or Deduct	Per 100 L.F.	5.90	3.40	2.40	1.90	1.45	1.35	1.10	1.00	0.85
Story Hgt. Adj., Add or Deduct	Per 1 Ft.	2.30	1.90	1.50	1.40	1.25	1.35	1.25	1.10	1.10

For Basement, add $35.65 per square foot of basement area

The above costs were calculated using the basic specifications shown on the facing page. These costs should be adjusted where necessary for design alternatives and owner's requirements. Reported completed project costs, for this type of structure, range from $130.60 to $228.35 per S.F.

Common additives

Description	Unit	$ Cost	Description	Unit	$ Cost
Bar, Front bar	L.F.	410	Laundry Equipment		
Back bar	L.F.	325	Folders, blankets & sheets, king size	Each	70,500
Booth, Upholstered, custom, straight	L.F.	237 - 435	Ironers, 110" single roll	Each	37,400
"L" or "U" shaped	L.F.	245 - 415	Combination washer & extractor 50#	Each	13,200
Closed Circuit Surveillance, One station			125#	Each	36,300
Camera and monitor	Each	1950	Sauna, Prefabricated, complete		
For additional camera stations, add	Each	1050	6' x 4'	Each	6225
Directory Boards, Plastic, glass covered			6' x 6'	Each	7325
30" x 20"	Each	660	6' x 9'	Each	8975
36" x 48"	Each	1475	8' x 8'	Each	10,600
Aluminum, 24" x 18"	Each	620	10' x 12'	Each	15,100
48" x 32"	Each	1050	Smoke Detectors		
48" x 60"	Each	2175	Ceiling type	Each	220
Elevators, Electric passenger, 10 stops			Duct type	Each	550
3500# capacity	Each	418,000	Sound System		
5000# capacity	Each	425,500	Amplifier, 250 watts	Each	1975
Additional stop, add	Each	12,700	Speaker, ceiling or wall	Each	206
Emergency Lighting, 25 watt, battery operated			Trumpet	Each	395
Lead battery	Each	305	TV Antenna, Master system, 12 outlet	Outlet	277
Nickel cadmium	Each	850	30 outlet	Outlet	216
			100 outlet	Outlet	205

Important: See the Reference Section for Location Factors.

Model costs calculated for a 15 story building with 10' story height and 450,000 square feet of floor area

Hotel, 8-24 Story

				Unit	Unit Cost	Cost Per S.F.	% Of Sub-Total
A. SUBSTRUCTURE							
1010	Standard Foundations	CIP concrete pile caps		S.F. Ground	5.85	.39	
1020	Special Foundations	Steel H-piles, concrete grade beams		S.F. Ground	120	8.02	
1030	Slab on Grade	4" reinforced concrete with vapor barrier and granular base		S.F. Slab	5.18	.34	5.9%
2010	Basement Excavation	Site preparation for slab, piles and grade beams		S.F. Ground	.18	.01	
2020	Basement Walls	4' foundation wall		L.F. Wall	79	.18	
B. SHELL							
B10 Superstructure							
1010	Floor Construction	Open web steel joists, slab form, concrete, columns		S.F. Floor	20.99	19.59	13.2%
1020	Roof Construction	Metal deck, open web steel joists, beams, columns		S.F. Roof	9.30	.62	
B20 Exterior Enclosure							
2010	Exterior Walls	N/A		—	—	—	
2020	Exterior Windows	Glass and metal curtain walls	100% of wall	Each	54	14.50	9.7%
2030	Exterior Doors	Glass and metal doors and entrances		Each	3227	.24	
B30 Roofing							
3010	Roof Coverings	Built-up tar and gravel with flashing; perlite/EPS composite insulation		S.F. Roof	6	.40	0.3%
3020	Roof Openings	N/A		—	—	—	
C. INTERIORS							
1010	Partitions	Gypsum board and sound deadening board, steel studs	9 S.F. Floor/L.F. Partition	S.F. Partition	6.72	5.97	
1020	Interior Doors	Single leaf hollow metal	90 S.F. Floor/Door	Each	1212	13.47	
1030	Fittings	N/A		—	—	—	
2010	Stair Construction	Concrete filled metal pan		Flight	14,150	2.87	23.0%
3010	Wall Finishes	20% paint, 75% vinyl cover, 5% ceramic tile		S.F. Surface	1.92	3.42	
3020	Floor Finishes	80% carpet tile, 10% vinyl composition tile, 10% ceramic tile		S.F. Floor	5.37	5.37	
3030	Ceiling Finishes	Gypsum board on resilient channel		S.F. Ceiling	3.98	3.98	
D. SERVICES							
D10 Conveying							
1010	Elevators & Lifts	One geared freight, six geared passenger elevators		Each	505,500	6.74	4.4%
1020	Escalators & Moving Walks	N/A		—	—	—	
D20 Plumbing							
2010	Plumbing Fixtures	Kitchen, toilet and service fixtures, supply and drainage	1 Fixture/165 S.F. Floor	Each	4257	25.80	
2020	Domestic Water Distribution	Electric water heater		S.F. Floor	3.23	3.23	19.1%
2040	Rain Water Drainage	Roof drains		S.F. Roof	2.10	.14	
D30 HVAC							
3010	Energy Supply	Gas fired hot water, wall fin radiation		S.F. Floor	2.60	2.60	
3020	Heat Generating Systems	N/A		—	—	—	
3030	Cooling Generating Systems	Chilled water, fan coil units		S.F. Floor	13.15	13.15	10.3%
3050	Terminal & Package Units	N/A		—	—	—	
3090	Other HVAC Sys. & Equipment	N/A		—	—	—	
D40 Fire Protection							
4010	Sprinklers	Sprinkler system, light hazard		S.F. Floor	4	4	4.9%
4020	Standpipes	Standpipes, fire pump and hose systems		S.F. Floor	3.48	3.48	
D50 Electrical							
5010	Electrical Service/Distribution	6000 ampere service, panel board and feeders		S.F. Floor	1.31	1.31	
5020	Lighting & Branch Wiring	High efficiency fluorescent fixtures, receptacles, switches, A.C. and misc. power		S.F. Floor	8.40	8.40	9.3%
5030	Communications & Security	Addressable alarm systems, internet wiring, communications systems and emergency lighting		S.F. Floor	4.13	4.13	
5090	Other Electrical Systems	Emergency generator, 500 kW		S.F. Floor	.36	.36	
E. EQUIPMENT & FURNISHINGS							
1010	Commercial Equipment	N/A		—	—	—	
1020	Institutional Equipment	N/A		—	—	—	
1030	Vehicular Equipment	N/A		—	—	—	0.0%
1090	Other Equipment	N/A		—	—	—	
F. SPECIAL CONSTRUCTION							
1020	Integrated Construction	N/A		—	—	—	0.0%
1040	Special Facilities	N/A		—	—	—	
G. BUILDING SITEWORK	**N/A**						

		Sub-Total	152.71	100%
CONTRACTOR FEES (General Requirements: 10%, Overhead: 5%, Profit: 10%)		25%	38.19	
ARCHITECT FEES		6%	11.45	
		Total Building Cost	**202.35**	

Costs per square foot of floor area

Exterior Wall	S.F. Area	5500	12000	20000	30000	40000	60000	80000	100000	145000
	L.F. Perimeter	310	330	400	490	530	590	680	760	920
Face Brick with Concrete Block Back-up	Steel Frame	476.55	360.35	327.40	311.15	299.60	286.95	281.70	278.30	273.65
	R/Conc. Frame	476.05	359.80	326.85	310.60	298.95	286.30	281.15	277.75	273.00
Stucco on Concrete Block	Steel Frame	444.30	344.75	316.10	301.95	292.15	281.45	277.05	274.15	270.15
	R/Conc. Frame	443.45	343.95	315.30	301.10	291.35	280.65	276.20	273.35	269.35
Reinforced Concrete	Steel Frame	451.95	348.45	318.85	304.20	293.95	282.80	278.20	275.20	271.00
	R/Conc. Frame	451.10	347.70	318.00	303.35	293.15	282.00	277.35	274.35	270.20
Perimeter Adj., Add or Deduct	Per 100 L.F.	51.05	23.40	14.05	9.40	6.90	4.70	3.50	2.85	1.90
Story Hgt. Adj., Add or Deduct	Per 1 Ft.	12.20	5.95	4.35	3.60	2.85	2.10	1.85	1.65	1.35

For Basement, add *$ 49.30 per square foot of basement area*

The above costs were calculated using the basic specifications shown on the facing page. These costs should be adjusted where necessary for design alternatives and owner's requirements. Reported completed project costs, for this type of structure, range from $ 190.05 to $ 504.70 per S.F.

Common additives

Description	Unit	$ Cost	Description	Unit	$ Cost
Clock System			Emergency Lighting, 25 watt, battery operated		
20 room	Each	19,200	Nickel cadmium	Each	850
50 room	Each	45,000	Flagpoles, Complete		
Closed Circuit Surveillance, One station			Aluminum, 20' high	Each	1750
Camera and monitor	Each	1950	40' High	Each	3875
For additional camera stations, add	Each	1050	70' High	Each	10,600
Elevators, Hydraulic passenger, 2 stops			Fiberglass, 23' High	Each	1325
1500# capacity	Each	66,000	39'-5" High	Each	3525
2500# capacity	Each	68,900	59' High	Each	7200
3500# capacity	Each	72,900	Laundry Equipment		
Additional stop, add	Each	8375	Folders, blankets, & sheets	Each	70,500
Emergency Generators, Complete system, gas			Ironers, 110" single roll	Each	37,400
15 KW	Each	18,400	Combination washer & extractor, 50#	Each	13,200
70 KW	Each	35,000	125#	Each	36,300
85 KW	Each	38,900	Safe, Office type, 1 hour rating		
115 KW	Each	75,000	30" x 18" x 18"	Each	2500
170 KW	Each	98,000	60" x 36" x 18", double door	Each	8075
Diesel, 50 KW	Each	29,300	Sound System		
100 KW	Each	41,800	Amplifier, 250 watts	Each	1975
150 KW	Each	50,500	Speaker, ceiling or wall	Each	206
350 KW	Each	79,500	Trumpet	Each	395

Important: See the Reference Section for Location Factors.

Model costs calculated for a 3 story building with 12' story height and 40,000 square feet of floor area

Jail

				Unit	Unit Cost	Cost Per S.F.	% Of Sub-Total
A. SUBSTRUCTURE							
1010	Standard Foundations	Poured concrete; strip and spread footings		S.F. Ground	2.13	.71	
1020	Special Foundations	N/A		—	—	—	
1030	Slab on Grade	4" reinforced concrete with vapor barrier and granular base		S.F. Slab	5.18	1.72	1.6%
2010	Basement Excavation	Site preparation for slab and trench for foundation wall and footing		S.F. Ground	.32	.11	
2020	Basement Walls	4' foundation wall		L.F. Wall	71	.94	
B. SHELL							
B10 Superstructure							
1010	Floor Construction	Concrete slab with metal deck and beams		S.F. Floor	29.30	19.53	12.1%
1020	Roof Construction	Concrete slab with metal deck and beams		S.F. Roof	22.80	7.60	
B20 Exterior Enclosure							
2010	Exterior Walls	Face brick with reinforced concrete block backup	85% of wall	S.F. Wall	32.46	13.16	
2020	Exterior Windows	Bullet resisting and metal horizontal pivoted	15% of wall	Each	158	11.28	11.0%
2030	Exterior Doors	Metal		Each	6375	.16	
B30 Roofing							
3010	Roof Coverings	Built-up tar and gravel with flashing; perlite/EPS composite insulation		S.F. Roof	6.18	2.06	0.9%
3020	Roof Openings	Roof hatches		S.F. Roof	.09	.03	
C. INTERIORS							
1010	Partitions	Concrete block	45 S.F. of Floor/L.F. Partition	S.F. Partition	10.13	2.70	
1020	Interior Doors	Hollow metal fire doors	930 S.F. Floor/Door	Each	1212	1.30	
1030	Fittings	N/A		—	—	—	
2010	Stair Construction	Concrete filled metal pan		Flight	16,950	2.12	6.1%
3010	Wall Finishes	Paint		S.F. Surface	4.24	2.26	
3020	Floor Finishes	70% vinyl composition tile, 20% carpet, 10% ceramic tile		S.F. Floor	3.31	3.31	
3030	Ceiling Finishes	Mineral tile on zee runners	30% of area	S.F. Ceiling	6.86	2.06	
D. SERVICES							
D10 Conveying							
1010	Elevators & Lifts	One hydraulic passenger elevator		Each	166,400	4.16	1.9%
1020	Escalators & Moving Walks	N/A		—	—	—	
D20 Plumbing							
2010	Plumbing Fixtures	Toilet and service fixtures, supply and drainage	1 Fixture/78 S.F. Floor	Each	4004	51.33	
2020	Domestic Water Distribution	Electric water heater		S.F. Floor	3.94	3.94	25.1%
2040	Rain Water Drainage	Roof drains		S.F. Roof	2.97	.99	
D30 HVAC							
3010	Energy Supply	N/A		—	—	—	
3020	Heat Generating Systems	Included in D3050		—	—	—	
3030	Cooling Generating Systems	N/A		—	—	—	7.1%
3050	Terminal & Package Units	Rooftop multizone unit systems		S.F. Floor	15.81	15.81	
3090	Other HVAC Sys. & Equipment	N/A		—	—	—	
D40 Fire Protection							
4010	Sprinklers	Sprinkler system, light hazard		S.F. Floor	2.92	2.92	1.6%
4020	Standpipes	Standpipes and hose systems		S.F. Floor	.58	.58	
D50 Electrical							
5010	Electrical Service/Distribution	600 ampere service, panel board and feeders		S.F. Floor	1.06	1.06	
5020	Lighting & Branch Wiring	Fluorescent fixtures, receptacles, switches, A.C. and misc. power		S.F. Floor	9.75	9.75	7.1%
5030	Communications & Security	Addressable alarm systems, intercom, and emergency lighting		S.F. Floor	4.32	4.32	
5090	Other Electrical Systems	Emergency generator, 80 kW		S.F. Floor	.76	.76	
E. EQUIPMENT & FURNISHINGS							
1010	Commercial Equipment	N/A		—	—	—	
1020	Institutional Equipment	Prefabricated cells, visitor cubicles		S.F. Floor	57.31	57.31	25.6%
1030	Vehicular Equipment	N/A		—	—	—	
1090	Other Equipment	N/A		—	—	—	
F. SPECIAL CONSTRUCTION							
1020	Integrated Construction	N/A		—	—	—	0.0%
1040	Special Facilities	N/A		—	—	—	
G. BUILDING SITEWORK	**N/A**						

			Sub-Total	223.98	100%
CONTRACTOR FEES (General Requirements: 10%, Overhead: 5%, Profit: 10%)			25%	56.02	
ARCHITECT FEES			7%	19.60	
		Total Building Cost		**299.60**	

Costs per square foot of floor area

Exterior Wall	S.F. Area	1000	2000	3000	4000	5000	10000	15000	20000	25000
	L.F. Perimeter	126	179	219	253	283	400	490	568	632
Decorative Concrete Block	Steel Frame	290.10	244.30	226.75	217.05	210.80	196.55	190.75	187.60	185.30
	Bearing Walls	290.80	244.55	226.80	217.00	210.65	196.25	190.35	187.10	184.80
Face Brick with Concrete Block Back-up	Steel Frame	324.00	269.40	247.80	235.80	227.95	209.65	202.10	197.95	194.85
	Bearing Walls	321.15	266.15	244.35	232.25	224.35	205.85	198.20	193.95	190.90
Metal Sandwich Panel	Steel Frame	295.80	249.55	231.75	222.00	215.65	201.20	195.30	192.05	189.80
Precast Concrete Panel	Bearing Walls	286.15	241.25	224.10	214.65	208.55	194.75	189.15	186.05	183.90
Perimeter Adj., Add or Deduct	Per 100 L.F.	55.10	27.55	18.35	13.80	11.00	5.50	3.70	2.75	2.25
Story Hgt. Adj., Add or Deduct	Per 1 Ft.	3.70	2.60	2.10	1.85	1.65	1.15	1.00	0.85	0.75

For Basement, add $32.60 per square foot of basement area

The above costs were calculated using the basic specifications shown on the facing page. These costs should be adjusted where necessary for design alternatives and owner's requirements. Reported completed project costs, for this type of structure, range from $105.80 to $258.60 per S.F.

Common additives

Description	Unit	$ Cost
Closed Circuit Surveillance, One station		
Camera and monitor	Each	1950
For additional camera stations, add	Each	1050
Emergency Lighting, 25 watt, battery operated		
Lead battery	Each	305
Nickel cadmium	Each	850
Laundry Equipment		
Dryers, coin operated 30 lb.	Each	3750
Double stacked	Each	8125
50 lb.	Each	4025
Dry cleaner 20 lb.	Each	43,100
30 lb.	Each	63,500
Washers, coin operated	Each	1600
Washer/extractor 20 lb.	Each	7025
30 lb.	Each	11,200
50 lb.	Each	13,200
75 lb.	Each	25,300
Smoke Detectors		
Ceiling type	Each	220
Duct type	Each	550

Important: See the Reference Section for Location Factors.

Model costs calculated for a 1 story building with 12' story height and 3,000 square feet of floor area

				Unit	Unit Cost	Cost Per S.F.	% Of Sub-Total
A. SUBSTRUCTURE							
1010	Standard Foundations	Poured concrete; strip and spread footings		S.F. Ground	3.27	3.27	
1020	Special Foundations	N/A		—	—	—	
1030	Slab on Grade	5" reinforced concrete with vapor barrier and granular base		S.F. Slab	5.59	5.59	8.6%
2010	Basement Excavation	Site preparation for slab and trench for foundation wall and footing		S.F. Ground	.55	.55	
2020	Basement Walls	4' Foundation wall		L.F. Wall	71	5.16	
B. SHELL							
B10 Superstructure							
1010	Floor Construction	N/A		—	—	—	
1020	Roof Construction	Metal deck, open web steel joists, beams, columns		S.F. Roof	10.15	10.15	6.0 %
B20 Exterior Enclosure							
2010	Exterior Walls	Decorative concrete block	90% of wall	S.F. Wall	15.35	12.10	
2020	Exterior Windows	Store front	10% of wall	Each	41.85	3.67	10.2%
2030	Exterior Doors	Double aluminum and glass		Each	4700	1.57	
B30 Roofing							
3010	Roof Coverings	Built-up tar and gravel with flashing; perlite/EPS composite insulation		S.F. Roof	9.67	9.67	5.7%
3020	Roof Openings	N/A		—	—	—	
C. INTERIORS							
1010	Partitions	Gypsum board on metal studs	60 S.F. Floor/L.F. Partition	S.F. Partition	25.02	4.17	
1020	Interior Doors	Single leaf wood	750 S.F. Floor/Door	Each	622	.83	
1030	Fittings	N/A		—	—	—	
2010	Stair Construction	N/A		—	—	—	6.4%
3010	Wall Finishes	Paint		S.F. Surface	.81	.27	
3020	Floor Finishes	Vinyl composition tile		S.F. Floor	1.92	1.92	
3030	Ceiling Finishes	Fiberglass board on exposed grid system		S.F. Ceiling	3.69	3.69	
D. SERVICES							
D10 Conveying							
1010	Elevators & Lifts	N/A		—	—	—	
1020	Escalators & Moving Walks	N/A		—	—	—	0.0 %
D20 Plumbing							
2010	Plumbing Fixtures	Toilet and service fixtures, supply and drainage	1 Fixture/600 S.F. Floor	Each	6492	10.82	
2020	Domestic Water Distribution	Gas fired hot water heater		S.F. Floor	48.20	48.20	36.0%
2040	Rain Water Drainage	Roof drains		S.F. Roof	1.97	1.97	
D30 HVAC							
3010	Energy Supply	N/A		—	—	—	
3020	Heat Generating Systems	Included in D3050		—	—	—	
3030	Cooling Generating Systems	N/A		—	—	—	6.0 %
3050	Terminal & Package Units	Rooftop single zone unit systems		S.F. Floor	10.10	10.10	
3090	Other HVAC Sys. & Equipment	N/A		—	—	—	
D40 Fire Protection							
4010	Sprinklers	Sprinkler, ordinary hazard		S.F. Floor	5.09	5.09	3.0%
4020	Standpipes	N/A		—	—	—	
D50 Electrical							
5010	Electrical Service/Distribution	200 ampere service, panel board and feeders		S.F. Floor	4.15	4.15	
5020	Lighting & Branch Wiring	High efficiency fluorescent fixtures, receptacles, switches, A.C. and misc. power		S.F. Floor	23	23	18.1%
5030	Communications & Security	Addressable alarm systems and emergency lighting		S.F. Floor	3.06	3.06	
5090	Other Electrical Systems	Emergency generator, 7.5 kW		S.F. Floor	.51	.51	
E. EQUIPMENT & FURNISHINGS							
1010	Commercial Equipment	N/A		—	—	—	
1020	Institutional Equipment	N/A		—	—	—	
1030	Vehicular Equipment	N/A		—	—	—	0.0 %
1090	Other Equipment	N/A		—	—	—	
F. SPECIAL CONSTRUCTION							
1020	Integrated Construction	N/A		—	—	—	
1040	Special Facilities	N/A		—	—	—	0.0 %
G. BUILDING SITEWORK	**N/A**						

		Sub-Total	169.51	100%
CONTRACTOR FEES (General Requirements: 10%, Overhead: 5%, Profit: 10%)		25%	42.41	
ARCHITECT FEES		7%	14.83	
		Total Building Cost	226.75	

Costs per square foot of floor area

Exterior Wall	S.F. Area	7000	10000	13000	16000	19000	22000	25000	28000	31000
	L.F. Perimeter	240	300	336	386	411	435	472	510	524
Face Brick with Concrete Block Back-up	R/Conc. Frame	189.40	180.00	172.00	168.40	163.85	160.50	158.75	157.35	155.10
	Steel Frame	190.30	180.95	172.95	169.35	164.80	161.40	159.65	158.35	156.05
Limestone with Concrete Block	R/Conc. Frame	233.50	217.00	202.95	196.60	188.70	182.80	179.70	177.40	173.45
	Steel Frame	229.05	214.85	202.10	196.65	189.25	183.75	181.00	178.90	175.10
Precast Concrete Panels	R/Conc. Frame	203.90	191.05	180.60	175.80	170.00	165.75	163.40	161.65	158.80
	Steel Frame	204.80	192.00	181.60	176.70	171.00	166.70	164.35	162.60	159.75
Perimeter Adj., Add or Deduct	Per 100 L.F.	22.90	16.15	12.35	10.10	8.50	7.30	6.45	5.75	5.20
Story Hgt. Adj., Add or Deduct	Per 1 Ft.	3.45	3.05	2.65	2.45	2.25	1.95	1.85	1.90	1.70

For Basement, add $46.70 per square foot of basement area

The above costs were calculated using the basic specifications shown on the facing page. These costs should be adjusted where necessary for design alternatives and owner's requirements. Reported completed project costs, for this type of structure, range from $101.65 to $260.90 per S.F.

Common additives

Description	Unit	$ Cost
Carrels Hardwood	Each	855 - 1625
Closed Circuit Surveillance, One station		
Camera and monitor	Each	1950
For additional camera stations, add	Each	1050
Elevators, Hydraulic passenger, 2 stops		
1500# capacity	Each	66,000
2500# capacity	Each	68,900
3500# capacity	Each	72,900
Emergency Lighting, 25 watt, battery operated		
Lead battery	Each	305
Nickel cadmium	Each	850
Flagpoles, Complete		
Aluminum, 20' high	Each	1750
40' high	Each	3875
70' high	Each	10,600
Fiberglass, 23' high	Each	1325
39'-5" high	Each	3525
59' high	Each	7200

Description	Unit	$ Cost
Library Furnishings		
Bookshelf, 90" high, 10" shelf double face	L.F.	202
single face	L.F.	173
Charging desk, built-in with counter		
Plastic laminated top	L.F.	410
Reading table, laminated		
top 60" x 36"	Each	675

Model costs calculated for a 2 story building with 14' story height and 22,000 square feet of floor area

				Unit	Unit Cost	Cost Per S.F.	% Of Sub-Total
A.	**SUBSTRUCTURE**						
1010	Standard Foundations	Poured concrete; strip and spread footings		S.F. Ground	3.78	1.89	
1020	Special Foundations	N/A		—	—	—	
1030	Slab on Grade	4" reinforced concrete with vapor barrier and granular base		S.F. Slab	5.18	2.60	5.2%
2010	Basement Excavation	Site preparation for slab and trench for foundation wall and footing		S.F. Ground	.32	.16	
2020	Basement Walls	4' foundation wall		L.F. Wall	79	1.56	
B.	**SHELL**						
	B10 Superstructure						
1010	Floor Construction	Concrete waffle slab		S.F. Floor	26.18	13.09	18.9%
1020	Roof Construction	Concrete waffle slab		S.F. Roof	18.70	9.35	
	B20 Exterior Enclosure						
2010	Exterior Walls	Face brick with concrete block backup	90% of wall	S.F. Wall	31.49	15.69	
2020	Exterior Windows	Window wall	10% of wall	Each	52	2.87	16.1%
2030	Exterior Doors	Double aluminum and glass, single leaf hollow metal		Each	5650	.52	
	B30 Roofing						
3010	Roof Coverings	Single ply membrane, EPDM, fully adhered; perlite/EPS composite insulation		S.F. Roof	5.26	2.63	2.2%
3020	Roof Openings	Roof hatches		S.F. Roof	.08	.04	
C.	**INTERIORS**						
1010	Partitions	Gypsum board on metal studs	30 S.F. Floor/L.F. Partition	S.F. Partition	11.43	4.57	
1020	Interior Doors	Single leaf wood	300 S.F. Floor/Door	Each	622	2.07	
1030	Fittings	N/A		—	—	—	
2010	Stair Construction	Concrete filled metal pan		Flight	9050	.83	15.6%
3010	Wall Finishes	Paint		S.F. Surface	.83	.66	
3020	Floor Finishes	50% carpet, 50% vinyl tile		S.F. Floor	3.55	3.55	
3030	Ceiling Finishes	Mineral fiber on concealed zee bars		S.F. Ceiling	6.86	6.86	
D.	**SERVICES**						
	D10 Conveying						
1010	Elevators & Lifts	One hydraulic passenger elevator		Each	86,900	3.95	3.3%
1020	Escalators & Moving Walks	N/A		—	—	—	
	D20 Plumbing						
2010	Plumbing Fixtures	Toilet and service fixtures, supply and drainage	1 Fixture/1835 S.F. Floor	Each	7377	4.02	
2020	Domestic Water Distribution	Gas fired water heater		S.F. Floor	1.29	1.29	5.0%
2040	Rain Water Drainage	Roof drains		S.F. Roof	1.18	.59	
	D30 HVAC						
3010	Energy Supply	N/A		—	—	—	
3020	Heat Generating Systems	Included in D3050		—	—	—	
3030	Cooling Generating Systems	N/A		—	—	—	18.0 %
3050	Terminal & Package Units	Multizone unit, gas heating, electric cooling		S.F. Floor	21.35	21.35	
3090	Other HVAC Sys. & Equipment	N/A		—	—	—	
	D40 Fire Protection						
4010	Sprinklers	Wet pipe sprinkler system		S.F. Floor	3.07	3.07	3.4%
4020	Standpipes	Standpipe		S.F. Floor	.92	.92	
	D50 Electrical						
5010	Electrical Service/Distribution	400 ampere service, panel board and feeders		S.F. Floor	1.21	1.21	
5020	Lighting & Branch Wiring	High efficiency fluorescent fixtures, receptacles, switches, A.C. and misc. power		S.F. Floor	11.21	11.21	12.4%
5030	Communications & Security	Addressable alarm systems, internet wiring, and emergency lighting		S.F. Floor	2.05	2.05	
5090	Other Electrical Systems	Emergency generator, 7.5 kW, Uninterruptible power supply		S.F. Floor	.26	.26	
E.	**EQUIPMENT & FURNISHINGS**						
1010	Commercial Equipment	N/A		—	—	—	
1020	Institutional Equipment	N/A		—	—	—	0.0 %
1030	Vehicular Equipment	N/A		—	—	—	
1090	Other Equipment	N/A		—	—	—	
F.	**SPECIAL CONSTRUCTION**						
1020	Integrated Construction	N/A		—	—	—	0.0 %
1040	Special Facilities	N/A		—	—	—	
G.	**BUILDING SITEWORK**	**N/A**					

	Sub-Total	118.86	100%
CONTRACTOR FEES (General Requirements: 10%, Overhead: 5%, Profit: 10%)	25%	29.75	
ARCHITECT FEES	8%	11.89	
Total Building Cost		**160.50**	

Costs per square foot of floor area

Exterior Wall	S.F. Area	4000	5500	7000	8500	10000	11500	13000	14500	16000
	L.F. Perimeter	280	320	380	440	453	503	510	522	560
Face Brick with Concrete Block Back-up	Steel Joists	217.80	208.65	205.45	203.35	198.65	197.40	194.20	191.90	191.10
	Wood Truss	224.20	215.15	211.85	209.75	205.10	203.80	200.60	198.30	197.55
Stucco on Concrete Block	Steel Joists	206.30	199.10	196.55	194.85	191.20	190.20	187.70	186.00	185.35
	Wood Truss	212.75	205.55	202.95	201.30	197.65	196.65	194.20	192.40	191.80
Brick Veneer	Wood Truss	217.00	208.70	205.70	203.80	199.55	198.40	195.50	193.40	192.70
Wood Siding	Wood Frame	209.10	202.55	200.15	198.60	195.30	194.35	192.15	190.55	189.95
Perimeter Adj., Add or Deduct	Per 100 L.F.	16.90	12.35	9.65	7.95	6.75	5.90	5.20	4.70	4.25
Story Hgt. Adj., Add or Deduct	Per 1 Ft.	3.45	2.90	2.65	2.55	2.25	2.20	1.90	1.75	1.75
For Basement, add $30.40 per square foot of basement area										

The above costs were calculated using the basic specifications shown on the facing page. These costs should be adjusted where necessary for design alternatives and owner's requirements. Reported completed project costs, for this type of structure, range from $85.65 to $220.35 per S.F.

Common additives

Description	Unit	$ Cost	Description	Unit	$ Cost
Cabinets, Hospital, base			Directory Boards, Plastic, glass covered		
Laminated plastic	L.F.	400	30" x 20"	Each	660
Stainless steel	L.F.	640	36" x 48"	Each	1475
Counter top, laminated plastic	L.F.	77	Aluminum, 24" x 18"	Each	620
Stainless steel	L.F.	189	36" x 24"	Each	725
For drop-in sink, add	Each	1050	48" x 32"	Each	1050
Nurses station, door type			48" x 60"	Each	2175
Laminated plastic	L.F.	445	Heat Therapy Unit		
Enameled steel	L.F.	430	Humidified, 26" x 78" x 28"	Each	3850
Stainless steel	L.F.	745	Smoke Detectors		
Wall cabinets, laminated plastic	L.F.	289	Ceiling type	Each	220
Enameled steel	L.F.	340	Duct type	Each	550
Stainless steel	L.F.	600	Tables, Examining, vinyl top		
			with base cabinets	Each	1450 - 6650
			Utensil Washer, Sanitizer	Each	12,500
			X-Ray, Mobile	Each	18,300 - 84,500

Important: See the Reference Section for Location Factors.

Model costs calculated for a 1 story building with 10' story height and 7,000 square feet of floor area

Medical Office, 1 Story

					Unit	Unit Cost	Cost Per S.F.	% Of Sub-Total

A. SUBSTRUCTURE

				Unit	Unit Cost	Cost Per S.F.	% Of Sub-Total
1010	Standard Foundations	Poured concrete; strip and spread footings		S.F. Ground	2.09	2.09	
1020	Special Foundations	N/A		—	—	—	
1030	Slab on Grade	4" reinforced concrete with vapor barrier and granular base		S.F. Slab	5.18	5.18	7.6%
2010	Basement Excavation	Site preparation for slab and trench for foundation wall and footing		S.F. Ground	.32	.32	
2020	Basement Walls	4' foundation wall		L.F. Wall	79	4.29	

B. SHELL

B10 Superstructure

				Unit	Unit Cost	Cost Per S.F.	% Of Sub-Total
1010	Floor Construction	N/A		—	—	—	4.2%
1020	Roof Construction	Plywood on wood trusses		S.F. Roof	6.57	6.57	

B20 Exterior Enclosure

				Unit	Unit Cost	Cost Per S.F.	% Of Sub-Total
2010	Exterior Walls	Face brick with concrete block backup	70% of wall	S.F. Wall	31.50	11.97	
2020	Exterior Windows	Wood double hung	30% of wall	Each	630	6.03	13.1%
2030	Exterior Doors	Aluminum and glass doors and entrance with transoms		Each	1425	2.44	

B30 Roofing

				Unit	Unit Cost	Cost Per S.F.	% Of Sub-Total
3010	Roof Coverings	Asphalt shingles with flashing (Pitched); rigid fiber glass insulation, gutters		S.F. Roof	4.51	4.51	2.9%
3020	Roof Openings	N/A		—	—	—	

C. INTERIORS

				Unit	Unit Cost	Cost Per S.F.	% Of Sub-Total
1010	Partitions	Gypsum bd. & sound deadening bd. on wood studs w/insul.	6 S.F. Floor/L.F. Partition	S.F. Partition	8.76	11.68	
1020	Interior Doors	Single leaf wood	60 S.F. Floor/Door	Each	622	10.37	
1030	Fittings	N/A		—	—	—	
2010	Stair Construction	N/A		—	—	—	24.8%
3010	Wall Finishes	50% paint, 50% vinyl wall covering		S.F. Surface	1.35	3.59	
3020	Floor Finishes	50% carpet, 50% vinyl composition tile		S.F. Floor	6.87	6.87	
3030	Ceiling Finishes	Mineral fiber tile on concealed zee bars		S.F. Ceiling	5.98	5.98	

D. SERVICES

D10 Conveying

				Unit	Unit Cost	Cost Per S.F.	% Of Sub-Total
1010	Elevators & Lifts	N/A		—	—	—	0.0%
1020	Escalators & Moving Walks	N/A		—	—	—	

D20 Plumbing

				Unit	Unit Cost	Cost Per S.F.	% Of Sub-Total
2010	Plumbing Fixtures	Toilet, exam room and service fixtures, supply and drainage	1 Fixture/195 S.F. Floor	Each	4155	21.31	
2020	Domestic Water Distribution	Gas fired water heater		S.F. Floor	3.26	3.26	16.4%
2040	Rain Water Drainage	Roof drains		S.F. Roof	.99	.99	

D30 HVAC

				Unit	Unit Cost	Cost Per S.F.	% Of Sub-Total
3010	Energy Supply	N/A		—	—	—	
3020	Heat Generating Systems	Included in D3050		—	—	—	
3030	Cooling Generating Systems	N/A		—	—	—	9.7%
3050	Terminal & Package Units	Multizone unit, gas heating, electric cooling		S.F. Floor	15.15	15.15	
3090	Other HVAC Sys. & Equipment	N/A		—	—	—	

D40 Fire Protection

				Unit	Unit Cost	Cost Per S.F.	% Of Sub-Total
4010	Sprinklers	Wet pipe sprinkler system		S.F. Floor	4.66	4.66	3.7%
4020	Standpipes	Standpipe		S.F. Floor	1.15	1.15	

D50 Electrical

				Unit	Unit Cost	Cost Per S.F.	% Of Sub-Total
5010	Electrical Service/Distribution	200 ampere service, panel board and feeders		S.F. Floor	1.66	1.66	
5020	Lighting & Branch Wiring	High efficiency fluorescent fixtures, receptacles, switches, A.C. and misc. power		S.F. Floor	7.54	7.54	11.5%
5030	Communications & Security	Alarm systems, internet and phone wiring, intercom system, and emergency lighting		S.F. Floor	7.80	7.80	
5090	Other Electrical Systems	Emergency generator, 7.5 kW		S.F. Floor	.93	.93	

E. EQUIPMENT & FURNISHINGS

				Unit	Unit Cost	Cost Per S.F.	% Of Sub-Total
1010	Commercial Equipment	N/A		—	—	—	
1020	Institutional Equipment	Exam room casework and countertops		S.F. Floor	9.16	9.16	5.9%
1030	Vehicular Equipment	N/A		—	—	—	
1090	Other Equipment	N/A		—	—	—	

F. SPECIAL CONSTRUCTION

				Unit	Unit Cost	Cost Per S.F.	% Of Sub-Total
1020	Integrated Construction	N/A		—	—	—	0.0%
1040	Special Facilities	N/A		—	—	—	

G. BUILDING SITEWORK   N/A

			Sub-Total	155.50	100%
CONTRACTOR FEES (General Requirements: 10%, Overhead: 5%, Profit: 10%)		25%	38.86		
ARCHITECT FEES		9%	17.49		

Total Building Cost   211.85

Costs per square foot of floor area

Exterior Wall	S.F. Area	4000	5500	7000	8500	10000	11500	13000	14500	16000
	L.F. Perimeter	180	210	240	270	286	311	336	361	386
Face Brick with Concrete Block Back-up	Steel Joists	260.35	248.20	241.30	236.75	232.00	229.40	227.35	225.70	224.45
	Wood Joists	264.90	252.80	245.85	241.40	236.60	233.95	231.95	230.35	229.05
Stucco on Concrete Block	Steel Joists	247.85	237.55	231.70	227.90	224.00	221.75	220.05	218.70	217.65
	Wood Joists	252.75	242.55	236.65	232.85	228.90	226.70	225.00	223.60	222.55
Brick Veneer	Wood Frame	257.90	246.85	240.55	236.40	232.10	229.75	227.95	226.40	225.25
Wood Siding	Wood Frame	246.60	237.25	231.95	228.40	224.90	222.95	221.45	220.20	219.20
Perimeter Adj., Add or Deduct	Per 100 L.F.	30.15	21.95	17.25	14.25	12.05	10.45	9.30	8.35	7.55
Story Hgt. Adj., Add or Deduct	Per 1 Ft.	4.40	3.75	3.35	3.15	2.80	2.65	2.55	2.50	2.35

For Basement, add $33.75 per square foot of basement area

The above costs were calculated using the basic specifications shown on the facing page. These costs should be adjusted where necessary for design alternatives and owner's requirements. Reported completed project costs, for this type of structure, range from $87.95 to $300.40 per S.F.

Common additives

Description	Unit	$ Cost
Cabinets, Hospital, base		
Laminated plastic	L.F.	400
Stainless steel	L.F.	640
Counter top, laminated plastic	L.F.	77
Stainless steel	L.F.	189
For drop-in sink, add	Each	1050
Nurses station, door type		
Laminated plastic	L.F.	445
Enameled steel	L.F.	430
Stainless steel	L.F.	745
Wall cabinets, laminated plastic	L.F.	289
Enameled steel	L.F.	340
Stainless steel	L.F.	600
Elevators, Hydraulic passenger, 2 stops		
1500# capacity	Each	66,000
2500# capacity	Each	68,900
3500# capacity	Each	72,900

Description	Unit	$ Cost
Directory Boards, Plastic, glass covered		
30" x 20"	Each	660
36" x 48"	Each	1475
Aluminum, 24" x 18"	Each	620
36" x 24"	Each	725
48" x 32"	Each	1050
48" x 60"	Each	2175
Emergency Lighting, 25 watt, battery operated		
Lead battery	Each	305
Nickel cadmium	Each	850
Heat Therapy Unit		
Humidified, 26" x 78" x 28"	Each	3850
Smoke Detectors		
Ceiling type	Each	220
Duct type	Each	550
Tables, Examining, vinyl top		
with base cabinets	Each	1450 - 6650
Utensil Washer, Sanitizer	Each	12,500
X-Ray, Mobile	Each	18,300 - 84,500

Important: See the Reference Section for Location Factors.

Model costs calculated for a 2 story building with 10' story height and 7,000 square feet of floor area

Medical Office, 2 Story

				Unit	Unit Cost	Cost Per S.F.	% Of Sub-Total
A.	**SUBSTRUCTURE**						
1010	Standard Foundations	Poured concrete; strip and spread footings		S.F. Ground	3.34	1.67	
1020	Special Foundations	N/A		—	—	—	
1030	Slab on Grade	4" reinforced concrete with vapor barrier and granular base		S.F. Slab	5.18	2.60	4.2%
2010	Basement Excavation	Site preparation for slab and trench for foundation wall and footing		S.F. Ground	.18	.09	
2020	Basement Walls	4' foundation wall		L.F. Wall	79	2.71	
B.	**SHELL**						
	B10 Superstructure						
1010	Floor Construction	Open web steel joists, slab form, concrete, columns		S.F. Floor	13.26	6.63	5.7%
1020	Roof Construction	Metal deck, open web steel joists, beams, columns		S.F. Roof	6.06	3.03	
	B20 Exterior Enclosure						
2010	Exterior Walls	Stucco on concrete block		S.F. Wall	18.19	8.73	
2020	Exterior Windows	Outward projecting metal	30% of wall	Each	524	7.19	10.3%
2030	Exterior Doors	Aluminum and glass doors with transoms		Each	5650	1.61	
	B30 Roofing						
3010	Roof Coverings	Built-up tar and gravel with flashing; perlite/EPS composite insulation		S.F. Roof	7.22	3.61	2.2%
3020	Roof Openings	Roof hatches		S.F. Roof	.26	.13	
C.	**INTERIORS**						
1010	Partitions	Gypsum bd. & acous. insul. on metal studs	6 S.F. Floor/L.F. Partition	S.F. Partition	7.24	9.65	
1020	Interior Doors	Single leaf wood	60 S.F. Floor/Door	Each	622	10.37	
1030	Fittings	N/A		—	—	—	
2010	Stair Construction	Concrete filled metal pan		Flight	14,150	4.04	23.8%
3010	Wall Finishes	45% paint, 50% vinyl wall coating, 5% ceramic tile		S.F. Surface	1.35	3.59	
3020	Floor Finishes	50% carpet, 50% vinyl composition tile		S.F. Floor	6.87	6.87	
3030	Ceiling Finishes	Mineral fiber tile on concealed zee bars		S.F. Ceiling	5.98	5.98	
D.	**SERVICES**						
	D10 Conveying						
1010	Elevators & Lifts	One hydraulic hospital elevator		Each	99,890	14.27	8.4%
1020	Escalators & Moving Walks	N/A		—	—	—	
	D20 Plumbing						
2010	Plumbing Fixtures	Toilet, exam room and service fixtures, supply and drainage	1 Fixture/160 S.F. Floor	Each	3899	24.37	
2020	Domestic Water Distribution	Gas fired water heater		S.F. Floor	3.26	3.26	16.9%
2040	Rain Water Drainage	Roof drains		S.F. Roof	2.12	1.06	
	D30 HVAC						
3010	Energy Supply	N/A		—	—	—	
3020	Heat Generating Systems	Included in D3050		—	—	—	
3030	Cooling Generating Systems	N/A		—	—	—	8.9 %
3050	Terminal & Package Units	Multizone unit, gas heating, electric cooling		S.F. Floor	15.15	15.15	
3090	Other HVAC Sys. & Equipment	N/A		—	—	—	
	D40 Fire Protection						
4010	Sprinklers	Wet pipe sprinkler system		S.F. Floor	3.74	3.74	3.2%
4020	Standpipes	Standpipe		S.F. Floor	1.72	1.72	
	D50 Electrical						
5010	Electrical Service/Distribution	400 ampere service, panel board and feeders		S.F. Floor	2.55	2.55	
5020	Lighting & Branch Wiring	High efficiency fluorescent fixtures, receptacles, switches, A.C. and misc. power		S.F. Floor	7.54	7.54	11.1%
5030	Communications & Security	Alarm system, internet and phone wiring, intercom system, and emergency lighting		S.F. Floor	7.80	7.80	
5090	Other Electrical Systems	Emergency generator, 7.5 kW		S.F. Floor	.93	.93	
E.	**EQUIPMENT & FURNISHINGS**						
1010	Commercial Equipment	N/A		—	—	—	
1020	Institutional Equipment	Exam room casework and contertops		S.F. Floor	9.16	9.16	5.4 %
1030	Vehicular Equipment	N/A		—	—	—	
1090	Other Equipment	N/A		—	—	—	
F.	**SPECIAL CONSTRUCTION**						
1020	Integrated Construction	N/A		—	—	—	0.0 %
1040	Special Facilities	N/A		—	—	—	
G.	**BUILDING SITEWORK**	**N/A**					

			Sub-Total	170.05	100%
CONTRACTOR FEES (General Requirements: 10%, Overhead: 5%, Profit: 10%)			25%	42.52	
ARCHITECT FEES			9%	19.13	

Total Building Cost	**231.70**

Costs per square foot of floor area

Exterior Wall	S.F. Area	2000	3000	4000	6000	8000	10000	12000	14000	16000
	L.F. Perimeter	240	260	280	380	480	560	580	660	740
Brick Veneer	Wood Frame	203.65	178.55	166.10	159.35	156.00	152.80	147.80	146.65	145.90
Aluminum Siding	Wood Frame	186.35	166.05	155.95	150.20	147.30	144.75	140.85	139.80	139.25
Wood Siding	Wood Frame	185.85	165.70	155.65	150.00	147.05	144.55	140.65	139.65	139.00
Wood Shingles	Wood Frame	189.65	168.45	157.80	151.90	148.95	146.25	142.10	141.05	140.45
Precast Concrete Block	Wood Truss	185.15	165.25	155.25	149.60	146.70	144.20	140.35	139.35	138.75
Brick on Concrete Block	Wood Truss	213.15	185.40	171.60	164.40	160.70	157.25	151.65	150.40	149.55
Perimeter Adj., Add or Deduct	Per 100 L.F.	29.10	19.45	14.50	9.70	7.30	5.85	4.80	4.20	3.60
Story Hgt. Adj., Add or Deduct	Per 1 Ft.	4.75	3.45	2.75	2.55	2.40	2.25	1.95	1.85	1.85

For Basement, add $23.20 per square foot of basement area

The above costs were calculated using the basic specifications shown on the facing page. These costs should be adjusted where necessary for design alternatives and owner's requirements. Reported completed project costs, for this type of structure, range from $70.95 to $336.00 per S.F.

Common additives

Description	Unit	$ Cost
Closed Circuit Surveillance, One station		
Camera and monitor	Each	1950
For additional camera stations, add	Each	1050
Emergency Lighting, 25 watt, battery operated		
Lead battery	Each	305
Nickel cadmium	Each	850
Laundry Equipment		
Dryer, gas, 16 lb. capacity	Each	945
30 lb. capacity	Each	3750
Washer, 4 cycle	Each	1175
Commercial	Each	1600
Sauna, Prefabricated, complete		
6' x 4'	Each	6225
6' x 6'	Each	7325
6' x 9'	Each	8975
8' x 8'	Each	10,600
8' x 10'	Each	11,900
10' x 12'	Each	15,100
Smoke Detectors		
Ceiling type	Each	220
Duct type	Each	550

Description	Unit	$ Cost
Swimming Pools, Complete, gunite	S.F.	85 - 105
TV Antenna, Master system, 12 outlet	Outlet	277
30 outlet	Outlet	216
100 outlet	Outlet	205

Important: See the Reference Section for Location Factors.

Model costs calculated for a 1 story building with 9' story height and 8,000 square feet of floor area

				Unit	Unit Cost	Cost Per S.F.	% Of Sub-Total
A.	**SUBSTRUCTURE**						
1010	Standard Foundations	Poured concrete; strip and spread footings		S.F. Ground	3.17	3.17	
1020	Special Foundations	N/A		—	—	—	
1030	Slab on Grade	4" reinforced concrete with vapor barrier and granular base		S.F. Slab	5.18	5.18	12.7%
2010	Basement Excavation	Site preparation for slab and trench for foundation wall and footing		S.F. Ground	.32	.32	
2020	Basement Walls	4' foundation wall		L.F. Wall	75	6.17	
B.	**SHELL**						
	B10 Superstructure						
1010	Floor Construction	N/A		—	—	—	
1020	Roof Construction	Plywood on wood trusses		S.F. Roof	6.57	6.57	5.6%
	B20 Exterior Enclosure						
2010	Exterior Walls	Face brick on wood studs with sheathing, insulation and paper	80% of wall	S.F. Wall	24.26	10.48	
2020	Exterior Windows	Wood double hung	20% of wall	Each	504	3.89	15.7%
2030	Exterior Doors	Wood solid core		Each	1425	3.92	
	B30 Roofing						
3010	Roof Coverings	Asphalt shingles with flashing (pitched); rigid fiberglass insulation		S.F. Roof	4.33	4.33	3.7%
3020	Roof Openings	N/A		—	—	—	
C.	**INTERIORS**						
1010	Partitions	Gypsum bd. and sound deadening bd. on wood studs	9 S.F. Floor/L.F. Partition	S.F. Partition	9.36	8.32	
1020	Interior Doors	Single leaf hollow core wood	300 S.F. Floor/Door	Each	543	1.81	
1030	Fittings	N/A		—	—	—	
2010	Stair Construction	N/A		—	—	—	24.1%
3010	Wall Finishes	90% paint, 10% ceramic tile		S.F. Surface	1.44	2.56	
3020	Floor Finishes	85% carpet, 15% ceramic tile		S.F. Floor	11.17	11.17	
3030	Ceiling Finishes	Painted gypsum board on furring		S.F. Ceiling	4.21	4.21	
D.	**SERVICES**						
	D10 Conveying						
1010	Elevators & Lifts	N/A		—	—	—	
1020	Escalators & Moving Walks	N/A		—	—	—	0.0%
	D20 Plumbing						
2010	Plumbing Fixtures	Toilet and service fixtures, supply and drainage	1 Fixture/90 S.F. Floor	Each	1897	21.08	
2020	Domestic Water Distribution	Gas fired water heater		S.F. Floor	1.77	1.77	20.9%
2040	Rain Water Drainage	Roof Drains		S.F. Roof	1.48	1.48	
	D30 HVAC						
3010	Energy Supply	N/A		—	—	—	
3020	Heat Generating Systems	Included in D3050		—	—	—	
3030	Cooling Generating Systems	N/A		—	—	—	2.3%
3050	Terminal & Package Units	Through the wall electric heating and cooling units		S.F. Floor	2.63	2.63	
3090	Other HVAC Sys. & Equipment	N/A		—	—	—	
	D40 Fire Protection						
4010	Sprinklers	Wet pipe sprinkler system		S.F. Floor	4.66	4.66	4.9%
4020	Standpipes	Standpipe, wet, Class III		S.F. Floor	1.00	1.00	
	D50 Electrical						
5010	Electrical Service/Distribution	200 ampere service, panel board and feeders		S.F. Floor	1.39	1.39	
5020	Lighting & Branch Wiring	Fluorescent fixtures, receptacles, switches and misc. power		S.F. Floor	6.91	6.91	9.4%
5030	Communications & Security	Addressable alarm systems		S.F. Floor	2.65	2.65	
5090	Other Electrical Systems	N/A		—	—	—	
E.	**EQUIPMENT & FURNISHINGS**						
1010	Commercial Equipment	Laundry equipment		S.F. Floor	.96	.96	
1020	Institutional Equipment	N/A		—	—	—	
1030	Vehicular Equipment	N/A		—	—	—	0.8%
1090	Other Equipment	N/A		—	—	—	
F.	**SPECIAL CONSTRUCTION**						
1020	Integrated Construction	N/A		—	—	—	0.0%
1040	Special Facilities	N/A		—	—	—	
G.	**BUILDING SITEWORK**	**N/A**					

		Sub-Total	116.63	100%
CONTRACTOR FEES (General Requirements: 10%, Overhead: 5%, Profit: 10%)		25%	29.16	
ARCHITECT FEES		7%	10.21	

Total Building Cost	**156**

Costs per square foot of floor area

Exterior Wall	S.F. Area	25000	37000	49000	61000	73000	81000	88000	96000	104000
	L.F. Perimeter	433	593	606	720	835	911	978	1054	1074
Decorative Concrete Block	Wood Joists	166.30	163.20	158.10	156.90	156.15	155.75	155.50	155.20	154.35
	Precast Conc.	179.30	175.15	169.45	167.95	167.00	166.50	166.15	165.80	164.85
Stucco on Concrete Block	Wood Joists	168.35	163.75	157.80	156.15	155.05	154.50	154.15	153.75	152.80
	Precast Conc.	180.35	175.80	169.80	168.15	167.10	166.60	166.20	165.80	164.85
Wood Siding	Wood Frame	162.35	159.50	155.05	154.00	153.30	152.90	152.70	152.45	151.70
Brick Veneer	Wood Frame	171.45	166.65	159.95	158.25	157.15	156.55	156.15	155.70	154.65
Perimeter Adj., Add or Deduct	Per 100 L.F.	4.85	3.20	2.45	1.95	1.65	1.50	1.35	1.20	1.10
Story Hgt. Adj., Add or Deduct	Per 1 Ft.	1.75	1.55	1.20	1.15	1.10	1.10	1.05	1.05	1.00

For Basement, add $30.45 per square foot of basement area

The above costs were calculated using the basic specifications shown on the facing page. These costs should be adjusted where necessary for design alternatives and owner's requirements. Reported completed project costs, for this type of structure, range from $65.35 to $336.00 per S.F.

Common additives

Description	Unit	$ Cost	Description	Unit	$ Cost
Closed Circuit Surveillance, One station			Sauna, Prefabricated, complete		
Camera and monitor	Each	1950	6' x 4'	Each	6225
For additional camera station, add	Each	1050	6' x 6'	Each	7325
Elevators, Hydraulic passenger, 2 stops			6' x 9'	Each	8975
1500# capacity	Each	66,000	8' x 8'	Each	10,600
2500# capacity	Each	68,900	8' x 10'	Each	11,900
3500# capacity	Each	72,900	10' x 12'	Each	15,100
Additional stop, add	Each	8375	Smoke Detectors		
Emergency Lighting, 25 watt, battery operated			Ceiling type	Each	220
Lead battery	Each	305	Duct type	Each	550
Nickel cadmium	Each	850	Swimming Pools, Complete, gunite	S.F.	85 - 105
Laundry Equipment			TV Antenna, Master system, 12 outlet	Outlet	277
Dryer, gas, 16 lb. capacity	Each	945	30 outlet	Outlet	216
30 lb. capacity	Each	3750	100 outlet	Outlet	205
Washer, 4 cycle	Each	1175			
Commercial	Each	1600			

Model costs calculated for a 3 story building with 9' story height and 49,000 square feet of floor area

Motel, 2-3 Story

				Unit	Unit Cost	Cost Per S.F.	% Of Sub-Total
A. SUBSTRUCTURE							
1010	Standard Foundations	Poured concrete; strip and spread footings		S.F. Ground	1.47	.49	
1020	Special Foundations	N/A		—	—	—	
1030	Slab on Grade	4" reinforced concrete with vapor barrier and granular base		S.F. Slab	5.18	1.72	3.3%
2010	Basement Excavation	Site preparation for slab and trench for foundation wall and footing		S.F. Ground	.18	.06	
2020	Basement Walls	4' foundation wall		L.F. Wall	79	1.96	
B. SHELL							
	B10 Superstructure						
1010	Floor Construction	Precast concrete plank		S.F. Floor	10.07	6.71	7.9%
1020	Roof Construction	Precast concrete plank		S.F. Roof	10.05	3.35	
	B20 Exterior Enclosure						
2010	Exterior Walls	Decorative concrete block	85% of wall	S.F. Wall	17.40	4.94	
2020	Exterior Windows	Aluminum sliding	15% of wall	Each	584	1.95	10.6%
2030	Exterior Doors	Aluminum and glass doors and entrance with transom		Each	2000	6.61	
	B30 Roofing						
3010	Roof Coverings	Built-up tar and gravel with flashing; perlite/EPS composite insulation		S.F. Roof	6.60	2.20	1.8%
3020	Roof Openings	Roof hatches		S.F. Roof	.12	.04	
C. INTERIORS							
1010	Partitions	Concrete block	7 S.F. Floor/L.F. Partition	S.F. Partition	18.78	21.46	
1020	Interior Doors	Wood hollow core	70 S.F. Floor/Door	Each	543	7.76	
1030	Fittings	N/A		—	—	—	
2010	Stair Construction	Concrete filled metal pan		Flight	14,150	3.46	40.0%
3010	Wall Finishes	90% paint, 10% ceramic tile		S.F. Surface	1.44	3.29	
3020	Floor Finishes	85% carpet, 5% vinyl composition tile, 10% ceramic tile		S.F. Floor	11.17	11.17	
3030	Ceiling Finishes	Textured finish		S.F. Ceiling	4.02	4.02	
D. SERVICES							
	D10 Conveying						
1010	Elevators & Lifts	Two hydraulic passenger elevators		Each	107,555	4.39	3.4%
1020	Escalators & Moving Walks	N/A		—	—	—	
	D20 Plumbing						
2010	Plumbing Fixtures	Toilet and service fixtures, supply and drainage	1 Fixture/180 S.F. Floor	Each	4239	23.55	
2020	Domestic Water Distribution	Gas fired water heater		S.F. Floor	1.24	1.24	19.7%
2040	Rain Water Drainage	Roof drains		S.F. Roof	1.32	.44	
	D30 HVAC						
3010	Energy Supply	N/A		—	—	—	
3020	Heat Generating Systems	Included in D3050		—	—	—	
3030	Cooling Generating Systems	N/A		—	—	—	1.9 %
3050	Terminal & Package Units	Through the wall electric heating and cooling units		S.F. Floor	2.38	2.38	
3090	Other HVAC Sys. & Equipment	N/A		—	—	—	
	D40 Fire Protection						
4010	Sprinklers	Sprinklers, wet, light hazard		S.F. Floor	2.92	2.92	2.5%
4020	Standpipes	Standpipe, wet, Class III		S.F. Floor	.23	.23	
	D50 Electrical						
5010	Electrical Service/Distribution	800 ampere service, panel board and feeders		S.F. Floor	1.69	1.69	
5020	Lighting & Branch Wiring	Fluorescent fixtures, receptacles, switches and misc. power		S.F. Floor	7.58	7.58	8.8%
5030	Communications & Security	Addressable alarm systems and emergency lighting		S.F. Floor	1.84	1.84	
5090	Other Electrical Systems	Emergency generator, 7.5 kW		S.F. Floor	.13	.13	
E. EQUIPMENT & FURNISHINGS							
1010	Commercial Equipment	Commercial laundry equipment		S.F. Floor	.31	.31	
1020	Institutional Equipment	N/A		—	—	—	
1030	Vehicular Equipment	N/A		—	—	—	0.2%
1090	Other Equipment	N/A		—	—	—	
F. SPECIAL CONSTRUCTION							
1020	Integrated Construction	N/A		—	—	—	0.0 %
1040	Special Facilities	N/A		—	—	—	
G. BUILDING SITEWORK	**N/A**						

	Sub-Total		127.89	**100%**
CONTRACTOR FEES (General Requirements: 10%, Overhead: 5%, Profit: 10%)		25%	31.97	
ARCHITECT FEES		6%	9.59	
	Total Building Cost		169.45	

Costs per square foot of floor area

Exterior Wall	S.F. Area	9000	10000	12000	13000	14000	15000	16000	18000	20000
	L.F. Perimeter	385	410	440	460	480	500	510	547	583
Decorative Concrete Block	Steel Joists	177.45	173.10	165.20	162.55	160.25	158.20	155.90	152.95	150.50
Painted Concrete Block	Steel Joists	169.40	165.35	158.35	155.90	153.75	151.90	149.95	147.15	145.00
Face Brick on Conc. Block	Steel Joists	192.35	187.40	178.00	174.90	172.20	169.85	167.05	163.50	160.70
Precast Concrete Panels	Steel Joists	204.00	198.55	188.05	184.55	181.55	178.90	175.70	171.80	168.65
Tilt-up Panels	Steel Joists	174.50	170.30	162.75	160.15	157.90	155.95	153.75	150.85	148.55
Metal Sandwich Panels	Steel Joists	173.50	169.30	161.80	159.30	157.05	155.10	152.95	150.10	147.75
Perimeter Adj., Add or Deduct	Per 100 L.F.	9.00	8.05	6.80	6.20	5.75	5.40	5.10	4.45	4.05
Story Hgt. Adj., Add or Deduct	Per 1 Ft.	1.35	1.20	1.20	1.05	1.05	1.00	1.00	0.90	0.90
Basement—Not Applicable										

The above costs were calculated using the basic specifications shown on the facing page. These costs should be adjusted where necessary for design alternatives and owner's requirements. Reported completed project costs, for this type of structure, range from $83.60 to $218.70 per S.F.

Common additives

Description	Unit	$ Cost
Emergency Lighting, 25 watt, battery operated		
Lead battery	Each	305
Nickel cadmium	Each	850
Seating		
Auditorium chair, all veneer	Each	265
Veneer back, padded seat	Each	335
Upholstered, spring seat	Each	320
Classroom, movable chair & desk	Set	81 - 171
Lecture hall, pedestal type	Each	257 - 750
Smoke Detectors		
Ceiling type	Each	220
Duct type	Each	550
Sound System		
Amplifier, 250 watts	Each	1975
Speaker, ceiling or wall	Each	206
Trumpet	Each	395

Important: See the Reference Section for Location Factors.

Model costs calculated for a 1 story building with 20' story height and 12,000 square feet of floor area

				Unit	Unit Cost	Cost Per S.F.	% Of Sub-Total
A.	**SUBSTRUCTURE**						
1010	Standard Foundations	Poured concrete; strip and spread footings		S.F. Ground	1.42	1.42	
1020	Special Foundations	N/A		—	—	—	
1030	Slab on Grade	4" reinforced concrete with vapor barrier and granular base		S.F. Slab	5.18	5.18	10.8%
2010	Basement Excavation	Site preparation for slab and trench for foundation wall and footing		S.F. Ground	3.81	3.81	
2020	Basement Walls	4' foundation wall		L.F. Wall	79	2.90	
B.	**SHELL**						
	B10 Superstructure						
1010	Floor Construction	Open web steel joists, slab form, concrete	mezzanine 2250 S.F.	S.F. Floor	11.34	2.12	10.2%
1020	Roof Construction	Metal deck on open web steel joists		S.F. Roof	10.45	10.45	
	B20 Exterior Enclosure						
2010	Exterior Walls	Decorative concrete block	100% of wall	S.F. Wall	19.39	14.22	
2020	Exterior Windows	N/A		—	—	—	12.5%
2030	Exterior Doors	Sliding mallfront aluminum and glass and hollow metal		Each	2533	1.27	
	B30 Roofing						
3010	Roof Coverings	Built-up tar and gravel with flashing; perlite/EPS composite insulation		S.F. Roof	6.07	6.07	4.9%
3020	Roof Openings	N/A		—	—	—	
C.	**INTERIORS**						
1010	Partitions	Concrete block	40 S.F. Floor/L.F. Partition	S.F. Partition	9.10	3.64	
1020	Interior Doors	Single leaf hollow metal	705 S.F. Floor/Door	Each	1212	1.72	
1030	Fittings	Toilet partitions		S.F. Floor	.60	.60	
2010	Stair Construction	Concrete filled metal pan		Flight	16,950	2.82	20.7%
3010	Wall Finishes	Paint		S.F. Surface	4.73	3.78	
3020	Floor Finishes	Carpet 50%, ceramic tile 5%	50% of area	S.F. Floor	12.32	6.16	
3030	Ceiling Finishes	Mineral fiber tile on concealed zee runners suspended		S.F. Ceiling	6.86	6.86	
D.	**SERVICES**						
	D10 Conveying						
1010	Elevators & Lifts	N/A		—	—	—	0.0 %
1020	Escalators & Moving Walks	N/A		—	—	—	
	D20 Plumbing						
2010	Plumbing Fixtures	Toilet and service fixtures, supply and drainage	1 Fixture/500 S.F. Floor	Each	4680	9.36	
2020	Domestic Water Distribution	Gas fired water heater		S.F. Floor	.44	.44	9.0%
2040	Rain Water Drainage	Roof drains		S.F. Roof	1.27	1.27	
	D30 HVAC						
3010	Energy Supply	N/A		—	—	—	
3020	Heat Generating Systems	Included in D3050		—	—	—	
3030	Cooling Generating Systems	N/A		—	—	—	8.3 %
3050	Terminal & Package Units	Single zone rooftop unit, gas heating, electric cooling		S.F. Floor	10.27	10.27	
3090	Other HVAC Sys. & Equipment	N/A		—	—	—	
	D40 Fire Protection						
4010	Sprinklers	Wet pipe sprinkler system		S.F. Floor	4.20	4.20	4.1%
4020	Standpipes	Standpipe		S.F. Floor	.82	.82	
	D50 Electrical						
5010	Electrical Service/Distribution	400 ampere service, panel board and feeders		S.F. Floor	1.87	1.87	
5020	Lighting & Branch Wiring	High efficiency fluorescent fixtures, receptacles, switches, A.C. and misc. power		S.F. Floor	6.46	6.46	8.7%
5030	Communications & Security	Addressable alarm systems, sound system and emergency lighting		S.F. Floor	2.26	2.26	
5090	Other Electrical Systems	Emergency generator, 7.5 kW		S.F. Floor	.18	.18	
E.	**EQUIPMENT & FURNISHINGS**						
1010	Commercial Equipment	N/A		—	—	—	
1020	Institutional Equipment	Projection equipment, screen		S.F. Floor	8.97	8.97	10.8 %
1030	Vehicular Equipment	N/A		—	—	—	
2010	Other Equipment	Movie theater seating		S.F. Floor	4.42	4.42	
F.	**SPECIAL CONSTRUCTION**						
1020	Integrated Construction	N/A		—	—	—	0.0 %
1040	Special Facilities	N/A		—	—	—	
G.	**BUILDING SITEWORK**	**N/A**					

	Sub-Total	123.54	**100%**
CONTRACTOR FEES (General Requirements: 10%, Overhead: 5%, Profit: 10%)	25%	30.85	
ARCHITECT FEES	7%	10.81	

Total Building Cost	165.20

Costs per square foot of floor area

Exterior Wall	S.F. Area	10000	15000	20000	25000	30000	35000	40000	45000	50000
	L.F. Perimeter	286	370	453	457	513	568	624	680	735
Precast Concrete Panels	Bearing Walls	224.80	215.70	211.05	203.10	200.65	198.90	197.55	196.55	195.75
	Steel Frame	232.50	223.40	218.75	210.80	208.35	206.60	205.25	204.25	203.40
Face Brick with Concrete Block Back-up	Bearing Walls	214.25	206.60	202.70	196.40	194.35	192.90	191.80	190.95	190.25
	Steel Joists	222.85	215.25	211.35	205.00	203.00	201.55	200.40	199.60	198.90
Stucco on Concrete Block	Bearing Walls	205.35	198.95	195.70	190.70	189.00	187.85	186.95	186.25	185.70
	Steel Joists	214.05	207.65	204.35	199.35	197.70	196.55	195.60	194.95	194.40
Perimeter Adj., Add or Deduct	Per 100 L.F.	16.35	10.85	8.20	6.55	5.45	4.65	4.10	3.65	3.20
Story Hgt. Adj., Add or Deduct	Per 1 Ft.	3.95	3.45	3.15	2.55	2.40	2.25	2.15	2.10	1.95
For Basement, add $31.60 per square foot of basement area										

The above costs were calculated using the basic specifications shown on the facing page. These costs should be adjusted where necessary for design alternatives and owner's requirements. Reported completed project costs, for this type of structure, range from $89.75 to $225.05 per S.F.

Common additives

Description	Unit	$ Cost	Description	Unit	$ Cost
Beds, Manual	Each	865 - 2750	Kitchen Equipment, cont.		
Elevators, Hydraulic passenger, 2 stops			Ice cube maker, 50 lb. per day	Each	1925
1500# capacity	Each	66,000	Range with 1 oven	Each	3325
2500# capacity	Each	68,900	Laundry Equipment		
3500# capacity	Each	72,900	Dryer, gas, 16 lb. capacity	Each	945
Emergency Lighting, 25 watt, battery operated			30 lb. capacity	Each	3750
Lead battery	Each	305	Washer, 4 cycle	Each	1175
Nickel cadmium	Each	850	Commercial	Each	1600
Intercom System, 25 station capacity			Nurses Call System		
Master station	Each	3125	Single bedside call station	Each	330
Intercom outlets	Each	183	Pillow speaker	Each	272
Handset	Each	500	Refrigerator, Prefabricated, walk-in		
Kitchen Equipment			7'-6" high, 6' x 6'	S.F.	190
Broiler	Each	4175	10' x 10'	S.F.	150
Coffee urn, twin 6 gallon	Each	2625	12' x 14'	S.F.	133
Cooler, 6 ft. long	Each	4000	12' x 20'	S.F.	117
Dishwasher, 10-12 racks per hr.	Each	5300	TV Antenna, Master system, 12 outlet	Outlet	277
Food warmer	Each	560	30 outlet	Outlet	216
Freezer, 44 C.F., reach-in	Each	4150	100 outlet	Outlet	205
			Whirlpool Bath, Mobile, 18" x 24" x 60"	Each	5325
			X-Ray, Mobile	Each	18,300 - 84,500

Important: See the Reference Section for Location Factors.

Model costs calculated for a 2 story building with 10' story height and 25,000 square feet of floor area

Nursing Home

			Unit	Unit Cost	Cost Per S.F.	% Of Sub-Total
A. SUBSTRUCTURE						
1010	Standard Foundations	Poured concrete; strip and spread footings	S.F. Ground	1.40	.70	
1020	Special Foundations	N/A	—	—	—	
1030	Slab on Grade	4" reinforced concrete with vapor barrier and granular base	S.F. Slab	5.18	2.60	3.5%
2010	Basement Excavation	Site preparation for slab and trench for foundation wall and footing	S.F. Ground	.32	.16	
2020	Basement Walls	4' foundation wall	L.F. Wall	71	1.68	
B. SHELL						
	B10 Superstructure					
1010	Floor Construction	Pre-cast double tees with concrete topping	S.F. Floor	12.88	6.44	8.2%
1020	Roof Construction	Pre-cast double tees	S.F. Roof	11.08	5.54	
	B20 Exterior Enclosure					
2010	Exterior Walls	Precast concrete panels	S.F. Wall	47.17	14.66	
2020	Exterior Windows	Wood double hung 15% of wall	Each	584	2.13	11.8%
2030	Exterior Doors	Double aluminum & glass doors, single leaf hollow metal	Each	2270	.45	
	B30 Roofing					
3010	Roof Coverings	Built-up tar and gravel with flashing; perlite/EPS composite insulation	S.F. Roof	6.54	3.27	2.3%
3020	Roof Openings	Roof hatches	S.F. Roof	.08	.04	
C. INTERIORS						
1010	Partitions	Gypsum board on metal studs 8 S.F. Floor/L.F. Partition	S.F. Partition	7.07	7.07	
1020	Interior Doors	Single leaf wood 80 S.F. Floor/Door	Each	622	7.78	
1030	Fittings	N/A	—	—	—	
2010	Stair Construction	Concrete stair	Flight	6400	1.53	17.9%
3010	Wall Finishes	50% vinyl wall coverings, 45% paint, 5% ceramic tile	S.F. Surface	1.66	3.32	
3020	Floor Finishes	95% vinyl tile, 5% ceramic tile	S.F. Floor	2.48	2.48	
3030	Ceiling Finishes	Painted gypsum board	S.F. Ceiling	3.98	3.98	
D. SERVICES						
	D10 Conveying					
1010	Elevators & Lifts	One hydraulic hospital elevator	Each	100,000	4	2.7%
1020	Escalators & Moving Walks	N/A	—	—	—	
	D20 Plumbing					
2010	Plumbing Fixtures	Kitchen, toilet and service fixtures, supply and drainage 1 Fixture/230 S.F. Floor	Each	9711	42.22	
2020	Domestic Water Distribution	Oil fired water heater	S.F. Floor	1.29	1.29	30.1%
2040	Rain Water Drainage	Roof drains	S.F. Roof	1.12	.56	
	D30 HVAC					
3010	Energy Supply	Oil fired hot water, wall fin radiation	S.F. Floor	8.80	8.80	
3020	Heat Generating Systems	N/A	—	—	—	
3030	Cooling Generating Systems	N/A	—	—	—	11.9%
3050	Terminal & Package Units	Split systems with air cooled condensing units	S.F. Floor	8.56	8.56	
3090	Other HVAC Sys. & Equipment	N/A	—	—	—	
	D40 Fire Protection					
4010	Sprinklers	Sprinkler, light hazard	S.F. Floor	3.07	3.07	2.4%
4020	Standpipes	Standpipe	S.F. Floor	.42	.42	
	D50 Electrical					
5010	Electrical Service/Distribution	800 ampere service, panel board and feeders	S.F. Floor	1.72	1.72	
5020	Lighting & Branch Wiring	High efficiency fluorescent fixtures, receptacles, switches, A.C. and misc. power	S.F. Floor	10.01	10.01	9.3%
5030	Communications & Security	Addressable alarm systems and emergency lighting	S.F. Floor	1.38	1.38	
5090	Other Electrical Systems	Emergency generator, 15 kW	S.F. Floor	.55	.55	
E. EQUIPMENT & FURNISHINGS						
1010	Commercial Equipment	N/A	—	—	—	
1020	Institutional Equipment	N/A	—	—	—	
1030	Vehicular Equipment	N/A	—	—	—	0.0 %
1090	Other Equipment	N/A	—	—	—	
F. SPECIAL CONSTRUCTION						
1020	Integrated Construction	N/A	—	—	—	0.0 %
1040	Special Facilities	N/A	—	—	—	
G. BUILDING SITEWORK	**N/A**					

		Sub-Total	146.41	100%
CONTRACTOR FEES (General Requirements: 10%, Overhead: 5%, Profit: 10%)		25%	36.56	
ARCHITECT FEES		11%	20.13	

Total Building Cost	**203.10**

Costs per square foot of floor area

Exterior Wall	S.F. Area	2000	3000	5000	7000	9000	12000	15000	20000	25000
	L.F. Perimeter	220	260	320	360	420	480	520	640	700
Wood Siding	Wood Truss	207.85	188.35	170.95	162.20	158.35	153.80	150.45	148.40	146.00
Brick Veneer	Wood Truss	256.75	231.50	208.70	197.05	192.00	185.95	181.45	178.80	175.55
Brick on Block	Wood Truss	196.30	169.55	145.25	132.80	127.50	121.00	116.10	113.20	109.65
Brick on Block	Steel Roof Deck	252.05	224.05	198.50	185.35	179.80	173.00	167.85	164.80	161.10
EIFS on Metal Studs	Steel Roof Deck	233.70	210.65	189.90	179.25	174.70	169.30	165.15	162.70	159.75
Tiltup Concrete Panel	Steel Roof Deck	235.50	211.60	190.10	179.15	174.45	168.80	164.60	162.00	159.05
Perimeter Adj., Add or Deduct	Per 100 L.F.	23.30	15.55	9.35	6.70	5.15	3.90	3.10	2.30	1.85
Story Hgt. Adj., Add or Deduct	Per 1 Ft.	2.70	2.20	1.65	1.30	1.20	1.00	0.90	0.75	0.70

For Basement, add $32.05 per square foot of basement area

The above costs were calculated using the basic specifications shown on the facing page. These costs should be adjusted where necessary for design alternatives and owner's requirements. Reported completed project costs, for this type of structure, range from $68.55 to $265.00 per S.F.

Common additives

Description	Unit	$ Cost
Closed circuit surveillance, one station		
Camera and monitor	Each	1950
For additional camera stations, add	Each	1050
Directory boards, plastic, glass covered		
30" x 20"	Each	660
36" x 48"	Each	1475
Aluminum, 24" x 18"	Each	620
36" x 24"	Each	725
48" x 32"	Each	1050
48" x 60"	Each	2175
Electronic, wall mounted	S.F.	980
Free standing	S.F.	1100
Pedestal access floor system w/ plastic laminate cover		
Computer room, less than 6000 SF	S.F.	11.95
Greater than 6000 SF	S.F.	11.20
Office, greater than 6000 S.F.	S.F.	8

Description	Unit	$ Cost
Security access systems		
Metal detectors, wand type	Each	98
Walk-through portal type, single-zone	Each	4325
Multi-zone	Each	5400
X-ray equipment		
Desk top, for mail, small packages	Each	3850
Conveyer type, including monitor, minimum	Each	17,500
Maximum	Each	31,100
Explosive detection equipment		
Hand held, battery operated	Each	28,100
Walk-through portal type	Each	47,900
Uninterruptible power supply, 15 kVA/12.75 kW	kW	1735

Important: See the Reference Section for Location Factors.

Model costs calculated for a 1 story building with 12' story height and 7,000 square feet of floor area

				Unit	Unit Cost	Cost Per S.F.	% Of Sub-Total
A. SUBSTRUCTURE							
1010	Standard Foundations	Poured concrete; strip and spread footings		S.F. Ground	2.75	2.75	
1020	Special Foundations	N/A		—	—	—	
1030	Slab on Grade	4" reinforced concrete with vapor barrier and granular base		S.F. Slab	5.18	5.18	9.3%
2010	Basement Excavation	Site preparation for slab and trench for foundation wall and footing		S.F. Ground	.32	.32	
2020	Basement Walls	4' foundation wall		L.F. Wall	150	4.27	
B. SHELL							
	B10 Superstructure						
1010	Floor Construction	N/A		—	—	—	7.6 %
1020	Roof Construction	Steel joists, girders & deck on columns		S.F. Roof	10.17	10.17	
	B20 Exterior Enclosure						
2010	Exterior Walls	E.I.F.S. on metal studs		S.F. Wall	16.79	8.29	
2020	Exterior Windows	Aluminum outward projecting	20% of wall	Each	735	3.94	11.3%
2030	Exterior Doors	Aluminum and glass, hollow metal		Each	3453	2.96	
	B30 Roofing						
3010	Roof Coverings	Single ply membrane, loose laid and ballasted		S.F. Roof	9.06	9.06	6.9%
3020	Roof Openings	Roof hatch		S.F. Roof	.17	.17	
C. INTERIORS							
1010	Partitions	Gypsum board on metal studs	20 S.F. Floor/L.F. Partition	S.F. Partition	8.58	4.29	
1020	Interior Doors	Single leaf hollow metal	200 S.F. Floor/Door	Each	1212	6.07	
1030	Fittings	Toilet partitions		S.F. Floor	.51	.51	
2010	Stair Construction	N/A		—	—	—	20.6%
3010	Wall Finishes	60% vinyl wall covering, 40% paint		S.F. Surface	1.45	1.45	
3020	Floor Finishes	60% carpet, 30% vinyl composition tile, 10% ceramic tile		S.F. Floor	8.44	8.44	
3030	Ceiling Finishes	Mineral fiber tile on concealed zee bars		S.F. Ceiling	6.86	6.86	
D. SERVICES							
	D10 Conveying						
1010	Elevators & Lifts	N/A		—	—	—	0.0 %
1020	Escalators & Moving Walks	N/A		—	—	—	
	D20 Plumbing						
2010	Plumbing Fixtures	Toilet and service fixtures, supply and drainage	1 Fixture/1320 S.F. Floor	Each	8567	6.49	
2020	Domestic Water Distribution	Gas fired water heater		S.F. Floor	1.31	1.31	6.4%
2040	Rain Water Drainage	Roof drains		S.F. Roof	.81	.81	
	D30 HVAC						
3010	Energy Supply	N/A		—	—	—	
3020	Heat Generating Systems	Included in D3050		—	—	—	
3030	Cooling Generating Systems	N/A		—	—	—	15.9 %
3050	Terminal & Package Units	Multizone unit gas heating, electric cooling		S.F. Floor	21.25	21.25	
3090	Other HVAC Sys. & Equipment	N/A		—	—	—	
	D40 Fire Protection						
4010	Sprinklers	Sprinkler system, light hazard		S.F. Floor	3.51	3.51	3.6%
4020	Standpipes	Standpipes and hose systems		S.F. Floor	1.37	1.37	
	D50 Electrical						
5010	Electrical Service/Distribution	400 ampere service, panel board and feeders		S.F. Floor	5	5	
5020	Lighting & Branch Wiring	High efficiency fluorescent fixtures, receptacles, switches, A.C. and misc. power		S.F. Floor	12.22	12.22	18.3%
5030	Communications & Security	Addressable alarm systems, internet and phone wiring, and emergency lighting		S.F. Floor	7.13	7.13	
5090	Other Electrical Systems	Emergency generator, 7.5 kW		S.F. Floor	.22	.22	
E. EQUIPMENT & FURNISHINGS							
1010	Commercial Equipment	N/A		—	—	—	
1020	Institutional Equipment	N/A		—	—	—	0.0 %
1030	Vehicular Equipment	N/A		—	—	—	
1090	Other Equipment	N/A		—	—	—	
F. SPECIAL CONSTRUCTION							
1020	Integrated Construction	N/A		—	—	—	0.0 %
1040	Special Facilities	N/A		—	—	—	
G. BUILDING SITEWORK	**N/A**						

		Sub-Total	134.04	100%
CONTRACTOR FEES (General Requirements: 10%, Overhead: 5%, Profit: 10%)		25%	33.48	
ARCHITECT FEES		7%	11.73	
	Total Building Cost		**179.25**	

Costs per square foot of floor area

Exterior Wall	S.F. Area	5000	8000	12000	16000	20000	35000	50000	65000	80000
	L.F. Perimeter	220	260	310	330	360	440	490	548	580
Face Brick with Concrete Block Back-up	Wood Joists	247.10	216.65	199.15	186.75	180.30	167.40	161.10	157.90	155.30
	Steel Joists	251.15	220.70	203.15	190.75	184.30	171.45	165.15	161.95	159.35
Glass and Metal Curtain Wall	Steel Frame	295.40	254.80	231.50	214.50	205.75	188.10	179.40	175.05	171.35
	R/Conc. Frame	291.20	251.15	228.05	211.25	202.55	185.10	176.50	172.10	168.45
Wood Siding	Wood Frame	196.05	175.05	163.00	154.90	150.60	142.25	138.15	136.15	134.50
Brick Veneer	Wood Frame	218.35	191.50	176.10	165.35	159.75	148.60	143.15	140.40	138.20
Perimeter Adj., Add or Deduct	Per 100 L.F.	38.80	24.20	16.10	12.10	9.70	5.55	3.90	3.00	2.45
Story Hgt. Adj., Add or Deduct	Per 1 Ft.	6.35	4.65	3.70	2.90	2.55	1.80	1.45	1.25	1.10
For Basement, add $35.85 per square foot of basement area										

The above costs were calculated using the basic specifications shown on the facing page. These costs should be adjusted where necessary for design alternatives and owner's requirements. Reported completed project costs, for this type of structure, range from $71.70 to $277.85 per S.F.

Common additives

Description	Unit	$ Cost
Closed circuit surveillance, one station		
Camera and monitor	Each	1950
For additional camera stations, add	Each	1050
Directory boards, plastic, glass covered		
30" x 20"	Each	660
36" x 48"	Each	1475
Aluminum, 24" x 18"	Each	620
36" x 24"	Each	725
48" x 32"	Each	1050
48" x 60"	Each	2175
Electronic, wall mounted	S.F.	980
Free standing	S.F.	1100
Escalators, 10' rise, 32" wide, glass balustrade	Each	137,400
Metal balustrade	Each	143,400
48" wide, glass balustrade	Each	143,400
Metal balustrade	Each	150,900
Pedestal access floor system w/ plastic laminate cover		
Computer room, less than 6000 SF	S.F.	11.95
Greater than 6000 SF	S.F.	11.20
Office, greater than 6000 S.F.	S.F.	8

Description	Unit	$ Cost
Security access systems		
Metal detectors, wand type	Each	98
Walk-through portal type, single-zone	Each	4325
Multi-zone	Each	5400
X-ray equipment		
Desk top, for mail, small packages	Each	3850
Conveyer type, including monitor, minimum	Each	17,500
Maximum	Each	31,100
Explosive detection equipment		
Hand held, battery operated	Each	28,100
Walk-through portal type	Each	47,900
Uninterruptible power supply, 15 kVA/12.75 kW	kW	1735

Important: See the Reference Section for Location Factors.

Model costs calculated for a 3 story building with 12' story height and 20,000 square feet of floor area

				Unit	Unit Cost	Cost Per S.F.	% Of Sub-Total
A.	**SUBSTRUCTURE**						
1010	Standard Foundations	Poured concrete; strip and spread footings		S.F. Ground	7.41	2.47	
1020	Special Foundations	N/A		—	—	—	
1030	Slab on Grade	4" reinforced concrete with vapor barrier and granular base		S.F. Slab	5.18	1.72	4.3%
2010	Basement Excavation	Site preparation for slab and trench for foundation wall and footing		S.F. Ground	.18	.06	
2020	Basement Walls	4' foundation wall		L.F. Wall	75	1.67	
B.	**SHELL**						
	B10 Superstructure						
1010	Floor Construction	Open web steel joists, slab form, concrete, columns		S.F. Floor	18.68	12.45	10.8%
1020	Roof Construction	Metal deck, open web steel joists, columns		S.F. Roof	7.23	2.41	
	B20 Exterior Enclosure						
2010	Exterior Walls	Face brick with concrete block backup	80% of wall	S.F. Wall	32.45	16.82	
2020	Exterior Windows	Aluminum outward projecting	20% of wall	Each	735	4.14	16.0%
2030	Exterior Doors	Aluminum and glass, hollow metal		Each	3453	1.04	
	B30 Roofing						
3010	Roof Coverings	Built-up tar and gravel with flashing; perlite/EPS composite		S.F. Roof	6.96	2.32	1.7%
3020	Roof Openings	N/A		—	—	—	
C.	**INTERIORS**						
1010	Partitions	Gypsum board on metal studs	20 S.F. Floor/L.F. Partition	S.F. Partition	9.65	3.86	
1020	Interior Doors	Single leaf hollow metal	200 S.F. Floor/Door	Each	1212	6.07	
1030	Fittings	Toilet partitions		S.F. Floor	1.06	1.06	
2010	Stair Construction	Concrete filled metal pan		Flight	14,150	4.96	23.5%
3010	Wall Finishes	60% vinyl wall covering, 40% paint		S.F. Surface	1.48	1.18	
3020	Floor Finishes	60% carpet, 30% vinyl composition tile, 10% ceramic tile		S.F. Floor	8.44	8.44	
3030	Ceiling Finishes	Mineral fiber tile on concealed zee bars		S.F. Ceiling	6.86	6.86	
D.	**SERVICES**						
	D10 Conveying						
1010	Elevators & Lifts	Two hydraulic passenger elevators		Each	118,100	11.81	8.6%
1020	Escalators & Moving Walks	N/A		—	—	—	
	D20 Plumbing						
2010	Plumbing Fixtures	Toilet and service fixtures, supply and drainage	1 Fixture/1320 S.F. Floor	Each	5016	3.80	
2020	Domestic Water Distribution	Gas fired water heater		S.F. Floor	.45	.45	3.6%
2040	Rain Water Drainage	Roof drains		S.F. Roof	1.95	.65	
	D30 HVAC						
3010	Energy Supply	N/A		—	—	—	
3020	Heat Generating Systems	Included in D3050		—	—	—	
3030	Cooling Generating Systems	N/A		—	—	—	12.5 %
3050	Terminal & Package Units	Multizone unit gas heating, electric cooling		S.F. Floor	17.25	17.25	
3090	Other HVAC Sys. & Equipment	N/A		—	—	—	
	D40 Fire Protection						
4010	Sprinklers	Wet pipe sprinkler system		S.F. Floor	3.65	3.65	3.3%
4020	Standpipes	Standpipes and hose systems		S.F. Floor	.88	.88	
	D50 Electrical						
5010	Electrical Service/Distribution	1000 ampere service, panel board and feeders		S.F. Floor	3.84	3.84	
5020	Lighting & Branch Wiring	High efficiency fluorescent fixtures, receptacles, switches, A.C. and misc. power		S.F. Floor	11.73	11.73	15.8%
5030	Communications & Security	Addressable alarm systems, internet and phone wiring, and emergency lighting		S.F. Floor	5.97	5.97	
5090	Other Electrical Systems	Emergency generator, 7.5 kW, uninterruptible power supply		S.F. Floor	.24	.24	
E.	**EQUIPMENT & FURNISHINGS**						
1010	Commercial Equipment	N/A		—	—	—	
1020	Institutional Equipment	N/A		—	—	—	0.0 %
1030	Vehicular Equipment	N/A		—	—	—	
1090	Other Equipment	N/A		—	—	—	
F.	**SPECIAL CONSTRUCTION**						
1020	Integrated Construction	N/A		—	—	—	0.0 %
1040	Special Facilities	N/A		—	—	—	
G.	**BUILDING SITEWORK**	**N/A**					

		Sub-Total	137.80	**100%**
CONTRACTOR FEES (General Requirements: 10%, Overhead: 5%, Profit: 10%)		25%	34.44	
ARCHITECT FEES		7%	12.06	

Total Building Cost	**184.30**

Costs per square foot of floor area

Exterior Wall	S.F. Area	20000	40000	60000	80000	100000	150000	200000	250000	300000
	L.F. Perimeter	260	360	400	420	460	520	600	640	700
Precast Concrete Panel	Steel Frame	235.70	204.10	187.05	176.90	172.15	164.00	160.55	157.50	155.85
	R/Conc. Frame	233.70	201.90	184.85	174.65	169.85	161.65	158.20	155.10	153.45
Face Brick with Concrete Block Back-up	Steel Frame	218.15	191.15	177.15	168.85	164.95	158.35	155.50	153.00	151.70
	R/Conc. Frame	213.60	188.00	174.50	166.55	162.75	156.25	153.60	151.15	149.85
Limestone Panel Concrete Block Back-up	Steel Frame	261.05	221.55	200.00	187.10	181.05	170.65	166.35	162.45	160.35
	R/Conc. Frame	258.65	219.20	197.60	184.70	178.65	168.30	163.95	160.05	157.95
Perimeter Adj., Add or Deduct	Per 100 L.F.	32.55	16.30	10.85	8.10	6.50	4.30	3.20	2.55	2.20
Story Hgt. Adj., Add or Deduct	Per 1 Ft.	6.75	4.70	3.45	2.75	2.40	1.80	1.55	1.30	1.25

For Basement, add $38.65 per square foot of basement area

The above costs were calculated using the basic specifications shown on the facing page. These costs should be adjusted where necessary for design alternatives and owner's requirements. Reported completed project costs, for this type of structure, range from $80.70 to $237.35 per S.F.

Common additives

Description	Unit	$ Cost
Closed circuit surveillance, one station		
Camera and monitor	Each	1950
For additional camera stations, add	Each	1050
Directory boards, plastic, glass covered		
30" x 20"	Each	660
36" x 48"	Each	1475
Aluminum, 24" x 18"	Each	620
36" x 24"	Each	725
48" x 32"	Each	1050
48" x 60"	Each	2175
Electronic, wall mounted	S.F.	980
Free standing	S.F.	1100
Escalators, 10' rise, 32" wide, glass balustrade	Each	137,400
Metal balustrade	Each	143,400
48" wide, glass balustrade	Each	143,400
Metal balustrade	Each	150,900
Pedestal access floor system w/ plastic laminate cover		
Computer room, less than 6000 SF	S.F.	11.95
Greater than 6000 SF	S.F.	11.20
Office, greater than 6000 S.F.	S.F.	8

Description	Unit	$ Cost
Security access systems		
Metal detectors, wand type	Each	98
Walk-through portal type, single-zone	Each	4325
Multi-zone	Each	5400
X-ray equipment		
Desk top, for mail, small packages	Each	3850
Conveyer type, including monitor, minimum	Each	17,500
Maximum	Each	31,100
Explosive detection equipment		
Hand held, battery operated	Each	28,100
Walk-through portal type	Each	47,900
Uninterruptible power supply, 15 kVA/12.75 kW	kW	1735

Important: See the Reference Section for Location Factors.

Model costs calculated for a 8 story building with 12' story height and 80,000 square feet of floor area

				Unit	Unit Cost	Cost Per S.F.	% Of Sub-Total
A.	**SUBSTRUCTURE**						
1010	Standard Foundations	Poured concrete; strip and spread footings		S.F. Ground	11.92	1.49	
1020	Special Foundations	N/A		—	—	—	
1030	Slab on Grade	4" reinforced concrete with vapor barrier and granular base		S.F. Slab	5.18	.65	2.0%
2010	Basement Excavation	Site preparation for slab and trench for foundation wall and footing		S.F. Ground	.32	.04	
2020	Basement Walls	4' foundation wall		L.F. Wall	79	.53	
B.	**SHELL**						
	B10 Superstructure						
1010	Floor Construction	Concrete slab with metal deck and beams		S.F. Floor	24.06	21.05	16.4%
1020	Roof Construction	Metal deck, open web steel joists, interior columns		S.F. Roof	7.04	.88	
	B20 Exterior Enclosure						
2010	Exterior Walls	Precast concrete panels	80% of wall	S.F. Wall	47.67	19.22	
2020	Exterior Windows	Vertical pivoted steel	20% of wall	Each	584	3.93	17.5%
2030	Exterior Doors	Double aluminum and glass doors and entrance with transoms		Each	4154	.26	
	B30 Roofing						
3010	Roof Coverings	Built-up tar and gravel with flashing; perlite/EPS composite insulation		S.F. Roof	6.08	.76	0.6%
3020	Roof Openings	N/A		—	—	—	
C.	**INTERIORS**						
1010	Partitions	Gypsum board on metal studs	30 S.F. Floor/L.F. Partition	S.F. Partition	9.54	3.18	
1020	Interior Doors	Single leaf hollow metal	400 S.F. Floor/Door	Each	1212	3.03	
1030	Fittings	Toilet Partitions		S.F. Floor	.70	.70	
2010	Stair Construction	Concrete filled metal pan		Flight	14,150	3.01	19.6%
3010	Wall Finishes	60% vinyl wall covering, 40% paint		S.F. Surface	1.46	.97	
3020	Floor Finishes	60% carpet, 30% vinyl composition tile, 10% ceramic tile		S.F. Floor	8.44	8.44	
3030	Ceiling Finishes	Mineral fiber tile on concealed zee bars		S.F. Ceiling	6.86	6.86	
D.	**SERVICES**						
	D10 Conveying						
1010	Elevators & Lifts	Four geared passenger elevators		Each	304,000	15.20	11.4%
1020	Escalators & Moving Walks	N/A		—	—	—	
	D20 Plumbing						
2010	Plumbing Fixtures	Toilet and service fixtures, supply and drainage	1 Fixture/1370 S.F. Floor	Each	3685	2.69	
2020	Domestic Water Distribution	Gas fired water heater		S.F. Floor	.56	.56	2.7%
2040	Rain Water Drainage	Roof drains		S.F. Roof	2.32	.29	
	D30 HVAC						
3010	Energy Supply	N/A		—	—	—	
3020	Heat Generating Systems	Included in D3050		—	—	—	
3030	Cooling Generating Systems	N/A		—	—	—	12.9 %
3050	Terminal & Package Units	Multizone unit gas heating, electric cooling		S.F. Floor	17.25	17.25	
3090	Other HVAC Sys. & Equipment	N/A		—	—	—	
	D40 Fire Protection						
4010	Sprinklers	Wet pipe sprinkler system		S.F. Floor	2.85	2.85	2.9%
4020	Standpipes	Standpipes and hose systems		S.F. Floor	1.02	1.02	
	D50 Electrical						
5010	Electrical Service/Distribution	1600 ampere service, panel board and feeders		S.F. Floor	1.48	1.48	
5020	Lighting & Branch Wiring	High efficiency fluorescent fixtures, receptacles, switches, A.C. and misc. power		S.F. Floor	11.63	11.63	14.0%
5030	Communications & Security	Addressable alarm systems, internet and phone wiring, emergency lighting		S.F. Floor	4.48	4.48	
5090	Other Electrical Systems	Emergency generator, 100 kW, uninterruptible power supply		S.F. Floor	1.07	1.07	
E.	**EQUIPMENT & FURNISHINGS**						
1010	Commercial Equipment	N/A		—	—	—	
1020	Institutional Equipment	N/A		—	—	—	
1030	Vehicular Equipment	N/A		—	—	—	0.0 %
1090	Other Equipment	N/A		—	—	—	
F.	**SPECIAL CONSTRUCTION**						
1020	Integrated Construction	N/A		—	—	—	
1040	Special Facilities	N/A		—	—	—	0.0 %
G.	**BUILDING SITEWORK**	**N/A**					

		Sub-Total	133.52	**100%**
CONTRACTOR FEES (General Requirements: 10%, Overhead: 5%, Profit: 10%)		25%	33.37	
ARCHITECT FEES		6%	10.01	
		Total Building Cost	**176.90**	

Costs per square foot of floor area

Exterior Wall	S.F. Area	120000	145000	170000	200000	230000	260000	400000	600000	800000
	L.F. Perimeter	420	450	470	490	510	530	600	730	820
Double Glazed Heat Absorbing Tinted Plate Glass Panels	Steel Frame	202.25	193.85	187.00	180.60	175.95	172.30	161.70	155.90	152.25
	R/Conc. Frame	192.75	184.60	177.85	171.60	167.00	163.45	153.00	147.35	143.70
Face Brick with Concrete Block Back-up	Steel Frame	174.25	169.05	164.95	161.05	158.25	156.10	149.75	146.20	144.05
	R/Conc. Frame	195.30	190.25	186.25	182.50	179.75	177.65	171.50	168.05	165.95
Precast Concrete Panel With Exposed Aggregate	Steel Frame	181.25	175.30	170.45	166.00	162.70	160.15	152.75	148.65	146.10
	R/Conc. Frame	171.85	166.00	161.35	156.95	153.75	151.30	144.05	140.05	137.55
Perimeter Adj., Add or Deduct	Per 100 L.F.	13.95	11.60	9.90	8.40	7.35	6.55	4.20	2.85	2.10
Story Hgt. Adj., Add or Deduct	Per 1 Ft.	5.65	5.05	4.50	4.00	3.60	3.35	2.45	2.00	1.65

For Basement, add $38.65 per square foot of basement area

The above costs were calculated using the basic specifications shown on the facing page. These costs should be adjusted where necessary for design alternatives and owner's requirements. Reported completed project costs, for this type of structure, range from $101.25 to $247.05 per S.F.

Common additives

Description	Unit	$ Cost
Closed circuit surveillance, one station		
Camera and monitor	Each	1950
For additional camera stations, add	Each	1050
Directory boards, plastic, glass covered		
30" x 20"	Each	660
36" x 48"	Each	1475
Aluminum, 24" x 18"	Each	620
36" x 24"	Each	725
48" x 32"	Each	1050
48" x 60"	Each	2175
Electronic, wall mounted	S.F.	980
Free standing	S.F.	1100
Escalators, 10' rise, 32" wide, glass balustrade	Each	137,400
Metal balustrade	Each	143,400
48" wide, glass balustrade	Each	143,400
Metal balustrade	Each	150,900
Pedestal access floor system w/ plastic laminate cover		
Computer room, less than 6000 SF	S.F.	11.95
Greater than 6000 SF	S.F.	11.20
Office, greater than 6000 S.F.	S.F.	8

Description	Unit	$ Cost
Security access systems		
Metal detectors, wand type	Each	98
Walk-through portal type, single-zone	Each	4325
Multi-zone	Each	5400
X-ray equipment		
Desk top, for mail, small packages	Each	3850
Conveyer type, including monitor, minimum	Each	17,500
Maximum	Each	31,100
Explosive detection equipment		
Hand held, battery operated	Each	28,100
Walk-through portal type	Each	47,900
Uninterruptible power supply, 15 kVA/12.75 kW	kW	1735

Model costs calculated for a 16 story building with 10' story height and 260,000 square feet of floor area

				Unit	Unit Cost	Cost Per S.F.	% Of Sub-Total
A. SUBSTRUCTURE							
1010	Standard Foundations	CIP concrete pile caps		S.F. Ground	9.60	.60	
1020	Special Foundations	Steel H-piles, concrete grade beams		S.F. Ground	65	4.09	
1030	Slab on Grade	4" reinforced concrete with vapor barrier and granular base		S.F. Slab	5.18	.33	4.2%
2010	Basement Excavation	Site preparation for slab, piles and grade beams		S.F. Ground	.32	.02	
2020	Basement Walls	4' foundation wall		L.F. Wall	79	.38	
B. SHELL							
B10 Superstructure							
1010	Floor Construction	Concrete slab, metal deck, beams		S.F. Floor	27.65	25.92	20.3%
1020	Roof Construction	Metal deck, open web steel joists, beams, columns		S.F. Roof	8.48	.53	
B20 Exterior Enclosure							
2010	Exterior Walls	N/A		—	—	—	
2020	Exterior Windows	Double glazed heat absorbing, tinted plate glass wall panels	100% of wall	Each	73	23.93	18.9 %
2030	Exterior Doors	Double aluminum & glass doors		Each	6264	.67	
B30 Roofing							
3010	Roof Coverings	Single ply membrane, fully adhered; perlite/EPS composite insulation		S.F. Roof	5.44	.34	0.3%
3020	Roof Openings	N/A		—	—	—	
C. INTERIORS							
1010	Partitions	Gypsum board on metal studs	30 S.F. Floor/L.F. Partition	S.F. Partition	10.65	2.84	
1020	Interior Doors	Single leaf hollow metal	400 S.F. Floor/Door	Each	1212	3.03	
1030	Fittings	Toilet partitions		S.F. Floor	.41	.41	
2010	Stair Construction	Concrete filled metal pan		Flight	16,950	2.29	16.1%
3010	Wall Finishes	60% vinyl wall covering, 40% paint		S.F. Surface	1.44	.77	
3020	Floor Finishes	60% carpet tile, 30% vinyl composition tile, 10% ceramic tile		S.F. Floor	4.76	4.76	
3030	Ceiling Finishes	Mineral fiber tile on concealed zee bars		S.F. Ceiling	6.86	6.86	
D. SERVICES							
D10 Conveying							
1010	Elevators & Lifts	Four geared passenger elevators		Each	494,650	7.61	5.9%
1020	Escalators & Moving Walks	N/A		—	—	—	
D20 Plumbing							
2010	Plumbing Fixtures	Toilet and service fixtures, supply and drainage	1 Fixture/1345 S.F. Floor	Each	5501	4.09	
2020	Domestic Water Distribution	Oil fired water heater		S.F. Floor	.32	.32	3.5%
2040	Rain Water Drainage	Roof drains		S.F. Roof	2.88	.18	
D30 HVAC							
3010	Energy Supply	N/A		—	—	—	
3020	Heat Generating Systems	Boiler, heat exchanger and fans		Each	461,975	2.47	
3030	Cooling Generating Systems	Chilled water, fan coil units		S.F. Floor	16.01	16.01	14.2 %
3050	Terminal & Package Units	N/A		—	—	—	
3090	Other HVAC Sys. & Equipment	N/A		—	—	—	
D40 Fire Protection							
4010	Sprinklers	Sprinkler system, light hazard		S.F. Floor	2.76	2.76	2.5%
4020	Standpipes	Standpipes and hose systems		S.F. Floor	.51	.51	
D50 Electrical							
5010	Electrical Service/Distribution	2400 ampere service, panel board and feeders		S.F. Floor	.93	.93	
5020	Lighting & Branch Wiring	High efficiency fluorescent fixtures, receptacles, switches, A.C. and misc. power		S.F. Floor	11.51	11.51	14.1%
5030	Communications & Security	Addressable alarm systems, internet and phone wiring, emergency lighting		S.F. Floor	5.34	5.34	
5090	Other Electrical Systems	Emergency generator, 200 kW, uninterruptible power supply		S.F. Floor	.56	.56	
E. EQUIPMENT & FURNISHINGS							
1010	Commercial Equipment	N/A		—	—	—	
1020	Institutional Equipment	N/A		—	—	—	
1030	Vehicular Equipment	N/A		—	—	—	0.0 %
1090	Other Equipment	N/A		—	—	—	
F. SPECIAL CONSTRUCTION							
1020	Integrated Construction	N/A		—	—	—	0.0 %
1040	Special Facilities	N/A		—	—	—	
G. BUILDING SITEWORK	**N/A**						

Sub-Total	130.06	**100%**
CONTRACTOR FEES (General Requirements: 10%, Overhead: 5%, Profit: 10%)	25%	32.49
ARCHITECT FEES	6%	9.75
Total Building Cost	**172.30**	

Costs per square foot of floor area

Exterior Wall	S.F. Area	5000	7500	10000	12500	15000	17500	20000	22500	25000
	L.F. Perimeter	284	350	410	450	490	530	570	600	632
Brick Veneer	Wood Truss	433.45	420.35	413.20	407.35	403.50	400.75	398.65	396.55	395.00
Brick Veneer	Steel Frame	438.55	425.90	419.10	413.50	409.85	407.20	405.20	403.25	401.80
Decorative Concrete Block	Steel Frame	440.20	427.25	420.25	414.60	410.80	408.10	406.05	404.00	402.50
Decorative Concrete Block	Bearing Wall	429.55	421.25	416.85	413.40	411.10	409.50	408.25	407.05	406.20
Tiltup Concrete Panel	Steel Frame	433.40	421.75	415.45	410.35	407.00	404.55	402.70	400.90	399.60
EIFS on Concrete Block	Steel Frame	436.80	422.80	415.30	409.55	405.70	403.00	400.90	398.95	397.45
Perimeter Adj., Add or Deduct	Per 100 L.F.	19.35	12.85	9.65	7.75	6.45	5.50	4.85	4.30	3.85
Story Hgt. Adj., Add or Deduct	Per 1 Ft.	2.90	2.40	2.05	1.80	1.65	1.50	1.45	1.40	1.25

For Basement, add $30.40 per square foot of basement area

The above costs were calculated using the basic specifications shown on the facing page. These costs should be adjusted where necessary for design alternatives and owner's requirements. Reported completed project costs, for this type of structure, range from $177.30 to $919.40 per S.F.

Common additives

Description	Unit	$ Cost
Cabinets, base, door units, metal	L.F.	305
Drawer units	L.F.	600
Tall storage cabinets, open, 7' high	L.F.	570
With glazed doors	L.F.	845
Wall cabinets, metal, 12-1/2" deep, open	L.F.	228
With doors	L.F.	415
Counter top, laminated plastic, no backsplash	L.F.	77
Stainless steel counter top	L.F.	189
For drop-in stainless 43" x 21" sink, add	Ea.	1050
Nurse Call Station		
Single bedside call station	Ea.	330
Ceiling speaker station	Ea.	152
Emergency call station	Ea.	157
Pillow speaker	Ea.	272
Standard call button	Ea.	173
Master control station for 20 stations	Total	6250

Description	Unit	$ Cost
Directory Boards, Plastic, glass covered		
Plastic, glass covered, 30" x 20"	Ea.	660
36" x 48"	Ea.	1475
Building directory, alum., black felt panels, 1 door, 24" x 18"	Ea.	620
36" x 24"	Ea.	725
48" x 32"	Ea.	1050
48" x 60"	Ea.	2175
Tables, Examining, vinyl top		
with base cabinets	Ea.	1450
Utensil washer-sanitizer	Ea.	12,500
X-ray, mobile, minimum	Ea.	18,300

Important: See the Reference Section for Location Factors.

Model costs calculated for a 1 story building with 15'-4" story height and 12,500 square feet of floor area

Outpatient Surgery Center

				Unit	Unit Cost	Cost Per S.F.	% Of Sub-Total
A. SUBSTRUCTURE							
1010	Standard Foundations	Poured concrete; strip and spread footings		S.F. Ground	1.94	1.94	
1020	Special Foundations	N/A		—	—	—	
1030	Slab on Grade	4" reinforced concrete with vapor barrier and granular base		S.F. Slab	6.36	6.36	3.8%
2010	Basement Excavation	Site preparation for slab and trench for foundation wall and footing		S.F. Ground	.32	.32	
2020	Basement Walls	4' Foundation wall		L.F. Wall	79	2.84	
B. SHELL							
B10 Superstructure							
1010	Floor Construction	N/A		—	—	—	2.5 %
1020	Roof Construction	Steel Joists and girders on columns		S.F. Roof	7.62	7.62	
B20 Exterior Enclosure							
2010	Exterior Walls	Brick veneer	85% of wall	S.F. Wall	18.70	8.77	
2020	Exterior Windows	Aluminum frame, fixed	15% of wall	Each	815	4.49	6.1%
2030	Exterior Doors	Aluminum and glass entry; steel flush doors		Each	6569	5.25	
B30 Roofing							
3010	Roof Coverings	Single-ply membrane, loose laid and ballasted; polyisocyanurate insulation		S.F. Roof	6.29	6.29	2.1%
3020	Roof Openings	Roof hatches		S.F. Roof	.10	.10	
C. INTERIORS							
1010	Partitions	Gypsum board on metal studs	7 S.F. Floor/L.F. Partition	S.F. Partition	7.36	10.52	
1020	Interior Doors	Single and double leaf fire doors	200 S.F. Floor/Door	Each	1340	6.70	
1030	Fittings	Stainless steel toilet partitions, steel lockers		S.F. Floor	1.61	1.61	
2010	Stair Construction	N/A		—	—	—	12.9%
3010	Wall Finishes	80% Paint, 10% vinyl wall covering, 10% ceramic tile		S.F. Surface	1.95	5.58	
3020	Floor Finishes	10% Carpet, 80% vinyl sheet, 10% ceramic tile		S.F. Floor	10.38	10.38	
3030	Ceiling Finishes	Mineral fiber tile and plastic coated tile on suspension system		S.F. Ceiling	3.98	3.99	
D. SERVICES							
D10 Conveying							
1010	Elevators & Lifts	N/A		—	—	—	0.0 %
1020	Escalators & Moving Walks	N/A		—	—	—	
D20 Plumbing							
2010	Plumbing Fixtures	Medical, patient & specialty fixtures, supply and drainage	1 fixture/350 S.F. Floor	Each	3483	9.95	
2020	Domestic Water Distribution	Electric water heater		S.F. Floor	1.57	1.57	4.6%
2040	Rain Water Drainage	Roof drains		S.F. Roof	2.26	2.26	
D30 HVAC							
3010	Energy Supply	Hot water reheat for surgery		S.F. Floor	12.21	12.21	
3020	Heat Generating Systems	Boiler		Each	56,725	6.17	
3030	Cooling Generating Systems	N/A		—	—	—	36.8%
3050	Terminal & Package Units	N/A		—	—	—	
3090	Other HVAC Sys. & Equipment	Rooftop AC, CAV, roof vents, heat exchanger, surgical air curtains		S.F. Floor	92	92.47	
D40 Fire Protection							
4010	Sprinklers	Wet pipe sprinkler system		S.F. Floor	3.51	3.51	1.7%
4020	Standpipes	Standpipe		S.F. Floor	1.53	1.53	
D50 Electrical							
5010	Electrical Service/Distribution	1200 ampere service, panel board and feeders		S.F. Floor	7.16	7.16	
5020	Lighting & Branch Wiring	T-8 fluorescent light fixtures, receptacles, switches, A.C. and misc. power		S.F. Floor	16.60	16.60	13.5%
5030	Communications & Security	Addressable alarm system, internet wiring, communications system, emergency lighting		S.F. Floor	9.91	9.91	
5090	Other Electrical Systems	Emergency generator, 300 kW with fuel tank		S.F. Floor	6.89	6.89	
E. EQUIPMENT & FURNISHINGS							
1010	Commercial Equipment	N/A		—	—	—	
1020	Institutional Equipment	Medical gas system, cabinetry, scrub sink, sterilizer		S.F. Floor	47.19	47.19	16.0 %
1030	Vehicular Equipment	N/A		—	—	—	
1090	Other Equipment	Ice maker, washing machine		S.F. Floor	.99	.99	
F. SPECIAL CONSTRUCTION & DEMOLITION							
1020	Integrated Construction	N/A		—	—	—	0.0 %
1040	Special Facilities	N/A		—	—	—	
G. BUILDING SITEWORK	**N/A**						

	Sub-Total	301.17	**100%**
CONTRACTOR FEES (General Requirements: 10%, Overhead: 5%, Profit: 10%)	25%	75.30	
ARCHITECT FEES	9%	33.88	
	Total Building Cost	410.35	

Costs per square foot of floor area

Exterior Wall	S.F. Area	7000	9000	11000	13000	15000	17000	19000	21000	23000
	L.F. Perimeter	240	280	303	325	354	372	397	422	447
Limestone with Concrete Block Back-up	Bearing Walls	277.75	259.65	244.10	233.20	226.30	219.40	214.90	211.25	208.35
	R/Conc. Frame	290.40	272.90	258.20	247.85	241.35	234.80	230.55	227.05	224.30
Face Brick with Concrete Block Back-up	Bearing Walls	239.00	224.35	212.80	204.75	199.45	194.45	191.10	188.30	186.10
	R/Conc. Frame	259.85	245.20	233.65	225.60	220.30	215.30	211.90	209.15	207.00
Decorative Concrete Block	Bearing Walls	227.35	213.75	203.40	196.20	191.35	187.00	183.90	181.50	179.50
	R/Conc. Frame	249.25	235.65	225.30	218.10	213.25	208.90	205.85	203.40	201.40
Perimeter Adj., Add or Deduct	Per 100 L.F.	36.95	28.65	23.40	19.85	17.25	15.15	13.65	12.35	11.20
Story Hgt. Adj., Add or Deduct	Per 1 Ft.	6.65	6.00	5.30	4.80	4.60	4.25	4.05	3.85	3.70
For Basement, add $30.40 per square foot of basement area										

The above costs were calculated using the basic specifications shown on the facing page. These costs should be adjusted where necessary for design alternatives and owner's requirements. Reported completed project costs, for this type of structure, range from $116.35 to $303.45 per S.F.

Common additives

Description	Unit	$ Cost
Cells Prefabricated, 5'-6' wide,		
7'-8' high, 7'-8' deep	Each	12,600
Elevators, Hydraulic passenger, 2 stops		
1500# capacity	Each	66,000
2500# capacity	Each	68,900
3500# capacity	Each	72,900
Emergency Lighting, 25 watt, battery operated		
Lead battery	Each	305
Nickel cadmium	Each	850
Flagpoles, Complete		
Aluminum, 20' high	Each	1750
40' high	Each	3875
70' high	Each	10,600
Fiberglass, 23' high	Each	1325
39'-5" high	Each	3525
59' high	Each	7200

Description	Unit	$ Cost
Lockers, Steel, Single tier, 60" to 72"	Opening	218 - 355
2 tier, 60" or 72" total	Opening	112 - 160
5 tier, box lockers	Opening	70 - 86
Locker bench, lam. maple top only	L.F.	25.50
Pedestals, steel pipe	Each	72.50
Safe, Office type, 1 hour rating		
30" x 18" x 18"	Each	2500
60" x 36" x 18", double door	Each	8075
Shooting Range, Incl. bullet traps, target provisions, and contols, not incl. structural shell	Each	50,000
Smoke Detectors		
Ceiling type	Each	220
Duct type	Each	550
Sound System		
Amplifier, 250 watts	Each	1975
Speaker, ceiling or wall	Each	206
Trumpet	Each	395

Model costs calculated for a 2 story building with 12' story height and 11,000 square feet of floor area

				Unit	Unit Cost	Cost Per S.F.	% Of Sub-Total
A.	**SUBSTRUCTURE**						
1010	Standard Foundations	Poured concrete; strip and spread footings		S.F. Ground	2.82	1.41	
1020	Special Foundations	N/A		—	—	—	
1030	Slab on Grade	4" reinforced concrete with vapor barrier and granular base		S.F. Slab	5.18	2.60	3.9%
2010	Basement Excavation	Site preparation for slab and trench for foundation wall and footing		S.F. Ground	.32	.16	
2020	Basement Walls	4' foundation wall		L.F. Wall	79	2.90	
B.	**SHELL**						
	B10 Superstructure						
1010	Floor Construction	Open web steel joists, slab form, concrete		S.F. Floor	12	6	4.7%
1020	Roof Construction	Metal deck on open web steel joists		S.F. Roof	4.80	2.40	
	B20 Exterior Enclosure						
2010	Exterior Walls	Limestone with concrete block backup	80% of wall	S.F. Wall	67	35.17	
2020	Exterior Windows	Metal horizontal sliding	20% of wall	Each	1125	9.92	26.3%
2030	Exterior Doors	Hollow metal		Each	2694	1.97	
	B30 Roofing						
3010	Roof Coverings	Built-up tar and gravel with flashing; perlite/EPS composite insulation		S.F. Roof	7	3.50	2.0%
3020	Roof Openings	N/A		—	—	—	
C.	**INTERIORS**						
1010	Partitions	Concrete block	20 S.F. Floor/L.F. Partition	S.F. Partition	9.12	4.56	
1020	Interior Doors	Single leaf kalamein fire door	200 S.F. Floor/Door	Each	1212	6.07	
1030	Fittings	Toilet partitions		S.F. Floor	.89	.89	
2010	Stair Construction	Concrete filled metal pan		Flight	16,950	3.08	16.0%
3010	Wall Finishes	90% paint, 10% ceramic tile		S.F. Surface	3.69	3.69	
3020	Floor Finishes	70% vinyl composition tile, 20% carpet tile, 10% ceramic tile		S.F. Floor	3.55	3.55	
3030	Ceiling Finishes	Mineral fiber tile on concealed zee bars		S.F. Ceiling	6.86	6.86	
D.	**SERVICES**						
	D10 Conveying						
1010	Elevators & Lifts	One hydraulic passenger elevator		Each	83,380	7.58	4.2%
1020	Escalators & Moving Walks	N/A		—	—	—	
	D20 Plumbing						
2010	Plumbing Fixtures	Toilet and service fixtures, supply and drainage	1 Fixture/580 S.F. Floor	Each	4739	8.17	
2020	Domestic Water Distribution	Oil fired water heater		S.F. Floor	3.59	3.59	7.6%
2040	Rain Water Drainage	Roof drains		S.F. Roof	3.64	1.82	
	D30 HVAC						
3010	Energy Supply	N/A		—	—	—	
3020	Heat Generating Systems	N/A		—	—	—	
3030	Cooling Generating Systems	N/A		—	—	—	11.9 %
3050	Terminal & Package Units	Multizone HVAC air cooled system		S.F. Floor	21.25	21.25	
3090	Other HVAC Sys. & Equipment	N/A		—	—	—	
	D40 Fire Protection						
4010	Sprinklers	Wet pipe sprinkler system		S.F. Floor	3.74	3.74	2.6%
4020	Standpipes	Standpipe		S.F. Floor	.96	.96	
	D50 Electrical						
5010	Electrical Service/Distribution	400 ampere service, panel board and feeders		S.F. Floor	2.06	2.06	
5020	Lighting & Branch Wiring	T-8 fluorescent fixtures, receptacles, switches, A.C. and misc. power		S.F. Floor	12	12	12.0%
5030	Communications & Security	Addressable alarm systems, internet wiring, intercom and emergency lighting		S.F. Floor	7.13	7.13	
5090	Other Electrical Systems	Emergency generator, 15 kW		S.F. Floor	.24	.24	
E.	**EQUIPMENT & FURNISHINGS**						
1010	Commercial Equipment	N/A		—	—	—	
1020	Institutional Equipment	Lockers, detention rooms, cells, gasoline dispensers		S.F. Floor	13.41	13.41	8.9 %
1030	Vehicular Equipment	Gasoline dispenser system		S.F. Floor	2.45	2.45	
1090	Other Equipment	N/A		—	—	—	
F.	**SPECIAL CONSTRUCTION**						
1020	Integrated Construction	N/A		—	—	—	0.0 %
1040	Special Facilities	N/A		—	—	—	
G.	**BUILDING SITEWORK**	**N/A**					

			Sub-Total	179.13	100%
CONTRACTOR FEES (General Requirements: 10%, Overhead: 5%, Profit: 10%)			25%	44.82	
ARCHITECT FEES			9%	20.15	
		Total Building Cost		**244.10**	

Costs per square foot of floor area

Exterior Wall	S.F. Area	600	750	925	1150	1450	1800	2100	2600	3200
	L.F. Perimeter	100	112	128	150	164	182	200	216	240
Steel Siding on Wood Studs	Wood Truss	73.10	67.80	64.30	61.80	56.80	53.40	51.65	48.25	45.95
Aluminum Siding	Wood Truss	71.85	67.15	64.15	61.85	57.55	54.55	53.05	50.10	48.15
Wood Board Siding	Wood Truss	74.10	69.20	65.95	63.60	59.05	55.90	54.30	51.20	49.10
Plywood Siding	Wood Truss	71.50	66.85	63.80	61.55	57.30	54.30	52.85	49.90	47.95
Perimeter Adj., Add or Deduct	Per 100 L.F.	43.80	35.00	28.45	22.85	18.15	14.55	12.50	10.10	8.20
Story Hgt. Adj., Add or Deduct	Per 1 Ft.	3.15	2.75	2.65	2.40	2.15	1.85	1.75	1.55	1.40
Basement—Not Applicable										

The above costs were calculated using the basic specifications shown on the facing page. These costs should be adjusted where necessary for design alternatives and owner's requirements. Reported completed project costs, for this type of structure, range from $37.20 to $148.45 per S.F.

Common additives

Description	Unit	$ Cost
Slab on Grade		
4" thick, non-industrial, non-reinforced	S.F.	4.66
Reinforced	S.F.	5.18
Light industrial, non-reinforced	S.F.	5.84
Reinforced	S.F.	6.36
Industrial, non-reinforced	S.F.	10
Reinforced	S.F.	10.56
6" thick, non-industrial, non-reinforced	S.F.	5.44
Reinforced	S.F.	6.16
Light industrial, non-reinforced	S.F.	6.65
Reinforced	S.F.	7.67
Heavy industrial, non-reinforced	S.F.	12.83
Reinforced	S.F.	13.56
Insulation		
Fiberboard, low density, 1" thick, R2.78	S.F.	1.25
Foil on reinforced scrim, single bubble air space, R8.8	C.S.F.	68
Vinyl faced fiberglass, 1-1/2" thick, R5	S.F.	0.78
Vinyl/scrim/foil (VSF), 1-1/2" thick, R5	S.F.	0.93

Description	Unit	$ Cost
Roof ventilation, ridge vent, aluminum	L.F.	7.35
Mushroom vent, aluminum	Each	119
Gutter, aluminum, 5" K type, enameled	L.F.	8.25
Steel, half round or box, 5", enameled	L.F.	7.60
Wood, clear treated cedar, fir or hemlock, 4" x 5"	L.F.	20.50
Downspout, aluminum, 3" x 4", enameled	L.F.	7.80
Steel, round, 4" diameter	L.F.	7.35

Important: See the Reference Section for Location Factors.

Model costs calculated for a 1 story building with 14' story height and 1,800 square feet of floor area

Post Frame Barn

				Unit	Unit Cost	Cost Per S.F.	% Of Sub-Total
A. SUBSTRUCTURE							
1010	Standard Foundations	Included in B2010 Exterior Walls		—	—	—	
1020	Special Foundations	N/A		—	—	—	
1030	Slab on Grade	N/A		—	—	—	0.0%
2010	Basement Excavation	N/A		—	—	—	
2020	Basement Walls	N/A		—	—	—	
B. SHELL							
	B10 Superstructure						
1010	Floor Construction	N/A		—	—	—	
1020	Roof Construction	Wood truss, 4/12 pitch		S.F. Roof	5.82	5.82	14.7 %
	B20 Exterior Enclosure						
2010	Exterior Walls	2 x 6 Wood framing, 6 x 6 wood posts in concrete; steel siding panels		S.F. Wall	11.56	15.55	
2020	Exterior Windows	Steel windows, fixed		Each	523	4.11	60.2%
2030	Exterior Doors	Steel overhead, hollow metal doors		Each	1863	4.13	
	B30 Roofing						
3010	Roof Coverings	Metal panel roof		S.F. Roof	4.27	4.27	10.8%
3020	Roof Openings	N/A		—	—	—	
C. INTERIORS							
1010	Partitions	N/A		—	—	—	
1020	Interior Doors	N/A		—	—	—	
1030	Fittings	N/A		—	—	—	
2010	Stair Construction	N/A		—	—	—	0.0 %
3010	Wall Finishes	N/A		—	—	—	
3020	Floor Finishes	N/A		—	—	—	
3030	Ceiling Finishes	N/A		—	—	—	
D. SERVICES							
	D10 Conveying						
1010	Elevators & Lifts	N/A		—	—	—	0.0 %
1020	Escalators & Moving Walks	N/A		—	—	—	
	D20 Plumbing						
2010	Plumbing Fixtures	N/A		—	—	—	
2020	Domestic Water Distribution	N/A		—	—	—	0.0 %
2040	Rain Water Drainage	N/A		—	—	—	
	D30 HVAC						
3010	Energy Supply	N/A		—	—	—	
3020	Heat Generating Systems	N/A		—	—	—	
3030	Cooling Generating Systems	N/A		—	—	—	0.0 %
3050	Terminal & Package Units	N/A		—	—	—	
3090	Other HVAC Sys. & Equipment	N/A		—	—	—	
	D40 Fire Protection						
4010	Sprinklers	N/A		—	—	—	0.0 %
4020	Standpipes	N/A		—	—	—	
	D50 Electrical						
5010	Electrical Service/Distribution	60 Ampere service, panel board & feeders		S.F. Floor	1.93	1.93	
5020	Lighting & Branch Wiring	High bay fixtures, receptacles & switches		S.F. Floor	3.73	3.73	14.3%
5030	Communications & Security	N/A		—	—	—	
5090	Other Electrical Systems	N/A		—	—	—	
E. EQUIPMENT & FURNISHINGS							
1010	Commercial Equipment	N/A		—	—	—	
1020	Institutional Equipment	N/A		—	—	—	0.0 %
1030	Vehicular Equipment	N/A		—	—	—	
1090	Other Equipment	N/A		—	—	—	
F. SPECIAL CONSTRUCTION & DEMOLITION							
1020	Integrated Construction	N/A		—	—	—	0.0 %
1040	Special Facilities	N/A		—	—	—	
G. BUILDING SITEWORK	**N/A**						

		Sub-Total	39.54	100%
CONTRACTOR FEES (General Requirements: 10%, Overhead: 5%, Profit: 10%)		25%	9.91	
ARCHITECT FEES		8%	3.95	

Total Building Cost	**53.40**

Costs per square foot of floor area

Exterior Wall	S.F. Area	5000	7000	9000	11000	13000	15000	17000	19000	21000
	L.F. Perimeter	300	380	420	468	486	513	540	580	620
Face Brick with Concrete Block Back-up	Steel Frame	170.20	160.65	151.30	145.95	140.25	136.50	133.70	132.05	130.75
	Bearing Walls	167.65	158.15	148.75	143.45	137.70	134.00	131.15	129.60	128.25
Limestone with Concrete Block Back-up	Steel Frame	201.40	188.85	175.55	168.05	159.65	154.30	150.15	147.95	146.10
	Bearing Walls	198.60	186.10	172.75	165.30	156.85	151.50	147.40	145.20	143.35
Decorative Concrete Block	Steel Frame	158.25	149.90	142.00	137.55	132.80	129.75	127.35	126.00	124.90
	Bearing Walls	155.75	147.35	139.50	134.95	130.25	127.15	124.85	123.50	122.35
Perimeter Adj., Add or Deduct	Per 100 L.F.	18.25	13.05	10.20	8.35	7.00	6.10	5.35	4.90	4.35
Story Hgt. Adj., Add or Deduct	Per 1 Ft.	2.90	2.55	2.25	2.05	1.75	1.65	1.45	1.55	1.40
For Basement, add $29.60 per square foot of basement area										

The above costs were calculated using the basic specifications shown on the facing page. These costs should be adjusted where necessary for design alternatives and owner's requirements. Reported completed project costs, for this type of structure, range from $96.50 to $250.50 per S.F.

Common additives

Description	Unit	$ Cost
Closed Circuit Surveillance, One station		
Camera and monitor	Each	1950
For additional camera stations, add	Each	1050
Emergency Lighting, 25 watt, battery operated		
Lead battery	Each	305
Nickel cadmium	Each	850
Flagpoles, Complete		
Aluminum, 20' high	Each	1750
40' high	Each	3875
70' high	Each	10,600
Fiberglass, 23' high	Each	1325
39'-5" high	Each	3525
59' high	Each	7200

Description	Unit	$ Cost
Mail Boxes, Horizontal, key lock, 15" x 6" x 5"	Each	55
Double 15" x 12" x 5"	Each	99.50
Quadruple 15" x 12" x 10"	Each	136
Vertical, 6" x 5" x 15", aluminum	Each	49.50
Bronze	Each	60.50
Steel, enameled	Each	49.50
Scales, Dial type, 5 ton cap.		
8' x 6' platform	Each	11,600
9' x 7' platform	Each	14,600
Smoke Detectors		
Ceiling type	Each	220
Duct type	Each	550

Model costs calculated for a 1 story building with 14' story height and 13,000 square feet of floor area

					Unit	Unit Cost	Cost Per S.F.	% Of Sub-Total
A. SUBSTRUCTURE								
1010	Standard Foundations	Poured concrete; strip and spread footings			S.F. Ground	1.98	1.98	
1020	Special Foundations	N/A			—	—	—	
1030	Slab on Grade	4" reinforced concrete with vapor barrier and granular base			S.F. Slab	5.18	5.18	11.0%
2010	Basement Excavation	Site preparation for slab and trench for foundation wall and footing			S.F. Ground	.32	.32	
2020	Basement Walls	4' foundation wall			L.F. Wall	79	3.84	
B. SHELL								
B10 Superstructure								
1010	Floor Construction	Steel column fireproofing			L.F. Column	33.61	.10	
1020	Roof Construction	Metal deck, open web steel joists, columns			S.F. Roof	9.10	9.10	8.9%
B20 Exterior Enclosure								
2010	Exterior Walls	Face brick with concrete block backup		80% of wall	S.F. Wall	32.46	13.59	
2020	Exterior Windows	Double strength window glass		20% of wall	Each	735	3.35	17.7%
2030	Exterior Doors	Double aluminum & glass, single aluminum, hollow metal, steel overhead			Each	2682	1.24	
B30 Roofing								
3010	Roof Coverings	Built-up tar and gravel with flashing; perlite/EPS composite insulation			S.F. Roof	6.28	6.28	6.1%
3020	Roof Openings	N/A			—	—	—	
C. INTERIORS								
1010	Partitions	Concrete block		15 S.F. Floor/L.F. Partition	S.F. Partition	9.11	7.29	
1020	Interior Doors	Single leaf hollow metal		150 S.F. Floor/Door	Each	1212	8.08	
1030	Fittings	Toilet partitions, cabinets, shelving, lockers			S.F. Floor	1.39	1.39	
2010	Stair Construction	N/A			—	—	—	23.4%
3010	Wall Finishes	Paint			S.F. Surface	2.29	3.66	
3020	Floor Finishes	50% vinyl tile, 50% paint			S.F. Floor	1.92	1.92	
3030	Ceiling Finishes	Mineral fiber tile on concealed zee bars		25% of area	S.F. Ceiling	6.86	1.72	
D. SERVICES								
D10 Conveying								
1010	Elevators & Lifts	N/A			—	—	—	0.0 %
1020	Escalators & Moving Walks	N/A			—	—	—	
D20 Plumbing								
2010	Plumbing Fixtures	Toilet and service fixtures, supply and drainage		1 Fixture/1180 S.F. Floor	Each	2419	2.05	
2020	Domestic Water Distribution	Gas fired water heater			S.F. Floor	.41	.41	3.7%
2040	Rain Water Drainage	Roof drains			S.F. Roof	1.33	1.33	
D30 HVAC								
3010	Energy Supply	N/A			—	—	—	
3020	Heat Generating Systems	Included in D3050			—	—	—	
3030	Cooling Generating Systems	N/A			—	—	—	8.4 %
3050	Terminal & Package Units	Single zone, gas heating, electric cooling			S.F. Floor	8.60	8.60	
3090	Other HVAC Sys. & Equipment	N/A			—	—	—	
D40 Fire Protection								
4010	Sprinklers	Wet pipe sprinkler system			S.F. Floor	4.53	4.53	5.2%
4020	Standpipes	Standpipe			S.F. Floor	.86	.86	
D50 Electrical								
5010	Electrical Service/Distribution	600 ampere service, panel board and feeders			S.F. Floor	2.56	2.56	
5020	Lighting & Branch Wiring	T-8 fluorescent fixtures, receptacles, switches, A.C. and misc. power			S.F. Floor	8.88	8.88	15.6%
5030	Communications & Security	Addressable alarm systems, internet wiring and emergency lighting			S.F. Floor	4.51	4.51	
5090	Other Electrical Systems	Emergency generator, 15 kW			S.F. Floor	.14	.14	
E. EQUIPMENT & FURNISHINGS								
1010	Commercial Equipment	N/A			—	—	—	
1020	Institutional Equipment	N/A			—	—	—	0.0 %
1030	Vehicular Equipment	N/A			—	—	—	
1090	Other Equipment	N/A			—	—	—	
F. SPECIAL CONSTRUCTION								
1020	Integrated Construction	N/A			—	—	—	0.0 %
1040	Special Facilities	N/A			—	—	—	
G. BUILDING SITEWORK	**N/A**							

							Sub-Total	102.91	100%
CONTRACTOR FEES (General Requirements: 10%, Overhead: 5%, Profit: 10%)							25%	25.76	
ARCHITECT FEES							9%	11.58	

Total Building Cost   **140.25**

Costs per square foot of floor area

Exterior Wall	S.F. Area	5000	10000	15000	21000	25000	30000	40000	50000	60000
	L.F. Perimeter	287	400	500	600	660	700	834	900	1000
Face Brick with Concrete Block Back-up	Steel Frame	243.10	209.65	197.20	188.65	184.85	179.75	175.40	170.70	168.50
	Bearing Walls	229.40	201.65	191.15	183.70	180.30	175.60	171.75	167.35	165.30
Concrete Block	Steel Frame	209.30	186.05	177.55	171.80	169.30	165.95	163.15	160.15	158.65
Brick Veneer	Steel Frame	232.75	202.40	191.15	183.45	180.05	175.50	171.65	167.50	165.50
Galvanized Steel Siding	Steel Frame	198.30	178.10	170.75	165.85	163.65	160.95	158.45	156.00	154.75
Metal Sandwich Panel	Steel Frame	250.30	214.65	201.35	192.20	188.10	182.65	178.00	173.00	170.60
Perimeter Adj., Add or Deduct	Per 100 L.F.	30.05	15.00	10.05	7.15	5.95	5.00	3.75	3.05	2.50
Story Hgt. Adj., Add or Deduct	Per 1 Ft.	6.10	4.25	3.50	3.00	2.80	2.45	2.20	1.95	1.80
For Basement, add $28.50 per square foot of basement area										

The above costs were calculated using the basic specifications shown on the facing page. These costs should be adjusted where necessary for design alternatives and owner's requirements. Reported completed project costs, for this type of structure, range from $91.45 to $285.10 per S.F.

Common additives

Description	Unit	$ Cost
Bar, Front Bar	L.F.	410
Back Bar	L.F.	325
Booth, Upholstered, custom straight	L.F.	237 - 435
"L" or "U" shaped	L.F.	245 - 415
Bleachers, Telescoping, manual		
To 15 tier	Seat	134 - 187
21-30 tier	Seat	294 - 355
Courts		
Ceiling	Court	9650
Floor	Court	16,200
Walls	Court	27,700
Emergency Lighting, 25 watt, battery operated		
Lead battery	Each	305
Nickel cadmium	Each	850
Kitchen Equipment		
Broiler	Each	4175
Cooler, 6 ft. long, reach-in	Each	4000
Dishwasher, 10-12 racks per hr.	Each	5300
Food warmer, counter 1.2 KW	Each	560
Freezer, reach-in, 44 C.F.	Each	4150
Ice cube maker, 50 lb. per day	Each	1925

Description	Unit	$ Cost
Lockers, Steel, single tier, 60" or 72"	Opening	218 - 355
2 tier, 60" or 72" total	Opening	112 - 160
5 tier, box lockers	Opening	70 - 86
Locker bench, lam. maple top only	L.F.	25.50
Pedestals, steel pipe	Each	72.50
Sauna, Prefabricated, complete		
6' x 4'	Each	6225
6' x 9'	Each	8975
8' x 8'	Each	10,600
8' x 10'	Each	11,900
10' x 12'	Each	15,100
Sound System		
Amplifier, 250 watts	Each	1975
Speaker, ceiling or wall	Each	206
Trumpet	Each	395
Steam Bath, Complete, to 140 C.F.	Each	2975
To 300 C.F.	Each	3350
To 800 C.F.	Each	10,600
To 2500 C.F.	Each	13,300

Important: See the Reference Section for Location Factors.

Model costs calculated for a 2 story building with 12' story height and 30,000 square feet of floor area

Racquetball Court

					Unit	Unit Cost	Cost Per S.F.	% Of Sub-Total
A. SUBSTRUCTURE								
1010	Standard Foundations	Poured concrete; strip and spread footings			S.F. Ground	3.42	1.71	
1020	Special Foundations	N/A			—	—	—	
1030	Slab on Grade	5" reinforced concrete with vapor barrier and granular base			S.F. Slab	6.78	4.41	6.1%
2010	Basement Excavation	Site preparation for slab and trench for foundation wall and footing			S.F. Ground	.32	.21	
2020	Basement Walls	4' foundation wall			L.F. Wall	79	1.84	
B. SHELL								
	B10 Superstructure							
1010	Floor Construction	Open web steel joists, slab form, concrete, columns	50% of area		S.F. Floor	18.36	9.18	10.6%
1020	Roof Construction	Metal deck on open web steel joists, columns			S.F. Roof	9.92	4.96	
	B20 Exterior Enclosure							
2010	Exterior Walls	Face brick with concrete block backup	95% of wall		S.F. Wall	31.50	16.76	
2020	Exterior Windows	Storefront	5% of wall		S.F. Window	100	2.80	15.0%
2030	Exterior Doors	Aluminum and glass and hollow metal			Each	3108	.42	
	B30 Roofing							
3010	Roof Coverings	Built-up tar and gravel with flashing; perlite/EPS composite insulation			S.F. Roof	7.86	3.93	3.0%
3020	Roof Openings	Roof hatches			S.F. Roof	.12	.06	
C. INTERIORS								
1010	Partitions	Concrete block, gypsum board on metal studs	25 S.F. Floor/L.F. Partition		S.F. Partition	13.98	5.59	
1020	Interior Doors	Single leaf hollow metal	810 S.F. Floor/Door		Each	1212	1.50	
1030	Fittings	Toilet partitions			S.F. Floor	.36	.36	
2010	Stair Construction	Concrete filled metal pan			Flight	14,150	1.42	13.0%
3010	Wall Finishes	Paint			S.F. Surface	1.41	1.13	
3020	Floor Finishes	80% carpet tile, 20% ceramic tile	50% of floor area		S.F. Floor	6.28	3.14	
3030	Ceiling Finishes	Mineral fiber tile on concealed zee bars	60% of area		S.F. Ceiling	6.86	4.12	
D. SERVICES								
	D10 Conveying							
1010	Elevators & Lifts	N/A			—	—	—	0.0 %
1020	Escalators & Moving Walks	N/A			—	—	—	
	D20 Plumbing							
2010	Plumbing Fixtures	Kitchen, bathroom and service fixtures, supply and drainage	1 Fixture/1000 S.F. Floor		Each	3990	3.99	
2020	Domestic Water Distribution	Gas fired water heater			S.F. Floor	3.22	3.22	5.7%
2040	Rain Water Drainage	Roof drains			S.F. Roof	.76	.38	
	D30 HVAC							
3010	Energy Supply	N/A			—	—	—	
3020	Heat Generating Systems	Included in D3050			—	—	—	
3030	Cooling Generating Systems	N/A			—	—	—	21.8 %
3050	Terminal & Package Units	Multizone unit, gas heating, electric cooling			S.F. Floor	29.05	29.05	
3090	Other HVAC Sys. & Equipment	N/A			—	—	—	
	D40 Fire Protection							
4010	Sprinklers	Sprinklers, light hazard			S.F. Floor	3.30	3.30	2.7%
4020	Standpipes	Standpipe			S.F. Floor	.35	.35	
	D50 Electrical							
5010	Electrical Service/Distribution	400 ampere service, panel board and feeders			S.F. Floor	.93	.93	
5020	Lighting & Branch Wiring	Fluorescent and high intensity discharge fixtures, receptacles, switches, A.C. and misc. power			S.F. Floor	8.01	8.01	7.6%
5030	Communications & Security	Addressable alarm systems and emergency lighting			S.F. Floor	1.07	1.07	
5090	Other Electrical Systems	Emergency generator, 15 kW			S.F. Floor	.06	.06	
E. EQUIPMENT & FURNISHINGS								
1010	Commercial Equipment	N/A			—	—	—	
1020	Institutional Equipment	N/A			—	—	—	
1030	Vehicular Equipment	N/A			—	—	—	14.5 %
1090	Other Equipment	Courts, sauna baths			S.F. Floor	19.24	19.24	
F. SPECIAL CONSTRUCTION								
1020	Integrated Construction	N/A			—	—	—	0.0 %
1040	Special Facilities	N/A			—	—	—	
G. BUILDING SITEWORK	**N/A**							

		Sub-Total	133.14	100%
CONTRACTOR FEES (General Requirements: 10%, Overhead: 5%, Profit: 10%)		25%	33.30	
ARCHITECT FEES		8%	13.31	

Total Building Cost	**179.75**

Costs per square foot of floor area

Exterior Wall	S.F. Area	5000	6000	7000	8000	9000	10000	11000	12000	13000
	L.F. Perimeter	286	320	353	386	397	425	454	460	486
Face Brick with Concrete Block Back-up	Steel Joists	189.95	185.40	182.15	179.70	175.70	173.90	172.60	169.85	168.80
	Wood Joists	191.35	186.60	183.15	180.55	176.40	174.50	173.15	170.20	169.15
Stucco on Concrete Block	Steel Joists	191.65	187.55	184.55	182.30	178.80	177.15	176.00	173.55	172.70
	Wood Joists	180.10	176.15	173.25	171.10	167.70	166.15	165.05	162.70	161.80
Limestone with Concrete Block Back-up	Steel Joists	229.10	222.45	217.50	213.80	207.50	204.80	202.85	198.40	196.90
	Wood Joists	218.40	211.85	207.00	203.45	197.25	194.60	192.70	188.35	186.80
Perimeter Adj., Add or Deduct	Per 100 L.F.	16.75	14.00	11.95	10.45	9.30	8.45	7.60	6.95	6.45
Story Hgt. Adj., Add or Deduct	Per 1 Ft.	2.90	2.75	2.55	2.45	2.25	2.20	2.15	1.90	1.90

For Basement, add $30.00 per square foot of basement area

The above costs were calculated using the basic specifications shown on the facing page. These costs should be adjusted where necessary for design alternatives and owner's requirements. Reported completed project costs, for this type of structure, range from $79.85 to $247.20 per S.F.

Common additives

Description	Unit	$ Cost
Carrels Hardwood	Each	855 - 1625
Emergency Lighting, 25 watt, battery operated		
Lead battery	Each	305
Nickel cadmium	Each	850
Flagpoles, Complete		
Aluminum, 20' high	Each	1750
40' high	Each	3875
70' high	Each	10,600
Fiberglass, 23' high	Each	1325
39'-5" high	Each	3525
59' high	Each	7200
Gym Floor, Incl. sleepers and finish, maple	S.F.	14.15
Intercom System, 25 Station capacity		
Master station	Each	3125
Intercom outlets	Each	183
Handset	Each	500

Description	Unit	$ Cost
Lockers, Steel, single tier, 60" to 72"	Opening	218 - 355
2 tier, 60" to 72" total	Opening	112 - 160
5 tier, box lockers	Opening	70 - 86
Locker bench, lam. maple top only	L.F.	25.50
Pedestals, steel pipe	Each	72.50
Seating		
Auditorium chair, all veneer	Each	265
Veneer back, padded seat	Each	335
Upholstered, spring seat	Each	320
Classroom, movable chair & desk	Set	81 - 171
Lecture hall, pedestal type	Each	257 - 750
Smoke Detectors		
Ceiling type	Each	220
Duct type	Each	550
Sound System		
Amplifier, 250 watts	Each	1975
Speaker, ceiling or wall	Each	206
Trumpet	Each	395
Swimming Pools, Complete, gunite	S.F.	85 - 105

Model costs calculated for a 1 story building with 12' story height and 10,000 square feet of floor area

				Unit	Unit Cost	Cost Per S.F.	% Of Sub-Total
A. SUBSTRUCTURE							
1010	Standard Foundations	Poured concrete; strip and spread footings		S.F. Ground	2.38	2.38	
1020	Special Foundations	N/A		—	—	—	
1030	Slab on Grade	4" reinforced concrete with vapor barrier and granular base		S.F. Slab	5.18	5.18	8.8%
2010	Basement Excavation	Site preparation for slab and trench for foundation wall and footing		S.F. Ground	.32	.32	
2020	Basement Walls	4' foundation wall		L.F. Wall	79	3.36	
B. SHELL							
	B10 Superstructure						
1010	Floor Construction	N/A		—	—	—	
1020	Roof Construction	Metal deck, open web steel joists, beams, interior columns		S.F. Roof	9.21	9.21	7.2%
	B20 Exterior Enclosure						
2010	Exterior Walls	Face brick with concrete block backup	85% of wall	S.F. Wall	32.46	14.07	
2020	Exterior Windows	Window wall	15% of wall	Each	46.90	3.58	15.8%
2030	Exterior Doors	Double aluminum and glass		Each	6325	2.54	
	B30 Roofing						
3010	Roof Coverings	Built-up tar and gravel with flashing; perlite/EPS composite insulation		S.F. Roof	6.28	6.28	4.9%
3020	Roof Openings	N/A		—	—	—	
C. INTERIORS							
1010	Partitions	Concrete block	8 S.F. Floor/S.F. Partition	S.F. Partition	9.11	11.39	
1020	Interior Doors	Single leaf hollow metal	700 S.F. Floor/Door	Each	1212	1.73	
1030	Fittings	Toilet partitions		S.F. Floor	1.10	1.10	
2010	Stair Construction	N/A		—	—	—	26.3%
3010	Wall Finishes	Paint		S.F. Surface	2.35	5.87	
3020	Floor Finishes	50% vinyl tile, 50% carpet		S.F. Floor	6.62	6.62	
3030	Ceiling Finishes	Mineral fiber tile on concealed zee bars		S.F. Ceiling	6.86	6.86	
D. SERVICES							
	D10 Conveying						
1010	Elevators & Lifts	N/A		—	—	—	
1020	Escalators & Moving Walks	N/A		—	—	—	0.0%
	D20 Plumbing						
2010	Plumbing Fixtures	Toilet and service fixtures, supply and drainage	1 Fixture/455 S.F. Floor	Each	3758	8.26	
2020	Domestic Water Distribution	Gas fired water heater		S.F. Floor	1.56	1.56	8.5%
2040	Rain Water Drainage	Roof drains		S.F. Roof	1.05	1.05	
	D30 HVAC						
3010	Energy Supply	N/A		—	—	—	
3020	Heat Generating Systems	N/A		—	—	—	
3030	Cooling Generating Systems	N/A		—	—	—	19.4%
3050	Terminal & Package Units	Multizone HVAC air cooled system		S.F. Floor	24.70	24.70	
3090	Other HVAC Sys. & Equipment	N/A		—	—	—	
	D40 Fire Protection						
4010	Sprinklers	Sprinkler, light hazard		S.F. Floor	3.51	3.51	3.2%
4020	Standpipes	Standpipe		S.F. Floor	.55	.55	
	D50 Electrical						
5010	Electrical Service/Distribution	200 ampere service, panel board and feeders		S.F. Floor	1.13	1.13	
5020	Lighting & Branch Wiring	T-8 fluorescent fixtures, receptacles, switches, A.C. and misc. power		S.F. Floor	2.83	2.83	5.9%
5030	Communications & Security	Addressable alarm systems, partial internet wiring and emergency lighting		S.F. Floor	3.46	3.46	
5090	Other Electrical Systems	Emergency generator, 15 kW		S.F. Floor	.09	.09	
E. EQUIPMENT & FURNISHINGS							
1010	Commercial Equipment	N/A		—	—	—	
1020	Institutional Equipment	N/A		—	—	—	
1030	Vehicular Equipment	N/A		—	—	—	0.0%
1090	Other Equipment	N/A		—	—	—	
F. SPECIAL CONSTRUCTION							
1020	Integrated Construction	N/A		—	—	—	0.0%
1040	Special Facilities	N/A		—	—	—	
G. BUILDING SITEWORK	**N/A**						

		Sub-Total	127.63	**100%**
CONTRACTOR FEES (General Requirements: 10%, Overhead: 5%, Profit: 10%)			25%	31.91
ARCHITECT FEES			9%	14.36
		Total Building Cost	**173.90**	

Costs per square foot of floor area

Exterior Wall	S.F. Area	2000	2800	3500	4200	5000	5800	6500	7200	8000
	L.F. Perimeter	180	212	240	268	300	314	336	344	368
Wood Siding	Wood Frame	240.80	228.15	221.75	217.65	214.30	210.10	208.15	205.50	204.05
Brick Veneer	Wood Frame	253.60	238.95	231.55	226.75	222.85	217.85	215.50	212.30	210.60
Face Brick with Concrete Block Back-up	Wood Joists	261.90	245.95	237.90	232.65	228.40	222.80	220.30	216.65	214.85
	Steel Joists	258.25	240.50	231.65	225.80	221.15	214.80	212.00	207.95	205.90
Stucco on Concrete Block	Wood Joists	247.65	233.95	227.00	222.50	218.90	214.25	212.10	209.10	207.55
	Steel Joists	235.35	221.60	214.70	210.25	206.55	201.95	199.80	196.80	195.25
Perimeter Adj., Add or Deduct	Per 100 L.F.	27.90	19.95	16.05	13.25	11.15	9.65	8.60	7.75	7.00
Story Hgt. Adj., Add or Deduct	Per 1 Ft.	2.85	2.40	2.25	2.05	1.95	1.70	1.65	1.50	1.45

For Basement, add $34.70 per square foot of basement area

The above costs were calculated using the basic specifications shown on the facing page. These costs should be adjusted where necessary for design alternatives and owner's requirements. Reported completed project costs, for this type of structure, range from $123.75 to $288.40 per S.F.

Common additives

Description	Unit	$ Cost	Description	Unit	$ Cost
Bar, Front Bar	L.F.	410	Fireplace, Brick, not incl. chimney or foundation		
Back bar	L.F.	325	30" x 29" opening	Each	3050
Booth, Upholstered, custom straight	L.F.	237 - 435	Chimney, standard brick		
"L" or "U" shaped	L.F.	245 - 415	Single flue, 16" x 20"	V.L.F.	103
Cupola, Stock unit, redwood			20" x 20"	V.L.F.	125
30" square, 37" high, aluminum roof	Each	715	2 Flue, 20" x 24"	V.L.F.	150
Copper roof	Each	400	20" x 32"	V.L.F.	191
Fiberglass, 5'-0" base, 63" high	Each	3625 - 4500	Kitchen Equipment		
6'-0" base, 63" high	Each	5000 - 7275	Broiler	Each	4175
Decorative Wood Beams, Non load bearing			Coffee urn, twin 6 gallon	Each	2625
Rough sawn, 4" x 6"	L.F.	8.45	Cooler, 6 ft. long	Each	4000
4" x 8"	L.F.	9.55	Dishwasher, 10-12 racks per hr.	Each	5300
4" x 10"	L.F.	11.40	Food warmer, counter, 1.2 KW	Each	560
4" x 12"	L.F.	12.95	Freezer, 44 C.F., reach-in	Each	4150
8" x 8"	L.F.	13.75	Ice cube maker, 50 lb. per day	Each	1925
Emergency Lighting, 25 watt, battery operated			Range with 1 oven	Each	3325
Lead battery	Each	305	Refrigerators, Prefabricated, walk-in		
Nickel cadmium	Each	850	7'-6" high, 6' x 6'	S.F.	190
			10' x 10'	S.F.	150
			12' x 14'	S.F.	133
			12' x 20'	S.F.	117

Important: See the Reference Section for Location Factors.

Model costs calculated for a 1 story building with 12' story height and 5,000 square feet of floor area

					Unit	Unit Cost	Cost Per S.F.	% Of Sub-Total
A. SUBSTRUCTURE								
1010	Standard Foundations	Poured concrete; strip and spread footings			S.F. Ground	2.59	2.59	
1020	Special Foundations	N/A			—	—	—	
1030	Slab on Grade	4" reinforced concrete with vapor barrier and granular base			S.F. Slab	5.18	5.18	8.2%
2010	Basement Excavation	Site preparation for slab and trench for foundation wall and footing			S.F. Ground	.55	.55	
2020	Basement Walls	4' foundation wall			L.F. Wall	79	4.74	
B. SHELL								
B10 Superstructure								
1010	Floor Construction	Wood columns			S.F. Floor	.33	.83	5.1%
1020	Roof Construction	Plywood on wood truss (pitched)			S.F. Roof	6.55	7.34	
B20 Exterior Enclosure								
2010	Exterior Walls	Cedar siding on wood studs with insulation	70% of wall		S.F. Wall	11.55	5.82	
2020	Exterior Windows	Storefront windows	30% of wall		Each	42.70	9.23	12.7%
2030	Exterior Doors	Aluminum and glass doors and entrance with transom			Each	5272	5.28	
B30 Roofing								
3010	Roof Coverings	Cedar shingles with flashing (pitched); rigid fiberglass insulation			S.F. Roof	19.35	8.16	5.1%
3020	Roof Openings	Skylights			S.F. Roof	.09	.09	
C. INTERIORS								
1010	Partitions	Gypsum board on wood studs	25 S.F. Floor/L.F. Partition		S.F. Partition	9.98	3.99	
1020	Interior Doors	Hollow core wood	250 S.F. Floor/Door		Each	543	2.17	
1030	Fittings	Toilet partitions			S.F. Floor	.70	.70	
2010	Stair Construction	N/A			—	—	—	15.2%
3010	Wall Finishes	75% paint, 25% ceramic tile			S.F. Surface	2.39	1.91	
3020	Floor Finishes	65% carpet, 35% quarry tile			S.F. Floor	8.78	8.78	
3030	Ceiling Finishes	Mineral fiber tile on concealed zee bars			S.F. Ceiling	6.86	6.86	
D. SERVICES								
D10 Conveying								
1010	Elevators & Lifts	N/A			—	—	—	0.0 %
1020	Escalators & Moving Walks	N/A			—	—	—	
D20 Plumbing								
2010	Plumbing Fixtures	Kitchen, bathroom and service fixtures, supply and drainage	1 Fixture/355 S.F. Floor		Each	4033	11.36	
2020	Domestic Water Distribution	Gas fired water heater			S.F. Floor	8.25	8.25	13.4%
2040	Rain Water Drainage	Roof drains			S.F. Roof	1.79	1.79	
D30 HVAC								
3010	Energy Supply	N/A			—	—	—	
3020	Heat Generating Systems	Included in D3050			—	—	—	
3030	Cooling Generating Systems	N/A			—	—	—	22.6 %
3050	Terminal & Package Units	Multizone unit, gas heating, electric cooling; kitchen ventilation			S.F. Floor	36.23	36.23	
3090	Other HVAC Sys. & Equipment	N/A			—	—	—	
D40 Fire Protection								
4010	Sprinklers	Sprinklers, light hazard and ordinary hazard kitchen			S.F. Floor	8.32	8.32	6.4%
4020	Standpipes	Standpipe			S.F. Floor	1.92	1.92	
D50 Electrical								
5010	Electrical Service/Distribution	400 ampere service, panel board and feeders			S.F. Floor	4.76	4.76	
5020	Lighting & Branch Wiring	T-8 fluorescent fixtures, receptacles, switches, A.C. and misc. power			S.F. Floor	9.50	9.50	11.3%
5030	Communications & Security	Addressable alarm systems and emergency lighting			S.F. Floor	3.51	3.51	
5090	Other Electrical Systems	Emergency generator, 15 kW			S.F. Floor	.36	.36	
E. EQUIPMENT & FURNISHINGS								
1010	Commercial Equipment	N/A			—	—	—	
1020	Institutional Equipment	N/A			—	—	—	0.0 %
1030	Vehicular Equipment	N/A			—	—	—	
1090	Other Equipment	N/A			—	—	—	
F. SPECIAL CONSTRUCTION								
1020	Integrated Construction	N/A			—	—	—	0.0 %
1040	Special Facilities	N/A			—	—	—	
G. BUILDING SITEWORK	**N/A**							

			Sub-Total	160.22	**100%**
CONTRACTOR FEES (General Requirements: 10%, Overhead: 5%, Profit: 10%)			25%	40.06	
ARCHITECT FEES			7%	14.02	

Total Building Cost	**214.30**

Costs per square foot of floor area

Exterior Wall	S.F. Area	2000	2800	3500	4000	5000	5800	6500	7200	8000
	L.F. Perimeter	180	212	240	260	300	314	336	344	368
Face Brick with Concrete Block Back-up	Bearing Walls	237.95	227.50	222.25	219.60	215.95	211.65	209.85	207.05	205.70
	Steel Frame	241.20	230.70	225.45	222.90	219.20	214.95	213.10	210.25	208.95
Concrete Block With Stucco	Bearing Walls	225.15	216.55	212.25	210.20	207.15	203.70	202.20	199.85	198.75
	Steel Frame	229.40	221.00	216.80	214.70	211.75	208.35	206.85	204.60	203.55
Wood Siding	Wood Frame	218.40	210.80	206.95	205.10	202.40	199.35	198.00	195.90	194.95
Brick Veneer	Steel Frame	235.00	225.75	221.10	218.80	215.55	211.75	210.10	207.60	206.45
Perimeter Adj., Add or Deduct	Per 100 L.F.	36.55	26.10	20.90	18.35	14.60	12.65	11.30	10.15	9.20
Story Hgt. Adj., Add or Deduct	Per 1 Ft.	4.80	4.00	3.65	3.50	3.25	2.90	2.80	2.50	2.50
Basement—Not Applicable										

The above costs were calculated using the basic specifications shown on the facing page. These costs should be adjusted where necessary for design alternatives and owner's requirements. Reported completed project costs, for this type of structure, range from $123.50 to $240.80 per S.F.

Common additives

Description	Unit	$ Cost	Description	Unit	$ Cost
Bar, Front Bar	L.F.	410	Refrigerators, Prefabricated, walk-in		
Back bar	L.F.	325	7'-6" High, 6' x 6'	S.F.	190
Booth, Upholstered, custom straight	L.F.	237 - 435	10' x 10'	S.F.	150
"L" or "U" shaped	L.F.	245 - 415	12' x 14'	S.F.	133
Drive-up Window	Each	8650 - 12,600	12' x 20'	S.F.	117
Emergency Lighting, 25 watt, battery operated			Serving		
Lead battery	Each	305	Counter top (Stainless steel)	L.F.	189
Nickel cadmium	Each	850	Base cabinets	L.F.	400 - 640
Kitchen Equipment			Sound System		
Broiler	Each	4175	Amplifier, 250 watts	Each	1975
Coffee urn, twin 6 gallon	Each	2625	Speaker, ceiling or wall	Each	206
Cooler, 6 ft. long	Each	4000	Trumpet	Each	395
Dishwasher, 10-12 racks per hr.	Each	5300	Storage		
Food warmer, counter, 1.2 KW	Each	560	Shelving	S.F.	8.85
Freezer, 44 C.F., reach-in	Each	4150	Washing		
Ice cube maker, 50 lb. per day	Each	1925	Stainless steel counter	L.F.	189
Range with 1 oven	Each	3325			

Important: See the Reference Section for Location Factors.

Model costs calculated for a 1 story building with 10' story height and 4,000 square feet of floor area

				Unit	Unit Cost	Cost Per S.F.	% Of Sub-Total
A. SUBSTRUCTURE							
1010	Standard Foundations	Poured concrete; strip and spread footings		S.F. Ground	2.50	2.50	
1020	Special Foundations	N/A		—	—	—	
1030	Slab on Grade	4" reinforced concrete with vapor barrier and granular base		S.F. Slab	5.18	5.18	8.2%
2010	Basement Excavation	Site preparation for slab and trench for foundation wall and footing		S.F. Ground	.55	.55	
2020	Basement Walls	4' foundation wall		L.F. Wall	79	5.14	
B. SHELL							
B10 Superstructure							
1010	Floor Construction	N/A		—	—	—	
1020	Roof Construction	Metal deck on open web steel joists		S.F. Roof	7.69	7.69	4.7 %
B20 Exterior Enclosure							
2010	Exterior Walls	Face brick with concrete block backup	70% of wall	S.F. Wall	32.44	14.76	
2020	Exterior Windows	Window wall	30% of wall	Each	48.55	9.47	19.7%
2030	Exterior Doors	Aluminum and glass		Each	3943	7.89	
B30 Roofing							
3010	Roof Coverings	Built-up tar and gravel with flashing; perlite/EPS composite insulation		S.F. Roof	6.68	6.68	4.4%
3020	Roof Openings	Roof hatches		S.F. Roof	.51	.51	
C. INTERIORS							
1010	Partitions	Gypsum board on metal studs	25 S.F. Floor/L.F. Partition	S.F. Partition	6	2.16	
1020	Interior Doors	Hollow core wood	1000 S.F. Floor/Door	Each	543	.55	
1030	Fittings	N/A		—	—	—	
2010	Stair Construction	N/A		—	—	—	16.8%
3010	Wall Finishes	Paint		S.F. Surface	2.85	2.05	
3020	Floor Finishes	Quarry tile		S.F. Floor	15.68	15.68	
3030	Ceiling Finishes	Mineral fiber tile on concealed zee bars		S.F. Ceiling	6.86	6.86	
D. SERVICES							
D10 Conveying							
1010	Elevators & Lifts	N/A		—	—	—	
1020	Escalators & Moving Walks	N/A		—	—	—	0.0 %
D20 Plumbing							
2010	Plumbing Fixtures	Kitchen, bathroom and service fixtures, supply and drainage	1 Fixture/400 S.F. Floor	Each	4176	10.44	
2020	Domestic Water Distribution	Gas fired water heater		S.F. Floor	5.89	5.89	11.3%
2040	Rain Water Drainage	Roof drains		S.F. Roof	1.99	1.99	
D30 HVAC							
3010	Energy Supply	N/A		—	—	—	
3020	Heat Generating Systems	Included in D3050		—	—	—	
3030	Cooling Generating Systems	N/A		—	—	—	11.9 %
3050	Terminal & Package Units	Multizone unit, gas heating, electric cooling; kitchen ventilation		S.F. Floor	19.35	19.35	
3090	Other HVAC Sys. & Equipment	N/A		—	—	—	
D40 Fire Protection							
4010	Sprinklers	Sprinklers, light hazard		S.F. Floor	8.32	8.32	6.3%
4020	Standpipes	Standpipe		S.F. Floor	2	2	
D50 Electrical							
5010	Electrical Service/Distribution	400 ampere service, panel board and feeders		S.F. Floor	4.91	4.91	
5020	Lighting & Branch Wiring	T-8 fluorescent fixtures, receptacles, switches, A.C. and misc. power		S.F. Floor	10.28	10.28	12.3%
5030	Communications & Security	Addressable alarm systems and emergency lighting		S.F. Floor	4.39	4.39	
5090	Other Electrical Systems	Emergency generator, 7.5 kW		S.F. Floor	.42	.42	
E. EQUIPMENT & FURNISHINGS							
1010	Commercial Equipment	N/A		—	—	—	
1020	Institutional Equipment	N/A		—	—	—	
1030	Vehicular Equipment	N/A		—	—	—	4.3 %
1090	Other Equipment	Walk-in refrigerator		S.F. Floor	7.03	7.03	
F. SPECIAL CONSTRUCTION							
1020	Integrated Construction	N/A		—	—	—	0.0 %
1040	Special Facilities	N/A		—	—	—	
G. BUILDING SITEWORK	**N/A**						

	Sub-Total	162.69	**100%**
CONTRACTOR FEES (General Requirements: 10%, Overhead: 5%, Profit: 10%)	25%	40.64	
ARCHITECT FEES	8%	16.27	
Total Building Cost		**219.60**	

Costs per square foot of floor area

Exterior Wall	S.F. Area	10000	15000	20000	25000	30000	35000	40000	45000	50000
	L.F. Perimeter	450	500	600	700	740	822	890	920	966
Face Brick with Concrete Block Back-up	Steel Frame	197.80	180.85	175.90	172.90	168.10	166.35	164.60	162.00	160.40
	Lam. Wood Truss	196.75	180.60	175.80	173.00	168.40	166.75	165.10	162.60	161.05
Concrete Block	Steel Frame	195.20	185.50	182.65	180.85	178.15	177.10	176.15	174.65	173.70
	Lam. Wood Truss	174.50	164.85	161.95	160.15	157.45	156.40	155.45	153.95	153.05
Galvanized Steel Siding	Steel Frame	184.30	174.70	171.85	170.10	167.45	166.45	165.50	163.95	163.05
Metal Sandwich Panel	Steel Joists	193.85	181.85	178.25	176.10	172.70	171.45	170.20	168.35	167.20
Perimeter Adj., Add or Deduct	Per 100 L.F.	14.00	9.35	7.00	5.65	4.70	4.05	3.50	3.15	2.80
Story Hgt. Adj., Add or Deduct	Per 1 Ft.	2.20	1.60	1.50	1.35	1.20	1.20	1.10	1.00	0.95
Basement—Not Applicable										

The above costs were calculated using the basic specifications shown on the facing page. These costs should be adjusted where necessary for design alternatives and owner's requirements. Reported completed project costs, for this type of structure, range from $73.05 to $227.30 per S.F.

Common additives

Description	Unit	$ Cost
Bar, Front Bar	L.F.	410
Back bar	L.F.	325
Booth, Upholstered, custom straight	L.F.	237 - 435
"L" or "U" shaped	L.F.	245 - 415
Bleachers, Telescoping, manual		
To 15 tier	Seat	134 - 187
16-20 tier	Seat	277 - 340
21-30 tier	Seat	294 - 355
For power operation, add	Seat	54 - 84.50
Emergency Lighting, 25 watt, battery operated		
Lead battery	Each	305
Nickel cadmium	Each	850
Lockers, Steel, single tier, 60" or 72"	Opening	218 - 355
2 tier, 60" or 72" total	Opening	112 - 160
5 tier, box lockers	Opening	70 - 86
Locker bench, lam. maple top only	L.F.	25.50
Pedestals, steel pipe	Each	72.50

Description	Unit	$ Cost
Rink		
Dasher boards & top guard	Each	186,000
Mats, rubber	S.F.	21.50
Score Board	Each	13,000 - 17,700

Important: See the Reference Section for Location Factors.

Rink, Hockey/Indoor Soccer

Model costs calculated for a 1 story building with 24' story height and 30,000 square feet of floor area

				Unit	Unit Cost	Cost Per S.F.	% Of Sub-Total
A. SUBSTRUCTURE							
1010	Standard Foundations	Poured concrete; strip and spread footings		S.F. Ground	1.73	1.73	
1020	Special Foundations	N/A		—	—	—	
1030	Slab on Grade	6" reinforced concrete with vapor barrier and granular base		S.F. Slab	6.16	6.16	8.0%
2010	Basement Excavation	Site preparation for slab and trench for foundation wall and footing		S.F. Ground	.18	.18	
2020	Basement Walls	4' foundation wall		L.F. Wall	75	2.53	
B. SHELL							
B10 Superstructure							
1010	Floor Construction	Wide flange beams and columns		S.F. Floor	28.13	19.69	27.1%
1020	Roof Construction	Metal deck on steel joist		S.F. Roof	16.41	16.41	
B20 Exterior Enclosure							
2010	Exterior Walls	Concrete block	95% of wall	S.F. Wall	14.35	8.07	
2020	Exterior Windows	Store front	5% of wall	Each	46.90	1.39	7.7%
2030	Exterior Doors	Aluminum and glass, hollow metal, overhead		Each	3228	.86	
B30 Roofing							
3010	Roof Coverings	Elastomeric neoprene membrane with flashing; perlite/EPS composite insulation		S.F. Roof	4.84	4.84	3.8%
3020	Roof Openings	Roof hatches		S.F. Roof	.24	.24	
C. INTERIORS							
1010	Partitions	Concrete block	140 S.F. Floor/L.F. Partition	S.F. Partition	9.10	.78	
1020	Interior Doors	Hollow metal	2500 S.F. Floor/Door	Each	1212	.49	
1030	Fittings	N/A		—	—	—	
2010	Stair Construction	N/A		—	—	—	6.3%
3010	Wall Finishes	Paint		S.F. Surface	12.31	2.11	
3020	Floor Finishes	80% rubber mat, 20% paint	50% of floor area	S.F. Floor	8.58	4.29	
3030	Ceiling Finishes	Mineral fiber tile on concealed zee bar	10% of area	S.F. Ceiling	6.86	.69	
D. SERVICES							
D10 Conveying							
1010	Elevators & Lifts	N/A		—	—	—	0.0%
1020	Escalators & Moving Walks	N/A		—	—	—	
D20 Plumbing							
2010	Plumbing Fixtures	Toilet and service fixtures, supply and drainage	1 Fixture/1070 S.F. Floor	Each	6238	5.83	
2020	Domestic Water Distribution	Oil fired water heater		S.F. Floor	3.22	3.22	6.8%
2040	Rain Water Drainage	N/A		—	—	—	
D30 HVAC							
3010	Energy Supply	Oil fired hot water, unit heaters	10% of area	S.F. Floor	.90	.90	
3020	Heat Generating Systems	N/A		—	—	—	
3030	Cooling Generating Systems	N/A		—	—	—	10.6%
3050	Terminal & Package Units	Single zone, electric cooling	90% of area	S.F. Floor	13.28	13.28	
3090	Other HVAC Sys. & Equipment	N/A		—	—	—	
D40 Fire Protection							
4010	Sprinklers	N/A		—	—	—	0.5%
4020	Standpipes	Standpipe		S.F. Floor	.63	.63	
D50 Electrical							
5010	Electrical Service/Distribution	600 ampere service, panel board and feeders		S.F. Floor	1.59	1.59	
5020	Lighting & Branch Wiring	High intensity discharge and fluorescent fixtures, receptacles, switches, A.C. and misc. power		S.F. Floor	7.74	7.74	8.7%
5030	Communications & Security	Addressable alarm systems, emergency lighting and public address		S.F. Floor	2.04	2.04	
5090	Other Electrical Systems	Emergency generator		S.F. Floor	.22	.22	
E. EQUIPMENT & FURNISHINGS							
1010	Commercial Equipment	N/A		—	—	—	
1020	Institutional Equipment	N/A		—	—	—	
1030	Vehicular Equipment	N/A		—	—	—	0.0%
1090	Other Equipment	N/A		—	—	—	
F. SPECIAL CONSTRUCTION							
1020	Integrated Construction	N/A		—	—	—	20.5%
1040	Special Facilities	Dasher boards and rink (including ice making system)		S.F. Floor	27.27	27.27	
G. BUILDING SITEWORK	**N/A**						

		Sub-Total	133.18	100%
CONTRACTOR FEES (General Requirements: 10%, Overhead: 5%, Profit: 10%)		25%	33.32	
ARCHITECT FEES		7%	11.65	
	Total Building Cost		**178.15**	

Costs per square foot of floor area

Exterior Wall	S.F. Area	25000	30000	35000	40000	45000	50000	55000	60000	65000
	L.F. Perimeter	900	1050	1200	1350	1510	1650	1800	1970	2100
Face Brick with Concrete Block Back-up	Steel Frame	180.30	178.75	177.60	176.70	176.35	175.55	175.15	175.15	174.45
	Bearing Walls	172.90	171.30	170.20	169.30	**168.90**	168.15	167.70	167.75	167.05
Stucco on Concrete Block	Steel Frame	173.20	171.75	170.80	170.05	169.65	169.05	168.65	168.65	168.00
	Bearing Walls	165.75	164.35	163.40	162.60	162.30	161.60	161.25	161.25	160.65
Decorative Concrete Block	Steel Frame	172.30	170.95	169.95	169.15	168.80	168.20	167.80	167.80	167.15
	Bearing Walls	167.80	166.40	165.40	164.60	164.30	163.60	163.20	163.25	162.65
Perimeter Adj., Add or Deduct	Per 100 L.F.	4.45	3.70	3.20	2.80	2.50	2.25	2.00	1.90	1.70
Story Hgt. Adj., Add or Deduct	Per 1 Ft.	1.65	1.60	1.55	1.60	1.55	1.50	1.45	1.50	1.50

For Basement, add $25.70 per square foot of basement area

The above costs were calculated using the basic specifications shown on the facing page. These costs should be adjusted where necessary for design alternatives and owner's requirements. Reported completed project costs, for this type of structure, range from $85.35 to $217.10 per S.F.

Common additives

Description	Unit	$ Cost
Bleachers, Telescoping, manual		
To 15 tier	Seat	134 - 187
16-20 tier	Seat	277 - 340
21-30 tier	Seat	294 - 355
For power operation, add	Seat	54 - 84.50
Carrels Hardwood	Each	855 - 1625
Clock System		
20 room	Each	19,200
50 room	Each	45,000
Emergency Lighting, 25 watt, battery operated		
Lead battery	Each	305
Nickel cadmium	Each	850
Flagpoles, Complete		
Aluminum, 20' high	Each	1750
40' high	Each	3875
Fiberglass, 23' high	Each	1325
39'-5" high	Each	3525
Kitchen Equipment		
Broiler	Each	4175
Cooler, 6 ft. long, reach-in	Each	4000

Description	Unit	$ Cost
Kitchen Equipment, cont.		
Dishwasher, 10-12 racks per hr.	Each	5300
Food warmer, counter, 1.2 KW	Each	560
Freezer, 44 C.F., reach-in	Each	4150
Ice cube maker, 50 lb. per day	Each	1925
Range with 1 oven	Each	3325
Lockers, Steel, single tier, 60" to 72"	Opening	218 - 355
2 tier, 60" to 72" total	Opening	112 - 160
5 tier, box lockers	Opening	70 - 86
Locker bench, lam. maple top only	L.F.	25.50
Pedestals, steel pipe	Each	72.50
Seating		
Auditorium chair, all veneer	Each	265
Veneer back, padded seat	Each	335
Upholstered, spring seat	Each	320
Classroom, movable chair & desk	Set	81 - 171
Lecture hall, pedestal type	Each	257 - 750
Sound System		
Amplifier, 250 watts	Each	1975
Speaker, ceiling or wall	Each	206
Trumpet	Each	395

Model costs calculated for a 1 story building with 15' story height and 45,000 square feet of floor area

School, Elementary

				Unit	Unit Cost	Cost Per S.F.	% Of Sub-Total
A. SUBSTRUCTURE							
1010	Standard Foundations	Poured concrete; strip and spread footings		S.F. Ground	4.91	4.91	
1020	Special Foundations	N/A		—	—	—	
1030	Slab on Grade	4" reinforced concrete with vapor barrier and granular base		S.F. Slab	5.18	5.18	11.7%
2010	Basement Excavation	Site preparation for slab and trench for foundation wall and footing		S.F. Ground	.18	.18	
2020	Basement Walls	4' foundation wall		L.F. Wall	79	4.51	
B. SHELL							
B10 Superstructure							
1010	Floor Construction	N/A		—	—	—	
1020	Roof Construction	Metal deck on open web steel joists		S.F. Roof	4.37	4.37	3.5 %
B20 Exterior Enclosure							
2010	Exterior Walls	Face brick with concrete block backup	70% of wall	S.F. Wall	32.44	11.43	
2020	Exterior Windows	Steel outward projecting	25% of wall	Each	735	4.82	13.4%
2030	Exterior Doors	Metal and glass	5% of wall	Each	3780	.67	
B30 Roofing							
3010	Roof Coverings	Single-ply membrane with flashing; polyisocyanurate insulation		S.F. Roof	6.99	6.99	5.5%
3020	Roof Openings	N/A		—	—	—	
C. INTERIORS							
1010	Partitions	Concrete block	20 S.F. Floor/L.F. Partition	S.F. Partition	9.64	4.82	
1020	Interior Doors	Single leaf kalamein fire doors	700 S.F. Floor/Door	Each	1212	1.73	
1030	Fittings	Toilet partitions		S.F. Floor	1.74	1.74	
2010	Stair Construction	N/A		—	—	—	20.0%
3010	Wall Finishes	75% paint, 15% glazed coating, 10% ceramic tile		S.F. Surface	4.57	4.57	
3020	Floor Finishes	65% vinyl composition tile, 25% carpet, 10% terrazzo		S.F. Floor	5.59	5.59	
3030	Ceiling Finishes	Mineral fiber tile on concealed zee bars		S.F. Ceiling	6.86	6.86	
D. SERVICES							
D10 Conveying							
1010	Elevators & Lifts	N/A		—	—	—	0.0 %
1020	Escalators & Moving Walks	N/A		—	—	—	
D20 Plumbing							
2010	Plumbing Fixtures	Kitchen, bathroom and service fixtures, supply and drainage	1 Fixture/625 S.F. Floor	Each	7813	12.50	
2020	Domestic Water Distribution	Gas fired water heater		S.F. Floor	.63	.63	11.4%
2040	Rain Water Drainage	Roof drains		S.F. Roof	1.24	1.24	
D30 HVAC							
3010	Energy Supply	Oil fired hot water, wall fin radiation		S.F. Floor	10.44	10.44	
3020	Heat Generating Systems	N/A		—	—	—	
3030	Cooling Generating Systems	N/A		—	—	—	19.5%
3050	Terminal & Package Units	Split systems with air cooled condensing units		S.F. Floor	14.20	14.20	
3090	Other HVAC Sys. & Equipment	N/A		—	—	—	
D40 Fire Protection							
4010	Sprinklers	Sprinklers, light hazard		S.F. Floor	2.86	2.86	2.6%
4020	Standpipes	Standpipe		S.F. Floor	.37	.37	
D50 Electrical							
5010	Electrical Service/Distribution	800 ampere service, panel board and feeders		S.F. Floor	1.16	1.16	
5020	Lighting & Branch Wiring	High efficiency fluorescent fixtures, receptacles, switches, A.C. and misc. power		S.F. Floor	10.56	10.56	12.3%
5030	Communications & Security	Addressable alarm systems, internet wiring, communications systems and emergency lighting		S.F. Floor	3.69	3.69	
5090	Other Electrical Systems	Emergency generator, 15 kW		S.F. Floor	.09	.09	
E. EQUIPMENT & FURNISHINGS							
1010	Commercial Equipment	N/A		—	—	—	
1020	Institutional Equipment	Chalkboards		S.F. Floor	.19	.19	0.2 %
1030	Vehicular Equipment	N/A		—	—	—	
1090	Other Equipment	N/A		—	—	—	
F. SPECIAL CONSTRUCTION							
1020	Integrated Construction	N/A		—	—	—	0.0 %
1040	Special Facilities	N/A		—	—	—	
G. BUILDING SITEWORK	**N/A**						

		Sub-Total	126.30	100%
CONTRACTOR FEES (General Requirements: 10%, Overhead: 5%, Profit: 10%)		25%	31.55	
ARCHITECT FEES		7%	11.05	

Total Building Cost	**168.90**

Costs per square foot of floor area

Exterior Wall	S.F. Area	50000	70000	90000	110000	130000	150000	170000	190000	210000
	L.F. Perimeter	850	1140	1420	1700	1980	2280	2560	2840	3120
Face Brick with Concrete Block Back-up	Steel Frame	170.10	167.40	165.70	164.65	163.90	163.65	163.15	162.85	162.50
	R/Conc. Frame	175.15	171.35	169.00	167.60	166.55	166.05	165.50	165.05	164.65
Decorative Concrete Block	Steel Frame	165.90	162.45	160.30	159.00	158.05	157.60	157.05	156.65	156.25
	R/Conc. Frame	166.30	162.80	160.70	159.40	158.45	158.00	157.45	157.00	156.65
Limestone with Concrete Block Back-up	Steel Frame	190.00	185.50	182.65	180.90	179.65	179.15	178.40	177.80	177.30
	R/Conc. Frame	196.70	192.20	189.35	187.60	186.40	185.85	185.10	184.50	184.00
Perimeter Adj., Add or Deduct	Per 100 L.F.	4.05	2.95	2.30	1.90	1.55	1.30	1.25	1.05	1.05
Story Hgt. Adj., Add or Deduct	Per 1 Ft.	2.05	2.00	1.90	1.90	1.85	1.80	1.85	1.75	1.85
For Basement, add $33.60 per square foot of basement area										

The above costs were calculated using the basic specifications shown on the facing page. These costs should be adjusted where necessary for design alternatives and owner's requirements. Reported completed project costs, for this type of structure, range from $99.20 to $233.60 per S.F.

Common additives

Description	Unit	$ Cost		Description	Unit	$ Cost
Bleachers, Telescoping, manual				Kitchen Equipment		
To 15 tier	Seat	134 - 187		Broiler	Each	4175
16-20 tier	Seat	277 - 340		Cooler, 6 ft. long, reach-in	Each	4000
21-30 tier	Seat	294 - 355		Dishwasher, 10-12 racks per hr.	Each	5300
For power operation, add	Seat	54 - 84.50		Food warmer, counter, 1.2 KW	Each	560
Carrels Hardwood	Each	855 - 1625		Freezer, 44 C.F., reach-in	Each	4150
Clock System				Lockers, Steel, single tier, 60" or 72"	Opening	218 - 355
20 room	Each	19,200		2 tier, 60" or 72" total	Opening	112 - 160
50 room	Each	45,000		5 tier, box lockers	Opening	70 - 86
Elevators, Hydraulic passenger, 2 stops				Locker bench, lam. maple top only	L.F.	25.50
1500# capacity	Each	66,000		Pedestals, steel pipe	Each	72.50
2500# capacity	Each	68,900		Seating		
Emergency Lighting, 25 watt, battery operated				Auditorium chair, all veneer	Each	265
Lead battery	Each	305		Veneer back, padded seat	Each	335
Nickel cadmium	Each	850		Upholstered, spring seat	Each	320
Flagpoles, Complete				Classroom, movable chair & desk	Set	81 - 171
Aluminum, 20' high	Each	1750		Lecture hall, pedestal type	Each	257 - 750
40' high	Each	3875		Sound System		
Fiberglass, 23' high	Each	1325		Amplifier, 250 watts	Each	1975
39'-5" high	Each	3525		Speaker, ceiling or wall	Each	206
				Trumpet	Each	395

Model costs calculated for a 2 story building with 15' story height and 130,000 square feet of floor area

School, High, 2-3 Story

					Unit	Unit Cost	Cost Per S.F.	% Of Sub-Total
A. SUBSTRUCTURE								
1010	Standard Foundations	Poured concrete; strip and spread footings			S.F. Ground	2.64	1.32	
1020	Special Foundations	N/A			—	—	—	
1030	Slab on Grade	4" reinforced concrete with vapor barrier and granular base			S.F. Slab	5.18	2.60	4.2%
2010	Basement Excavation	Site preparation for slab and trench for foundation wall and footing			S.F. Ground	.18	.09	
2020	Basement Walls	4' foundation wall			L.F. Wall	79	1.20	
B. SHELL								
	B10 Superstructure							
1010	Floor Construction	Concrete slab without drop panel, concrete columns			S.F. Floor	21.48	10.74	14.6%
1020	Roof Construction	Concrete slab without drop panel			S.F. Roof	14.82	7.41	
	B20 Exterior Enclosure							
2010	Exterior Walls	Face brick with concrete block backup	75% of wall		S.F. Wall	32.45	11.12	
2020	Exterior Windows	Window wall	25% of wall		Each	71	8.11	15.9%
2030	Exterior Doors	Metal and glass			Each	2265	.51	
	B30 Roofing							
3010	Roof Coverings	Single-ply membrane and standing seam metal; polyisocyanurate insulation			S.F. Roof	10.88	5.44	4.4%
3020	Roof Openings	Roof hatches			S.F. Roof	.10	.05	
C. INTERIORS								
1010	Partitions	Concrete block	25 S.F. Floor/L.F. Partition		S.F. Partition	12.23	5.87	
1020	Interior Doors	Single leaf kalamein fire doors	700 S.F. Floor/Door		Each	1212	1.73	
1030	Fittings	Toilet partitions, chalkboards			S.F. Floor	1.22	1.22	
2010	Stair Construction	Concrete filled metal pan			Flight	14,150	.65	19.9%
3010	Wall Finishes	75% paint, 15% glazed coating, 10% ceramic tile			S.F. Surface	3.45	3.31	
3020	Floor Finishes	70% vinyl composition tile, 20% carpet, 10% terrazzo			S.F. Floor	5.14	5.14	
3030	Ceiling Finishes	Mineral fiber tile on concealed zee bars			S.F. Ceiling	6.86	6.86	
D. SERVICES								
	D10 Conveying							
1010	Elevators & Lifts	One hydraulic passenger elevator			Each	83,200	.64	0.5%
1020	Escalators & Moving Walks	N/A			—	—	—	
	D20 Plumbing							
2010	Plumbing Fixtures	Kitchen, bathroom and service fixtures, supply and drainage	1 Fixture/860 S.F. Floor		Each	4481	5.21	
2020	Domestic Water Distribution	Gas fired water heater			S.F. Floor	1.09	1.09	5.6%
2040	Rain Water Drainage	Roof drains			S.F. Roof	1.38	.69	
	D30 HVAC							
3010	Energy Supply	Oil fired hot water, wall fin radiation			S.F. Floor	5.02	5.02	
3020	Heat Generating Systems	N/A			—	—	—	
3030	Cooling Generating Systems	Chilled water, cooling tower systems			S.F. Floor	17.43	17.43	18.0%
3050	Terminal & Package Units	N/A			—	—	—	
3090	Other HVAC Sys. & Equipment	N/A			—	—	—	
	D40 Fire Protection							
4010	Sprinklers	Sprinklers, light hazard			S.F. Floor	2.50	2.50	2.3%
4020	Standpipes	Standpipe, wet, Class III			S.F. Floor	.32	.32	
	D50 Electrical							
5010	Electrical Service/Distribution	2000 ampere service, panel board and feeders			S.F. Floor	1.13	1.13	
5020	Lighting & Branch Wiring	High efficiency fluorescent fixtures, receptacles, switches, A.C. and misc. power			S.F. Floor	9.81	9.81	12.4%
5030	Communications & Security	Addressable alarm systems, internet wiring, communications systems and emergency lighting			S.F. Floor	3.98	3.98	
5090	Other Electrical Systems	Emergency generator, 250 kW			S.F. Floor	.51	.51	
E. EQUIPMENT & FURNISHINGS								
1010	Commercial Equipment	N/A			—	—	—	
1020	Institutional Equipment	Laboratory casework and counters			S.F. Floor	1.90	1.90	2.3 %
1030	Vehicular Equipment	N/A			—	—	—	
1090	Other Equipment	Built-in athletic equipment , bleachers and scoreboard			S.F. Floor	.93	.93	
F. SPECIAL CONSTRUCTION								
1020	Integrated Construction	N/A			—	—	—	0.0 %
1040	Special Facilities	N/A			—	—	—	
G. BUILDING SITEWORK	**N/A**							

						Sub-Total	124.53	100%
	CONTRACTOR FEES (General Requirements: 10%, Overhead: 5%, Profit: 10%)					25%	31.12	
	ARCHITECT FEES					7%	10.90	

	Total Building Cost	**166.55**

Costs per square foot of floor area

Exterior Wall	S.F. Area	50000	65000	80000	95000	110000	125000	140000	155000	170000
	L.F. Perimeter	850	1060	1280	1490	1700	1920	2140	2340	2560
Face Brick with Concrete Block Back-up	Steel Frame	174.50	172.30	171.20	170.20	169.45	169.05	168.80	168.25	168.10
	Bearing Walls	166.90	164.65	163.55	162.55	161.85	161.50	161.20	160.65	160.50
Concrete Block Stucco Face	Steel Frame	166.20	164.25	163.30	162.45	161.80	161.50	161.20	160.85	160.65
	Bearing Walls	159.50	157.65	156.65	155.80	155.15	154.85	154.60	154.15	154.00
Decorative Concrete Block	Steel Frame	167.90	165.95	164.95	164.05	163.45	163.10	162.80	162.40	162.20
	Bearing Walls	159.80	157.85	156.80	156.00	155.35	155.00	154.75	154.30	154.10
Perimeter Adj., Add or Deduct	Per 100 L.F.	3.65	2.75	2.25	1.90	1.70	1.50	1.25	1.15	1.10
Story Hgt. Adj., Add or Deduct	Per 1 Ft.	1.85	1.75	1.65	1.65	1.70	1.70	1.60	1.65	1.65

For Basement, add $34.95 per square foot of basement area

The above costs were calculated using the basic specifications shown on the facing page. These costs should be adjusted where necessary for design alternatives and owner's requirements. Reported completed project costs, for this type of structure, range from $93.65 to $220.05 per S.F.

Common additives

Description	Unit	$ Cost		Description	Unit	$ Cost
Bleachers, Telescoping, manual				Kitchen Equipment		
To 15 tier	Seat	134 - 187		Broiler	Each	4175
16-20 tier	Seat	277 - 340		Cooler, 6 ft. long, reach-in	Each	4000
21-30 tier	Seat	294 - 355		Dishwasher, 10-12 racks per hr.	Each	5300
For power operation, add	Seat	54 - 84.50		Food warmer, counter, 1.2 KW	Each	560
Carrels Hardwood	Each	855 - 1625		Freezer, 44 C.F., reach-in	Each	4150
Clock System				Lockers, Steel, single tier, 60" to 72"	Opening	218 - 355
20 room	Each	19,200		2 tier, 60" to 72" total	Opening	112 - 160
50 room	Each	45,000		5 tier, box lockers	Opening	70 - 86
Elevators, Hydraulic passenger, 2 stops				Locker bench, lam. maple top only	L.F.	25.50
1500# capacity	Each	66,000		Pedestals, steel pipe	Each	72.50
2500# capacity	Each	68,900		Seating		
Emergency Lighting, 25 watt, battery operated				Auditorium chair, all veneer	Each	265
Lead battery	Each	305		Veneer back, padded seat	Each	335
Nickel cadmium	Each	850		Upholstered, spring seat	Each	320
Flagpoles, Complete				Classroom, movable chair & desk	Set	81 - 171
Aluminum, 20' high	Each	1750		Lecture hall, pedestal type	Each	257 - 750
40' high	Each	3875		Sound System		
Fiberglass, 23' high	Each	1325		Amplifier, 250 watts	Each	1975
39'-5" high	Each	3525		Speaker, ceiling or wall	Each	206
				Trumpet	Each	395

Model costs calculated for a 2 story building with 15' story height and 110,000 square feet of floor area

				Unit	Unit Cost	Cost Per S.F.	% Of Sub-Total
A. SUBSTRUCTURE							
1010	Standard Foundations	Poured concrete; strip and spread footings		S.F. Ground	2.40	1.20	
1020	Special Foundations	N/A		—	—	—	
1030	Slab on Grade	4" reinforced concrete with vapor barrier and granular base		S.F. Slab	5.18	2.60	4.0%
2010	Basement Excavation	Site preparation for slab and trench for foundation wall and footing		S.F. Ground	.18	.09	
2020	Basement Walls	4' foundation wall		L.F. Wall	79	1.22	
B. SHELL							
B10 Superstructure							
1010	Floor Construction	Open web steel joists, slab form, concrete, columns		S.F. Floor	28	14	15.7%
1020	Roof Construction	Metal deck, open web steel joists, columns		S.F. Roof	11.74	5.87	
B20 Exterior Enclosure							
2010	Exterior Walls	Face brick with concrete block backup	75% of wall	S.F. Wall	32.44	11.28	
2020	Exterior Windows	Window wall	25% of wall	Each	54	6.22	14.3%
2030	Exterior Doors	Double aluminum & glass		Each	2389	.65	
B30 Roofing							
3010	Roof Coverings	Single-ply membrane and standing seam metal; polyisocyanurate insulation		S.F. Roof	11.06	5.53	4.4%
3020	Roof Openings	Roof hatches		S.F. Roof	.04	.02	
C. INTERIORS							
1010	Partitions	Concrete block	20 S.F. Floor/L.F. Partition	S.F. Partition	9.12	5.47	
1020	Interior Doors	Single leaf kalamein fire doors	750 S.F. Floor/Door	Each	1212	1.62	
1030	Fittings	Toilet partitions, chalkboards		S.F. Floor	1.31	1.31	
2010	Stair Construction	Concrete filled metal pan		Flight	14,150	.77	22.4%
3010	Wall Finishes	50% paint, 40% glazed coatings, 10% ceramic tile		S.F. Surface	4.13	4.96	
3020	Floor Finishes	50% vinyl composition tile, 30% carpet, 20% terrrazzo		S.F. Floor	7.35	7.35	
3030	Ceiling Finishes	Mineral fiberboard on concealed zee bars		S.F. Ceiling	6.86	6.86	
D. SERVICES							
D10 Conveying							
1010	Elevators & Lifts	One hydraulic passenger elevator		Each	83,600	.76	0.6%
1020	Escalators & Moving Walks	N/A		—	—	—	
D20 Plumbing							
2010	Plumbing Fixtures	Kitchen, toilet and service fixtures, supply and drainage	1 Fixture/1170 S.F. Floor	Each	6517	5.57	
2020	Domestic Water Distribution	Gas fired water heater		S.F. Floor	.65	.65	5.5%
2040	Rain Water Drainage	Roof drains		S.F. Roof	1.40	.70	
D30 HVAC							
3010	Energy Supply	N/A		—	—	—	
3020	Heat Generating Systems	Included in D3050		—	—	—	
3030	Cooling Generating Systems	N/A		—	—	—	15.9 %
3050	Terminal & Package Units	Multizone unit, gas heating, electric cooling		S.F. Floor	20.20	20.20	
3090	Other HVAC Sys. & Equipment	N/A		—	—	—	
D40 Fire Protection							
4010	Sprinklers	Sprinklers, light hazard	10% of area	S.F. Floor	2.50	2.50	2.3%
4020	Standpipes	Standpipe, wet, Class III		S.F. Floor	.38	.38	
D50 Electrical							
5010	Electrical Service/Distribution	1600 ampere service, panel board and feeders		S.F. Floor	1.07	1.07	
5020	Lighting & Branch Wiring	High efficiency fluorescent fixtures, receptacles, switches, A.C. and misc. power		S.F. Floor	9.81	9.81	12.3%
5030	Communications & Security	Addressable alarm systems, internet wiring, communications systems and emergency lighting		S.F. Floor	4.28	4.28	
5090	Other Electrical Systems	Emergency generator, 100 kW		S.F. Floor	.38	.38	
E. EQUIPMENT & FURNISHINGS							
1010	Commercial Equipment	N/A		—	—	—	
1020	Institutional Equipment	Laboratory casework and counters		S.F. Floor	2.24	2.24	2.7 %
1030	Vehicular Equipment	N/A		—	—	—	
1090	Other Equipment	Bleachers, basketball backstops		S.F. Floor	1.12	1.12	
F. SPECIAL CONSTRUCTION							
1020	Integrated Construction	N/A		—	—	—	0.0 %
1040	Special Facilities	N/A		—	—	—	
G. BUILDING SITEWORK	**N/A**						

	Sub-Total	126.68	100%
CONTRACTOR FEES (General Requirements: 10%, Overhead: 5%, Profit: 10%)	25%	31.69	
ARCHITECT FEES	7%	11.08	
Total Building Cost		**169.45**	

Costs per square foot of floor area

Exterior Wall	S.F. Area	20000	30000	40000	50000	60000	70000	80000	90000	100000
	L.F. Perimeter	440	590	740	900	1050	1200	1360	1510	1660
Face Brick with Concrete Block Back-up	Steel Frame	181.65	175.60	172.55	171.05	169.85	168.90	168.50	167.95	167.45
	Bearing Walls	175.55	169.50	166.60	165.05	163.85	162.95	162.50	162.00	161.45
Decorative Concrete Block	Steel Frame	167.75	163.10	160.85	159.65	158.70	158.00	157.70	157.30	156.90
	Bearing Walls	162.00	157.35	155.10	153.90	153.00	152.25	151.95	151.55	151.15
Steel Siding on Steel Studs	Steel Frame	162.35	158.30	156.30	155.30	154.45	153.80	153.55	153.15	152.85
Metal Sandwich Panel	Steel Frame	184.55	178.15	175.00	173.40	172.10	171.10	170.65	170.10	169.60
Perimeter Adj., Add or Deduct	Per 100 L.F.	9.30	6.15	4.65	3.75	3.10	2.65	2.30	2.00	1.85
Story Hgt. Adj., Add or Deduct	Per 1 Ft.	2.20	2.00	1.90	1.85	1.80	1.75	1.75	1.75	1.70
For Basement, add $33.90 per square foot of basement area										

The above costs were calculated using the basic specifications shown on the facing page. These costs should be adjusted where necessary for design alternatives and owner's requirements. Reported completed project costs, for this type of structure, range from $84.05 to $240.15 per S.F.

Common additives

Description	Unit	$ Cost	Description	Unit	$ Cost
Carrels Hardwood	Each	855 - 1625	Flagpoles, Complete		
Clock System			Aluminum, 20' high	Each	1750
20 room	Each	19,200	40' high	Each	3875
50 room	Each	45,000	Fiberglass, 23' high	Each	1325
Directory Boards, Plastic, glass covered			39'-5" high	Each	3525
30" x 20"	Each	660	Seating		
36" x 48"	Each	1475	Auditorium chair, all veneer	Each	265
Aluminum, 24" x 18"	Each	620	Veneer back, padded seat	Each	335
36" x 24"	Each	725	Upholstered, spring seat	Each	320
48" x 32"	Each	1050	Classroom, movable chair & desk	Set	81 - 171
48" x 60"	Each	2175	Lecture hall, pedestal type	Each	257 - 750
Elevators, Hydraulic passenger, 2 stops			Shops & Workroom:		
1500# capacity	Each	66,000	Benches, metal	Each	790
2500# capacity	Each	68,900	Parts bins 6'-3" high, 3' wide, 12" deep, 72 bins	Each	690
3500# capacity	Each	72,900	Shelving, metal 1' x 3'	S.F.	12.20
Emergency Lighting, 25 watt, battery operated			Wide span 6' wide x 24" deep	S.F.	8.85
Lead battery	Each	305	Sound System		
Nickel cadmium	Each	850	Amplifier, 250 watts	Each	1975
			Speaker, ceiling or wall	Each	206
			Trumpet	Each	395

Model costs calculated for a 2 story building with 16' story height and 40,000 square feet of floor area

				Unit	Unit Cost	Cost Per S.F.	% Of Sub-Total
A. SUBSTRUCTURE							
1010	Standard Foundations	Poured concrete; strip and spread footings		S.F. Ground	2.74	1.37	
1020	Special Foundations	N/A		—	—	—	
1030	Slab on Grade	5" reinforced concrete with vapor barrier and granular base		S.F. Slab	12.82	6.41	7.5%
2010	Basement Excavation	Site preparation for slab and trench for foundation wall and footing		S.F. Ground	.18	.09	
2020	Basement Walls	4' foundation wall		L.F. Wall	79	1.46	
B. SHELL							
B10 Superstructure							
1010	Floor Construction	Open web steel joists, slab form, concrete, beams, columns		S.F. Floor	18.40	9.20	9.9%
1020	Roof Construction	Metal deck, open web steel joists, beams, columns		S.F. Roof	6.36	3.18	
B20 Exterior Enclosure							
2010	Exterior Walls	Face brick with concrete block backup	85% of wall	S.F. Wall	32.45	16.33	
2020	Exterior Windows	Tubular aluminum framing with insulated glass	15% of wall	Each	54	4.77	17.4%
2030	Exterior Doors	Metal and glass doors		Each	2858	.58	
B30 Roofing							
3010	Roof Coverings	Single-ply membrane with polyisocyanurate insulation		S.F. Roof	6.38	3.19	2.7%
3020	Roof Openings	Roof hatches		S.F. Roof	.34	.17	
C. INTERIORS							
1010	Partitions	Concrete block	20 S.F. Floor/L.F. Partition	S.F. Partition	9.12	5.47	
1020	Interior Doors	Single leaf kalamein fire doors	600 S.F. Floor/Door	Each	1212	2.02	
1030	Fittings	Toilet partitions, chalkboards		S.F. Floor	1.44	1.44	
2010	Stair Construction	Concrete filled metal pan		Flight	14,150	1.42	20.1%
3010	Wall Finishes	50% paint, 40% glazed coating, 10% ceramic tile		S.F. Surface	4.55	5.46	
3020	Floor Finishes	70% vinyl composition tile, 20% carpet, 10% terrazzo		S.F. Floor	5.14	5.14	
3030	Ceiling Finishes	Mineral fiber tile on concealed zee bars	60% of area	S.F. Ceiling	6.86	4.12	
D. SERVICES							
D10 Conveying							
1010	Elevators & Lifts	One hydraulic passenger elevator		Each	83,600	2.09	1.7%
1020	Escalators & Moving Walks	N/A		—	—	—	
D20 Plumbing							
2010	Plumbing Fixtures	Toilet and service fixtures, supply and drainage	1 Fixture/700 S.F. Floor	Each	3304	4.72	
2020	Domestic Water Distribution	Gas fired water heater		S.F. Floor	1.03	1.03	5.3%
2040	Rain Water Drainage	Roof drains		S.F. Roof	1.82	.91	
D30 HVAC							
3010	Energy Supply	Oil fired hot water, wall fin radiation		S.F. Floor	9.88	9.88	
3020	Heat Generating Systems	N/A		—	—	—	
3030	Cooling Generating Systems	Chilled water, cooling tower systems		S.F. Floor	14.85	14.85	19.9%
3050	Terminal & Package Units	N/A		—	—	—	
3090	Other HVAC Sys. & Equipment	N/A		—	—	—	
D40 Fire Protection							
4010	Sprinklers	Sprinklers, light hazard		S.F. Floor	3.07	3.07	3.2%
4020	Standpipes	Standpipe, wet, Class III		S.F. Floor	.88	.88	
D50 Electrical							
5010	Electrical Service/Distribution	800 ampere service, panel board and feeders		S.F. Floor	1.30	1.30	
5020	Lighting & Branch Wiring	High efficiency fluorescent fixtures, receptacles, switches, A.C. and misc. power		S.F. Floor	10.60	10.60	12.1%
5030	Communications & Security	Addressable alarm systems, internet wiring, communications systems and emergency lighting		S.F. Floor	3.11	3.11	
5090	Other Electrical Systems	Emergency generator, 11.5 kW		S.F. Floor	.12	.12	
E. EQUIPMENT & FURNISHINGS							
1010	Commercial Equipment	N/A		—	—	—	
1020	Institutional Equipment	Stainless steel countertops		S.F. Floor	.16	.16	0.1%
1030	Vehicular Equipment	N/A		—	—	—	
1090	Other Equipment	N/A		—	—	—	
F. SPECIAL CONSTRUCTION							
1020	Integrated Construction	N/A		—	—	—	0.0%
1040	Special Facilities	N/A		—	—	—	
G. BUILDING SITEWORK	**N/A**						

			Sub-Total	124.54	100%
	CONTRACTOR FEES (General Requirements: 10%, Overhead: 5%, Profit: 10%)		25%	31.16	
	ARCHITECT FEES		7%	10.90	

Total Building Cost	**166.60**

Costs per square foot of floor area

Exterior Wall	S.F. Area	1000	2000	3000	4000	6000	8000	10000	12000	15000
	L.F. Perimeter	126	179	219	253	310	358	400	438	490
Wood Siding	Wood Frame	150.00	127.25	118.15	113.00	107.25	103.95	101.75	100.10	98.40
Face Brick Veneer	Wood Frame	177.65	146.30	133.25	125.80	117.25	112.30	108.95	106.55	103.90
Stucco on Concrete Block	Steel Frame	172.90	144.55	132.85	126.25	118.65	114.25	111.35	109.20	106.85
	Bearing Walls	168.75	140.40	128.70	122.10	114.50	110.15	107.20	105.05	102.75
Metal Sandwich Panel	Steel Frame	188.60	155.95	142.30	134.50	125.50	120.25	116.80	114.20	111.45
Precast Concrete	Steel Frame	226.15	182.15	163.35	152.50	139.95	132.60	127.60	124.00	120.05
Perimeter Adj., Add or Deduct	Per 100 L.F.	36.55	18.35	12.20	9.15	6.05	4.55	3.60	3.05	2.45
Story Hgt. Adj., Add or Deduct	Per 1 Ft.	2.50	1.85	1.50	1.30	1.00	0.90	0.75	0.75	0.65

For Basement, add $25.60 per square foot of basement area

The above costs were calculated using the basic specifications shown on the facing page. These costs should be adjusted where necessary for design alternatives and owner's requirements. Reported completed project costs, for this type of structure, range from $69.60 to $241.35 per S.F.

Common additives

Description	Unit	$ Cost	Description	Unit	$ Cost
Check Out Counter			Refrigerated Food Cases, cont.		
Single belt	Each	3375	Frozen food, chest type, 12' long	Each	9150
Double belt	Each	4825	Glass door reach-in, 5 door	Each	15,700
Emergency Lighting, 25 watt, battery operated			Island case 12' long, single deck	Each	8300
Lead battery	Each	305	Multi deck	Each	9650
Nickel cadmium	Each	850	Meat cases, 12' long, single deck	Each	8300
Refrigerators, Prefabricated, walk-in			Multi deck	Each	11,700
7'-6" high, 6' x 6'	S.F.	190	Produce, 12' long single deck	Each	7825
10' x 10'	S.F.	150	Multi deck	Each	9875
12' x 14'	S.F.	133	Safe, Office type, 1 hour rating		
12' x 20'	S.F.	117	30" x 18" x 18"	Each	2500
Refrigerated Food Cases			60" x 36" x 18", double door	Each	8075
Dairy, multi deck, 12' long	Each	12,700	Smoke Detectors		
Delicatessen case, single deck, 12' long	Each	8525	Ceiling type	Each	220
Multi deck, 18 S.F. shelf display	Each	7875	Duct type	Each	550
Freezer, self-contained chest type, 30 C.F.	Each	9250	Sound System		
Glass door upright, 78 C.F.	Each	13,500	Amplifier, 250 watts	Each	1975
			Speaker, ceiling or wall	Each	206
			Trumpet	Each	395

Important: See the Reference Section for Location Factors.

Model costs calculated for a 1 story building with 12' story height and 4,000 square feet of floor area

				Unit	Unit Cost	Cost Per S.F.	% Of Sub-Total
A.	**SUBSTRUCTURE**						
1010	Standard Foundations	Poured concrete; strip and spread footings		S.F. Ground	1.64	1.64	
1020	Special Foundations	N/A		—	—	—	
1030	Slab on Grade	4" reinforced concrete with vapor barrier and granular base		S.F. Slab	5.18	5.18	14.0%
2010	Basement Excavation	Site preparation for slab and trench for foundation wall and footing		S.F. Ground	.55	.55	
2020	Basement Walls	4' foundation wall		L.F. Wall	71	4.47	
B.	**SHELL**						
	B10	**Superstructure**					
1010	Floor Construction	N/A		—	—	—	7.8 %
1020	Roof Construction	Wood truss with plywood sheathing		S.F. Roof	6.57	6.57	
	B20	**Exterior Enclosure**					
2010	Exterior Walls	Wood siding on wood studs, insulated	80% of wall	S.F. Wall	9.29	5.64	
2020	Exterior Windows	Storefront	20% of wall	Each	43.60	4.44	14.5%
2030	Exterior Doors	Double aluminum and glass, solid core wood		Each	2942	2.20	
	B30	**Roofing**					
3010	Roof Coverings	Asphalt shingles; rigid fiberglass insulation		S.F. Roof	5.15	5.15	6.1%
3020	Roof Openings	N/A		—	—	—	
C.	**INTERIORS**						
1010	Partitions	Gypsum board on wood studs	60 S.F. Floor/L.F. Partition	S.F. Partition	12.72	2.12	
1020	Interior Doors	Single leaf wood, hollow metal	1300 S.F. Floor/Door	Each	1230	.94	
1030	Fittings	N/A		—	—	—	
2010	Stair Construction	N/A		—	—	—	13.6%
3010	Wall Finishes	Paint		S.F. Surface	1.17	.39	
3020	Floor Finishes	Vinyl composition tile		S.F. Floor	2.02	2.02	
3030	Ceiling Finishes	Mineral fiber tile on wood furring		S.F. Ceiling	5.98	5.98	
D.	**SERVICES**						
	D10	**Conveying**					
1010	Elevators & Lifts	N/A		—	—	—	0.0 %
1020	Escalators & Moving Walks	N/A		—	—	—	
	D20	**Plumbing**					
2010	Plumbing Fixtures	Toilet and service fixtures, supply and drainage	1 Fixture/1000 S.F. Floor	Each	4090	4.09	
2020	Domestic Water Distribution	Gas fired water heater		S.F. Floor	1.33	1.33	6.4%
2040	Rain Water Drainage	N/A		—	—	—	
	D30	**HVAC**					
3010	Energy Supply	N/A		—	—	—	
3020	Heat Generating Systems	Included in D3050		—	—	—	
3030	Cooling Generating Systems	N/A		—	—	—	9.8 %
3050	Terminal & Package Units	Single zone rooftop unit, gas heating, electric cooling		S.F. Floor	8.32	8.32	
3090	Other HVAC Sys. & Equipment	N/A		—	—	—	
	D40	**Fire Protection**					
4010	Sprinklers	Sprinkler, ordinary hazard		S.F. Floor	4.66	4.66	7.9%
4020	Standpipes	Standpipe		S.F. floor	2.00	2.00	
	D50	**Electrical**					
5010	Electrical Service/Distribution	200 ampere service, panel board and feeders		S.F. Floor	3.10	3.10	
5020	Lighting & Branch Wiring	High efficiency fluorescent fixtures, receptacles, switches, A.C. and misc. power		S.F. Floor	10.58	10.58	19.9%
5030	Communications & Security	Addressable alarm systems and emergency lighting		S.F. Floor	2.75	2.75	
5090	Other Electrical Systems	Emergency generator, 7.5 kW		S.F. Floor	.38	.38	
E.	**EQUIPMENT & FURNISHINGS**						
1010	Commercial Equipment	N/A		—	—	—	
1020	Institutional Equipment	N/A		—	—	—	
1030	Vehicular Equipment	N/A		—	—	—	0.0 %
1090	Other Equipment	N/A		—	—	—	
F.	**SPECIAL CONSTRUCTION**						
1020	Integrated Construction	N/A		—	—	—	
1040	Special Facilities	N/A		—	—	—	0.0 %
G.	**BUILDING SITEWORK**	**N/A**					
					Sub-Total	84.50	**100%**
	CONTRACTOR FEES (General Requirements: 10%, Overhead: 5%, Profit: 10%)				25%	21.11	
	ARCHITECT FEES				7%	7.39	
					Total Building Cost	113	

Costs per square foot of floor area

Exterior Wall	S.F. Area	50000	65000	80000	95000	110000	125000	140000	155000	170000
	L.F. Perimeter	920	1065	1167	1303	1333	1433	1533	1633	1733
Face Brick with Concrete Block Back-up	R/Conc. Frame	123.10	120.70	118.70	117.65	115.90	115.25	114.65	114.15	113.80
	Steel Frame	107.80	105.40	103.40	102.30	100.60	99.95	99.30	98.90	98.50
Decorative Concrete Block	R/Conc. Frame	117.80	115.95	114.45	113.60	112.35	111.90	111.50	111.10	110.85
	Steel Joists	103.55	101.55	100.00	99.10	97.80	97.25	96.80	96.40	96.10
Precast Concrete Panels	R/Conc. Frame	122.10	119.85	117.95	116.95	115.40	114.80	114.25	113.85	113.45
	Steel Joists	106.30	104.05	102.15	101.20	99.60	99.00	98.40	98.00	97.65
Perimeter Adj., Add or Deduct	Per 100 L.F.	1.85	1.45	1.10	0.95	0.90	0.75	0.65	0.60	0.55
Story Hgt. Adj., Add or Deduct	Per 1 Ft.	1.00	0.85	0.75	0.70	0.65	0.60	0.55	0.55	0.55

For Basement, add $25.15 per square foot of basement area

The above costs were calculated using the basic specifications shown on the facing page. These costs should be adjusted where necessary for design alternatives and owner's requirements. Reported completed project costs, for this type of structure, range from $60.10 to $149.45 per S.F.

Common additives

Description	Unit	$ Cost
Closed Circuit Surveillance, One station		
Camera and monitor	Each	1950
For additional camera stations, add	Each	1050
Directory Boards, Plastic, glass covered		
30" x 20"	Each	660
36" x 48"	Each	1475
Aluminum, 24" x 18"	Each	620
36" x 24"	Each	725
48" x 32"	Each	1050
48" x 60"	Each	2175
Emergency Lighting, 25 watt, battery operated		
Lead battery	Each	305
Nickel cadmium	Each	850
Safe, Office type, 1 hour rating		
30" x 18" x 18"	Each	2500
60" x 36" x 18", double door	Each	8075
Sound System		
Amplifier, 250 watts	Each	1975
Speaker, ceiling or wall	Each	206
Trumpet	Each	395

Important: See the Reference Section for Location Factors.

Model costs calculated for a 1 story building with 14' story height and 110,000 square feet of floor area

				Unit	Unit Cost	Cost Per S.F.	% Of Sub-Total
A.	**SUBSTRUCTURE**						
1010	Standard Foundations	Poured concrete; strip and spread footings		S.F. Ground	.89	.89	
1020	Special Foundations	N/A		—	—	—	
1030	Slab on Grade	4" reinforced concrete with vapor barrier and granular base		S.F. Slab	5.18	5.18	8.4%
2010	Basement Excavation	Site preparation for slab and trench for foundation wall and footing		S.F. Ground	.30	.30	
2020	Basement Walls	4' foundation wall		L.F. Wall	79	.95	
B.	**SHELL**						
	B10 Superstructure						
1010	Floor Construction	Cast-in-place concrete columns		L.F. Column	72	.92	25.7%
1020	Roof Construction	Precast concrete beam and plank, concrete columns		S.F. Roof	21.59	21.59	
	B20 Exterior Enclosure						
2010	Exterior Walls	Face brick with concrete block backup	90% of wall	S.F. Wall	31.50	4.81	
2020	Exterior Windows	Storefront	10% of wall	Each	65	1.11	7.2%
2030	Exterior Doors	Sliding electric operated entrance, hollow metal		Each	7607	.42	
	B30 Roofing						
3010	Roof Coverings	Built-up tar and gravel with flashing; perlite/EPS composite insulation		S.F. Roof	5.19	5.19	6.0%
3020	Roof Openings	Roof hatches		S.F. Roof	.05	.05	
C.	**INTERIORS**						
1010	Partitions	Gypsum board on metal studs	60 S.F. Floor/L.F. Partition	S.F. Partition	10.02	1.67	
1020	Interior Doors	Single leaf hollow metal	600 S.F. Floor/Door	Each	1212	2.02	
1030	Fittings	N/A		—	—	—	
2010	Stair Construction	N/A		—	—	—	24.8%
3010	Wall Finishes	Paint		S.F. Surface	.81	.27	
3020	Floor Finishes	50% ceramic tile, 50% carpet tile		S.F. Floor	13.26	13.26	
3030	Ceiling Finishes	Mineral fiber board on exposed grid system, suspended		S.F. Ceiling	4.44	4.44	
D.	**SERVICES**						
	D10 Conveying						
1010	Elevators & Lifts	N/A		—	—	—	0.0%
1020	Escalators & Moving Walks	N/A		—	—	—	
	D20 Plumbing						
2010	Plumbing Fixtures	Toilet and service fixtures, supply and drainage	1 Fixture/4075 S.F. Floor	Each	6642	1.63	
2020	Domestic Water Distribution	Gas fired water heater		S.F. Floor	.37	.37	3.2%
2040	Rain Water Drainage	Roof drains		S.F. Roof	.79	.79	
	D30 HVAC						
3010	Energy Supply	N/A		—	—	—	
3020	Heat Generating Systems	Included in D3050		—	—	—	
3030	Cooling Generating Systems	N/A		—	—	—	9.0%
3050	Terminal & Package Units	Single zone unit gas heating, electric cooling		S.F. Floor	7.90	7.90	
3090	Other HVAC Sys. & Equipment	N/A		—	—	—	
	D40 Fire Protection						
4010	Sprinklers	Sprinklers, light hazard		S.F. Floor	2.86	2.86	3.6%
4020	Standpipes	Standpipe		S.F. Floor	.25	.25	
	D50 Electrical						
5010	Electrical Service/Distribution	1200 ampere service, panel board and feeders		S.F. Floor	.81	.81	
5020	Lighting & Branch Wiring	High efficiency fluorescent fixtures, receptacles, switches, A.C. and misc. power		S.F. Floor	8.52	8.52	12.1%
5030	Communications & Security	Addressable alarm systems, internet wiring and emergency lighting		S.F. Floor	1.25	1.25	
5090	Other Electrical Systems	Emergency generator, 7.5 kW		S.F. Floor	.04	.04	
E.	**EQUIPMENT & FURNISHINGS**						
1010	Commercial Equipment	N/A		—	—	—	
1020	Institutional Equipment	N/A		—	—	—	0.0%
1030	Vehicular Equipment	N/A		—	—	—	
1090	Other Equipment	N/A		—	—	—	
F.	**SPECIAL CONSTRUCTION**						
1020	Integrated Construction	N/A		—	—	—	0.0%
1040	Special Facilities	N/A		—	—	—	
G.	**BUILDING SITEWORK**	**N/A**					

		Sub-Total	87.49	100%
CONTRACTOR FEES (General Requirements: 10%, Overhead: 5%, Profit: 10%)		25%	21.85	
ARCHITECT FEES		6%	6.56	
Total Building Cost			**115.90**	

Costs per square foot of floor area

Exterior Wall	S.F. Area	50000	65000	80000	95000	110000	125000	140000	155000	170000
	L.F. Perimeter	533	593	670	715	778	840	871	923	976
Face Brick with Concrete Block Back-up	Steel Frame	151.85	145.95	142.85	139.90	138.15	136.75	135.20	134.20	133.40
	R/Conc. Frame	150.50	144.65	141.50	138.50	136.75	135.40	133.80	132.85	132.00
Face Brick on Steel Studs	Steel Frame	147.30	142.05	139.25	136.65	135.10	133.85	132.50	131.65	130.90
	R/Conc. Frame	148.40	143.20	140.35	137.75	136.20	135.00	133.60	132.75	132.05
Precast Concrete Panels Exposed Aggregate	Steel Frame	158.15	151.35	147.80	144.40	142.40	140.80	138.95	137.80	136.80
	R/Conc. Frame	158.70	151.95	148.40	144.95	142.95	141.35	139.45	138.35	137.40
Perimeter Adj., Add or Deduct	Per 100 L.F.	4.95	3.90	3.10	2.60	2.25	2.00	1.75	1.65	1.40
Story Hgt. Adj., Add or Deduct	Per 1 Ft.	1.50	1.30	1.15	1.10	1.00	0.95	0.85	0.90	0.80
For Basement, add $42.65 per square foot of basement area										

The above costs were calculated using the basic specifications shown on the facing page. These costs should be adjusted where necessary for design alternatives and owner's requirements. Reported completed project costs, for this type of structure, range from $62.05 to $166.10 per S.F.

Common additives

Description	Unit	$ Cost	Description	Unit	$ Cost
Closed Circuit Surveillance, One station			Escalators, Metal		
Camera and monitor	Each	1950	32" wide, 10' story height	Each	143,400
For additional camera stations, add	Each	1050	20' story height	Each	174,500
Directory Boards, Plastic, glass covered			48" wide, 10' story height	Each	150,900
30" x 20"	Each	660	20' story height	Each	184,000
36" x 48"	Each	1475	Glass		
Aluminum, 24" x 18"	Each	620	32" wide, 10' story height	Each	137,400
36" x 24"	Each	725	20' story height	Each	167,000
48" x 32"	Each	1050	48" wide, 10' story height	Each	143,400
48" x 60"	Each	2175	20' story height	Each	175,500
Elevators, Hydraulic passenger, 2 stops			Safe, Office type, 1 hour rating		
1500# capacity	Each	66,000	30" x 18" x 18"	Each	2500
2500# capacity	Each	68,900	60" x 36" x 18", double door	Each	8075
3500# capacity	Each	72,900	Sound System		
Additional stop, add	Each	8375	Amplifier, 250 watts	Each	1975
Emergency Lighting, 25 watt, battery operated			Speaker, ceiling or wall	Each	206
Lead battery	Each	305	Trumpet	Each	395
Nickel cadmium	Each	850			

Important: See the Reference Section for Location Factors.

Model costs calculated for a 3 story building with 16' story height and 95,000 square feet of floor area

				Unit	Unit Cost	Cost Per S.F.	% Of Sub-Total
A. SUBSTRUCTURE							
1010	Standard Foundations	Poured concrete; strip and spread footings		S.F. Ground	3.63	1.21	
1020	Special Foundations	N/A		—	—	—	
1030	Slab on Grade	4" reinforced concrete with vapor barrier and granular base		S.F. Slab	5.18	1.72	3.4%
2010	Basement Excavation	Site preparation for slab and trench for foundation wall and footing		S.F. Ground	.18	.06	
2020	Basement Walls	4' foundation wall		L.F. Wall	75	.64	
B. SHELL							
B10 Superstructure							
1010	Floor Construction	Concrete slab with metal deck and beams, steel columns		S.F. Floor	29.34	19.56	21.8%
1020	Roof Construction	Metal deck, open web steel joists, beams, columns		S.F. Roof	10.32	3.44	
B20 Exterior Enclosure							
2010	Exterior Walls	Face brick with concrete block backup	90% of wall	S.F. Wall	31.49	10.24	
2020	Exterior Windows	Storefront	10% of wall	Each	46.40	1.68	12.9%
2030	Exterior Doors	Revolving and sliding panel, mall-front		Each	11,848	1.75	
B30 Roofing							
3010	Roof Coverings	Built-up tar and gravel with flashing; perlite/EPS composite insulation		S.F. Roof	5.70	1.90	1.9%
3020	Roof Openings	Roof hatches		S.F. Roof	.21	.07	
C. INTERIORS							
1010	Partitions	Gypsum board on metal studs	60 S.F. Floor/L.F. Partition	S.F. Partition	6	1	
1020	Interior Doors	Single leaf hollow metal	600 S.F. Floor/Door	Each	1212	2.02	
1030	Fittings	N/A		—	—	—	
2010	Stair Construction	Concrete filled metal pan		Flight	19,725	2.08	24.8%
3010	Wall Finishes	70% paint, 20% vinyl wall covering, 10% ceramic tile		S.F. Surface	4.74	1.58	
3020	Floor Finishes	50% carpet tile, 40% marble tile, 10% terrazzo		S.F. Floor	12.60	12.60	
3030	Ceiling Finishes	Mineral fiber tile on concealed zee bars		S.F. Ceiling	6.86	6.86	
D. SERVICES							
D10 Conveying							
1010	Elevators & Lifts	One hydraulic passenger, one hydraulic freight		Each	312,550	3.29	9.1%
1020	Escalators & Moving Walks	Four escalators		Each	150,100	6.32	
D20 Plumbing							
2010	Plumbing Fixtures	Toilet and service fixtures, supply and drainage	1 Fixture/2570 S.F. Floor	Each	4472	1.74	
2020	Domestic Water Distribution	Gas fired water heater		S.F. Floor	.43	.43	2.5%
2040	Rain Water Drainage	Roof drains		S.F. Roof	1.56	.52	
D30 HVAC							
3010	Energy Supply	N/A		—	—	—	
3020	Heat Generating Systems	Included in D3050		—	—	—	
3030	Cooling Generating Systems	N/A		—	—	—	7.5 %
3050	Terminal & Package Units	Multizone rooftop unit, gas heating, electric cooling		S.F. Floor	7.90	7.90	
3090	Other HVAC Sys. & Equipment	N/A		—	—	—	
D40 Fire Protection							
4010	Sprinklers	Sprinklers, light hazard		S.F. Floor	2.39	2.39	3.7%
4020	Standpipes	Standpipe		S.F. Floor	1.56	1.56	
D50 Electrical							
5010	Electrical Service/Distribution	1200 ampere service, panel board and feeders		S.F. Floor	1.31	1.31	
5020	Lighting & Branch Wiring	High efficiency fluorescent fixtures, receptacles, switches, A.C. and misc. power		S.F. Floor	8.94	8.94	12.3%
5030	Communications & Security	Addressable alarm systems, internet wiring and emergency lighting		S.F. Floor	2.46	2.46	
5090	Other Electrical Systems	Emergency generator, 50 kW		S.F. Floor	.31	.31	
E. EQUIPMENT & FURNISHINGS							
1010	Commercial Equipment	N/A		—	—	—	
1020	Institutional Equipment	N/A		—	—	—	
1030	Vehicular Equipment	N/A		—	—	—	0.0 %
1090	Other Equipment	N/A		—	—	—	
F. SPECIAL CONSTRUCTION							
1020	Integrated Construction	N/A		—	—	—	
1040	Special Facilities	N/A		—	—	—	0.0 %
G. BUILDING SITEWORK	**N/A**						

		Sub-Total	105.58	100%
CONTRACTOR FEES (General Requirements: 10%, Overhead: 5%, Profit: 10%)		25%	26.40	
ARCHITECT FEES		6%	7.92	

Total Building Cost **139.90**

Costs per square foot of floor area

Exterior Wall	S.F. Area	4000	6000	8000	10000	12000	15000	18000	20000	22000
	L.F. Perimeter	260	340	360	410	440	490	540	565	594
Split Face Concrete Block	Steel Joists	146.90	134.00	122.70	117.85	113.55	109.45	106.75	105.15	103.85
Stucco on Concrete Block	Steel Joists	144.20	131.70	120.85	116.15	112.00	108.10	105.50	103.90	102.70
EIFS on Metal Studs	Steel Joists	146.60	133.70	122.50	117.65	113.35	109.30	106.60	105.00	103.75
Face Brick on Concrete Block	Steel Joists	162.50	147.60	133.45	127.70	122.30	117.30	113.95	111.85	110.30
Painted Reinforced Concrete	Steel Joists	154.80	140.90	128.15	122.85	117.95	113.40	110.40	108.55	107.15
Tilt-up Concrete Panels	Steel Joists	141.60	129.35	119.00	114.50	110.50	106.75	104.25	102.80	101.65
Perimeter Adj., Add or Deduct	Per 100 L.F.	16.10	10.75	8.05	6.50	5.35	4.30	3.60	3.15	2.95
Story Hgt. Adj., Add or Deduct	Per 1 Ft.	2.05	1.80	1.40	1.30	1.15	1.10	0.95	0.80	0.90
For Basement, add $35.45 per square foot of basement area										

The above costs were calculated using the basic specifications shown on the facing page. These costs should be adjusted where necessary for design alternatives and owner's requirements. Reported completed project costs, for this type of structure, range from $60.55 to $210.60 per S.F.

Common additives

Description	Unit	$ Cost
Emergency Lighting, 25 watt, battery operated		
Lead battery	Each	305
Nickel cadmium	Each	850
Safe, Office type, 1 hour rating		
30" x 18" x 18"	Each	2500
60" x 36" x 18", double door	Each	8075
Smoke Detectors		
Ceiling type	Each	220
Duct type	Each	550
Sound System		
Amplifier, 250 watts	Each	1975
Speaker, ceiling or wall	Each	206
Trumpet	Each	395

Model costs calculated for a 1 story building with 14' story height and 8,000 square feet of floor area

Store, Retail

				Unit	Unit Cost	Cost Per S.F.	% Of Sub-Total
A. SUBSTRUCTURE							
1010	Standard Foundations	Poured concrete; strip and spread footings		S.F. Ground	1.86	1.86	
1020	Special Foundations	N/A		—	—	—	
1030	Slab on Grade	4" reinforced concrete with vapor barrier and granular base		S.F. Slab	5.18	5.18	12.0%
2010	Basement Excavation	Site preparation for slab and trench for foundation wall and footing		S.F. Ground	.32	.32	
2020	Basement Walls	4' foundation wall		L.F. Wall	79	3.56	
B. SHELL							
B10 Superstructure							
1010	Floor Construction	N/A		—	—	—	
1020	Roof Construction	Metal deck, open web steel joists, beams, interior columns		S.F. Roof	7.36	7.36	8.1%
B20 Exterior Enclosure							
2010	Exterior Walls	Decorative concrete block	90% of wall	S.F. Wall	17.41	9.87	
2020	Exterior Windows	Storefront windows	10% of wall	Each	46.40	2.93	14.7%
2030	Exterior Doors	Sliding entrance door and hollow metal service doors		Each	2268	.57	
B30 Roofing							
3010	Roof Coverings	Built-up tar and gravel with flashing; perlite/EPS composite insulation		S.F. Roof	6.38	6.38	7.2%
3020	Roof Openings	Roof hatches		S.F. Roof	.12	.12	
C. INTERIORS							
1010	Partitions	Gypsum board on metal studs	60 S.F. Floor/L.F. Partition	S.F. Partition	6	1	
1020	Interior Doors	Single leaf hollow metal	600 S.F. Floor/Door	Each	1212	2.02	
1030	Fittings	N/A		—	—	—	
2010	Stair Construction	N/A		—	—	—	15.4%
3010	Wall Finishes	Paint		S.F. Surface	6.27	2.09	
3020	Floor Finishes	Vinyl tile		S.F. Floor	2.02	2.02	
3030	Ceiling Finishes	Mineral fiber tile on concealed zee bars		S.F. Ceiling	6.86	6.86	
D. SERVICES							
D10 Conveying							
1010	Elevators & Lifts	N/A		—	—	—	
1020	Escalators & Moving Walks	N/A		—	—	—	0.0%
D20 Plumbing							
2010	Plumbing Fixtures	Toilet and service fixtures, supply and drainage	1 Fixture/890 S.F. Floor	Each	3071	3.45	
2020	Domestic Water Distribution	Gas fired water heater		S.F. Floor	5.13	5.13	11.2%
2040	Rain Water Drainage	Roof drains		S.F. Roof	1.61	1.61	
D30 HVAC							
3010	Energy Supply	N/A		—	—	—	
3020	Heat Generating Systems	Included in D3050		—	—	—	
3030	Cooling Generating Systems	N/A		—	—	—	8.7%
3050	Terminal & Package Units	Single zone unit, gas heating, electric cooling		S.F. Floor	7.90	7.90	
3090	Other HVAC Sys. & Equipment	N/A		—	—	—	
D40 Fire Protection							
4010	Sprinklers	Wet pipe sprinkler system		S.F. Floor	4.53	4.53	6.1%
4020	Standpipes	Standpipe, wet, Class III		S.F. Floor	1	1	
D50 Electrical							
5010	Electrical Service/Distribution	400 ampere service, panel board and feeders		S.F. Floor	2.84	2.84	
5020	Lighting & Branch Wiring	High efficiency fluorescent fixtures, receptacles, switches, A.C. and misc. power		S.F. Floor	10.43	10.43	16.7%
5030	Communications & Security	Addressable alarm systems and emergency lighting		S.F. Floor	1.64	1.64	
5090	Other Electrical Systems	Emergency generator, 15 kW		S.F. Floor	.23	.23	
E. EQUIPMENT & FURNISHINGS							
1010	Commercial Equipment	N/A		—	—	—	
1020	Institutional Equipment	N/A		—	—	—	
1030	Vehicular Equipment	N/A		—	—	—	0.0%
1090	Other Equipment	N/A		—	—	—	
F. SPECIAL CONSTRUCTION							
1020	Integrated Construction	N/A		—	—	—	
1040	Special Facilities	N/A		—	—	—	0.0%
G. BUILDING SITEWORK	**N/A**						

			Sub-Total	90.90	100%
CONTRACTOR FEES (General Requirements: 10%, Overhead: 5%, Profit: 10%)			25%	22.71	
ARCHITECT FEES			8%	9.09	
		Total Building Cost		**122.70**	

215

Costs per square foot of floor area

Exterior Wall	S.F. Area	12000	16000	20000	26000	32000	38000	44000	52000	60000
	L.F. Perimeter	450	510	570	650	716	780	840	920	1020
Face Brick with Concrete Block Back-up	Steel Frame	139.45	129.20	123.15	117.00	112.65	109.65	107.35	105.05	103.80
	Bearing Walls	137.30	127.05	120.95	114.85	110.50	107.50	105.15	102.90	101.60
Stucco on Concrete Block	Steel Frame	128.60	120.00	114.90	109.85	106.20	103.75	101.80	99.95	98.90
	Bearing Walls	129.15	120.50	115.45	110.35	106.70	104.25	102.30	100.50	99.40
Precast Concrete Panels	Steel Frame	148.50	136.90	130.05	123.10	118.10	114.65	111.95	109.35	107.90
Metal Sandwich Panels	Steel Frame	139.45	129.20	123.15	117.05	112.70	109.65	107.35	105.05	103.80
Perimeter Adj., Add or Deduct	Per 100 L.F.	10.05	7.55	5.95	4.65	3.80	3.15	2.75	2.30	2.00
Story Hgt. Adj., Add or Deduct	Per 1 Ft.	2.05	1.80	1.60	1.40	1.30	1.15	1.05	1.00	0.90
For Basement, add $25.75 per square foot of basement area										

The above costs were calculated using the basic specifications shown on the facing page. These costs should be adjusted where necessary for design alternatives and owner's requirements. Reported completed project costs, for this type of structure, range from $70.75 to $202.65 per S.F.

Common additives

Description	Unit	$ Cost	Description	Unit	$ Cost
Check Out Counter			Refrigerated Food Cases, cont.		
Single belt	Each	3375	Frozen food, chest type, 12' long	Each	9150
Double belt	Each	4825	Glass door reach in, 5 door	Each	15,700
Scanner, registers, guns & memory 2 lanes	Each	17,200	Island case 12' long, single deck	Each	8300
10 lanes	Each	163,500	Multi deck	Each	9650
Power take away	Each	5675	Meat case 12' long, single deck	Each	8300
Emergency Lighting, 25 watt, battery operated			Multi deck	Each	11,700
Lead battery	Each	305	Produce, 12' long single deck	Each	7825
Nickel cadmium	Each	850	Multi deck	Each	9875
Refrigerators, Prefabricated, walk-in			Safe, Office type, 1 hour rating		
7'-6" High, 6' x 6'	S.F.	190	30" x 18" x 18"	Each	2500
10' x 10'	S.F.	150	60" x 36" x 18", double door	Each	8075
12' x 14'	S.F.	133	Smoke Detectors		
12' x 20'	S.F.	117	Ceiling type	Each	220
Refrigerated Food Cases			Duct type	Each	550
Dairy, multi deck, 12' long	Each	12,700	Sound System		
Delicatessen case, single deck, 12' long	Each	8525	Amplifier, 250 watts	Each	1975
Multi deck, 18 S.F. shelf display	Each	7875	Speaker, ceiling or wall	Each	206
Freezer, self-contained chest type, 30 C.F.	Each	9250	Trumpet	Each	395
Glass door upright, 78 C.F.	Each	13,500			

Important: See the Reference Section for Location Factors.

Model costs calculated for a 1 story building with 18' story height and 44,000 square feet of floor area

Supermarket

				Unit	Unit Cost	Cost Per S.F.	% Of Sub-Total
A. SUBSTRUCTURE							
1010	Standard Foundations	Poured concrete; strip and spread footings		S.F. Ground	1.11	1.11	
1020	Special Foundations	N/A		—	—	—	
1030	Slab on Grade	4" reinforced concrete with vapor barrier and granular base		S.F. Slab	5.18	5.18	10.2%
2010	Basement Excavation	Site preparation for slab and trench for foundation wall and footing		S.F. Ground	.18	.18	
2020	Basement Walls	4' foundation wall		L.F. Wall	79	1.51	
B. SHELL							
B10 Superstructure							
1010	Floor Construction	N/A		—	—	—	10.1 %
1020	Roof Construction	Metal deck, open web steel joists, beams, interior columns		S.F. Roof	7.95	7.95	
B20 Exterior Enclosure							
2010	Exterior Walls	Face brick with concrete block backup	85% of wall	S.F. Wall	32.46	9.48	
2020	Exterior Windows	Storefront windows	15% of wall	Each	74	3.80	18.6%
2030	Exterior Doors	Sliding entrance doors with electrical operator, hollow metal		Each	5273	1.33	
B30 Roofing							
3010	Roof Coverings	Built-up tar and gravel with flashing; perlite/EPS composite insulation		S.F. Roof	5.45	5.45	7.1%
3020	Roof Openings	Roof hatches		S.F. Roof	.12	.12	
C. INTERIORS							
1010	Partitions	50% concrete block, 50% gypsum board on metal studs	50 S.F. Floor/L.F. Partition	S.F. Partition	7.58	1.82	
1020	Interior Doors	Single leaf hollow metal	2000 S.F. Floor/Door	Each	1212	.61	
1030	Fittings	N/A		—	—	—	
2010	Stair Construction	N/A		—	—	—	17.5%
3010	Wall Finishes	Paint		S.F. Surface	5.15	2.47	
3020	Floor Finishes	Vinyl composition tile		S.F. Floor	2.02	2.02	
3030	Ceiling Finishes	Mineral fiber tile on concealed zee bars		S.F. Ceiling	6.86	6.86	
D. SERVICES							
D10 Conveying							
1010	Elevators & Lifts	N/A		—	—	—	0.0 %
1020	Escalators & Moving Walks	N/A		—	—	—	
D20 Plumbing							
2010	Plumbing Fixtures	Toilet and service fixtures, supply and drainage	1 Fixture/1820 S.F. Floor	Each	2148	1.18	
2020	Domestic Water Distribution	Gas fired water heater		S.F. Floor	.41	.41	3.3%
2040	Rain Water Drainage	Roof drains		S.F. Roof	.97	.97	
D30 HVAC							
3010	Energy Supply	N/A		—	—	—	
3020	Heat Generating Systems	Included in D3050		—	—	—	
3030	Cooling Generating Systems	N/A		—	—	—	6.6 %
3050	Terminal & Package Units	Single zone rooftop unit, gas heating, electric cooling		S.F. Floor	5.17	5.17	
3090	Other HVAC Sys. & Equipment	N/A		—	—	—	
D40 Fire Protection							
4010	Sprinklers	Sprinklers, light hazard		S.F. Floor	3.51	3.51	5.1%
4020	Standpipes	Standpipe		S.F. Floor	.52	.52	
D50 Electrical							
5010	Electrical Service/Distribution	1600 ampere service, panel board and feeders		S.F. Floor	2.69	2.69	
5020	Lighting & Branch Wiring	High efficiency fluorescent fixtures, receptacles, switches, A.C. and misc. power		S.F. Floor	12.04	12.04	21.6%
5030	Communications & Security	Addressable alarm systems, partial internet wiring and emergency lighting		S.F. Floor	2.19	2.19	
5090	Other Electrical Systems	Emergency generator, 15 kW		S.F. Floor	.05	.05	
E. EQUIPMENT & FURNISHINGS							
1010	Commercial Equipment	N/A		—	—	—	
1020	Institutional Equipment	N/A		—	—	—	0.0 %
1030	Vehicular Equipment	N/A		—	—	—	
1090	Other Equipment	N/A		—	—	—	
F. SPECIAL CONSTRUCTION							
1020	Integrated Construction	N/A		—	—	—	0.0 %
1040	Special Facilities	N/A		—	—	—	
G. BUILDING SITEWORK	**N/A**						

	Sub-Total	78.62	100%
CONTRACTOR FEES (General Requirements: 10%, Overhead: 5%, Profit: 10%)	25%	19.65	
ARCHITECT FEES	7%	6.88	
Total Building Cost		**105.15**	

Costs per square foot of floor area

Exterior Wall	S.F. Area	10000	16000	20000	22000	24000	26000	28000	30000	32000
	L.F. Perimeter	420	510	600	640	660	684	706	740	737
Face Brick with Concrete Block Back-up	Wood Truss	258.75	239.80	236.05	234.25	231.40	229.20	227.20	226.20	223.35
	Precast Conc.	295.60	272.40	267.90	265.70	262.15	259.50	257.00	255.75	252.20
Metal Sandwich Panel	Wood Truss	262.60	244.30	240.05	238.15	235.65	233.65	231.90	230.85	228.55
Precast Concrete Panel	Precast Conc.	297.50	270.85	265.10	262.50	258.65	255.70	253.00	251.55	247.90
Painted Concrete Block	Wood Frame	256.85	243.15	240.35	239.05	237.00	235.45	234.00	233.30	231.30
	Precast Conc.	279.35	257.20	252.30	250.10	246.95	244.55	242.35	241.10	238.15
Perimeter Adj., Add or Deduct	Per 100 L.F.	16.90	10.60	8.40	7.70	7.00	6.50	6.05	5.60	5.25
Story Hgt. Adj., Add or Deduct	Per 1 Ft.	2.10	1.60	1.50	1.50	1.40	1.30	1.30	1.20	1.15

For Basement, add $38.20 per square foot of basement area

The above costs were calculated using the basic specifications shown on the facing page. These costs should be adjusted where necessary for design alternatives and owner's requirements. Reported completed project costs, for this type of structure, range from $111.75 to $328.55 per S.F.

Common additives

Description	Unit	$ Cost	Description	Unit	$ Cost
Bleachers, Telescoping, manual			Sauna, Prefabricated, complete		
To 15 tier	Seat	134 - 187	6' x 4'	Each	6225
16-20 tier	Seat	277 - 340	6' x 6'	Each	7325
21-30 tier	Seat	294 - 355	6' x 9'	Each	8975
For power operation, add	Seat	54 - 84.50	8' x 8'	Each	10,600
Emergency Lighting, 25 watt, battery operated			8' x 10'	Each	11,900
Lead battery	Each	305	10' x 12'	Each	15,100
Nickel cadmium	Each	850	Sound System		
Lockers, Steel, single tier, 60" or 72"	Opening	218 - 355	Amplifier, 250 watts	Each	1975
2 tier, 60" or 72" total	Opening	112 - 160	Speaker, ceiling or wall	Each	206
5 tier, box lockers	Opening	70 - 86	Trumpet	Each	395
Locker bench, lam. maple top only	L.F.	25.50	Steam Bath, Complete, to 140 C.F.	Each	2975
Pedestal, steel pipe	Each	72.50	To 300 C.F.	Each	3350
Pool Equipment			To 800 C.F.	Each	10,600
Diving stand, 3 meter	Each	15,600	To 2500 C.F.	Each	13,300
1 meter	Each	7875			
Diving board, 16' aluminum	Each	4700			
Fiberglass	Each	3525			
Lifeguard chair, fixed	Each	3950			
Portable	Each	3025			
Lights, underwater, 12 volt, 300 watt	Each	945			

Important: See the Reference Section for Location Factors.

Model costs calculated for a 1 story building with 24' story height and 20,000 square feet of floor area

Swimming Pool, Enclosed

				Unit	Unit Cost	Cost Per S.F.	% Of Sub-Total
A. SUBSTRUCTURE							
1010	Standard Foundations	Poured concrete; strip and spread footings		S.F. Ground	1.60	1.60	
1020	Special Foundations	N/A		—	—	—	
1030	Slab on Grade	4" reinforced concrete with vapor barrier and granular base		S.F. Slab	5.18	5.18	15.5%
2010	Basement Excavation	Site preparation for slab and trench for foundation wall and footing		S.F. Ground	10.75	10.75	
2020	Basement Walls	4' foundation wall		L.F. Wall	79	9.49	
B. SHELL							
B10 Superstructure							
1010	Floor Construction	N/A		—	—	—	
1020	Roof Construction	Wood deck on laminated wood truss		S.F. Roof	18.42	18.42	10.5 %
B20 Exterior Enclosure							
2010	Exterior Walls	Face brick with concrete block backup	80% of wall	S.F. Wall	31.49	18.14	
2020	Exterior Windows	Outward projecting steel	20% of wall	Each	46.40	6.69	14.6%
2030	Exterior Doors	Double aluminum and glass, hollow metal		Each	2658	.66	
B30 Roofing							
3010	Roof Coverings	Asphalt shingles with flashing; perlite/EPS composite insulation		S.F. Roof	2.40	2.40	1.4%
3020	Roof Openings	N/A		—	—	—	
C. INTERIORS							
1010	Partitions	Concrete block	100 S.F. Floor/L.F. Partition	S.F. Partition	9.10	.91	
1020	Interior Doors	Single leaf hollow metal	1000 S.F. Floor/Door	Each	1212	1.22	
1030	Fittings	Toilet partitions		S.F. Floor	.18	.18	
2010	Stair Construction	N/A		—	—	—	11.0%
3010	Wall Finishes	Acrylic glazed coating		S.F. Surface	11.45	2.29	
3020	Floor Finishes	70% terrazzo, 30% ceramic tile		S.F. Floor	13.75	13.75	
3030	Ceiling Finishes	Mineral fiber tile on concealed zee bars	20% of area	S.F. Ceiling	4.44	.89	
D. SERVICES							
D10 Conveying							
1010	Elevators & Lifts	N/A		—	—	—	
1020	Escalators & Moving Walks	N/A		—	—	—	0.0 %
D20 Plumbing							
2010	Plumbing Fixtures	Toilet and service fixtures, supply and drainage	1 Fixture/210 S.F. Floor	Each	1214	5.78	
2020	Domestic Water Distribution	Gas fired water heater		S.F. Floor	4.12	4.12	5.7%
2040	Rain Water Drainage	N/A		—	—	—	
D30 HVAC							
3010	Energy Supply	Terminal unit heaters	10% of area	S.F. Floor	.90	.90	
3020	Heat Generating Systems	N/A		—	—	—	
3030	Cooling Generating Systems	N/A		—	—	—	8.1%
3050	Terminal & Package Units	Single zone unit gas heating, electric cooling		S.F. Floor	13.28	13.28	
3090	Other HVAC Sys. & Equipment	N/A		—	—	—	
D40 Fire Protection							
4010	Sprinklers	Sprinklers, light hazard		S.F. Floor	1.76	1.76	
4020	Standpipes	Standpipe		S.F. Floor	.52	.52	1.3%
D50 Electrical							
5010	Electrical Service/Distribution	400 ampere service, panel board and feeders		S.F. Floor	1.19	1.19	
5020	Lighting & Branch Wiring	High efficiency fluorescent fixtures, receptacles, switches, A.C. and misc. power		S.F. Floor	20.73	20.73	
5030	Communications & Security	Addressable alarm systems and emergency lighting		S.F. Floor	1.03	1.03	13.2%
5090	Other Electrical Systems	Emergency generator, 15 kW		S.F. Floor	.09	.09	
E. EQUIPMENT & FURNISHINGS							
1010	Commercial Equipment	N/A		—	—	—	
1020	Institutional Equipment	N/A		—	—	—	
1030	Vehicular Equipment	N/A		—	—	—	18.8 %
1090	Other Equipment	Swimming pool		S.F. Floor	32.87	32.87	
F. SPECIAL CONSTRUCTION							
1020	Integrated Construction	N/A		—	—	—	
1040	Special Facilities	N/A		—	—	—	0.0 %
G. BUILDING SITEWORK	**N/A**						

		Sub-Total	174.84	100%
CONTRACTOR FEES (General Requirements: 10%, Overhead: 5%, Profit: 10%)			25%	43.73	
ARCHITECT FEES			8%	17.48	

Total Building Cost	**236.05**

Costs per square foot of floor area

Exterior Wall	S.F. Area	2000	3000	4000	5000	6000	7000	8000	9000	10000
	L.F. Perimeter	180	220	260	286	320	353	368	397	425
Face Brick with Concrete Block Back-up	Steel Frame	225.55	200.75	188.40	178.30	172.75	168.70	163.50	161.00	158.85
	Bearing Walls	216.10	194.00	182.95	173.65	168.65	165.00	160.05	157.75	155.75
Limestone with Concrete Block Back-up	Steel Frame	261.55	231.90	217.05	204.25	197.50	192.50	185.60	182.35	179.70
	Bearing Walls	261.70	229.35	213.20	199.60	192.30	186.95	179.70	176.35	173.50
Decorative Concrete Block	Steel Frame	209.90	188.00	177.10	168.35	163.50	159.95	155.50	153.25	151.45
	Bearing Walls	205.25	183.40	172.45	163.70	158.85	155.30	150.85	148.65	146.80
Perimeter Adj., Add or Deduct	Per 100 L.F.	48.65	32.50	24.35	19.40	16.25	13.90	12.15	10.75	9.75
Story Hgt. Adj., Add or Deduct	Per 1 Ft.	5.70	4.70	4.15	3.55	3.40	3.25	2.90	2.75	2.65
For Basement, add $37.00 per square foot of basement area										

The above costs were calculated using the basic specifications shown on the facing page. These costs should be adjusted where necessary for design alternatives and owner's requirements. Reported completed project costs, for this type of structure, range from $98.75 to $356.40 per S.F.

Common additives

Description	Unit	$ Cost
Emergency Lighting, 25 watt, battery operated		
Lead battery	Each	305
Nickel cadmium	Each	850
Smoke Detectors		
Ceiling type	Each	220
Duct type	Each	550
Emergency Generators, complete system, gas		
15 kw	Each	18,400
85 kw	Each	38,900
170 kw	Each	98,000
Diesel, 50 kw	Each	29,300
150 kw	Each	50,500
350 kw	Each	79,500

Important: See the Reference Section for Location Factors.

Model costs calculated for a 1 story building with 12' story height and 5,000 square feet of floor area

				Unit	Unit Cost	Cost Per S.F.	% Of Sub-Total
A.	**SUBSTRUCTURE**						
1010	Standard Foundations	Poured concrete; strip and spread footings		S.F. Ground	2.73	2.73	
1020	Special Foundations	N/A		—	—	—	
1030	Slab on Grade	4" reinforced concrete with vapor barrier and granular base		S.F. Slab	5.18	5.18	10.1%
2010	Basement Excavation	Site preparation for slab and trench for foundation wall and footing		S.F. Ground	.55	.55	
2020	Basement Walls	4' foundation wall		L.F. Wall	79	4.52	
B.	**SHELL**						
	B10 Superstructure						
1010	Floor Construction	N/A		—	—	—	7.9 %
1020	Roof Construction	Metal deck, open web steel joists, beams, columns		S.F. Roof	10.15	10.15	
	B20 Exterior Enclosure						
2010	Exterior Walls	Face brick with concrete block backup	80% of wall	S.F. Wall	32.45	17.82	
2020	Exterior Windows	Outward projecting steel	20% of wall	Each	1720	11.81	24.3%
2030	Exterior Doors	Single aluminum glass with transom		Each	3905	1.57	
	B30 Roofing						
3010	Roof Coverings	Built-up tar and gravel with flashing; perlite/EPS composite insulation		S.F. Roof	6.81	6.81	5.3%
3020	Roof Openings						
C.	**INTERIORS**						
1010	Partitions	Double layer gypsum board on metal studs	15 S.F. Floor/L.F. Partition	S.F. Partition	5.99	3.99	
1020	Interior Doors	Single leaf hollow metal	150 S.F. Floor/Door	Each	1212	8.08	
1030	Fittings	Toilet partitions		S.F. Floor	1.10	1.10	
2010	Stair Construction	N/A		—	—	—	24.2%
3010	Wall Finishes	Paint		S.F. Surface	2.15	2.86	
3020	Floor Finishes	90% carpet, 10% terrazzo		S.F. Floor	10.72	10.72	
3030	Ceiling Finishes	Fiberglass board on exposed grid system, suspended		S.F. Ceiling	4.35	4.35	
D.	**SERVICES**						
	D10 Conveying						
1010	Elevators & Lifts	N/A		—	—	—	0.0 %
1020	Escalators & Moving Walks	N/A		—	—	—	
	D20 Plumbing						
2010	Plumbing Fixtures	Kitchen, toilet and service fixtures, supply and drainage	1 Fixture/715 S.F. Floor	Each	3825	5.35	
2020	Domestic Water Distribution	Gas fired water heater		S.F. Floor	2.30	2.30	6.8%
2040	Rain Water Drainage	Roof drains		S.F. Roof	1.11	1.11	
	D30 HVAC						
3010	Energy Supply	N/A		—	—	—	
3020	Heat Generating Systems	Included in D3050		—	—	—	
3030	Cooling Generating Systems	N/A		—	—	—	7.2 %
3050	Terminal & Package Units	Single zone unit, gas heating, electric cooling		S.F. Floor	9.30	9.30	
3090	Other HVAC Sys. & Equipment	N/A		—	—	—	
	D40 Fire Protection						
4010	Sprinklers	Wet pipe sprinkler system		S.F. Floor	5.09	5.09	5.5%
4020	Standpipes	Standpipe		S.F. Floor	1.92	1.92	
	D50 Electrical						
5010	Electrical Service/Distribution	200 ampere service, panel board and feeders		S.F. Floor	1.97	1.97	
5020	Lighting & Branch Wiring	High efficiency fluorescent fixtures, receptacles, switches, A.C. and misc. power		S.F. Floor	6.13	6.13	8.7%
5030	Communications & Security	Addressable alarm systems and emergency lighting		S.F. Floor	2.88	2.88	
5090	Other Electrical Systems	Emergency generator, 15 kW		S.F. Floor	.19	.19	
E.	**EQUIPMENT & FURNISHINGS**						
1010	Commercial Equipment	N/A		—	—	—	
1020	Institutional Equipment	N/A		—	—	—	0.0 %
1030	Vehicular Equipment	N/A		—	—	—	
1090	Other Equipment	N/A		—	—	—	
F.	**SPECIAL CONSTRUCTION**						
1020	Integrated Construction	N/A		—	—	—	0.0 %
1040	Special Facilities	N/A		—	—	—	
G.	**BUILDING SITEWORK**	**N/A**					

		Sub-Total	128.48	100%
CONTRACTOR FEES (General Requirements: 10%, Overhead: 5%, Profit: 10%)		25%	32.15	
ARCHITECT FEES		11%	17.67	

Total Building Cost	**178.30**

Costs per square foot of floor area

Exterior Wall	S.F. Area	5000	6500	8000	9500	11000	14000	17500	21000	24000
	L.F. Perimeter	300	360	386	396	435	510	550	620	680
Face Brick with Concrete Block Back-up	Steel Joists	157.55	151.50	144.30	138.05	135.55	132.05	127.35	125.30	124.10
	Wood Joists	157.25	150.95	143.65	137.20	134.65	131.05	126.25	124.15	122.90
Stone with Concrete Block Back-up	Steel Joists	164.40	157.80	149.85	142.80	140.10	136.25	130.95	128.65	127.35
	Wood Joists	164.05	157.30	149.15	141.95	139.20	135.25	129.80	127.55	126.15
Brick Veneer	Wood Frame	147.95	142.25	135.85	130.20	127.95	124.80	120.60	118.75	117.65
E.I.F.S.	Wood Frame	135.95	130.95	125.60	120.95	119.00	116.30	112.85	111.30	110.40
Perimeter Adj., Add or Deduct	Per 100 L.F.	15.85	12.15	9.90	8.30	7.20	5.70	4.55	3.80	3.30
Story Hgt. Adj., Add or Deduct	Per 1 Ft.	2.90	2.65	2.30	2.00	1.95	1.75	1.50	1.40	1.30
For Basement, add $30.45 per square foot of basement area										

The above costs were calculated using the basic specifications shown on the facing page. These costs should be adjusted where necessary for design alternatives and owner's requirements. Reported completed project costs, for this type of structure, range from $93.35 to $245.60 per S.F.

Common additives

Description	Unit	$ Cost		Description	Unit	$ Cost
Directory Boards, Plastic, glass covered				Smoke Detectors		
30" x 20"	Each	660		Ceiling type	Each	220
36" x 48"	Each	1475		Duct type	Each	550
Aluminum, 24" x 18"	Each	620		Vault Front, Door & frame		
36" x 24"	Each	725		1 Hour test, 32" x 78"	Opening	5275
48" x 32"	Each	1050		2 Hour test, 32" door	Opening	5650
48" x 60"	Each	2175		40" door	Opening	6600
Emergency Lighting, 25 watt, battery operated				4 Hour test, 32" door	Opening	6000
Lead battery	Each	305		40" door	Opening	7175
Nickel cadmium	Each	850		Time lock movement; two movement	Each	2800
Flagpoles, Complete						
Aluminum, 20' high	Each	1750				
40' high	Each	3875				
70' high	Each	10,600				
Fiberglass, 23' high	Each	1325				
39'-5" high	Each	3525				
59' high	Each	7200				
Safe, Office type, 1 hour rating						
30" x 18" x 18"	Each	2500				
60" x 36" x 18", double door	Each	8075				

Model costs calculated for a 1 story building with 12' story height and 11,000 square feet of floor area

				Unit	Unit Cost	Cost Per S.F.	% Of Sub-Total
A. SUBSTRUCTURE							
1010	Standard Foundations	Poured concrete; strip and spread footings		S.F. Ground	1.67	1.67	
1020	Special Foundations	N/A		—	—	—	
1030	Slab on Grade	4" reinforced concrete with vapor barrier and granular base		S.F. Slab	5.18	5.18	10.4%
2010	Basement Excavation	Site preparation for slab and trench for foundation wall and footing		S.F. Ground	.32	.32	
2020	Basement Walls	4' foundation wall		L.F. Wall	79	3.13	
B. SHELL							
B10 Superstructure							
1010	Floor Construction	N/A		—	—	—	
1020	Roof Construction	Metal deck, open web steel joists, beams, interior columns		S.F. Roof	7.69	7.69	7.7 %
B20 Exterior Enclosure							
2010	Exterior Walls	Face brick with concrete block backup	70% of wall	S.F. Wall	32.45	10.78	
2020	Exterior Windows	Metal outward projecting	30% of wall	Each	735	4.55	16.5%
2030	Exterior Doors	Metal and glass with transom		Each	2845	1.04	
B30 Roofing							
3010	Roof Coverings	Built-up tar and gravel with flashing; perlite/EPS composite insulation		S.F. Roof	6.37	6.37	6.4%
3020	Roof Openings	Roof hatches		S.F. Roof			
C. INTERIORS							
1010	Partitions	Gypsum board on metal studs	20 S.F. Floor/L.F. Partition	S.F. Partition	8.90	4.45	
1020	Interior Doors	Wood solid core	200 S.F. Floor/Door	Each	622	3.12	
1030	Fittings	Toilet partitions		S.F. Floor	.48	.48	
2010	Stair Construction	N/A		—	—	—	22.5%
3010	Wall Finishes	90% paint, 10% ceramic tile		S.F. Surface	1.44	1.44	
3020	Floor Finishes	70% carpet tile, 15% terrazzo, 15% vinyl composition tile		S.F. Floor	6.08	6.08	
3030	Ceiling Finishes	Mineral fiber tile on concealed zee bars		S.F. Ceiling	6.86	6.86	
D. SERVICES							
D10 Conveying							
1010	Elevators & Lifts	N/A		—	—	—	0.0 %
1020	Escalators & Moving Walks	N/A		—	—	—	
D20 Plumbing							
2010	Plumbing Fixtures	Kitchen, toilet and service fixtures, supply and drainage	1 Fixture/500 S.F. Floor	Each	2430	4.86	
2020	Domestic Water Distribution	Gas fired water heater		S.F. Floor	1.46	1.46	7.9%
2040	Rain Water Drainage	Roof drains		S.F. Roof	1.50	1.50	
D30 HVAC							
3010	Energy Supply	N/A		—	—	—	
3020	Heat Generating Systems	Included in D3050		—	—	—	
3030	Cooling Generating Systems	N/A		—	—	—	8.6 %
3050	Terminal & Package Units	Multizone unit, gas heating, electric cooling		S.F. Floor	8.60	8.60	
3090	Other HVAC Sys. & Equipment	N/A		—	—	—	
D40 Fire Protection							
4010	Sprinklers	Wet pipe sprinkler system		S.F. Floor	3.51	3.51	4.3%
4020	Standpipes	Standpipe, wet, Class III		S.F. Floor	.73	.73	
D50 Electrical							
5010	Electrical Service/Distribution	400 ampere service, panel board and feeders		S.F. Floor	2.06	2.06	
5020	Lighting & Branch Wiring	High efficiency fluorescent fixtures, receptacles, switches, A.C. and misc. power		S.F. Floor	10.22	10.22	15.8%
5030	Communications & Security	Addressable alarm systems, internet wiring, and emergency lighting		S.F. Floor	3.23	3.23	
5090	Other Electrical Systems	Emergency generator, 15 kW		S.F. Floor	.16	.16	
E. EQUIPMENT & FURNISHINGS							
1010	Commercial Equipment	N/A		—	—	—	
1020	Institutional Equipment	N/A		—	—	—	0.0 %
1030	Vehicular Equipment	N/A		—	—	—	
1090	Other Equipment	N/A		—	—	—	
F. SPECIAL CONSTRUCTION							
1020	Integrated Construction	N/A		—	—	—	0.0 %
1040	Special Facilities	N/A		—	—	—	
G. BUILDING SITEWORK	**N/A**						

		Sub-Total	99.49	100%
CONTRACTOR FEES (General Requirements: 10%, Overhead: 5%, Profit: 10%)		25%	24.87	
ARCHITECT FEES		9%	11.19	

Total Building Cost   135.55

Costs per square foot of floor area

Exterior Wall	S.F. Area	8000	10000	12000	15000	18000	24000	28000	35000	40000
	L.F. Perimeter	206	233	260	300	320	360	393	451	493
Face Brick with Concrete Block Back-up	Steel Frame	210.60	199.30	191.65	184.00	176.70	167.60	164.20	160.05	158.05
	R/Conc. Frame	217.30	205.95	198.35	190.65	183.40	174.35	170.90	166.75	164.70
Stone with Concrete Block Back-up	Steel Frame	206.70	196.35	189.35	182.30	**175.45**	166.80	163.55	159.70	157.75
	R/Conc. Frame	226.10	213.90	205.80	197.50	189.50	179.50	175.70	171.20	168.90
Limestone with Concrete Block Back-up	Steel Frame	230.40	216.30	206.85	197.15	187.40	175.10	170.60	165.15	162.45
	R/Conc. Frame	237.05	222.95	213.50	203.85	194.05	181.75	177.25	171.80	169.10
Perimeter Adj., Add or Deduct	Per 100 L.F.	24.40	19.45	16.25	12.95	10.90	8.20	6.95	5.60	4.85
Story Hgt. Adj., Add or Deduct	Per 1 Ft.	3.75	3.30	3.15	2.85	2.55	2.20	2.00	1.90	1.80

For Basement, add $29.65 per square foot of basement area

The above costs were calculated using the basic specifications shown on the facing page. These costs should be adjusted where necessary for design alternatives and owner's requirements. Reported completed project costs, for this type of structure, range from $83.70 to $245.60 per S.F.

Common additives

Description	Unit	$ Cost
Directory Boards, Plastic, glass covered		
30" x 20"	Each	660
36" x 48"	Each	1475
Aluminum, 24" x 18"	Each	620
36" x 24"	Each	725
48" x 32"	Each	1050
48" x 60"	Each	2175
Elevators, Hydraulic passenger, 2 stops		
1500# capacity	Each	66,000
2500# capacity	Each	68,900
3500# capacity	Each	72,900
Additional stop, add	Each	8375
Emergency Lighting, 25 watt, battery operated		
Lead battery	Each	305
Nickel cadmium	Each	850

Description	Unit	$ Cost
Flagpoles, Complete		
Aluminum, 20' high	Each	1750
40' high	Each	3875
70' high	Each	10,600
Fiberglass, 23' high	Each	1325
39'-5" high	Each	3525
59' high	Each	7200
Safe, Office type, 1 hour rating		
30" x 18" x 18"	Each	2500
60" x 36" x 18", double door	Each	8075
Smoke Detectors		
Ceiling type	Each	220
Duct type	Each	550
Vault Front, Door & frame		
1 Hour test, 32" x 78"	Opening	5275
2 Hour test, 32" door	Opening	5650
40" door	Opening	6600
4 Hour test, 32" door	Opening	6000
40" door	Opening	7175
Time lock movement; two movement	Each	2800

Important: See the Reference Section for Location Factors.

Model costs calculated for a 3 story building with 12' story height and 18,000 square feet of floor area

					Unit	Unit Cost	Cost Per S.F.	% Of Sub-Total
A.	**SUBSTRUCTURE**							
1010	Standard Foundations	Poured concrete; strip and spread footings			S.F. Ground	3.99	1.33	
1020	Special Foundations	N/A			—	—	—	
1030	Slab on Grade	4" reinforced concrete with vapor barrier and granular base			S.F. Slab	5.18	1.72	3.6%
2010	Basement Excavation	Site preparation for slab and trench for foundation wall and footing			S.F. Ground	.18	.06	
2020	Basement Walls	4' foundation wall			L.F. Wall	75	1.56	
B.	**SHELL**							
	B10 Superstructure							
1010	Floor Construction	Open web steel joists, slab form, concrete, wide flange steel columns			S.F. Floor	25.71	17.14	15.5%
1020	Roof Construction	Metal deck, open web steel joists, beams, interior columns			S.F. Roof	8.43	2.81	
	B20 Exterior Enclosure							
2010	Exterior Walls	Stone with concrete block backup	70% of wall		S.F. Wall	42.46	19.02	
2020	Exterior Windows	Metal outward projecting	10% of wall		Each	1125	2.60	17.4%
2030	Exterior Doors	Metal and glass with transoms			Each	2658	.75	
	B30 Roofing							
3010	Roof Coverings	Built-up tar and gravel with flashing; perlite/EPS composite insulation			S.F. Roof	6.33	2.11	1.6%
3020	Roof Openings	N/A			—	—	—	
C.	**INTERIORS**							
1010	Partitions	Gypsum board on metal studs	20 S.F. Floor/L.F. Partition		S.F. Partition	9.92	4.96	
1020	Interior Doors	Wood solid core	200 S.F. Floor/Door		Each	622	3.12	
1030	Fittings	Toilet partitions			S.F. Floor	.29	.29	
2010	Stair Construction	Concrete filled metal pan			Flight	16,950	5.65	22.1%
3010	Wall Finishes	90% paint, 10% ceramic tile			S.F. Surface	1.44	1.44	
3020	Floor Finishes	70% carpet tile, 15% terrazzo, 15% vinyl composition tile			S.F. Floor	6.08	6.08	
3030	Ceiling Finishes	Mineral fiber tile on concealed zee bars			S.F. Ceiling	6.86	6.86	
D.	**SERVICES**							
	D10 Conveying							
1010	Elevators & Lifts	Two hydraulic elevators			Each	118,080	13.12	10.2%
1020	Escalators & Moving Walks	N/A			—	—	—	
	D20 Plumbing							
2010	Plumbing Fixtures	Toilet and service fixtures, supply and drainage	1 Fixture/1385 S.F. Floor		Each	6468	4.67	
2020	Domestic Water Distribution	Gas fired water heater			S.F. Floor	.63	.63	4.6%
2040	Rain Water Drainage	Roof drains			S.F. Roof	1.71	.57	
	D30 HVAC							
3010	Energy Supply	N/A			—	—	—	
3020	Heat Generating Systems	Included in D3050			—	—	—	
3030	Cooling Generating Systems	N/A			—	—	—	6.7 %
3050	Terminal & Package Units	Multizone unit, gas heating, electric cooling			S.F. Floor	8.60	8.60	
3090	Other HVAC Sys. & Equipment	N/A			—	—	—	
	D40 Fire Protection							
4010	Sprinklers	Sprinklers, light hazard			S.F. Floor	3.23	3.23	4.9%
4020	Standpipes	Standpipe, wet, Class III			S.F. Floor	3.02	3.02	
	D50 Electrical							
5010	Electrical Service/Distribution	400 ampere service, panel board and feeders			S.F. Floor	1.76	1.76	
5020	Lighting & Branch Wiring	High efficiency fluorescent fixtures, receptacles, switches, A.C. and misc. power			S.F. Floor	12.10	12.10	13.5%
5030	Communications & Security	Addressable alarm systems, internet wiring, and emergency lighting			S.F. Floor	3.36	3.36	
5090	Other Electrical Systems	Emergency generator, 15 kW			S.F. Floor	.20	.20	
E.	**EQUIPMENT & FURNISHINGS**							
1010	Commercial Equipment	N/A			—	—	—	
1020	Institutional Equipment	N/A			—	—	—	
1030	Vehicular Equipment	N/A			—	—	—	0.0 %
1090	Other Equipment	N/A			—	—	—	
F.	**SPECIAL CONSTRUCTION**							
1020	Integrated Construction	N/A			—	—	—	0.0 %
1040	Special Facilities	N/A			—	—	—	
G.	**BUILDING SITEWORK**	**N/A**						

		Sub-Total	128.76	**100%**
CONTRACTOR FEES (General Requirements: 10%, Overhead: 5%, Profit: 10%)		25%	32.20	
ARCHITECT FEES		9%	14.49	
		Total Building Cost	175.45	

Costs per square foot of floor area

Exterior Wall	S.F. Area	3500	5000	7500	9000	11000	12500	15000	20000	25000
	L.F. Perimeter	240	290	350	390	430	460	510	580	650
E.I.F.S.	Wood Truss	195.25	185.45	176.35	173.50	170.35	168.70	166.55	163.25	161.25
	Steel Joists	180.75	170.60	161.10	158.10	154.80	153.00	150.80	147.30	145.20
Wood Siding	Wood Truss	175.50	165.80	156.80	154.00	150.90	149.25	147.15	143.90	141.90
Brick Veneer	Wood Truss	185.90	174.60	163.85	160.55	156.85	154.80	152.30	148.30	145.85
Brick on Block	Wood Truss	188.00	175.60	163.60	160.00	155.80	153.50	150.70	146.15	143.40
Tilt Up Concrete	Steel Joists	173.65	164.50	155.95	153.30	150.40	148.75	146.80	143.75	141.90
Perimeter Adj., Add or Deduct	Per 100 L.F.	13.90	9.75	6.45	5.40	4.45	3.85	3.25	2.40	1.90
Story Hgt. Adj., Add or Deduct	Per 1 Ft.	1.50	1.30	1.00	0.90	0.85	0.75	0.75	0.65	0.55

For Basement, add $32.05 per square foot of basement area

The above costs were calculated using the basic specifications shown on the facing page. These costs should be adjusted where necessary for design alternatives and owner's requirements. Reported completed project costs, for this type of structure, range from $68.55 to $265.00 per S.F.

Common additives

Description	Unit	$ Cost
Closed circuit surveillance, one station		
Camera and monitor	Each	1950
For additional camera stations, add	Each	1050
Directory boards, plastic, glass covered		
30" x 20"	Each	660
36" x 48"	Each	1475
Aluminum, 24" x 18"	Each	620
36" x 24"	Each	725
48" x 32"	Each	1050
48" x 60"	Each	2175
Electronic, wall mounted	S.F.	980
Free standing	S.F.	1100
Kennel fencing		
Kennel fencing, 1-1/2" mesh, 6' long, 3'-6" wide, 6'-2" high	Each	775
12' long	Each	1050
Top covers, 1-1/2" mesh, 6' long	Each	210
12' long	Each	285

Description	Unit	$ Cost
Kennel doors		
2 way, swinging type, 13" x 19" opening	Opng.	200
17" x 29" opening	Opng.	222
9" x 9" opening, electronic with accessories	Opng.	259

Model costs calculated for a 1 story building with 12' story height and 7,500 square feet of floor area

Veterinary Hospital

				Unit	Unit Cost	Cost Per S.F.	% Of Sub-Total
A. SUBSTRUCTURE							
1010	Standard Foundations	Poured concrete; strip and spread footings		S.F. Ground	4.36	4.36	
1020	Special Foundations	N/A		—	—	—	
1030	Slab on Grade	4" and 5" reinforced concrete with vapor barrier and granular base		S.F. Slab	10.77	5.22	12.0%
2010	Basement Excavation	Site preparation for slab and trench for foundation wall and footing		S.F. Ground	.55	.55	
2020	Basement Walls	4' foundation wall		L.F. Wall	79	3.69	
B. SHELL							
B10 Superstructure							
1010	Floor Construction	6 x 6 wood columns		S.F. Floor	.22	.22	8.0%
1020	Roof Construction	Wood roof truss with plywood sheathing		S.F. Roof	8.99	8.99	
B20 Exterior Enclosure							
2010	Exterior Walls	Cedar bevel siding on wood studs, insulated	85% of wall	S.F. Wall	13.34	6.35	
2020	Exterior Windows	Wood double hung and awning	15% of wall	Each	1146	3.86	10.7%
2030	Exterior Doors	Aluminum and glass		Each	2670	2.14	
B30 Roofing							
3010	Roof Coverings	Ashphalt shingles with flashing (pitched); al. gutters and downspouts		S.F. Roof	4.64	4.64	4.0%
3020	Roof Openings	N/A		—	—	—	
C. INTERIORS							
1010	Partitions	Gypsum board on wood studs	8 S.F. Floor/L.F. Partition	S.F. Partition	4.81	6.01	
1020	Interior Doors	Solid core wood	180 S.F. Floor/Door	Each	689	3.83	
1030	Fittings	Lockers		Each	.43	.43	
2010	Stair Construction	N/A		—	—	—	18.2%
3010	Wall Finishes	Paint on drywall		S.F. Surface	.98	2.45	
3020	Floor Finishes	90% vinyl composition tile, 5% ceramic tile, 5% sealed concrete		S.F. Floor	2.62	2.62	
3030	Ceiling Finishes	90% acoustic tile, 10% gypsum board		S.F. Ceiling	5.60	5.61	
D. SERVICES							
D10 Conveying							
1010	Elevators & Lifts	N/A		—	—	—	0.0 %
1020	Escalators & Moving Walks	N/A		—	—	—	
D20 Plumbing							
2010	Plumbing Fixtures	Toilet and service fixtures, supply and drainage	1 Fixture/1075 S.F. Floor	Each	1882	3.36	
2020	Domestic Water Distribution	Electric water heater		S.F. Floor	1.10	1.10	3.9%
2040	Rain Water Drainage	N/A		—	—	—	
D30 HVAC							
3010	Energy Supply	N/A		—	—	—	
3020	Heat Generating Systems	N/A		—	—	—	
3030	Cooling Generating Systems	N/A		—	—	—	6.0 %
3050	Terminal & Package Units	Split system with air cooled condensing units		S.F. Floor	6.85	6.85	
3090	Other HVAC Sys. & Equipment	N/A					
D40 Fire Protection							
4010	Sprinklers	Wet pipe sprinkler system		S.F. Floor	3.51	3.51	3.1%
4020	Standpipes	N/A		—	—	—	
D50 Electrical							
5010	Electrical Service/Distribution	800 ampere service, panel board and feeders		S.F. Floor	7.17	7.17	
5020	Lighting & Branch Wiring	High efficiency fluorescent fixtures, receptacles, switches, A.C. and misc. power		S.F. Floor	11.96	11.96	20.8%
5030	Communications & Security	Addressable alarm systems, internet wiring, and emergency lighting		S.F. Floor	4.74	4.74	
5090	Other Electrical Systems	Emergency generator, 100 kW		S.F. Floor	.05	.05	
E. EQUIPMENT & FURNISHINGS							
1010	Commercial Equipment	N/A		—	—	—	
1020	Institutional Equipment	Eye wash station		S.F. Floor	13.52	13.52	13.4 %
1030	Vehicular Equipment	N/A		—	—	—	
1090	Other Equipment	Countertop, plastic laminate		S.F.Floor	1.85	1.85	
F. SPECIAL CONSTRUCTION & DEMOLITION							
1020	Integrated Construction	N/A		—	—	—	0.0 %
1040	Special Facilities	N/A		—	—	—	
G. BUILDING SITEWORK	**N/A**						

		Sub-Total	115.08	100%
CONTRACTOR FEES (General Requirements: 10%, Overhead: 5%, Profit: 10%)		25%	28.77	
ARCHITECT FEES		9%	12.95	
	Total Building Cost		**156.80**	

Costs per square foot of floor area

Exterior Wall	S.F. Area	10000	15000	20000	25000	30000	35000	40000	50000	60000
	L.F. Perimeter	410	500	600	640	700	766	833	966	1000
Tiltup Concrete Panels	Steel Frame	115.95	106.15	101.55	97.00	94.55	92.90	91.70	89.95	87.60
Brick with Block Back-up	Bearing Walls	137.60	122.80	116.05	108.80	104.85	102.25	100.40	97.70	93.65
Concrete Block	Steel Frame	118.85	111.00	107.45	103.65	101.55	100.15	99.15	97.75	95.55
	Bearing Walls	115.25	107.35	103.70	99.80	97.70	96.30	95.25	93.85	91.60
Galvanized Steel Siding	Steel Frame	127.20	118.90	115.15	111.05	108.90	107.40	106.30	104.85	102.50
Metal Sandwich Panels	Steel Frame	128.65	118.95	114.65	109.75	107.10	105.40	104.10	102.35	99.60
Perimeter Adj., Add or Deduct	Per 100 L.F.	7.25	4.80	3.65	3.00	2.45	2.10	1.85	1.45	1.15
Story Hgt. Adj., Add or Deduct	Per 1 Ft.	0.85	0.70	0.65	0.55	0.50	0.45	0.40	0.45	0.30
For Basement, add $28.50 per square foot of basement area										

The above costs were calculated using the basic specifications shown on the facing page. These costs should be adjusted where necessary for design alternatives and owner's requirements. Reported completed project costs, for this type of structure, range from $37.20 to $148.45 per S.F.

Common additives

Description	Unit	$ Cost	Description	Unit	$ Cost
Dock Leveler, 10 ton cap.			Sound System		
6' x 8'	Each	6975	Amplifier, 250 watts	Each	1975
7' x 8'	Each	7850	Speaker, ceiling or wall	Each	206
Emergency Lighting, 25 watt, battery operated			Trumpet	Each	395
Lead battery	Each	305	Yard Lighting, 20' aluminum pole	Each	3075
Nickel cadmium	Each	850	with 400 watt		
Fence, Chain link, 6' high			high pressure sodium		
9 ga. wire	L.F.	28	fixture.		
6 ga. wire	L.F.	30			
Gate	Each	370			
Flagpoles, Complete					
Aluminum, 20' high	Each	1750			
40' high	Each	3875			
70' high	Each	10,600			
Fiberglass, 23' high	Each	1325			
39'-5" high	Each	3525			
59' high	Each	7200			
Paving, Bituminous					
Wearing course plus base course	S.Y.	8.05			
Sidewalks, Concrete 4" thick	S.F.	4.28			

Important: See the Reference Section for Location Factors.

Model costs calculated for a 1 story building with 24' story height and 30,000 square feet of floor area

Warehouse

				Unit	Unit Cost	Cost Per S.F.	% Of Sub-Total
A. SUBSTRUCTURE							
1010	Standard Foundations	Poured concrete; strip and spread footings		S.F. Ground	1.42	1.42	
1020	Special Foundations	N/A		—	—		
1030	Slab on Grade	5" reinforced concrete with vapor barrier and granular base		S.F. Slab	12.82	12.82	23.0%
2010	Basement Excavation	Site preparation for slab and trench for foundation wall and footing		S.F. Ground	.18	.18	
2020	Basement Walls	4' foundation wall		L.F. Wall	79	2.40	
B. SHELL							
	B10 Superstructure						
1010	Floor Construction	Mezzanine: open web steel joists, slab form, concrete beams, columns	10% of area	S.F. Floor	21.30	2.13	12.8%
1020	Roof Construction	Metal deck, open web steel joists, beams, columns		S.F. Roof	7.23	7.23	
	B20 Exterior Enclosure						
2010	Exterior Walls	Concrete block	95% of wall	S.F. Wall	12.35	6.57	
2020	Exterior Windows	N/A		—	—	—	10.6%
2030	Exterior Doors	Steel overhead, hollow metal	5% of wall	Each	3117	1.15	
	B30 Roofing						
3010	Roof Coverings	Built-up tar and gravel with flashing; perlite/EPS composite insulation		S.F. Roof	5.60	5.60	8.1%
3020	Roof Openings	Roof hatches and skylight		S.F. Roof	.33	.33	
C. INTERIORS							
1010	Partitions	Concrete block (office and washrooms)	100 S.F. Floor/L.F. Partition	S.F. Partition	9	.72	
1020	Interior Doors	Single leaf hollow metal	5000 S.F. Floor/Door	Each	1212	.24	
1030	Fittings	N/A		—	—	—	
2010	Stair Construction	Steel gate with rails		Flight	13,375	.90	9.1%
3010	Wall Finishes	Paint		S.F. Surface	13.88	2.22	
3020	Floor Finishes	90% hardener, 10% vinyl composition tile		S.F. Floor	1.87	1.87	
3030	Ceiling Finishes	Suspended mineral tile on zee channels in office area	10% of area	S.F. Ceiling	6.86	.69	
D. SERVICES							
	D10 Conveying						
1010	Elevators & Lifts	N/A		—	—	—	0.0%
1020	Escalators & Moving Walks	N/A		—	—	—	
	D20 Plumbing						
2010	Plumbing Fixtures	Toilet and service fixtures, supply and drainage	1 Fixture/2500 S.F. Floor	Each	3625	1.45	
2020	Domestic Water Distribution	Gas fired water heater		S.F. Floor	.38	.38	4.8%
2040	Rain Water Drainage	Roof drains		S.F. Roof	1.68	1.68	
	D30 HVAC						
3010	Energy Supply	N/A		—	—	—	
3020	Heat Generating Systems	Ventilation with heat		Each	119,500	4.30	
3030	Cooling Generating Systems	N/A		—	—	—	7.2%
3050	Terminal & Package Units	Single zone unit gas, heating, electric cooling	10% of area	S.F. Floor	.96	.96	
3090	Other HVAC Sys. & Equipment	N/A		—	—	—	
	D40 Fire Protection						
4010	Sprinklers	Sprinklers, ordinary hazard		S.F. Floor	3.85	3.85	8.2%
4020	Standpipes	Standpipe		S.F. Floor	2.12	2.12	
	D50 Electrical						
5010	Electrical Service/Distribution	200 ampere service, panel board and feeders		S.F. Floor	.56	.56	
5020	Lighting & Branch Wiring	High efficiency fluorescent fixtures, receptacles, switches, A.C. and misc. power		S.F. Floor	5.99	5.99	12.6%
5030	Communications & Security	Addressable alarm systems		S.F. Floor	2.67	2.67	
5090	Other Electrical Systems	N/A		—	—	—	
E. EQUIPMENT & FURNISHINGS							
1010	Commercial Equipment	N/A		—	—	—	
1020	Institutional Equipment	N/A		—	—	—	
1030	Vehicular Equipment	Dock boards, dock levelers		S.F. Floor	2.62	2.62	3.6%
1090	Other Equipment	N/A		—	—	—	
F. SPECIAL CONSTRUCTION							
1020	Integrated Construction	N/A		—	—	—	0.0%
1040	Special Facilities	N/A		—	—	—	
G. BUILDING SITEWORK	**N/A**						

			Sub-Total	73.05	**100%**
CONTRACTOR FEES (General Requirements: 10%, Overhead: 5%, Profit: 10%)			25%	18.26	
ARCHITECT FEES			7%	6.39	
		Total Building Cost		**97.70**	

Costs per square foot of floor area

Exterior Wall	S.F. Area	2000	3000	5000	8000	12000	20000	30000	50000	100000
	L.F. Perimeter	180	220	300	420	580	900	1300	2100	4100
Concrete Block	Steel Frame	117.25	107.05	98.95	94.30	91.75	89.80	88.75	87.95	87.35
	R/Conc. Frame	121.70	115.00	109.70	106.70	105.05	103.75	103.05	102.50	102.15
Metal Sandwich Panel	Steel Frame	111.25	103.70	97.70	94.30	92.40	90.90	90.10	89.55	89.10
Tilt-up Concrete Panel	Steel and Concrete	91.85	85.30	80.15	77.25	75.65	74.30	73.65	73.15	72.80
Precast Concrete Panel	Steel Frame	121.50	111.75	103.95	99.60	97.15	95.20	94.20	93.40	92.80
	R/Conc. Frame	142.60	133.40	126.10	121.95	119.70	117.80	116.90	116.20	115.65
Perimeter Adj., Add or Deduct	Per 100 L.F.	14.15	9.45	5.65	3.60	2.45	1.40	1.00	0.60	0.25
Story Hgt. Adj., Add or Deduct	Per 1 Ft.	0.70	0.55	0.45	0.45	0.40	0.35	0.35	0.35	0.30
Basement—Not Applicable										

The above costs were calculated using the basic specifications shown on the facing page. These costs should be adjusted where necessary for design alternatives and owner's requirements. Reported completed project costs, for this type of structure, range from $35.60 to $204.20 per S.F.

Common additives

Description	Unit	$ Cost
Dock Leveler, 10 ton cap.		
6' x 8'	Each	6975
7' x 8'	Each	7850
Emergency Lighting, 25 watt, battery operated		
Lead battery	Each	305
Nickel cadmium	Each	850
Fence, Chain link, 6' high		
9 ga. wire	L.F.	28
6 ga. wire	L.F.	30
Gate	Each	370
Flagpoles, Complete		
Aluminum, 20' high	Each	1750
40' high	Each	3875
70' high	Each	10,600
Fiberglass, 23' high	Each	1325
39'-5" high	Each	3525
59' high	Each	7200
Paving, Bituminous		
Wearing course plus base course	S.Y.	8.05
Sidewalks, Concrete 4" thick	S.F.	4.28

Description	Unit	$ Cost
Sound System		
Amplifier, 250 watts	Each	1975
Speaker, ceiling or wall	Each	206
Trumpet	Each	395
Yard Lighting, 20' aluminum pole with 400 watt high pressure sodium fixture	Each	3075

Warehouse, Self Storage

Model costs calculated for a 1 story building with 12' story height and 20,000 square feet of floor area

				Unit	Unit Cost	Cost Per S.F.	% Of Sub-Total
A. SUBSTRUCTURE							
1010	Standard Foundations	Poured concrete; strip and spread footings		S.F. Ground	1.91	1.91	
1020	Special Foundations	N/A		—	—	—	
1030	Slab on Grade	4" reinforced concrete with vapor barrier and granular base		S.F. Slab	6.36	6.36	17.5%
2010	Basement Excavation	Site preparation for slab and trench for foundation wall and footing		S.F. Ground	.32	.32	
2020	Basement Walls	4' foundation wall		L.F. Wall	71	3.18	
B. SHELL							
	B10 Superstructure						
1010	Floor Construction	N/A		—	—	—	9.7%
1020	Roof Construction	Metal deck, open web steel joists, beams, columns		S.F. Roof	6.54	6.54	
	B20 Exterior Enclosure						
2010	Exterior Walls	Concrete block	53% of wall	S.F. Wall	40.85	11.69	
2020	Exterior Windows	Aluminum projecting	1% of wall	Each	820	.08	26.3%
2030	Exterior Doors	Steel overhead, hollow metal	46% of wall	Each	1451	5.88	
	B30 Roofing						
3010	Roof Coverings	Metal panel roof; perlite/EPS composite insulation		S.F. Roof	20.60	20.60	30.7%
3020	Roof Openings	N/A		—	—	—	
C. INTERIORS							
1010	Partitions	Metal panels on metal studs	11 S.F. Floor/L.F. Partition	S.F. Partition	1.11	1.21	
1020	Interior Doors	Single leaf hollow metal	20,000 S.F. Floor/Door	Each	600	.06	
1030	Fittings	N/A		—	—	—	
2010	Stair Construction	N/A		—	—	—	1.9%
3010	Wall Finishes	N/A		—	—	—	
3020	Floor Finishes	N/A		—	—	—	
3030	Ceiling Finishes	N/A		—	—	—	
D. SERVICES							
	D10 Conveying						
1010	Elevators & Lifts	N/A		—	—	—	0.0%
1020	Escalators & Moving Walks	N/A		—	—	—	
	D20 Plumbing						
2010	Plumbing Fixtures	Toilet and service fixtures, supply and drainage	1 Fixture/10,000 S.F. Floor	Each	2600	.26	
2020	Domestic Water Distribution	Gas fired water heater		S.F. Floor	.27	.27	0.8%
2040	Rain Water Drainage	N/A		—	—	—	
	D30 HVAC						
3010	Energy Supply	N/A		—	—	—	
3020	Heat Generating Systems	N/A		—	—	—	
3030	Cooling Generating Systems	N/A		—	—	—	2.1%
3050	Terminal & Package Units	Single zone rooftop unit		S.F. Floor	1.41	1.41	
3090	Other HVAC Sys. & Equipment	N/A		—	—	—	
	D40 Fire Protection						
4010	Sprinklers	N/A		—	—	—	0.0%
4020	Standpipes	N/A		—	—	—	
	D50 Electrical						
5010	Electrical Service/Distribution	60 ampere service, panel board and feeders		S.F. Floor	.42	.42	
5020	Lighting & Branch Wiring	High bay fixtures, receptacles, switches and misc. power		S.F. Floor	5.64	5.64	11.0%
5030	Communications & Security	Addressable alarm systems		S.F. Floor	1.31	1.31	
5090	Other Electrical Systems	N/A		—	—	—	
E. EQUIPMENT & FURNISHINGS							
1010	Commercial Equipment	N/A		—	—	—	
1020	Institutional Equipment	N/A		—	—	—	
1030	Vehicular Equipment	N/A		—	—	—	0.0%
1090	Other Equipment	N/A		—	—	—	
F. SPECIAL CONSTRUCTION							
1020	Integrated Construction	N/A		—	—	—	0.0%
1040	Special Facilities	N/A		—	—	—	
G. BUILDING SITEWORK	**N/A**						

		Sub-Total	67.14	100%
CONTRACTOR FEES (General Requirements: 10%, Overhead: 5%, Profit: 10%)		25%	16.78	
ARCHITECT FEES		7%	5.88	
	Total Building Cost		**89.80**	

Deterioration occurs within the structure itself and is determined by the observation of both materials and equipment.

Curable deterioration can be remedied either by maintenance, repair or replacement, within prudent economic limits.

Incurable deterioration that has progressed to the point of actually affecting the structural integrity of the structure, making repair or replacement not economically feasible.

Actual Versus Observed Age

The observed age of a structure refers to the age that the structure appears to be. Periodic maintenance, remodeling, and renovation all tend to reduce the amount of deterioration that has taken place, thereby decreasing the observed age. Actual age, on the other hand, relates solely to the year that the structure was built.

The Depreciation Table shown here relates to the observed age.

Obsolescence arises from conditions either occurring within the structure (functional) or caused by factors outside the limits of the structure (economic).

Functional obsolescence is any inadequacy caused by outmoded design, dated construction materials, or oversized or undersized areas, all of which cause excessive operation costs.

Incurable is so costly as not to economically justify the capital expenditure required to correct the deficiency.

Economic obsolescence is caused by factors outside the limits of the structure. The prime causes of economic obsolescence are:

- zoning and environmental laws
- government legislation
- negative neighborhood influences
- business climate
- proximity to transportation facilities

Depreciation Table
Commercial/Industrial/Institutional

	Building Material		
Observed Age (Years)	**Frame**	**Masonry On Wood**	**Masonry On Masonry Or Steel**
1	1%	0%	0%
2	2	1	0
3	3	2	1
4	4	3	2
5	6	5	3
10	20	15	8
15	25	20	15
20	30	25	20
25	35	30	25
30	40	35	30
35	45	40	35
40	50	45	40
45	55	50	45
50	60	55	50
55	65	60	55
60	70	65	60

Assemblies Section

Table of Contents

Table of Contents

Introduction to the Assemblies Section

This section of the manual contains the technical description and installed cost of all components used in the commercial/industrial/institutional section. In most cases, there is an illustration of the component to aid in identification.

All Costs Include:

Materials purchased in lot sizes typical of normal construction projects with discounts appropriate to established contractors.

Installation work performed by union labor working under standard conditions at a normal pace.

Installing Contractor's Overhead & Profit

Standard installing contractor's overhead & profit are included in the assemblies costs.

These assemblies costs contain no provisions for General Contractor's overhead and profit or Architect's fees, nor do they include premiums for labor or materials, or savings which may be realized under certain economic situations.

The assemblies tables are arranged by physical size (dimensions) of the component wherever possible.

How RSMeans Assemblies Data Works

Assemblies estimating provides a fast and reasonably accurate way to develop construction costs. An assembly is the grouping of individual work items, with appropriate quantities, to provide a cost for a major construction component in a convenient unit of measure.

An assemblies estimate is often used during early stages of design development to compare the cost impact of various design alternatives on total building cost.

Assemblies estimates are also used as an efficient tool to verify construction estimates.

Assemblies estimates do not require a completed design or detailed drawings. Instead, they are based on the general size of the structure and other known parameters of the project. The degree of accuracy of an assemblies estimate generally within +/- 15%.

NARRATIVE FORMAT

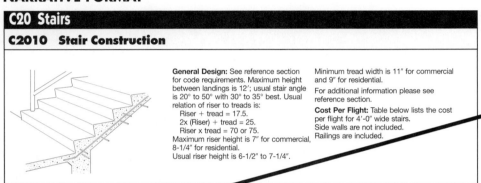

C20 Stairs

C2010 Stair Construction

General Design: See reference section for code requirements. Maximum height between landings is 12'; usual stair angle is 20° to 50° with 30° to 35° best. Usual relation of riser to treads is:
  Riser + tread = 17.5.
  2x (Riser) + tread = 25.
  Riser x tread = 70 or 75.
Maximum riser height is 7" for commercial, 8-1/4" for residential.
Usual riser height is 6-1/2" to 7-1/4".

Minimum tread width is 11" for commercial and 9" for residential.
For additional information please see reference section.

Cost Per Flight: Table below lists the cost per flight for 4'-0" wide stairs. Side walls are not included. Railings are included.

RSMeans assemblies are identified by a **unique 12-character identifier**. The assemblies are numbered using UNIFORMAT II, ASTM Standard E1557. The first 6 characters represent this system to Level 4. The last 6 characters represent further breakdown by RSMeans in order to arrange items in understandable groups of similar tasks. Line numbers are consistent across all RSMeans publications, so a line number in any RSMeans assemblies data set will always refer to the same work.

System Components	QUANTITY	UNIT	COST PER FLIGHT		
			MAT.	INST.	TOTAL
SYSTEM C2010 110 0560					
STAIRS, C.I.P. CONCRETE WITH LANDING, 12 RISERS					
Concrete in place, free standing stairs not incl. safety treads	48.000	L.F.	283.20	1,917.12	2,200.32
Concrete in place, free standing stair landing	32.000	S.F.	153.28	527.36	680.64
Cast alum nosing insert, abr surface, pre-drilled, 3" wide x 4' long	12.000	Ea.	768	196.20	964.20
Industrial railing, welded, 2 rail 3'-6" high 1-1/2" pipe	18.000	L.F.	522	214.56	736.56
Wall railing with returns, steel pipe	17.000	L.F.	282.20	202.64	484.84
TOTAL			2,008.68	3,057.88	5,066.56

Information is available in the Reference Section to assist the estimator with estimating procedures, alternate pricing methods, and additional technical information.

The **Reference Box** indicates the exact location of this information in the Reference Section at the back of the book. The "R" stands for "reference," and the remaining characters are the RSMeans line numbers.

C2010 110	Stairs		COST PER FLIGHT		
			MAT.	INST.	TOTAL
0470	Stairs, C.I.P. concrete, w/o landing, 12 risers, w/o nosing	RC2010 -100	1,075	2,350	3,425
0480	With nosing		1,850	2,550	4,400
0550	W/landing, 12 risers, w/o nosing		1,250	2,875	4,125
0560	With nosing		2,000	3,050	5,050
0570	16 risers, w/o nosing		1,525	3,575	5,100
0580	With nosing		2,550	3,850	6,400
0590	20 risers, w/o nosing		1,800	4,325	6,125
0600	With nosing		3,075	4,650	7,725
0610	24 risers, w/o nosing		2,075	5,050	7,125
0620	With nosing		3,600	5,450	9,050
0630	Steel, grate type w/nosing & rails, 12 risers, w/o landing		5,525	1,225	6,750
0640	With landing		7,225	1,675	8,900
0660	16 risers, with landing		9,075	2,075	11,150
0680	20 risers, with landing		10,900	2,475	13,375
0700	24 risers, with landing		12,700	2,875	15,575
0710	Concrete fill metal pan & picket rail, 12 risers, w/o landing		7,200	1,225	8,425
0720			9,475	1,825	11,300
0740	16 risers, with landing		11,900	2,250	14,150
0760	20 risers, with landing		14,300	2,650	16,950
0780	24 risers, with landing		16,700	3,025	19,725
0790	Cast iron tread & pipe rail, 12 risers, w/o landing		7,200	1,225	8,425
0800	With landing		9,475	1,825	11,300
0820	16 risers, with landing		11,900	2,250	14,150

RSMeans assemblies descriptions appear in two formats: narrative and table. **Narrative descriptions** are shown in a hierarchical structure to make them readable. In order to read a complete description, read up through the indents to the top of the section. Include everything that is above and to the left that is not contradicted by information below.

For supplemental customizable square foot estimating forms, visit: **http://www.reedconstructiondata.com/rsmeans/sf-costs**

Most assemblies consist of three major elements: a graphic, the system components, and the cost data itself. The **Graphic** is a visual representation showing the typical appearance of the assembly in question, frequently accompanied by additional explanatory technical information describing the class of items. The **System Components** is a listing of the individual tasks that make up the assembly, including the quantity and unit of measure for each item, along with the cost of material and installation. The **Assemblies Data** below lists prices for other similar systems with dimensional and/or size variations.

All RSMeans assemblies costs represent the cost for the installing contractor. An allowance for profit has been added to all material, labor, and equipment rental costs. A markup for labor burdens, including workers' compensation, fixed overhead, and business overhead, is included with installation costs.

The data published in RSMeans print books represents a "national average" cost. This data should be modified to the project location using the **City Cost Indexes** or **Location Factors** tables found in the Reference Section in the back of the book.

TABLE FORMAT

A20 Basement Construction

A2020 Basement Walls

The Foundation Bearing Wall System includes: forms up to 16' high (four uses); 3,000 p.s.i. concrete placed and vibrated; and form removal with breaking form ties and patching walls. The wall systems list walls from 6" to 16" thick and are designed with minimum reinforcement.

Excavation and backfill are not included.

Please see the reference section for further design and cost information.

All RSMeans assemblies data includes a typical **Unit of Measure** used for estimating that item. For instance, for continuous footings or foundation walls the unit is linear feet (L.F.). For spread footings the unit is each (Ea.). The estimator needs to take special care that the unit in the data matches the unit in the takeoff. Abbreviations and unit conversions can be found in the Reference Section.

System components are listed separately to detail what is included in the development of the total system price.

System Components				COST PER L.F.		
		QUANTITY	UNIT	MAT.	INST.	TOTAL
SYSTEM A2020 110 1500						
FOUNDATION WALL, CAST IN PLACE, DIRECT CHUTE, 4' HIGH, 6" THICK						
Formwork		8.000	SFCA	5.28	43.20	48.48
Reinforcing		3.300	Lb.	1.82	1.39	3.21
Unloading & sorting reinforcing		3.300	Lb.		.08	.08
Concrete, 3,000 psi		.074	C.Y.	7.92		7.92
Place concrete, direct chute		.074	C.Y.		2.28	2.28
Finish walls, break ties and patch voids, one side		4.000	S.F.	.16	3.76	3.92
	TOTAL			15.18	50.71	65.89

Table descriptions work in a similar fashion, except that if there is a blank in the column at a particular line number, read up to the description above in the same column.

A2020 110			**Walls, Cast in Place**					
	WALL HEIGHT (FT.)	PLACING METHOD	CONCRETE (C.Y. per L.F.)	REINFORCING (LBS. per L.F.)	WALL THICKNESS (IN.)	COST PER L.F.		
						MAT.	INST.	TOTAL
1500	4'	direct chute	.074	3.3	6	15.20	50.50	65.70
1520			.099	4.8	8	18.65	52	70.65
1540			.123	6.0	10	22	53	75
1560			.148	7.2	12	25	54	79
1580	RA2020 -210		.173	8.1	14	28.50	55	83.50
1600			.197	9.44	16	31.50	56	87.50
1700	4'	pumped	.074	3.3	6	15.20	51.50	66.70
1720			.099	4.8	8	18.65	54	72.65
1740			.123	6.0	10	22	55	77
1760			.148	7.2	12	25	56	81
1780			.173	8.1	14	28.50	57.50	86
1800			.197	9.44	16	31.50	58.50	90
3000	6'	direct chute	.111	4.95	6	23	76	99
3020			.149	7.20	8	28	78	106
3040			.184	9.00	10	33	79.50	112.50
3060			.222	10.8	12	38	81	119
3080			.260	12.15	14	42.50	82.50	125
3100			.300	14.39	16	48	85	133

How RSMeans Assemblies Data Works (continued)

Sample Estimate

This sample demonstrates the elements of an estimate, including a tally of the RSMeans data lines. Published assemblies costs include all markups for labor burden and profit for the installing contractor. This estimate adds a summary of the markups applied by a general contractor on the installing contractors' work. These figures represent the total cost to the owner. The RSMeans location factor is added at the bottom of the estimate to adjust the cost of the work to a specific location

Work performed: The body of the estimate shows the RSMeans data selected, including line numbers, a brief description of each item, its takeoff quantity and unit, and the total installed cost, including the installing contractor's overhead and profit.

Location Factor: RSMeans published data is based on national average costs.
If necessary, adjust the total cost of the project using a location factor from the "Location Factor" table or the "City Cost Indexes" table, found in the Reference Section. Use location factors if the work is general, covering the work of multiple trades. If the work is by a single trade (e.g. masonry) use the more specific data found in the City Cost Indexes.

To adjust costs by location factors, multiply the base cost by the factor and divide by 100.

Contingency: A factor for contingency may be added to any estimate to represent the cost of unknowns that may occur between the time that the estimate is performed and the time the project is constructed. The amount of the allowance will depend on the stage of design at which the estimate is done, and the contractor's assessment of the risk invloved.

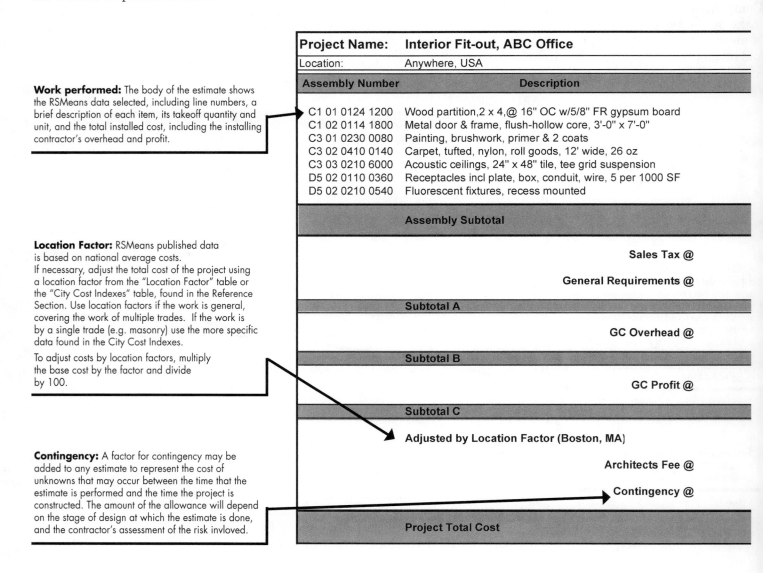

Project Name:	Interior Fit-out, ABC Office
Location:	Anywhere, USA
Assembly Number	**Description**
C1 01 0124 1200	Wood partition,2 x 4,@ 16" OC w/5/8" FR gypsum board
C1 02 0114 1800	Metal door & frame, flush-hollow core, 3'-0" x 7'-0"
C3 01 0230 0080	Painting, brushwork, primer & 2 coats
C3 02 0410 0140	Carpet, tufted, nylon, roll goods, 12' wide, 26 oz
C3 03 0210 6000	Acoustic ceilings, 24" x 48" tile, tee grid suspension
D5 02 0110 0360	Receptacles incl plate, box, conduit, wire, 5 per 1000 SF
D5 02 0210 0540	Fluorescent fixtures, recess mounted
Assembly Subtotal	
	Sales Tax @
	General Requirements @
Subtotal A	
	GC Overhead @
Subtotal B	
	GC Profit @
Subtotal C	
Adjusted by Location Factor (Boston, MA)	
	Architects Fee @
	Contingency @
Project Total Cost	

This estimate is based on an interactive spreadsheet.
A copy of this spreadsheet is located on the RSMeans website at
http://www.reedconstructiondata.com/rsmeans/extras/546011.
You are free to download it and adjust it to your methodology.

Date:		
Qty.	**Unit**	**Subtotal**
600.000	S.F.	$2,682.00
2.000	Ea.	$2,550.00
600.000	S.F.	$696.00
240.000	S.F.	$1,708.80
200.000	S.F.	$870.00
6.000	S.F.	$14.76
200.000	S.F.	$1,580.00
		$10,101.56
5 %		$ 252.54
7 %		$ 707.11
		$11,061.21
5 %		$ 553.06
		$11,614.27
5 %		$ 580.71
		$12,194.98
118.1		$ 14,402.27
8 %		$ 1,152.18
15 %		$ 2,160.34
		$ 17,714.80

Sales Tax: If the work is subject to state or local sales taxes, the amount must be added to the estimate. In a conceptual estimate it can be assumed that one half of the total represents material costs. Therefore, apply the sales tax rate to 50% of the assembly subtotal.

General Requirements: This item covers project-wide needs provided by the General Contractor. These items vary by project, but may include temporary facilities and utilities, security, testing, project cleanup, etc. In assemblies estimates a percentage is used, typically between 5% and 15% of project cost.

General Contractor Overhead: This entry represents the General Contractor's markup on all work to cover project administration costs.

General Contractor Profit: This entry represents the GC's profit on all work performed. The value included here can vary widely by project, and is influenced by the GC's perception of the project's financial risk and market conditions.

Architects Fee: If appropriate, add the design cost to the project estimate. These fees vary based on project complexity and size. Typical design and engineering fees can be found in the reference section.

Example

Assemblies costs can be used to calculate the cost for each component of a building and, accumulated with appropriate markups, produce a cost for the complete structure. Components from a model building with similar characteristics in conjunction with Assemblies components may be used to calculate costs for a complete structure.

Example:

Outline Specifications

General	Building size—60' x 100', 8 suspended floors, 12' floor to floor, 4' high parapet above roof, full basement 11'-8" floor to floor, bay size 25' x 30', ceiling heights - 9' in office area and 8' in core area.
A Substructure	Concrete, spread and strip footings, concrete walls waterproofed, 4" slab on grade.
B Shell	
B10 Superstructure	Columns, wide flange; 3 hr. fire rated; Floors, composite steel frame & deck with concrete slab; Roof, steel beams, open web joists and deck.
B20 Exterior Enclosure	Walls; North, East & West, brick & lightweight concrete block with 2" cavity insulation, 25% window; South, 8" lightweight concrete block insulated, 10% window. Doors, aluminum & glass. Windows, aluminum, 3'-0" x 5'-4", insulating glass.
B30 Roofing	Tar & gravel, 2" rigid insulation, R 12.5.
C Interiors	Core—6" lightweight concrete block, full height with plaster on exposed faces to ceiling height.
	Corridors—1st & 2nd floor; 3-5/8" steel studs with F.R. gypsum board, full height.
	Exterior Wall—plaster, ceiling height.
	Fittings—Toilet accessories, directory boards.
	Doors—hollow metal.
	Wall Finishes—lobby, mahogany paneling on furring, remainder plaster & gypsum board, paint.
	Floor Finishes—1st floor lobby, corridors & toilet rooms, terrazzo; remainder concrete, tenant developed. 2nd thru 8th, toilet rooms, ceramic tile; office & corridor, carpet.
	Ceiling Finishes—24" x 48" fiberglass board on Tee grid.
D Services	
D10 Conveying	2-2500 lb. capacity, 200 F.P.M. geared elevators, 9 stops.
D20 Plumbing	Fixtures, see sketch.
	Roof Drains—2-4" C.I. pipe.
D30 HVAC	Heating—fin tube radiation, forced hot water.
	Air Conditioning—chilled water with water cooled condenser.
D40 Fire Protection	Fire Protection—4" standpipe, 9 hose cabinets.
D50 Electrical	Lighting, 1st thru 8th, 15 fluorescent fixtures/1000 S.F., 3 Watts/S.F.
	Basement, 10 fluorescent fixtures/1000 S.F., 2 Watts/S.F.
	Receptacles, 1st thru 8th, 16.5/1000 S.F., 2 Watts/S.F.
	Basement, 10 receptacles/1000 S.F., 1.2 Watts/S.F.
	Air Conditioning, 4 Watts/S.F.
	Miscellaneous Connections, 1.2 Watts/S.F.
	Elevator Power, 2-10 H.P., 230 volt motors.
	Wall Switches, 2/1000 S.F.
	Service, panel board & feeder, 2000 Amp.
	Fire Detection System, pull stations, signals, smoke and heat detectors.
	Emergency Lighting Generator, 30 KW.
E Equipment & Furnishings	NA.
F Special Construction	NA.
G Building Sitework	NA.

Front Elevation

Basement Plan

Ground Floor Plan

Typical Floor Plan

Example

If spread footing & column sizes are unknown, develop approximate loads as follows. Enter tables with these loads to determine costs.

Approximate loads/S.F. for roof & floors.
Roof. Assume 40 psf superimposed load.
Steel joists, beams & deck.
Table B1020 112—Line 3900

Superimposed Load Ranges

Apartments & Residential Structures	65 to	75 psf
Assembly Areas & Retail Stores	110 to	125 psf
Commercial & Manufacturing	150 to	250 psf
Offices	75 to	100 psf

B1020 112		Steel Joists, Beams, & Deck on Columns						
	BAY SIZE (FT.)	SUPERIMPOSED LOAD (P.S.F.)	DEPTH (IN.)	TOTAL LOAD (P.S.F.)	COLUMN ADD	COST PER S.F.		
						MAT.	INST.	TOTAL
3500	25x30	20	22	40		5.05	1.58	6.63
3600					columns	1.14	.27	1.41
3900		40	25	60		6.15	1.88	8.03
4000					columns	1.37	.33	1.70

B1010 256		Composite Beams, Deck & Slab						
	BAY SIZE (FT.)	SUPERIMPOSED LOAD (P.S.F.)	SLAB THICKNESS (IN.)	TOTAL DEPTH (FT.-IN.)	TOTAL LOAD (P.S.F.)	COST PER S.F.		
						MAT.	INST.	TOTAL
3400	25x30	40	5-1/2	1 - 11-1/2	83	11.35	5.65	17
3600		75	5-1/2	1 - 11-1/2	119	12.25	5.75	18
3900		125	5-1/2	1 - 11-1/2	170	14.35	6.50	20.85
4000		200	6-1/4	2 - 6-1/4	252	17.70	7.35	25.05

Floors–Total load, 119 psf.
Interior foundation load.
Roof

[(25' x 30' x 60 psf) + 8 floors x (25' x 30' x 119 psf)] x 1/1000 lb./Kip =	759 Kips
Approximate Footing Loads, Interior footing =	759 Kips
Exterior footing (1/2 bay) 759 k x .6 =	455 Kips
Corner footing (1/4 bay) 759 k x .45 =	342 Kips
[Factors to convert Interior load to Exterior & Corner loads]	
Approximate average Column load 759 k/2 =	379 Kips

PRELIMINARY ESTIMATE

PROJECT	**Office Building**	TOTAL SITE AREA			
BUILDING TYPE		OWNER			
LOCATION		ARCHITECT			
DATE OF CONSTRUCTION		ESTIMATED CONSTRUCTION PERIOD			

BRIEF DESCRIPTION **Building size: 60' x 100' - 8 structural floors - 12' floor to floor**

4' high parapet above roof, full basement 11' - 8" floor to floor, bay size

25' x 30', ceiling heights - 9' in office area + 8' in core area

TYPE OF PLAN		TYPE OF CONSTRUCTION			
QUALITY		BUILDING CAPACITY			

Floor				**Wall Area**				
Below Grade Levels				Foundation Walls	L.F.		Ht.	S.F.
Area		S.F.		Frost Walls	L.F.		Ht.	S.F.
Area		S.F.		Exterior Closure		Total		S.F.
Total Area		S.F.		Comment				
Ground Floor				Fenestration		%		S.F.
Area		S.F.				%		S.F.
Area		S.F.		Exterior Wall		%		S.F.
Total Area		S.F.				%		S.F.
Supported Levels				**Site Work**				
Area				Parking		S.F. (For		Cars)
Area		S.F.		Access Roads		L.F. (X		Ft. Wide)
Area		S.F.		Sidewalk		L.F. (X		Ft. Wide)
Area		S.F.		Landscaping		S.F. (		% Unbuilt Site)
Area		S.F.		**Building Codes**				
Total Area		S.F.		City			Country	
Miscellaneous				National			Other	
Area		S.F.		**Loading**				
Area		S.F.		Roof	psf	Ground Floor		psf
Area		S.F.		Supported Floors	psf	Corridor		psf
Area		S.F.		Balcony	psf	Partition, allow		psf
Total Area		S.F.		Miscellaneous				psf
Net Finished Area		S.F.		Live Load Reduction				
Net Floor Area		S.F.		Wind				
Gross Floor Area	**54,000**	S.F.		Earthquake			Zone	
Roof				Comment				
Total Area		S.F.		Soil Type				
Comments				Bearing Capacity				K.S.F.
				Frost Depth				Ft.
Volume				**Frame**				
Depth of Floor System				Type		Bay Spacing		
Minimum		In.		Foundation				
Maximum		In.		Special				
Foundation Wall Height		Ft.		Substructure				
Floor to Floor Height		Ft.		Comment				
Floor to Ceiling Height		Ft.		Superstructure, Vertical				
Subgrade Volume		C.F.		Fireproofing		☐ Columns		Hrs.
Above Grade Volume		C.F.		☐ Girders	Hrs.	☐ Beams		Hrs.
Total Building Volume	**648,000**	C.F.		☐ Floor	Hrs.	☐ None		

NUMBER		QTY.	UNIT	TOTAL COST UNIT	TOTAL	COST PER S.F.
A	**SUBSTRUCTURE**					
A1010 210 7900	Corner Footings 8'-6" SQ. x 27"	4	Ea.	1,855.00	7,420	
8010	Exterior 9' -6" SQ. x 30"	8		2,475.00	19,800	
8300	Interior 12" SQ.	3		4,550.00	13,650	
A1010 110 2700	Strip 2' Wide x 1' Thick					
	320 L.F. [(4 x 8.5) + 8 x 9.5)] =	320	L.F.	38.60	12,352	
A2020 110 7262	Foundation Wall 12' High, 1' Thick	320		248.00	79,360	
A1010 320 2800	Foundation Waterproofing	320		21.35	6,832	
A1030 120 2240	4", non-industrial, reinf. slab on grade	6000	S.F.	5.11	30,660	
A2010 110 3440	Building Excavation + Backfill	6000	S.F.	6.48	38,880	
3500	(Interpolated ; 12' Between					
4620	8' and 16' ; 6,000 Between					
4680	4,000 and 10,000 S.F.					
	Total				208,954	3.87
B	**SHELL**					
B10	**Superstructure**					
B1010 208 5800	Columns: Load 379K Use 400K					
	Exterior 12 x 96' + Interior 3 x 108'	1476	V.L.F.	142.45	210,256	
B1010 720 3650	Column Fireproofing - Interior 4 sides					
3700	Exterior 1/2 (Interpolated for 12")	900	V.L.F	33.61	30,249	
B1010 256 3600	Floors: Composite steel + lt. wt. conc.	48000	S.F.	18.00	864,000	
B1020 112 3900	Roof: Open web joists, beams + decks	6000	S.F.	8.03	48,180	
	Total				1,152,685	21.35
B20	**Exterior Enclosure**					
B2010 134 1200	4" Brick + 6" Block - Insulated					
	75% NE + W Walls 220' x 100' x .75	16500	S.F.	29.45	485,925	
B2010 110 3410	8" Block - Insulated 90%					
	100' x 100' x .9	9000	S.F.	11.28	101,520	
B2030 110 6950	Double Aluminum + Glass Door	1	Pr.	5,575.00	5,575	
6300	Single	2	Ea.	2,725.00	5,450	
B2020 106 8800	Aluminum windows - Insul. glass					
	(5,500 S.F. + 1,000 S.F.)/(3 x 5.33)	406	Ea.	901.00	365,806	
B2010 105 8480	Precast concrete coping	320	L.F.	41.05	13,136	
	Total				977,412	18.10
B30	**Roofing**					
B3010 320 4020	Insulation: Perlite/Polyisocyanurate	6000	S.F.	1.69	10,140	
B3010 105 1400	Roof: 4 Ply T & G composite	6000		2.69	16,140	
B3010 430 0300	Flashing: Aluminum - fabric backed	640		3.07	1,965	
B3020 210 0500	Roof hatch	1	Ea.	1,237.00	1,237	
	Total				29,482	0.55

ASSEMBLY NUMBER		QTY.	UNIT	TOTAL COST		COST PER S.F.
				UNIT	TOTAL	
C	INTERIORS					
C1010 104 5500	Core partitions: 6" Lt. Wt. concrete					
	block 288 L.F. x 11.5' x 8 Floors	26,449	S.F.	9.01	238,305	
C1010 144 0920	Plaster: [(196 L.F. x 8') + (144 L.F. x 9')] 8					
	+ Ext. Wall [(320 x 9 x 8 Floors) - 6500]	39,452	S.F.	3.03	119,540	
C1010 126 5400	Corridor partitions 1st + 2nd floors					
	steel studs + F.R. gypsum board	2,530	S.F.	3.89	9,842	
C1020 102 2600	Interior doors	66	Ea.	1,194.00	78,804	
	Wall finishes - Lobby: Mahogany w/furring					
C1010 128 0652	Furring: 150 L.F. x 9' + Ht.	1,350	S.F.	1.48	1,998	
C1030 710 0120	Towel Dispenser	16	Ea.	80.50	1,288	
C1030 710 0140	Grab Bar	32	Ea.	54.50	1,744	
C1030 710 0160	Mirror	31	Ea.	226.00	7,006	
C1030 710 0180	Toilet Tissue Dispenser	31	Ea.	36.85	1,142	
C1030 510 0140	Directory Board	8	Ea.	319.00	2,552	
C2010 110 0780	Stairs: Steel w/conc. fill	18	Flt.	19,075.00	343,350	
C3010 230 1662	Mahogany paneling	1,350	S.F.	5.50	7,425	
C3010 230 0080	Paint, primer + 2 coats finish, on plaster or gyp. board					
	39,452 + [(220' x 9' x 2) - 1350] + 72	42,134	S.F.	1.12	47,190	
	Floor Finishes					
C3020 410 1100	1st floor lobby + terrazzo	2,175	S.F.	18.90	41,108	
	Remainder: Concrete - Tenant finished					
C3020 410 0060	2nd - 8th: Carpet - 4725 S.F. x 7	33,075	S.F.	4.11	135,938	
C3020 410 1720	Ceramic Tile - 300 S.F. x 7	2,100	S.F.	10.78	22,638	
C3030 240 2780	Suspended Ceiling Tiles- 5200 S.F. x 8	41,600	S.F.	2.09	86,944	
C3030 240 3260	Suspended Ceiling Grid- 5200 S.F. x 8	41,600	S.F.	1.45	60,320	
C1030 110 0700	Toilet partitions	31	Ea.	700.00	21,700	
0760	Handicapped addition	16	Ea.	330.00	5,280	
	Total				1,234,114	22.85
D	SERVICES					
D10	Conveying					
	2 Elevators - 2500# 200'/min. Geared					
D1010 140 1600	5 Floors $159,300					
1800	15 Floors $ 382,500					
	Diff. $223,200/10 = 22,320/Flr.					
	$159,300 + (4 x 22,320) = $248,580/Ea.	2	Ea.	248,580.00	497,160	
	Total				497,160	9.21
D20	Plumbing					
D2010 310 1600	Plumbing - Lavatories	31	Ea.	1,345.00	41,695	
D2010 440 4340	-Service Sink	8	Ea.	3,900.00	31,200	
D2010 210 2000	-Urinals	8	Ea.	1,425.00	11,400	

ASSEMBLY NUMBER		QTY.	UNIT	TOTAL COST		COST PER S.F.
				UNIT	TOTAL	
D2010 110 2080	-Water Closets	31	Ea.	2,595.00	80,445	
	Water Control, Waste Vent Piping	45%			74,133	
D2040 210 4200	- Roof Drains, 4" C.I.	2	Ea.	1,690.00	3,380	
4240	- Pipe, 9 Flr. x 12' x 12' Ea.	216	L.F.	38.50	8,316	
	Total				250,569	4.64
D30	HVAC					
D3010 520 2000	10,000 S.F. @ $10.65/S.F. Interpolate					
2040	100,000 S.F. @ $4.51/S.F.	48,000	S.F.	8.06	386,880	
	Cooling - Chilled Water, Air Cool, Cond.					
D3030 115 4000	40,000 S.F. @14.35/S.F. Interpolate	48,000	S.F.	14.89	714,720	
4040	60,000 S.F. @15.25/S.F.					
	Total				1,101,600	20.40
D40	Fire Protection					
D4020 310 0560	Wet Stand Pipe: 4" x 10' 1st floor	12/10	Ea.	7,900.00	9,480	
0580	Additional Floors: (12'/10 x 8)	9.6	Ea.	2,225.00	21,360	
D4020 410 8400	Cabinet Assembly	9	Ea.	1,040.00	9,360	
	Total				40,200	0.74
D50	Electrical					
D5020 210 0280	Office Lighting, 15/1000 S.F. - 3 Watts/S.F.	48,000	S.F.	7.95	381,600	
0240	Basement Lighting, 10/1000 S.F. - 2 Watts/S.F.	6,000	S.F.	5.29	31,740	
D5020 110 0640	Office Receptacles 16.5/1000 S.F. - 2 Watts/S.F.	48,000	S.F.	4.63	222,240	
0560	Basement Receptacles 10/1000 S.F. - 1.2 Watts/S.F.	6,000	S.F.	3.51	21,060	
D5020 140 0280	Central A.C. - 4 Watts/S.F.	48,000	S.F.	0.61	29,280	
D5020 135 0320	Misc. Connections - 1.2 Watts/S.F.	48,000	S.F.	0.32	15,360	
D5020 145 0680	Elevator Motor Power - 10 H.P.	2	Ea.	3,175.00	6,350	
D5020 130 0280	Wall Switches - 2/1000 S.F.	54,000	S.F.	0.47	25,380	
D5010 120 0560	2000 Amp Service	1	Ea.	41,700.00	41,700	
D5010 240 0400	Switchgear	1	Ea.	56,400.00	56,400	
D5010 230 0560	Feeder	50	L.F.	635.00	31,750	
D5030 910 0400	Fire Detection System - 50 Detectors	1	Ea.	34,500.00	34,500	
D5090 210 0320	Emergency Generator - 30 kW	30	kW	622.00	18,660	
	Total				916,020	16.96
E	EQUIPMENT & FURNISHINGS					
G	BUILDING SITEWORK					
	Miscellaneous					

Preliminary Estimate Cost Summary

PROJECT: Office Building		TOTAL AREA	54,000 S.F.	SHEET NO.
LOCATION		TOTAL VOLUME	648,000 C.F.	ESTIMATE NO.
ARCHITECT		COST PER S.F.		DATE
OWNER		COST PER C.F.		NO OF STORIES
QUANTITIES BY		EXTENSIONS BY		CHECKED BY

DIV	DESCRIPTION	SUBTOTAL COST	COST/S.F.	PERCENTAGE
A	SUBSTRUCTURE	208,954	$ 3.87	
B10	SHELL: SUPERSTRUCTURE	1,152,685	$ 21.35	
B20	SHELL: EXTERIOR ENCLOSURE	977,412	$ 18.10	
B30	SHELL: ROOFING	29,482	$ 0.55	
C	INTERIORS	1,234,114	$ 22.85	
D10	SERVICES: CONVEYING	497,160	$ 9.21	
D20	SERVICES: PLUMBING	250,569	$ 4.64	
D30	SERVICES: HVAC	1,101,600	$ 20.40	
D40	SERVICES: FIRE PROTECTION	40,200	$ 0.74	
D50	SERVICES: ELECTRICAL	916,020	$ 16.96	
E	EQUIPMENT & FURNISHINGS			
F	SPECIAL CONSTRUCTION & DEMO.			
G	BUILDING SITEWORK			
	BUILDING SUBTOTAL	$ 6,408,196		$ 6,408,196

Sales Tax % x Subtotal /2	N/A	$	-
General Conditions 10 % x Subtotal		$	640,820
	Subtotal "A"	$	7,049,015
Overhead 5 % x Subtotal "A"		$	352,451
	Subtotal "B"	$	7,401,466
Profit 10% x Subtotal "B"		$	740,147
	Subtotal "C"	$	8,141,613
Location Factor % x Subtotal "C"	N/A Localized Cost		
Architects Fee 6.5% x Localized Cost =		$	529,205
Contingency x Localized Cost =	N/A	$	-
	Project Total Cost	$	8,670,817
Square Foot Cost $8,670,817/ 54,000 S.F. =	S.F. Cost	$	160.57
Cubic Foot Cost $8,670,817/ 648,000 C.F. =	C.F. Cost	$	13.38

A1010 Standard Foundations

The Strip Footing System includes: excavation; hand trim; all forms needed for footing placement; forms for 2″ x 6″ keyway (four uses); dowels; and 3,000 p.s.i. concrete.

The footing size required varies for different soils. Soil bearing capacities are listed for 3 KSF and 6 KSF. Depths of the system range from 8″ and deeper. Widths range from 16″ and wider. Smaller strip footings may not require reinforcement.

Please see the reference section for further design and cost information.

A1010 110	Strip Footings	COST PER L.F.		
		MAT.	INST.	TOTAL
2100	Strip footing, load 2.6 KLF, soil capacity 3 KSF, 16″ wide x 8″ deep plain	8.20	13.65	21.85
2300	Load 3.9 KLF, soil capacity, 3 KSF, 24″ wide x 8″ deep, plain	10	15	25
2500	Load 5.1 KLF, soil capacity 3 KSF, 24″ wide x 12″ deep, reinf.	16	22.50	38.50
2700	Load 11.1 KLF, soil capacity 6 KSF, 24″ wide x 12″ deep, reinf.	16	22.50	38.50
2900	Load 6.8 KLF, soil capacity 3 KSF, 32″ wide x 12″ deep, reinf.	19.30	24.50	43.80
3100	Load 14.8 KLF, soil capacity 6 KSF, 32″ wide x 12″ deep, reinf.	19.30	24.50	43.80
3300	Load 9.3 KLF, soil capacity 3 KSF, 40″ wide x 12″ deep, reinf.	22.50	26.50	49
3500	Load 18.4 KLF, soil capacity 6 KSF, 40″ wide x 12″ deep, reinf.	22.50	27	49.50
4500	Load 10 KLF, soil capacity 3 KSF, 48″ wide x 16″ deep, reinf.	31.50	34	65.50
4700	Load 22 KLF, soil capacity 6 KSF, 48″ wide, 16″ deep, reinf.	32.50	34.50	67
5700	Load 15 KLF, soil capacity 3 KSF, 72″ wide x 20″ deep, reinf.	54.50	48	102.50
5900	Load 33 KLF, soil capacity 6 KSF, 72″ wide x 20″ deep, reinf.	58	51.50	109.50

A1010 Standard Foundations

Dampproofing

Spread Footing

Strip Footing

A1010 210	Spread Footings	COST EACH		
		MAT.	INST.	TOTAL
7090	Spread footings, 3000 psi concrete, chute delivered			
7100	Load 25K, soil capacity 3 KSF, 3'-0" sq. x 12" deep	56	118	174
7150	Load 50K, soil capacity 3 KSF, 4'-6" sq. x 12" deep	120	203	323
7200	Load 50K, soil capacity 6 KSF, 3'-0" sq. x 12" deep	56	118	174
7250	Load 75K, soil capacity 3 KSF, 5'-6" sq. x 13" deep	191	286	477
7300	Load 75K, soil capacity 6 KSF, 4'-0" sq. x 12" deep	97	175	272
7350	Load 100K, soil capacity 3 KSF, 6'-0" sq. x 14" deep	241	345	586
7410	Load 100K, soil capacity 6 KSF, 4'-6" sq. x 15" deep	147	239	386
7450	Load 125K, soil capacity 3 KSF, 7'-0" sq. x 17" deep	385	495	880
7500	Load 125K, soil capacity 6 KSF, 5'-0" sq. x 16" deep	190	287	477
7550	Load 150K, soil capacity 3 KSF 7'-6" sq. x 18" deep	465	575	1,040
7610	Load 150K, soil capacity 6 KSF, 5'-6" sq. x 18" deep	254	360	614
7650	Load 200K, soil capacity 3 KSF, 8'-6" sq. x 20" deep	660	760	1,420
7700	Load 200K, soil capacity 6 KSF, 6'-0" sq. x 20" deep	335	445	780
7750	Load 300K, soil capacity 3 KSF, 10'-6" sq. x 25" deep	1,225	1,250	2,475
7810	Load 300K, soil capacity 6 KSF, 7'-6" sq. x 25" deep	635	750	1,385
7850	Load 400K, soil capacity 3 KSF, 12'-6" sq. x 28" deep	1,925	1,850	3,775
7900	Load 400K, soil capacity 6 KSF, 8'-6" sq. x 27" deep	880	970	1,850
8010	Load 500K, soil capacity 6 KSF, 9'-6" sq. x 30" deep	1,225	1,275	2,500
8100	Load 600K, soil capacity 6 KSF, 10'-6" sq. x 33" deep	1,650	1,625	3,275
8200	Load 700K, soil capacity 6 KSF, 11'-6" sq. x 36" deep	2,100	1,975	4,075
8300	Load 800K, soil capacity 6 KSF, 12'-0" sq. x 37" deep	2,350	2,200	4,550
8400	Load 900K, soil capacity 6 KSF, 13'-0" sq. x 39" deep	2,925	2,600	5,525
8500	Load 1000K, soil capacity 6 KSF, 13'-6" sq. x 41" deep	3,300	2,900	6,200

A1010 320	Foundation Dampproofing	COST PER L.F.		
		MAT.	INST.	TOTAL
1000	Foundation dampproofing, bituminous, 1 coat, 4' high	.92	4.39	5.31
1400	8' high	1.84	8.75	10.59
1800	12' high	2.76	13.60	16.36
2000	2 coats, 4' high	1.84	5.45	7.29
2400	8' high	3.68	10.85	14.53
2800	12' high	5.50	16.70	22.20
3000	Asphalt with fibers, 1/16" thick, 4' high	1.36	5.45	6.81
3400	8' high	2.72	10.85	13.57
3800	12' high	4.08	16.70	20.78
4000	1/8" thick, 4' high	2.40	6.45	8.85
4400	8' high	4.80	12.95	17.75
4800	12' high	7.20	19.85	27.05
5000	Asphalt coated board and mastic, 1/4" thick, 4' high	4.08	5.85	9.93
5400	8' high	8.15	11.75	19.90

A10 Foundations

A1010 Standard Foundations

A1010 320	Foundation Dampproofing	COST PER L.F.		
		MAT.	INST.	TOTAL
5800	12' high	12.25	18.05	30.30
6000	1/2" thick, 4' high	6.10	8.15	14.25
6400	8' high	12.25	16.30	28.55
6800	12' high	18.35	25	43.35
7000	Cementitious coating, on walls, 1/8" thick coating, 4' high	2	8.45	10.45
7400	8' high	4	16.95	20.95
7800	12' high	6	25.50	31.50
8000	Cementitious/metallic slurry, 4 coat, 1/2"thick, 2' high	.57	8.50	9.07
8400	4' high	1.14	17	18.14
8800	6' high	1.71	25.50	27.21

A1030 Slab on Grade

Fibre Expansion Joint

Wire Mesh Reinforcing

Concrete Slab

Vapor Barrier

Compacted Gravel

Reinforced Slab on Grade

A Slab on Grade system includes fine grading; 6″ of compacted gravel; vapor barrier; 3500 p.s.i. concrete; bituminous fiber expansion joint; all necessary edge forms 4 uses; steel trowel finish; and sprayed on membrane curing compound. Wire mesh reinforcing used in all reinforced slabs.

Non-industrial slabs are for foot traffic only with negligible abrasion. Light industrial slabs are for pneumatic wheels and light abrasion. Industrial slabs are for solid rubber wheels and moderate abrasion. Heavy industrial slabs are for steel wheels and severe abrasion.

A1030 120	Plain & Reinforced	COST PER S.F.		
		MAT.	INST.	TOTAL
2220	Slab on grade, 4″ thick, non industrial, non reinforced	2.01	2.65	4.66
2240	Reinforced	2.17	3.01	5.18
2260	Light industrial, non reinforced	2.59	3.25	5.84
2280	Reinforced	2.75	3.61	6.36
2300	Industrial, non reinforced	3.25	6.75	10
2320	Reinforced	3.41	7.15	10.56
3340	5″ thick, non industrial, non reinforced	2.34	2.73	5.07
3360	Reinforced	2.50	3.09	5.59
3380	Light industrial, non reinforced	2.93	3.33	6.26
3400	Reinforced	3.09	3.69	6.78
3420	Heavy industrial, non reinforced	4.22	8.05	12.27
3440	Reinforced	4.32	8.50	12.82
4460	6″ thick, non industrial, non reinforced	2.77	2.67	5.44
4480	Reinforced	3	3.16	6.16
4500	Light industrial, non reinforced	3.38	3.27	6.65
4520	Reinforced	3.74	3.93	7.67
4540	Heavy industrial, non reinforced	4.68	8.15	12.83
4560	Reinforced	4.91	8.65	13.56
5580	7″ thick, non industrial, non reinforced	3.12	2.76	5.88
5600	Reinforced	3.42	3.28	6.70
5620	Light industrial, non reinforced	3.73	3.36	7.09
5640	Reinforced	4.03	3.88	7.91
5660	Heavy industrial, non reinforced	5	8.05	13.05
5680	Reinforced	5.25	8.55	13.80
6700	8″ thick, non industrial, non reinforced	3.45	2.81	6.26
6720	Reinforced	3.70	3.25	6.95
6740	Light industrial, non reinforced	4.07	3.41	7.48
6760	Reinforced	4.32	3.85	8.17
6780	Heavy industrial, non reinforced	5.40	8.15	13.55
6800	Reinforced	5.75	8.60	14.35

A2010 Basement Excavation

Clay 1/2:1

Sand & Gravel 1:1

Line of Excavation

In general, the following items are accounted for in the table below.

Costs:
1) Excavation for building or other structure to depth and extent indicated.
2) Backfill compacted in place
3) Haul of excavated waste
4) Replacement of unsuitable backfill material with bank run gravel.

A2010 110	Building Excavation & Backfill	COST PER S.F.		
		MAT.	INST.	TOTAL
2220	Excav & fill, 1000 S.F. 4' sand, gravel, or common earth, on site storage		1.02	1.02
2240	Off site storage		1.55	1.55
2260	Clay excavation, bank run gravel borrow for backfill	2.65	2.39	5.04
2280	8' deep, sand, gravel, or common earth, on site storage		6.20	6.20
2300	Off site storage		13.75	13.75
2320	Clay excavation, bank run gravel borrow for backfill	10.75	11.20	21.95
2340	16' deep, sand, gravel, or common earth, on site storage		17	17
2350	Off site storage		34	34
2360	Clay excavation, bank run gravel borrow for backfill	30	29	59
3380	4000 S.F., 4' deep, sand, gravel, or common earth, on site storage		.55	.55
3400	Off site storage		1.09	1.09
3420	Clay excavation, bank run gravel borrow for backfill	1.35	1.20	2.55
3440	8' deep, sand, gravel, or common earth, on site storage		4.39	4.39
3460	Off site storage		7.70	7.70
3480	Clay excavation, bank run gravel borrow for backfill	4.90	6.70	11.60
3500	16' deep, sand, gravel, or common earth, on site storage		10.55	10.55
3520	Off site storage		21	21
3540	Clay, excavation, bank run gravel borrow for backfill	13.10	15.95	29.05
4560	10,000 S.F., 4' deep, sand gravel, or common earth, on site storage		.32	.32
4580	Off site storage		.65	.65
4600	Clay excavation, bank run gravel borrow for backfill	.82	.73	1.55
4620	8' deep, sand, gravel, or common earth, on site storage		3.81	3.81
4640	Off site storage		5.80	5.80
4660	Clay excavation, bank run gravel borrow for backfill	2.97	5.25	8.22
4680	16' deep, sand, gravel, or common earth, on site storage		8.65	8.65
4700	Off site storage		14.75	14.75
4720	Clay excavation, bank run gravel borrow for backfill	7.85	11.95	19.80
5740	30,000 S.F., 4' deep, sand, gravel, or common earth, on site storage		.18	.18
5760	Off site storage		.38	.38
5780	Clay excavation, bank run gravel borrow for backfill	.48	.43	.91
5860	16' deep, sand, gravel, or common earth, on site storage		7.40	7.40
5880	Off site storage		10.75	10.75
5900	Clay excavation, bank run gravel borrow for backfill	4.37	9.30	13.67
6910	100,000 S.F., 4' deep, sand, gravel, or common earth, on site storage		.10	.10
6940	8' deep, sand, gravel, or common earth, on site storage		3.21	3.21
6970	16' deep, sand, gravel, or common earth, on site storage		6.75	6.75
6980	Off site storage		8.50	8.50
6990	Clay excavation, bank run gravel borrow for backfill	2.36	7.70	10.06

A2020 Basement Walls

The Foundation Bearing Wall System includes: forms up to 16' high (four uses); 3,000 p.s.i. concrete placed and vibrated; and form removal with breaking form ties and patching walls. The wall systems list walls from 6" to 16" thick and are designed with minimum reinforcement.

Excavation and backfill are not included.

Please see the reference section for further design and cost information.

A2020 110		Walls, Cast in Place						
	WALL HEIGHT (FT.)	PLACING METHOD	CONCRETE (C.Y. per L.F.)	REINFORCING (LBS. per L.F.)	WALL THICKNESS (IN.)	COST PER L.F.		
						MAT.	INST.	TOTAL
1500	4'	direct chute	.074	3.3	6	15.20	50.50	65.70
1520			.099	4.8	8	18.65	52	70.65
1540			.123	6.0	10	22	53	75
1560			.148	7.2	12	25	54	79
1580			.173	8.1	14	28.50	55	83.50
1600			.197	9.44	16	31.50	56	87.50
3000	6'	direct chute	.111	4.95	6	23	76	99
3020			.149	7.20	8	28	78	106
3040			.184	9.00	10	33	79.50	112.50
3060			.222	10.8	12	38	81	119
5000	8'	direct chute	.148	6.6	6	30.50	101	131.50
5020			.199	9.6	8	37.50	104	141.50
5040			.250	12	10	44	106	150
5061			.296	14.39	12	50.50	108	158.50
6020	10'	direct chute	.248	12	8	46.50	130	176.50
6040			.307	14.99	10	54.50	132	186.50
6061			.370	17.99	12	63	135	198
7220	12'	pumped	.298	14.39	8	56	161	217
7240			.369	17.99	10	65.50	164	229.50
7260			.444	21.59	12	75.50	170	245.50
9220	16'	pumped	.397	19.19	8	75	214	289
9240			.492	23.99	10	87.50	219	306.50
9260			.593	28.79	12	101	226	327

General: It is desirable for purposes of consistency and simplicity to maintain constant column sizes throughout building height. To do this, concrete strength may be varied (higher strength concrete at lower stories and lower strength concrete at upper stories), as well as varying the amount of reinforcing.

The table provides probably minimum column sizes with related costs and weight per lineal foot of story height.

B1010 203			C.I.P. Column, Square Tied					
	LOAD (KIPS)	STORY HEIGHT (FT.)	COLUMN SIZE (IN.)	COLUMN WEIGHT (P.L.F.)	CONCRETE STRENGTH (PSI)	COST PER V.L.F.		
						MAT.	INST.	TOTAL
0640	100	10	10	96	4000	11.25	47	58.25
0680		12	10	97	4000	11.40	47	58.40
0700		14	12	142	4000	14.85	57	71.85
0840	200	10	12	140	4000	16.25	59	75.25
0860		12	12	142	4000	16.50	59.50	76
0900		14	14	196	4000	19.05	66	85.05
0920	300	10	14	192	4000	20	67.50	87.50
0960		12	14	194	4000	20.50	68	88.50
0980		14	16	253	4000	23.50	75.50	99
1020	400	10	16	248	4000	25	77.50	102.50
1060		12	16	251	4000	25.50	78.50	104
1080		14	16	253	4000	26	79	105
1200	500	10	18	315	4000	29.50	86.50	116
1250		12	20	394	4000	35.50	98.50	134
1300		14	20	397	4000	36	100	136
1350	600	10	20	388	4000	37	101	138
1400		12	20	394	4000	38	102	140
1600		14	20	397	4000	38.50	103	141.50
3400	900	10	24	560	4000	50	125	175
3800		12	24	567	4000	51	127	178
4000		14	24	571	4000	52	128	180
7300	300	10	14	192	6000	21	67	88
7500		12	14	194	6000	21.50	67.50	89
7600		14	14	196	6000	21.50	68	89.50
8000	500	10	16	248	6000	26.50	77.50	104
8050		12	16	251	6000	27	78.50	105.50
8100		14	16	253	6000	27.50	79	106.50
8200	600	10	18	315	6000	31	87.50	118.50
8300		12	18	319	6000	31.50	88.50	120
8400		14	18	321	6000	32	89	121
8800	800	10	20	388	6000	38	98.50	136.50
8900		12	20	394	6000	38.50	100	138.50
9000		14	20	397	6000	39.50	101	140.50

B10 Superstructure

B1010 Floor Construction

B1010 203 — C.I.P. Column, Square Tied

	LOAD (KIPS)	STORY HEIGHT (FT.)	COLUMN SIZE (IN.)	COLUMN WEIGHT (P.L.F.)	CONCRETE STRENGTH (PSI)	COST PER V.L.F.		
						MAT.	INST.	TOTAL
9100	900	10	20	388	6000	41	103	144
9300		12	20	394	6000	41.50	104	145.50
9600		14	20	397	6000	42.50	106	148.50

B1010 204 — C.I.P. Column, Square Tied-Minimum Reinforcing

	LOAD (KIPS)	STORY HEIGHT (FT.)	COLUMN SIZE (IN.)	COLUMN WEIGHT (P.L.F.)	CONCRETE STRENGTH (PSI)	COST PER V.L.F.		
						MAT.	INST.	TOTAL
9913	150	10-14	12	135	4000	13.45	55	68.45
9918	300	10-14	16	240	4000	21	72	93
9924	500	10-14	20	375	4000	31.50	93	124.50
9930	700	10-14	24	540	4000	43	115	158
9936	1000	10-14	28	740	4000	56.50	141	197.50
9942	1400	10-14	32	965	4000	70	159	229
9948	1800	10-14	36	1220	4000	87	185	272

B10 Superstructure

B1010 Floor Construction

Concentric Load

Eccentric Load

General: Data presented here is for plant produced members transported 50 miles to 100 miles to the site and erected.

Design and pricing assumptions:
Normal wt. concrete, f'c = 5 KSI

Main reinforcement, fy = 60 KSI
Ties, fy = 40 KSI

Minimum design eccentricity, 0.1t.

Concrete encased structural steel haunches are assumed where practical; otherwise galvanized rebar haunches are assumed.

Base plates are integral with columns.

Foundation anchor bolts, nuts and washers are included in price.

B1010 206	Tied, Concentric Loaded Precast Concrete Columns							
	LOAD (KIPS)	STORY HEIGHT (FT.)	COLUMN SIZE (IN.)	COLUMN WEIGHT (P.L.F.)	LOAD LEVELS	COST PER V.L.F.		
						MAT.	INST.	TOTAL
0560	100	10	12x12	164	2	165	13.85	178.85
0570		12	12x12	162	2	160	11.55	171.55
0580		14	12x12	161	2	159	11.35	170.35
0590	150	10	12x12	166	3	159	11.25	170.25
0600		12	12x12	169	3	159	10.10	169.10
0610		14	12x12	162	3	159	9.95	168.95
0620	200	10	12x12	168	4	159	12.05	171.05
0630		12	12x12	170	4	159	10.90	169.90
0640		14	12x12	220	4	159	10.80	169.80

B1010 207	Tied, Eccentric Loaded Precast Concrete Columns							
	LOAD (KIPS)	STORY HEIGHT (FT.)	COLUMN SIZE (IN.)	COLUMN WEIGHT (P.L.F.)	LOAD LEVELS	COST PER V.L.F.		
						MAT.	INST.	TOTAL
1130	100	10	12x12	161	2	157	13.85	170.85
1140		12	12x12	159	2	157	11.55	168.55
1150		14	12x12	159	2	156	11.35	167.35
1390	600	10	18x18	385	4	230	12.05	242.05
1400		12	18x18	380	4	230	10.90	240.90
1410		14	18x18	375	4	229	10.80	239.80
1480	800	10	20x20	490	4	209	12.05	221.05
1490		12	20x20	480	4	209	10.90	219.90
1500		14	20x20	475	4	210	10.80	220.80

B1010 Floor Construction

 (A) Wide Flange

 (B) Pipe

 (C) Pipe, Concrete Filled

 (G) Rectangular Tube, Concrete Filled

B1010 208				Steel Columns			
	LOAD (KIPS)	UNSUPPORTED HEIGHT (FT.)	WEIGHT (P.L.F.)	SIZE (IN.)	TYPE	COST PER V.L.F.		
						MAT.	INST.	TOTAL
1000	25	10	13	4	A	22.50	10.65	33.15
1020			7.58	3	B	13.10	10.65	23.75
1040			15	3-1/2	C	16	10.65	26.65
1120			20	4x3	G	18.45	10.65	29.10
1200		16	16	5	A	25.50	8	33.50
1220			10.79	4	B	17.30	8	25.30
1240			36	5-1/2	C	24	8	32
1320			64	8x6	G	37	8	45
1600	50	10	16	5	A	27.50	10.65	38.15
1620			14.62	5	B	25.50	10.65	36.15
1640			24	4-1/2	C	19.05	10.65	29.70
1720			28	6x3	G	24.50	10.65	35.15
1800		16	24	8	A	38.50	8	46.50
1840			36	5-1/2	C	24	8	32
1920			64	8x6	G	37	8	45
2000		20	28	8	A	42.50	8	50.50
2040			49	6-5/8	C	29.50	8	37.50
2120			64	8x6	G	35	8	43
2200	75	10	20	6	A	34.50	10.65	45.15
2240			36	4-1/2	C	48	10.65	58.65
2320			35	6x4	G	27.50	10.65	38.15
2400		16	31	8	A	49.50	8	57.50
2440			49	6-5/8	C	31.50	8	39.50
2520			64	8x6	G	37	8	45
2600		20	31	8	A	47	8	55
2640			81	8-5/8	C	45	8	53
2720			64	8x6	G	35	8	43

B1010 Floor Construction

B1010 208				Steel Columns				

	LOAD (KIPS)	UNSUPPORTED HEIGHT (FT.)	WEIGHT (P.L.F.)	SIZE (IN.)	TYPE	COST PER V.L.F.		
						MAT.	INST.	TOTAL
2800	100	10	24	8	A	41.50	10.65	52.15
2840			35	4-1/2	C	48	10.65	58.65
2920			46	8x4	G	33.50	10.65	44.15
3000		16	31	8	A	49.50	8	57.50
3040			56	6-5/8	C	46.50	8	54.50
3120			64	8x6	G	37	8	45
3200		20	40	8	A	60.50	8	68.50
3240			81	8-5/8	C	45	8	53
3320			70	8x6	G	50.50	8	58.50
3400	125	10	31	8	A	53.50	10.65	64.15
3440			81	8	C	51	10.65	61.65
3520			64	8x6	G	40	10.65	50.65
3600		16	40	8	A	64	8	72
3640			81	8	C	47.50	8	55.50
3720			64	8x6	G	37	8	45
3800		20	48	8	A	73	8	81
3840			81	8	C	45	8	53
3920			60	8x6	G	50.50	8	58.50
4000	150	10	35	8	A	60.50	10.65	71.15
4040			81	8-5/8	C	51	10.65	61.65
4120			64	8x6	G	40	10.65	50.65
4200		16	45	10	A	72	8	80
4240			81	8-5/8	C	47.50	8	55.50
4320			70	8x6	G	53.50	8	61.50
4400		20	49	10	A	74.50	8	82.50
4440			123	10-3/4	C	64	8	72
4520			86	10x6	G	50	8	58
4600	200	10	45	10	A	78	10.65	88.65
4640			81	8-5/8	C	51	10.65	61.65
4720			70	8x6	G	57.50	10.65	68.15
4800		16	49	10	A	78.50	8	86.50
4840			123	10-3/4	C	67	8	75
4920			85	10x6	G	62	8	70
5200	300	10	61	14	A	106	10.65	116.65
5240			169	12-3/4	C	89	10.65	99.65
5320			86	10x6	G	86	10.65	96.65
5400		16	72	12	A	115	8	123
5440			169	12-3/4	C	83	8	91
5600		20	79	12	A	120	8	128
5640			169	12-3/4	C	78.50	8	86.50
5800	400	10	79	12	A	137	10.65	147.65
5840			178	12-3/4	C	117	10.65	127.65
6000		16	87	12	A	139	8	147
6040			178	12-3/4	C	108	8	116
6400	500	10	99	14	A	171	10.65	181.65
6600		16	109	14	A	175	8	183
6800		20	120	12	A	182	8	190
7000	600	10	120	12	A	208	10.65	218.65
7200		16	132	14	A	211	8	219
7400		20	132	14	A	200	8	208
7600	700	10	136	12	A	235	10.65	245.65
7800		16	145	14	A	232	8	240
8000		20	145	14	A	220	8	228

B10 Superstructure

B1010 Floor Construction

B1010 208		Steel Columns						

	LOAD (KIPS)	UNSUPPORTED HEIGHT (FT.)	WEIGHT (P.L.F.)	SIZE (IN.)	TYPE	COST PER V.L.F.		
						MAT.	INST.	TOTAL
8200	800	10	145	14	A	251	10.65	261.65
8300		16	159	14	A	255	8	263
8400		20	176	14	A	267	8	275
8800	900	10	159	14	A	275	10.65	285.65
8900		16	176	14	A	282	8	290
9000		20	193	14	A	293	8	301
9100	1000	10	176	14	A	305	10.65	315.65
9200		16	193	14	A	310	8	318
9300		20	211	14	A	320	8	328

B1010 Floor Construction

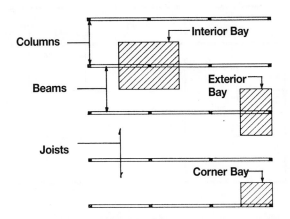

Columns — Interior Bay

Beams

Exterior Bay

Joists

Corner Bay

Description: Table below lists costs per S.F. of bay size for wood columns of various sizes and unsupported heights and the maximum allowable total load per S.F. per bay size.

B1010 210					Wood Columns			
	NOMINAL COLUMN SIZE (IN.)	BAY SIZE (FT.)	UNSUPPORTED HEIGHT (FT.)	MATERIAL (BF per M.S.F.)	TOTAL LOAD (P.S.F.)	COST PER S.F.		
						MAT.	INST.	TOTAL
1000	4 x 4	10 x 8	8	133	100	.18	.25	.43
1050			10	167	60	.23	.31	.54
1200		10 x 10	8	106	80	.14	.20	.34
1250			10	133	50	.18	.25	.43
1400		10 x 15	8	71	50	.10	.13	.23
1450			10	88	30	.12	.16	.28
1600		15 x 15	8	47	30	.06	.09	.15
1650			10	59	15	.08	.11	.19
2000	6 x 6	10 x 15	8	160	230	.23	.27	.50
2050			10	200	210	.29	.34	.63
2200		15 x 15	8	107	150	.15	.18	.33
2250			10	133	140	.19	.23	.42
2400		15 x 20	8	80	110	.11	.14	.25
2450			10	100	100	.14	.17	.31
2600		20 x 20	8	60	80	.09	.10	.19
2650			10	75	70	.11	.13	.24
2800		20 x 25	8	48	60	.07	.08	.15
2850			10	60	50	.09	.10	.19
3400	8 x 8	20 x 20	8	107	160	.16	.17	.33
3450			10	133	160	.20	.21	.41
3600		20 x 25	8	85	130	.11	.16	.27
3650			10	107	130	.14	.20	.34
3800		25 x 25	8	68	100	.11	.10	.21
3850			10	85	100	.14	.13	.27
4200	10 x 10	20 x 25	8	133	210	.21	.20	.41
4250			10	167	210	.27	.25	.52
4400		25 x 25	8	107	160	.17	.16	.33
4450			10	133	160	.21	.20	.41
4700	12 x 12	20 x 25	8	192	310	.32	.26	.58
4750			10	240	310	.40	.33	.73
4900		25 x 25	8	154	240	.26	.21	.47
4950			10	192	240	.32	.26	.58

B1010 Floor Construction

"T" Shaped Precast Beams

"L" Shaped Precast Beams

B1010 214	"T" Shaped Precast Beams							
	SPAN (FT.)	SUPERIMPOSED LOAD (K.L.F.)	SIZE W X D (IN.)	BEAM WEIGHT (P.L.F.)	TOTAL LOAD (K.L.F.)	COST PER L.F.		
						MAT.	INST.	TOTAL
2300	15	2.8	12x16	260	3.06	176	20	196
2500		8.37	12x28	515	8.89	209	22	231
8900	45	3.34	12x60	1165	4.51	335	13	348
9900		6.5	24x60	1915	8.42	365	18.15	383.15

B1010 215	"L" Shaped Precast Beams							
	SPAN (FT.)	SUPERIMPOSED LOAD (K.L.F.)	SIZE W X D (IN.)	BEAM WEIGHT (P.L.F.)	TOTAL LOAD (K.L.F.)	COST PER L.F.		
						MAT.	INST.	TOTAL
2250	15	2.58	12x16	230	2.81	165	20	185
2400		5.92	12x24	370	6.29	182	20	202
4000	25	2.64	12x28	435	3.08	193	13.10	206.10
4450		6.44	18x36	790	7.23	226	15.60	241.60
5300	30	2.80	12x36	565	3.37	204	11.85	215.85
6400		8.66	24x44	1245	9.90	270	17	287

B1010 Floor Construction

Cast in Place Floor Slab, One Way
General: Solid concrete slabs of uniform depth reinforced for flexure in one direction and for temperature and shrinkage in the other direction.

Cast in Place Beam & Slab, One Way
General: Solid concrete one way slab cast monolithically with reinforced concrete beams and girders.

B1010 217 — Cast in Place Slabs, One Way

	SLAB DESIGN & SPAN (FT.)	SUPERIMPOSED LOAD (P.S.F.)	THICKNESS (IN.)	TOTAL LOAD (P.S.F.)		COST PER S.F. MAT.	INST.	TOTAL
2500	Single 8	40	4	90		3.87	8	11.87
2600		75	4	125		3.90	8.70	12.60
2700		125	4-1/2	181		4.07	8.75	12.82
2800		200	5	262		4.53	9.05	13.58
3000	Single 10	40	4	90		4.05	8.60	12.65
3100		75	4	125		4.05	8.60	12.65
3200		125	5	188		4.50	8.80	13.30
3300		200	7-1/2	293		6	9.90	15.90
3500	Single 15	40	5-1/2	90		4.86	8.70	13.56
3600		75	6-1/2	156		5.40	9.15	14.55
3700		125	7-1/2	219		5.90	9.40	15.30
3800		200	8-1/2	306		6.40	9.65	16.05
4000	Single 20	40	7-1/2	115		5.85	9.20	15.05
4100		75	9	200		6.40	9.15	15.55
4200		125	10	250		7.05	9.55	16.60
4300		200	10	324		7.60	9.95	17.55

B1010 219 — Cast in Place Beam & Slab, One Way

	BAY SIZE (FT.)	SUPERIMPOSED LOAD (P.S.F.)	MINIMUM COL. SIZE (IN.)	SLAB THICKNESS (IN.)	TOTAL LOAD (P.S.F.)	COST PER S.F. MAT.	INST.	TOTAL
5000	20x25	40	12	5-1/2	121	4.97	11	15.97
5200		125	16	5-1/2	215	5.90	12.60	18.50
5500	25x25	40	12	6	129	5.25	10.80	16.05
5700		125	18	6	227	6.75	13.25	20
7500	30x35	40	16	8	158	6.60	12.20	18.80
7600		75	18	8	196	7.05	12.60	19.65
7700		125	22	8	254	7.95	13.95	21.90
7800		200	26	8	332	8.70	14.45	23.15
8000	35x35	40	16	9	169	7.40	12.55	19.95
8200		75	20	9	213	8	13.70	21.70
8400		125	24	9	272	8.85	14.20	23.05
8600		200	26	9	355	9.80	15.20	25

B1010 Floor Construction

General: Solid concrete two way slab cast monolithically with reinforced concrete support beams and girders.

General: Flat Slab: Solid uniform depth concrete two way slabs with drop panels at columns and no column capitals.

B1010 220 — Cast in Place Beam & Slab, Two Way

	BAY SIZE (FT.)	SUPERIMPOSED LOAD (P.S.F.)	MINIMUM COL. SIZE (IN.)	SLAB THICKNESS (IN.)	TOTAL LOAD (P.S.F.)	COST PER S.F.		
						MAT.	INST.	TOTAL
4000	20 x 25	40	12	7	141	5.80	11.05	16.85
4300		75	14	7	181	6.65	12.10	18.75
4500		125	16	7	236	6.80	12.40	19.20
5100	25 x 25	40	12	7-1/2	149	6.05	11.20	17.25
5200		75	16	7-1/2	185	6.60	12	18.60
5300		125	18	7-1/2	250	7.20	13	20.20
7600	30 x 35	40	16	10	188	8.10	12.70	20.80
7700		75	18	10	225	8.60	13.15	21.75
8000		125	22	10	282	9.60	14.15	23.75
8500	35 x 35	40	16	10-1/2	193	8.70	12.95	21.65
8600		75	20	10-1/2	233	9.15	13.40	22.55
9000		125	24	10-1/2	287	10.25	14.40	24.65

B1010 222 — Cast in Place Flat Slab with Drop Panels

	BAY SIZE (FT.)	SUPERIMPOSED LOAD (P.S.F.)	MINIMUM COL. SIZE (IN.)	SLAB & DROP (IN.)	TOTAL LOAD (P.S.F.)	COST PER S.F.		
						MAT.	INST.	TOTAL
1960	20 x 20	40	12	7 - 3	132	5.10	8.85	13.95
1980		75	16	7 - 4	168	5.40	9.10	14.50
2000		125	18	7 - 6	221	6.05	9.40	15.45
3200	25 x 25	40	12	8-1/2 - 5-1/2	154	6	9.30	15.30
4000		125	20	8-1/2 - 8-1/2	243	6.90	9.95	16.85
4400		200	24	9 - 8-1/2	329	7.25	10.20	17.45
5000	25 x 30	40	14	9-1/2 - 7	168	6.60	9.60	16.20
5200		75	18	9-1/2 - 7	203	7.05	10	17.05
5600		125	22	9-1/2 - 8	256	7.40	10.20	17.60
6400	30 x 30	40	14	10-1/2 - 7-1/2	182	7.15	9.85	17
6600		75	18	10-1/2 - 7-1/2	217	7.65	10.25	17.90
6800		125	22	10-1/2 - 9	269	8.05	10.50	18.55
7400	30 x 35	40	16	11-1/2 - 9	196	7.80	10.25	18.05
7900		75	20	11-1/2 - 9	231	8.40	10.65	19.05
8000		125	24	11-1/2 - 11	284	8.75	10.95	19.70
9000	35 x 35	40	16	12 - 9	202	8.05	10.35	18.40
9400		75	20	12 - 11	240	8.70	10.80	19.50
9600		125	24	12 - 11	290	9	11	20

B1010 Floor Construction

Multispan Joist Slab

General: Flat Plates: Solid uniform depth concrete two way slab without drops or interior beams. Primary design limit is shear at columns.

General: Combination of thin concrete slab and monolithic ribs at uniform spacing to reduce dead weight and increase rigidity.

B1010 223		Cast in Place Flat Plate						
	BAY SIZE (FT.)	SUPERIMPOSED LOAD (P.S.F.)	MINIMUM COL. SIZE (IN.)	SLAB THICKNESS (IN.)	TOTAL LOAD (P.S.F.)	COST PER S.F.		
						MAT.	INST.	TOTAL
3000	15 x 20	40	14	7	127	4.81	8.50	13.31
3400		75	16	7-1/2	169	5.15	8.65	13.80
3600		125	22	8-1/2	231	5.70	8.90	14.60
3800		175	24	8-1/2	281	5.75	8.90	14.65
4200	20 x 20	40	16	7	127	4.83	8.50	13.33
4400		75	20	7-1/2	175	5.20	8.65	13.85
4600		125	24	8-1/2	231	5.70	8.85	14.55
5000		175	24	8-1/2	281	5.75	8.90	14.65
5600	20 x 25	40	18	8-1/2	146	5.65	8.90	14.55
6000		75	20	9	188	5.85	8.95	14.80
6400		125	26	9-1/2	244	6.35	9.25	15.60
6600		175	30	10	300	6.60	9.35	15.95
7000	25 x 25	40	20	9	152	5.85	8.95	14.80
7400		75	24	9-1/2	194	6.20	9.20	15.40
7600		125	30	10	250	6.65	9.40	16.05

B1010 226		Cast in Place Multispan Joist Slab						
	BAY SIZE (FT.)	SUPERIMPOSED LOAD (P.S.F.)	MINIMUM COL. SIZE (IN.)	RIB DEPTH (IN.)	TOTAL LOAD (P.S.F.)	COST PER S.F.		
						MAT.	INST.	TOTAL
2000	15 x 15	40	12	8	115	6.10	10.45	16.55
2100		75	12	8	150	6.15	10.45	16.60
2200		125	12	8	200	6.30	10.55	16.85
2300		200	14	8	275	6.45	10.95	17.40
2600	15 x 20	40	12	8	115	6.25	10.40	16.65
2800		75	12	8	150	6.35	11.05	17.40
3000		125	14	8	200	6.60	11.20	17.80
3300		200	16	8	275	6.90	11.40	18.30
3600	20 x 20	40	12	10	120	6.40	10.30	16.70
3900		75	14	10	155	6.65	10.90	17.55
4000		125	16	10	205	6.70	11.10	17.80
4100		200	18	10	280	7.05	11.60	18.65
6200	30 x 30	40	14	14	131	7.10	10.75	17.85
6400		75	18	14	166	7.30	11.10	18.40
6600		125	20	14	216	7.75	11.70	19.45
6700		200	24	16	297	8.30	12.15	20.45

B1010 Floor Construction

Waffle Slab

B1010 227	Cast in Place Waffle Slab							
	BAY SIZE (FT.)	SUPERIMPOSED LOAD (P.S.F.)	MINIMUM COL. SIZE (IN.)	RIB DEPTH (IN.)	TOTAL LOAD (P.S.F.)	COST PER S.F.		
						MAT.	INST.	TOTAL
3900	20 x 20	40	12	8	144	7.80	10.30	18.10
4000		75	12	8	179	7.95	10.45	18.40
4100		125	16	8	229	8.10	10.55	18.65
4200		200	18	8	304	8.60	10.95	19.55
4400	20 x 25	40	12	8	146	7.95	10.40	18.35
4500		75	14	8	181	8.15	10.55	18.70
4600		125	16	8	231	8.35	10.70	19.05
4700		200	18	8	306	8.75	11	19.75
4900	25 x 25	40	12	10	150	8.20	10.50	18.70
5000		75	16	10	185	8.45	10.70	19.15
5300		125	18	10	235	8.70	10.90	19.60
5500		200	20	10	310	9	11.10	20.10
5700	25 x 30	40	14	10	154	8.35	10.60	18.95
5800		75	16	10	189	8.60	10.80	19.40
5900		125	18	10	239	8.85	11	19.85
6000		200	20	12	329	9.75	11.45	21.20
6400	30 x 30	40	14	12	169	8.90	10.80	19.70
6500		75	18	12	204	9.15	11	20.15
6600		125	20	12	254	9.30	11.10	20.40
6700		200	24	12	329	10.15	11.70	21.85
6900	30 x 35	40	16	12	169	9.10	10.90	20
7000		75	18	12	204	9.10	10.90	20
7100		125	22	12	254	9.60	11.25	20.85
7200		200	26	14	334	10.70	12	22.70
7400	35 x 35	40	16	14	174	9.65	11.20	20.85
7500		75	20	14	209	9.85	11.40	21.25
7600		125	24	14	259	10.10	11.55	21.65
7700		200	26	16	346	10.90	12.15	23.05
8000	35 x 40	40	18	14	176	9.90	11.40	21.30
8300		75	22	14	211	10.25	11.70	21.95
8500		125	26	16	271	10.80	11.95	22.75
8750		200	30	20	372	11.85	12.60	24.45
9200	40 x 40	40	18	14	176	10.25	11.70	21.95
9400		75	24	14	211	10.70	12	22.70
9500		125	26	16	271	10.95	12.10	23.05
9700	40 x 45	40	20	16	186	10.65	11.85	22.50
9800		75	24	16	221	11.10	12.15	23.25
9900		125	28	16	271	11.30	12.30	23.60

B1010 Floor Construction

Precast Plank with No Topping

Precast Plank with 2″ Concrete Topping

B1010 229					Precast Plank with No Topping				
	SPAN (FT.)	SUPERIMPOSED LOAD (P.S.F.)	TOTAL DEPTH (IN.)	DEAD LOAD (P.S.F.)	TOTAL LOAD (P.S.F.)	COST PER S.F.			
						MAT.	INST.	TOTAL	
0720	10	40	4	50	90	6.65	3.40	10.05	
0750		75	6	50	125	7.15	2.92	10.07	
0770		100	6	50	150	7.15	2.92	10.07	
0800	15	40	6	50	90	7.15	2.92	10.07	
0820		75	6	50	125	7.15	2.92	10.07	
0850		100	6	50	150	7.15	2.92	10.07	
0950	25	40	6	50	90	7.15	2.92	10.07	
0970		75	8	55	130	7.50	2.56	10.06	
1000		100	8	55	155	7.50	2.56	10.06	
1200	30	40	8	55	95	7.50	2.56	10.06	
1300		75	8	55	130	7.50	2.56	10.06	
1400		100	10	70	170	8.15	2.27	10.42	
1500	40	40	10	70	110	8.15	2.27	10.42	
1600		75	12	70	145	8.95	2.04	10.99	
1700	45	40	12	70	110	8.95	2.04	10.99	

B1010 230					Precast Plank with 2″ Concrete Topping				
	SPAN (FT.)	SUPERIMPOSED LOAD (P.S.F.)	TOTAL DEPTH (IN.)	DEAD LOAD (P.S.F.)	TOTAL LOAD (P.S.F.)	COST PER S.F.			
						MAT.	INST.	TOTAL	
2000	10	40	6	75	115	7.60	5.70	13.30	
2100		75	8	75	150	8.10	5.20	13.30	
2200		100	8	75	175	8.10	5.20	13.30	
2500	15	40	8	75	115	8.10	5.20	13.30	
2600		75	8	75	150	8.10	5.20	13.30	
2700		100	8	75	175	8.10	5.20	13.30	
3100	25	40	8	75	115	8.10	5.20	13.30	
3200		75	8	75	150	8.10	5.20	13.30	
3300		100	10	80	180	8.45	4.84	13.29	
3400	30	40	10	80	120	8.45	4.84	13.29	
3500		75	10	80	155	8.45	4.84	13.29	
3600		100	10	80	180	8.45	4.84	13.29	
4000	40	40	12	95	135	9.10	4.55	13.65	
4500		75	14	95	170	9.90	4.32	14.22	
5000	45	40	14	95	135	9.90	4.32	14.22	

B1010 Floor Construction

Precast Double "T" Beams with No Topping

Precast Double "T" Beams with 2″ Topping

Most widely used for moderate span floors and moderate and long span roofs. At shorter spans, they tend to be competitive with hollow core slabs. They are also used as wall panels.

B1010 234			Precast Double "T" Beams with No Topping					
	SPAN (FT.)	SUPERIMPOSED LOAD (P.S.F.)	DBL. "T" SIZE D (IN.) W (FT.)	CONCRETE "T" TYPE	TOTAL LOAD (P.S.F.)	COST PER S.F.		
						MAT.	INST.	TOTAL
4300	50	30	20x8	Lt. Wt.	66	9.60	1.55	11.15
4400		40	20x8	Lt. Wt.	76	9.65	1.42	11.07
4500		50	20x8	Lt. Wt.	86	9.75	1.70	11.45
4600		75	20x8	Lt. Wt.	111	9.80	1.94	11.74
5600	70	30	32x10	Lt. Wt.	78	9.60	1.07	10.67
5750		40	32x10	Lt. Wt.	88	9.70	1.30	11
5900		50	32x10	Lt. Wt.	98	9.75	1.49	11.24
6000		75	32x10	Lt. Wt.	123	9.85	1.80	11.65
6100		100	32x10	Lt. Wt.	148	10.05	2.44	12.49
6200	80	30	32x10	Lt. Wt.	78	9.75	1.49	11.24
6300		40	32x10	Lt. Wt.	88	9.85	1.80	11.65
6400		50	32x10	Lt. Wt.	98	9.95	2.12	12.07

B1010 235			Precast Double "T" Beams With 2″ Topping					
	SPAN (FT.)	SUPERIMPOSED LOAD (P.S.F.)	DBL. "T" SIZE D (IN.) W (FT.)	CONCRETE "T" TYPE	TOTAL LOAD (P.S.F.)	COST PER S.F.		
						MAT.	INST.	TOTAL
7100	40	30	18x8	Reg. Wt.	120	8.35	3.11	11.46
7200		40	20x8	Reg. Wt.	130	9.75	2.98	12.73
7300		50	20x8	Reg. Wt.	140	9.85	3.26	13.11
7400		75	20x8	Reg. Wt.	165	9.90	3.39	13.29
7500		100	20x8	Reg. Wt.	190	10	3.79	13.79
7550	50	30	24x8	Reg. Wt.	120	9.85	3.12	12.97
7600		40	24x8	Reg. Wt.	130	9.90	3.22	13.12
7750		50	24x8	Reg. Wt.	140	9.90	3.25	13.15
7800		75	24x8	Reg. Wt.	165	10.05	3.64	13.69
7900		100	32x10	Reg. Wt.	189	9.90	2.98	12.88

B1010 Floor Construction

Precast Beam and Plank with No Topping

Precast Beam and Plank with 2″ Topping

B1010 236 — Precast Beam & Plank with No Topping

	BAY SIZE (FT.)	SUPERIMPOSED LOAD (P.S.F.)	PLANK THICKNESS (IN.)	TOTAL DEPTH (IN.)	TOTAL LOAD (P.S.F.)	COST PER S.F. MAT.	INST.	TOTAL
6200	25x25	40	8	28	118	20.50	3.41	23.91
6400		75	8	36	158	20.50	3.41	23.91
6500		100	8	36	183	20.50	3.41	23.91
7000	25x30	40	8	36	110	18.70	3.31	22.01
7200		75	10	36	159	19.60	3.02	22.62
7400		100	12	36	188	20	2.74	22.74
7600	30x30	40	8	36	121	18.95	3.20	22.15
8000		75	10	44	140	18.85	3.21	22.06
8250		100	12	52	206	21.50	2.71	24.21
8500	30x35	40	12	44	135	18.90	2.61	21.51
8750		75	12	52	176	19.85	2.62	22.47
9000	35x35	40	12	52	141	19.85	2.55	22.40
9250		75	12	60	181	20.50	2.55	23.05
9500	35x40	40	12	52	137	19.85	2.55	22.40

B1010 238 — Precast Beam & Plank with 2″ Topping

	BAY SIZE (FT.)	SUPERIMPOSED LOAD (P.S.F.)	PLANK THICKNESS (IN.)	TOTAL DEPTH (IN.)	TOTAL LOAD (P.S.F.)	COST PER S.F. MAT.	INST.	TOTAL
4300	20x20	40	6	22	135	22	6	28
4400		75	6	24	173	23	6.05	29.05
4500		100	6	28	200	23	6.05	29.05
4600	20x25	40	6	26	134	21	5.90	26.90
5000		75	8	30	177	21	5.45	26.45
5200		100	8	30	202	21	5.45	26.45
5400	25x25	40	6	38	143	21.50	5.60	27.10
5600		75	8	38	183	16.90	5.30	22.20
6000		100	8	46	216	16.70	4.89	21.59
6200	25x30	40	8	38	144	19.45	5.10	24.55
6400		75	10	46	200	20	4.77	24.77
6600		100	10	46	225	20.50	4.78	25.28
7000	30x30	40	8	46	150	20	5	25
7200		75	10	54	181	21	5	26
7600		100	10	54	231	21	5	26
7800	30x35	40	10	54	166	20	4.61	24.61
8000		75	12	54	200	20.50	4.36	24.86

B1010 Floor Construction

General: Beams and double tees priced here are for plant produced prestressed members transported to the site and erected.

The 2″ structural topping is applied after the beams and double tees are in place and is reinforced with W.W.F.

With Topping

B1010 239	Precast Double "T" & 2″ Topping on Precast Beams							
	BAY SIZE (FT.)	SUPERIMPOSED LOAD (P.S.F.)	DEPTH (IN.)		TOTAL LOAD (P.S.F.)	COST PER S.F.		
						MAT.	INST.	TOTAL
3000	25x30	40	38		130	21	4.24	25.24
3100		75	38		168	21	4.24	25.24
3300		100	46		196	21	4.20	25.20
3600	30x30	40	46		150	21.50	4.10	25.60
3750		75	46		174	21.50	4.10	25.60
4000		100	54		203	22.50	4.09	26.59
4100	30x40	40	46		136	18.85	3.68	22.53
4300		75	54		173	19.40	3.66	23.06
4400		100	62		204	20	3.68	23.68
4600	30x50	40	54		138	17.65	3.41	21.06
4800		75	54		181	17.45	3.46	20.91
5000		100	54		219	17.45	3.15	20.60
5200	30x60	40	62		151	17.15	3.08	20.23
5400		75	62		192	16.75	3.13	19.88
5600		100	62		215	16.90	3.13	20.03
5800	35x40	40	54		139	19.65	3.62	23.27
6000		75	62		179	19.80	3.49	23.29
6250		100	62		212	19.95	3.62	23.57
6500	35x50	40	62		142	18.20	3.40	21.60
6750		75	62		186	18.25	3.51	21.76
7300		100	62		231	18.35	3.20	21.55
7600	35x60	40	54		154	17.65	2.96	20.61
7750		75	54		179	17.75	2.97	20.72
8000		100	62		224	18.20	3.02	21.22
8250	40x40	40	62		145	21.50	3.63	25.13
8400		75	62		187	20.50	3.67	24.17
8750		100	62		223	21.50	3.74	25.24
9000	40x50	40	62		151	19	3.42	22.42
9300		75	62		193	18.55	2.54	21.09
9800	40x60	40	62		164	21	2.45	23.45

B1010 Floor Construction

General: The following table is based upon structural wide flange (WF) beam and girder framing. Non-composite action is assumed between WF framing and decking. Deck costs not included.

The deck spans the short direction. The steel beams and girders are fireproofed with sprayed fiber fireproofing.

No columns included in price.

B1010 241		W Shape Beams & Girders						
	BAY SIZE (FT.) BEAM X GIRD	SUPERIMPOSED LOAD (P.S.F.)	STEEL FRAMING DEPTH (IN.)	FIREPROOFING (S.F. PER S.F.)	TOTAL LOAD (P.S.F.)	COST PER S.F.		
						MAT.	INST.	TOTAL
1350	15x20	40	12	.535	50	4.98	1.92	6.90
1400		40	16	.65	90	6.50	2.47	8.97
1450		75	18	.694	125	8.60	3.10	11.70
1500		125	24	.796	175	11.85	4.31	16.16
2000	20x20	40	12	.55	50	5.60	2.10	7.70
2050		40	14	.579	90	7.65	2.74	10.39
2100		75	16	.672	125	9.15	3.26	12.41
2150		125	16	.714	175	10.90	3.96	14.86
2900	20x25	40	16	.53	50	6.15	2.25	8.40
2950		40	18	.621	96	9.70	3.39	13.09
3000		75	18	.651	131	11.20	3.85	15.05
3050		125	24	.77	200	14.75	5.20	19.95
5200	25x25	40	18	.486	50	6.70	2.39	9.09
5300		40	18	.592	96	10	3.43	13.43
5400		75	21	.668	131	12.05	4.12	16.17
5450		125	24	.738	191	15.90	5.50	21.40
6300	25x30	40	21	.568	50	7.65	2.73	10.38
6350		40	21	.694	90	10.35	3.62	13.97
6400		75	24	.776	125	13.30	4.57	17.87
6450		125	30	.904	175	16	5.70	21.70
7700	30x30	40	21	.52	50	8.20	2.85	11.05
7750		40	24	.629	103	12.60	4.25	16.85
7800		75	30	.715	138	15	5	20
7850		125	36	.822	206	19.75	6.75	26.50
8250	30x35	40	21	.508	50	9.35	3.18	12.53
8300		40	24	.651	109	13.80	4.62	18.42
8350		75	33	.732	150	16.75	5.55	22.30
8400		125	36	.802	225	21	7.10	28.10
9550	35x35	40	27	.560	50	9.65	3.33	12.98
9600		40	36	.706	109	17.65	5.80	23.45
9650		75	36	.750	150	18.25	6	24.25
9820		125	36	.797	225	24.50	8.20	32.70

B1010 Floor Construction

Description: Table below lists costs for light gage CEE or PUNCHED DOUBLE joists to suit the span and loading with the minimum thickness subfloor required by the joist spacing.

B1010 244		Light Gauge Steel Floor Systems						
	SPAN (FT.)	SUPERIMPOSED LOAD (P.S.F.)	FRAMING DEPTH (IN.)	FRAMING SPAC. (IN.)	TOTAL LOAD (P.S.F.)	COST PER S.F.		
						MAT.	INST.	TOTAL
1500	15	40	8	16	54	2.71	2.18	4.89
1550			8	24	54	2.73	2.23	4.96
1600		65	10	16	80	3.50	2.66	6.16
1650			10	24	80	3.36	2.59	5.95
1700		75	10	16	90	3.50	2.66	6.16
1750			10	24	90	3.36	2.59	5.95
1800		100	10	16	116	4.55	3.38	7.93
1850			10	24	116	3.36	2.59	5.95
1900		125	10	16	141	4.55	3.38	7.93
1950			10	24	141	3.36	2.59	5.95
2500	20	40	8	16	55	3.11	2.47	5.58
2550			8	24	55	2.73	2.23	4.96
2600		65	8	16	80	3.60	2.82	6.42
2650			10	24	80	3.37	2.61	5.98
2700		75	10	16	90	3.31	2.52	5.83
2750			12	24	90	3.79	2.27	6.06
2800		100	10	16	115	3.31	2.52	5.83
2850			10	24	116	4.35	2.94	7.29
2900		125	12	16	142	5.20	2.86	8.06
2950			12	24	141	4.73	3.12	7.85
3500	25	40	10	16	55	3.50	2.66	6.16
3550			10	24	55	3.36	2.59	5.95
3600		65	10	16	81	4.55	3.38	7.93
3650			12	24	81	4.21	2.46	6.67
3700		75	12	16	92	5.20	2.86	8.06
3750			12	24	91	4.73	3.12	7.85
3800		100	12	16	117	5.85	3.16	9.01
3850		125	12	16	143	6.65	4.17	10.82
4500	30	40	12	16	57	5.20	2.86	8.06
4550			12	24	56	4.19	2.45	6.64
4600		65	12	16	82	5.80	3.15	8.95
4650		75	12	16	92	5.80	3.15	8.95

B10 Superstructure

B1010 Floor Construction

Description: The table below lists costs per S.F. for a floor system on bearing walls using open web steel joists, galvanized steel slab form and 2-1/2″ concrete slab reinforced with welded wire fabric. Costs of the bearing walls are not included.

B1010 246		Deck & Joists on Bearing Walls						
	SPAN (FT.)	SUPERIMPOSED LOAD (P.S.F.)	JOIST SPACING FT. - IN.	DEPTH (IN.)	TOTAL LOAD (P.S.F.)	COST PER S.F.		
						MAT.	INST.	TOTAL
1050	20	40	2-0	14-1/2	83	5.65	3.82	9.47
1070		65	2-0	16-1/2	109	6.15	4.06	10.21
1100		75	2-0	16-1/2	119	6.15	4.06	10.21
1120		100	2-0	18-1/2	145	6.20	4.07	10.27
1150		125	1-9	18-1/2	170	6.75	4.32	11.07
1170	25	40	2-0	18-1/2	84	6.20	4.07	10.27
1200		65	2-0	20-1/2	109	6.40	4.16	10.56
1220		75	2-0	20-1/2	119	6.80	4.34	11.14
1250		100	2-0	22-1/2	145	6.95	4.39	11.34
1270		125	1-9	22-1/2	170	7.90	4.79	12.69
1300	30	40	2-0	22-1/2	84	6.95	3.95	10.90
1320		65	2-0	24-1/2	110	7.55	4.12	11.67
1350		75	2-0	26-1/2	121	7.80	4.20	12
1370		100	2-0	26-1/2	146	8.35	4.35	12.70
1400		125	2-0	24-1/2	172	9.80	4.79	14.59
1420	35	40	2-0	26-1/2	85	9.25	4.62	13.87
1450		65	2-0	28-1/2	111	9.55	4.72	14.27
1470		75	2-0	28-1/2	121	9.55	4.72	14.27
1500		100	1-11	28-1/2	147	10.40	4.97	15.37
1520		125	1-8	28-1/2	172	11.45	5.30	16.75
1550	5/8″ gyp. fireproof.							
1560	On metal furring, add					.76	2.85	3.61

B10 Superstructure

B1010 Floor Construction

The table below lists costs for a floor system on exterior bearing walls and interior columns and beams using open web steel joists, galvanized steel slab form, 2-1/2″ concrete slab reinforced with welded wire fabric.

Columns are sized to accommodate one floor plus roof loading but costs for columns are from suspended floor to slab on grade only. Costs of the bearing walls are not included.

Steel Joists and Concrete Slab on Walls and Beams

B1010 248		Steel Joists on Beam & Wall						
	BAY SIZE (FT.)	SUPERIMPOSED LOAD (P.S.F.)	DEPTH (IN.)	TOTAL LOAD (P.S.F.)	COLUMN ADD	COST PER S.F.		
						MAT.	INST.	TOTAL
1720	25x25	40	23	84		7.80	4.37	12.17
1730					columns	.41	.12	.53
1750	25x25	65	29	110		8.20	4.52	12.72
1760					columns	.41	.12	.53
1770	25x25	75	26	120		9.15	4.86	14.01
1780					columns	.47	.14	.61
1800	25x25	100	29	145		10.30	5.30	15.60
1810					columns	.47	.14	.61
1820	25x25	125	29	170		10.85	5.50	16.35
1830					columns	.53	.15	.68
2020	30x30	40	29	84		9	4.38	13.38
2030					columns	.37	.10	.47
2050	30x30	65	29	110		10.20	4.73	14.93
2060					columns	.37	.10	.47
2070	30x30	75	32	120		10.50	4.82	15.32
2080					columns	.42	.13	.55
2100	30x30	100	35	145		11.60	5.15	16.75
2110					columns	.49	.14	.63
2120	30x30	125	35	172		13.45	5.70	19.15
2130					columns	.59	.18	.77
2170	30x35	40	29	85		10.15	4.72	14.87
2180					columns	.35	.10	.45
2200	30x35	65	29	111		12.15	5.30	17.45
2210					columns	.40	.12	.52
2220	30x35	75	32	121		12.10	5.30	17.40
2230					columns	.41	.13	.54
2250	30x35	100	35	148		12.35	5.35	17.70
2260					columns	.51	.15	.66
2270	30x35	125	38	173		13.85	5.80	19.65
2280					columns	.51	.15	.66
2320	35x35	40	32	85		10.50	4.82	15.32
2330					columns	.36	.10	.46
2350	35x35	65	35	111		12.80	5.50	18.30
2360					columns	.43	.13	.56
2370	35x35	75	35	121		13.10	5.60	18.70
2380					columns	.43	.13	.56
2400	35x35	100	38	148		13.65	5.75	19.40
2410					columns	.53	.15	.68

B1010 Floor Construction

B1010 248		Steel Joists on Beam & Wall						
	BAY SIZE (FT.)	SUPERIMPOSED LOAD (P.S.F.)	DEPTH (IN.)	TOTAL LOAD (P.S.F.)	COLUMN ADD	COST PER S.F.		
						MAT.	INST.	TOTAL
2460 2475	5/8 gyp. fireproof. On metal furring, add					.76	2.85	3.61

B1010   Floor Construction

The table below lists costs per S.F. for a floor system on steel columns and beams using open web steel joists, galvanized steel slab form, 2-1/2″ concrete slab reinforced with welded wire fabric.

Columns are sized to accommodate one floor plus roof loading but costs for columns are from floor to grade only.

Steel Joists and Concrete Slab on Steel Columns and Beams

B1010 250	BAY SIZE (FT.)	SUPERIMPOSED LOAD (P.S.F.)	DEPTH (IN.)	TOTAL LOAD (P.S.F.)	COLUMN ADD	COST PER S.F. MAT.	COST PER S.F. INST.	COST PER S.F. TOTAL
2350	15x20	40	17	83		8.10	4.41	12.51
2400					column	1.35	.40	1.75
2450	15x20	65	19	108		9.05	4.68	13.73
2500					column	1.35	.40	1.75
2550	15x20	75	19	119		9.35	4.83	14.18
2600					column	1.48	.43	1.91
2650	15x20	100	19	144		9.95	5.05	15
2700					column	1.48	.43	1.91
2750	15x20	125	19	170		10.50	5.15	15.65
2800					column	1.97	.57	2.54
2850	20x20	40	19	83		8.85	4.61	13.46
2900					column	1.11	.33	1.44
2950	20x20	65	23	109		9.75	4.93	14.68
3000					column	1.48	.43	1.91
3100	20x20	75	26	119		10.35	5.10	15.45
3200					column	1.48	.43	1.91
3400	20x20	100	23	144		10.75	5.25	16
3450					column	1.48	.43	1.91
3500	20x20	125	23	170		11.95	5.65	17.60
3600					column	1.77	.52	2.29
3700	20x25	40	44	83		9.50	4.84	14.34
3800					column	1.18	.34	1.52
3900	20x25	65	26	110		10.40	5.15	15.55
4000					column	1.18	.34	1.52
4100	20x25	75	26	120		10.85	5.30	16.15
4200					column	1.42	.42	1.84
4300	20x25	100	26	145		11.50	5.55	17.05
4400					column	1.42	.42	1.84
4500	20x25	125	29	170		12.80	6.05	18.85
4600					column	1.65	.48	2.13
4700	25x25	40	23	84		10.30	5.05	15.35
4800					column	1.13	.33	1.46
4900	25x25	65	29	110		10.90	5.25	16.15
5000					column	1.13	.33	1.46
5100	25x25	75	26	120		12.05	5.70	17.75
5200					column	1.32	.38	1.70

B1010 Floor Construction

B1010 250	Steel Joists, Beams & Slab on Columns

	BAY SIZE (FT.)	SUPERIMPOSED LOAD (P.S.F.)	DEPTH (IN.)	TOTAL LOAD (P.S.F.)	COLUMN ADD	COST PER S.F.		
						MAT.	INST.	TOTAL
5300	25x25	100	29	145		13.40	6.20	19.60
5400					column	1.32	.38	1.70
5500	25x25	125	32	170		14.15	6.45	20.60
5600					column	1.46	.43	1.89
5700	25x30	40	29	84		11.80	5.15	16.95
5800					column	1.10	.32	1.42
5900	25x30	65	29	110		12.30	5.30	17.60
6000					column	1.10	.32	1.42
6050	25x30	75	29	120		12.65	5.45	18.10
6100					column	1.22	.36	1.58
6150	25x30	100	29	145		13.75	5.75	19.50
6200					column	1.22	.36	1.58
6250	25x30	125	32	170		15.45	6.25	21.70
6300					column	1.40	.41	1.81
6350	30x30	40	29	84		11.85	5.15	17
6400					column	1.02	.29	1.31
6500	30x30	65	29	110		13.45	5.65	19.10
6600					column	1.02	.29	1.31
6700	30x30	75	32	120		13.75	5.75	19.50
6800					column	1.17	.34	1.51
6900	30x30	100	35	145		15.25	6.15	21.40
7000					column	1.36	.40	1.76
7100	30x30	125	35	172		17.45	6.80	24.25
7200					column	1.52	.45	1.97
7300	30x35	40	29	85		13.30	5.60	18.90
7400					column	.87	.25	1.12
7500	30x35	65	29	111		15.55	6.30	21.85
7600					column	1.12	.33	1.45
7700	30x35	75	32	121		15.55	6.30	21.85
7800					column	1.15	.33	1.48
7900	30x35	100	35	148		16.05	6.40	22.45
8000					column	1.40	.41	1.81
8100	30x35	125	38	173		17.80	6.95	24.75
8200					column	1.43	.42	1.85
8300	35x35	40	32	85		13.70	5.75	19.45
8400					column	1	.29	1.29
8500	35x35	65	35	111		16.35	6.50	22.85
8600					column	1.20	.36	1.56
9300	35x35	75	38	121		16.80	6.60	23.40
9400					column	1.20	.36	1.56
9500	35x35	100	38	148		18.10	7.05	25.15
9600					column	1.48	.43	1.91
9750	35x35	125	41	173		18.70	7.20	25.90
9800					column	1.51	.45	1.96
9810	5/8 gyp. fireproof.							
9815	On metal furring, add					.76	2.85	3.61

B10 Superstructure

B1010 Floor Construction

Composite Beam & Cast in Place Slab

General: Composite construction of wide flange beams and concrete slabs is most efficiently used when loads are heavy and spans are moderately long. It is stiffer with less deflection than non-composite construction of similar depth and spans.

B1010 252		Composite Beam & Cast in Place Slab						
	BAY SIZE (FT.)	SUPERIMPOSED LOAD (P.S.F.)	SLAB THICKNESS (IN.)	TOTAL DEPTH (FT.-IN.)	TOTAL LOAD (P.S.F.)	COST PER S.F.		
						MAT.	INST.	TOTAL
3800	20x25	40	4	1 - 8	94	9.90	10.60	20.50
3900		75	4	1 - 8	130	11.60	11.30	22.90
4000		125	4	1 - 10	181	13.55	12.05	25.60
4100		200	5	2 - 2	272	17.70	13.65	31.35
4200	25x25	40	4-1/2	1 - 8-1/2	99	10.35	10.75	21.10
4300		75	4-1/2	1 - 10-1/2	136	12.55	11.65	24.20
4400		125	5-1/2	2 - 0-1/2	200	14.90	12.55	27.45
4500		200	5-1/2	2 - 2-1/2	278	18.60	13.95	32.55
4600	25x30	40	4-1/2	1 - 8-1/2	100	11.60	11.20	22.80
4700		75	4-1/2	1 - 10-1/2	136	13.70	12	25.70
4800		125	5-1/2	2 - 2-1/2	202	16.80	13.20	30
4900		200	5-1/2	2 - 5-1/2	279	20	14.45	34.45
5000	30x30	40	4	1 - 8	95	11.65	11.25	22.90
5200		75	4	2 - 1	131	13.55	11.85	25.40
5400		125	4	2 - 4	183	16.85	13.10	29.95
5600		200	5	2 - 10	274	21.50	14.85	36.35
5800	30x35	40	4	2 - 1	95	12.35	11.45	23.80
6000		75	4	2 - 4	139	15	12.35	27.35
6250		125	4	2 - 4	185	19.60	13.95	33.55
6500		200	5	2 - 11	276	24	15.60	39.60
8000	35x35	40	4	2 - 1	96	13.05	11.65	24.70
8250		75	4	2 - 4	133	15.60	12.60	28.20
8500		125	4	2 - 10	185	18.80	13.75	32.55
8750		200	5	3 - 5	276	24.50	15.75	40.25
9000	35x40	40	4	2 - 4	97	14.55	12.15	26.70
9250		75	4	2 - 4	134	17.10	12.95	30.05
9500		125	4	2 - 7	186	21	14.35	35.35
9750		200	5	3 - 5	278	27	16.55	43.55

B1010 Floor Construction

Wide Flange, Composite Deck & Slab

The table below lists costs per S.F. for floors using steel beams and girders, composite steel deck, concrete slab reinforced with W.W.F. and sprayed fiber fireproofing (non- asbestos) on the steel beams and girders and on the steel deck.

B1010 254		W Shape, Composite Deck, & Slab						
	BAY SIZE (FT.) BEAM X GIRD	SUPERIMPOSED LOAD (P.S.F.)	SLAB THICKNESS (IN.)	TOTAL DEPTH (FT.-IN.)	TOTAL LOAD (P.S.F.)	COST PER S.F.		
						MAT.	INST.	TOTAL
0540	15x20	40	5	1-7	89	11.45	6.50	17.95
0560		75	5	1-9	125	12.65	6.90	19.55
0580		125	5	1-11	176	14.85	7.55	22.40
0600		200	5	2-2	254	18.30	8.65	26.95
0700	20x20	40	5	1-9	90	12.50	6.80	19.30
0720		75	5	1-9	126	14.20	7.35	21.55
0740		125	5	1-11	177	16.15	7.95	24.10
0760		200	5	2-5	255	19.55	8.95	28.50
0780	20x25	40	5	1-9	90	12.70	6.75	19.45
0800		75	5	1-9	127	15.95	7.80	23.75
0820		125	5	1-11	180	19.10	8.50	27.60
0840		200	5	2-5	256	21.50	9.60	31.10
0940	25x25	40	5	1-11	91	14.45	7.25	21.70
0960		75	5	2-5	178	16.70	8	24.70
0980		125	5	2-5	181	20.50	8.90	29.40
1000		200	5-1/2	2-8-1/2	263	23.50	10.30	33.80
1400	25x30	40	5	2-5	91	14.35	7.45	21.80
1500		75	5	2-5	128	17.55	8.30	25.85
1600		125	5	2-8	180	20	9.20	29.20
1700		200	5	2-11	259	25	10.60	35.60
2200	30x30	40	5	2-2	92	16	7.90	23.90
2300		75	5	2-5	129	18.75	8.80	27.55
2400		125	5	2-11	182	22.50	9.90	32.40
2500		200	5	3-2	263	30.50	12.40	42.90
2600	30x35	40	5	2-5	94	17.45	8.40	25.85
2700		75	5	2-11	131	20.50	9.25	29.75
2800		125	5	3-2	183	24	10.35	34.35
2900		200	5-1/2	3-5-1/2	268	29.50	11.95	41.45
3800	35x35	40	5	2-8	94	17.95	8.35	26.30
3900		75	5	2-11	131	21	9.30	30.30
4000		125	5	3-5	184	25	10.65	35.65
4100		200	5-1/2	3-5-1/2	270	32.50	12.60	45.10

B1010   Floor Construction

The table below lists costs per S.F. for floors using composite steel beams with welded shear studs, composite steel deck and light weight concrete slab reinforced with W.W.F.

Composite Beam, Deck & Slab

B1010 256		Composite Beams, Deck & Slab						
	BAY SIZE (FT.)	SUPERIMPOSED LOAD (P.S.F.)	SLAB THICKNESS (IN.)	TOTAL DEPTH (FT.-IN.)	TOTAL LOAD (P.S.F.)	COST PER S.F.		
						MAT.	INST.	TOTAL
2400	20x25	40	5-1/2	1 - 5-1/2	80	11.85	6.10	17.95
2500		75	5-1/2	1 - 9-1/2	115	12.35	6.10	18.45
2750		125	5-1/2	1 - 9-1/2	167	15.20	7.10	22.30
2900		200	6-1/4	1 - 11-1/2	251	17.10	7.70	24.80
3000	25x25	40	5-1/2	1 - 9-1/2	82	11.65	5.80	17.45
3100		75	5-1/2	1 - 11-1/2	118	13	5.90	18.90
3200		125	5-1/2	2 - 2-1/2	169	13.60	6.35	19.95
3300		200	6-1/4	2 - 6-1/4	252	18.45	7.45	25.90
3400	25x30	40	5-1/2	1 - 11-1/2	83	11.90	5.75	17.65
3600		75	5-1/2	1 - 11-1/2	119	12.85	5.85	18.70
3900		125	5-1/2	1 - 11-1/2	170	15	6.60	21.60
4000		200	6-1/4	2 - 6-1/4	252	18.55	7.45	26
4200	30x30	40	5-1/2	1 - 11-1/2	81	11.95	5.90	17.85
4400		75	5-1/2	2 - 2-1/2	116	12.95	6.15	19.10
4500		125	5-1/2	2 - 5-1/2	168	15.80	6.95	22.75
4700		200	6-1/4	2 - 9-1/4	252	18.95	8.05	27
4900	30x35	40	5-1/2	2 - 2-1/2	82	12.55	6.10	18.65
5100		75	5-1/2	2 - 5-1/2	117	13.75	6.25	20
5300		125	5-1/2	2 - 5-1/2	169	16.30	7.10	23.40
5500		200	6-1/4	2 - 9-1/4	254	19.15	8.05	27.20
5750	35x35	40	5-1/2	2 - 5-1/2	84	13.40	6.15	19.55
6000		75	5-1/2	2 - 5-1/2	121	15.35	6.60	21.95
7000		125	5-1/2	2 - 8-1/2	170	18.15	7.55	25.70
7200		200	5-1/2	2 - 11-1/2	254	21	8.35	29.35
7400	35x40	40	5-1/2	2 - 5-1/2	85	14.85	6.60	21.45
7600		75	5-1/2	2 - 5-1/2	121	16.15	6.85	23
8000		125	5-1/2	2 - 5-1/2	171	18.60	7.65	26.25
9000		200	5-1/2	2 - 11-1/2	255	22.50	8.70	31.20

B1010 Floor Construction

Description: Table B1010 258 lists S.F. costs for steel deck and concrete slabs for various spans.

Description: Table B1010 261 lists the S.F. costs for wood joists and a minimum thickness plywood subfloor.

Description: Table B1010 264 lists the S.F. costs, total load, and member sizes, for various bay sizes and loading conditions.

Metal Deck/Concrete Fill

Wood Joist

Wood Beam and Joist

B1010 258		Metal Deck/Concrete Fill						
	SUPERIMPOSED LOAD (P.S.F.)	DECK SPAN (FT.)	DECK GAGE DEPTH	SLAB THICKNESS (IN.)	TOTAL LOAD (P.S.F.)	COST PER S.F.		
						MAT.	INST.	TOTAL
0900	125	6	22 1-1/2	4	164	2.83	2.47	5.30
0920		7	20 1-1/2	4	164	3.11	2.62	5.73
0950		8	20 1-1/2	4	165	3.11	2.62	5.73
0970		9	18 1-1/2	4	165	3.66	2.62	6.28
1000		10	18 2	4	165	3.56	2.76	6.32
1020		11	18 3	5	169	4.40	3	7.40

B1010 261	Wood Joist	COST PER S.F.		
		MAT.	INST.	TOTAL
2900	2"x8", 12" O.C.	1.62	1.89	3.51
2950	16" O.C.	1.39	1.62	3.01
3000	24" O.C.	1.37	1.49	2.86
3300	2"x10", 12" O.C.	1.99	2.15	4.14
3350	16" O.C.	1.66	1.81	3.47
3400	24" O.C.	1.54	1.61	3.15
3700	2"x12", 12" O.C.	2.28	2.17	4.45
3750	16" O.C.	1.87	1.82	3.69
3800	24" O.C.	1.69	1.62	3.31
4100	2"x14", 12" O.C.	2.76	2.37	5.13
4150	16" O.C.	2.24	1.98	4.22
4200	24" O.C.	1.94	1.73	3.67
7101	Note: Subfloor cost is included in these prices.			

B1010 264		Wood Beam & Joist						
	BAY SIZE (FT.)	SUPERIMPOSED LOAD (P.S.F.)	GIRDER BEAM (IN.)	JOISTS (IN.)	TOTAL LOAD (P.S.F.)	COST PER S.F.		
						MAT.	INST.	TOTAL
2000	15x15	40	8 x 12 4 x 12	2 x 6 @ 16	53	9	4.36	13.36
2050		75	8 x 16 4 x 16	2 x 8 @ 16	90	11.90	4.77	16.67
2100		125	12 x 16 6 x 16	2 x 8 @ 12	144	18.45	6.10	24.55
2150		200	14 x 22 12 x 16	2 x 10 @ 12	227	37	9.35	46.35
3000	20x20	40	10 x 14 10 x 12	2 x 8 @ 16	63	11.05	4.39	15.44
3050		75	12 x 16 8 x 16	2 x 10 @ 16	102	15.80	4.92	20.72
3100		125	14 x 22 12 x 16	2 x 10 @ 12	163	31	7.80	38.80

B10 Superstructure

B1010 Floor Construction

Steel Column Fireproofing

"T" Shaped
Precast Beams

"L" Shaped
Precast Beams

B1010 720						Steel Column Fireproofing		
	ENCASEMENT SYSTEM	COLUMN SIZE (IN.)	THICKNESS (IN.)	FIRE RATING (HRS.)	WEIGHT (P.L.F.)	COST PER V.L.F.		
						MAT.	INST.	TOTAL
3000	Concrete	8	1	1	110	6.65	35	41.65
3300		14	1	1	258	10.90	51	61.90
3400			2	3	325	13	58	71
3450	Gypsum board	8	1/2	2	8	3.25	21.50	24.75
3550	1 layer	14	1/2	2	18	3.48	23	26.48
3600	Gypsum board	8	1	3	14	4.33	27.50	31.83
3650	1/2" fire rated	10	1	3	17	4.61	29	33.61
3700	2 layers	14	1	3	22	4.76	30	34.76
3750	Gypsum board	8	1-1/2	3	23	5.60	34.50	40.10
3800	1/2" fire rated	10	1-1/2	3	27	6.25	38	44.25
3850	3 layers	14	1-1/2	3	35	6.90	42	48.90
3900	Sprayed fiber	8	1-1/2	2	6.3	3.93	6.85	10.78
3950	Direct application		2	3	8.3	5.40	9.45	14.85
4050		10	1-1/2	2	7.9	4.75	8.25	13
4200		14	1-1/2	2	10.8	5.90	10.25	16.15

B1020 Roof Construction

The table below lists prices per S.F. for roof rafters and sheathing by nominal size and spacing. Sheathing is 5/16" CDX for 12" and 16" spacing and 3/8" CDX for 24" spacing.

Factors for Converting Inclined to Horizontal

Roof Slope	Approx. Angle	Factor	Roof Slope	Approx. Angle	Factor
Flat	0°	1.000	12 in 12	45.0°	1.414
1 in 12	4.8°	1.003	13 in 12	47.3°	1.474
2 in 12	9.5°	1.014	14 in 12	49.4°	1.537
3 in 12	14.0°	1.031	15 in 12	51.3°	1.601
4 in 12	18.4°	1.054	16 in 12	53.1°	1.667
5 in 12	22.6°	1.083	17 in 12	54.8°	1.734
6 in 12	26.6°	1.118	18 in 12	56.3°	1.803
7 in 12	30.3°	1.158	19 in 12	57.7°	1.873
8 in 12	33.7°	1.202	20 in 12	59.0°	1.943
9 in 12	36.9°	1.250	21 in 12	60.3°	2.015
10 in 12	39.8°	1.302	22 in 12	61.4°	2.088
11 in 12	42.5°	1.357	23 in 12	62.4°	2.162

B1020 102	Wood/Flat or Pitched	COST PER S.F.		
		MAT.	INST.	TOTAL
2500	Flat rafter, 2"x4", 12" O.C.	1.06	1.69	2.75
2550	16" O.C.	.96	1.45	2.41
2600	24" O.C.	.81	1.26	2.07
2900	2"x6", 12" O.C.	1.31	1.69	3
2950	16" O.C.	1.14	1.45	2.59
3000	24" O.C.	.93	1.26	2.19
3300	2"x8", 12" O.C.	1.58	1.83	3.41
3350	16" O.C.	1.35	1.56	2.91
3400	24" O.C.	1.07	1.33	2.40
3700	2"x10", 12" O.C.	1.95	2.09	4.04
3750	16" O.C.	1.62	1.75	3.37
3800	24" O.C.	1.24	1.45	2.69
4100	2"x12", 12" O.C.	2.24	2.11	4.35
4150	16" O.C.	1.83	1.76	3.59
4200	24" O.C.	1.39	1.46	2.85
4500	2"x14", 12" O.C.	2.72	3.08	5.80
4550	16" O.C.	2.20	2.50	4.70
4600	24" O.C.	1.64	1.96	3.60
4900	3"x6", 12" O.C.	2.27	1.79	4.06
4950	16" O.C.	1.86	1.53	3.39
5000	24" O.C.	1.42	1.31	2.73
5300	3"x8", 12" O.C.	3.33	2	5.33
5350	16" O.C.	2.65	1.68	4.33
5400	24" O.C.	1.94	1.41	3.35
5700	3"x10", 12" O.C.	4	2.27	6.27
5750	16" O.C.	3.17	1.89	5.06
5800	24" O.C.	2.28	1.55	3.83
6100	3"x12", 12" O.C.	4.67	2.75	7.42
6150	16" O.C.	3.67	2.25	5.92
6200	24" O.C.	2.61	1.78	4.39
7001	Wood truss, 4 in 12 slope, 24" O.C., 24' to 29' span	3.27	2.55	5.82
7100	30' to 43' span	4.02	2.55	6.57
7200	44' to 60' span	4.21	2.55	6.76

B10 Superstructure

B1020 Roof Construction

The table below lists the cost per S.F. for a roof system with steel columns, beams and deck using open web steel joists and 1-1/2″ galvanized metal deck. Perimeter of system is supported on bearing walls.

Fireproofing is not included. Costs/S.F. are based on a building 4 bays long and 4 bays wide.

Column costs are additive. Costs for the bearing walls are not included.

Steel Joists, Beams and Deck on Bearing Walls

B1020 108		Steel Joists, Beams & Deck on Columns & Walls						
	BAY SIZE (FT.)	SUPERIMPOSED LOAD (P.S.F.)	DEPTH (IN.)	TOTAL LOAD (P.S.F.)	COLUMN ADD	COST PER S.F.		
						MAT.	INST.	TOTAL
3000	25x25	20	18	40		4.13	1.43	5.56
3100					columns	.38	.09	.47
3200		30	22	50		4.43	1.52	5.95
3300					columns	.51	.12	.63
3400		40	20	60		4.92	1.68	6.60
3500					columns	.51	.12	.63
3600	25x30	20	22	40		4.38	1.38	5.76
3700					columns	.43	.10	.53
3800		30	20	50		4.96	1.54	6.50
3900					columns	.43	.10	.53
4000		40	25	60		5.15	1.58	6.73
4100					columns	.51	.12	.63
4200	30x30	20	25	42		4.82	1.48	6.30
4300					columns	.35	.09	.44
4400		30	22	52		5.25	1.61	6.86
4500					columns	.43	.10	.53
4600		40	28	62		5.50	1.66	7.16
4700					columns	.43	.10	.53
4800	30x35	20	22	42		5	1.54	6.54
4900					columns	.36	.09	.45
5000		30	28	52		5.30	1.61	6.91
5100					columns	.36	.09	.45
5200		40	25	62		5.70	1.72	7.42
5300					columns	.43	.10	.53
5400	35x35	20	28	42		5	1.55	6.55
5500					columns	.31	.08	.39

B1020 Roof Construction

Description: The table below lists the cost per S.F. for a roof system with steel columns, beams, and deck, using open web steel joists and 1-1/2" galvanized metal deck.

Column costs are additive. Fireproofing is not included.

Steel Joists, Beams and Deck on Columns

B1020 112		Steel Joists, Beams, & Deck on Columns						
	BAY SIZE (FT.)	SUPERIMPOSED LOAD (P.S.F.)	DEPTH (IN.)	TOTAL LOAD (P.S.F.)	COLUMN ADD	COST PER S.F.		
						MAT.	INST.	TOTAL
1100	15x20	20	16	40		4.08	1.43	5.51
1200					columns	2.21	.53	2.74
1500		40	18	60		4.56	1.58	6.14
1600					columns	2.21	.53	2.74
2300	20x25	20	18	40		4.65	1.59	6.24
2400					columns	1.33	.31	1.64
2700		40	20	60		5.25	1.77	7.02
2800					columns	1.77	.42	2.19
2900	25x25	20	18	40		5.45	1.80	7.25
3000					columns	1.06	.25	1.31
3100		30	22	50		5.65	1.84	7.49
3200					columns	1.42	.34	1.76
3300		40	20	60		6.35	2.04	8.39
3400					columns	1.42	.34	1.76
3500	25x30	20	22	40		5.35	1.61	6.96
3600					columns	1.18	.28	1.46
3900		40	25	60		6.50	1.91	8.41
4000					columns	1.42	.34	1.76
4100	30x30	20	25	42		6.10	1.79	7.89
4200					columns	.98	.24	1.22
4300		30	22	52		6.70	1.96	8.66
4400					columns	1.18	.28	1.46
4500		40	28	62		7.05	2.04	9.09
4600					columns	1.18	.28	1.46
5300	35x35	20	28	42		6.65	1.94	8.59
5400					columns	.87	.20	1.07
5500		30	25	52		7.45	2.14	9.59
5600					columns	1.01	.24	1.25
5700		40	28	62		8.05	2.29	10.34
5800					columns	1.12	.26	1.38

B10 Superstructure

B1020 Roof Construction

Description: The table below lists the cost per S.F. for a roof system using open web steel joists and 1-1/2″ galvanized metal deck. The system is assumed supported on bearing walls or other suitable support. Costs for the supports are not included.

B1020 116	Steel Joists & Deck on Bearing Walls							
	BAY SIZE (FT.)	SUPERIMPOSED LOAD (P.S.F.)	DEPTH (IN.)	TOTAL LOAD (P.S.F.)		COST PER S.F.		
						MAT.	INST.	TOTAL
1100	20	20	13-1/2	40		2.59	1.04	3.63
1200		30	15-1/2	50		2.65	1.06	3.71
1300		40	15-1/2	60		2.83	1.14	3.97
1400	25	20	17-1/2	40		2.84	1.15	3.99
1500		30	17-1/2	50		3.06	1.24	4.30
1600		40	19-1/2	60		3.11	1.26	4.37
1700	30	20	19-1/2	40		3.10	1.26	4.36
1800		30	21-1/2	50		3.17	1.28	4.45
1900		40	23-1/2	60		3.41	1.38	4.79
2000	35	20	23-1/2	40		3.38	1.16	4.54
2100		30	25-1/2	50		3.49	1.19	4.68
2200		40	25-1/2	60		3.71	1.25	4.96
2300	40	20	25-1/2	41		3.76	1.27	5.03
2400		30	25-1/2	51		3.99	1.34	5.33
2500		40	25-1/2	61		4.13	1.38	5.51
2600	45	20	27-1/2	41		4.53	1.50	6.03
2700		30	31-1/2	51		4.79	1.58	6.37
2800		40	31-1/2	61		5.05	1.66	6.71
2900	50	20	29-1/2	42		5.10	1.67	6.77
3000		30	31-1/2	52		5.55	1.81	7.36
3100		40	31-1/2	62		5.90	1.92	7.82
3200	60	20	37-1/2	42		6	2.15	8.15
3300		30	37-1/2	52		6.65	2.38	9.03
3400		40	37-1/2	62		6.65	2.38	9.03
3500	70	20	41-1/2	42		6.65	2.37	9.02
3600		30	41-1/2	52		7.10	2.53	9.63
3700		40	41-1/2	64		8.60	3.06	11.66
3800	80	20	45-1/2	44		8.35	2.97	11.32
3900		30	45-1/2	54		8.35	2.97	11.32
4000		40	45-1/2	64		9.25	3.28	12.53
4400	100	20	57-1/2	44		8.25	2.82	11.07
4500		30	57-1/2	54		9.85	3.35	13.20
4600		40	57-1/2	65		10.85	3.69	14.54
4700	125	20	69-1/2	44		9.85	3.34	13.19
4800		30	69-1/2	56		11.45	3.87	15.32
4900		40	69-1/2	67		13	4.40	17.40

B1020 Roof Construction

Description: The table below lists costs for a roof system supported on exterior bearing walls and interior columns. Costs include bracing, joist girders, open web steel joists and 1-1/2″ galvanized metal deck.

Column costs are additive. Fireproofing is not included.

Costs/S.F. are based on a building 4 bays long and 4 bays wide. Costs for bearing walls are not included.

B1020 120	Steel Joists, Joist Girders & Deck on Columns & Walls							
	BAY SIZE (FT.) GIRD X JOISTS	SUPERIMPOSED LOAD (P.S.F.)	DEPTH (IN.)	TOTAL LOAD (P.S.F.)	COLUMN ADD	COST PER S.F.		
						MAT.	INST.	TOTAL
2350	30x35	20	32-1/2	40		3.79	1.44	5.23
2400					columns	.36	.09	.45
2550		40	36-1/2	60		4.24	1.59	5.83
2600					columns	.42	.10	.52
3000	35x35	20	36-1/2	40		4.13	1.78	5.91
3050					columns	.31	.08	.39
3200		40	36-1/2	60		4.64	1.95	6.59
3250					columns	.40	.10	.50
3300	35x40	20	36-1/2	40		4.20	1.62	5.82
3350					columns	.32	.08	.40
3500		40	36-1/2	60		4.71	1.79	6.50
3550					columns	.35	.09	.44
3900	40x40	20	40-1/2	41		4.87	1.81	6.68
3950					columns	.31	.08	.39
4100		40	40-1/2	61		5.45	1.98	7.43
4150					columns	.31	.08	.39
5100	45x50	20	52-1/2	41		5.50	2.06	7.56
5150					columns	.22	.05	.27
5300		40	52-1/2	61		6.70	2.44	9.14
5350					columns	.32	.08	.40
5400	50x45	20	56-1/2	41		5.30	2.01	7.31
5450					columns	.22	.05	.27
5600		40	56-1/2	61		6.85	2.50	9.35
5650					columns	.30	.07	.37
5700	50x50	20	56-1/2	42		5.95	2.24	8.19
5750					columns	.22	.05	.27
5900		40	59	64		6.85	2.73	9.58
5950					columns	.31	.08	.39
6300	60x50	20	62-1/2	43		6	2.28	8.28
6350					columns	.35	.09	.44
6500		40	71	65		7.05	2.82	9.87
6550					columns	.47	.11	.58

B1020 Roof Construction

Description: The table below lists the cost per S.F. for a roof system supported on columns. Costs include joist girders, open web steel joists and 1-1/2″ galvanized metal deck. Column costs are additive.

Fireproofing is not included. Costs/S.F. are based on a building 4 bays long and 4 bays wide.

Costs for wind columns are not included.

B1020 124	Steel Joists & Joist Girders on Columns							
	BAY SIZE (FT.) GIRD X JOISTS	SUPERIMPOSED LOAD (P.S.F.)	DEPTH (IN.)	TOTAL LOAD (P.S.F.)	COLUMN ADD	COST PER S.F.		
						MAT.	INST.	TOTAL
2000	30x30	20	17-1/2	40		3.83	1.72	5.55
2050					columns	1.18	.28	1.46
2100		30	17-1/2	50		4.16	1.86	6.02
2150					columns	1.18	.28	1.46
2200		40	21-1/2	60		4.26	1.90	6.16
2250					columns	1.18	.28	1.46
2300	30x35	20	32-1/2	40		4.07	1.61	5.68
2350					columns	1.01	.24	1.25
2500		40	36-1/2	60		4.58	1.78	6.36
2550					columns	1.18	.28	1.46
3200	35x40	20	36-1/2	40		5.30	2.17	7.47
3250					columns	.88	.21	1.09
3400		40	36-1/2	60		5.95	2.39	8.34
3450					columns	.98	.23	1.21
3800	40x40	20	40-1/2	41		5.35	2.18	7.53
3850					columns	.85	.20	1.05
4000		40	40-1/2	61		6	2.40	8.40
4050					columns	.85	.20	1.05
4100	40x45	20	40-1/2	41		5.75	2.48	8.23
4150					columns	.76	.18	.94
4200		30	40-1/2	51		6.25	2.63	8.88
4250					columns	.76	.18	.94
4300		40	40-1/2	61		7	2.85	9.85
4350					columns	.86	.20	1.06
5000	45x50	20	52-1/2	41		6.10	2.50	8.60
5050					columns	.61	.14	.75
5200		40	52-1/2	61		7.45	2.94	10.39
5250					columns	.78	.19	.97
5300	50x45	20	56-1/2	41		5.90	2.45	8.35
5350					columns	.61	.14	.75
5500		40	56-1/2	61		7.65	3	10.65
5550					columns	.78	.19	.97
5600	50x50	20	56-1/2	42		6.60	2.69	9.29
5650					columns	.62	.15	.77
5800		40	59	64		5.95	2.57	8.52
5850					columns	.86	.20	1.06

B2010 Exterior Walls

Cast in Place Concrete Wall

Flat Precast Concrete Wall

Table B2010 101 describes a concrete wall system for exterior closure. There are several types of wall finishes priced from plain finish to a finish with 3/4″ rustication strip.

In Table B2010 103, precast concrete wall panels are either solid or insulated. Prices below are based on delivery within 50 miles of a plant. Small jobs can double the prices below. For large, highly repetitive jobs, deduct up to 15%.

B2010 101	Cast In Place Concrete	COST PER S.F.		
		MAT.	INST.	TOTAL
2100	Conc wall reinforced, 8' high, 6″ thick, plain finish, 3000 PSI	4.56	15.85	20.41
2700	Aged wood liner, 3000 PSI	5.75	18	23.75
3000	Sand blast light 1 side, 3000 PSI	5.40	18.05	23.45
3700	3/4″ bevel rustication strip, 3000 PSI	4.73	16.80	21.53
4000	8″ thick, plain finish, 3000 PSI	5.40	16.25	21.65
4100	4000 PSI	5.50	16.25	21.75
4300	Rub concrete 1 side, 3000 PSI	5.40	19.15	24.55
4550	8″ thick, aged wood liner, 3000 PSI	6.55	18.40	24.95
4750	Sand blast light 1 side, 3000 PSI	6.20	18.45	24.65
5300	3/4″ bevel rustication strip, 3000 PSI	5.55	17.20	22.75
5600	10″ thick, plain finish, 3000 PSI	6.15	16.60	22.75
5900	Rub concrete 1 side, 3000 PSI	6.15	19.50	25.65
6200	Aged wood liner, 3000 PSI	7.35	18.80	26.15
6500	Sand blast light 1 side, 3000 PSI	7	18.80	25.80
7100	3/4″ bevel rustication strip, 3000 PSI	6.35	17.55	23.90
7400	12″ thick, plain finish, 3000 PSI	7.10	17.05	24.15
7700	Rub concrete 1 side, 3000 PSI	7.10	19.95	27.05
8000	Aged wood liner, 3000 PSI	8.30	19.25	27.55
8300	Sand blast light 1 side, 3000 PSI	7.95	19.25	27.20
8900	3/4″ bevel rustication strip, 3000 PSI	7.30	18	25.30

B2010 102			Flat Precast Concrete					
	THICKNESS (IN.)	PANEL SIZE (FT.)	FINISHES	RIGID INSULATION (IN)	TYPE	COST PER S.F.		
						MAT.	INST.	TOTAL
3000	4	5x18	smooth gray	none	low rise	12.95	7.80	20.75
3050		6x18				10.80	6.55	17.35
3100		8x20				21	3.16	24.16
3150		12x20				19.90	2.97	22.87
3200	6	5x18	smooth gray	2	low rise	13.85	8.35	22.20
3250		6x18				11.75	7.05	18.80
3300		8x20				22.50	3.81	26.31
3350		12x20				20	3.49	23.49
3400	8	5x18	smooth gray	2	low rise	29	4.79	33.79
3450		6x18				27.50	4.57	32.07
3500		8x20				25	4.21	29.21
3550		12x20				22.50	3.86	26.36

B20 Exterior Enclosure

B2010 Exterior Walls

B2010 102 — Flat Precast Concrete

	THICKNESS (IN.)	PANEL SIZE (FT.)	FINISHES	RIGID INSULATION (IN)	TYPE	COST PER S.F.		
						MAT.	INST.	TOTAL
3600	4	4x8	white face	none	low rise	48.50	4.52	53.02
3650		8x8				36.50	3.38	39.88
3700		10x10				32	2.97	34.97
3750		20x10				29	2.69	31.69
3800	5	4x8	white face	none	low rise	49.50	4.61	54.11
3850		8x8				37.50	3.48	40.98
3900		10x10				33.50	3.08	36.58
3950		20x20				30.50	2.82	33.32
4000	6	4x8	white face	none	low rise	51.50	4.77	56.27
4050		8x8				39	3.62	42.62
4100		10x10				34.50	3.20	37.70
4150		20x10				31.50	2.93	34.43
4200	6	4x8	white face	2	low rise	52.50	5.40	57.90
4250		8x8				40	4.24	44.24
4300		10x10				35.50	3.82	39.32
4350		20x10				31.50	2.93	34.43
4400	7	4x8	white face	none	low rise	53	4.89	57.89
4450		8x8				40.50	3.76	44.26
4500		10x10				36.50	3.37	39.87
4550		20x10				33.50	3.08	36.58
4600	7	4x8	white face	2	low rise	54	5.50	59.50
4650		8x8				41.50	4.38	45.88
4700		10x10				37.50	3.99	41.49
4750		20x10				34.50	3.70	38.20
4800	8	4x8	white face	none	low rise	54	5	59
4850		8x8				41.50	3.85	45.35
4900		10x10				37.50	3.46	40.96
4950		20x10				34.50	3.19	37.69
5000	8	4x8	white face	2	low rise	55	5.60	60.60
5050		8x8				42.50	4.47	46.97
5100		10x10				38.50	4.08	42.58
5150		20x10				35.50	3.81	39.31

B2010 103 — Fluted Window or Mullion Precast Concrete

	THICKNESS (IN.)	PANEL SIZE (FT.)	FINISHES	RIGID INSULATION (IN)	TYPE	COST PER S.F.		
						MAT.	INST.	TOTAL
5200	4	4x8	smooth gray	none	high rise	29.50	17.85	47.35
5250		8x8				21	12.80	33.80
5300		10x10				40	6.05	46.05
5350		20x10				35.50	5.30	40.80
5400	5	4x8	smooth gray	none	high rise	30	18.20	48.20
5450		8x8				22	13.25	35.25
5500		10x10				42	6.30	48.30
5550		20x10				37	5.55	42.55
5600	6	4x8	smooth gray	none	high rise	31	18.65	49.65
5650		8x8				22.50	13.65	36.15
5700		10x10				43	6.45	49.45
5750		20x10				38.50	5.75	44.25
5800	6	4x8	smooth gray	2	high rise	32	19.25	51.25
5850		8x8				23.50	14.30	37.80
5900		10x10				44	7.05	51.05
5950		20x10				39.50	6.35	45.85

B2010 Exterior Walls

B2010 103 — Fluted Window or Mullion Precast Concrete

	THICKNESS (IN.)	PANEL SIZE (FT.)	FINISHES	RIGID INSULATION (IN)	TYPE	COST PER S.F.		
						MAT.	INST.	TOTAL
6000	7	4x8	smooth gray	none	high rise	31.50	19.05	50.55
6050		8x8				23.50	14.05	37.55
6100		10x10				45	6.75	51.75
6150		20x10				39.50	5.90	45.40
6200	7	4x8	smooth gray	2	high rise	32.50	19.65	52.15
6250		8x8				24.50	14.65	39.15
6300		10x10				46	7.35	53.35
6350		20x10				40.50	6.55	47.05
6400	8	4x8	smooth gray	none	high rise	32	19.35	51.35
6450		8x8				24	14.40	38.40
6500		10x10				46.50	6.95	53.45
6550		20x10				41.50	6.20	47.70
6600	8	4x8	smooth gray	2	high rise	33	20	53
6650		8x8				25	15.05	40.05
6700		10x10				47.50	7.60	55.10
6750		20x10				42.50	6.85	49.35

B2010 104 — Ribbed Precast Concrete

	THICKNESS (IN.)	PANEL SIZE (FT.)	FINISHES	RIGID INSULATION(IN.)	TYPE	COST PER S.F.		
						MAT.	INST.	TOTAL
6800	4	4x8	aggregate	none	high rise	34.50	17.85	52.35
6850		8x8				25.50	13.05	38.55
6900		10x10				43	4.74	47.74
6950		20x10				38	4.18	42.18
7000	5	4x8	aggregate	none	high rise	35	18.20	53.20
7050		8x8				26	13.40	39.40
7100		10x10				45	4.91	49.91
7150		20x10				40	4.39	44.39
7200	6	4x8	aggregate	none	high rise	36	18.55	54.55
7250		8x8				26.50	13.80	40.30
7300		10x10				46.50	5.05	51.55
7350		20x10				41.50	4.55	46.05
7400	6	4x8	aggregate	2	high rise	37	19.20	56.20
7450		8x8				28	14.40	42.40
7500		10x10				47.50	5.70	53.20
7550		20x10				42.50	5.15	47.65
7600	7	4x8	aggregate	none	high rise	36.50	19	55.50
7650		8x8				27.50	14.15	41.65
7700		10x10				47.50	5.25	52.75
7750		20x10				42.50	4.68	47.18
7800	7	4x8	aggregate	2	high rise	38	19.60	57.60
7850		8x8				28.50	14.75	43.25
7900		10x10				49	5.85	54.85
7950		20x10				43.50	5.30	48.80
8000	8	4x8	aggregate	none	high rise	37.50	19.25	56.75
8050		8x8				28	14.55	42.55
8100		10x10				49.50	5.40	54.90
8150		20x10				44.50	4.90	49.40
8200	8	4x8	aggregate	2	high rise	38.50	19.90	58.40
8250		8x8				29.50	15.20	44.70
8300		10x10				50.50	6.05	56.55
8350		20x10				45.50	5.50	51

B20 Exterior Enclosure

B2010 Exterior Walls

| B2010 105 | | | | | | | Precast Concrete Specialties | | |

	TYPE	SIZE				COST PER L.F.		
						MAT.	INST.	TOTAL
8400	Coping, precast	6" wide				20	13.05	33.05
8450	Stock units	10" wide				21.50	14	35.50
8460		12" wide				23.50	15.05	38.55
8480		14" wide				25	16.30	41.30
8500	Window sills	6" wide				13.50	14	27.50
8550	Precast	10" wide				17.60	16.30	33.90
8600		14" wide				17.60	19.55	37.15
8610								

B2010 Exterior Walls

The advantage of tilt up construction is in the low cost of forms and placing of concrete and reinforcing. Tilt up has been used for several types of buildings, including warehouses, stores, offices, and schools. The panels are cast in forms on the ground, or floor slab. Most jobs use 5-1/2″ thick solid reinforced concrete panels.

Design Assumptions:
Conc. f'c = 3000 psi
Reinf. fy = 60,000

B2010 106	Tilt-Up Concrete Panel	COST PER S.F.		
		MAT.	INST.	TOTAL
3200	Tilt up conc panels, broom finish, 5-1/2″ thick, 3000 PSI	4.05	6.05	10.10
3250	5000 PSI	4.12	5.95	10.07
3300	6″ thick, 3000 PSI	4.44	6.20	10.64
3350	5000 PSI	4.55	6.10	10.65
3400	7-1/2″ thick, 3000 PSI	5.65	6.45	12.10
3450	5000 PSI	5.80	6.35	12.15
3500	8″ thick, 3000 PSI	6.05	6.60	12.65
3550	5000 PSI	6.25	6.50	12.75
3700	Steel trowel finish, 5-1/2″ thick, 3000 PSI	4.03	6.15	10.18
3750	5000 PSI	4.12	6.05	10.17
3800	6″ thick, 3000 PSI	4.44	6.30	10.74
3850	5000 PSI	4.55	6.20	10.75
3900	7-1/2″ thick, 3000 PSI	5.65	6.50	12.15
3950	5000 PSI	5.80	6.40	12.20
4000	8″ thick, 3000 PSI	6.05	6.65	12.70
4050	5000 PSI	6.25	6.55	12.80
4200	Exp. aggregate finish, 5-1/2″ thick, 3000 PSI	4.41	6.20	10.61
4250	5000 PSI	4.50	6.10	10.60
4300	6″ thick, 3000 PSI	4.82	6.35	11.17
4350	5000 PSI	4.93	6.25	11.18
4400	7-1/2″ thick, 3000 PSI	6	6.60	12.60
4450	5000 PSI	6.20	6.45	12.65
4500	8″ thick, 3000 PSI	6.45	6.75	13.20
4550	5000 PSI	6.65	6.60	13.25
4600	Exposed aggregate & vert. rustication 5-1/2″ thick, 3000 PSI	6.65	7.65	14.30
4650	5000 PSI	6.75	7.55	14.30
4700	6″ thick, 3000 PSI	7.05	7.80	14.85
4750	5000 PSI	7.15	7.70	14.85
4800	7-1/2″ thick, 3000 PSI	8.25	8	16.25
4850	5000 PSI	8.45	7.90	16.35
4900	8″ thick, 3000 PSI	8.70	8.15	16.85
4950	5000 PSI	8.85	8.05	16.90
5000	Vertical rib & light sandblast, 5-1/2″ thick, 3000 PSI	6.65	10	16.65
5100	6″ thick, 3000 PSI	7.10	10.15	17.25
5200	7-1/2″ thick, 3000 PSI	8.30	10.40	18.70
5300	8″ thick, 3000 PSI	8.70	10.55	19.25
6000	Broom finish w/2″ polystyrene insulation, 6″ thick, 3000 PSI	3.84	7.45	11.29
6100	Broom finish 2″ fiberplank insulation, 6″ thick, 3000 PSI	4.33	7.35	11.68
6200	Exposed aggregate w/2″ polystyrene insulation, 6″ thick, 3000 PSI	4.11	7.45	11.56
6300	Exposed aggregate 2″ fiberplank insulation, 6″ thick, 3000 PSI	4.60	7.35	11.95

B2010 Exterior Walls

Exterior concrete block walls are defined in the following terms; structural reinforcement, weight, percent solid, size, strength and insulation. Within each of these categories, two to four variations are shown. No costs are included for brick shelf or relieving angles.

B2010 109	Concrete Block Wall - Regular Weight							
	TYPE	**SIZE (IN.)**	**STRENGTH (P.S.I.)**	**CORE FILL**		**COST PER S.F.**		
						MAT.	**INST.**	**TOTAL**
1200	Hollow	4x8x16	2,000	none		1.85	6	7.85
1250			4,500	none		2.25	6	8.25
1400		8x8x16	2,000	perlite		4.43	7.30	11.73
1410				styrofoam		3.98	6.90	10.88
1440				none		2.73	6.90	9.63
1450			4,500	perlite		5.10	7.30	12.40
1460				styrofoam		4.63	6.90	11.53
1490				none		3.38	6.90	10.28
2000	75% solid	4x8x16	2,000	none		2.23	6.10	8.33
2100		6x8x16	2,000	perlite		3.67	6.75	10.42
2500	Solid	4x8x16	2,000	none		2.11	6.20	8.31
2700		8x8x16	2,000	none		3.88	7.20	11.08

B2010 110	Concrete Block Wall - Lightweight							
	TYPE	**SIZE (IN.)**	**WEIGHT (P.C.F.)**	**CORE FILL**		**COST PER S.F.**		
						MAT.	**INST.**	**TOTAL**
3100	Hollow	8x4x16	105	perlite		4.27	6.90	11.17
3110				styrofoam		3.82	6.50	10.32
3200		4x8x16	105	none		2	5.90	7.90
3250			85	none		2.48	5.75	8.23
3300		6x8x16	105	perlite		3.96	6.65	10.61
3310				styrofoam		4.06	6.30	10.36
3340				none		2.81	6.30	9.11
3400		8x8x16	105	perlite		5.10	7.15	12.25
3410				styrofoam		4.63	6.75	11.38
3440				none		3.38	6.75	10.13
3450			85	perlite		5.35	7	12.35
4000	75% solid	4x8x16	105	none		2.88	5.95	8.83
4050			85	none		3.05	5.80	8.85
4100		6x8x16	105	perlite		4.68	6.55	11.23
4500	Solid	4x8x16	105	none		1.97	6.15	8.12
4700		8x8x16	105	none		3.92	7.10	11.02

B2010 Exterior Walls

B2010 111   Reinforced Concrete Block Wall - Regular Weight

	TYPE	SIZE (IN.)	STRENGTH (P.S.I.)	VERT. REINF & GROUT SPACING		COST PER S.F. MAT.	INST.	TOTAL
5200	Hollow	4x8x16	2,000	#4 @ 48"		2	6.65	8.65
5300		6x8x16	2,000	#4 @ 48"		2.81	7.15	9.96
5330				#5 @ 32"		3.02	7.40	10.42
5340				#5 @ 16"		3.49	8.35	11.84
5350			4,500	#4 @ 28"		3.03	7.15	10.18
5390				#5 @ 16"		3.71	8.35	12.06
5400		8x8x16	2,000	#4 @ 48"		3.11	7.60	10.71
5430				#5 @ 32"		3.28	8	11.28
5440				#5 @ 16"		3.82	9.15	12.97
5450			4,500	#4 @ 48"		3.69	7.70	11.39
5490				#5 @ 16"		4.47	9.15	13.62
5500		12x8x16	2,000	#4 @ 48"		4.51	9.80	14.31
5540				#5 @ 16"		5.55	11.30	16.85
6100	75% solid	6x8x16	2,000	#4 @ 48"		3.23	7.10	10.33
6140				#5 @ 16"		3.68	8.10	11.78
6150			4,500	#4 @ 48"		3.52	7.10	10.62
6190				#5 @ 16"		3.97	8.10	12.07
6200		8x8x16	2,000	#4 @ 48"		3.43	7.65	11.08
6230				#5 @ 32"		3.60	7.90	11.50
6240				#5 @ 16"		3.91	8.85	12.76
6250			4,500	#4 @ 48"		4.28	7.65	11.93
6280				#5 @ 32"		4.45	7.90	12.35
6290				#5 @ 16"		4.76	8.85	13.61
6500	Solid-double Wythe	2-4x8x16	2,000	#4 @ 48" E.W.		5.10	14.05	19.15
6530				#5 @ 16" E.W.		5.80	14.95	20.75
6550			4,500	#4 @ 48" E.W.		7.05	13.95	21
6580				#5 @ 16" E.W.		7.80	14.85	22.65

B2010 112   Reinforced Concrete Block Wall - Lightweight

	TYPE	SIZE (IN.)	WEIGHT (P.C.F.)	VERT REINF. & GROUT SPACING		COST PER S.F. MAT.	INST.	TOTAL
7100	Hollow	8x4x16	105	#4 @ 48"		2.88	7.30	10.18
7140				#5 @ 16"		3.66	8.75	12.41
7150			85	#4 @ 48"		4.15	7.15	11.30
7190				#5 @ 16"		4.93	8.60	13.53
7400		8x8x16	105	#4 @ 48"		3.69	7.55	11.24
7440				#5 @ 16"		4.47	9	13.47
7450			85	#4 @ 48"		3.96	7.40	11.36
7490				#5 @ 16"		4.74	8.85	13.59
7800		8x8x24	105	#4 @ 48"		2.85	8.10	10.95
7840				#5 @ 16"		3.63	9.55	13.18
7850			85	#4 @ 48"		6.40	6.95	13.35
7890				#5 @ 16"		7.15	8.40	15.55
8100	75% solid	6x8x16	105	#4 @ 48"		4.24	6.90	11.14
8130				#5 @ 32"		4.40	7.15	11.55
8150			85	#4 @ 48"		4.41	6.75	11.16
8180				#5 @ 32"		4.57	7	11.57
8200		8x8x16	105	#4 @ 48"		5.05	7.50	12.55
8230				#5 @ 32"		5.25	7.75	13
8250			85	#4 @ 48"		4.63	7.30	11.93
8280				#5 @ 32"		4.80	7.55	12.35

B20 Exterior Enclosure

B2010 Exterior Walls

B2010 112	Reinforced Concrete Block Wall - Lightweight

	TYPE	SIZE (IN.)	WEIGHT (P.C.F.)	VERT REINF. & GROUT SPACING		COST PER S.F. MAT.	COST PER S.F. INST.	COST PER S.F. TOTAL
8500	Solid-double	2-4x8x16	105	#4 @ 48"		4.80	13.95	18.75
8530	Wythe		105	#5 @ 16"		5.55	14.85	20.40
8600		2-6x8x16	105	#4 @ 48"		7	14.90	21.90
8630				#5 @ 16"		7.75	15.80	23.55
8650			85	#4 @ 48"		8.65	14.30	22.95

B2010 Exterior Walls

Exterior split ribbed block walls are defined in the following terms; structural reinforcement, weight, percent solid, size, number of ribs and insulation. Within each of these categories two to four variations are shown. No costs are included for brick shelf or relieving angles. Costs include control joints every 20′ and horizontal reinforcing.

B2010 113			Split Ribbed Block Wall - Regular Weight					
	TYPE	SIZE (IN.)	RIBS	CORE FILL		COST PER S.F.		
						MAT.	INST.	TOTAL
1220	Hollow	4x8x16	4	none		3.86	7.45	11.31
1250			8	none		4.19	7.45	11.64
1280			16	none		4.52	7.55	12.07
1430		8x8x16	8	perlite		7.35	8.85	16.20
1440				styrofoam		6.90	8.45	15.35
1450				none		5.65	8.45	14.10
1530		12x8x16	8	perlite		9.35	11.75	21.10
1540				styrofoam		8.75	10.95	19.70
1550				none		6.55	10.95	17.50
2120	75% solid	4x8x16	4	none		4.85	7.55	12.40
2150			8	none		5.30	7.55	12.85
2180			16	none		5.70	7.65	13.35
2520	Solid	4x8x16	4	none		5.50	7.65	13.15
2550			8	none		6.50	7.65	14.15
2580			16	none		6.50	7.75	14.25

B2010 115			Reinforced Split Ribbed Block Wall - Regular Weight					
	TYPE	SIZE (IN.)	RIBS	VERT. REINF. & GROUT SPACING		COST PER S.F.		
						MAT.	INST.	TOTAL
5200	Hollow	4x8x16	4	#4 @ 48″		4.01	8.10	12.11
5230			8	#4 @ 48″		4.34	8.10	12.44
5260			16	#4 @ 48″		4.67	8.20	12.87
5430		8x8x16	8	#4 @ 48″		5.95	9.25	15.20
5440				#5 @ 32″		6.20	9.55	15.75
5450				#5 @ 16″		6.70	10.70	17.40
5530		12x8x16	8	#4 @ 48″		7	11.80	18.80
5540				#5 @ 32″		7.30	12.10	19.40
5550				#5 @ 16″		8.05	13.30	21.35
6230	75% solid	6x8x16	8	#4 @ 48″		6.25	8.55	14.80
6240				#5 @ 32″		6.40	8.80	15.20
6250				#5 @ 16″		6.70	9.55	16.25
6330		8x8x16	8	#4 @ 48″		7.20	9.25	16.45
6340				#5 @ 32″		7.35	9.50	16.85
6350				#5 @ 16″		7.65	10.45	18.10

B2010 Exterior Walls

Exterior split face block walls are defined in the following terms; structural reinforcement, weight, percent solid, size, scores and insulation. Within each of these categories two to four variations are shown. No costs are included for brick shelf or relieving angles. Costs include control joints every 20′ and horizontal reinforcing.

B2010 117			Split Face Block Wall - Regular Weight					
	TYPE	SIZE (IN.)	SCORES	CORE FILL		COST PER S.F.		
						MAT.	INST.	TOTAL
1200	Hollow	8x4x16	0	perlite		7.10	9.75	16.85
1210				styrofoam		6.65	9.35	16
1240				none		5.40	9.35	14.75
1250			1	perlite		7.45	9.75	17.20
1260				styrofoam		7	9.35	16.35
1290				none		5.75	9.35	15.10
1300		12x4x16	0	perlite		9.35	11.10	20.45
1310				styrofoam		8.75	10.30	19.05
1340				none		6.55	10.30	16.85
1350			1	perlite		9.85	11.10	20.95
1360				styrofoam		9.25	10.30	19.55
1390				none		7.05	10.30	17.35
1400		4x8x16	0	none		3.53	7.35	10.88
1450			1	none		3.82	7.45	11.27
1500		6x8x16	0	perlite		5.25	8.50	13.75
1510				styrofoam		5.35	8.15	13.50
1540				none		4.09	8.15	12.24
1550			1	perlite		5.60	8.60	14.20
1560				styrofoam		5.70	8.25	13.95
1590				none		4.43	8.25	12.68
1600		8x8x16	0	perlite		6.40	9.10	15.50
1610				styrofoam		5.95	8.70	14.65
1640				none		4.72	8.70	13.42
1650			1	perlite		6.80	9.25	16.05
1660				styrofoam		6.35	8.85	15.20
1690				none		5.10	8.85	13.95
1700		12x8x16	0	perlite		8.45	11.95	20.40
1710				styrofoam		7.85	11.15	19
1740				none		5.65	11.15	16.80
1750			1	perlite		8.90	12.20	21.10
1760				styrofoam		8.30	11.40	19.70
1790				none		6.10	11.40	17.50

B2010 Exterior Walls

B2010 117	Split Face Block Wall - Regular Weight

	TYPE	SIZE (IN.)	SCORES	CORE FILL		COST PER S.F. MAT.	INST.	TOTAL
1800	75% solid	8x4x16	0	perlite		7.40	9.70	17.10
1840				none		6.55	9.50	16.05
1850			1	perlite		7.90	9.70	17.60
1890				none		7.05	9.50	16.55
2000		4x8x16	0	none		4.41	7.45	11.86
2050			1	none		4.80	7.55	12.35
2400	Solid	8x4x16	0	none		7.30	9.70	17
2450			1	none		7.90	9.70	17.60
2900		12x8x16	0	none		7.90	11.60	19.50
2950			1	none		8.60	11.80	20.40

B2010 119	Reinforced Split Face Block Wall - Regular Weight

	TYPE	SIZE (IN.)	SCORES	VERT. REINF. & GROUT SPACING		COST PER S.F. MAT.	INST.	TOTAL
5200	Hollow	8x4x16	0	#4 @ 48"		5.70	10.15	15.85
5210				#5 @ 32"		5.95	10.45	16.40
5240				#5 @ 16"		6.45	11.60	18.05
5250			1	#4 @ 48"		6.05	10.15	16.20
5260				#5 @ 32"		6.30	10.45	16.75
5290				#5 @ 16"		6.80	11.60	18.40
5700		12x8x16	0	#4 @ 48"		6.10	12	18.10
5710				#5 @ 32"		6.40	12.30	18.70
5740				#5 @ 16"		7.15	13.50	20.65
5750			1	#4 @ 48"		6.55	12.25	18.80
5760				#5 @ 32"		6.85	12.55	19.40
5790				#5 @ 16"		7.60	13.75	21.35
6000	75% solid	8x4x16	0	#4 @ 48"		6.70	10.15	16.85
6010				#5 @ 32"		6.85	10.40	17.25
6040				#5 @ 16"		7.15	11.35	18.50
6050			1	#4 @ 48"		7.20	10.15	17.35
6060				#5 @ 32"		7.35	10.40	17.75
6090				#5 @ 16"		7.65	11.35	19
6100		12x4x16	0	#4 @ 48"		8.20	11.20	19.40
6110				#5 @ 32"		8.45	11.45	19.90
6140				#5 @ 16"		8.85	12.20	21.05
6150			1	#4 @ 48"		8.75	11.20	19.95
6160				#5 @ 32"		9	11.45	20.45
6190				#5 @ 16"		9.45	12.40	21.85
6700	Solid-double Wythe	2-4x8x16	0	#4 @ 48" E.W.		10.90	16.75	27.65
6710				#5 @ 32" E.W.		11.15	16.90	28.05
6740				#5 @ 16" E.W.		11.60	17.65	29.25
6750			1	#4 @ 48" E.W.		11.75	16.95	28.70
6760				#5 @ 32" E.W.		12.05	17.10	29.15
6790				#5 @ 16" E.W.		12.50	17.85	30.35
6800		2-6x8x16	0	#4 @ 48" E.W.		12.35	18.50	30.85
6810				#5 @ 32" E.W.		12.60	18.65	31.25
6840				#5 @ 16" E.W.		13.10	19.40	32.50
6850			1	#4 @ 48" E.W.		13.35	18.80	32.15
6860				#5 @ 32" E.W.		13.60	18.95	32.55
6890				#5 @ 16" E.W.		14.10	19.70	33.80

B2010 Exterior Walls

Exterior solid brick walls are defined in the following terms; structural reinforcement, type of brick, thickness, and bond. Nine different types of face bricks are presented, with single wythes shown in four different bonds. Shelf angles are included in system components. These walls do not include ties and as such are not tied to a backup wall.

B2010 125	Solid Brick Walls - Single Wythe							
	TYPE	THICKNESS (IN.)	BOND			COST PER S.F.		
						MAT.	INST.	TOTAL
1000	Common	4	running			4.41	13.55	17.96
1010			common			5.25	15.50	20.75
1050			Flemish			5.80	18.85	24.65
1100			English			6.50	19.95	26.45
1150	Standard	4	running			5.70	14.05	19.75
1160			common			6.70	16.20	22.90
1200			Flemish			7.40	19.35	26.75
1250			English			8.30	20.50	28.80
1300	Glazed	4	running			19.65	14.60	34.25
1310			common			23.50	17	40.50
1350			Flemish			26.50	20.50	47
1400			English			29.50	22	51.50
1600	Economy	4	running			8.05	10.75	18.80
1610			common			9.50	12.30	21.80
1650			Flemish			10.50	14.60	25.10
1700			English			21	15.50	36.50
1900	Fire	4-1/2	running			12.60	12.30	24.90
1910			common			15	14.05	29.05
1950			Flemish			16.65	17	33.65
2000			English			18.70	17.85	36.55
2050	King	3-1/2	running			4.30	11.05	15.35
2060			common			5	12.70	17.70
2100			Flemish			6	15.20	21.20
2150			English			6.65	15.85	22.50
2200	Roman	4	running			9.40	12.70	22.10
2210			common			11.25	14.60	25.85
2250			Flemish			12.40	17.40	29.80
2300			English			14	18.85	32.85
2350	Norman	4	running			6.50	10.50	17
2360			common			7.65	11.95	19.60
2400			Flemish			17.50	14.35	31.85
2450			English			9.50	15.20	24.70

B2010 Exterior Walls

B2010 125	Solid Brick Walls - Single Wythe								

	TYPE	THICKNESS (IN.)	BOND			COST PER S.F.		
						MAT.	INST.	TOTAL
2500	Norwegian	4	running			6.40	9.35	15.75
2510			common			7.55	10.65	18.20
2550			Flemish			8.35	12.70	21.05
2600			English			9.35	13.35	22.70

B2010 126	Solid Brick Walls - Double Wythe								

	TYPE	THICKNESS (IN.)	COLLAR JOINT THICKNESS (IN.)			COST PER S.F.		
						MAT.	INST.	TOTAL
4100	Common	8	3/4			8.35	23	31.35
4150	Standard	8	1/2			10.75	23.50	34.25
4200	Glazed	8	1/2			38.50	24.50	63
4250	Engineer	8	1/2			10	21	31
4300	Economy	8	3/4			15.50	17.85	33.35
4350	Double	8	3/4			15.40	14.40	29.80
4400	Fire	8	3/4			24.50	21	45.50
4450	King	7	3/4			7.95	18.25	26.20
4500	Roman	8	1			18.25	21.50	39.75
4550	Norman	8	3/4			12.40	17.45	29.85
4600	Norwegian	8	3/4			12.25	15.45	27.70
4650	Utility	8	3/4			12.35	13.35	25.70
4700	Triple	8	3/4			8.70	12	20.70
4750	SCR	12	3/4			16.30	17.90	34.20
4800	Norwegian	12	3/4			14.50	15.80	30.30
4850	Jumbo	12	3/4			12.05	13.80	25.85

B2010 127	Reinforced Brick Walls								

	TYPE	THICKNESS (IN.)	WYTHE	REINF. & SPACING		COST PER S.F.		
						MAT.	INST.	TOTAL
7200	Common	4"	1	#4 @ 48"vert		4.61	14.05	18.66
7220				#5 @ 32"vert		4.76	14.25	19.01
7230				#5 @ 16"vert		5.05	14.90	19.95
7700	King	9"	2	#4 @ 48" E.W.		8.35	19.15	27.50
7720				#5 @ 32" E.W.		8.60	19.30	27.90
7730				#5 @ 16" E.W.		9.10	19.85	28.95
7800	Common	10"	2	#4 @ 48" E.W.		8.75	23.50	32.25
7820				#5 @ 32" E.W.		9	23.50	32.50
7830				#5 @ 16" E.W.		9.50	24.50	34
7900	Standard	10"	2	#4 @ 48" E.W.		11.15	24.50	35.65
7920				#5 @ 32" E.W.		11.40	24.50	35.90
7930				#5 @ 16" E.W.		11.90	25	36.90
8100	Economy	10"	2	#4 @ 48" E.W.		15.90	18.75	34.65
8120				#5 @ 32" E.W.		16.15	18.90	35.05
8130				#5 @ 16" E.W.		16.65	19.45	36.10
8200	Double	10"	2	#4 @ 48" E.W.		15.80	15.30	31.10
8220				#5 @ 32" E.W.		16.05	15.45	31.50
8230				#5 @ 16" E.W.		16.55	16	32.55
8300	Fire	10"	2	#4 @ 48" E.W.		25	21.50	46.50
8320				#5 @ 32" E.W.		25	21.50	46.50
8330				#5 @ 16" E.W.		25.50	22.50	48
8400	Roman	10"	2	#4 @ 48" E.W.		18.65	22	40.65
8420				#5 @ 32" E.W.		18.90	22.50	41.40
8430				#5 @ 16" E.W.		19.40	23	42.40

B2010 127				Reinforced Brick Walls				
	TYPE	THICKNESS (IN.)	WYTHE	REINF. & SPACING		COST PER S.F.		
						MAT.	INST.	TOTAL
8500	Norman	10"	2	#4 @ 48" E.W.		12.80	18.35	31.15
8520				#5 @ 32" E.W.		13.05	18.50	31.55
8530				#5 @ 16" E.W.		13.55	19.05	32.60
8600	Norwegian	10"	2	#4 @ 48" E.W.		12.65	16.35	29
8620				#5 @ 32" E.W.		12.90	16.50	29.40
8630				#5 @ 16" E.W.		13.40	17.05	30.45
8800	Triple	10"	2	#4 @ 48" E.W.		9.10	12.90	22
8820				#5 @ 32" E.W.		9.35	13.05	22.40
8830				#5 @ 16" E.W.		9.85	13.60	23.45

B2010 Exterior Walls

The table below lists costs per S.F. for stone veneer walls on various backup using different stone.

Stone Veneer

B2010 128	Stone Veneer	COST PER S.F.		
		MAT.	INST.	TOTAL
2000	Ashlar veneer, 4", 2"x4" stud backup, 16" O.C., 8' high, low priced stone	14.30	21.50	35.80
2100	Metal stud backup, 8' high, 16" O.C.	15.10	21.50	36.60
2150	24" O.C.	14.75	21	35.75
2200	Conc. block backup, 4" thick	15.60	25.50	41.10
2300	6" thick	16.25	25.50	41.75
2350	8" thick	16.45	26	42.45
2400	10" thick	16.95	27.50	44.45
2500	12" thick	17.80	29.50	47.30
3100	High priced stone, wood stud backup, 10' high, 16" O.C.	21.50	24.50	46
3200	Metal stud backup, 10' high, 16" O.C.	22	25	47
3250	24" O.C.	21.50	24.50	46
3300	Conc. block backup, 10' high, 4" thick	22.50	28.50	51
3350	6" thick	23	28.50	51.50
3400	8" thick	23.50	29.50	53
3450	10" thick	24	30.50	54.50
3500	12" thick	24.50	32.50	57
4000	Indiana limestone 2" thick, sawn finish, wood stud backup, 10' high, 16" O.C.	40	13.15	53.15
4100	Metal stud backup, 10' high, 16" O.C.	40.50	13.45	53.95
4150	24" O.C.	40.50	12.95	53.45
4200	Conc. block backup, 4" thick	41	17.20	58.20
4250	6" thick	42	17.45	59.45
4300	8" thick	42	18	60
4350	10" thick	42.50	19.30	61.80
4400	12" thick	43.50	21	64.50
4450	2" thick, smooth finish, wood stud backup, 8' high, 16" O.C.	40	13.15	53.15
4550	Metal stud backup, 8' high, 16" O.C.	40.50	13.25	53.75
4600	24" O.C.	40.50	12.75	53.25
4650	Conc. block backup, 4" thick	41	17	58
4700	6" thick	42	17.25	59.25
4750	8" thick	42	17.80	59.80
4800	10" thick	42.50	19.30	61.80
4850	12" thick	43.50	21	64.50
5350	4" thick, smooth finish, wood stud backup, 8' high, 16" O.C.	46.50	13.15	59.65
5450	Metal stud backup, 8' high, 16" O.C.	47	13.45	60.45
5500	24" O.C.	47	12.95	59.95
5550	Conc. block backup, 4" thick	47.50	17.20	64.70
5600	6" thick	48.50	17.45	65.95
5650	8" thick	48.50	18	66.50

B20 Exterior Enclosure

B2010 Exterior Walls

B2010 128	Stone Veneer	COST PER S.F.		
		MAT.	INST.	TOTAL
5700	10" thick	49	19.30	68.30
5750	12" thick	50	21	71
6000	Granite, gray or pink, 2" thick, wood stud backup, 8' high, 16" O.C.	24	26.50	50.50
6100	Metal studs, 8' high, 16" O.C.	24.50	27	51.50
6150	24" O.C.	24.50	26.50	51
6200	Conc. block backup, 4" thick	25	30.50	55.50
6250	6" thick	26	31	57
6300	8" thick	26	31.50	57.50
6350	10" thick	26.50	32.50	59
6400	12" thick	27.50	34.50	62
6900	4" thick, wood stud backup, 8' high, 16" O.C.	38.50	30	68.50
7000	Metal studs, 8' high, 16" O.C.	39	30.50	69.50
7050	24" O.C.	39	30	69
7100	Conc. block backup, 4" thick	39.50	34	73.50
7150	6" thick	40.50	34.50	75
7200	8" thick	40.50	35	75.50
7250	10" thick	41	36	77
7300	12" thick	42	38	80

B2010 Exterior Walls

Exterior brick veneer/stud backup walls are defined in the following terms; type of brick and studs, stud spacing and bond. All systems include a brick shelf, ties to the backup and all necessary dampproofing and insulation.

B2010 129 — Brick Veneer/Wood Stud Backup

	FACE BRICK	STUD BACKUP	STUD SPACING (IN.)	BOND	FACE	MAT.	INST.	TOTAL
1100	Standard	2x4-wood	16	running		7.25	16.65	23.90
1120				common		8.30	18.80	27.10
1140				Flemish		9	22	31
1160				English		9.90	23	32.90
1400		2x6-wood	16	running		7.50	16.75	24.25
1420				common		8.50	18.90	27.40
1440				Flemish		9.20	22	31.20
1460				English		10.10	23.50	33.60
1700	Glazed	2x4-wood	16	running		21	17.20	38.20
1720				common		25	19.60	44.60
1740				Flemish		28	23	51
1760				English		31	24.50	55.50
2300	Engineer	2x4-wood	16	running		6.90	14.90	21.80
2320				common		7.80	16.65	24.45
2340				Flemish		8.45	19.60	28.05
2360				English		9.30	20.50	29.80
2900	Roman	2x4-wood	16	running		11	15.30	26.30
2920				common		12.85	17.20	30.05
2940				Flemish		14	20	34
2960				English		15.60	21.50	37.10
4100	Norwegian	2x4-wood	16	running		8	11.95	19.95
4120				common		9.15	13.25	22.40
4140				Flemish		9.95	15.30	25.25
4160				English		10.95	15.95	26.90

B2010 130 — Brick Veneer/Metal Stud Backup

	FACE BRICK	STUD BACKUP	STUD SPACING (IN.)	BOND	FACE	MAT.	INST.	TOTAL
5100	Standard	25ga.x6"NLB	24	running		6.85	16.25	23.10
5120				common		7.90	18.40	26.30
5140				Flemish		8.25	21	29.25
5160				English		9.50	23	32.50
5200		20ga.x3-5/8"NLB	16	running		7	16.95	23.95
5220				common		8.05	19.10	27.15
5240				Flemish		8.75	22.50	31.25
5260				English		9.65	23.50	33.15

B2010 Exterior Walls

B2010 130	Brick Veneer/Metal Stud Backup

	FACE BRICK	STUD BACKUP	STUD SPACING (IN.)	BOND	FACE	COST PER S.F.		
						MAT.	INST.	TOTAL
5400	Standard	16ga.x3-5/8"LB	16	running		7.75	17.20	24.95
5420				common		8.75	19.35	28.10
5440				Flemish		9.45	22.50	31.95
5460				English		10.35	23.50	33.85
5500		16ga.x3-5/8"LB	24	running		7.40	16.70	24.10
5520				common		8.40	18.85	27.25
5540				Flemish		9.10	22	31.10
5560				English		10	23	33
5700	Glazed	25ga.x6"NLB	24	running		21	16.80	37.80
5720				common		24.50	19.20	43.70
5740				Flemish		27.50	23	50.50
5760				English		30.50	24	54.50
5800		20ga.x3-5/8"NLB	24	running		21	17	38
5820				common		24.50	19.40	43.90
5840				Flemish		27.50	23	50.50
5860				English		30.50	24.50	55
6000		16ga.x3-5/8"LB	16	running		21.50	17.75	39.25
6020				common		25.50	20	45.50
6040				Flemish		28.50	23.50	52
6060				English		31.50	25	56.50
6300	Engineer	25ga.x6"NLB	24	running		6.50	14.50	21
6320				common		7.40	16.25	23.65
6340				Flemish		8.05	19.20	27.25
6360				English		8.90	20	28.90
6400		20ga.x3-5/8"NLB	16	running		6.65	15.20	21.85
6420				common		7.55	16.95	24.50
6440				Flemish		8.20	19.90	28.10
6460				English		9.05	21	30.05
6900	Roman	25ga.x6"NLB	24	running		10.60	14.90	25.50
6920				common		12.45	16.75	29.20
6940				Flemish		13.60	19.60	33.20
6960				English		15.20	21	36.20
7000		20ga.x3-5/8"NLB	16	running		10.75	15.60	26.35
7020				common		12.60	17.50	30.10
7040				Flemish		13.75	20.50	34.25
7060				English		15.35	22	37.35
7500	Norman	25ga.x6"NLB	24	running		7.70	12.70	20.40
7520				common		8.85	14.15	23
7540				Flemish		18.70	16.55	35.25
7560				English		10.70	17.40	28.10
7600		20ga.x3-5/8"NLB	24	running		7.70	12.90	20.60
7620				common		8.85	14.35	23.20
7640				Flemish		18.70	16.75	35.45
7660				English		10.70	17.60	28.30
8100	Norwegian	25ga.x6"NLB	24	running		7.60	11.55	19.15
8120				common		8.75	12.85	21.60
8140				Flemish		9.55	14.90	24.45
8160				English		10.55	15.55	26.10
8400		16ga.x3-5/8"LB	16	running		8.45	12.50	20.95
8420				common		9.60	13.80	23.40
8440				Flemish		10.40	15.85	26.25
8460				English		11.40	16.50	27.90

B2010 Exterior Walls

Brick Face Composite Wall

B2010 132		Brick Face Composite Wall - Double Wythe							
	FACE BRICK	BACKUP MASONRY	BACKUP THICKNESS (IN.)	BACKUP CORE FILL		COST PER S.F.		
						MAT.	INST.	TOTAL
1000	Standard	common brick	4	none		9.35	25.50	34.85
1040		SCR brick	6	none		13.45	23	36.45
1080		conc. block	4	none		7.90	20.50	28.40
1120			6	perlite		9.75	21.50	31.25
1160				styrofoam		9.85	21	30.85
1200			8	perlite		10.45	22	32.45
1240				styrofoam		10	21.50	31.50
1520		glazed block	4	none		16.65	22	38.65
1560			6	perlite		18.65	22.50	41.15
1600				styrofoam		18.75	22.50	41.25
1640			8	perlite		19.60	23.50	43.10
1680				styrofoam		19.15	23	42.15
1720		clay tile	4	none		12.45	19.90	32.35
1760			6	none		14.65	20.50	35.15
1800			8	none		16.70	21	37.70
1840		glazed tile	4	none		19.25	26	45.25
1880								
2000	Glazed	common brick	4	none		23.50	26	49.50
2040		SCR brick	6	none		27.50	23.50	51
2080		conc. block	4	none		22	21	43
2200			8	perlite		24.50	22.50	47
2240				styrofoam		24	22	46
2280		L.W. block	4	none		22	21	43
2400			8	perlite		25	22.50	47.50
2520		glazed block	4	none		30.50	22.50	53
2560			6	perlite		32.50	23.50	56
2600				styrofoam		32.50	23	55.50
2640			8	perlite		33.50	24	57.50
2680				styrofoam		33	23.50	56.50

B2010 Exterior Walls

B2010 132 — Brick Face Composite Wall - Double Wythe

	FACE BRICK	BACKUP MASONRY	BACKUP THICKNESS (IN.)	BACKUP CORE FILL		COST PER S.F.		
						MAT.	INST.	TOTAL
2720	Glazed	clay tile	4	none		26.50	20.50	47
2760			6	none		28.50	21	49.50
2800			8	none		30.50	21.50	52
2840		glazed tile	4	none		33	26.50	59.50
3000	Engineer	common brick	4	none		8.95	24	32.95
3040		SCR brick	6	none		13.05	21	34.05
3080		conc. block	4	none		7.50	18.95	26.45
3200			8	perlite		10.05	20	30.05
3280		L.W. block	4	none		5.95	13.40	19.35
3320			6	perlite		9.60	19.55	29.15
3520		glazed block	4	none		16.25	20.50	36.75
3560			6	perlite		18.25	21	39.25
3600				styrofoam		18.35	20.50	38.85
3640			8	perlite		19.20	21.50	40.70
3680				styrofoam		18.75	21	39.75
3720		clay tile	4	none		12.05	18.15	30.20
3800			8	none		16.30	19.40	35.70
3840		glazed tile	4	none		18.85	24.50	43.35
4000	Roman	common brick	4	none		13.05	24.50	37.55
4040		SCR brick	6	none		17.20	21.50	38.70
4080		conc. block	4	none		11.65	19.35	31
4120			6	perlite		13.45	20	33.45
4200			8	perlite		14.20	20.50	34.70
5000	Norman	common brick	4	none		10.15	22	32.15
5040		SCR brick	6	none		14.30	19.40	33.70
5280		L.W. block	4	none		8.90	17.05	25.95
5320			6	perlite		10.85	17.85	28.70
5720		clay tile	4	none		13.30	16.35	29.65
5760			6	none		15.50	16.90	32.40
5840		glazed tile	4	none		20	22.50	42.50
6000	Norwegian	common brick	4	none		10.05	21	31.05
6040		SCR brick	6	none		14.20	18.25	32.45
6080		conc. block	4	none		8.65	16	24.65
6120			6	perlite		10.45	16.75	27.20
6160				styrofoam		10.55	16.45	27
6200			8	perlite		11.20	17.25	28.45
6520		glazed block	4	none		17.40	17.45	34.85
6560			6	perlite		19.40	18.05	37.45
7000	Utility	common brick	4	none		10.15	19.95	30.10
7040		SCR brick	6	none		14.30	17.15	31.45
7080		conc. block	4	none		8.75	14.90	23.65
7120			6	perlite		10.55	15.65	26.20
7160				styrofoam		10.65	15.35	26
7200			8	perlite		11.30	16.15	27.45
7240				styrofoam		10.85	15.75	26.60
7280		L.W. block	4	none		8.90	14.80	23.70
7320			6	perlite		10.80	15.50	26.30
7520		glazed block	4	none		17.50	16.35	33.85
7560			6	perlite		19.50	16.95	36.45
7720		clay tile	4	none		13.30	14.10	27.40
7760			6	none		15.50	14.65	30.15
7840		glazed tile	4	none		20	20.50	40.50

B20 Exterior Enclosure

B2010 Exterior Walls

B2010 133		Brick Face Composite Wall - Triple Wythe						

	FACE BRICK	MIDDLE WYTHE	INSIDE MASONRY	TOTAL THICKNESS (IN.)		COST PER S.F.		
						MAT.	INST.	TOTAL
8000	Standard	common brick	standard brick	12		14.20	37	51.20
8100		4" conc. brick	standard brick	12		13.60	38	51.60
8120		4" conc. brick	common brick	12		12.30	37.50	49.80
8200	Glazed	common brick	standard brick	12		28	37.50	65.50
8300		4" conc. brick	standard brick	12		27.50	38.50	66
8320		4" conc. brick	glazed brick	12		26.50	38	64.50

Brick Face Cavity Wall

B2010 134	Brick Face Cavity Wall							
	FACE BRICK	**BACKUP MASONRY**	**TOTAL THICKNESS (IN.)**	**CAVITY INSULATION**		**COST PER S.F.**		
						MAT.	**INST.**	**TOTAL**
1000	Standard	4" common brick	10	polystyrene		9.25	26	35.25
1020				none		9	25	34
1040		6" SCR brick	12	polystyrene		13.40	23	36.40
1060				none		13.10	22.50	35.60
1080		4" conc. block	10	polystyrene		7.80	20.50	28.30
1100				none		7.55	20	27.55
1120		6" conc. block	12	polystyrene		8.50	21	29.50
1140				none		8.25	20.50	28.75
1160		4" L.W. block	10	polystyrene		7.95	20.50	28.45
1180				none		7.70	19.95	27.65
1200		6" L.W. block	12	polystyrene		8.75	21	29.75
1220				none		8.50	20.50	29
1240		4" glazed block	10	polystyrene		16.55	22	38.55
1260				none		16.30	21.50	37.80
1280		6" glazed block	12	polystyrene		16.60	22	38.60
1300				none		16.30	21.50	37.80
1320		4" clay tile	10	polystyrene		12.35	19.95	32.30
1340				none		12.10	19.25	31.35
1360		4" glazed tile	10	polystyrene		19.15	26.50	45.65
1380				none		18.90	25.50	44.40
1500	Glazed	4" common brick	10	polystyrene		23	26.50	49.50
1520				none		23	25.50	48.50
1580		4" conc. block	10	polystyrene		22	21.50	43.50
1600				none		21.50	20.50	42
1660		4" L.W. block	10	polystyrene		22	21	43
1680				none		21.50	20.50	42
1740		4" glazed block	10	polystyrene		30.50	22.50	53
1760				none		30.50	22	52.50
1820		4" clay tile	10	polystyrene		26.50	20.50	47
1840				none		26	19.80	45.80
1860		4" glazed tile	10	polystyrene		28.50	21	49.50
1880				none		28.50	20.50	49
2000	Engineer	4" common brick	10	polystyrene		8.85	24	32.85
2020				none		8.60	23.50	32.10
2080		4" conc. block	10	polystyrene		7.40	19	26.40
2100				none		7.15	18.30	25.45

B2010 134				Brick Face Cavity Wall				
	FACE BRICK	BACKUP MASONRY	TOTAL THICKNESS (IN.)	CAVITY INSULATION		COST PER S.F.		
						MAT.	INST.	TOTAL
2160	Engineer	4" L.W. block	10	polystyrene		7.55	18.90	26.45
2180				none		7.30	18.20	25.50
2240		4" glazed block	10	polystyrene		16.15	20.50	36.65
2260				none		15.90	19.75	35.65
2320		4" clay tile	10	polystyrene		11.95	18.20	30.15
2340				none		11.70	17.50	29.20
2360		4" glazed tile	10	polystyrene		18.75	24.50	43.25
2380				none		18.50	24	42.50
2500	Roman	4" common brick	10	polystyrene		12.95	24.50	37.45
2520				none		12.70	24	36.70
2580		4" conc. block	10	polystyrene		11.55	19.40	30.95
2600				none		11.25	18.70	29.95
2660		4" L.W. block	10	polystyrene		11.70	19.30	31
2680				none		11.40	18.60	30
2740		4" glazed block	10	polystyrene		20.50	21	41.50
2760				none		20	20	40
2820		4" clay tile	10	polystyrene		16.10	18.60	34.70
2840				none		15.80	17.90	33.70
2860		4" glazed tile	10	polystyrene		23	25	48
2880				none		22.50	24.50	47
3000	Norman	4" common brick	10	polystyrene		10.05	22	32.05
3020				none		9.80	21.50	31.30
3080		4" conc. block	10	polystyrene		8.65	17.20	25.85
3100				none		8.35	16.50	24.85
3160		4" L.W. block	10	polystyrene		8.80	17.10	25.90
3180				none		8.50	16.40	24.90
3240		4" glazed block	10	polystyrene		17.40	18.65	36.05
3260				none		17.10	17.95	35.05
3320		4" clay tile	10	polystyrene		13.20	16.40	29.60
3340				none		12.90	15.70	28.60
3360		4" glazed tile	10	polystyrene		20	22.50	42.50
3380				none		19.70	22	41.70
3500	Norwegian	4" common brick	10	polystyrene		9.95	21	30.95
3520				none		9.70	20.50	30.20
3580		4" conc. block	10	polystyrene		8.55	16.05	24.60
3600				none		8.25	15.35	23.60
3660		4" L.W. block	10	polystyrene		8.70	15.95	24.65
3680				none		8.40	15.25	23.65
3740		4" glazed block	10	polystyrene		17.30	17.50	34.80
3760				none		17	16.80	33.80
3820		4" clay tile	10	polystyrene		13.10	15.25	28.35
3840				none		12.80	14.55	27.35
3860		4" glazed tile	10	polystyrene		19.90	21.50	41.40
3880				none		19.60	21	40.60
4000	Utility	4" common brick	10	polystyrene		10.05	20	30.05
4020				none		9.80	19.30	29.10
4080		4" conc. block	10	polystyrene		8.65	14.95	23.60
4100				none		8.35	14.25	22.60
4160		4" L.W. block	10	polystyrene		8.80	14.85	23.65
4180				none		8.50	14.15	22.65
4240		4" glazed block	10	polystyrene		17.40	16.40	33.80
4260				none		17.10	15.70	32.80

B2010 Exterior Walls

B2010 134	Brick Face Cavity Wall							

	FACE BRICK	BACKUP MASONRY	TOTAL THICKNESS (IN.)	CAVITY INSULATION		COST PER S.F.		
						MAT.	INST.	TOTAL
4320	Utility	4" clay tile	10	polystyrene		13.20	14.15	27.35
4340				none		12.90	13.45	26.35
4360		4" glazed tile	10	polystyrene		20	20.50	40.50
4380				none		19.70	19.80	39.50

B2010 135	Brick Face Cavity Wall - Insulated Backup							

	FACE BRICK	BACKUP MASONRY	TOTAL THICKNESS (IN.)	BACKUP CORE FILL		COST PER S.F.		
						MAT.	INST.	TOTAL
5100	Standard	6" conc. block	10	perlite		9.35	21	30.35
5120				styrofoam		9.45	20.50	29.95
5180		6" L.W. block	10	perlite		9.65	20.50	30.15
5200				styrofoam		9.75	20.50	30.25
5260		6" glazed block	10	perlite		18.30	22	40.30
5280				styrofoam		18.40	22	40.40
5340		6" clay tile	10	none		14.30	19.80	34.10
5360		8" clay tile	12	none		16.35	20.50	36.85
5600	Glazed	6" conc. block	10	perlite		23.50	21.50	45
5620				styrofoam		23.50	21	44.50
5680		6" L.W. block	10	perlite		23.50	21	44.50
5700				styrofoam		23.50	21	44.50
5760		6" glazed block	10	perlite		32.50	22.50	55
5780				styrofoam		32.50	22.50	55
5840		6" clay tile	10	none		28.50	20.50	49
5860		8" clay tile	8	none		30.50	21	51.50
6100	Engineer	6" conc. block	10	perlite		9	19.10	28.10
6120				styrofoam		9.10	18.75	27.85
6180		6" L.W. block	10	perlite		9.25	18.95	28.20
6200				styrofoam		9.35	18.60	27.95
6260		6" glazed block	10	perlite		17.90	20.50	38.40
6280				styrofoam		18	20	38
6340		6" clay tile	10	none		13.90	18.05	31.95
6360		8" clay tile	12	none		16	18.75	34.75
6600	Roman	6" conc. block	10	perlite		13.10	19.50	32.60
6620				styrofoam		13.20	19.15	32.35
6680		6" L.W. block	10	perlite		13.35	19.35	32.70
6700				styrofoam		13.45	19	32.45
6760		6" glazed block	10	perlite		22	21	43
6780				styrofoam		22	20.50	42.50
6840		6" clay tile	10	none		18	18.45	36.45
6860		8" clay tile	12	none		20	19.15	39.15
7100	Norman	6" conc. block	10	perlite		10.20	17.30	27.50
7120				styrofoam		10.30	16.95	27.25
7180		6" L.W. block	10	perlite		10.45	17.15	27.60
7200				styrofoam		10.55	16.80	27.35
7260		6" glazed block	10	perlite		19.10	18.60	37.70
7280				styrofoam		19.20	18.25	37.45
7340		6" clay tile	10	none		15.10	16.25	31.35
7360		8" clay tile	12	none		17.20	16.95	34.15
7600	Norwegian	6" conc. block	10	perlite		10.10	16.15	26.25
7620				styrofoam		10.20	15.80	26
7680		6" L.W. block	10	perlite		10.35	16	26.35
7700				styrofoam		10.45	15.65	26.10

B2010 Exterior Walls

B2010 135	Brick Face Cavity Wall - Insulated Backup

	FACE BRICK	BACKUP MASONRY	TOTAL THICKNESS (IN.)	BACKUP CORE FILL		COST PER S.F.		
						MAT.	INST.	TOTAL
7760	Norwegian	6" glazed block	10	perlite		19	17.45	36.45
7780				styrofoam		19.10	17.10	36.20
7840		6" clay tile	10	none		15	15.10	30.10
7860		8" clay tile	12	none		17.10	15.80	32.90
8100	Utility	6" conc. block	10	perlite		10.20	15.05	25.25
8120				styrofoam		10.30	14.70	25
8180		6" L.W. block	10	perlite		10.45	14.90	25.35
8200				styrofoam		10.55	14.55	25.10
8220		8" L.W. block	12	perlite		11.60	15.35	26.95
8240				styrofoam		11.15	14.95	26.10
8260		6" glazed block	10	perlite		19.10	16.35	35.45
8280				styrofoam		19.20	16	35.20
8300		8" glazed block	12	perlite		20	16.90	36.90
8340		6" clay tile	10	none		15.10	14	29.10
8360		8" clay tile	12	none		17.20	14.70	31.90

B2010 Exterior Walls

Glass Block

B2010 140	Glass Block	COST PER S.F.		
		MAT.	INST.	TOTAL
2300	Glass block 4" thick, 6"x6" plain, under 1,000 S.F.	30.50	24	54.50
2400	1,000 to 5,000 S.F.	30	21	51
2500	Over 5,000 S.F.	29	19.75	48.75
2600	Solar reflective, under 1,000 S.F.	43	32.50	75.50
2700	1,000 to 5,000 S.F.	42	28.50	70.50
2800	Over 5,000 S.F.	41	26.50	67.50
3500	8"x8" plain, under 1,000 S.F.	18.20	18.10	36.30
3600	1,000 to 5,000 S.F.	17.85	15.65	33.50
3700	Over 5,000 S.F.	17.30	14.15	31.45
3800	Solar reflective, under 1,000 S.F.	25.50	24.50	50
3900	1,000 to 5,000 S.F.	25	21	46
4000	Over 5,000 S.F.	24	18.80	42.80
5000	12"x12" plain, under 1,000 S.F.	25	16.80	41.80
5100	1,000 to 5,000 S.F.	24.50	14.15	38.65
5200	Over 5,000 S.F.	24	12.90	36.90
5300	Solar reflective, under 1,000 S.F.	35	22.50	57.50
5400	1,000 to 5,000 S.F.	34.50	18.80	53.30
5600	Over 5,000 S.F.	34	17.05	51.05
5800	3" thinline, 6"x6" plain, under 1,000 S.F.	22.50	24	46.50
5900	Over 5,000 S.F.	21.50	19.75	41.25
6000	Solar reflective, under 1,000 S.F.	31.50	32.50	64
6100	Over 5,000 S.F.	30.50	26.50	57
6200	8"x8" plain, under 1,000 S.F.	12.65	18.10	30.75
6300	Over 5,000 S.F.	12.35	14.15	26.50
6400	Solar reflective, under 1,000 S.F.	17.60	24.50	42.10
6500	Over 5,000 S.F.	17.20	18.80	36

B2010 Exterior Walls

Metal Siding

The table below lists costs for metal siding of various descriptions, not including the steel frame, or the structural steel, of a building. Costs are per S.F. including all accessories and insulation.

B2010 146	Metal Siding Panel	COST PER S.F.		
		MAT.	INST.	TOTAL
1400	Metal siding aluminum panel, corrugated, .024" thick, natural	2.95	3.98	6.93
1450	Painted	3.13	3.98	7.11
1500	.032" thick, natural	3.30	3.98	7.28
1550	Painted	3.89	3.98	7.87
1600	Ribbed 4" pitch, .032" thick, natural	3.23	3.98	7.21
1650	Painted	3.93	3.98	7.91
1700	.040" thick, natural	3.72	3.98	7.70
1750	Painted	4.26	3.98	8.24
1800	.050" thick, natural	4.19	3.98	8.17
1850	Painted	4.75	3.98	8.73
1900	8" pitch panel, .032" thick, natural	3.08	3.82	6.90
1950	Painted	3.76	3.83	7.59
2000	.040" thick, natural	3.57	3.84	7.41
2050	Painted	4.09	3.88	7.97
2100	.050" thick, natural	4.04	3.86	7.90
2150	Painted	4.62	3.89	8.51
3000	Steel, corrugated or ribbed, 29 Ga., .0135" thick, galvanized	2.26	3.63	5.89
3050	Colored	3.21	3.69	6.90
3100	26 Ga., .0179" thick, galvanized	2.34	3.65	5.99
3150	Colored	3.32	3.71	7.03
3200	24 Ga., .0239" thick, galvanized	3	3.66	6.66
3250	Colored	3.74	3.72	7.46
3300	22 Ga., .0299" thick, galvanized	3.01	3.68	6.69
3350	Colored	4.53	3.74	8.27
3400	20 Ga., .0359" thick, galvanized	3.01	3.68	6.69
3450	Colored	4.82	3.95	8.77
4100	Sandwich panels, factory fab., 1" polystyrene, steel core, 26 Ga., galv.	6.20	5.70	11.90
4200	Colored, 1 side	7.60	5.70	13.30
4300	2 sides	9.75	5.70	15.45
4400	2" polystyrene, steel core, 26 Ga., galvanized	7.35	5.70	13.05
4500	Colored, 1 side	8.75	5.70	14.45
4600	2 sides	10.90	5.70	16.60
4700	22 Ga., baked enamel exterior	14.40	6.05	20.45
4800	Polyvinyl chloride exterior	15.15	6.05	21.20
5100	Textured aluminum, 4' x 8' x 5/16" plywood backing, single face	4.24	3.47	7.71
5200	Double face	5.75	3.47	9.22

B2010 Exterior Walls

The table below lists costs per S.F. for exterior walls with wood siding. A variety of systems are presented using both wood and metal studs at 16" and 24" O.C.

Exterior Wall Siding

B2010 148	Panel, Shingle & Lap Siding	COST PER S.F.		
		MAT.	INST.	TOTAL
1400	Wood siding w/2"x4"studs, 16"O.C., insul. wall, 5/8"text 1-11 fir plywood	3.18	5.65	8.83
1450	5/8" text 1-11 cedar plywood	4.47	5.45	9.92
1500	1" x 4" vert T.&G. redwood	4.59	6.50	11.09
1600	1" x 8" vert T.&G. redwood	6.30	5.70	12
1650	1" x 5" rabbetted cedar bev. siding	6.85	5.90	12.75
1700	1" x 6" cedar drop siding	6.95	5.95	12.90
1750	1" x 12" rough sawn cedar	3.16	5.70	8.86
1800	1" x 12" sawn cedar, 1" x 4" battens	6.20	5.35	11.55
1850	1" x 10" redwood shiplap siding	6.35	5.50	11.85
1900	18" no. 1 red cedar shingles, 5-1/2" exposed	4.77	7.45	12.22
1950	6" exposed	4.56	7.20	11.76
2000	6-1/2" exposed	4.34	6.95	11.29
2100	7" exposed	4.13	6.70	10.83
2150	7-1/2" exposed	3.92	6.45	10.37
3000	8" wide aluminum siding	4.21	5.05	9.26
3150	8" plain vinyl siding	2.75	5.15	7.90
3250	8" insulated vinyl siding	3.08	5.70	8.78
3300				
3400	2" x 6" studs, 16" O.C., insul. wall, w/ 5/8" text 1-11 fir plywood	3.49	5.80	9.29
3500	5/8" text 1-11 cedar plywood	4.78	5.80	10.58
3600	1" x 4" vert T.&G. redwood	4.90	6.70	11.60
3700	1" x 8" vert T.&G. redwood	6.60	5.85	12.45
3800	1" x 5" rabbetted cedar bev siding	7.15	6.05	13.20
3900	1" x 6" cedar drop siding	7.25	6.10	13.35
4000	1" x 12" rough sawn cedar	3.47	5.85	9.32
4200	1" x 12" sawn cedar, 1" x 4" battens	6.55	5.50	12.05
4500	1" x 10" redwood shiplap siding	6.65	5.65	12.30
4550	18" no. 1 red cedar shingles, 5-1/2" exposed	5.10	7.60	12.70
4600	6" exposed	4.87	7.35	12.22
4650	6-1/2" exposed	4.65	7.10	11.75
4700	7" exposed	4.44	6.90	11.34
4750	7-1/2" exposed	4.23	6.65	10.88
4800	8" wide aluminum siding	4.52	5.25	9.77
4850	8" plain vinyl siding	3.06	5.30	8.36
4900	8" insulated vinyl siding	3.39	5.85	9.24
4910				
5000	2" x 6" studs, 24" O.C., insul. wall, 5/8" text 1-11, fir plywood	3.34	5.45	8.79
5050	5/8" text 1-11 cedar plywood	4.63	5.45	10.08
5100	1" x 4" vert T.&G. redwood	4.75	6.35	11.10
5150	1" x 8" vert T.&G. redwood	6.45	5.50	11.95

B2010 Exterior Walls

B2010 148	Panel, Shingle & Lap Siding	COST PER S.F.		
		MAT.	INST.	TOTAL
5200	1" x 5" rabbetted cedar bev siding	7	5.70	12.70
5250	1" x 6" cedar drop siding	7.10	5.75	12.85
5300	1" x 12" rough sawn cedar	3.32	5.50	8.82
5400	1" x 12" sawn cedar, 1" x 4" battens	6.40	5.15	11.55
5450	1" x 10" redwood shiplap siding	6.50	5.30	11.80
5500	18" no. 1 red cedar shingles, 5-1/2" exposed	4.93	7.25	12.18
5550	6" exposed	4.72	7.05	11.77
5650	7" exposed	4.29	6.55	10.84
5700	7-1/2" exposed	4.08	6.30	10.38
5750	8" wide aluminum siding	4.37	4.89	9.26
5800	8" plain vinyl siding	2.91	4.97	7.88
5850	8" insulated vinyl siding	3.24	5.50	8.74
5900	3-5/8" metal studs, 16 Ga., 16" O.C. insul.wall, 5/8"text 1-11 fir plywood	3.99	5.95	9.94
5950	5/8" text 1-11 cedar plywood	5.30	5.95	11.25
6000	1" x 4" vert T.&G. redwood	5.40	6.80	12.20
6050	1" x 8" vert T.&G. redwood	7.10	6	13.10
6100	1" x 5" rabbetted cedar bev siding	7.65	6.20	13.85
6150	1" x 6" cedar drop siding	7.75	6.25	14
6200	1" x 12" rough sawn cedar	3.97	6	9.97
6250	1" x 12" sawn cedar, 1" x 4" battens	7.05	5.65	12.70
6300	1" x 10" redwood shiplap siding	7.15	5.80	12.95
6350	18" no. 1 red cedar shingles, 5-1/2" exposed	5.60	7.75	13.35
6500	6" exposed	5.35	7.50	12.85
6550	6-1/2" exposed	5.15	7.25	12.40
6600	7" exposed	4.94	7	11.94
6650	7-1/2" exposed	4.94	6.85	11.79
6700	8" wide aluminum siding	5	5.35	10.35
6750	8" plain vinyl siding	3.56	5.25	8.81
6800	8" insulated vinyl siding	3.89	5.80	9.69
7000	3-5/8" metal studs, 16 Ga. 24" O.C. insul wall, 5/8" text 1-11 fir plywood	3.64	5.30	8.94
7050	5/8" text 1-11 cedar plywood	4.93	5.30	10.23
7100	1" x 4" vert T.&G. redwood	5.05	6.35	11.40
7150	1" x 8" vert T.&G. redwood	6.75	5.50	12.25
7200	1" x 5" rabbetted cedar bev siding	7.30	5.70	13
7250	1" x 6" cedar drop siding	7.40	5.75	13.15
7300	1" x 12" rough sawn cedar	3.62	5.55	9.17
7350	1" x 12" sawn cedar 1" x 4" battens	6.70	5.15	11.85
7400	1" x 10" redwood shiplap siding	6.80	5.30	12.10
7450	18" no. 1 red cedar shingles, 5-1/2" exposed	5.45	7.55	13
7500	6" exposed	5.25	7.30	12.55
7550	6-1/2" exposed	5	7.05	12.05
7600	7" exposed	4.80	6.80	11.60
7650	7-1/2" exposed	4.38	6.30	10.68
7700	8" wide aluminum siding	4.67	4.90	9.57
7750	8" plain vinyl siding	3.21	4.98	8.19
7800	8" insul. vinyl siding	3.54	5.55	9.09

B2010 Exterior Walls

The table below lists costs for some typical stucco walls including all the components as demonstrated in the component block below. Prices are presented for backup walls using wood studs, metal studs and CMU.

Exterior Stucco Wall

B2010 151	Stucco Wall	COST PER S.F.		
		MAT.	INST.	TOTAL
2100	Cement stucco, 7/8" th., plywood sheathing, stud wall, 2" x 4", 16" O.C.	2.67	9.20	11.87
2200	24" O.C.	2.57	8.95	11.52
2300	2" x 6", 16" O.C.	2.98	9.40	12.38
2400	24" O.C.	2.83	9.05	11.88
2500	No sheathing, metal lath on stud wall, 2" x 4", 16" O.C.	1.81	8	9.81
2600	24" O.C.	1.71	7.75	9.46
2700	2" x 6", 16" O.C.	2.12	8.20	10.32
2800	24" O.C.	1.97	7.85	9.82
2900	1/2" gypsum sheathing, 3-5/8" metal studs, 16" O.C.	2.85	8.35	11.20
2950	24" O.C.	2.50	7.90	10.40
3000	Cement stucco, 5/8" th., 2 coats on std. CMU block, 8"x 16", 8" thick	3.14	11.15	14.29
3100	10" thick	4.34	11.45	15.79
3200	12" thick	4.49	13.15	17.64
3300	Std. light Wt. block 8" x 16", 8" thick	3.76	11	14.76
3400	10" thick	4.40	11.25	15.65
3500	12" thick	4.61	12.90	17.51
3600	3 coat stucco, self furring metal lath 3.4 Lb/SY, on 8" x 16", 8" thick	3.18	11.80	14.98
3700	10" thick	4.38	12.10	16.48
3800	12" thick	4.53	13.80	18.33
3900	Lt. Wt. block, 8" thick	3.80	11.65	15.45
4000	10" thick	4.44	11.90	16.34
4100	12" thick	4.65	13.55	18.20

B2010 152	E.I.F.S.	COST PER S.F.		
		MAT.	INST.	TOTAL
5100	E.I.F.S., plywood sheathing, stud wall, 2" x 4", 16" O.C., 1" EPS	3.79	9.30	13.09
5110	2" EPS	4.05	9.30	13.35
5120	3" EPS	4.32	9.30	13.62
5130	4" EPS	4.58	9.30	13.88
5140	2" x 6", 16" O.C., 1" EPS	4.10	9.45	13.55
5150	2" EPS	4.36	9.45	13.81
5160	3" EPS	4.63	9.45	14.08
5170	4" EPS	4.89	9.45	14.34
5180	Cement board sheathing, 3-5/8" metal studs, 16" O.C., 1" EPS	4.55	11.50	16.05
5190	2" EPS	4.81	11.50	16.31
5200	3" EPS	5.10	11.50	16.60
5210	4" EPS	6.10	11.50	17.60
5220	6" metal studs, 16" O.C., 1" EPS	5	11.55	16.55
5230	2" EPS	5.25	11.55	16.80

B2010 Exterior Walls

B2010 152	E.I.F.S.	COST PER S.F.		
		MAT.	**INST.**	**TOTAL**
5240	3" EPS	5.55	11.55	17.10
5250	4" EPS	6.55	11.55	18.10
5260	CMU block, 8" x 8" x 16", 1" EPS	4.81	13.20	18.01
5270	2" EPS	5.05	13.20	18.25
5280	3" EPS	5.35	13.20	18.55
5290	4" EPS	5.60	13.20	18.80
5300	8" x 10" x 16", 1" EPS	6	13.50	19.50
5310	2" EPS	6.25	13.50	19.75
5320	3" EPS	6.55	13.50	20.05
5330	4" EPS	6.80	13.50	20.30
5340	8" x 12" x 16", 1" EPS	6.15	15.20	21.35
5350	2" EPS	6.40	15.20	21.60
5360	3" EPS	6.70	15.20	21.90
5370	4" EPS	6.95	15.20	22.15

Description: The table below lists costs, $/S.F., for channel girts with sag rods and connector angles top and bottom for various column spacings, building heights and wind loads. Additive costs are shown for wind columns.

How to Use this Table: Add the cost of girts, sag rods, and angles to the framing costs of steel buildings clad in metal or composition siding. If the column spacing is in excess of the column spacing shown, use intermediate wind columns. Additive costs are shown under "Wind Columns."

B2010 154		Metal Siding Support						
	BLDG. HEIGHT (FT.)	WIND LOAD (P.S.F.)	COL. SPACING (FT.)		INTERMEDIATE COLUMNS	COST PER S.F.		
						MAT.	INST.	TOTAL
3000	18	20	20			2.37	4	6.37
3100					wind cols.	1.22	.29	1.51
3200		20	25			2.59	4.05	6.64
3300					wind cols.	.98	.23	1.21
3400		20	30			2.86	4.12	6.98
3500					wind cols.	.82	.19	1.01
3600		20	35			3.15	4.19	7.34
3700					wind cols.	.70	.16	.86
3800		30	20			2.62	4.06	6.68
3900					wind cols.	1.22	.29	1.51
4000		30	25			2.86	4.12	6.98
4100					wind cols.	.98	.23	1.21
4200		30	30			3.16	4.19	7.35
4300					wind cols.	1.11	.26	1.37
4600	30	20	20			2.30	2.81	5.11
4700					wind cols.	1.73	.41	2.14
4800		20	25			2.59	2.88	5.47
4900					wind cols.	1.38	.33	1.71
5000		20	30			2.91	2.96	5.87
5100					wind cols.	1.36	.33	1.69
5200		20	35			3.25	3.04	6.29
5300					wind cols.	1.35	.32	1.67
5400		30	20			2.62	2.89	5.51
5500					wind cols.	2.04	.49	2.53
5600		30	25			2.90	2.95	5.85
5700					wind cols.	1.95	.47	2.42
5800		30	30			3.27	3.04	6.31
5900					wind cols.	1.83	.44	2.27

B2010 Exterior Walls

B2010 160	Pole Barn Exterior Wall	COST PER S.F.		
		MAT.	INST.	TOTAL
2000	Pole barn exterior wall, pressure treated pole in concrete, 8' O.C.			
3000	Steel siding, 8' eave	4.57	9.35	13.92
3050	10' eave	4.41	8.60	13.01
3100	12' eave	4.21	7.85	12.06
3150	14' eave	4.11	7.45	11.56
3200	16' eave	4.02	7.10	11.12
4000	Aluminum siding, 8' eave	3.95	9.35	13.30
4050	10' eave	3.73	8.45	12.18
4100	12' eave	3.59	7.85	11.44
4150	14' eave	3.49	7.45	10.94
4200	16' eave	3.40	7.10	10.50
5000	Wood siding, 8' eave	4.36	8.25	12.61
5050	10' eave	4.14	7.30	11.44
5100	12' eave	4	6.70	10.70
5150	14' eave	3.90	6.30	10.20
5200	16' eave	3.81	5.95	9.76
6000	Plywood siding, 8' eave	3.54	8.20	11.74
6050	10' eave	3.32	7.30	10.62
6100	12' eave	3.18	6.70	9.88
6150	14' eave	3.08	6.25	9.33
6200	16' eave	2.99	5.90	8.89

B2020 Exterior Windows

The table below lists window systems by material, type and size. Prices between sizes listed can be interpolated with reasonable accuracy. Prices include frame, hardware, and casing as illustrated in the component block below.

B2020 102		Wood Windows						
	MATERIAL	TYPE	GLAZING	SIZE	DETAIL	COST PER UNIT		
						MAT.	INST.	TOTAL
3000	Wood	double hung	std. glass	2'-8" x 4'-6"		249	240	489
3050				3'-0" x 5'-6"		325	275	600
3100			insul. glass	2'-8" x 4'-6"		264	240	504
3150				3'-0" x 5'-6"		355	275	630
3200		sliding	std. glass	3'-4" x 2'-7"		330	200	530
3250				4'-4" x 3'-3"		345	219	564
3300				5'-4" x 6'-0"		435	263	698
3350			insul. glass	3'-4" x 2'-7"		400	234	634
3400				4'-4" x 3'-3"		415	254	669
3450				5'-4" x 6'-0"		525	300	825
3500		awning	std. glass	2'-10" x 1'-9"		231	120	351
3600				4'-4" x 2'-8"		370	116	486
3700			insul. glass	2'-10" x 1'-9"		282	138	420
3800				4'-4" x 2'-8"		450	127	577
3900		casement	std. glass	1'-10" x 3'-2"	1 lite	335	158	493
3950				4'-2" x 4'-2"	2 lite	595	210	805
4000				5'-11" x 5'-2"	3 lite	900	270	1,170
4050				7'-11" x 6'-3"	4 lite	1,275	325	1,600
4100				9'-11" x 6'-3"	5 lite	1,650	375	2,025
4150			insul. glass	1'-10" x 3'-2"	1 lite	335	158	493
4200				4'-2" x 4'-2"	2 lite	610	210	820
4250				5'-11" x 5'-2"	3 lite	945	270	1,215
4300				7'-11" x 6'-3"	4 lite	1,350	325	1,675
4350				9'-11" x 6'-3"	5 lite	1,750	375	2,125
4400		picture	std. glass	4'-6" x 4'-6"		470	271	741
4450				5'-8" x 4'-6"		530	300	830
4500			insul. glass	4'-6" x 4'-6"		570	315	885
4550				5'-8" x 4'-6"		650	350	1,000
4600		fixed bay	std. glass	8' x 5'		1,625	500	2,125
4650				9'-9" x 5'-4"		1,175	700	1,875
4700			insul. glass	8' x 5'		2,150	500	2,650
4750				9'-9" x 5'-4"		1,275	700	1,975
4800		casement bay	std. glass	8' x 5'		1,625	575	2,200
4850			insul. glass	8' x 5'		1,825	575	2,400
4900		vert. bay	std. glass	8' x 5'		1,825	575	2,400
4950			insul. glass	8' x 5'		1,925	575	2,500

B2020 Exterior Windows

B2020 104				Steel Windows				
	MATERIAL	TYPE	GLAZING	SIZE	DETAIL	COST PER UNIT		
						MAT.	INST.	TOTAL
5000	Steel	double hung	1/4" tempered	2'-8" x 4'-6"		880	191	1,071
5050				3'-4" x 5'-6"		1,350	291	1,641
5100			insul. glass	2'-8" x 4'-6"		895	218	1,113
5150				3'-4" x 5'-6"		1,375	335	1,710
5200		horz. pivoted	std. glass	2' x 2'		269	63.50	332.50
5250				3' x 3'		605	143	748
5300				4' x 4'		1,075	254	1,329
5350				6' x 4'		1,625	380	2,005
5400			insul. glass	2' x 2'		274	73	347
5450				3' x 3'		615	164	779
5500				4' x 4'		1,100	291	1,391
5550				6' x 4'		1,650	435	2,085
5600		picture window	std. glass	3' x 3'		380	143	523
5650				6' x 4'		1,025	380	1,405
5700			insul. glass	3' x 3'		390	164	554
5750				6' x 4'		1,050	435	1,485
5800		industrial security	std. glass	2'-9" x 4'-1"		810	178	988
5850				4'-1" x 5'-5"		1,600	350	1,950
5900			insul. glass	2'-9" x 4'-1"		825	204	1,029
5950				4'-1" x 5'-5"		1,625	400	2,025
6000		comm. projected	std. glass	3'-9" x 5'-5"		1,325	325	1,650
6050				6'-9" x 4'-1"		1,800	440	2,240
6100			insul. glass	3'-9" x 5'-5"		1,350	370	1,720
6150				6'-9" x 4'-1"		1,825	500	2,325
6200		casement	std. glass	4'-2" x 4'-2"	2 lite	1,175	276	1,451
6250			insul. glass	4'-2" x 4'-2"		1,200	315	1,515
6300			std. glass	5'-11" x 5'-2"	3 lite	2,475	485	2,960
6350			insul. glass	5'-11" x 5'-2"		2,500	555	3,055

B2020 106				Aluminum Windows				
	MATERIAL	TYPE	GLAZING	SIZE	DETAIL	COST PER UNIT		
						MAT.	INST.	TOTAL
6400	Aluminum	double hung	insul. glass	2'-0" x 4'-0"		385	143	528
6450				4'-5" x 5'-3"		435	179	614
6500			insul. glass	3'-1" x 3'-2"		460	172	632
6550				4'-5" x 5'-3"		520	215	735
6600		sliding	std. glass	3' x 2'		239	143	382
6650				5' x 3'		365	159	524
6700				8' x 4'		385	238	623
6750				9' x 5'		585	355	940
6800			insul. glass	3' x 2'		255	143	398
6850				5' x 3'		425	159	584
6900				8' x 4'		620	238	858
6950				9' x 5'		935	355	1,290
7000		single hung	std. glass	2' x 3'		226	143	369
7050				2'-8" x 6'-8"		400	179	579
7100				3'-4" x 5'-0"		325	159	484
7150			insul. glass	2' x 3'		274	143	417
7200				2'-8" x 6'-8"		515	179	694
7250				3'-4" x 5'		360	159	519
7300		double hung	std. glass	2' x 3'		310	95.50	405.50
7350				2'-8" x 6'-8"		920	283	1,203

B2020 Exterior Windows

B2020 106		Aluminum Windows						

	MATERIAL	TYPE	GLAZING	SIZE	DETAIL	COST PER UNIT		
						MAT.	INST.	TOTAL
7400	Aluminum	double hung		3'-4" x 5'		860	265	1,125
7450			insul. glass	2' x 3'		320	109	429
7500				2'-8" x 6'-8"		945	325	1,270
7550				3'-4" x 5'-0"		885	305	1,190
7600		casement	std. glass	3'-1" x 3'-2"		273	155	428
7650				4'-5" x 5'-3"		660	375	1,035
7700			insul. glass	3'-1" x 3'-2"		287	177	464
7750				4'-5" x 5'-3"		695	430	1,125
7800		hinged swing	std. glass	3' x 4'		570	191	761
7850				4' x 5'		955	320	1,275
7900			insul. glass	3' x 4'		590	218	808
7950				4' x 5'		980	365	1,345
8200		picture unit	std. glass	2'-0" x 3'-0"		175	95.50	270.50
8250				2'-8" x 6'-8"		520	283	803
8300				3'-4" x 5'-0"		485	265	750
8350			insul. glass	2'-0" x 3'-0"		183	109	292
8400				2'-8" x 6'-8"		545	325	870
8450				3'-4" x 5'-0"		510	305	815
8500		awning type	std. glass	3'-0" x 3'-0"	2 lite	460	102	562
8550				3'-0" x 4'-0"	3 lite	540	143	683
8600				3'-0" x 5'-4"	4 lite	650	143	793
8650				4'-0" x 5'-4"	4 lite	715	159	874
8700			insul. glass	3'-0" x 3'-0"	2 lite	490	102	592
8750				3'-0" x 4'-0"	3 lite	620	143	763
8800				3'-0" x 5'-4"	4 lite	765	143	908
8850				4'-0" x 5'-4"	4 lite	855	159	1,014
9051								
9052								

B2020 210	Tubular Aluminum Framing	COST/S.F. OPNG.		
		MAT.	INST.	TOTAL
1100	Alum flush tube frame, for 1/4" glass, 1-3/4"x4", 5'x6'opng, no inter horiz	13.80	10.95	24.75
1150	One intermediate horizontal	18.85	13.10	31.95
1200	Two intermediate horizontals	24	15.20	39.20
1250	5' x 20' opening, three intermediate horizontals	12.70	9.40	22.10
1400	1-3/4" x 4-1/2", 5' x 6' opening, no intermediate horizontals	15.95	10.95	26.90
1450	One intermediate horizontal	21	13.10	34.10
1500	Two intermediate horizontals	26.50	15.20	41.70
1550	5' x 20' opening, three intermediate horizontals	14.30	9.40	23.70
1700	For insulating glass, 2"x4-1/2", 5'x6' opening, no intermediate horizontals	16.25	11.60	27.85
1750	One intermediate horizontal	21	13.85	34.85
1800	Two intermediate horizontals	26	16.10	42.10
1850	5' x 20' opening, three intermediate horizontals	14.15	9.95	24.10
2000	Thermal break frame, 2-1/4"x4-1/2", 5'x6'opng, no intermediate horizontals	16.90	11.75	28.65
2050	One intermediate horizontal	22.50	14.45	36.95
2100	Two intermediate horizontals	28	17.10	45.10
2150	5' x 20' opening, three intermediate horizontals	15.30	10.40	25.70

B2020 Exterior Windows

Glazing Panel

Plate Glass

Spandrel Glass

Polycarbonate

Sandwich Panel

The table below lists costs of curtain wall and spandrel panels per S.F. Costs do not include structural framing used to hang the panels.

B2020 220	Curtain Wall Panels	COST PER S.F.		
		MAT.	INST.	TOTAL
1000	Glazing panel, insulating, 1/2" thick, 2 lites 1/8" float, clear	10.60	11.05	21.65
1100	Tinted	14.85	11.05	25.90
1200	5/8" thick units, 2 lites 3/16" float, clear	14.85	11.65	26.50
1300	Tinted	14.90	11.65	26.55
1400	1" thick units, 2 lites, 1/4" float, clear	17.80	14	31.80
1500	Tinted	25.50	14	39.50
1600	Heat reflective film inside	30	12.35	42.35
1700	Light and heat reflective glass, tinted	34	12.35	46.35
2000	Plate glass, 1/4" thick, clear	6.20	8.75	14.95
2050	Tempered	9.20	8.75	17.95
2100	Tinted	9.70	8.75	18.45
2200	3/8" thick, clear	10.80	14	24.80
2250	Tempered	17.80	14	31.80
2300	Tinted	17	14	31
2400	1/2" thick, clear	18.30	19.05	37.35
2450	Tempered	26.50	19.05	45.55
2500	Tinted	30	19.05	49.05
2600	3/4" thick, clear	38.50	30	68.50
2650	Tempered	44.50	30	74.50
3000	Spandrel glass, panels, 1/4" plate glass insul w/fiberglass, 1" thick	18.25	8.75	27
3100	2" thick	21.50	8.75	30.25
3200	Galvanized steel backing, add	6.35		6.35
3300	3/8" plate glass, 1" thick	31	8.75	39.75
3400	2" thick	34.50	8.75	43.25
4000	Polycarbonate, masked, clear or colored, 1/8" thick	11	6.15	17.15
4100	3/16" thick	13.20	6.35	19.55
4200	1/4" thick	14.30	6.75	21.05
4300	3/8" thick	24.50	7	31.50
5000	Facing panel, textured alum., 4' x 8' x 5/16" plywood backing, sgl face	4.24	3.47	7.71
5100	Double face	5.75	3.47	9.22
5200	4' x 10' x 5/16" plywood backing, single face	4.53	3.47	8
5300	Double face	6.10	3.47	9.57
5400	4' x 12' x 5/16" plywood backing, single face	4.21	3.47	7.68
5500	Sandwich panel, 22 Ga. galv., both sides 2" insulation, enamel exterior	14.40	6.05	20.45
5600	Polyvinylidene fluoride exterior finish	15.15	6.05	21.20
5700	26 Ga., galv. both sides, 1" insulation, colored 1 side	7.60	5.70	13.30
5800	Colored 2 sides	9.75	5.70	15.45

B2030 Exterior Doors

Costs are listed for exterior door systems by material, type and size. Prices between sizes listed can be interpolated with reasonable accuracy. Prices are per opening for a complete door system including frame.

B2030 110			Glazed Doors, Steel or Aluminum					
	MATERIAL	TYPE	DOORS	SPECIFICATION	OPENING	COST PER OPNG.		
						MAT.	INST.	TOTAL
5600	St. Stl. & glass	revolving	stock unit	manual oper.	6'-0" x 7'-0"	43,700	9,475	53,175
5650				auto Cntrls.	6'-10" x 7'-0"	60,000	10,100	70,100
5700	Bronze		stock unit	manual oper.	6'-10" x 7'-0"	51,500	19,100	70,600
5750				auto Cntrls.	6'-10" x 7'-0"	67,500	19,700	87,200
5800	St. Stl. & glass	balanced	standard	economy	3'-0" x 7'-0"	9,650	1,600	11,250
5850				premium	3'-0" x 7'-0"	16,900	2,100	19,000
6300	Alum. & glass	w/o transom	narrow stile	w/panic Hrdwre.	3'-0" x 7'-0"	1,725	1,050	2,775
6350				dbl. door, Hrdwre.	6'-0" x 7'-0"	2,975	1,725	4,700
6400			wide stile	hdwre.	3'-0" x 7'-0"	2,025	1,025	3,050
6450				dbl. door, Hdwre.	6'-0" x 7'-0"	3,925	2,050	5,975
6500			full vision	hdwre.	3'-0" x 7'-0"	2,550	1,625	4,175
6550				dbl. door, Hdwre.	6'-0" x 7'-0"	3,575	2,325	5,900
6600			non-standard	hdwre.	3'-0" x 7'-0"	2,200	1,025	3,225
6650				dbl. door, Hdwre.	6'-0" x 7'-0"	4,400	2,050	6,450
6700			bronze fin.	hdwre.	3'-0" x 7'-0"	1,600	1,025	2,625
6750				dbl. door, Hrdwre.	6'-0" x 7'-0"	3,225	2,050	5,275
6800			black fin.	hdwre.	3'-0" x 7'-0"	2,325	1,025	3,350
6850				dbl. door, Hdwre.	6'-0" x 7'-0"	4,650	2,050	6,700
6900		w/transom	narrow stile	hdwre.	3'-0" x 10'-0"	2,275	1,175	3,450
6950				dbl. door, Hdwre.	6'-0" x 10'-0"	3,575	2,075	5,650
7000			wide stile	hdwre.	3'-0" x 10'-0"	2,600	1,425	4,025
7050				dbl. door, Hdwre.	6'-0" x 10'-0"	3,925	2,450	6,375
7100			full vision	hdwre.	3'-0" x 10'-0"	2,875	1,575	4,450
7150				dbl. door, Hdwre.	6'-0" x 10'-0"	4,300	2,725	7,025

B2030 Exterior Doors

| B2030 110 | Glazed Doors, Steel or Aluminum |

	MATERIAL	TYPE	DOORS	SPECIFICATION	OPENING	COST PER OPNG.		
						MAT.	INST.	TOTAL
7200	Alum. & glass	W/transom	non-standard	hdwre.	3'-0" x 10'-0"	2,275	1,100	3,375
7250				dbl. door, Hdwre.	6'-0" x 10'-0"	4,550	2,200	6,750
7300			bronze fin.	hdwre.	3'-0" x 10'-0"	1,700	1,100	2,800
7350				dbl. door, Hdwre.	6'-0" x 10'-0"	3,375	2,200	5,575
7400			black fin.	hdwre.	3'-0" x 10'-0"	2,400	1,100	3,500
7450				dbl. door, Hdwre.	6'-0" x 10'-0"	4,800	2,200	7,000
7500		revolving	stock design	minimum	6'-10" x 7'-0"	22,300	3,800	26,100
7550				average	6'-0" x 7'-0"	26,900	4,775	31,675
7600				maximum	6'-10" x 7'-0"	46,100	6,350	52,450
7650				min., automatic	6'-10" x 7'-0"	38,500	4,425	42,925
7700				avg., automatic	6'-10" x 7'-0"	43,100	5,400	48,500
7750				max., automatic	6'-10" x 7'-0"	62,500	6,975	69,475
7800		balanced	standard	economy	3'-0" x 7'-0"	7,000	1,600	8,600
7850				premium	3'-0" x 7'-0"	8,550	2,050	10,600
7900		mall front	sliding panels	alum. fin.	16'-0" x 9'-0"	3,650	805	4,455
7950					24'-0" x 9'-0"	5,175	1,500	6,675
8000				bronze fin.	16'-0" x 9'-0"	4,250	940	5,190
8050					24'-0" x 9'-0"	6,025	1,750	7,775
8100			fixed panels	alum. fin.	48'-0" x 9'-0"	9,650	1,175	10,825
8150				bronze fin.	48'-0" x 9'-0"	11,200	1,375	12,575
8200		sliding entrance	5' x 7' door	electric oper.	12'-0" x 7'-6"	9,025	1,500	10,525
8250		sliding patio	temp. glass	economy	6'-0" x 7'-0"	1,825	277	2,102
8300			temp. glass	economy	12'-0" x 7'-0"	4,275	370	4,645
8350				premium	6'-0" x 7'-0"	2,750	415	3,165
8400					12'-0" x 7'-0"	6,425	555	6,980

B2030 Exterior Doors

Exterior Door System

The table below lists exterior door systems by material, type and size. Prices between sizes listed can be interpolated with reasonable accuracy. Prices are per opening for a complete door system including frame and required hardware.

Wood doors in this table are designed with wood frames and metal doors with hollow metal frames. Depending upon quality the total material and installation cost of a wood door frame is about the same as a hollow metal frame.

B2030 210 — Wood Doors

	MATERIAL	TYPE	DOORS	SPECIFICATION	OPENING	COST PER OPNG. MAT.	INST.	TOTAL
2350	Birch	solid core	single door	hinged	2'-6" x 6'-8"	1,125	300	1,425
2400					2'-6" x 7'-0"	1,125	300	1,425
2450					2'-8" x 7'-0"	1,100	300	1,400
2500					3'-0" x 7'-0"	1,125	300	1,425
2550			double door	hinged	2'-6" x 6'-8"	2,100	550	2,650
2600					2'-6" x 7'-0"	2,100	550	2,650
2650					2'-8" x 7'-0"	2,075	550	2,625
2700					3'-0" x 7'-0"	1,950	480	2,430
2750	Wood	combination	storm & screen	hinged	3'-0" x 6'-8"	335	74	409
2800					3'-0" x 7'-0"	370	81.50	451.50
2850		overhead	panels, H.D.	manual oper.	8'-0" x 8'-0"	1,125	555	1,680
2900					10'-0" x 10'-0"	1,625	615	2,240
2950					12'-0" x 12'-0"	2,225	740	2,965
3000					14'-0" x 14'-0"	3,550	850	4,400
3050					20'-0" x 16'-0"	5,850	1,700	7,550
3100				electric oper.	8'-0" x 8'-0"	2,150	830	2,980
3150					10'-0" x 10'-0"	2,650	890	3,540
3200					12'-0" x 12'-0"	3,250	1,025	4,275
3250					14'-0" x 14'-0"	4,575	1,125	5,700
3300					20'-0" x 16'-0"	7,050	2,250	9,300

B2030 220 — Steel Doors

	MATERIAL	TYPE	DOORS	SPECIFICATION	OPENING	COST PER OPNG. MAT.	INST.	TOTAL
3350	Steel 18 Ga.	hollow metal	1 door w/frame	no label	2'-6" x 7'-0"	1,600	310	1,910
3400					2'-8" x 7'-0"	1,600	310	1,910
3450					3'-0" x 7'-0"	1,600	310	1,910
3500					3'-6" x 7'-0"	1,650	325	1,975
3550					4'-0" x 8'-0"	1,950	330	2,280
3600			2 doors w/frame	no label	5'-0" x 7'-0"	3,125	575	3,700
3650					5'-4" x 7'-0"	3,125	575	3,700
3700					6'-0" x 7'-0"	3,125	575	3,700
3750					7'-0" x 7'-0"	3,225	610	3,835
3800					8'-0" x 8'-0"	3,775	610	4,385
3850			1 door w/frame	"A" label	2'-6" x 7'-0"	1,925	375	2,300

B2030 Exterior Doors

| B2030 220 | | | | Steel Doors | | | |

	MATERIAL	TYPE	DOORS	SPECIFICATION	OPENING	COST PER OPNG.		
						MAT.	INST.	TOTAL
3900	Steel 18 Ga.	hollow metal	1 door w/frame	"A" label	2'-8" x 7'-0"	1,925	380	2,305
3950					3'-0" x 7'-0"	1,925	380	2,305
4000					3'-6" x 7'-0"	2,075	390	2,465
4050					4'-0" x 8'-0"	2,275	405	2,680
4100			2 doors w/frame		5'-0" x 7'-0"	3,700	695	4,395
4150					5'-4" x 7'-0"	3,700	705	4,405
4200					6'-0" x 7'-0"	3,700	705	4,405
4250					7'-0" x 7'-0"	4,025	720	4,745
4300					8'-0" x 8'-0"	4,025	720	4,745
4350	Steel 24 Ga.	overhead	sectional	manual oper.	8'-0" x 8'-0"	890	555	1,445
4400					10'-0" x 10'-0"	1,200	615	1,815
4450					12'-0" x 12'-0"	1,450	740	2,190
4500					20'-0" x 14'-0"	4,100	1,575	5,675
4550				electric oper.	8'-0" x 8'-0"	1,925	830	2,755
4600					10'-0" x 10'-0"	2,225	890	3,115
4650					12'-0" x 12'-0"	2,475	1,025	3,500
4700					20'-0" x 14'-0"	5,300	2,125	7,425
4750	Steel		rolling	manual oper.	8'-0" x 8'-0"	1,225	895	2,120
4800					10'-0" x 10'-0"	2,025	1,025	3,050
4850					12'-0" x 12'-0"	2,050	1,200	3,250
4900					14'-0" x 14'-0"	3,325	1,775	5,100
4950					20'-0" x 12'-0"	2,325	1,600	3,925
5000					20'-0" x 16'-0"	3,825	2,375	6,200
5050				electric oper.	8'-0" x 8'-0"	2,400	1,175	3,575
5100					10'-0" x 10'-0"	3,200	1,300	4,500
5150					12'-0" x 12'-0"	3,225	1,475	4,700
5200					14'-0" x 14'-0"	4,500	2,050	6,550
5250					20'-0" x 12'-0"	3,475	1,875	5,350
5300					20'-0" x 16'-0"	4,975	2,650	7,625
5350				fire rated	10'-0" x 10'-0"	2,250	1,300	3,550
5400			rolling grille	manual oper.	10'-0" x 10'-0"	2,750	1,425	4,175
5450					15'-0" x 8'-0"	3,175	1,775	4,950
5500		vertical lift	1 door w/frame	motor operator	16'-0" x 16'-0"	23,700	5,525	29,225
5550					32'-0" x 24'-0"	51,000	3,675	54,675

| B2030 230 | | | | Aluminum Doors | | | |

	MATERIAL	TYPE	DOORS	SPECIFICATION	OPENING	COST PER OPNG.		
						MAT.	INST.	TOTAL
6000	Aluminum	combination	storm & screen	hinged	3'-0" x 6'-8"	330	79	409
6050					3'-0" x 7'-0"	365	87	452
6100		overhead	rolling grille	manual oper.	12'-0" x 12'-0"	4,400	2,500	6,900
6150				motor oper.	12'-0" x 12'-0"	5,775	2,800	8,575
6200	Alum. & Fbrgls.		heavy duty	manual oper.	12'-0" x 12'-0"	2,975	740	3,715
6250				electric oper.	12'-0" x 12'-0"	4,000	1,025	5,025

B3010 Roof Coverings

Built Up Ply

Multiple ply roofing is the most popular covering for minimum pitch roofs.

B3010 105	Built-Up	COST PER S.F.		
		MAT.	**INST.**	**TOTAL**
1200	Asphalt flood coat w/gravel; not incl. insul, flash., nailers			
1300				
1400	Asphalt base sheets & 3 plies #15 asphalt felt, mopped	1.08	1.79	2.87
1500	On nailable deck	1.09	1.87	2.96
1600	4 plies #15 asphalt felt, mopped	1.45	1.96	3.41
1700	On nailable deck	1.28	2.07	3.35
1800	Coated glass base sheet, 2 plies glass (type IV), mopped	1.14	1.79	2.93
1900	For 3 plies	1.37	1.96	3.33
2000	On nailable deck	1.28	2.07	3.35
2300	4 plies glass fiber felt (type IV), mopped	1.68	1.96	3.64
2400	On nailable deck	1.51	2.07	3.58
2500	Organic base sheet & 3 plies #15 organic felt, mopped	1.14	1.98	3.12
2600	On nailable deck	1.12	2.07	3.19
2700	4 plies #15 organic felt, mopped	1.40	1.79	3.19
2750				
2800	Asphalt flood coat, smooth surface, not incl. insul, flash., nailers			
2900	Asphalt base sheet & 3 plies #15 asphalt felt, mopped	1.12	1.64	2.76
3000	On nailable deck	1.03	1.71	2.74
3100	Coated glass fiber base sheet & 2 plies glass fiber felt, mopped	1.09	1.58	2.67
3200	On nailable deck	1.03	1.64	2.67
3300	For 3 plies, mopped	1.32	1.71	3.03
3400	On nailable deck	1.23	1.79	3.02
3700	4 plies glass fiber felt (type IV), mopped	1.55	1.71	3.26
3800	On nailable deck	1.46	1.79	3.25
3900	Organic base sheet & 3 plies #15 organic felt, mopped	1.15	1.64	2.79
4000	On nailable decks	1.07	1.71	2.78
4100	4 plies #15 organic felt, mopped	1.35	1.79	3.14
4200	Coal tar pitch with gravel surfacing			
4300	4 plies #15 tarred felt, mopped	2.56	1.87	4.43
4400	3 plies glass fiber felt (type IV), mopped	2.11	2.07	4.18
4500	Coated glass fiber base sheets 2 plies glass fiber felt, mopped	2.13	2.07	4.20
4600	On nailable decks	1.85	2.18	4.03
4800	3 plies glass fiber felt (type IV), mopped	2.89	1.87	4.76
4900	On nailable decks	2.61	1.96	4.57

B3010 Roof Coverings

Full Adhered Single Ply Membrane

Preformed Metal

Flat Seam

Batten Seam

Standing Seam

B3010 120	Single Ply Membrane	COST PER S.F.		
		MAT.	INST.	TOTAL
1000	CSPE (Chlorosulfonated polyethylene), 35 mils, fully adhered	2.14	.98	3.12
2000	EPDM (Ethylene propylene diene monomer), 45 mils, fully adhered	1.25	.98	2.23
4000	Modified bit., SBS modified, granule surface cap sheet, mopped, 150 mils	.72	1.96	2.68
4500	APP modified, granule surface cap sheet, torched, 180 mils	.98	1.28	2.26
6000	Reinforced PVC, 48 mils, loose laid and ballasted with stone	1.24	.50	1.74
6200	Fully adhered with adhesive	1.55	.98	2.53

B3010 130	Preformed Metal Roofing	COST PER S.F.		
		MAT.	INST.	TOTAL
0200	Corrugated roofing, aluminum, mill finish, .0175" thick, .272 P.S.F.	1.09	1.81	2.90
0250	.0215" thick, .334 P.S.F.	1.43	1.81	3.24

B3010 135	Formed Metal	COST PER S.F.		
		MAT.	INST.	TOTAL
1000	Batten seam, formed copper roofing, 3" min slope, 16 oz., 1.2 P.S.F.	13.35	6	19.35
1100	18 oz., 1.35 P.S.F.	14.85	6.60	21.45
2000	Zinc copper alloy, 3" min slope, .020" thick, .88 P.S.F.	12.85	5.50	18.35
3000	Flat seam, copper, 1/4" min. slope, 16 oz., 1.2 P.S.F.	9.75	5.50	15.25
3100	18 oz., 1.35 P.S.F.	10.85	5.75	16.60
5000	Standing seam, copper, 2-1/2" min. slope, 16 oz., 1.25 P.S.F.	10.60	5.10	15.70
5100	18 oz., 1.40 P.S.F.	11.85	5.50	17.35
6000	Zinc copper alloy, 2-1/2" min. slope, .020" thick, .87 P.S.F.	12.35	5.50	17.85
6100	.032" thick, 1.39 P.S.F.	19.60	6	25.60

B3010 Roof Coverings

Shingle and Tile

Shingles and tiles are practical in applications where the roof slope is more than 3-1/2″ per foot of rise. Lines 1100 through 6000 list the various materials and the weight per square foot.

Roof Edges

The table below lists the costs for various types of roof perimeter edge treatments.

Roof edge systems include the cost per L.F. for a 2″ x 8″ treated wood nailer fastened at 4′-0″ O.C. and a diagonally cut 4″ x 6″ treated wood cant.

Roof edge and base flashing are assumed to be made from the same material.

B3010 140	Shingle & Tile	COST PER S.F.		
		MAT.	INST.	TOTAL
1095	Asphalt roofing			
1100	Strip shingles, 4″ slope, inorganic class A 210-235 lb./sq.	.95	1.09	2.04
1150	Organic, class C, 235-240 lb./sq.	1.19	1.18	2.37
1200	Premium laminated multi-layered, class A, 260-300 lb./sq.	1.75	1.62	3.37
1545	Metal roofing			
1550	Alum., shingles, colors, 3″ min slope, .019″ thick, 0.4 PSF	2.76	1.20	3.96
1850	Steel, colors, 3″ min slope, 26 gauge, 1.0 PSF	4.02	2.45	6.47
2795	Slate roofing			
2800	Slate roofing, 4″ min. slope, shingles, 3/16″ thick, 8.0 PSF	5.35	3.05	8.40
3495	Wood roofing			
3500	4″ min slope, cedar shingles, 16″ x 5″, 5″ exposure 1.6 PSF	2.82	2.38	5.20
4000	Shakes, 18″, 8-1/2″ exposure, 2.8 PSF	2.33	2.94	5.27
5095	Tile roofing			
5100	Aluminum, mission, 3″ min slope, .019″ thick, 0.65 PSF	9.05	2.30	11.35
6000	Clay, Americana, 3″ minimum slope, 8 PSF	5.40	2.68	8.08

B3010 Roof Coverings

B3010 320	Roof Deck Rigid Insulation	COST PER S.F.		
		MAT.	INST.	TOTAL
0100	Fiberboard low density, 1/2" thick, R1.39			
0150	1" thick R2.78	.55	.70	1.25
0300	1 1/2" thick R4.17	.84	.70	1.54
0350	2" thick R5.56	1.11	.70	1.81
0370	Fiberboard high density, 1/2" thick R1.3	.29	.61	.90
0380	1" thick R2.5	.57	.70	1.27
0390	1 1/2" thick R3.8	.86	.70	1.56
0410	Fiberglass, 3/4" thick R2.78	.62	.61	1.23
0450	15/16" thick R3.70	.81	.61	1.42
0550	1-5/16" thick R5.26	1.39	.61	2
0650	2 7/16" thick R10	1.69	.70	2.39
1510	Polyisocyanurate 2#/CF density, 1" thick	.55	.49	1.04
1550	1 1/2" thick	.70	.53	1.23
1600	2" thick	.87	.57	1.44
1650	2 1/2" thick	1.11	.59	1.70
1700	3" thick	1.34	.61	1.95
1750	3 1/2" thick	2.15	.61	2.76
1800	Tapered for drainage	.63	.49	1.12
1810	Expanded polystyrene, 1#/CF density, 3/4" thick R2.89	.20	.47	.67
1820	2" thick R7.69	.53	.53	1.06
1830	Extruded polystyrene, 15 PSI compressive strength, 1" thick, R5	.53	.47	1
1835	2" thick R10	.68	.53	1.21
1840	3" thick R15	1.38	.61	1.99
2550	40 PSI compressive strength, 1" thick R5	.56	.47	1.03
2600	2" thick R10	1.06	.53	1.59
2650	3" thick R15	1.54	.61	2.15
2700	4" thick R20	2.04	.61	2.65
2750	Tapered for drainage	.80	.49	1.29
2810	60 PSI compressive strength, 1" thick R5	.78	.48	1.26
2850	2" thick R10	1.50	.54	2.04
2900	Tapered for drainage	1.02	.49	1.51
3050	Composites with 2" EPS			
3060	1" Fiberboard	1.46	.60	2.06
3070	7/16" oriented strand board	1.18	.70	1.88
3080	1/2" plywood	1.45	.70	2.15
3090	1" perlite	1.20	.70	1.90
4000	Composites with 1-1/2" polyisocyanurate			
4010	1" fiberboard	1.24	.70	1.94
4020	1" perlite	1.11	.68	1.79
4030	7/16" oriented strand board	.98	.70	1.68

B3010 420			Roof Edges			COST PER L.F.		
	EDGE TYPE	DESCRIPTION	SPECIFICATION	FACE HEIGHT		MAT.	INST.	TOTAL
1000	Aluminum	mill finish	.050" thick	4"		13.35	9.80	23.15
1100				6"		13.85	10.15	24
1300		duranodic	.050" thick	4"		13.90	9.80	23.70
1400				6"		14.40	10.15	24.55
1600		painted	.050" thick	4"		14.25	9.80	24.05
1700				6"		15.65	10.15	25.80
2000	Copper	plain	16 oz.	4"		49.50	9.80	59.30
2100				6"		49.50	10.15	59.65
2300			20 oz.	4"		64	11.30	75.30
2400				6"		56.50	10.90	67.40
2700	Sheet Metal	galvanized	20 Ga.	4"		16.20	11.75	27.95
2800				6"		16.20	11.75	27.95

B3010 Roof Coverings

B3010 420 — Roof Edges

	EDGE TYPE	DESCRIPTION	SPECIFICATION	FACE HEIGHT		COST PER L.F.		
						MAT.	INST.	TOTAL
3000	Sheet metal	galvanized	24 Ga.	4"		13.35	9.80	23.15
3100				6"		13.35	9.80	23.15

B3010 430 — Flashing

	MATERIAL	BACKING	SIDES	SPECIFICATION	QUANTITY	COST PER S.F.		
						MAT.	INST.	TOTAL
0040	Aluminum	none		.019"		1.18	3.55	4.73
0050				.032"		1.38	3.55	4.93
0300		fabric	2	.004"		1.55	1.56	3.11
0400		mastic		.004"		1.55	1.56	3.11
0700	Copper	none		16 oz.	<500 lbs.	8.35	4.48	12.83
0800				24 oz.	<500 lbs.	13.35	4.90	18.25
3500	PVC black	none		.010"		.23	1.81	2.04
3700				.030"		.39	1.81	2.20
4200	Neoprene			1/16"		2.45	1.81	4.26
4500	Stainless steel	none		.015"	<500 lbs.	6.40	4.48	10.88
4600	Copper clad				>2000 lbs.	6.10	3.32	9.42
5000	Plain			32 ga.		4.40	3.32	7.72

B3010   Roof Coverings

B3010 610 — Gutters

	SECTION	MATERIAL	THICKNESS	SIZE	FINISH	COST PER L.F. MAT.	INST.	TOTAL
0050	Box	aluminum	.027"	5"	enameled	2.84	5.40	8.24
0100					mill	2.91	5.40	8.31
0200			.032"	5"	enameled	3.24	5.40	8.64
0500		copper	16 Oz.	4"	lead coated	21	5.40	26.40
0600					mill	13.85	5.40	19.25
1000		steel galv.	28 Ga.	5"	enameled	2.22	5.40	7.62
1200			26 Ga.	5"	mill	2.40	5.40	7.80
1800		vinyl		4"	colors	1.28	5.05	6.33
1900				5"	colors	1.71	5.05	6.76
2300		hemlock or fir		4"x5"	treated	14.85	5.55	20.40
3000	Half round	copper	16 Oz.	4"	lead coated	15.20	5.40	20.60
3100					mill	10.75	5.40	16.15
3600		steel galv.	28 Ga.	5"	enameled	2.22	5.40	7.62
4102		stainless steel		5"	mill	9.80	5.40	15.20
5000		vinyl		4"	white	1.19	5.05	6.24

B3010 620 — Downspouts

	MATERIALS	SECTION	SIZE	FINISH	THICKNESS	COST PER V.L.F. MAT.	INST.	TOTAL
0100	Aluminum	rectangular	2"x3"	embossed mill	.020"	1.07	3.42	4.49
0150				enameled	.020"	1.45	3.42	4.87
0250			3"x4"	enameled	.024"	3.17	4.65	7.82
0300		round corrugated	3"	enameled	.020"	1.89	3.42	5.31
0350			4"	enameled	.025"	2.92	4.65	7.57
0500	Copper	rectangular corr.	2"x3"	mill	16 Oz.	9.85	3.42	13.27
0600		smooth		mill	16 Oz.	16.65	3.42	20.07
0700		rectangular corr.	3"x4"	mill	16 Oz.	12.95	4.49	17.44
1300	Steel		2"x3"	galvanized	28 Ga.	2.13	3.42	5.55
1350				epoxy coated	24 Ga.	2.42	3.42	5.84
1400		smooth		galvanized	28 Ga.	3.33	3.42	6.75
1450		rectangular corr.	3"x4"	galvanized	28 Ga.	2	4.49	6.49
1500				epoxy coated	24 Ga.	3.03	4.49	7.52
1550		smooth		galvanized	28 Ga.	3.37	4.49	7.86
1600		round corrugated	2"	galvanized	28 Ga.	2.29	3.42	5.71
1650			3"	galvanized	28 Ga.	2.29	3.42	5.71
1700			4"	galvanized	28 Ga.	2.87	4.49	7.36
1750			5"	galvanized	28 Ga.	4.04	5	9.04
2552	S.S. tubing sch.5	rectangular	3"x4"	mill		53	4.49	57.49

B3010 630 — Gravel Stop

	MATERIALS	SECTION	SIZE	FINISH	THICKNESS	COST PER L.F. MAT.	INST.	TOTAL
5100	Aluminum	extruded	4"	mill	.050"	6.60	4.49	11.09
5200			4"	duranodic	.050"	7.15	4.49	11.64
5300			8"	mill	.050"	8	5.20	13.20
5400			8"	duranodic	.050"	8.80	5.20	14
6000			12"-2 pc.	duranodic	.050"	11.05	6.50	17.55
6100	Stainless	formed	6"	mill	24 Ga.	10.80	4.82	15.62

B30 Roofing

B3020 Roof Openings

Roof Hatch

Smoke Hatch

Skylight

B3020 110	Skylights	COST PER S.F.		
		MAT.	INST.	TOTAL
5100	Skylights, plastic domes, insul curbs, nom. size to 10 S.F., single glaze	29	13.60	42.60
5200	Double glazing	33	16.75	49.75
5300	10 S.F. to 20 S.F., single glazing	28	5.50	33.50
5400	Double glazing	28	6.90	34.90
5500	20 S.F. to 30 S.F., single glazing	21.50	4.68	26.18
5600	Double glazing	26.50	5.50	32
5700	30 S.F. to 65 S.F., single glazing	19.70	3.56	23.26
5800	Double glazing	27	4.68	31.68
6000	Sandwich panels fiberglass, 1-9/16" thick, 2 S.F. to 10 S.F.	18.90	10.85	29.75
6100	10 S.F. to 18 S.F.	17.05	8.20	25.25
6200	2-3/4" thick, 25 S.F. to 40 S.F.	27.50	7.35	34.85
6300	40 S.F. to 70 S.F.	22.50	6.60	29.10

B3020 210	Hatches	COST PER OPNG.		
		MAT.	INST.	TOTAL
0200	Roof hatches with curb and 1" fiberglass insulation, 2'-6"x3'-0", aluminum	1,175	217	1,392
0300	Galvanized steel 165 lbs.	705	217	922
0400	Primed steel 164 lbs.	695	217	912
0500	2'-6"x4'-6" aluminum curb and cover, 150 lbs.	1,000	242	1,242
0600	Galvanized steel 220 lbs.	975	242	1,217
0650	Primed steel 218 lbs.	980	242	1,222
0800	2'x6"x8'-0" aluminum curb and cover, 260 lbs.	1,975	330	2,305
0900	Galvanized steel, 360 lbs.	1,975	330	2,305
0950	Primed steel 358 lbs.	1,575	330	1,905
1200	For plexiglass panels, add to the above	485		485
2100	Smoke hatches, unlabeled not incl. hand winch operator, 2'-6"x3', galv	855	263	1,118
2200	Plain steel, 160 lbs.	840	263	1,103
2400	2'-6"x8'-0",galvanized steel, 360 lbs.	2,150	360	2,510
2500	Plain steel, 350 lbs.	1,725	360	2,085
3000	4'-0"x8'-0", double leaf low profile, aluminum cover, 359 lb.	3,075	272	3,347
3100	Galvanized steel 475 lbs.	2,625	272	2,897
3200	High profile, aluminum cover, galvanized curb, 361 lbs.	2,950	272	3,222

C1010 Partitions

The Concrete Block Partition Systems are defined by weight and type of block, thickness, type of finish and number of sides finished. System components include joint reinforcing on alternate courses and vertical control joints.

C1010 102			Concrete Block Partitions - Regular Weight					
	TYPE	THICKNESS (IN.)	TYPE FINISH	SIDES FINISHED		COST PER S.F.		
						MAT.	INST.	TOTAL
1000	Hollow	4	none	0		1.85	6	7.85
1010			gyp. plaster 2 coat	1		2.23	8.75	10.98
1020				2		2.61	11.60	14.21
1200			portland - 3 coat	1		2.12	9.20	11.32
1400			5/8" drywall	1		2.40	7.90	10.30
1500		6	none	0		2.55	6.45	9
1510			gyp. plaster 2 coat	1		2.93	9.20	12.13
1520				2		3.31	12.05	15.36
1700			portland - 3 coat	1		2.82	9.65	12.47
1900			5/8" drywall	1		3.10	8.35	11.45
1910				2		3.65	10.20	13.85
2000		8	none	0		2.73	6.90	9.63
2010			gyp. plaster 2 coat	1		3.11	9.65	12.76
2020				2		3.49	12.45	15.94
2200			portland - 3 coat	1		3	10.10	13.10
2400			5/8" drywall	1		3.28	8.75	12.03
2410				2		3.83	10.65	14.48
2500		10	none	0		3.23	7.25	10.48
2510			gyp. plaster 2 coat	1		3.61	10	13.61
2520				2		3.99	12.80	16.79
2700			portland - 3 coat	1		3.50	10.45	13.95
2900			5/8" drywall	1		3.78	9.10	12.88
2910				2		4.33	11	15.33
3000	Solid	2	none	0		1.60	5.95	7.55
3010			gyp. plaster	1		1.98	8.70	10.68
3020				2		2.36	11.55	13.91
3200			portland - 3 coat	1		1.87	9.15	11.02
3400			5/8" drywall	1		2.15	7.85	10
3410				2		2.70	9.70	12.40
3500		4	none	0		2.11	6.20	8.31
3510			gyp. plaster	1		2.62	9	11.62
3520				2		2.87	11.80	14.67
3700			portland - 3 coat	1		2.38	9.40	11.78
3900			5/8" drywall	1		2.66	8.10	10.76
3910				2		3.21	9.95	13.16

C10 Interior Construction

C1010 Partitions

C1010 102 — Concrete Block Partitions - Regular Weight

	TYPE	THICKNESS (IN.)	TYPE FINISH	SIDES FINISHED		COST PER S.F. MAT.	INST.	TOTAL
4000	Solid	6	none	0		2.15	6.70	8.85
4010			gyp. plaster	1		2.53	9.45	11.98
4020				2		2.91	12.30	15.21
4200			portland - 3 coat	1		2.42	9.90	12.32
4400			5/8" drywall	1		2.70	8.60	11.30
4410				2		3.25	10.45	13.70

C1010 104 — Concrete Block Partitions - Lightweight

	TYPE	THICKNESS (IN.)	TYPE FINISH	SIDES FINISHED		COST PER S.F. MAT.	INST.	TOTAL
5000	Hollow	4	none	0		2	5.90	7.90
5010			gyp. plaster	1		2.38	8.65	11.03
5020				2		2.76	11.50	14.26
5200			portland - 3 coat	1		2.27	9.10	11.37
5400			5/8" drywall	1		2.55	7.80	10.35
5410				2		3.10	9.65	12.75
5500		6	none	0		2.81	6.30	9.11
5520			gyp. plaster	2		3.57	11.90	15.47
5700			portland - 3 coat	1		3.08	9.50	12.58
5900			5/8" drywall	1		3.36	8.20	11.56
5910				2		3.91	10.05	13.96
6000		8	none	0		3.38	6.75	10.13
6010			gyp. plaster	1		3.76	9.50	13.26
6020				2		4.14	12.30	16.44
6200			portland - 3 coat	1		3.65	9.95	13.60
6400			5/8" drywall	1		3.93	8.60	12.53
6410				2		4.48	10.50	14.98
6500		10	none	0		4.02	7.05	11.07
6510			gyp. plaster	1		4.40	9.80	14.20
6520				2		4.78	12.60	17.38
6700			portland - 3 coat	1		4.29	10.25	14.54
6900			5/8" drywall	1		4.57	8.90	13.47
6910				2		5.10	10.80	15.90
7000	Solid	4	none	0		1.97	6.15	8.12
7010			gyp. plaster	1		2.35	8.90	11.25
7020				2		2.73	11.75	14.48
7200			portland - 3 coat	1		2.24	9.35	11.59
7400			5/8" drywall	1		2.52	8.05	10.57
7410				2		3.07	9.90	12.97
7500		6	none	0		3.10	6.60	9.70
7510			gyp. plaster	1		3.59	9.45	13.04
7520				2		3.86	12.20	16.06
7700			portland - 3 coat	1		3.37	9.80	13.17
7900			5/8" drywall	1		3.65	8.50	12.15
7910				2		4.20	10.35	14.55
8000		8	none	0		3.92	7.10	11.02
8010			gyp. plaster	1		4.30	9.85	14.15
8020				2		4.68	12.65	17.33
8200			portland - 3 coat	1		4.19	10.30	14.49
8400			5/8" drywall	1		4.47	8.95	13.42
8410				2		5	10.85	15.85

C1010 Partitions

Single Wythe

Double Wythe

C1010 120	Tile Partitions	COST PER S.F.		
		MAT.	INST.	TOTAL
1000	8W series 8"x16", 4" thick wall, reinf every 2 courses, glazed 1 side	14.80	7.25	22.05
1100	Glazed 2 sides	18.30	7.70	26
1200	Glazed 2 sides, using 2 wythes of 2" thick tile	22	13.90	35.90
1300	6" thick wall, horizontal reinf every 2 courses, glazed 1 side	25.50	7.55	33.05
1400	Glazed 2 sides, each face different color, 2" and 4" tile	26	14.20	40.20
1500	8" thick wall, glazed 2 sides using 2 wythes of 4" thick tile	29.50	14.50	44
1600	10" thick wall, glazed 2 sides using 1 wythe of 4" tile & 1 wythe of 6" tile	40.50	14.80	55.30
1700	Glazed 2 sides cavity wall, using 2 wythes of 4" thick tile	29.50	14.50	44
1800	12" thick wall, glazed 2 sides using 2 wythes of 6" thick tile	51	15.10	66.10
1900	Glazed 2 sides cavity wall, using 2 wythes of 4" thick tile	29.50	14.50	44
2100	6T series 5-1/3"x12" tile, 4" thick, non load bearing glazed one side,	13	11.35	24.35
2200	Glazed two sides	17.10	12.80	29.90
2300	Glazed two sides, using two wythes of 2" thick tile	19.10	22	41.10
2400	6" thick, glazed one side	19.40	11.90	31.30
2500	Glazed two sides	23	13.50	36.50
2600	Glazed two sides using 2" thick tile and 4" thick tile	22.50	22.50	45
2700	8" thick, glazed one side	25.50	13.90	39.40
2800	Glazed two sides using two wythes of 4" thick tile	26	22.50	48.50
2900	Glazed two sides using 6" thick tile and 2" thick tile	29	23	52
3000	10" thick cavity wall, glazed two sides using two wythes of 4" tile	26	22.50	48.50
3100	12" thick, glazed two sides using 4" thick tile and 8" thick tile	38.50	25.50	64
3200	2" thick facing tile, glazed one side, on 6" concrete block	13.45	13.20	26.65
3300	On 8" concrete block	13.60	13.60	27.20
3400	On 10" concrete block	14.10	13.90	28

C10 Interior Construction

C1010 Partitions

Wood Stud Framing

Metal Stud Framing

The Drywall Partitions/Stud Framing Systems are defined by type of drywall and number of layers, type and spacing of stud framing, and treatment on the opposite face. Components include taping and finishing.

Cost differences between regular and fire resistant drywall are negligible, and terminology is interchangeable. In some cases fiberglass insulation is included for additional sound deadening.

C1010 124		Drywall Partitions/Wood Stud Framing						
	FACE LAYER	BASE LAYER	FRAMING	OPPOSITE FACE	INSULATION	COST PER S.F.		
						MAT.	INST.	TOTAL
1200	5/8" FR drywall	none	2 x 4, @ 16" O.C.	same	0	1.16	3.31	4.47
1250				5/8" reg. drywall	0	1.14	3.31	4.45
1300				nothing	0	.79	2.21	3
1400		1/4" SD gypsum	2 x 4 @ 16" O.C.	same	1-1/2" fiberglass	2.41	5.10	7.51
1450				5/8" FR drywall	1-1/2" fiberglass	1.99	4.47	6.46
1500				nothing	1-1/2" fiberglass	1.62	3.37	4.99
1600		resil. channels	2 x 4 @ 16", O.C.	same	1-1/2" fiberglass	1.96	6.45	8.41
1650				5/8" FR drywall	1-1/2" fiberglass	1.77	5.15	6.92
1700				nothing	1-1/2" fiberglass	1.40	4.06	5.46
1800		5/8" FR drywall	2 x 4 @ 24" O.C.	same	0	1.70	4.19	5.89
1850				5/8" FR drywall	0	1.38	3.64	5.02
1900				nothing	0	1.01	2.54	3.55
2200			2 rows-2 x 4 16"O.C.	same	2" fiberglass	2.66	6.05	8.71
2250				5/8" FR drywall	2" fiberglass	2.34	5.50	7.84
2300				nothing	2" fiberglass	1.97	4.41	6.38
2400	5/8" WR drywall	none	2 x 4, @ 16" O.C.	same	0	1.44	3.31	4.75
2450				5/8" FR drywall	0	1.30	3.31	4.61
2500				nothing	0	.93	2.21	3.14
2600		5/8" FR drywall	2 x 4, @ 24" O.C.	same	0	1.98	4.19	6.17
2650				5/8" FR drywall	0	1.52	3.64	5.16
2700				nothing	0	1.15	2.54	3.69
2800	5/8" VF drywall	none	2 x 4, @ 16" O.C.	same	0	2.34	3.57	5.91
2850				5/8" FR drywall	0	1.75	3.44	5.19
2900				nothing	0	1.38	2.34	3.72
3000		5/8" FR drywall	2 x 4 , 24" O.C.	same	0	2.88	4.45	7.33
3050				5/8" FR drywall	0	1.97	3.77	5.74
3100				nothing	0	1.60	2.67	4.27
3200	1/2" reg drywall	3/8" reg drywall	2 x 4, @ 16" O.C.	same	0	1.54	4.41	5.95
3250				5/8" FR drywall	0	1.35	3.86	5.21
3300				nothing	0	.98	2.76	3.74

C1010 Partitions

C1010 126	Drywall Partitions/Metal Stud Framing

	FACE LAYER	BASE LAYER	FRAMING	OPPOSITE FACE	INSULATION	COST PER S.F.		
						MAT.	INST.	TOTAL
5200	5/8" FR drywall	none	1-5/8" @ 24" O.C.	same	0	.94	2.93	3.87
5250				5/8" reg. drywall	0	.92	2.93	3.85
5300				nothing	0	.57	1.83	2.40
5400			3-5/8" @ 24" O.C.	same	0	1.02	2.95	3.97
5450				5/8" reg. drywall	0	1	2.95	3.95
5500				nothing	0	.65	1.85	2.50
5600		1/4" SD gypsum	1-5/8" @ 24" O.C.	same	0	1.78	4.15	5.93
5650				5/8" FR drywall	0	1.36	3.54	4.90
5700				nothing	0	.99	2.44	3.43
5800			2-1/2" @ 24" O.C.	same	0	1.83	4.16	5.99
5850				5/8" FR drywall	0	1.41	3.55	4.96
5900				nothing	0	1.04	2.45	3.49
6000		5/8" FR drywall	2-1/2" @ 16" O.C.	same	0	1.86	4.72	6.58
6050				5/8" FR drywall	0	1.54	4.17	5.71
6100				nothing	0	1.17	3.07	4.24
6200			3-5/8" @ 24" O.C.	same	0	1.66	4.05	5.71
6250				5/8"FR drywall	3-1/2" fiberglass	1.75	3.91	5.66
6300				nothing	0	.97	2.40	3.37
6400	5/8" WR drywall	none	1-5/8" @ 24" O.C.	same	0	1.22	2.93	4.15
6450				5/8" FR drywall	0	1.08	2.93	4.01
6500				nothing	0	.71	1.83	2.54
6600			3-5/8" @ 24" O.C.	same	0	1.30	2.95	4.25
6650				5/8" FR drywall	0	1.16	2.95	4.11
6700				nothing	0	.79	1.85	2.64
6800		5/8" FR drywall	2-1/2" @ 16" O.C.	same	0	2.14	4.72	6.86
6850				5/8" FR drywall	0	1.68	4.17	5.85
6900				nothing	0	1.31	3.07	4.38
7000			3-5/8" @ 24" O.C.	same	0	1.94	4.05	5.99
7050				5/8"FR drywall	3-1/2" fiberglass	1.89	3.91	5.80
7100				nothing	0	1.11	2.40	3.51
7200	5/8" VF drywall	none	1-5/8" @ 24" O.C.	same	0	2.12	3.19	5.31
7250				5/8" FR drywall	0	1.53	3.06	4.59
7300				nothing	0	1.16	1.96	3.12
7400			3-5/8" @ 24" O.C.	same	0	2.20	3.21	5.41
7450				5/8" FR drywall	0	1.61	3.08	4.69
7500				nothing	0	1.24	1.98	3.22
7600		5/8" FR drywall	2-1/2" @ 16" O.C.	same	0	3.04	4.98	8.02
7650				5/8" FR drywall	0	2.13	4.30	6.43
7700				nothing	0	1.76	3.20	4.96
7800			3-5/8" @ 24" O.C.	same	0	2.84	4.31	7.15
7850				5/8"FR drywall	3-1/2" fiberglass	2.34	4.04	6.38
7900				nothing	0	1.56	2.53	4.09

C1010 Partitions

C1010 128	Drywall Components	COST PER S.F.		
		MAT.	INST.	TOTAL
0060	Metal studs, 24" O.C. including track, load bearing, 20 gage, 2-1/2"	.66	1.04	1.70
0080	3-5/8"	.78	1.06	1.84
0100	4"	.70	1.08	1.78
0120	6"	1.04	1.10	2.14
0140	Metal studs, 24" O.C. including track, load bearing, 18 gage, 2-1/2"	.66	1.04	1.70
0160	3-5/8"	.78	1.06	1.84
0180	4"	.70	1.08	1.78
0200	6"	1.04	1.10	2.14
0220	16 gage, 2-1/2"	.77	1.18	1.95
0240	3-5/8"	.91	1.21	2.12
0260	4"	.96	1.23	2.19
0280	6"	1.20	1.26	2.46
0300	Non load bearing, 25 gage, 1-5/8"	.20	.73	.93
0340	3-5/8"	.28	.75	1.03
0360	4"	.32	.75	1.07
0380	6"	.40	.76	1.16
0400	20 gage, 2-1/2"	.35	.92	1.27
0420	3-5/8"	.40	.94	1.34
0440	4"	.47	.94	1.41
0460	6"	.57	.95	1.52
0540	Wood studs including blocking, shoe and double top plate, 2"x4", 12" O.C.	.52	1.39	1.91
0560	16" O.C.	.42	1.11	1.53
0580	24" O.C.	.32	.89	1.21
0600	2"x6", 12" O.C.	.80	1.58	2.38
0620	16" O.C.	.65	1.23	1.88
0640	24" O.C.	.49	.96	1.45
0642	Furring one side only, steel channels, 3/4", 12" O.C.	.39	1.98	2.37
0644	16" O.C.	.35	1.76	2.11
0646	24" O.C.	.23	1.33	1.56
0647	1-1/2", 12" O.C.	.49	2.22	2.71
0648	16" O.C.	.44	1.94	2.38
0649	24" O.C.	.29	1.53	1.82
0650	Wood strips, 1" x 3", on wood, 12" O.C.	.38	1.01	1.39
0651	16" O.C.	.29	.76	1.05
0652	On masonry, 12" O.C.	.38	1.12	1.50
0653	16" O.C.	.29	.84	1.13
0654	On concrete, 12" O.C.	.38	2.13	2.51
0655	16" O.C.	.29	1.60	1.89
0665	Gypsum board, one face only, exterior sheathing, 1/2"	.45	.98	1.43
0680	Interior, fire resistant, 1/2"	.30	.55	.85
0700	5/8"	.32	.55	.87
0720	Sound deadening board 1/4"	.42	.61	1.03
0740	Standard drywall 3/8"	.25	.55	.80
0760	1/2"	.26	.55	.81
0780	5/8"	.30	.55	.85
0800	Tongue & groove coreboard 1"	.79	2.30	3.09
0820	Water resistant, 1/2"	.40	.55	.95
0840	5/8"	.46	.55	1.01
0860	Add for the following:, foil backing	.15		.15
0880	Fiberglass insulation, 3-1/2"	.41	.41	.82
0900	6"	.48	.41	.89
0920	Rigid insulation 1"	.53	.55	1.08
0940	Resilient furring @ 16" O.C.	.26	1.74	2
0960	Taping and finishing	.05	.55	.60
0980	Texture spray	.04	.58	.62
1000	Thin coat plaster	.12	.69	.81
1050	Sound wall framing, 2x6 plates, 2x4 staggered studs, 12" O.C.	.63	1.35	1.98

C1010 Partitions

Gypsum plaster and gypsum
lath on metal studs.

Gypsum plaster and diamond
metal lath on metal studs.

Gypsum plaster and gypsum
lath on wood studs.

Gypsum plaster and diamond
metal lath on wood studs.

C1010 140				Plaster Partitions/Metal Stud Framing				
	TYPE	FRAMING	LATH	OPPOSITE FACE		COST PER S.F.		
						MAT.	INST.	TOTAL
1000	2 coat gypsum	2-1/2" @ 16"O.C.	3/8" gypsum	same		2.55	8.05	10.60
1010				nothing		1.50	4.66	6.16
1100		3-1/4" @ 24"O.C.	1/2" gypsum	same		2.56	7.75	10.31
1110				nothing		1.47	4.33	5.80
1500	2 coat vermiculite	2-1/2" @ 16"O.C.	3/8" gypsum	same		2.41	8.75	11.16
1510				nothing		1.42	5	6.42
1600		3-1/4" @ 24"O.C.	1/2" gypsum	same		2.42	8.45	10.87
1610				nothing		1.39	4.68	6.07
2000	3 coat gypsum	2-1/2" @ 16"O.C.	3/8" gypsum	same		2.38	9.05	11.43
2010				nothing		1.41	5.20	6.61
2020			3.4 lb. diamond	same		2.44	9.05	11.49
2030				nothing		1.44	5.20	6.64
2040			2.75 lb. ribbed	same		2.07	9.05	11.12
2050				nothing		1.25	5.20	6.45
2100		3-1/4" @ 24"O.C.	1/2" gypsum	same		2.39	8.80	11.19
2110				nothing		1.38	4.86	6.24
2120			3.4 lb. ribbed	same		2.47	8.80	11.27
2130				nothing		1.42	4.86	6.28
3500	3 coat gypsum	2-1/2" @ 16"O.C.	3/8" gypsum	same		2.88	11.65	14.53
3510	W/med. Keenes			nothing		1.66	6.50	8.16
3520			3.4 lb. diamond	same		2.94	11.65	14.59
3530				nothing		1.69	6.50	8.19
3540			2.75 lb. ribbed	same		2.57	11.65	14.22
3550				nothing		1.50	6.50	8
3600		3-1/4" @ 24"O.C.	1/2" gypsum	same		2.89	11.35	14.24
3610				nothing		1.63	6.20	7.83
3620			3.4 lb. ribbed	same		2.97	11.35	14.32
3630				nothing		1.67	6.20	7.87

C10 Interior Construction

C1010 Partitions

| C1010 140 | Plaster Partitions/Metal Stud Framing | | | | | | | |

	TYPE	FRAMING	LATH	OPPOSITE FACE		COST PER S.F.		
						MAT.	INST.	TOTAL
4000	3 coat gypsum	2-1/2" @ 16"O.C.	3/8" gypsum	same		2.89	12.95	15.84
4010	W/hard Keenes			nothing		1.66	7.10	8.76
4020			3.4 lb. diamond	same		2.95	12.95	15.90
4030				nothing		1.69	7.10	8.79
4040			2.75 lb. ribbed	same		2.58	12.95	15.53
4050				nothing		1.50	7.10	8.60
4100		3-1/4" @ 24"O.C.	1/2" gypsum	same		2.90	12.65	15.55
4110				nothing		1.63	6.75	8.38
4120			3.4 lb. ribbed	same		2.98	12.65	15.63
4130				nothing		1.67	6.75	8.42

| C1010 142 | Plaster Partitions/Wood Stud Framing | | | | | | | |

	TYPE	FRAMING	LATH	OPPOSITE FACE		COST PER S.F.		
						MAT.	INST.	TOTAL
5000	2 coat gypsum	2"x4" @ 16"O.C.	3/8" gypsum	same		2.63	7.85	10.48
5010				nothing		1.58	4.56	6.14
5100		2"x4" @ 24"O.C.	1/2" gypsum	same		2.60	7.70	10.30
5110				nothing		1.51	4.38	5.89
5500	2 coat vermiculite	2"x4" @ 16"O.C.	3/8" gypsum	same		2.49	8.55	11.04
5510				nothing		1.50	4.91	6.41
5600		2"x4" @ 24"O.C.	1/2" gypsum	same		2.46	8.40	10.86
5610				nothing		1.43	4.73	6.16
6000	3 coat gypsum	2"x4" @ 16"O.C.	3/8" gypsum	same		2.46	8.90	11.36
6010				nothing		1.49	5.10	6.59
6020			3.4 lb. diamond	same		2.52	8.95	11.47
6030				nothing		1.52	5.15	6.67
6040			2.75 lb. ribbed	same		2.13	9	11.13
6050				nothing		1.32	5.15	6.47
6100		2"x4" @ 24"O.C.	1/2" gypsum	same		2.43	8.75	11.18
6110				nothing		1.42	4.91	6.33
6120			3.4 lb. ribbed	same		2.05	8.85	10.90
6130				nothing		1.23	4.95	6.18
7500	3 coat gypsum	2"x4" @ 16"O.C.	3/8" gypsum	same		2.96	11.45	14.41
7510	W/med Keenes			nothing		1.74	6.40	8.14
7520			3.4 lb. diamond	same		3.02	11.55	14.57
7530				nothing		1.77	6.45	8.22
7540			2.75 lb. ribbed	same		2.63	11.60	14.23
7550				nothing		1.57	6.45	8.02
7600		2"x4" @ 24"O.C.	1/2" gypsum	same		2.93	11.35	14.28
7610				nothing		1.67	6.25	7.92
7620			3.4 lb. ribbed	same		3.01	11.50	14.51
7630				nothing		1.71	6.30	8.01
8000	3 coat gypsum	2"x4" @ 16"O.C.	3/8" gypsum	same		2.97	12.75	15.72
8010	W/hard Keenes			nothing		1.74	7	8.74
8020			3.4 lb. diamond	same		3.03	12.85	15.88
8030				nothing		1.77	7.05	8.82
8040			2.75 lb. ribbed	same		2.64	12.90	15.54
8050				nothing		1.57	7.05	8.62
8100		2"x4" @ 24"O.C.	1/2" gypsum	same		2.94	12.65	15.59
8110				nothing		1.67	6.80	8.47
8120			3.4 lb. ribbed	same		3.02	12.80	15.82
8130				nothing		1.71	6.90	8.61

C1010 Partitions

C1010 144	Plaster Partition Components	COST PER S.F.		
		MAT.	INST.	TOTAL
0060	Metal studs, 16" O.C., including track, non load bearing, 25 gage, 1-5/8"	.34	1.13	1.47
0080	2-1/2"	.34	1.13	1.47
0100	3-1/4"	.38	1.15	1.53
0120	3-5/8"	.38	1.15	1.53
0140	4"	.43	1.16	1.59
0160	6"	.54	1.18	1.72
0180	Load bearing, 20 gage, 2-1/2"	.89	1.44	2.33
0200	3-5/8"	1.05	1.46	2.51
0220	4"	1.11	1.50	2.61
0240	6"	1.40	1.52	2.92
0260	16 gage 2-1/2"	1.05	1.63	2.68
0280	3-5/8"	1.26	1.68	2.94
0300	4"	1.32	1.70	3.02
0320	6"	1.65	1.73	3.38
0340	Wood studs, including blocking, shoe and double plate, 2"x4" , 12" O.C.	.52	1.39	1.91
0360	16" O.C.	.42	1.11	1.53
0380	24" O.C.	.32	.89	1.21
0400	2"x6" , 12" O.C.	.80	1.58	2.38
0420	16" O.C.	.65	1.23	1.88
0440	24" O.C.	.49	.96	1.45
0460	Furring one face only, steel channels, 3/4" , 12" O.C.	.39	1.98	2.37
0480	16" O.C.	.35	1.76	2.11
0500	24" O.C.	.23	1.33	1.56
0520	1-1/2" , 12" O.C.	.49	2.22	2.71
0540	16" O.C.	.44	1.94	2.38
0560	24"O.C.	.29	1.53	1.82
0580	Wood strips 1"x3", on wood., 12" O.C.	.38	1.01	1.39
0600	16"O.C.	.29	.76	1.05
0620	On masonry, 12" O.C.	.38	1.12	1.50
0640	16" O.C.	.29	.84	1.13
0660	On concrete, 12" O.C.	.38	2.13	2.51
0680	16" O.C.	.29	1.60	1.89
0700	Gypsum lath. plain or perforated, nailed to studs, 3/8" thick	.52	.61	1.13
0720	1/2" thick	.55	.65	1.20
0740	Clipped to studs, 3/8" thick	.52	.69	1.21
0760	1/2" thick	.55	.74	1.29
0780	Metal lath, diamond painted, nailed to wood studs, 2.5 lb.	.47	.61	1.08
0800	3.4 lb.	.55	.65	1.20
0820	Screwed to steel studs, 2.5 lb.	.47	.65	1.12
0840	3.4 lb.	.46	.69	1.15
0860	Rib painted, wired to steel, 2.75 lb	.36	.69	1.05
0880	3.4 lb	.59	.74	1.33
0900	4.0 lb	.69	.80	1.49
0910				
0920	Gypsum plaster, 2 coats	.42	2.67	3.09
0940	3 coats	.59	3.20	3.79
0960	Perlite or vermiculite plaster, 2 coats	.47	3.02	3.49
0980	3 coats	.77	3.79	4.56
1000	Stucco, 3 coats, 1" thick, on wood framing	.73	5.30	6.03
1020	On masonry	.28	4.15	4.43
1100	Metal base galvanized and painted 2-1/2" high	.66	1.94	2.60

C10 Interior Construction

C1010 Partitions

Folding Accordion

Folding Leaf

Movable and Borrow Lites

C1010 205	Partitions	COST PER S.F.		
		MAT.	**INST.**	**TOTAL**
0360	Folding accordion, vinyl covered, acoustical, 3 lb. S.F., 17 ft max. hgt	30	11.05	41.05
0380	5 lb. per S.F. 27 ft max height	42	11.65	53.65
0400	5.5 lb. per S.F., 17 ft. max height	49	12.30	61.30
0420	Commercial, 1.75 lb per S.F., 8 ft. max height	27.50	4.92	32.42
0440	2.0 Lb per S.F., 17 ft. max height	28.50	7.40	35.90
0460	Industrial, 4.0 lb. per S.F. 27 ft. max height	42.50	14.75	57.25
0480	Vinyl clad wood or steel, electric operation 6 psf	60	6.90	66.90
0500	Wood, non acoustic, birch or mahogany	32	3.69	35.69
0560	Folding leaf, aluminum framed acoustical 12 ft.high., 5.5 lb per S.F., min.	42.50	18.45	60.95
0580	Maximum	51.50	37	88.50
0600	6.5 lb. per S.F., minimum	45	18.45	63.45
0620	Maximum	55.50	37	92.50
0640	Steel acoustical, 7.5 per S.F., vinyl faced, minimum	63.50	18.45	81.95
0660	Maximum	77.50	37	114.50
0680	Wood acoustic type, vinyl faced to 18' high 6 psf, minimum	59.50	18.45	77.95
0700	Average	71	24.50	95.50
0720	Maximum	92	37	129
0740	Formica or hardwood faced, minimum	61.50	18.45	79.95
0760	Maximum	65.50	37	102.50
0780	Wood, low acoustical type to 12 ft. high 4.5 psf	44.50	22	66.50
0840	Demountable, trackless wall, cork finish, semi acous, 1-5/8" th, min	40.50	3.40	43.90
0860	Maximum	45	5.80	50.80
0880	Acoustic, 2" thick, minimum	38.50	3.63	42.13
0900	Maximum	59	4.92	63.92
0920	In-plant modular office system, w/prehung steel door			
0940	3" thick honeycomb core panels			
0960	12' x 12', 2 wall	13.60	.69	14.29
0970	4 wall	15.55	.92	16.47
0980	16' x 16', 2 wall	13.60	.49	14.09
0990	4 wall	9.55	.49	10.04
1000	Gypsum, demountable, 3" to 3-3/4" thick x 9' high, vinyl clad	7.05	2.55	9.60
1020	Fabric clad	17.55	2.78	20.33
1040	1.75 system, vinyl clad hardboard, paper honeycomb core panel			
1060	1-3/4" to 2-1/2" thick x 9' high	11.90	2.55	14.45
1080	Unitized gypsum panel system, 2" to 2-1/2" thick x 9' high			
1100	Vinyl clad gypsum	15.20	2.55	17.75
1120	Fabric clad gypsum	25	2.78	27.78
1140	Movable steel walls, modular system			
1160	Unitized panels, 48" wide x 9' high			
1180	Baked enamel, pre-finished	17.20	2.05	19.25
1200	Fabric clad	25	2.19	27.19

C10 Interior Construction

C1020 Interior Doors

Single Leaf **Sliding Entrance** **Rolling Overhead**

C1020 102	Special Doors	MAT.	INST.	TOTAL
		COST PER OPNG.		
2500	Single leaf, wood, 3'-0"x7'-0"x1 3/8", birch, solid core	415	207	622
2510	Hollow core	345	198	543
2530	Hollow core, lauan	335	198	533
2540	Louvered pine	480	198	678
2550	Paneled pine	515	198	713
2600	Hollow metal, comm. quality, flush, 3'-0"x7'-0"x1-3/8"	995	217	1,212
2650	3'-0"x10'-0" openings with panel	1,525	281	1,806
2700	Metal fire, comm. quality, 3'-0"x7'-0"x1-3/8"	1,150	226	1,376
2800	Kalamein fire, comm. quality, 3'-0"x7'-0"x1-3/4"	810	415	1,225
3200	Double leaf, wood, hollow core, 2 - 3'-0"x7'-0"x1-3/8"	590	380	970
3300	Hollow metal, comm. quality, B label, 2'-3'-0"x7'-0"x1-3/8"	2,175	475	2,650
3400	6'-0"x10'-0" opening, with panel	2,975	585	3,560
3500	Double swing door system, 12'-0"x7'-0", mill finish	8,825	3,125	11,950
3700	Black finish	8,950	3,225	12,175
3800	Sliding entrance door and system mill finish	10,300	2,650	12,950
3900	Bronze finish	11,300	2,875	14,175
4000	Black finish	11,800	2,975	14,775
4100	Sliding panel mall front, 16'x9' opening, mill finish	3,650	805	4,455
4200	Bronze finish	4,750	1,050	5,800
4300	Black finish	5,850	1,300	7,150
4400	24'x9' opening mill finish	5,175	1,500	6,675
4500	Bronze finish	6,725	1,950	8,675
4600	Black finish	8,275	2,400	10,675
4700	48'x9' opening mill finish	9,650	1,175	10,825
4800	Bronze finish	12,500	1,525	14,025
4900	Black finish	15,400	1,875	17,275
5000	Rolling overhead steel door, manual, 8' x 8' high	1,225	895	2,120
5100	10' x 10' high	2,025	1,025	3,050
5200	20' x 10' high	3,375	1,425	4,800
5300	12' x 12' high	2,050	1,200	3,250
5400	Motor operated, 8' x 8' high	2,400	1,175	3,575
5500	10' x 10' high	3,200	1,300	4,500
5600	20' x 10' high	4,550	1,700	6,250
5700	12' x 12' high	3,225	1,475	4,700
5800	Roll up grille, aluminum, manual, 10' x 10' high, mill finish	3,050	1,750	4,800
5900	Bronze anodized	4,900	1,750	6,650
6000	Motor operated, 10' x 10' high, mill finish	4,425	2,025	6,450
6100	Bronze anodized	6,275	2,025	8,300
6200	Steel, manual, 10' x 10' high	2,750	1,425	4,175
6300	15' x 8' high	3,175	1,775	4,950
6400	Motor operated, 10' x 10' high	4,125	1,700	5,825
6500	15' x 8' high	4,550	2,050	6,600

C10 Interior Construction

C1030 Fittings

Toilet Units

Entrance Screens

Urinal Screens

C1030 110	Toilet Partitions	COST PER UNIT		
		MAT.	INST.	TOTAL
0380	Toilet partitions, cubicles, ceiling hung, marble	1,775	500	2,275
0400	Painted metal	565	277	842
0420	Plastic laminate	605	277	882
0460	Stainless steel	1,100	277	1,377
0480	Handicap addition	475		475
0520	Floor and ceiling anchored, marble	2,250	400	2,650
0540	Painted metal	645	221	866
0560	Plastic laminate	860	221	1,081
0600	Stainless steel	1,225	221	1,446
0620	Handicap addition	345		345
0660	Floor mounted marble	1,325	335	1,660
0680	Painted metal	640	158	798
0700	Plastic laminate	535	158	693
0740	Stainless steel	1,525	158	1,683
0760	Handicap addition	345		345
0780	Juvenile deduction	45.50		45.50
0820	Floor mounted with headrail marble	1,250	335	1,585
0840	Painted metal	425	184	609
0860	Plastic laminate	825	184	1,009
0900	Stainless steel	1,150	184	1,334
0920	Handicap addition	410		410
0960	Wall hung, painted metal	700	158	858
1020	Stainless steel	1,825	158	1,983
1040	Handicap addition	410		410
1080	Entrance screens, floor mounted, 54" high, marble	830	112	942
1100	Painted metal	266	74	340
1140	Stainless steel	1,000	74	1,074
1300	Urinal screens, floor mounted, 24" wide, laminated plastic	198	138	336
1320	Marble	700	163	863
1340	Painted metal	253	138	391
1380	Stainless steel	655	138	793
1428	Wall mounted wedge type, painted metal	148	111	259
1460	Stainless steel	635	111	746
1500	Partitions, shower stall, single wall, painted steel, 2'-8" x 2'-8"	1,125	260	1,385
1510	Fiberglass, 2'-8" x 2'-8"	875	289	1,164
1520	Double wall, enameled steel, 2'-8" x 2'-8"	1,125	260	1,385
1530	Stainless steel, 2'-8" x 2'-8"	2,500	260	2,760
1560	Tub enclosure, sliding panels, tempered glass, aluminum frame	450	325	775
1570	Chrome/brass frame-deluxe	1,325	435	1,760

C10 Interior Construction

C1030 Fittings

C1030 310	Storage Specialties, EACH	COST EACH		
		MAT.	INST.	TOTAL
0200	Lockers, steel, single tier, 5' to 6' high, per opening, min.	171	46.50	217.50
0210	Maximum	300	54	354
0600	Shelving, metal industrial, braced, 3' wide, 1' deep	24.50	12.25	36.75
0610	2' deep	31.50	12.95	44.45

C1030 510	Identifying/Visual Aid Specialties, EACH	COST EACH		
		MAT.	INST.	TOTAL
0100	Control boards, magnetic, porcelain finish, framed, 24" x 18"	219	138	357
0110	96" x 48"	1,175	221	1,396
0120	Directory boards, outdoor, black plastic, 36" x 24"	840	555	1,395
0130	36" x 36"	970	740	1,710
0140	Indoor, economy, open faced, 18" x 24"	168	158	326
0500	Signs, interior electric exit sign, wall mounted, 6"	92	78.50	170.50
0510	Street, reflective alum., dbl. face, 4 way, w/bracket	254	37	291
0520	Letters, cast aluminum, 1/2" deep, 4" high	29	30.50	59.50
0530	1" deep, 10" high	61.50	30.50	92
0540	Plaques, cast aluminum, 20" x 30"	1,500	277	1,777
0550	Cast bronze, 36" x 48"	5,075	555	5,630

C1030 520	Identifying/Visual Aid Specialties, S.F.	COST PER S.F.		
		MAT.	INST.	TOTAL
0100	Bulletin board, cork sheets, no frame, 1/4" thick	2.28	3.82	6.10
0120	Aluminum frame, 1/4" thick, 3' x 5'	8.90	4.62	13.52
0200	Chalkboards, wall hung, alum. frame & chalktrough	14.40	2.45	16.85
0210	Wood frame & chalktrough	12.10	2.64	14.74
0220	Sliding board, one board with back panel	53.50	2.37	55.87
0230	Two boards with back panel	89	2.37	91.37
0240	Liquid chalk type, alum. frame & chalktrough	12.55	2.45	15
0250	Wood frame & chalktrough	33.50	2.45	35.95

C1030 710	Bath and Toilet Accessories, EACH	COST EACH		
		MAT.	INST.	TOTAL
0100	Specialties, bathroom accessories, st. steel, curtain rod, 5' long, 1" diam.	37	42.50	79.50
0120	Dispenser, towel, surface mounted	49	34.50	83.50
0140	Grab bar, 1-1/4" diam., 12" long	32	23	55
0160	Mirror, framed with shelf, 18" x 24"	205	27.50	232.50
0170	72" x 24"	330	92	422
0180	Toilet tissue dispenser, surface mounted, single roll	19.60	18.45	38.05
0200	Towel bar, 18" long	46	24	70
0300	Medicine cabinets, sliding mirror doors, 20" x 16" x 4-3/4", unlighted	137	79	216
0310	24" x 19" x 8-1/2", lighted	197	111	308

C1030 730	Bath and Toilet Accessories, L.F.	COST PER L.F.		
		MAT.	INST.	TOTAL
0100	Partitions, hospital curtain, ceiling hung, polyester oxford cloth	23	5.40	28.40
0110	Designer oxford cloth	19.45	6.85	26.30

C1030 830	Fabricated Cabinets & Counters, L.F.	COST PER L.F.		
		MAT.	INST.	TOTAL
0110	Household, base, hardwood, one top drawer & one door below x 12" wide	264	44.50	308.50
0120	Four drawer x 24" wide	330	49.50	379.50
0130	Wall, hardwood, 30" high with one door x 12" wide	214	50.50	264.50
0140	Two doors x 48" wide	500	60	560
0150	Counter top-laminated plastic, stock, economy	18.35	18.45	36.80
0160	Custom-square edge, 7/8" thick	16.30	41.50	57.80

C1030 Fittings

C1030 830	Fabricated Cabinets & Counters, L.F.	COST PER L.F.		
		MAT.	INST.	TOTAL
0170	School, counter, wood, 32" high	251	55.50	306.50
0180	Metal, 84" high	450	74	524

C1030 910	Other Fittings, EACH	COST EACH		
		MAT.	INST.	TOTAL
0500	Mail boxes, horizontal, rear loaded, aluminum, 5" x 6" x 15" deep	38.50	16.25	54.75
0510	Front loaded, aluminum, 10" x 12" x 15" deep	119	27.50	146.50
0520	Vertical, front loaded, aluminum, 15" x 5" x 6" deep	33	16.25	49.25
0530	Bronze, duranodic finish	44	16.25	60.25
0540	Letter slot, post office	116	69	185
0550	Mail counter, window, post office, with grille	605	277	882
0700	Turnstiles, one way, 4' arm, 46" diam., manual	1,800	221	2,021
0710	Electric	2,125	920	3,045
0720	3 arm, 5'-5" diam. & 7' high, manual	6,275	1,100	7,375
0730	Electric	8,475	1,850	10,325

C2010 Stair Construction

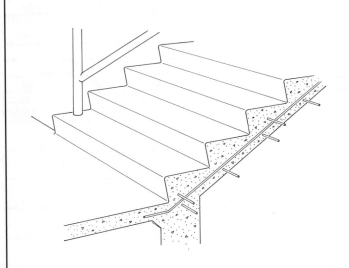

The table below lists the cost per flight for 4'-0" wide stairs. Side walls are not included. Railings are included in the prices.

C2010 110	Stairs	COST PER FLIGHT		
		MAT.	INST.	TOTAL
0470	Stairs, C.I.P. concrete, w/o landing, 12 risers, w/o nosing	1,075	2,350	3,425
0480	With nosing	1,850	2,550	4,400
0550	W/landing, 12 risers, w/o nosing	1,250	2,875	4,125
0560	With nosing	2,000	3,050	5,050
0570	16 risers, w/o nosing	1,525	3,575	5,100
0580	With nosing	2,550	3,850	6,400
0590	20 risers, w/o nosing	1,800	4,325	6,125
0600	With nosing	3,075	4,650	7,725
0610	24 risers, w/o nosing	2,075	5,050	7,125
0620	With nosing	3,600	5,450	9,050
0630	Steel, grate type w/nosing & rails, 12 risers, w/o landing	5,525	1,225	6,750
0640	With landing	7,225	1,675	8,900
0660	16 risers, with landing	9,075	2,075	11,150
0680	20 risers, with landing	10,900	2,475	13,375
0700	24 risers, with landing	12,700	2,875	15,575
0701				
0710	Concrete fill metal pan & picket rail, 12 risers, w/o landing	7,200	1,225	8,425
0720	With landing	9,475	1,825	11,300
0740	16 risers, with landing	11,900	2,250	14,150
0760	20 risers, with landing	14,300	2,650	16,950
0780	24 risers, with landing	16,700	3,025	19,725
0790	Cast iron tread & pipe rail, 12 risers, w/o landing	7,200	1,225	8,425
0800	With landing	9,475	1,825	11,300
1120	Wood, prefab box type, oak treads, wood rails 3'-6" wide, 14 risers	2,125	460	2,585
1150	Prefab basement type, oak treads, wood rails 3'-0" wide, 14 risers	960	114	1,074

C30 Interior Finishes

C3010 Wall Finishes

C3010 230	Paint & Covering	COST PER S.F.		
		MAT.	INST.	TOTAL
0060	Painting, interior on plaster and drywall, brushwork, primer & 1 coat	.13	.71	.84
0080	Primer & 2 coats	.21	.95	1.16
0100	Primer & 3 coats	.29	1.16	1.45
0120	Walls & ceilings, roller work, primer & 1 coat	.13	.48	.61
0140	Primer & 2 coats	.21	.61	.82
0160	Woodwork incl. puttying, brushwork, primer & 1 coat	.13	1.03	1.16
0180	Primer & 2 coats	.21	1.37	1.58
0200	Primer & 3 coats	.29	1.86	2.15
0260	Cabinets and casework, enamel, primer & 1 coat	.14	1.16	1.30
0280	Primer & 2 coats	.21	1.43	1.64
0300	Masonry or concrete, latex, brushwork, primer & 1 coat	.31	.97	1.28
0320	Primer & 2 coats	.40	1.39	1.79
0340	Addition for block filler	.23	1.19	1.42
0380	Fireproof paints, intumescent, 1/8" thick 3/4 hour	2.33	.95	3.28
0400	3/16" thick 1 hour	5.20	1.43	6.63
0420	7/16" thick 2 hour	6.70	3.32	10.02
0440	1-1/16" thick 3 hour	10.95	6.65	17.60
0500	Gratings, primer & 1 coat	.35	1.45	1.80
0600	Pipes over 12" diameter	.81	4.64	5.45
0700	Structural steel, brushwork, light framing 300-500 S.F./Ton	.10	1.80	1.90
0720	Heavy framing 50-100 S.F./Ton	.10	.90	1
0740	Spraywork, light framing 300-500 S.F./Ton	.11	.40	.51
0760	Heavy framing 50-100 S.F./Ton	.11	.44	.55
0800	Varnish, interior wood trim, no sanding sealer & 1 coat	.08	1.16	1.24
0820	Hardwood floor, no sanding 2 coats	.17	.25	.42
0840	Wall coatings, acrylic glazed coatings, minimum	.34	.88	1.22
0860	Maximum	.72	1.52	2.24
0880	Epoxy coatings, minimum	.44	.88	1.32
0900	Maximum	1.32	2.73	4.05
0940	Exposed epoxy aggregate, troweled on, 1/16" to 1/4" aggregate, minimum	.67	1.98	2.65
0960	Maximum	1.44	3.57	5.01
0980	1/2" to 5/8" aggregate, minimum	1.31	3.57	4.88
1000	Maximum	2.27	5.80	8.07
1020	1" aggregate, minimum	2.30	5.15	7.45
1040	Maximum	3.51	8.45	11.96
1060	Sprayed on, minimum	.62	1.57	2.19
1080	Maximum	1.14	3.20	4.34
1100	High build epoxy 50 mil, minimum	.74	1.19	1.93
1120	Maximum	1.25	4.89	6.14
1140	Laminated epoxy with fiberglass minimum	.79	1.57	2.36
1160	Maximum	1.43	3.20	4.63
1180	Sprayed perlite or vermiculite 1/16" thick, minimum	.30	.16	.46
1200	Maximum	.80	.73	1.53
1260	Wall coatings, vinyl plastic, minimum	.36	.63	.99
1280	Maximum	.89	1.93	2.82
1300	Urethane on smooth surface, 2 coats, minimum	.30	.41	.71
1320	Maximum	.64	.70	1.34
1340	3 coats, minimum	.39	.55	.94
1360	Maximum	.86	.99	1.85
1380	Ceramic-like glazed coating, cementitious, minimum	.52	1.06	1.58
1400	Maximum	.88	1.35	2.23
1420	Resin base, minimum	.36	.73	1.09
1440	Maximum	.58	1.41	1.99
1460	Wall coverings, aluminum foil	1.14	1.70	2.84
1500	Vinyl backing	6.10	1.94	8.04
1520	Cork tiles, 12"x12", light or dark, 3/16" thick	6.15	1.94	8.09
1540	5/16" thick	2.53	1.98	4.51
1560	Basketweave, 1/4" thick	6.45	1.94	8.39

C3010 Wall Finishes

C3010 230	Paint & Covering	COST PER S.F.		
		MAT.	INST.	TOTAL
1580	Natural, non-directional, 1/2" thick	8.70	1.94	10.64
1600	12"x36", granular, 3/16" thick	1.42	1.21	2.63
1620	1" thick	1.83	1.26	3.09
1640	12"x12", polyurethane coated, 3/16" thick	4.43	1.94	6.37
1660	5/16" thick	6.20	1.98	8.18
1661	Paneling, prefinished plywood, birch	1.51	2.63	4.14
1662	Mahogany, African	2.83	2.77	5.60
1663	Philippine (lauan)	1.21	2.21	3.42
1664	Oak or cherry	3.61	2.77	6.38
1665	Rosewood	5.15	3.46	8.61
1666	Teak	3.61	2.77	6.38
1667	Chestnut	5.35	2.95	8.30
1668	Pecan	2.31	2.77	5.08
1669	Walnut	5.85	2.77	8.62
1670	Wood board, knotty pine, finished	2.44	4.57	7.01
1671	Rough sawn cedar	3.52	4.57	8.09
1672	Redwood	5.80	4.57	10.37
1673	Aromatic cedar	2.56	4.90	7.46
1680	Cork wallpaper, paper backed, natural	2.21	.97	3.18
1700	Color	3.12	.97	4.09
1720	Gypsum based, fabric backed, minimum	.88	.58	1.46
1740	Average	1.38	.65	2.03
1760	Maximum	1.54	.73	2.27
1780	Vinyl wall covering, fabric back, light weight	.94	.73	1.67
1800	Medium weight	.91	.97	1.88
1820	Heavy weight	1.42	1.07	2.49
1840	Wall paper, double roll, solid pattern, avg. workmanship	.47	.73	1.20
1860	Basic pattern, avg. workmanship	1.07	.87	1.94
1880	Basic pattern, quality workmanship	2.13	1.07	3.20
1900	Grass cloths with lining paper, minimum	.73	1.17	1.90
1920	Maximum	2.93	1.33	4.26
1940	Ceramic tile, thin set, 4-1/4" x 4-1/4"	2.43	4.58	7.01
1942				
1960	12" x 12"	3.89	5.45	9.34

C3010 235	Paint Trim	COST PER L.F.		
		MAT.	INST.	TOTAL
2040	Painting, wood trim, to 6" wide, enamel, primer & 1 coat	.14	.58	.72
2060	Primer & 2 coats	.21	.74	.95
2080	Misc. metal brushwork, ladders	.71	5.80	6.51
2100	Pipes, to 4" dia.	.10	1.22	1.32
2120	6" to 8" dia.	.21	2.44	2.65
2140	10" to 12" dia.	.64	3.66	4.30
2160	Railings, 2" pipe	.25	2.90	3.15
2180	Handrail, single	.18	1.16	1.34
2185	Caulking & Sealants, Polyurethane, In place, 1 or 2 component, 1/2" X 1/4"	.36	1.88	2.24

C3020 Floor Finishes

C3020 410	Tile & Covering	COST PER S.F.		
		MAT.	INST.	TOTAL
0060	Carpet tile, nylon, fusion bonded, 18" x 18" or 24" x 24", 24 oz.	3.55	.67	4.22
0080	35 oz.	4.38	.67	5.05
0100	42 oz.	5.65	.67	6.32
0140	Carpet, tufted, nylon, roll goods, 12' wide, 26 oz.	6.35	.77	7.12
0160	36 oz.	9.45	.77	10.22
0180	Woven, wool, 36 oz.	16	.77	16.77
0200	42 oz.	16.55	.77	17.32
0220	Padding, add to above, minimum	.60	.36	.96
0240	Maximum	1.10	.36	1.46
0260	Composition flooring, acrylic, 1/4" thick	1.79	5.35	7.14
0280	3/8" thick	2.27	6.20	8.47
0300	Epoxy, minimum	3.03	4.13	7.16
0320	Maximum	4.37	5.70	10.07
0340	Epoxy terrazzo, minimum	6.45	6	12.45
0360	Maximum	9.80	8	17.80
0380	Mastic, hot laid, 1-1/2" thick, minimum	4.48	4.05	8.53
0400	Maximum	5.75	5.35	11.10
0420	Neoprene 1/4" thick, minimum	4.41	5.10	9.51
0440	Maximum	6.05	6.45	12.50
0460	Polyacrylate with ground granite 1/4", minimum	3.83	3.80	7.63
0480	Maximum	6.90	5.80	12.70
0500	Polyester with colored quart 2 chips 1/16", minimum	3.42	2.62	6.04
0520	Maximum	5.10	4.13	9.23
0540	Polyurethane with vinyl chips, minimum	8.25	2.62	10.87
0560	Maximum	11.90	3.24	15.14
0600	Concrete topping, granolithic concrete, 1/2" thick	.34	4.47	4.81
0620	1" thick	.69	4.58	5.27
0640	2" thick	1.38	5.25	6.63
0660	Heavy duty 3/4" thick, minimum	.47	6.90	7.37
0680	Maximum	.75	8.20	8.95
0700	For colors, add to above, minimum	.73	1.60	2.33
0720	Maximum	1.22	1.76	2.98
0740	Exposed aggregate finish, minimum	.27	.81	1.08
0760	Maximum	.83	1.09	1.92
0780	Abrasives, .25 P.S.F. add to above, minimum	.54	.60	1.14
0800	Maximum	.85	.60	1.45
0820	Dust on coloring, add, minimum	.73	.39	1.12
0840	Maximum	1.22	.81	2.03
0860	Floor coloring using 0.6 psf powdered color, 1/2" integral, minimum	5.45	4.47	9.92
0880	Maximum	5.90	4.47	10.37
0900	Dustproofing, add, minimum	.29	.27	.56
0920	Maximum	1.06	.39	1.45
0930	Paint	.40	1.39	1.79
0940	Hardeners, metallic add, minimum	.55	.60	1.15
0960	Maximum	1.65	.88	2.53
0980	Non-metallic, minimum	.22	.60	.82
1000	Maximum	.66	.88	1.54
1020	Integral topping and finish, 1:1:2 mix, 3/16" thick	.12	2.63	2.75
1040	1/2" thick	.31	2.78	3.09
1060	3/4" thick	.47	3.11	3.58
1080	1" thick	.62	3.51	4.13
1100	Terrazzo, minimum	3.58	15.95	19.53
1120	Maximum	6.85	8	14.85
1340	Cork tile, minimum	8.25	1.54	9.79
1360	Maximum	17.55	1.54	19.09
1380	Polyethylene, in rolls, minimum	3.82	1.76	5.58
1400	Maximum	7.35	1.76	9.11
1420	Polyurethane, thermoset, minimum	5.85	4.85	10.70

C3020 Floor Finishes

C3020 410	Tile & Covering	COST PER S.F.		
		MAT.	INST.	TOTAL
1440	Maximum	7.45	9.70	17.15
1460	Rubber, sheet goods, minimum	8.40	4.04	12.44
1480	Maximum	13.50	5.40	18.90
1500	Tile, minimum	7.90	1.21	9.11
1520	Maximum	15.35	1.76	17.11
1580	Vinyl, composition tile, minimum	.95	.97	1.92
1600	Maximum	1.05	.97	2.02
1620	Vinyl tile, minimum	3.74	.97	4.71
1640	Maximum	4.02	.97	4.99
1660	Sheet goods, minimum	4.51	1.94	6.45
1680	Maximum	7.85	2.42	10.27
1720	Tile, ceramic natural clay	6.40	4.75	11.15
1730	Marble, synthetic 12"x12"x5/8"	6.95	14.50	21.45
1740	Porcelain type, minimum	5.50	4.75	10.25
1760	Maximum	6.15	5.60	11.75
1800	Quarry tile, mud set, minimum	7.95	6.20	14.15
1820	Maximum	9.30	7.90	17.20
1840	Thin set, deduct		1.24	1.24
1850	Tile, natural stone, marble, in mortar bed, 12" x 12" x 3/8" thick	99.50	24	123.50
1860	Terrazzo precast, minimum	5.15	6.45	11.60
1880	Maximum	10.25	6.45	16.70
1900	Non-slip, minimum	21.50	16.30	37.80
1920	Maximum	22.50	21.50	44
2020	Wood block, end grain factory type, natural finish, 2" thick	4.48	4.43	8.91
2040	Fir, vertical grain, 1"x4", no finish, minimum	2.84	2.17	5.01
2060	Maximum	3.01	2.17	5.18
2080	Prefinished white oak, prime grade, 2-1/4" wide	5.35	3.25	8.60
2100	3-1/4" wide	5.95	2.99	8.94
2120	Maple strip, sanded and finished, minimum	5.70	4.73	10.43
2140	Maximum	5.90	4.73	10.63
2160	Oak strip, sanded and finished, minimum	3.59	4.73	8.32
2180	Maximum	5.30	4.73	10.03
2200	Parquetry, sanded and finished, minimum	6.95	4.94	11.89
2220	Maximum	5.40	7.05	12.45
2260	Add for sleepers on concrete, treated, 24" O.C., 1"x2"	1.73	2.83	4.56
2280	1"x3"	1.88	2.23	4.11
2300	2"x4"	.69	1.13	1.82
2340	Underlayment, plywood, 3/8" thick	.89	.74	1.63
2350	1/2" thick	1.03	.76	1.79
2360	5/8" thick	1.20	.79	1.99
2370	3/4" thick	1.36	.85	2.21
2380	Particle board, 3/8" thick	.41	.74	1.15
2390	1/2" thick	.45	.76	1.21
2400	5/8" thick	.55	.79	1.34
2410	3/4" thick	.75	.85	1.60
2420	Hardboard, 4' x 4', .215" thick	.63	.74	1.37

C3030 Ceiling Finishes

2 Coats of Plaster on Gypsum
Lath on Wood Furring

Fiberglass Board on
Exposed Suspended Grid System

Plaster and Metal Lath
on Metal Furring

C3030 105	Plaster Ceilings					COST PER S.F.		
	TYPE	LATH	FURRING	SUPPORT		MAT.	INST.	TOTAL
2400	2 coat gypsum	3/8" gypsum	1"x3" wood, 16" O.C.	wood		1.36	5.55	6.91
2500	Painted			masonry		1.36	5.65	7.01
2600				concrete		1.36	6.35	7.71
2700	3 coat gypsum	3.4# metal		wood		1.56	5.95	7.51
2800	Painted			masonry		1.56	6.05	7.61
2900				concrete		1.56	6.70	8.26
3000	2 coat perlite	3/8" gypsum		wood		1.41	5.80	7.21
3100	Painted			masonry		1.41	5.90	7.31
3200				concrete		1.41	6.60	8.01
3300	3 coat perlite	3.4# metal		wood		1.44	5.95	7.39
3400	Painted			masonry		1.44	6.05	7.49
3500				concrete		1.44	6.70	8.14
3600	2 coat gypsum	3/8" gypsum	3/4" CRC, 12" O.C.	1-1/2" CRC, 48"O.C.		1.46	6.60	8.06
3700	Painted		3/4" CRC, 16" O.C.	1-1/2" CRC, 48"O.C.		1.42	5.95	7.37
3800			3/4" CRC, 24" O.C.	1-1/2" CRC, 48"O.C.		1.30	5.45	6.75
3900	2 coat perlite		3/4" CRC, 12" O.C.	1-1/2" CRC, 48"O.C		1.51	7.10	8.61
4000	Painted		3/4" CRC, 16" O.C.	1-1/2" CRC, 48"O.C.		1.47	6.45	7.92
4100			3/4" CRC, 24" O.C.	1-1/2" CRC, 48"O.C.		1.35	5.95	7.30
4200	3 coat gypsum	3.4# metal	3/4" CRC, 12" O.C.	1-1/2" CRC, 36" O.C.		1.85	8.65	10.50
4300	Painted		3/4" CRC, 16" O.C.	1-1/2" CRC, 36" O.C.		1.81	8	9.81
4400			3/4" CRC, 24" O.C.	1-1/2" CRC, 36" O.C.		1.69	7.50	9.19
4500	3 coat perlite		3/4" CRC, 12" O.C.	1-1/2" CRC,36" O.C.		2.03	9.55	11.58
4600	Painted		3/4" CRC, 16" O.C.	1-1/2" CRC, 36" O.C.		1.99	8.95	10.94
4700			3/4" CRC, 24" O.C.	1-1/2" CRC, 36" O.C.		1.87	8.45	10.32

C3030 110	Drywall Ceilings					COST PER S.F.		
	TYPE	FINISH	FURRING	SUPPORT		MAT.	INST.	TOTAL
4800	1/2" F.R. drywall	painted and textured	1"x3" wood, 16" O.C.	wood		.81	3.40	4.21
4900				masonry		.81	3.51	4.32
5000				concrete		.81	4.18	4.99
5100	5/8" F.R. drywall			wood		.83	3.40	4.23
5200				masonry		.83	3.51	4.34
5300				concrete		.83	4.18	5.01
5400	1/2" F.R. drywall		7/8" resil. channels	24" O.C.		.68	3.30	3.98

C3030 Ceiling Finishes

C3030 110		Drywall Ceilings				COST PER S.F.		
	TYPE	FINISH	FURRING	SUPPORT		MAT.	INST.	TOTAL
5500	1/2" F.R. drywall	painted and textured	1"x2" wood	stud clips		.85	3.17	4.02
5602		painted	1-5/8" metal studs	24" O.C.		.72	2.94	3.66
5700	5/8" F.R. drywall	painted and textured				.74	2.94	3.68
5702								

C3030 140	Plaster Ceiling Components	COST PER S.F.		
		MAT.	INST.	TOTAL
0060	Plaster, gypsum incl. finish, 2 coats			
0080	3 coats	.59	3.55	4.14
0100	Perlite, incl. finish, 2 coats	.47	3.55	4.02
0120	3 coats	.77	4.44	5.21
0140	Thin coat on drywall	.12	.69	.81
0200	Lath, gypsum, 3/8" thick	.52	.85	1.37
0220	1/2" thick	.55	.89	1.44
0240	5/8" thick	.30	1.04	1.34
0260	Metal, diamond, 2.5 lb.	.47	.69	1.16
0280	3.4 lb.	.55	.74	1.29
0300	Flat rib, 2.75 lb.	.36	.69	1.05
0320	3.4 lb.	.59	.74	1.33
0440	Furring, steel channels, 3/4" galvanized , 12" O.C.	.39	2.22	2.61
0460	16" O.C.	.35	1.61	1.96
0480	24" O.C.	.23	1.11	1.34
0500	1-1/2" galvanized , 12" O.C.	.49	2.46	2.95
0520	16" O.C.	.44	1.79	2.23
0540	24" O.C.	.29	1.20	1.49
0560	Wood strips, 1"x3", on wood, 12" O.C.	.38	1.58	1.96
0580	16" O.C.	.29	1.19	1.48
0600	24" O.C.	.19	.79	.98
0620	On masonry, 12" O.C.	.38	1.73	2.11
0640	16" O.C.	.29	1.30	1.59
0660	24" O.C.	.19	.87	1.06
0680	On concrete, 12" O.C.	.38	2.63	3.01
0700	16" O.C.	.29	1.97	2.26
0720	24" O.C.	.19	1.32	1.51
0940	Paint on plaster or drywall, roller work, primer + 1 coat	.13	.48	.61
0960	Primer + 2 coats	.21	.61	.82

C3030 210		Acoustical Ceilings				COST PER S.F.		
	TYPE	TILE	GRID	SUPPORT		MAT.	INST.	TOTAL
5800	5/8" fiberglass board	24" x 48"	tee	suspended		2.03	1.66	3.69
5900		24" x 24"	tee	suspended		2.23	1.82	4.05
6000	3/4" fiberglass board	24" x 48"	tee	suspended		2.66	1.69	4.35
6100		24" x 24"	tee	suspended		2.86	1.85	4.71
6500	5/8" mineral fiber	12" x 12"	1"x3" wood, 12" O.C.	wood		2.56	3.42	5.98
6600				masonry		2.56	3.57	6.13
6700				concrete		2.56	4.47	7.03
6800	3/4" mineral fiber			wood		2.34	3.42	5.76
6900				masonry		2.34	3.42	5.76
7000				concrete		2.34	3.42	5.76
7100	3/4"mineral fiber on		25 ga. channels	runners		2.75	4.21	6.96
7102	5/8" F.R. drywall							

C30 Interior Finishes

C3030 Ceiling Finishes

C3030 210 — Acoustical Ceilings

	TYPE	TILE	GRID	SUPPORT		COST PER S.F.		
						MAT.	INST.	TOTAL
7200 7201 7202	5/8" plastic coated Mineral fiber	12" x 12"	25 ga. channels	adhesive backed		2.60	1.84	4.44
7300 7301 7302	3/4" plastic coated Mineral fiber	12" x 12"		adhesive backed		2.38	1.84	4.22
7400 7401 7402	3/4" mineral fiber	12" x 12"	conceal 2" bar & channels	suspended		2.70	4.16	6.86

C3030 240 — Acoustical Ceiling Components

		COST PER S.F.		
		MAT.	INST.	TOTAL
2480	Ceiling boards, eggcrate, acrylic, 1/2" x 1/2" x 1/2" cubes	2	1.11	3.11
2500	Polystyrene, 3/8" x 3/8" x 1/2" cubes	1.68	1.09	2.77
2520	1/2" x 1/2" x 1/2" cubes	2.24	1.11	3.35
2540	Fiberglass boards, plain, 5/8" thick	1.06	.89	1.95
2560	3/4" thick	1.69	.92	2.61
2580	Grass cloth faced, 3/4" thick	2.35	1.11	3.46
2600	1" thick	3.10	1.14	4.24
2620	Luminous panels, prismatic, acrylic	2.44	1.38	3.82
2640	Polystyrene	1.25	1.38	2.63
2660	Flat or ribbed, acrylic	4.25	1.38	5.63
2680	Polystyrene	2.92	1.38	4.30
2700	Drop pan, white, acrylic	6.25	1.38	7.63
2720	Polystyrene	5.20	1.38	6.58
2740	Mineral fiber boards, 5/8" thick, standard	.75	.82	1.57
2760	Plastic faced	1.16	1.38	2.54
2780	2 hour rating	1.27	.82	2.09
2800	Perforated aluminum sheets, .024 thick, corrugated painted	2.52	1.13	3.65
2820	Plain	4.16	1.11	5.27
3080	Mineral fiber, plastic coated, 12" x 12" or 12" x 24", 5/8" thick	2.18	1.84	4.02
3100	3/4" thick	1.96	1.84	3.80
3120	Fire rated, 3/4" thick, plain faced	1.49	1.84	3.33
3140	Mylar faced	1.41	1.84	3.25
3160	Add for ceiling primer	.14		.14
3180	Add for ceiling cement	.42		.42
3240	Suspension system, furring, 1" x 3" wood 12" O.C.	.38	1.58	1.96
3260	T bar suspension system, 2' x 4' grid	.74	.69	1.43
3280	2' x 2' grid	.94	.85	1.79
3300	Concealed Z bar suspension system 12" module	.86	1.06	1.92
3320	Add to above for 1-1/2" carrier channels 4' O.C.	.12	1.18	1.30
3340	Add to above for carrier channels for recessed lighting	.22	1.20	1.42

D1010 Elevators and Lifts

Hydraulic

Traction Geared

Traction Gearless

D1010 110	Hydraulic	COST EACH		
		MAT.	INST.	TOTAL
1300	Pass. elev., 1500 lb., 2 Floors, 100 FPM	48,600	17,400	66,000
1400	5 Floors, 100 FPM	89,000	46,400	135,400
1600	2000 lb., 2 Floors, 100 FPM	50,500	17,400	67,900
1700	5 floors, 100 FPM	91,000	46,400	137,400
1900	2500 lb., 2 Floors, 100 FPM	51,500	17,400	68,900
2000	5 floors, 100 FPM	92,000	46,400	138,400
2200	3000 lb., 2 Floors, 100 FPM	53,000	17,400	70,400
2300	5 floors, 100 FPM	93,500	46,400	139,900
2500	3500 lb., 2 Floors, 100 FPM	55,500	17,400	72,900
2600	5 floors, 100 FPM	96,000	46,400	142,400
2800	4000 lb., 2 Floors, 100 FPM	57,000	17,400	74,400
2900	5 floors, 100 FPM	97,500	46,400	143,900
3100	4500 lb., 2 Floors, 100 FPM	59,500	17,400	76,900
3200	5 floors, 100 FPM	100,000	46,400	146,400
4000	Hospital elevators, 3500 lb., 2 Floors, 100 FPM	74,500	17,400	91,900
4100	5 floors, 100 FPM	124,000	46,400	170,400
4300	4000 lb., 2 Floors, 100 FPM	74,500	17,400	91,900
4400	5 floors, 100 FPM	124,000	46,400	170,400
4600	4500 lb., 2 Floors, 100 FPM	81,000	17,400	98,400
4800	5 floors, 100 FPM	131,000	46,400	177,400
4900	5000 lb., 2 Floors, 100 FPM	84,500	17,400	101,900
5000	5 floors, 100 FPM	134,000	46,400	180,400
6700	Freight elevators (Class "B"), 3000 lb., 2 Floors, 50 FPM	118,500	22,200	140,700
6800	5 floors, 100 FPM	216,000	58,000	274,000
7000	4000 lb., 2 Floors, 50 FPM	125,500	22,200	147,700
7100	5 floors, 100 FPM	223,000	58,000	281,000
7500	10,000 lb., 2 Floors, 50 FPM	155,500	22,200	177,700
7600	5 floors, 100 FPM	253,000	58,000	311,000
8100	20,000 lb., 2 Floors, 50 FPM	206,500	22,200	228,700
8200	5 Floors, 100 FPM	304,000	58,000	362,000

D1010 140	Traction Geared Elevators	COST EACH		
		MAT.	INST.	TOTAL
1300	Passenger, 2000 Lb., 5 floors, 200 FPM	121,500	43,800	165,300
1500	15 floors, 350 FPM	288,500	132,500	421,000
1600	2500 Lb., 5 floors, 200 FPM	125,500	43,800	169,300
1800	15 floors, 350 FPM	292,500	132,500	425,000
1900	3000 Lb., 5 floors, 200 FPM	126,000	43,800	169,800
2100	15 floors, 350 FPM	293,000	132,500	425,500
2200	3500 Lb., 5 floors, 200 FPM	127,500	43,800	171,300
2400	15 floors, 350 FPM	294,500	132,500	427,000
2500	4000 Lb., 5 floors, 200 FPM	128,500	43,800	172,300
2700	15 floors, 350 FPM	295,500	132,500	428,000

D1010 Elevators and Lifts

D1010 140	Traction Geared Elevators	COST EACH		
		MAT.	INST.	TOTAL
2800	4500 Lb., 5 floors, 200 FPM	131,000	43,800	174,800
3000	15 floors, 350 FPM	298,000	132,500	430,500
3100	5000 Lb., 5 floors, 200 FPM	133,500	43,800	177,300
3300	15 floors, 350 FPM	300,500	132,500	433,000
4000	Hospital, 3500 Lb., 5 floors, 200 FPM	122,000	43,800	165,800
4200	15 floors, 350 FPM	302,000	132,500	434,500
4300	4000 Lb., 5 floors, 200 FPM	122,000	43,800	165,800
4500	15 floors, 350 FPM	302,000	132,500	434,500
4600	4500 Lb., 5 floors, 200 FPM	127,500	43,800	171,300
4800	15 floors, 350 FPM	308,000	132,500	440,500
4900	5000 Lb., 5 floors, 200 FPM	129,500	43,800	173,300
5100	15 floors, 350 FPM	310,000	132,500	442,500
6000	Freight, 4000 Lb., 5 floors, 50 FPM class 'B'	143,500	45,300	188,800
6200	15 floors, 200 FPM class 'B'	347,500	157,500	505,000
6300	8000 Lb., 5 floors, 50 FPM class 'B'	182,000	45,300	227,300
6500	15 floors, 200 FPM class 'B'	386,000	157,500	543,500
7000	10,000 Lb., 5 floors, 50 FPM class 'B'	193,000	45,300	238,300
7200	15 floors, 200 FPM class 'B'	566,500	157,500	724,000
8000	20,000 Lb., 5 floors, 50 FPM class 'B'	213,500	45,300	258,800
8200	15 floors, 200 FPM class 'B'	587,000	157,500	744,500

D1010 150	Traction Gearless Elevators	COST EACH		
		MAT.	INST.	TOTAL
1700	Passenger, 2500 Lb., 10 floors, 200 FPM	303,500	114,000	417,500
1900	30 floors, 600 FPM	662,000	291,500	953,500
2000	3000 Lb., 10 floors, 200 FPM	304,000	114,000	418,000
2200	30 floors, 600 FPM	663,000	291,500	954,500
2300	3500 Lb., 10 floors, 200 FPM	305,500	114,000	419,500
2500	30 floors, 600 FPM	664,000	291,500	955,500
2700	50 floors, 800 FPM	988,000	468,500	1,456,500
2800	4000 Lb., 10 floors, 200 FPM	306,500	114,000	420,500
3000	30 floors, 600 FPM	665,500	291,500	957,000
3200	50 floors, 800 FPM	989,000	468,500	1,457,500
3300	4500 Lb., 10 floors, 200 FPM	309,000	114,000	423,000
3500	30 floors, 600 FPM	667,500	291,500	959,000
3700	50 floors, 800 FPM	991,500	468,500	1,460,000
3800	5000 Lb., 10 floors, 200 FPM	311,500	114,000	425,500
4000	30 floors, 600 FPM	670,000	291,500	961,500
4200	50 floors, 800 FPM	994,000	468,500	1,462,500
6000	Hospital, 3500 Lb., 10 floors, 200 FPM	306,500	114,000	420,500
6200	30 floors, 600 FPM	694,500	291,500	986,000
6400	4000 Lb., 10 floors, 200 FPM	306,500	114,000	420,500
6600	30 floors, 600 FPM	694,500	291,500	986,000
6800	4500 Lb., 10 floors, 200 FPM	312,500	114,000	426,500
7000	30 floors, 600 FPM	700,500	291,500	992,000
7200	5000 Lb., 10 floors, 200 FPM	314,000	114,000	428,000
7400	30 floors, 600 FPM	702,000	291,500	993,500

D10 Conveying

D1090 Other Conveying Systems

D1090 510	Chutes	COST EACH		
		MAT.	INST.	TOTAL
0100	Chutes, linen or refuse incl. sprinklers, galv. steel, 18" diam., per floor	1,875	370	2,245
0110	Aluminized steel, 36" diam., per floor	2,900	465	3,365
0120	Mail, 8-3/4" x 3-1/2", aluminum & glass, per floor	715	260	975
0130	Bronze or stainless, per floor	1,100	289	1,389

Minimum Plumbing Fixture Requirements

Type of Building or Occupancy (2)	Water Closets (14) (Fixtures per Person)		Urinals (5,10) (Fixtures per Person)		Lavatories (Fixtures per Person)		Bathtubs or Showers (Fixtures per Person)	Drinking Fountains (Fixtures per Person) (3, 13)
	Male	Female	Male	Female	Male	Female		
Assembly Places- Theatres, Auditoriums, Convention Halls, etc.-for permanent employee use	1: 1 - 15 2: 16 - 35 3: 36 - 55	1: 1 - 15 2: 16 - 35 3: 36 - 55	0: 1 - 9 1: 10 - 50		1 per 40	1 per 40		
	Over 55, add 1 fixture for each additional 40 persons		Add one fixture for each additional 50 males					
Assembly Places- Theatres, Auditoriums, Convention Halls, etc. - for public use	1: 1 - 100 2: 101 - 200 3: 201 - 400	3: 1 - 50 4: 51 - 100 8: 101 - 200 11: 201 - 400	1: 1 - 100 2: 101 - 200 3: 201 - 400 4: 401 - 600		1: 1 - 200 2: 201 - 400 3: 401 - 750	1: 1 - 200 2: 201 - 400 3: 401 - 750		1: 1 - 150 2: 151 - 400 3: 401 - 750
	Over 400, add 1 fixture for each additional 500 males and 1 for each additional 125 females		Over 600, add 1 fixture for each additional 300 males		Over 750, add 1 fixture for each additional 500 persons			Over 750, add one fixture for each additional 500 persons
Dormitories (9) School or Labor	1 per 10	1 per 8	1 per 25		1 per 12	1 per 12	1 per 8	1 per 150 (12)
	Add 1 fixture for each additional 25 males (over 10) and 1 for each additional 20 females (over 8)		Over 150, add 1 fixture for each additional 50 males		Over 12 add 1 fixture for each additional 20 males and 1 for each 15 additional females		For females add 1 bathtub per 30. Over 150, add 1 per 20	
Dormitories- for Staff Use	1: 1 - 15 2: 16 - 35 3: 36 - 55	1: 1 - 15 3: 16 - 35 4: 36 - 55	1 per 50		1 per 40	1 per 40	1 per 8	
	Over 55, add 1 fixture for each additional 40 persons							
Dwellings: Single Dwelling Multiple Dwelling or Apartment House	1 per dwelling 1 per dwelling or apartment unit				1 per dwelling 1 per dwelling or apartment unit		1 per dwelling 1 per dwelling or apartment unit	
Hospital Waiting rooms	1 per room				1 per room			1 per 150 (12)
Hospitals- for employee use	1: 1 - 15 2: 16 - 35 3: 36 - 55	1: 1 - 15 3: 16 - 35 4: 36 - 55	0: 1 - 9 1: 10 - 50		1 per 40	1 per 40		
	Over 55, add 1 fixture for each additional 40 persons		Add 1 fixture for each additional 50 males					
Hospitals: Individual Room Ward Room	1 per room 1 per 8 patients				1 per room 1 per 10 patients		1 per room 1 per 20 patients	1 per 150 (12)
Industrial (6) Warehouses Workshops, Foundries and similar establishments- for employee use	1: 1 -10 2: 11 - 25 3: 26 - 50 4: 51 - 75 5: 76 - 100	1: 1 -10 2: 11 - 25 3: 26 - 50 4: 51 - 75 5: 76 - 100			Up to 100, per 10 persons Over 100, 1 per 15 persons (7, 8)		1 shower for each 15 persons exposed to excessive heat or to skin contamination with poisonous, infectious or irritating material	1 per 150 (12)
	Over 100, add 1 fixture for each additional 30 persons							
Institutional - Other than Hospitals or Penal Institutions (on each occupied floor)	1 per 25	1 per 20	0: 1 - 9 1: 10 - 50		1 per 10	1 per 10	1 per 8	1 per 150 (12)
			Add 1 fixture for each additional 50 males					
Institutional - Other than Hospitals or Penal Institutions (on each occupied floor)- for employee use	1: 1 - 15 2: 16 - 35 3: 36 - 55	1: 1 - 15 3: 16 - 35 4: 36 - 55	0: 1 - 9 1: 10 - 50		1 per 40	1 per 40	1 per 8	1 per 150 (12)
	Over 55, add 1 fixture for each additional 40 persons		Add 1 fixture for each additional 50 males					
Office or Public Buildings	1: 1 - 100 2: 101 - 200 3: 201 - 400	3: 1 - 50 4: 51 - 100 8: 101 - 200 11: 201 - 400	1: 1 - 100 2: 101 - 200 3: 201 - 400 4: 401 - 600		1: 1 - 200 2: 201 - 400 3: 401 - 750	1: 1 - 200 2: 201 - 400 3: 401 - 750		1 per 150 (12)
	Over 400, add 1 fixture for each additional 500 males and 1 for each additional 150 females		Over 600, add 1 fixture for each additional 300 males		Over 750, add 1 fixture for each additional 500 persons			
Office or Public Buildings - for employee use	1: 1 - 15 2: 16 - 35 3: 36 - 55	1: 1 - 15 3: 16 - 35 4: 36 - 55	0: 1 - 9 1: 10 - 50		1 per 40	1 per 40		
	Over 55, add 1 fixture for each additional 40 persons		Add 1 fixture for each additional 50 males					

D2010 Plumbing Fixtures

Minimum Plumbing Fixture Requirements

Type of Building or Occupancy	Water Closets (14) (Fixtures per Person)		Urinals (5, 10) (Fixtures per Person)		Lavatories (Fixtures per Person)		Bathtubs or Showers (Fixtures per Person)	Drinking Fountains (Fixtures per Person) (3, 13)
	Male	Female	Male	Female	Male	Female		
Penal Institutions - for employee use	1: 1 - 15 2: 16 - 35 3: 36 - 55 Over 55, add 1 fixture for each additional 40 persons	1: 1 - 15 3: 16 - 35 4: 36 - 55	0: 1 - 9 1: 10 - 50 Add 1 fixture for each additional 50 males		1 per 40	1 per 40		1 per 150 (12)
Penal Institutions - for prison use Cell	1 per cell				1 per cell			1 per cellblock floor
Exercise room	1 per exercise room		1 per exercise room		1 per exercise room			1 per exercise room
Restaurants, Pubs and Lounges (11)	1: 1 - 50 2: 51 - 150 3: 151 - 300 Over 300, add 1 fixture for each additional 200 persons	1: 1 - 50 2: 51 - 150 4: 151 - 300	1: 1 - 150 Over 150, add 1 fixture for each additional 150 males		1: 1 - 150 2: 151 - 200 3: 201 - 400 Over 400, add 1 fixture for each additional 400 persons	1: 1 - 150 2: 151 - 200 3: 201 - 400		
Schools - for staff use All Schools	1: 1 - 15 2: 16 - 35 3: 36 - 55 Over 55, add 1 fixture for each additional 40 persons	1: 1 - 15 3: 16 - 35 4: 36 - 55	1 per 50		1 per 40	1 per 40		
Schools - for student use: Nursery	1: 1 - 20 2: 21 - 50 Over 50, add 1 fixture for each additional 50 persons	1: 1 - 20 2: 21 - 50			1: 1 - 25 2: 26 - 50 Over 50, add 1 fixture for each additional 50 persons	1: 1 - 25 2: 26 - 50		1 per 150 (12)
Elementary	1 per 30	1 per 25	1 per 75		1 per 35	1 per 35		1 per 150 (12)
Secondary	1 per 40	1 per 30	1 per 35		1 per 40	1 per 40		1 per 150 (12)
Others (Colleges, Universities, Adult Centers, etc.	1 per 40	1 per 30	1 per 35		1 per 40	1 per 40		1 per 150 (12)
Worship Places Educational and Activities Unit	1 per 150	1 per 75	1 per 150		1 per 2 water closets			1 per 150 (12)
Worship Places Principal Assembly Place	1 per 150	1 per 75	1 per 150		1 per 2 water closets			1 per 150 (12)

Notes:

1. The figures shown are based upon one (1) fixture being the minimum required for the number of persons indicated or any fraction thereof.
2. Building categories not shown on this table shall be considered separately by the Administrative Authority.
3. Drinking fountains shall not be installed in toilet rooms.
4. Laundry trays. One (1) laundry tray or one (1) automatic washer standpipe for each dwelling unit or one (1) laundry trays or one (1) automatic washer standpipes, or combination thereof, for each twelve (12) apartments. Kitchen sinks, one (1) for each dwelling or apartment unit.
5. For each urinal added in excess of the minimum required, one water closet may be deducted. The number of water closets shall not be reduced to less than two-thirds (2/3) of the minimum requirement.
6. As required by ANSI Z4.1-1968, Sanitation in Places of Employment.
7. Where there is exposure to skin contamination with poisonous, infectious, or irritating materials, provide one (1) lavatory for each five (5) persons.
8. Twenty-four (24) lineal inches of wash sink or eighteen (18) inches of a circular basin, when provided with water outlets for such space shall be considered equivalent to one (1) lavatory.
9. Laundry trays, one (1) for each fifty (50) persons. Service sinks, one (1) for each hundred (100) persons.
10. General. In applying this schedule of facilities, consideration shall be given to the accessibility of the fixtures. Conformity purely on a numerical basis may not result in an installation suited to the need of the individual establishment. For example, schools should be provided with toilet facilities on each floor having classrooms.
    a. Surrounding materials, wall and floor space to a point two (2) feet in front of urinal lip and four (4) feet above the floor, and at least two (2) feet to each side of the urinal shall be lined with non-absorbent materials.
    b. Trough urinals shall be prohibited.
11. A restaurant is defined as a business which sells food to be consumed on the premises.
    a. The number of occupants for a drive-in restaurant shall be considered as equal to the number of parking stalls.
    b. Employee toilet facilities shall not to be included in the above restaurant requirements. Hand washing facilities shall be available in the kitchen for employees.
12. Where food is consumed indoors, water stations may be substituted for drinking fountains. Offices, or public buildings for use by more than six (6) persons shall have one (1) drinking fountain for the first one hundred fifty (150) persons and one additional fountain for each three hundred (300) persons thereafter.
13. There shall be a minimum of one (1) drinking fountain per occupied floor in schools, theaters, auditoriums, dormitories, offices of public building.
14. The total number of water closets for females shall be at least equal to the total number of water closets and urinals required for males.

D2010 Plumbing Fixtures

One Piece Wall Hung Water Closet

Wall Hung Urinal

Side By Side Water Closet Group

Floor Mount Water Closet

Stall Type Urinal

Back to Back Water Closet Group

D2010 110	Water Closet Systems	COST EACH		
		MAT.	INST.	TOTAL
1800	Water closet, vitreous china			
1840	Tank type, wall hung			
1880	Close coupled two piece	1,575	715	2,290
1920	Floor mount, one piece	1,525	760	2,285
1960	One piece low profile	920	760	1,680
2000	Two piece close coupled	645	760	1,405
2040	Bowl only with flush valve			
2080	Wall hung	1,900	810	2,710
2120	Floor mount	840	770	1,610
2122				
2160	Floor mount, ADA compliant with 18" high bowl	840	790	1,630

D2010 120	Water Closets, Group	COST EACH		
		MAT.	INST.	TOTAL
1760	Water closets, battery mount, wall hung, side by side, first closet	2,000	835	2,835
1800	Each additional water closet, add	1,900	790	2,690
3000	Back to back, first pair of closets	3,450	1,100	4,550
3100	Each additional pair of closets, back to back	3,400	1,100	4,500

D2010 210	Urinal Systems	COST EACH		
		MAT.	INST.	TOTAL
2000	Urinal, vitreous china, wall hung	625	800	1,425
2040	Stall type	1,375	960	2,335

D2010 Plumbing Fixtures

Systems are complete with trim and rough-in (supply, waste and vent) to connect to supply branches and waste mains.

Vanity Top

Wall Hung

Counter Top Single Bowl

Counter Top Double Bowl

D2010 310	Lavatory Systems	COST EACH		
		MAT.	INST.	TOTAL
1560	Lavatory w/trim, vanity top, PE on CI, 20" x 18", Vanity top by others.	800	715	1,515
1600	19" x 16" oval	630	715	1,345
1640	18" round	705	715	1,420
1680	Cultured marble, 19" x 17"	645	715	1,360
1720	25" x 19"	675	715	1,390
1760	Stainless, self-rimming, 25" x 22"	840	715	1,555
1800	17" x 22"	830	715	1,545
1840	Steel enameled, 20" x 17"	660	735	1,395
1880	19" round	630	735	1,365
1920	Vitreous china, 20" x 16"	730	750	1,480
1960	19" x 16"	735	750	1,485
2000	22" x 13"	735	750	1,485
2040	Wall hung, PE on CI, 18" x 15"	965	790	1,755
2080	19" x 17"	985	790	1,775
2120	20" x 18"	830	790	1,620
2160	Vitreous china, 18" x 15"	785	810	1,595
2200	19" x 17"	740	810	1,550
2240	24" x 20"	830	810	1,640
2300	20" x 27", handicap	1,100	875	1,975

D2010 410	Kitchen Sink Systems	COST EACH		
		MAT.	INST.	TOTAL
1720	Kitchen sink w/trim, countertop, PE on CI, 24"x21", single bowl	800	785	1,585
1760	30" x 21" single bowl	1,100	785	1,885
1800	32" x 21" double bowl	870	845	1,715
1840	42" x 21" double bowl	1,800	855	2,655
1880	Stainless steel, 19" x 18" single bowl	1,125	785	1,910
1920	25" x 22" single bowl	1,200	785	1,985
1960	33" x 22" double bowl	1,575	845	2,420
2000	43" x 22" double bowl	1,750	855	2,605
2040	44" x 22" triple bowl	2,025	890	2,915
2080	44" x 24" corner double bowl	1,450	855	2,305
2120	Steel, enameled, 24" x 21" single bowl	1,000	785	1,785
2160	32" x 21" double bowl	1,050	845	1,895
2240	Raised deck, PE on CI, 32" x 21", dual level, double bowl	970	1,075	2,045
2280	42" x 21" dual level, triple bowl	1,850	1,175	3,025

D20 Plumbing

D2010 Plumbing Fixtures

Laboratory Sink

Service Sink

Systems are complete with trim and rough-in (supply, waste and vent) to connect to supply branches and waste mains.

Single Compartment Sink

Double Compartment Sink

D2010 420	Laundry Sink Systems	COST EACH		
		MAT.	INST.	TOTAL
1740	Laundry sink w/trim, PE on CI, black iron frame			
1760	24" x 20", single compartment	965	770	1,735
1840	48" x 21" double compartment	1,450	835	2,285
1920	Molded stone, on wall, 22" x 21" single compartment	705	770	1,475
1960	45"x 21" double compartment	895	835	1,730
2040	Plastic, on wall or legs, 18" x 23" single compartment	665	755	1,420
2080	20" x 24" single compartment	665	755	1,420
2120	36" x 23" double compartment	730	810	1,540
2160	40" x 24" double compartment	830	810	1,640

D2010 430	Laboratory Sink Systems	COST EACH		
		MAT.	INST.	TOTAL
1580	Laboratory sink w/trim, stainless steel, single bowl,			
1590	Stainless steel, single bowl,			
1600	Double drainboard, 54" x 24" O.D.	2,075	1,000	3,075
1640	Single drainboard, 47" x 24"O.D.	1,525	1,000	2,525
1670	Stainless steel, double bowl,			
1680	70" x 24" O.D.	2,275	1,000	3,275
1750	Polyethylene, single bowl,			
1760	Flanged, 14-1/2" x 14-1/2" O.D.	495	905	1,400
1800	18-1/2" x 18-1/2" O.D.	595	905	1,500
1840	23-1/2" x 20-1/2" O.D.	670	905	1,575
1920	Polypropylene, cup sink, oval, 7" x 4" O.D.	370	795	1,165
1960	10" x 4-1/2" O.D.	395	795	1,190
1961				

D2010 Plumbing Fixtures

D2010 440	Service Sink Systems	COST EACH		
		MAT.	INST.	TOTAL
4260	Service sink w/trim, PE on CI, corner floor, 28" x 28", w/rim guard	2,575	1,000	3,575
4300	Wall hung w/rim guard, 22" x 18"	2,775	1,175	3,950
4340	24" x 20"	2,875	1,175	4,050
4380	Vitreous china, wall hung 22" x 20"	2,600	1,175	3,775
4383				

D2010 Plumbing Fixtures

Recessed Bathtub

Corner Bathtub

Systems are complete with trim and rough-in (supply, waste and vent) to connect to supply branches and waste mains.

Circular Wash Fountain

Semi-Circular Wash Fountain

D2010 510	Bathtub Systems	COST EACH		
		MAT.	INST.	TOTAL
2000	Bathtub, recessed, P.E. on Cl., 48" x 42"	3,425	890	4,315
2040	72" x 36"	3,525	990	4,515
2080	Mat bottom, 5' long	1,800	860	2,660
2120	5'-6" long	2,525	890	3,415
2160	Corner, 48" x 42"	3,425	860	4,285
2200	Formed steel, enameled, 4'-6" long	1,150	795	1,945

D2010 610	Group Wash Fountain Systems	COST EACH		
		MAT.	INST.	TOTAL
1740	Group wash fountain, precast terrazzo			
1760	Circular, 36" diameter	6,750	1,300	8,050
1800	54" diameter	8,225	1,425	9,650
1840	Semi-circular, 36" diameter	6,000	1,300	7,300
1880	54" diameter	7,400	1,425	8,825
1960	Stainless steel, circular, 36" diameter	6,750	1,200	7,950
2000	54" diameter	8,500	1,325	9,825
2040	Semi-circular, 36" diameter	5,975	1,200	7,175
2080	54" diameter	7,350	1,325	8,675
2160	Thermoplastic, circular, 36" diameter	5,075	980	6,055
2200	54" diameter	5,775	1,125	6,900
2240	Semi-circular, 36" diameter	4,075	980	5,055
2280	54" diameter	5,600	1,125	6,725

D2010 Plumbing Fixtures

Square Shower Stall

Corner Angle Shower Stall

Systems are complete with trim and rough-in (supply, waste and vent) to connect to supply branches and waste mains.

Wall Mounted, Low Back

Wall Mounted, No Back

D2010 710	Shower Systems	COST EACH		
		MAT.	INST.	TOTAL
1560	Shower, stall, baked enamel, molded stone receptor, 30" square	1,775	825	2,600
1600	32" square	1,775	835	2,610
1640	Terrazzo receptor, 32" square	1,975	835	2,810
1680	36" square	2,125	845	2,970
1720	36" corner angle	2,400	845	3,245
1800	Fiberglass one piece, three walls, 32" square	1,100	810	1,910
1840	36" square	1,125	810	1,935
1880	Polypropylene, molded stone receptor, 30" square	1,200	1,200	2,400
1920	32" square	1,200	1,200	2,400
1960	Built-in head, arm, bypass, stops and handles	121	310	431
2050	Shower, stainless steel panels, handicap			
2100	w/fixed and handheld head, control valves, grab bar, and seat	6,025	3,700	9,725
2500	Shower, group with six heads, thermostatic mix valves & balancing valve	5,450	910	6,360
2520	Five heads	4,525	825	5,350

D2010 810	Drinking Fountain Systems	COST EACH		
		MAT.	INST.	TOTAL
1740	Drinking fountain, one bubbler, wall mounted			
1760	Non recessed			
1800	Bronze, no back	1,525	470	1,995
1840	Cast iron, enameled, low back	1,375	470	1,845
1880	Fiberglass, 12" back	2,075	470	2,545
1920	Stainless steel, no back	1,525	470	1,995
1960	Semi-recessed, poly marble	1,400	470	1,870
2040	Stainless steel	1,625	470	2,095
2080	Vitreous china	1,350	470	1,820
2120	Full recessed, poly marble	2,125	470	2,595
2200	Stainless steel	2,050	470	2,520
2240	Floor mounted, pedestal type, aluminum	2,850	640	3,490
2320	Bronze	2,500	640	3,140
2360	Stainless steel	2,300	640	2,940

D2010 Plumbing Fixtures

Wall Hung Water Cooler

Floor Mounted Water Cooler

Systems are complete with trim and rough-in (supply, waste and vent) to connect to supply branches and waste mains.

D2010 820	Water Cooler Systems	MAT.	INST.	TOTAL
1840	Water cooler, electric, wall hung, 8.2 G.P.H.	1,175	610	1,785
1880	Dual height, 14.3 G.P.H.	1,450	625	2,075
1920	Wheelchair type, 7.5 G.P.H.	1,375	610	1,985
1960	Semi recessed, 8.1 G.P.H.	1,175	610	1,785
2000	Full recessed, 8 G.P.H.	2,500	650	3,150
2040	Floor mounted, 14.3 G.P.H.	1,325	525	1,850
2080	Dual height, 14.3 G.P.H.	1,650	640	2,290
2120	Refrigerated compartment type, 1.5 G.P.H.	1,925	525	2,450

D2010 Plumbing Fixtures

Two Fixture   Three Fixture

*Common wall is with adjacent bathroom.

D2010 920	Two Fixture Bathroom, Two Wall Plumbing	COST EACH		
		MAT.	INST.	TOTAL
1180	Bathroom, lavatory & water closet, 2 wall plumbing, stand alone	1,800	1,975	3,775
1200	Share common plumbing wall*	1,600	1,700	3,300

D2010 922	Two Fixture Bathroom, One Wall Plumbing	COST EACH		
		MAT.	INST.	TOTAL
2220	Bathroom, lavatory & water closet, one wall plumbing, stand alone	1,650	1,775	3,425
2240	Share common plumbing wall*	1,450	1,500	2,950

D2010 924	Three Fixture Bathroom, One Wall Plumbing	COST EACH		
		MAT.	INST.	TOTAL
1150	Bathroom, three fixture, one wall plumbing			
1160	Lavatory, water closet & bathtub			
1170	Stand alone	3,100	2,300	5,400
1180	Share common plumbing wall *	2,650	1,675	4,325

D2010 926	Three Fixture Bathroom, Two Wall Plumbing	COST EACH		
		MAT.	INST.	TOTAL
2130	Bathroom, three fixture, two wall plumbing			
2140	Lavatory, water closet & bathtub			
2160	Stand alone	3,125	2,325	5,450
2180	Long plumbing wall common *	2,800	1,850	4,650
3610	Lavatory, bathtub & water closet			
3620	Stand alone	3,425	2,650	6,075
3640	Long plumbing wall common *	3,175	2,400	5,575
4660	Water closet, corner bathtub & lavatory			
4680	Stand alone	4,775	2,375	7,150
4700	Long plumbing wall common *	4,275	1,775	6,050
6100	Water closet, stall shower & lavatory			
6120	Stand alone	3,650	2,650	6,300
6140	Long plumbing wall common *	3,450	2,450	5,900
7060	Lavatory, corner stall shower & water closet			
7080	Stand alone	4,000	2,350	6,350
7100	Short plumbing wall common *	3,200	1,575	4,775

D2010 Plumbing Fixtures

Four Fixture

Five Fixture

*Common wall is with adjacent bathroom.

D2010 928	Four Fixture Bathroom, Two Wall Plumbing	COST EACH		
		MAT.	INST.	TOTAL
1140	Bathroom, four fixture, two wall plumbing			
1150	Bathtub, water closet, stall shower & lavatory			
1160	Stand alone	4,600	2,550	7,150
1180	Long plumbing wall common *	4,100	1,950	6,050
2260	Bathtub, lavatory, corner stall shower & water closet			
2280	Stand alone	5,225	2,550	7,775
2320	Long plumbing wall common *	4,725	1,975	6,700
3620	Bathtub, stall shower, lavatory & water closet			
3640	Stand alone	5,100	3,200	8,300
3660	Long plumbing wall (opp. door) common *	4,575	2,600	7,175
4680	Bathroom, four fixture, three wall plumbing			
4700	Bathtub, stall shower, lavatory & water closet			
4720	Stand alone	5,925	3,525	9,450
4760	Long plumbing wall (opposite door) common *	5,750	3,250	9,000

D2010 932	Five Fixture Bathroom, Two Wall Plumbing	COST EACH		
		MAT.	INST.	TOTAL
1320	Bathroom, five fixture, two wall plumbing			
1340	Bathtub, water closet, stall shower & two lavatories			
1360	Stand alone	5,925	4,075	10,000
1400	One short plumbing wall common *	5,400	3,500	8,900
2360	Bathroom, five fixture, three wall plumbing			
2380	Water closet, bathtub, two lavatories & stall shower			
2400	Stand alone	6,550	4,075	10,625
2440	One short plumbing wall common *	6,025	3,500	9,525
4080	Bathroom, five fixture, one wall plumbing			
4100	Bathtub, two lavatories, corner stall shower & water closet			
4120	Stand alone	6,225	3,650	9,875
4160	Share common wall *	5,250	2,350	7,600

D2020   Domestic Water Distribution

Installation includes piping and fittings within 10′ of heater. Gas and oil fired heaters require vent piping (not included with these units).

Gas Fired

Oil Fired

D2020 220	Gas Fired Water Heaters - Residential Systems	COST EACH		
		MAT.	INST.	TOTAL
2200	Gas fired water heater, residential, 100°F rise			
2260	30 gallon tank, 32 GPH	1,725	1,400	3,125
2300	40 gallon tank, 32 GPH	2,075	1,575	3,650
2340	50 gallon tank, 63 GPH	2,100	1,575	3,675
2380	75 gallon tank, 63 GPH	3,000	1,750	4,750
2420	100 gallon tank, 63 GPH	3,175	1,825	5,000
2422				

D2020 230	Oil Fired Water Heaters - Residential Systems	COST EACH		
		MAT.	INST.	TOTAL
2200	Oil fired water heater, residential, 100°F rise			
2220	30 gallon tank, 103 GPH	1,900	1,300	3,200
2260	50 gallon tank, 145 GPH	2,450	1,450	3,900
2300	70 gallon tank, 164 GPH	3,475	1,625	5,100
2340	85 gallon tank, 181 GPH	10,700	1,675	12,375

D2020 Domestic Water Distribution

Electric

Systems below include piping and fittings within 10' of heater. Electric water heaters do not require venting. Gas fired heaters require vent piping (not included in these prices).

Gas Fired

D2020 240	Electric Water Heaters - Commercial Systems	COST EACH		
		MAT.	INST.	TOTAL
1800	Electric water heater, commercial, 100°F rise			
1820	50 gallon tank, 9 KW 37 GPH	5,075	1,175	6,250
1860	80 gal, 12 KW 49 GPH	6,800	1,450	8,250
1900	36 KW 147 GPH	9,225	1,550	10,775
1940	120 gal, 36 KW 147 GPH	9,875	1,675	11,550
1980	150 gal, 120 KW 490 GPH	28,600	1,800	30,400
2020	200 gal, 120 KW 490 GPH	30,000	1,850	31,850
2060	250 gal, 150 KW 615 GPH	33,800	2,150	35,950
2100	300 gal, 180 KW 738 GPH	43,200	2,275	45,475
2140	350 gal, 30 KW 123 GPH	27,600	2,450	30,050
2180	180 KW 738 GPH	37,700	2,450	40,150
2220	500 gal, 30 KW 123 GPH	35,600	2,900	38,500
2260	240 KW 984 GPH	60,500	2,900	63,400
2300	700 gal, 30 KW 123 GPH	29,400	3,300	32,700
2340	300 KW 1230 GPH	42,600	3,300	45,900
2380	1000 gal, 60 KW 245 GPH	37,100	4,600	41,700
2420	480 KW 1970 GPH	57,500	4,625	62,125
2460	1500 gal, 60 KW 245 GPH	75,500	5,700	81,200
2500	480 KW 1970 GPH	102,500	5,700	108,200

D2020 250	Gas Fired Water Heaters - Commercial Systems	COST EACH		
		MAT.	INST.	TOTAL
1760	Gas fired water heater, commercial, 100°F rise			
1780	75.5 MBH input, 63 GPH	4,975	1,800	6,775
1820	95 MBH input, 86 GPH	7,050	1,800	8,850
1860	100 MBH input, 91 GPH	7,250	1,875	9,125
1900	115 MBH input, 110 GPH	8,475	1,950	10,425
1980	155 MBH input, 150 GPH	10,100	2,175	12,275
2020	175 MBH input, 168 GPH	10,800	2,325	13,125
2060	200 MBH input, 192 GPH	11,200	2,650	13,850
2100	240 MBH input, 230 GPH	11,900	2,875	14,775
2140	300 MBH input, 278 GPH	13,500	3,275	16,775
2180	390 MBH input, 374 GPH	15,600	3,325	18,925
2220	500 MBH input, 480 GPH	20,700	3,550	24,250
2260	600 MBH input, 576 GPH	24,500	3,850	28,350

D2020 Domestic Water Distribution

Oil Fired

Oil fired water heater systems include piping and fittings within 10′ of heater. Oil fired heaters require vent piping (not included in these systems).

D2020 260	Oil Fired Water Heaters - Commercial Systems	COST EACH		
		MAT.	INST.	TOTAL
1800	Oil fired water heater, commercial, 100°F rise			
1820	140 gal., 140 MBH input, 134 GPH	21,700	1,525	23,225
1900	140 gal., 255 MBH input, 247 GPH	23,700	1,925	25,625
1940	140 gal., 270 MBH input, 259 GPH	28,900	2,175	31,075
1980	140 gal., 400 MBH input, 384 GPH	30,100	2,525	32,625
2060	140 gal., 720 MBH input, 691 GPH	32,000	2,650	34,650
2100	221 gal., 300 MBH input, 288 GPH	42,500	2,900	45,400
2140	221 gal., 600 MBH input, 576 GPH	47,100	2,925	50,025
2180	221 gal., 800 MBH input, 768 GPH	47,800	3,050	50,850
2220	201 gal., 1000 MBH input, 960 GPH	48,900	3,075	51,975
2260	201 gal., 1250 MBH input, 1200 GPH	49,800	3,150	52,950
2300	201 gal., 1500 MBH input, 1441 GPH	54,000	3,225	57,225
2340	411 gal., 600 MBH input, 576 GPH	54,500	3,300	57,800
2380	411 gal., 800 MBH input, 768 GPH	54,500	3,400	57,900
2420	411 gal., 1000 MBH input, 960 GPH	60,000	3,900	63,900
2460	411 gal., 1250 MBH input, 1200 GPH	61,000	4,025	65,025
2500	397 gal., 1500 MBH input, 1441 GPH	62,000	4,150	66,150
2540	397 gal., 1750 MBH input, 1681 GPH	66,500	4,300	70,800

D2040 Rain Water Drainage

Design Assumptions: Vertical conductor size is based on a maximum rate of rainfall of 4″ per hour. To convert roof area to other rates multiply "Max. S.F. Roof Area" shown by four and divide the result by desired local rate. The answer is the local roof area that may be handled by the indicated pipe diameter.

Basic cost is for roof drain, 10′ of vertical leader and 10′ of horizontal, plus connection to the main.

Pipe Dia.	Max. S.F. Roof Area	Gallons per Min.
2″	544	23
3″	1610	67
4″	3460	144
5″	6280	261
6″	10,200	424
8″	22,000	913

D2040 210	Roof Drain Systems	MAT.	INST.	TOTAL
		COST EACH		
1880	Roof drain, DWV PVC, 2″ diam., piping, 10′ high	267	675	942
1920	For each additional foot add	3.95	20.50	24.45
1960	3″ diam., 10′ high	360	790	1,150
2000	For each additional foot add	7.10	23	30.10
2040	4″ diam., 10′ high	445	880	1,325
2080	For each additional foot add	9.65	25	34.65
2120	5″ diam., 10′ high	1,325	1,025	2,350
2160	For each additional foot add	28.50	28	56.50
2200	6″ diam., 10′ high	1,350	1,125	2,475
2240	For each additional foot add	19.40	31	50.40
2280	8″ diam., 10′ high	2,950	1,925	4,875
2320	For each additional foot add	45	39	84
3940	C.I., soil, single hub, service wt., 2″ diam. piping, 10′ high	560	735	1,295
3980	For each additional foot add	10	19.20	29.20
4120	3″ diam., 10′ high	780	795	1,575
4160	For each additional foot add	14.05	20	34.05
4200	4″ diam., 10′ high	920	870	1,790
4240	For each additional foot add	18.60	22	40.60
4280	5″ diam., 10′ high	1,275	970	2,245
4320	For each additional foot add	25	25	50
4360	6″ diam., 10′ high	1,625	1,025	2,650
4400	For each additional foot add	31.50	26	57.50
4440	8″ diam., 10′ high	3,175	2,100	5,275
4480	For each additional foot add	48.50	43.50	92
6040	Steel galv. sch 40 threaded, 2″ diam. piping, 10′ high	730	715	1,445
6080	For each additional foot add	14.65	18.90	33.55
6120	3″ diam., 10′ high	1,425	1,025	2,450
6160	For each additional foot add	31	28	59
6200	4″ diam., 10′ high	2,050	1,325	3,375
6240	For each additional foot add	45.50	33.50	79
6280	5″ diam., 10′ high	2,325	1,100	3,425
6320	For each additional foot add	64.50	32.50	97
6360	6″ diam, 10′ high	2,775	1,425	4,200
6400	For each additional foot add	73	45	118
6440	8″ diam., 10′ high	4,375	2,050	6,425
6480	For each additional foot add	86	51	137

D20 Plumbing

D2090 Other Plumbing Systems

D2090 810	Piping - Installed - Unit Costs	COST PER L.F.		
		MAT.	INST.	TOTAL
0840	Cast iron, soil, B & S, service weight, 2" diameter	10	19.20	29.20
0860	3" diameter	14.05	20	34.05
0880	4" diameter	18.60	22	40.60
0900	5" diameter	25	25	50
0920	6" diameter	31.50	26	57.50
0940	8" diameter	48.50	43.50	92
0960	10" diameter	78	47.50	125.50
0980	12" diameter	111	53.50	164.50
1040	No hub, 1-1/2" diameter	9.90	17.05	26.95
1060	2" diameter	10.15	18.05	28.20
1080	3" diameter	14.05	18.90	32.95
1100	4" diameter	18.35	21	39.35
1120	5" diameter	24.50	22.50	47
1140	6" diameter	32	24	56
1160	8" diameter	55	37	92
1180	10" diameter	91.50	42	133.50
1220	Copper tubing, hard temper, solder, type K, 1/2" diameter	6.10	8.60	14.70
1260	3/4" diameter	10.75	9.10	19.85
1280	1" diameter	14.40	10.20	24.60
1300	1-1/4" diameter	17.90	12	29.90
1320	1-1/2" diameter	23.50	13.45	36.95
1340	2" diameter	36	16.80	52.80
1360	2-1/2" diameter	55	20	75
1380	3" diameter	77	22.50	99.50
1400	4" diameter	132	32	164
1480	5" diameter	292	38	330
1500	6" diameter	430	49.50	479.50
1520	8" diameter	720	55.50	775.50
1560	Type L, 1/2" diameter	4.66	8.30	12.96
1600	3/4" diameter	7.20	8.85	16.05
1620	1" diameter	10.55	9.90	20.45
1640	1-1/4" diameter	15.05	11.60	26.65
1660	1-1/2" diameter	19.30	12.95	32.25
1680	2" diameter	30.50	16	46.50
1700	2-1/2" diameter	47	19.50	66.50
1720	3" diameter	64	21.50	85.50
1740	4" diameter	111	31	142
1760	5" diameter	235	35.50	270.50
1780	6" diameter	335	47	382
1800	8" diameter	540	52.50	592.50
1840	Type M, 1/2" diameter	3.59	8	11.59
1880	3/4" diameter	5.45	8.60	14.05
1900	1" diameter	8.60	9.60	18.20
1920	1-1/4" diameter	12.60	11.20	23.80
1940	1-1/2" diameter	17	12.45	29.45
1960	2" diameter	27.50	15.30	42.80
1980	2-1/2" diameter	41.50	18.90	60.40
2000	3" diameter	56	21	77
2020	4" diameter	102	30.50	132.50
2060	6" diameter	335	45	380
2080	8" diameter	540	49.50	589.50
2120	Type DWV, 1-1/4" diameter	13.35	11.20	24.55
2160	1-1/2" diameter	16.75	12.45	29.20
2180	2" diameter	23	15.30	38.30
2200	3" diameter	43.50	21	64.50
2220	4" diameter	78	30.50	108.50
2240	5" diameter	213	33.50	246.50
2260	6" diameter	310	45	355

D2090 Other Plumbing Systems

D2090 810	Piping - Installed - Unit Costs	COST PER L.F.		
		MAT.	INST.	TOTAL
2280	8" diameter	675	49.50	724.50
2800	Plastic, PVC, DWV, schedule 40, 1-1/4" diameter	3.78	16	19.78
2820	1-1/2" diameter	3.43	18.70	22.13
2830	2" diameter	3.95	20.50	24.45
2840	3" diameter	7.10	23	30.10
2850	4" diameter	9.65	25	34.65
2890	6" diameter	19.40	31	50.40
3010	Pressure pipe 200 PSI, 1/2" diameter	2.97	12.45	15.42
3030	3/4" diameter	3.30	13.20	16.50
3040	1" diameter	3.56	14.60	18.16
3050	1-1/4" diameter	4.34	16	20.34
3060	1-1/2" diameter	4.67	18.70	23.37
3070	2" diameter	5.60	20.50	26.10
3080	2-1/2" diameter	10	21.50	31.50
3090	3" diameter	11.35	23	34.35
3100	4" diameter	18.25	25	43.25
3110	6" diameter	33	31	64
3120	8" diameter	50.50	39	89.50
4000	Steel, schedule 40, black, threaded, 1/2" diameter	3.25	10.65	13.90
4020	3/4" diameter	3.81	11	14.81
4030	1" diameter	5.60	12.70	18.30
4040	1-1/4" diameter	6.95	13.60	20.55
4050	1-1/2" diameter	10.45	15.15	25.60
4060	2" diameter	10.75	18.90	29.65
4070	2-1/2" diameter	16.80	24	40.80
4080	3" diameter	21.50	28	49.50
4090	4" diameter	32.50	33.50	66
4100	Grooved, 5" diameter	40.50	32.50	73
4110	6" diameter	44.50	45	89.50
4120	8" diameter	68	51	119
4130	10" diameter	85	60.50	145.50
4140	12" diameter	98	69.50	167.50
4200	Galvanized, threaded, 1/2" diameter	4.51	10.65	15.16
4220	3/4" diameter	5.35	11	16.35
4230	1" diameter	7.40	12.70	20.10
4240	1-1/4" diameter	9.35	13.60	22.95
4250	1-1/2" diameter	10.95	15.15	26.10
4260	2" diameter	14.65	18.90	33.55
4270	2-1/2" diameter	24.50	24	48.50
4280	3" diameter	31	28	59
4290	4" diameter	45.50	33.50	79
4300	Grooved, 5" diameter	64.50	32.50	97
4310	6" diameter	73	45	118
4320	8" diameter	86	51	137
4330	10" diameter	137	60.50	197.50
4340	12" diameter	168	69.50	237.50
5010	Flanged, black, 1" diameter	11.15	18.15	29.30
5020	1-1/4" diameter	12.80	19.80	32.60
5030	1-1/2" diameter	13.50	22	35.50
5040	2" diameter	16	28.50	44.50
5050	2-1/2" diameter	23	35	58
5060	3" diameter	27	40	67
5070	4" diameter	35.50	48.50	84
5080	5" diameter	55	60.50	115.50
5090	6" diameter	59	78	137
5100	8" diameter	91	102	193
5110	10" diameter	132	122	254
5120	12" diameter	166	138	304

D2090 Other Plumbing Systems

D2090 810	Piping - Installed - Unit Costs	COST PER L.F.		
		MAT.	INST.	TOTAL
5720	Grooved joints, black, 3/4" diameter	5.15	9.45	14.60
5730	1" diameter	4.91	10.65	15.56
5740	1-1/4" diameter	6.35	11.60	17.95
5750	1-1/2" diameter	7.20	13.20	20.40
5760	2" diameter	8.85	16.80	25.65
5770	2-1/2" diameter	15.35	21	36.35
5900	3" diameter	19.05	24	43.05
5910	4" diameter	26	27	53
5920	5" diameter	40.50	32.50	73
5930	6" diameter	44.50	45	89.50
5940	8" diameter	68	51	119
5950	10" diameter	85	60.50	145.50
5960	12" diameter	98	69.50	167.50

D3010 Energy Supply

Forced Hot Water Heating System

Fin Tube Radiation

Terminal Unit Heater

D3010 510	Apartment Building Heating - Fin Tube Radiation	COST PER S.F.		
		MAT.	INST.	TOTAL
1740	Heating systems, fin tube radiation, forced hot water			
1760	1,000 S.F. area, 10,000 C.F. volume	7.18	6.50	13.68
1800	10,000 S.F. area, 100,000 C.F. volume	3.48	3.93	7.41
1840	20,000 S.F. area, 200,000 C.F. volume	3.99	4.39	8.38
1880	30,000 S.F. area, 300,000 C.F. volume	3.82	4.25	8.07

D3010 520	Commercial Building Heating - Fin Tube Radiation	COST PER S.F.		
		MAT.	INST.	TOTAL
1940	Heating systems, fin tube radiation, forced hot water			
1960	1,000 S.F. bldg, one floor	18	14.70	32.70
2000	10,000 S.F., 100,000 C.F., total two floors	4.89	5.55	10.44
2040	100,000 S.F., 1,000,000 C.F., total three floors	1.89	2.48	4.37
2080	1,000,000 S.F., 10,000,000 C.F., total five floors	1.05	1.31	2.36

D3010 530	Commercial Bldg. Heating - Terminal Unit Heaters	COST PER S.F.		
		MAT.	INST.	TOTAL
1860	Heating systems, terminal unit heaters, forced hot water			
1880	1,000 S.F. bldg., one floor	17.70	13.50	31.20
1920	10,000 S.F. bldg., 100,000 C.F. total two floors	4.44	4.60	9.04
1960	100,000 S.F. bldg., 1,000,000 C.F. total three floors	1.93	2.27	4.20
2000	1,000,000 S.F. bldg., 10,000,000 C.F. total five floors	1.33	1.47	2.80

D3020 Heat Generating Systems

Small Electric Boiler Systems Considerations:

1. Terminal units are fin tube baseboard radiation rated at 720 BTU/hr with 200° water temperature or 820 BTU/hr steam.
2. Primary use being for residential or smaller supplementary areas, the floor levels are based on 7-1/2″ ceiling heights.
3. All distribution piping is copper for boilers through 205 MBH. All piping for larger systems is steel pipe.

Large Electric Boiler System Considerations:

1. Terminal units are all unit heaters of the same size. Quantities are varied to accommodate total requirements.
2. All air is circulated through the heaters a minimum of three times per hour.
3. As the capacities are adequate for commercial use, gross output rating by floor levels are based on 10′ ceiling height.
4. All distribution piping is black steel pipe.

D3020 102	Small Heating Systems, Hydronic, Electric Boilers	COST PER S.F.		
		MAT.	INST.	TOTAL
1100	Small heating systems, hydronic, electric boilers			
1120	Steam, 1 floor, 1480 S.F., 61 M.B.H.	8.18	8.38	16.56
1160	3,000 S.F., 123 M.B.H.	6.45	7.35	13.80
1200	5,000 S.F., 205 M.B.H.	5.75	6.80	12.55
1240	2 floors, 12,400 S.F., 512 M.B.H.	4.58	6.75	11.33
1280	3 floors, 24,800 S.F., 1023 M.B.H.	5	6.65	11.65
1320	34,750 S.F., 1,433 M.B.H.	4.62	6.45	11.07
1360	Hot water, 1 floor, 1,000 S.F., 41 M.B.H.	12.90	4.64	17.54
1400	2,500 S.F., 103 M.B.H.	9.85	8.35	18.20
1440	2 floors, 4,850 S.F., 205 M.B.H.	9.90	10.05	19.95
1480	3 floors, 9,700 S.F., 410 M.B.H.	10.95	10.35	21.30

D3020 104	Large Heating Systems, Hydronic, Electric Boilers	COST PER S.F.		
		MAT.	INST.	TOTAL
1230	Large heating systems, hydronic, electric boilers			
1240	9,280 S.F., 135 K.W., 461 M.B.H., 1 floor	4.05	3.14	7.19
1280	14,900 S.F., 240 K.W., 820 M.B.H., 2 floors	5.40	5.05	10.45
1320	18,600 S.F., 296 K.W., 1,010 M.B.H., 3 floors	5.35	5.50	10.85
1360	26,100 S.F., 420 K.W., 1,432 M.B.H., 4 floors	5.25	5.40	10.65
1400	39,100 S.F., 666 K.W., 2,273 M.B.H., 4 floors	4.49	4.51	9
1440	57,700 S.F., 900 K.W., 3,071 M.B.H., 5 floors	4.34	4.42	8.76
1480	111,700 S.F., 1,800 K.W., 6,148 M.B.H., 6 floors	3.85	3.80	7.65
1520	149,000 S.F., 2,400 K.W., 8,191 M.B.H., 8 floors	3.80	3.80	7.60
1560	223,300 S.F., 3,600 K.W., 12,283 M.B.H., 14 floors	4.06	4.30	8.36

D3020 Heat Generating Systems

Boiler Selection: The maximum allowable working pressures are limited by ASME "Code for Heating Boilers" to 15 PSI for steam and 160 PSI for hot water heating boilers, with a maximum temperature limitation of 250° F. Hot water boilers are generally rated for a working pressure of 30 PSI. High pressure boilers are governed by the ASME "Code for Power Boilers" which is used almost universally for boilers operating over 15 PSIG. High pressure boilers used for a combination of heating/process loads are usually designed for 150 PSIG.

Boiler ratings are usually indicated as either Gross or Net Output. The Gross Load is equal to the Net Load plus a piping and pickup allowance. When this allowance cannot be determined, divide the gross output rating by 1.25 for a value equal to or greater than the next heat loss requirement of the building.

Table below lists installed cost per boiler and includes insulating jacket, standard controls, burner and safety controls. Costs do not include piping or boiler base pad. Outputs are Gross.

D3020 106	Boilers, Hot Water & Steam	COST EACH		
		MAT.	INST.	TOTAL
0600	Boiler, electric, steel, hot water, 12 K.W., 41 M.B.H.	5,275	1,425	6,700
0620	30 K.W., 103 M.B.H.	5,775	1,550	7,325
0640	60 K.W., 205 M.B.H.	6,425	1,675	8,100
0660	120 K.W., 410 M.B.H.	6,800	2,050	8,850
0680	210 K.W., 716 M.B.H.	8,125	3,100	11,225
0700	510 K.W., 1,739 M.B.H.	23,500	5,775	29,275
0720	720 K.W., 2,452 M.B.H.	29,000	6,500	35,500
0740	1,200 K.W., 4,095 M.B.H.	36,900	7,475	44,375
0760	2,100 K.W., 7,167 M.B.H.	62,500	9,400	71,900
0780	3,600 K.W., 12,283 M.B.H.	98,500	15,900	114,400
0820	Steam, 6 K.W., 20.5 M.B.H.	4,175	1,550	5,725
0840	24 K.W., 81.8 M.B.H.	5,125	1,675	6,800
0860	60 K.W., 205 M.B.H.	7,100	1,850	8,950
0880	150 K.W., 512 M.B.H.	10,200	2,850	13,050
0900	510 K.W., 1,740 M.B.H.	34,800	7,050	41,850
0920	1,080 K.W., 3,685 M.B.H.	43,400	10,200	53,600
0940	2,340 K.W., 7,984 M.B.H.	92,500	15,900	108,400
0980	Gas, cast iron, hot water, 80 M.B.H.	1,775	1,775	3,550
1000	100 M.B.H.	2,075	1,925	4,000
1020	163 M.B.H.	2,825	2,600	5,425
1040	280 M.B.H.	4,025	2,875	6,900
1060	544 M.B.H.	9,875	5,125	15,000
1080	1,088 M.B.H.	14,900	6,500	21,400
1100	2,000 M.B.H.	21,600	10,200	31,800
1120	2,856 M.B.H.	31,100	13,000	44,100
1140	4,720 M.B.H.	86,500	17,900	104,400
1160	6,970 M.B.H.	104,500	29,200	133,700
1180	For steam systems under 2,856 M.B.H., add 8%			
1520	Oil, cast iron, hot water, 109 M.B.H.	2,400	2,175	4,575
1540	173 M.B.H.	3,050	2,600	5,650
1560	236 M.B.H.	3,875	3,050	6,925
1580	1,084 M.B.H.	10,700	6,900	17,600
1600	1,600 M.B.H.	13,700	9,925	23,625
1620	2,480 M.B.H.	21,000	12,700	33,700
1640	3,550 M.B.H.	27,100	15,200	42,300
1660	Steam systems same price as hot water			

D3020 Heat Generating Systems

Unit Heater

Fossil Fuel Boiler System Considerations:

1. Terminal units are horizontal unit heaters. Quantities are varied to accommodate total heat loss per building.
2. Unit heater selection was determined by their capacity to circulate the building volume a minimum of three times per hour in addition to the BTU output.
3. Systems shown are forced hot water. Steam boilers cost slightly more than hot water boilers. However, this is compensated for by the smaller size or fewer terminal units required with steam.
4. Floor levels are based on 10′ story heights.
5. MBH requirements are gross boiler output.

D3020 108	Heating Systems, Unit Heaters	COST PER S.F.		
		MAT.	INST.	TOTAL
1260	Heating systems, hydronic, fossil fuel, terminal unit heaters,			
1280	Cast iron boiler, gas, 80 M.B.H., 1,070 S.F. bldg.	10.75	10	20.75
1320	163 M.B.H., 2,140 S.F. bldg.	7.15	6.80	13.95
1360	544 M.B.H., 7,250 S.F. bldg.	5.30	4.81	10.11
1400	1,088 M.B.H., 14,500 S.F. bldg.	4.54	4.54	9.08
1440	3,264 M.B.H., 43,500 S.F. bldg.	3.62	3.42	7.04
1480	5,032 M.B.H., 67,100 S.F. bldg.	4.35	3.52	7.87
1520	Oil, 109 M.B.H., 1,420 S.F. bldg.	11.60	8.90	20.50
1560	235 M.B.H., 3,150 S.F. bldg.	7.60	6.35	13.95
1600	940 M.B.H., 12,500 S.F. bldg.	5.30	4.14	9.44
1640	1,600 M.B.H., 21,300 S.F. bldg.	4.89	3.96	8.85
1680	2,480 M.B.H., 33,100 S.F. bldg.	4.91	3.59	8.50
1720	3,350 M.B.H., 44,500 S.F. bldg.	4.56	3.65	8.21
1760	Coal, 148 M.B.H., 1,975 S.F. bldg.	10.25	5.80	16.05
1800	300 M.B.H., 4,000 S.F. bldg.	7.75	4.54	12.29
1840	2,360 M.B.H., 31,500 S.F. bldg.	5.35	3.65	9

D3020   Heat Generating Systems

Fin Tube Radiator

Fossil Fuel Boiler System Considerations:

1. Terminal units are commercial steel fin tube radiation. Quantities are varied to accommodate total heat loss per building.
2. Systems shown are forced hot water. Steam boilers cost slightly more than hot water boilers. However, this is compensated for by the smaller size or fewer terminal units required with steam.
3. Floor levels are based on 10' story heights.
4. MBH requirements are gross boiler output.

D3020 110	Heating System, Fin Tube Radiation	COST PER S.F.		
		MAT.	INST.	TOTAL
3230	Heating systems, hydronic, fossil fuel, fin tube radiation			
3240	Cast iron boiler, gas, 80 MBH, 1,070 S.F. bldg.	13.18	15.05	28.23
3280	169 M.B.H., 2,140 S.F. bldg.	8.25	9.50	17.75
3320	544 M.B.H., 7,250 S.F. bldg.	7.10	8.15	15.25
3360	1,088 M.B.H., 14,500 S.F. bldg.	6.40	8	14.40
3400	3,264 M.B.H., 43,500 S.F. bldg.	5.65	7	12.65
3440	5,032 M.B.H., 67,100 S.F. bldg.	6.40	7.15	13.55
3480	Oil, 109 M.B.H., 1,420 S.F. bldg.	15.85	16.40	32.25
3520	235 M.B.H., 3,150 S.F. bldg.	9.35	9.60	18.95
3560	940 M.B.H., 12,500 S.F. bldg.	7.20	7.60	14.80
3600	1,600 M.B.H., 21,300 S.F. bldg.	6.90	7.55	14.45
3640	2,480 M.B.H., 33,100 S.F. bldg.	6.95	7.20	14.15
3680	3,350 M.B.H., 44,500 S.F. bldg.	6.55	7.25	13.80
3720	Coal, 148 M.B.H., 1,975 S.F. bldg.	11.95	9.05	21
3760	300 M.B.H., 4,000 S.F. bldg.	9.40	7.80	17.20
3800	2,360 M.B.H., 31,500 S.F. bldg.	7.35	7.20	14.55
4080	Steel boiler, oil, 97 M.B.H., 1,300 S.F. bldg.	14.05	13.45	27.50
4120	315 M.B.H., 4,550 S.F. bldg.	6.20	6.85	13.05
4160	525 M.B.H., 7,000 S.F. bldg.	8.70	7.90	16.60
4200	1,050 M.B.H., 14,000 S.F. bldg.	7.60	7.90	15.50
4240	2,310 M.B.H., 30,800 S.F. bldg.	7.35	7.30	14.65
4280	3,150 M.B.H., 42,000 S.F. bldg.	7.10	7.35	14.45

D3030 Cooling Generating Systems

Chilled Water Supply & Return Piping

Air Cooled Water Chiller Unit

Roof

Insulate

Return

Fan Coil Unit

Supply

Finish Ceiling

*Cooling requirements would lead to a
choice of multiple chillers.

D3030 110	Chilled Water, Air Cooled Condenser Systems	COST PER S.F.		
		MAT.	INST.	TOTAL
1180	Packaged chiller, air cooled, with fan coil unit			
1200	Apartment corridors, 3,000 S.F., 5.50 ton	8.20	8.30	16.50
1240	6,000 S.F., 11.00 ton	6.90	6.65	13.55
1280	10,000 S.F., 18.33 ton	5.95	5.05	11
1320	20,000 S.F., 36.66 ton	4.33	3.65	7.98
1360	40,000 S.F., 73.33 ton	5.10	3.68	8.78
1440	Banks and libraries, 3,000 S.F., 12.50 ton	12.65	9.75	22.40
1480	6,000 S.F., 25.00 ton	11.55	7.95	19.50
1520	10,000 S.F., 41.66 ton	8.35	5.55	13.90
1560	20,000 S.F., 83.33 ton	8.55	5	13.55
1600	40,000 S.F., 167 ton*			
1680	Bars and taverns, 3,000 S.F., 33.25 ton	20.50	11.60	32.10
1720	6,000 S.F., 66.50 ton	16.90	8.90	25.80
1760	10,000 S.F., 110.83 ton	13.95	2.88	16.83
1800	20,000 S.F., 220 ton*			
1840	40,000 S.F., 440 ton*			
1920	Bowling alleys, 3,000 S.F., 17.00 ton	15.85	10.95	26.80
1960	6,000 S.F., 34.00 ton	11.40	7.70	19.10
2000	10,000 S.F., 56.66 ton	9.15	5.55	14.70
2040	20,000 S.F., 113.33 ton	10	5.10	15.10
2080	40,000 S.F., 227 ton*			
2160	Department stores, 3,000 S.F., 8.75 ton	11.20	9.25	20.45
2200	6,000 S.F., 17.50 ton	8.90	7.15	16.05
2240	10,000 S.F., 29.17 ton	6.70	5.15	11.85
2280	20,000 S.F., 58.33 ton	5.30	3.81	9.11
2320	40,000 S.F., 116.66 ton	6.60	3.82	10.42
2400	Drug stores, 3,000 S.F., 20.00 ton	17.90	11.35	29.25
2440	6,000 S.F., 40.00 ton	13.40	8.45	21.85
2480	10,000 S.F., 66.66 ton	13	7.20	20.20
2520	20,000 S.F., 133.33 ton	12.10	5.40	17.50
2560	40,000 S.F., 267 ton*			
2640	Factories, 2,000 S.F., 10.00 ton	11.10	9.35	20.45
2680	6,000 S.F., 20.00 ton	10	7.55	17.55
2720	10,000 S.F., 33.33 ton	7.25	5.30	12.55
2760	20,000 S.F., 66.66 ton	7.50	4.76	12.26
2800	40,000 S.F., 133.33 ton	7.15	3.92	11.07
2880	Food supermarkets, 3,000 S.F., 8.50 ton	11.05	9.20	20.25
2920	6,000 S.F., 17.00 ton	8.75	7.10	15.85
2960	10,000 S.F., 28.33 ton	6.40	5	11.40
3000	20,000 S.F., 56.66 ton	5.10	3.71	8.81

D3030 Cooling Generating Systems

D3030 110	Chilled Water, Air Cooled Condenser Systems	COST PER S.F.		
		MAT.	INST.	TOTAL
3040	40,000 S.F., 113.33 ton	6.30	3.82	10.12
3120	Medical centers, 3,000 S.F., 7.00 ton	9.95	8.95	18.90
3160	6,000 S.F., 14.00 ton	7.80	6.85	14.65
3200	10,000 S.F., 23.33 ton	6.95	5.25	12.20
3240	20,000 S.F., 46.66 ton	5.15	3.86	9.01
3280	40,000 S.F., 93.33 ton	5.80	3.74	9.54
3360	Offices, 3,000 S.F., 9.50 ton	10.25	9.10	19.35
3400	6,000 S.F., 19.00 ton	9.70	7.45	17.15
3440	10,000 S.F., 31.66 ton	7.05	5.25	12.30
3480	20,000 S.F., 63.33 ton	7.50	4.85	12.35
3520	40,000 S.F., 126.66 ton	7.05	3.96	11.01
3600	Restaurants, 3,000 S.F., 15.00 ton	14.15	10.15	24.30
3640	6,000 S.F., 30.00 ton	10.85	7.70	18.55
3680	10,000 S.F., 50.00 ton	8.80	5.65	14.45
3720	20,000 S.F., 100.00 ton	10.15	5.35	15.50
3760	40,000 S.F., 200 ton*			
3840	Schools and colleges, 3,000 S.F., 11.50 ton	12	9.60	21.60
3880	6,000 S.F., 23.00 ton	10.95	7.80	18.75
3920	10,000 S.F., 38.33 ton	7.90	5.45	13.35
3960	20,000 S.F., 76.66 ton	8.35	5.05	13.40
4000	40,000 S.F., 153 ton*			

D3030 Cooling Generating Systems

*Cooling requirements would lead to a choice of multiple chillers.

D3030 115	Chilled Water, Cooling Tower Systems	COST PER S.F.		
		MAT.	INST.	TOTAL
1300	Packaged chiller, water cooled, with fan coil unit			
1320	Apartment corridors, 4,000 S.F., 7.33 ton	7.33	8.10	15.43
1360	6,000 S.F., 11.00 ton	6.25	7	13.25
1400	10,000 S.F., 18.33 ton	5.95	5.20	11.15
1440	20,000 S.F., 26.66 ton	4.82	3.88	8.70
1480	40,000 S.F., 73.33 ton	5.05	3.97	9.02
1520	60,000 S.F., 110.00 ton	5.35	4.08	9.43
1600	Banks and libraries, 4,000 S.F., 16.66 ton	12.85	8.90	21.75
1640	6,000 S.F., 25.00 ton	10.85	7.85	18.70
1680	10,000 S.F., 41.66 ton	9.30	5.85	15.15
1720	20,000 S.F., 83.33 ton	8.40	5.60	14
1760	40,000 S.F., 166.66 ton	8.75	6.75	15.50
1800	60,000 S.F., 250.00 ton	8.95	7.15	16.10
1880	Bars and taverns, 4,000 S.F., 44.33 ton	22	11.25	33.25
1920	6,000 S.F., 66.50 ton	23	11.65	34.65
1960	10,000 S.F., 110.83 ton	18.45	9.25	27.70
2000	20,000 S.F., 221.66 ton	21	9.35	30.35
2040	40,000 S.F., 440 ton*			
2080	60,000 S.F., 660 ton*			
2160	Bowling alleys, 4,000 S.F., 22.66 ton	14.65	9.80	24.45
2200	6,000 S.F., 34.00 ton	13.10	8.65	21.75
2240	10,000 S.F., 56.66 ton	11.35	6.35	17.70
2280	20,000 S.F., 113.33 ton	10.45	5.90	16.35
2320	40,000 S.F., 226.66 ton	11.85	6.75	18.60
2360	60,000 S.F., 340 ton			
2440	Department stores, 4,000 S.F., 11.66 ton	8.35	8.80	17.15
2480	6,000 S.F., 17.50 ton	9.55	7.25	16.80
2520	10,000 S.F., 29.17 ton	7.20	5.40	12.60
2560	20,000 S.F., 58.33 ton	5.90	4.04	9.94
2600	40,000 S.F., 116.66 ton	6.35	4.19	10.54
2640	60,000 S.F., 175.00 ton	7.90	6.55	14.45
2720	Drug stores, 4,000 S.F., 26.66 ton	15.90	10.10	26
2760	6,000 S.F., 40.00 ton	14.70	8.65	23.35
2800	10,000 S.F., 66.66 ton	14.65	7.85	22.50
2840	20,000 S.F., 133.33 ton	11.95	6.20	18.15
2880	40,000 S.F., 266.67 ton	12.10	7.65	19.75
2920	60,000 S.F., 400 ton*			
3000	Factories, 4,000 S.F., 13.33 ton	11.10	8.50	19.60
3040	6,000 S.F., 20.00 ton	9.50	7.30	16.80
3080	10,000 S.F., 33.33 ton	7.85	5.60	13.45
3120	20,000 S.F., 66.66 ton	8.25	5.05	13.30
3160	40,000 S.F., 133.33 ton	7.10	4.32	11.42

D30 HVAC

D3030 Cooling Generating Systems

D3030 115	Chilled Water, Cooling Tower Systems	COST PER S.F.		
		MAT.	INST.	TOTAL
3200	60,000 S.F., 200.00 ton	8.05	6.90	14.95
3280	Food supermarkets, 4,000 S.F., 11.33 ton	8.25	8.75	17
3320	6,000 S.F., 17.00 ton	8.45	7.10	15.55
3360	10,000 S.F., 28.33 ton	7.05	5.40	12.45
3400	20,000 S.F., 56.66 ton	6.10	4.04	10.14
3440	40,000 S.F., 113.33 ton	6.30	4.18	10.48
3480	60,000 S.F., 170.00 ton	7.85	6.55	14.40
3560	Medical centers, 4.000 S.F., 9.33 ton	7.20	8.05	15.25
3600	6,000 S.F., 14.00 ton	8.35	6.95	15.30
3640	10,000 S.F., 23.33 ton	6.55	5.20	11.75
3680	20,000 S.F., 46.66 ton	5.45	3.95	9.40
3720	40,000 S.F., 93.33 ton	5.75	4.10	9.85
3760	60,000 S.F., 140.00 ton	6.55	6.60	13.15
3840	Offices, 4,000 S.F., 12.66 ton	10.75	8.40	19.15
3880	6,000 S.F., 19.00 ton	9.40	7.50	16.90
3920	10,000 S.F., 31.66 ton	7.80	5.65	13.45
3960	20,000 S.F., 63.33 ton	8.10	5.05	13.15
4000	40,000 S.F., 126.66 ton	7.20	6.45	13.65
4040	60,000 S.F., 190.00 ton	7.75	6.80	14.55
4120	Restaurants, 4,000 S.F., 20.00 ton	13.15	9.10	22.25
4160	6,000 S.F., 30.00 ton	11.85	8.10	19.95
4200	10,000 S.F., 50.00 ton	10.45	6.10	16.55
4240	20,000 S.F., 100.00 ton	9.80	5.90	15.70
4280	40,000 S.F., 200.00 ton	9.35	6.55	15.90
4320	60,000 S.F., 300.00 ton	9.95	7.45	17.40
4400	Schools and colleges, 4,000 S.F., 15.33 ton	12.15	8.75	20.90
4440	6,000 S.F., 23.00 ton	10.30	7.70	18
4480	10,000 S.F., 38.33 ton	8.80	5.70	14.50
4520	20,000 S.F., 76.66 ton	8.10	5.55	13.65
4560	40,000 S.F., 153.33 ton	8.25	6.60	14.85
4600	60,000 S.F., 230.00 ton	8.25	6.90	15.15

D3050 Terminal & Package Units

Return Duct — **Roof** — **Roof Top Unit**

Fan Coil Unit — **Return Grille** — **Supply Diff.** — **Finish Ceiling** — **Insulated Supply Duct**

*Above normal capacity

Rooftop, Single Zone System

D3050 150	Rooftop Single Zone Unit Systems	COST PER S.F.		
		MAT.	INST.	TOTAL
1260	Rooftop, single zone, air conditioner			
1280	Apartment corridors, 500 S.F., .92 ton	4.30	3.85	8.15
1320	1,000 S.F., 1.83 ton	4.28	3.84	8.12
1360	1500 S.F., 2.75 ton	2.73	3.17	5.90
1400	3,000 S.F., 5.50 ton	2.58	3.06	5.64
1440	5,000 S.F., 9.17 ton	2.79	2.78	5.57
1480	10,000 S.F., 18.33 ton	2.85	2.61	5.46
1560	Banks or libraries, 500 S.F., 2.08 ton	9.75	8.75	18.50
1600	1,000 S.F., 4.17 ton	6.20	7.20	13.40
1640	1,500 S.F., 6.25 ton	5.85	6.95	12.80
1680	3,000 S.F., 12.50 ton	6.35	6.30	12.65
1720	5,000 S.F., 20.80 ton	6.45	5.90	12.35
1760	10,000 S.F., 41.67 ton	5.30	5.85	11.15
1840	Bars and taverns, 500 S.F. 5.54 ton	14	11.65	25.65
1880	1,000 S.F., 11.08 ton	15.25	10	25.25
1920	1,500 S.F., 16.62 ton	14.90	9.25	24.15
1960	3,000 S.F., 33.25 ton	12.85	8.85	21.70
2000	5,000 S.F., 55.42 ton	11.60	8.85	20.45
2040	10,000 S.F., 110.83 ton*			
2080	Bowling alleys, 500 S.F., 2.83 ton	8.45	9.80	18.25
2120	1,000 S.F., 5.67 ton	8	9.45	17.45
2160	1,500 S.F., 8.50 ton	8.60	8.55	17.15
2200	3,000 S.F., 17.00 ton	8.45	8.20	16.65
2240	5,000 S.F., 28.33 ton	7.40	8	15.40
2280	10,000 S.F., 56.67 ton	6.75	8	14.75
2360	Department stores, 500 S.F., 1.46 ton	6.80	6.15	12.95
2400	1,000 S.F., 2.92 ton	4.34	5.05	9.39
2480	3,000 S.F., 8.75 ton	4.43	4.41	8.84
2520	5,000 S.F., 14.58 ton	4.34	4.22	8.56
2560	10,000 S.F., 29.17 ton	3.79	4.11	7.90
2640	Drug stores, 500 S.F., 3.33 ton	9.90	11.55	21.45
2680	1,000 S.F., 6.67 ton	9.40	11.10	20.50
2720	1,500 S.F., 10.00 ton	10.15	10.10	20.25
2760	3,000 S.F., 20.00 ton	10.35	9.45	19.80
2800	5,000 S.F., 33.33 ton	8.70	9.40	18.10
2840	10,000 S.F., 66.67 ton	7.95	9.40	17.35
2920	Factories, 500 S.F., 1.67 ton	7.80	7	14.80
3000	1,500 S.F., 5.00 ton	4.69	5.55	10.24
3040	3,000 S.F., 10.00 ton	5.05	5.05	10.10
3080	5,000 S.F., 16.67 ton	4.96	4.82	9.78

D3050 Terminal & Package Units

D3050 150	Rooftop Single Zone Unit Systems	COST PER S.F.		
		MAT.	INST.	TOTAL
3120	10,000 S.F., 33.33 ton	4.34	4.70	9.04
3200	Food supermarkets, 500 S.F., 1.42 ton	6.65	5.95	12.60
3240	1,000 S.F., 2.83 ton	4.18	4.89	9.07
3280	1,500 S.F., 4.25 ton	3.99	4.72	8.71
3320	3,000 S.F., 8.50 ton	4.31	4.28	8.59
3360	5,000 S.F., 14.17 ton	4.22	4.10	8.32
3400	10,000 S.F., 28.33 ton	3.69	3.99	7.68
3480	Medical centers, 500 S.F., 1.17 ton	5.45	4.91	10.36
3520	1,000 S.F., 2.33 ton	5.45	4.89	10.34
3560	1,500 S.F., 3.50 ton	3.47	4.04	7.51
3640	5,000 S.F., 11.67 ton	3.54	3.53	7.07
3680	10,000 S.F., 23.33 ton	3.62	3.32	6.94
3760	Offices, 500 S.F., 1.58 ton	7.40	6.65	14.05
3800	1,000 S.F., 3.17 ton	4.72	5.50	10.22
3840	1,500 S.F., 4.75 ton	4.47	5.25	9.72
3880	3,000 S.F., 9.50 ton	4.82	4.79	9.61
3920	5,000 S.F., 15.83 ton	4.71	4.59	9.30
3960	10,000 S.F., 31.67 ton	4.13	4.47	8.60
4000	Restaurants, 500 S.F., 2.50 ton	11.70	10.50	22.20
4040	1,000 S.F., 5.00 ton	7.05	8.35	15.40
4080	1,500 S.F., 7.50 ton	7.60	7.55	15.15
4120	3,000 S.F., 15.00 ton	7.45	7.25	14.70
4160	5,000 S.F., 25.00 ton	7.80	7.10	14.90
4200	10,000 S.F., 50.00 ton	5.95	7.05	13
4240	Schools and colleges, 500 S.F., 1.92 ton	8.95	8.05	17
4280	1,000 S.F., 3.83 ton	5.70	6.65	12.35
4360	3,000 S.F., 11.50 ton	5.85	5.80	11.65
4400	5,000 S.F., 19.17 ton	5.95	5.45	11.40

D30 HVAC

D3050 Terminal & Package Units

*Note A: Small single zone unit recommended.

*Note B: A combination of multizone units recommended.

D3050 155	Rooftop Multizone Unit Systems	COST PER S.F.		
		MAT.	INST.	TOTAL
1240	Rooftop, multizone, air conditioner			
1260	Apartment corridors, 1,500 S.F., 2.75 ton. See Note A.			
1280	3,000 S.F., 5.50 ton	10.70	4.93	15.63
1320	10,000 S.F., 18.30 ton	7.40	4.75	12.15
1360	15,000 S.F., 27.50 ton	7.25	4.73	11.98
1400	20,000 S.F., 36.70 ton	7.55	4.73	12.28
1440	25,000 S.F., 45.80 ton	7.05	4.77	11.82
1520	Banks or libraries, 1,500 S.F., 6.25 ton	24.50	11.20	35.70
1560	3,000 S.F., 12.50 ton	20.50	11	31.50
1600	10,000 S.F., 41.67 ton	16.05	10.85	26.90
1640	15,000 S.F., 62.50 ton	11.90	10.75	22.65
1680	20,000 S.F., 83.33 ton	11.90	10.75	22.65
1720	25,000 S.F., 104.00 ton	10.65	10.70	21.35
1800	Bars and taverns, 1,500 S.F., 16.62 ton	52.50	16.20	68.70
1840	3,000 S.F., 33.24 ton	43.50	15.60	59.10
1880	10,000 S.F., 110.83 ton	26	15.50	41.50
1920	15,000 S.F., 165 ton, See Note B.			
1960	20,000 S.F., 220 ton, See Note B.			
2000	25,000 S.F., 275 ton, See Note B.			
2080	Bowling alleys, 1,500 S.F., 8.50 ton	33	15.25	48.25
2120	3,000 S.F., 17.00 ton	28	14.95	42.95
2160	10,000 S.F., 56.70 ton	22	14.75	36.75
2200	15,000 S.F., 85.00 ton	16.20	14.65	30.85
2240	20,000 S.F., 113.00 ton	14.50	14.55	29.05
2280	25,000 S.F., 140.00 ton see Note B.			
2360	Department stores, 1,500 S.F., 4.37 ton, See Note A.			
2400	3,000 S.F., 8.75 ton	17	7.85	24.85
2440	10,000 S.F., 29.17 ton	12.05	7.55	19.60
2520	20,000 S.F., 58.33 ton	8.35	7.50	15.85
2560	25,000 S.F., 72.92 ton	8.35	7.55	15.90
2640	Drug stores, 1,500 S.F., 10.00 ton	39	17.95	56.95
2680	3,000 S.F., 20.00 ton	27	17.30	44.30
2720	10,000 S.F., 66.66 ton	19.05	17.20	36.25
2760	15,000 S.F., 100.00 ton	17.10	17.15	34.25
2800	20,000 S.F., 135 ton, See Note B.			
2840	25,000 S.F., 165 ton, See Note B.			
2920	Factories, 1,500 S.F., 5 ton, See Note A.			
3000	10,000 S.F., 33.33 ton	13.75	8.60	22.35

D3050 Terminal & Package Units

D3050 155	Rooftop Multizone Unit Systems	COST PER S.F.		
		MAT.	INST.	TOTAL
3040	15,000 S.F., 50.00 ton	12.85	8.65	21.50
3080	20,000 S.F., 66.66 ton	9.55	8.60	18.15
3120	25,000 S.F., 83.33 ton	9.55	8.60	18.15
3200	Food supermarkets, 1,500 S.F., 4.25 ton, See Note A.			
3240	3,000 S.F., 8.50 ton	16.55	7.65	24.20
3280	10,000 S.F., 28.33 ton	11.25	7.30	18.55
3320	15,000 S.F., 42.50 ton	10.90	7.35	18.25
3360	20,000 S.F., 56.67 ton	8.10	7.30	15.40
3400	25,000 S.F., 70.83 ton	8.10	7.30	15.40
3480	Medical centers, 1,500 S.F., 3.5 ton, See Note A.			
3520	3,000 S.F., 7.00 ton	13.60	6.30	19.90
3560	10,000 S.F., 23.33 ton	9.10	6.05	15.15
3600	15,000 S.F., 35.00 ton	9.65	6	15.65
3640	20,000 S.F., 46.66 ton	9	6.05	15.05
3680	25,000 S.F., 58.33 ton	6.65	6	12.65
3760	Offices, 1,500 S.F., 4.75 ton, See Note A.			
3800	3,000 S.F., 9.50 ton	18.50	8.50	27
3840	10,000 S.F., 31.66 ton	13.10	8.15	21.25
3920	20,000 S.F., 63.33 ton	9.05	8.15	17.20
3960	25,000 S.F., 79.16 ton	9.05	8.20	17.25
4000	Restaurants, 1,500 S.F., 7.50 ton	29	13.45	42.45
4040	3,000 S.F., 15.00 ton	24.50	13.20	37.70
4080	10,000 S.F., 50.00 ton	19.25	13	32.25
4120	15,000 S.F., 75.00 ton	14.30	12.90	27.20
4160	20,000 S.F., 100.00 ton	12.85	12.85	25.70
4200	25,000 S.F., 125 ton, See Note B.			
4240	Schools and colleges, 1,500 S.F., 5.75 ton	22.50	10.30	32.80
4320	10,000 S.F., 38.33 ton	14.75	9.95	24.70
4360	15,000 S.F., 57.50 ton	10.95	9.90	20.85
4400	20,000 S.F.,76.66 ton	10.95	9.90	20.85
4440	25,000 S.F., 95.83 ton	10.30	9.90	20.20

D3050 Terminal & Package Units

Self-Contained Water Cooled System

D3050 160	Self-contained, Water Cooled Unit Systems	COST PER S.F.		
		MAT.	INST.	TOTAL
1280	Self-contained, water cooled unit	3.43	3.12	6.55
1300	Apartment corridors, 500 S.F., .92 ton	3.45	3.10	6.55
1320	1,000 S.F., 1.83 ton	3.43	3.10	6.53
1360	3,000 S.F., 5.50 ton	2.87	2.61	5.48
1400	5,000 S.F., 9.17 ton	2.73	2.41	5.14
1440	10,000 S.F., 18.33 ton	3.84	2.16	6
1520	Banks or libraries, 500 S.F., 2.08 ton	7.45	3.20	10.65
1560	1,000 S.F., 4.17 ton	6.50	5.90	12.40
1600	3,000 S.F., 12.50 ton	6.20	5.50	11.70
1640	5,000 S.F., 20.80 ton	8.70	4.91	13.61
1680	10,000 S.F., 41.66 ton	7.60	4.91	12.51
1760	Bars and taverns, 500 S.F., 5.54 ton	16.30	5.40	21.70
1800	1,000 S.F., 11.08 ton	15.90	9.40	25.30
1840	3,000 S.F., 33.25 ton	20.50	7.45	27.95
1880	5,000 S.F., 55.42 ton	19.25	7.75	27
1920	10,000 S.F., 110.00 ton	18.70	7.65	26.35
2000	Bowling alleys, 500 S.F., 2.83 ton	10.15	4.35	14.50
2040	1,000 S.F., 5.66 ton	8.85	8.05	16.90
2080	3,000 S.F., 17.00 ton	11.85	6.70	18.55
2120	5,000 S.F., 28.33 ton	10.75	6.45	17.20
2160	10,000 S.F., 56.66 ton	10.10	6.60	16.70
2200	Department stores, 500 S.F., 1.46 ton	5.20	2.24	7.44
2240	1,000 S.F., 2.92 ton	4.55	4.14	8.69
2280	3,000 S.F., 8.75 ton	4.33	3.84	8.17
2320	5,000 S.F., 14.58 ton	6.10	3.44	9.54
2360	10,000 S.F., 29.17 ton	5.55	3.32	8.87
2440	Drug stores, 500 S.F., 3.33 ton	11.90	5.10	17
2480	1,000 S.F., 6.66 ton	10.40	9.45	19.85
2520	3,000 S.F., 20.00 ton	13.95	7.85	21.80
2560	5,000 S.F., 33.33 ton	14.65	7.75	22.40
2600	10,000 S.F., 66.66 ton	11.85	7.75	19.60
2680	Factories, 500 S.F., 1.66 ton	5.95	2.56	8.51
2720	1,000 S.F. 3.37 ton	5.25	4.78	10.03
2760	3,000 S.F., 10.00 ton	4.95	4.39	9.34
2800	5,000 S.F., 16.66 ton	6.95	3.94	10.89
2840	10,000 S.F., 33.33 ton	6.35	3.79	10.14

D3050 Terminal & Package Units

D3050 160	Self-contained, Water Cooled Unit Systems	COST PER S.F.		
		MAT.	INST.	TOTAL
2920	Food supermarkets, 500 S.F., 1.42 ton	5.05	2.17	7.22
2960	1,000 S.F., 2.83 ton	5.30	4.80	10.10
3000	3,000 S.F., 8.50 ton	4.42	4.02	8.44
3040	5,000 S.F., 14.17 ton	4.20	3.73	7.93
3080	10,000 S.F., 28.33 ton	5.40	3.21	8.61
3160	Medical centers, 500 S.F., 1.17 ton	4.17	1.79	5.96
3200	1,000 S.F., 2.33 ton	4.37	3.96	8.33
3240	3,000 S.F., 7.00 ton	3.65	3.32	6.97
3280	5,000 S.F., 11.66 ton	3.47	3.08	6.55
3320	10,000 S.F., 23.33 ton	4.88	2.76	7.64
3400	Offices, 500 S.F., 1.58 ton	5.65	2.43	8.08
3440	1,000 S.F., 3.17 ton	5.90	5.35	11.25
3480	3,000 S.F., 9.50 ton	4.70	4.18	8.88
3520	5,000 S.F., 15.83 ton	6.60	3.74	10.34
3560	10,000 S.F., 31.67 ton	6	3.60	9.60
3640	Restaurants, 500 S.F., 2.50 ton	8.95	3.84	12.79
3680	1,000 S.F., 5.00 ton	7.80	7.10	14.90
3720	3,000 S.F., 15.00 ton	7.45	6.60	14.05
3760	5,000 S.F., 25.00 ton	10.45	5.90	16.35
3800	10,000 S.F., 50.00 ton	6.55	5.40	11.95
3880	Schools and colleges, 500 S.F., 1.92 ton	6.85	2.94	9.79
3920	1,000 S.F., 3.83 ton	6	5.45	11.45
3960	3,000 S.F., 11.50 ton	5.70	5.05	10.75
4000	5,000 S.F., 19.17 ton	8	4.52	12.52
4040	10,000 S.F., 38.33 ton	7	4.52	11.52

D3050 Terminal & Package Units

Self-Contained Air Cooled System

D3050 165	Self-contained, Air Cooled Unit Systems	COST PER S.F.		
		MAT.	**INST.**	**TOTAL**
1300	Self-contained, air cooled unit			
1320	Apartment corridors, 500 S.F., .92 ton	4.70	4	8.70
1360	1,000 S.F., 1.83 ton	4.64	3.96	8.60
1400	3,000 S.F., 5.50 ton	3.95	3.72	7.67
1440	5,000 S.F., 9.17 ton	3.44	3.52	6.96
1480	10,000 S.F., 18.33 ton	3.58	3.23	6.81
1560	Banks or libraries, 500 S.F., 2.08 ton	10.15	5.10	15.25
1600	1,000 S.F., 4.17 ton	8.90	8.40	17.30
1640	3,000 S.F., 12.50 ton	7.85	8	15.85
1680	5,000 S.F., 20.80 ton	8.15	7.35	15.50
1720	10,000 S.F., 41.66 ton	8.40	7.20	15.60
1800	Bars and taverns, 500 S.F., 5.54 ton	21	11.15	32.15
1840	1,000 S.F., 11.08 ton	20.50	16.10	36.60
1880	3,000 S.F., 33.25 ton	22	13.95	35.95
1920	5,000 S.F., 55.42 ton	22.50	13.95	36.45
1960	10,000 S.F., 110.00 ton	22.50	13.90	36.40
2040	Bowling alleys, 500 S.F., 2.83 ton	13.90	7	20.90
2080	1,000 S.F., 5.66 ton	12.20	11.45	23.65
2120	3,000 S.F., 17.00 ton	11.05	10	21.05
2160	5,000 S.F., 28.33 ton	11.40	9.75	21.15
2200	10,000 S.F., 56.66 ton	11.65	9.75	21.40
2240	Department stores, 500 S.F., 1.46 ton	7.15	3.60	10.75
2280	1,000 S.F., 2.92 ton	6.25	5.90	12.15
2320	3,000 S.F., 8.75 ton	5.45	5.60	11.05
2360	5,000 S.F., 14.58 ton	5.45	5.60	11.05
2400	10,000 S.F., 29.17 ton	5.85	5	10.85
2480	Drug stores, 500 S.F., 3.33 ton	16.35	8.20	24.55
2520	1,000 S.F., 6.66 ton	14.30	13.45	27.75
2560	3,000 S.F., 20.00 ton	13.05	11.75	24.80
2600	5,000 S.F., 33.33 ton	13.40	11.50	24.90
2640	10,000 S.F., 66.66 ton	13.80	11.50	25.30
2720	Factories, 500 S.F., 1.66 ton	8.30	4.15	12.45
2760	1,000 S.F., 3.33 ton	7.20	6.75	13.95
2800	3,000 S.F., 10.00 ton	6.25	6.40	12.65
2840	5,000 S.F., 16.66 ton	6.45	5.85	12.30
2880	10,000 S.F., 33.33 ton	6.70	5.75	12.45
2960	Food supermarkets, 500 S.F., 1.42 ton	6.95	3.49	10.44
3000	1,000 S.F., 2.83 ton	7.15	6.10	13.25
3040	3,000 S.F., 8.50 ton	6.10	5.75	11.85
3080	5,000 S.F., 14.17 ton	5.30	5.45	10.75

D3050 Terminal & Package Units

D3050 165	Self-contained, Air Cooled Unit Systems	COST PER S.F.		
		MAT.	INST.	TOTAL
3120	10,000 S.F., 28.33 ton	5.70	4.88	10.58
3200	Medical centers, 500 S.F., 1.17 ton	5.75	2.88	8.63
3240	1,000 S.F., 2.33 ton	5.95	5.05	11
3280	3,000 S.F., 7.00 ton	5	4.72	9.72
3320	5,000 S.F., 16.66 ton	4.39	4.47	8.86
3360	10,000 S.F., 23.33 ton	4.55	4.11	8.66
3440	Offices, 500 S.F., 1.58 ton	7.80	3.91	11.71
3480	1,000 S.F., 3.16 ton	8.10	6.85	14.95
3520	3,000 S.F., 9.50 ton	5.95	6.10	12.05
3560	5,000 S.F., 15.83 ton	6.20	5.60	11.80
3600	10,000 S.F., 31.66 ton	6.35	5.45	11.80
3680	Restaurants, 500 S.F., 2.50 ton	12.20	6.15	18.35
3720	1,000 S.F., 5.00 ton	10.75	10.10	20.85
3760	3,000 S.F., 15.00 ton	9.40	9.60	19
3800	5,000 S.F., 25.00 ton	10	8.60	18.60
3840	10,000 S.F., 50.00 ton	10.50	8.60	19.10
3920	Schools and colleges, 500 S.F., 1.92 ton	9.40	4.73	14.13
3960	1,000 S.F., 3.83 ton	8.30	7.75	16.05
4000	3,000 S.F., 11.50 ton	7.20	7.35	14.55
4040	5,000 S.F., 19.17 ton	7.50	6.75	14.25
4080	10,000 S.F., 38.33 ton	7.75	6.60	14.35

D3050 Terminal & Package Units

*Cooling requirements would lead to more than one system.

D3050 170	Split Systems With Air Cooled Condensing Units	COST PER S.F.		
		MAT.	INST.	TOTAL
1260	Split system, air cooled condensing unit			
1280	Apartment corridors, 1,000 S.F., 1.83 ton	2.78	2.50	5.28
1320	2,000 S.F., 3.66 ton	2.06	2.52	4.58
1360	5,000 S.F., 9.17 ton	2.25	3.13	5.38
1400	10,000 S.F., 18.33 ton	2.60	3.44	6.04
1440	20,000 S.F., 36.66 ton	2.91	3.49	6.40
1520	Banks and libraries, 1,000 S.F., 4.17 ton	4.68	5.75	10.43
1560	2,000 S.F., 8.33 ton	5.10	7.15	12.25
1600	5,000 S.F., 20.80 ton	5.90	7.80	13.70
1640	10,000 S.F., 41.66 ton	6.60	7.95	14.55
1680	20,000 S.F., 83.32 ton	7.75	8.25	16
1760	Bars and taverns, 1,000 S.F., 11.08 ton	11.80	11.25	23.05
1800	2,000 S.F., 22.16 ton	15.95	13.50	29.45
1840	5,000 S.F., 55.42 ton	14.35	12.65	27
1880	10,000 S.F., 110.84 ton	16.80	13.20	30
1920	20,000 S.F., 220 ton*			
2000	Bowling alleys, 1,000 S.F., 5.66 ton	6.55	10.70	17.25
2040	2,000 S.F., 11.33 ton	6.95	9.70	16.65
2080	5,000 S.F., 28.33 ton	8.05	10.60	18.65
2120	10,000 S.F., 56.66 ton	9	10.80	19.80
2160	20,000 S.F., 113.32 ton	11.65	11.70	23.35
2320	Department stores, 1,000 S.F., 2.92 ton	3.30	3.94	7.24
2360	2,000 S.F., 5.83 ton	3.36	5.50	8.86
2400	5,000 S.F., 14.58 ton	3.56	4.99	8.55
2440	10,000 S.F., 29.17 ton	4.14	5.45	9.59
2480	20,000 S.F., 58.33 ton	4.62	5.55	10.17
2560	Drug stores, 1,000 S.F., 6.66 ton	7.70	12.55	20.25
2600	2,000 S.F., 13.32 ton	8.15	11.40	19.55
2640	5,000 S.F., 33.33 ton	10.55	12.70	23.25
2680	10,000 S.F., 66.66 ton	11.45	13.30	24.75
2720	20,000 S.F., 133.32 ton*			
2800	Factories, 1,000 S.F., 3.33 ton	3.78	4.49	8.27
2840	2,000 S.F., 6.66 ton	3.86	6.30	10.16
2880	5,000 S.F., 16.66 ton	4.73	6.25	10.98
2920	10,000 S.F., 33.33 ton	5.30	6.35	11.65
2960	20,000 S.F., 66.66 ton	5.75	6.65	12.40
3040	Food supermarkets, 1,000 S.F., 2.83 ton	3.21	3.81	7.02
3080	2,000 S.F., 5.66 ton	3.28	5.35	8.63
3120	5,000 S.F., 14.66 ton	3.47	4.85	8.32
3160	10,000 S.F., 28.33 ton	4.02	5.30	9.32
3200	20,000 S.F., 56.66 ton	4.49	5.40	9.89
3280	Medical centers, 1,000 S.F., 2.33 ton	2.90	3.08	5.98

D3050 Terminal & Package Units

D3050 170	Split Systems With Air Cooled Condensing Units	COST PER S.F.		
		MAT.	INST.	TOTAL
3320	2,000 S.F., 4.66 ton	2.69	4.40	7.09
3360	5,000 S.F., 11.66 ton	2.86	3.99	6.85
3400	10,000 S.F., 23.33 ton	3.31	4.37	7.68
3440	20,000 S.F., 46.66 ton	3.71	4.44	8.15
3520	Offices, 1,000 S.F., 3.17 ton	3.59	4.27	7.86
3560	2,000 S.F., 6.33 ton	3.66	5.95	9.61
3600	5,000 S.F., 15.83 ton	3.88	5.40	9.28
3640	10,000 S.F., 31.66 ton	4.49	5.95	10.44
3680	20,000 S.F., 63.32 ton	5.45	6.30	11.75
3760	Restaurants, 1,000 S.F., 5.00 ton	5.80	9.45	15.25
3800	2,000 S.F., 10.00 ton	6.15	8.60	14.75
3840	5,000 S.F., 25.00 ton	7.10	9.40	16.50
3880	10,000 S.F., 50.00 ton	7.95	9.55	17.50
3920	20,000 S.F., 100.00 ton	10.25	10.35	20.60
4000	Schools and colleges, 1,000 S.F., 3.83 ton	4.32	5.30	9.62
4040	2,000 S.F., 7.66 ton	4.70	6.55	11.25
4080	5,000 S.F., 19.17 ton	5.45	7.20	12.65
4120	10,000 S.F., 38.33 ton	6.10	7.30	13.40

D3090   Other HVAC Systems/Equip

Cast Iron Garage Exhaust System

Dual Exhaust System

D3090 320	Garage Exhaust Systems	COST PER BAY		
		MAT.	INST.	TOTAL
1040	Garage, single 3″ exhaust outlet, cars & light trucks, one bay	4,208	1,858	6,067
1060	Additional bays up to seven bays	995	500	1,495
1500	4″ outlet, trucks, one bay	4,250	1,850	6,100
1520	Additional bays up to six bays	1,025	500	1,525
1600	5″ outlet, diesel trucks, one bay	4,550	1,850	6,400
1650	Additional single bays up to six	1,450	590	2,040
1700	Two adjoining bays	4,550	1,850	6,400
2000	Dual exhaust, 3″ outlets, pair of adjoining bays	5,225	2,325	7,550
2100	Additional pairs of adjoining bays	1,575	590	2,165

Reference - System Classification

System Classification
Rules for installation of sprinkler systems vary depending on the classification of occupancy falling into one of three categories as follows:

Light Hazard Occupancy
The protection area allotted per sprinkler should not exceed 225 S.F., with the maximum distance between lines and sprinklers on lines being 15'. The sprinklers do not need to be staggered. Branch lines should not exceed eight sprinklers on either side of a cross main. Each large area requiring more than 100 sprinklers and without a sub-dividing partition should be supplied by feed mains or risers sized for ordinary hazard occupancy.
Maximum system area = 52,000 S.F.

Included in this group are:
Churches	Nursing Homes
Clubs	Offices
Educational	Residential
Hospitals	Restaurants
Institutional	Theaters and Auditoriums
Libraries	(except stages and prosceniums)
(except large stack rooms)	Unused Attics
Museums	

Ordinary Hazard Occupancy
The protection area allotted per sprinkler shall not exceed 130 S.F. of noncombustible ceiling and 130 S.F. of combustible ceiling. The maximum allowable distance between sprinkler lines and sprinklers on line is 15'. Sprinklers shall be staggered if the distance between heads exceeds 12'. Branch lines should not exceed eight sprinklers on either side of a cross main.
Maximum system area = 52,000 S.F.

Included in this group are:
Group 1	Group 2
Automotive Parking and Showrooms	Cereal Mills
Bakeries	Chemical Plants—Ordinary
Beverage manufacturing	Confectionery Products
Canneries	Distilleries
Dairy Products Manufacturing/Processing	Dry Cleaners
Electronic Plans	Feed Mills
Glass and Glass Products Manufacturing	Horse Stables
Laundries	Leather Goods Manufacturing
Restaurant Service Areas	Libraries—Large Stack Room Areas
	Machine Shops
	Metal Working
	Mercantile
	Paper and Pulp Mills
	Paper Process Plants
	Piers and Wharves
	Post Offices
	Printing and Publishing
	Repair Garages
	Stages
	Textile Manufacturing
	Tire Manufacturing
	Tobacco Products Manufacturing
	Wood Machining
	Wood Product Assembly

Extra Hazard Occupancy
The protection area allotted per sprinkler shall not exceed 100 S.F. of noncombustible ceiling and 100 S.F. of combustible ceiling. The maximum allowable distance between lines and between sprinklers on lines is 12'. Sprinklers on alternate lines shall be staggered if the distance between sprinklers on lines exceeds 8'. Branch lines should not exceed six sprinklers on either side of a cross main.
Maximum system area:
   Design by pipe schedule = 25,000 S.F.
   Design by hydraulic calculation = 40,000 S.F.

Included in this group are:
Group 1	Group 2
Aircraft hangars	Asphalt Saturating
Combustible Hydraulic Fluid Use Area	Flammable Liquids Spraying
Die Casting	Flow Coating
Metal Extruding	Manufactured/Modular Home
Plywood/Particle Board Manufacturing	Building Assemblies (where
Printing (inkls with flash points < 100 degrees F	finished enclosure is present and has combustible interiors)
Rubber Reclaiming, Compounding, Drying, Milling, Vulcanizing	Open Oil Quenching
Saw Mills	Plastics Processing
Textile Picking, Opening, Blending, Garnetting, Carding, Combing of Cotton, Synthetics, Wood Shoddy, or Burlap	Solvent Cleaning
Upholstering with Plastic Foams	Varnish and Paint Dipping

Reference - Sprinkler Systems (Automatic)

Sprinkler systems may be classified by type as follows:

1. **Wet Pipe System.** A system employing automatic sprinklers attached to a piping system containing water and connected to a water supply so that water discharges immediately from sprinklers opened by a fire.

2. **Dry Pipe System.** A system employing automatic sprinklers attached to a piping system containing air under pressure, the release of which as from the opening of sprinklers permits the water pressure to open a valve known as a "dry pipe valve". The water then flows into the piping system and out the opened sprinklers.

3. **Pre-Action System.** A system employing automatic sprinklers attached to a piping system containing air that may or may not be under pressure, with a supplemental heat responsive system of generally more sensitive characteristics than the automatic sprinklers themselves, installed in the same areas as the sprinklers; actuation of the heat responsive system, as from a fire, opens a valve which permits water to flow into the sprinkler piping system and to be discharged from any sprinklers which may be open.

4. **Deluge System.** A system employing open sprinklers attached to a piping system connected to a water supply through a valve which is opened by the operation of a heat responsive system installed in the same areas as the sprinklers. When this valve opens, water flows into the piping system and discharges from all sprinklers attached thereto.

5. **Combined Dry Pipe and Pre-Action Sprinkler System.** A system employing automatic sprinklers attached to a piping system containing air under pressure with a supplemental heat responsive system of generally more sensitive characteristics than the automatic sprinklers themselves, installed in the same areas as the sprinklers; operation of the heat responsive system, as from a fire, actuates tripping devices which open dry pipe valves simultaneously and without loss of air pressure in the system. Operation of the heat responsive system also opens approved air exhaust valves at the end of the feed main which facilitates the filling of the system with water which usually precedes the opening of sprinklers. The heat responsive system also serves as an automatic fire alarm system.

6. **Limited Water Supply System.** A system employing automatic sprinklers and conforming to these standards but supplied by a pressure tank of limited capacity.

7. **Chemical Systems.** Systems using halon, carbon dioxide, dry chemical or high expansion foam as selected for special requirements. Agent may extinguish flames by chemically inhibiting flame propagation, suffocate flames by excluding oxygen, interrupting chemical action of oxygen uniting with fuel or sealing and cooling the combustion center.

8. **Firecycle System.** Firecycle is a fixed fire protection sprinkler system utilizing water as its extinguishing agent. It is a time delayed, recycling, preaction type which automatically shuts the water off when heat is reduced below the detector operating temperature and turns the water back on when that temperature is exceeded. The system senses a fire condition through a closed circuit electrical detector system which controls water flow to the fire automatically. Batteries supply up to 90 hour emergency power supply for system operation. The piping system is dry (until water is required) and is monitored with pressurized air. Should any leak in the system piping occur, an alarm will sound, but water will not enter the system until heat is sensed by a firecycle detector.

Area coverage sprinkler systems may be laid out and fed from the supply in any one of several patterns as shown below. It is desirable, if possible, to utilize a central feed and achieve a shorter flow path from the riser to the furthest sprinkler. This permits use of the smallest sizes of pipe possible with resulting savings.

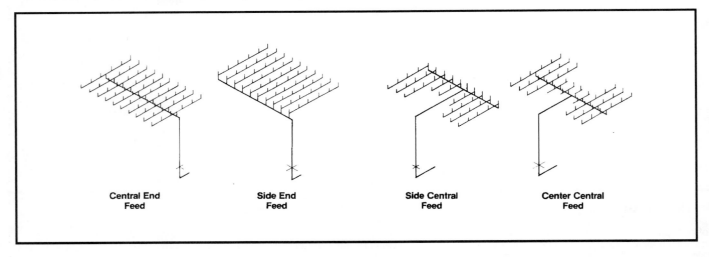

Central End Feed **Side End Feed** **Side Central Feed** **Center Central Feed**

D4010 Sprinklers

Dry Pipe System: A system employing automatic sprinklers attached to a piping system containing air under pressure, the release of which from the opening of sprinklers permits the water pressure to open a valve known as a "dry pipe valve". The water then flows into the piping system and out the opened sprinklers.

All areas are assumed to be open.

D4010 310	Dry Pipe Sprinkler Systems	COST PER S.F.		
		MAT.	INST.	TOTAL
0520	Dry pipe sprinkler systems, steel, black, sch. 40 pipe			
0530	Light hazard, one floor, 500 S.F.	9	5.95	14.95
0560	1000 S.F.	5.20	3.50	8.70
0580	2000 S.F.	4.91	3.50	8.41
0600	5000 S.F.	2.64	2.39	5.03
0620	10,000 S.F.	1.92	1.98	3.90
0640	50,000 S.F.	1.53	1.75	3.28
0660	Each additional floor, 500 S.F.	2.08	2.90	4.98
0680	1000 S.F.	1.80	2.40	4.20
0700	2000 S.F.	1.74	2.21	3.95
0720	5000 S.F.	1.45	1.90	3.35
0740	10,000 S.F.	1.36	1.75	3.11
0760	50,000 S.F.	1.22	1.55	2.77
1000	Ordinary hazard, one floor, 500 S.F.	9.20	6	15.20
1020	1000 S.F.	5.30	3.52	8.82
1040	2000 S.F.	5.05	3.65	8.70
1060	5000 S.F.	3.04	2.54	5.58
1080	10,000 S.F.	2.51	2.60	5.11
1100	50,000 S.F.	2.28	2.45	4.73
1140	Each additional floor, 500 S.F.	2.26	2.97	5.23
1160	1000 S.F.	2.07	2.67	4.74
1180	2000 S.F.	2.06	2.42	4.48
1200	5000 S.F.	1.92	2.08	4
1220	10,000 S.F.	1.78	2.05	3.83
1240	50,000 S.F.	1.66	1.78	3.44
1500	Extra hazard, one floor, 500 S.F.	12.65	7.45	20.10
1520	1000 S.F.	7.80	5.40	13.20
1540	2000 S.F.	5.75	4.70	10.45
1560	5000 S.F.	3.63	3.57	7.20
1580	10,000 S.F.	3.69	3.40	7.09
1600	50,000 S.F.	4	3.28	7.28
1660	Each additional floor, 500 S.F.	3.12	3.68	6.80
1680	1000 S.F.	3.06	3.50	6.56
1700	2000 S.F.	2.93	3.50	6.43
1720	5000 S.F.	2.51	3.07	5.58
1740	10,000 S.F.	2.82	2.80	5.62
1760	50,000 S.F.	2.90	2.69	5.59
2020	Grooved steel, black, sch. 40 pipe, light hazard, one floor, 2000 S.F.	4.64	3	7.64

D4010 Sprinklers

D4010 310	Dry Pipe Sprinkler Systems	COST PER S.F.		
		MAT.	INST.	TOTAL
2060	10,000 S.F.	1.76	1.73	3.49
2100	Each additional floor, 2000 S.F.	1.65	1.77	3.42
2150	10,000 S.F.	1.20	1.50	2.70
2200	Ordinary hazard, one floor, 2000 S.F.	4.88	3.20	8.08
2250	10,000 S.F.	2.24	2.19	4.43
2300	Each additional floor, 2000 S.F.	1.89	1.97	3.86
2350	10,000 S.F.	1.69	1.96	3.65
2400	Extra hazard, one floor, 2000 S.F.	5.50	4	9.50
2450	10,000 S.F.	3.14	2.84	5.98
2500	Each additional floor, 2000 S.F.	2.69	2.86	5.55
2550	10,000 S.F.	2.44	2.46	4.90
3050	Grooved steel, black, sch. 10 pipe, light hazard, one floor, 2000 S.F.	4.52	2.97	7.49
3100	10,000 S.F.	1.69	1.70	3.39
3150	Each additional floor, 2000 S.F.	1.53	1.74	3.27
3200	10,000 S.F.	1.13	1.47	2.60
3250	Ordinary hazard, one floor, 2000 S.F.	4.78	3.18	7.96
3300	10,000 S.F.	2.12	2.15	4.27
3350	Each additional floor, 2000 S.F.	1.79	1.95	3.74
3400	10,000 S.F.	1.57	1.92	3.49
3450	Extra hazard, one floor, 2000 S.F.	5.45	3.98	9.43
3500	10,000 S.F.	2.93	2.79	5.72
3550	Each additional floor, 2000 S.F.	2.61	2.84	5.45
3600	10,000 S.F.	2.32	2.43	4.75
4050	Copper tubing, type M, light hazard, one floor, 2000 S.F.	6.05	2.97	9.02
4100	10,000 S.F.	2.99	1.72	4.71
4150	Each additional floor, 2000 S.F.	3.09	1.78	4.87
4200	10,000 S.F.	2.43	1.50	3.93
4250	Ordinary hazard, one floor, 2000 S.F.	6.50	3.33	9.83
4300	10,000 S.F.	3.69	2.04	5.73
4350	Each additional floor, 2000 S.F.	4.04	2.07	6.11
4400	10,000 S.F.	3	1.78	4.78
4450	Extra hazard, one floor, 2000 S.F.	7.75	4.05	11.80
4500	10,000 S.F.	8.10	3.08	11.18
4550	Each additional floor, 2000 S.F.	4.92	2.91	7.83
4600	10,000 S.F.	6.15	2.68	8.83
5050	Copper tubing, type M, T-drill system, light hazard, one floor			
5060	2000 S.F.	5.95	2.78	8.73
5100	10,000 S.F.	2.62	1.43	4.05
5150	Each additional floor, 2000 S.F.	2.97	1.59	4.56
5200	10,000 S.F.	2.06	1.21	3.27
5250	Ordinary hazard, one floor, 2000 S.F.	5.90	2.85	8.75
5300	10,000 S.F.	3.37	1.81	5.18
5350	Each additional floor, 2000 S.F.	2.92	1.62	4.54
5400	10,000 S.F.	2.68	1.50	4.18
5450	Extra hazard, one floor, 2000 S.F.	6.70	3.35	10.05
5500	10,000 S.F.	5.85	2.26	8.11
5550	Each additional floor, 2000 S.F.	3.86	2.21	6.07
5600	10,000 S.F.	3.93	1.86	5.79

D40 Fire Protection

D4010 Sprinklers

Preaction System: A system employing automatic sprinklers attached to a piping system containing air that may or may not be under pressure, with a supplemental heat responsive system of generally more sensitive characteristics than the automatic sprinklers themselves, installed in the same areas as the sprinklers. Actuation of the heat responsive system, as from a fire, opens a valve which permits water to flow into the sprinkler piping system and to be discharged from those sprinklers which were opened by heat from the fire.

All areas are assumed to be opened.

D4010 350	Preaction Sprinkler Systems	COST PER S.F.		
		MAT.	INST.	TOTAL
0520	Preaction sprinkler systems, steel, black, sch. 40 pipe			
0530	Light hazard, one floor, 500 S.F.	8.95	4.76	13.71
0560	1000 S.F.	5.20	3.57	8.77
0580	2000 S.F.	4.76	3.49	8.25
0600	5000 S.F.	2.51	2.38	4.89
0620	10,000 S.F.	1.81	1.97	3.78
0640	50,000 S.F.	1.43	1.75	3.18
0660	Each additional floor, 500 S.F.	2.34	2.59	4.93
0680	1000 S.F.	1.81	2.39	4.20
0700	2000 S.F.	1.74	2.20	3.94
0720	5000 S.F.	1.32	1.89	3.21
0740	10,000 S.F.	1.25	1.74	2.99
0760	50,000 S.F.	1.23	1.61	2.84
1000	Ordinary hazard, one floor, 500 S.F.	9.20	5.15	14.35
1020	1000 S.F.	5.15	3.51	8.66
1040	2000 S.F.	5.15	3.66	8.81
1060	5000 S.F.	2.79	2.53	5.32
1080	10,000 S.F.	2.21	2.59	4.80
1100	50,000 S.F.	1.96	2.43	4.39
1140	Each additional floor, 500 S.F.	2.60	2.99	5.59
1160	1000 S.F.	1.78	2.41	4.19
1180	2000 S.F.	1.66	2.40	4.06
1200	5000 S.F.	1.80	2.25	4.05
1220	10,000 S.F.	1.66	2.36	4.02
1240	50,000 S.F.	1.55	2.10	3.65
1500	Extra hazard, one floor, 500 S.F.	12.35	6.60	18.95
1520	1000 S.F.	7.15	4.94	12.09
1540	2000 S.F.	5.25	4.67	9.92

D4010 Sprinklers

D4010 350	Preaction Sprinkler Systems	COST PER S.F.		
		MAT.	INST.	TOTAL
1560	5000 S.F.	3.44	3.85	7.29
1580	10,000 S.F.	3.33	3.76	7.09
1600	50,000 S.F.	3.54	3.66	7.20
1660	Each additional floor, 500 S.F.	3.13	3.68	6.81
1680	1000 S.F.	2.58	3.47	6.05
1700	2000 S.F.	2.44	3.47	5.91
1720	5000 S.F.	2.07	3.08	5.15
1740	10,000 S.F.	2.22	2.82	5.04
1760	50,000 S.F.	2.19	2.66	4.85
2020	Grooved steel, black, sch. 40 pipe, light hazard, one floor, 2000 S.F.	4.64	2.99	7.63
2060	10,000 S.F.	1.65	1.72	3.37
2100	Each additional floor of 2000 S.F.	1.65	1.76	3.41
2150	10,000 S.F.	1.09	1.49	2.58
2200	Ordinary hazard, one floor, 2000 S.F.	4.72	3.19	7.91
2250	10,000 S.F.	1.94	2.18	4.12
2300	Each additional floor, 2000 S.F.	1.73	1.96	3.69
2350	10,000 S.F.	1.39	1.95	3.34
2400	Extra hazard, one floor, 2000 S.F.	5	3.97	8.97
2450	10,000 S.F.	2.54	2.80	5.34
2500	Each additional floor, 2000 S.F.	2.20	2.83	5.03
2550	10,000 S.F.	1.80	2.42	4.22
3050	Grooved steel, black, sch. 10 pipe light hazard, one floor, 2000 S.F.	4.52	2.96	7.48
3100	10,000 S.F.	1.58	1.69	3.27
3150	Each additional floor, 2000 S.F.	1.53	1.73	3.26
3200	10,000 S.F.	1.02	1.46	2.48
3250	Ordinary hazard, one floor, 2000 S.F.	4.55	2.97	7.52
3300	10,000 S.F.	1.57	2.12	3.69
3350	Each additional floor, 2000 S.F.	1.63	1.94	3.57
3400	10,000 S.F.	1.27	1.91	3.18
3450	Extra hazard, one floor, 2000 S.F.	4.94	3.95	8.89
3500	10,000 S.F.	2.29	2.75	5.04
3550	Each additional floor, 2000 S.F.	2.12	2.81	4.93
3600	10,000 S.F.	1.68	2.39	4.07
4050	Copper tubing, type M, light hazard, one floor, 2000 S.F.	6.05	2.96	9.01
4100	10,000 S.F.	2.88	1.71	4.59
4150	Each additional floor, 2000 S.F.	3.10	1.77	4.87
4200	10,000 S.F.	2.07	1.47	3.54
4250	Ordinary hazard, one floor, 2000 S.F.	6.35	3.32	9.67
4300	10,000 S.F.	3.39	2.03	5.42
4350	Each additional floor, 2000 S.F.	3.11	1.83	4.94
4400	10,000 S.F.	2.45	1.63	4.08
4450	Extra hazard, one floor, 2000 S.F.	7.25	4.02	11.27
4500	10,000 S.F.	7.40	3.04	10.44
4550	Each additional floor, 2000 S.F.	4.43	2.88	7.31
4600	10,000 S.F.	5.50	2.64	8.14
5050	Copper tubing, type M, T-drill system, light hazard, one floor			
5060	2000 S.F.	5.95	2.77	8.72
5100	10,000 S.F.	2.51	1.42	3.93
5150	Each additional floor, 2000 S.F.	2.97	1.58	4.55
5200	10,000 S.F.	1.95	1.20	3.15
5250	Ordinary hazard, one floor, 2000 S.F.	5.75	2.84	8.59
5300	10,000 S.F.	3.07	1.80	4.87
5350	Each additional floor, 2000 S.F.	2.78	1.62	4.40
5400	10,000 S.F.	2.52	1.57	4.09
5450	Extra hazard, one floor, 2000 S.F.	6.20	3.32	9.52
5500	10,000 S.F.	5.15	2.22	7.37
5550	Each additional floor, 2000 S.F.	3.37	2.18	5.55
5600	10,000 S.F.	3.29	1.82	5.11

D40 Fire Protection

D4010 Sprinklers

Deluge System: A system employing open sprinklers attached to a piping system connected to a water supply through a valve which is opened by the operation of a heat responsive system installed in the same areas as the sprinklers. When this valve opens, water flows into the piping system and discharges from all sprinklers attached thereto.

D4010 370	Deluge Sprinkler Systems	COST PER S.F.		
		MAT.	INST.	TOTAL
0520	Deluge sprinkler systems, steel, black, sch. 40 pipe			
0530	Light hazard, one floor, 500 S.F.	27.50	4.81	32.31
0560	1000 S.F.	14.35	3.43	17.78
0580	2000 S.F.	9.65	3.50	13.15
0600	5000 S.F.	4.47	2.39	6.86
0620	10,000 S.F.	2.79	1.97	4.76
0640	50,000 S.F.	1.63	1.75	3.38
0660	Each additional floor, 500 S.F.	2.34	2.59	4.93
0680	1000 S.F.	1.81	2.39	4.20
0700	2000 S.F.	1.74	2.20	3.94
0720	5000 S.F.	1.32	1.89	3.21
0740	10,000 S.F.	1.25	1.74	2.99
0760	50,000 S.F.	1.23	1.61	2.84
1000	Ordinary hazard, one floor, 500 S.F.	28	5.50	33.50
1020	1000 S.F.	14.35	3.54	17.89
1040	2000 S.F.	10.05	3.67	13.72
1060	5000 S.F.	4.75	2.54	7.29
1080	10,000 S.F.	3.19	2.59	5.78
1100	50,000 S.F.	2.21	2.46	4.67
1140	Each additional floor, 500 S.F.	2.60	2.99	5.59
1160	1000 S.F.	1.78	2.41	4.19
1180	2000 S.F.	1.66	2.40	4.06
1200	5000 S.F.	1.67	2.07	3.74
1220	10,000 S.F.	1.60	2.09	3.69
1240	50,000 S.F.	1.46	1.92	3.38
1500	Extra hazard, one floor, 500 S.F.	30.50	6.65	37.15
1520	1000 S.F.	16.75	5.15	21.90
1540	2000 S.F.	10.15	4.68	14.83
1560	5000 S.F.	5.10	3.55	8.65
1580	10,000 S.F.	4.10	3.45	7.55
1600	50,000 S.F.	4.13	3.36	7.49
1660	Each additional floor, 500 S.F.	3.13	3.68	6.81

D4010 370	Deluge Sprinkler Systems	COST PER S.F.		
		MAT.	INST.	TOTAL
1680	1000 S.F.	2.58	3.47	6.05
1700	2000 S.F.	2.44	3.47	5.91
1720	5000 S.F.	2.07	3.08	5.15
1740	10,000 S.F.	2.30	2.93	5.23
1760	50,000 S.F.	2.31	2.84	5.15
2000	Grooved steel, black, sch. 40 pipe, light hazard, one floor			
2020	2000 S.F.	9.55	3	12.55
2060	10,000 S.F.	2.65	1.74	4.39
2100	Each additional floor, 2,000 S.F.	1.65	1.76	3.41
2150	10,000 S.F.	1.09	1.49	2.58
2200	Ordinary hazard, one floor, 2000 S.F.	4.72	3.19	7.91
2250	10,000 S.F.	2.92	2.18	5.10
2300	Each additional floor, 2000 S.F.	1.73	1.96	3.69
2350	10,000 S.F.	1.39	1.95	3.34
2400	Extra hazard, one floor, 2000 S.F.	9.90	3.98	13.88
2450	10,000 S.F.	3.54	2.82	6.36
2500	Each additional floor, 2000 S.F.	2.20	2.83	5.03
2550	10,000 S.F.	1.80	2.42	4.22
3000	Grooved steel, black, sch. 10 pipe, light hazard, one floor			
3050	2000 S.F.	8.85	2.83	11.68
3100	10,000 S.F.	2.56	1.69	4.25
3150	Each additional floor, 2000 S.F.	1.53	1.73	3.26
3200	10,000 S.F.	1.02	1.46	2.48
3250	Ordinary hazard, one floor, 2000 S.F.	9.50	3.18	12.68
3300	10,000 S.F.	2.55	2.12	4.67
3350	Each additional floor, 2000 S.F.	1.63	1.94	3.57
3400	10,000 S.F.	1.27	1.91	3.18
3450	Extra hazard, one floor, 2000 S.F.	9.85	3.96	13.81
3500	10,000 S.F.	3.27	2.75	6.02
3550	Each additional floor, 2000 S.F.	2.12	2.81	4.93
3600	10,000 S.F.	1.68	2.39	4.07
4000	Copper tubing, type M, light hazard, one floor			
4050	2000 S.F.	10.95	2.97	13.92
4100	10,000 S.F.	3.86	1.71	5.57
4150	Each additional floor, 2000 S.F.	3.09	1.77	4.86
4200	10,000 S.F.	2.07	1.47	3.54
4250	Ordinary hazard, one floor, 2000 S.F.	11.20	3.33	14.53
4300	10,000 S.F.	4.37	2.03	6.40
4350	Each additional floor, 2000 S.F.	3.11	1.83	4.94
4400	10,000 S.F.	2.45	1.63	4.08
4450	Extra hazard, one floor, 2000 S.F.	12.15	4.03	16.18
4500	10,000 S.F.	8.45	3.06	11.51
4550	Each additional floor, 2000 S.F.	4.43	2.88	7.31
4600	10,000 S.F.	5.50	2.64	8.14
5000	Copper tubing, type M, T-drill system, light hazard, one floor			
5050	2000 S.F.	10.85	2.78	13.63
5100	10,000 S.F.	3.49	1.42	4.91
5150	Each additional floor, 2000 S.F.	2.61	1.60	4.21
5200	10,000 S.F.	1.95	1.20	3.15
5250	Ordinary hazard, one floor, 2000 S.F.	10.65	2.85	13.50
5300	10,000 S.F.	4.05	1.80	5.85
5350	Each additional floor, 2000 S.F.	2.76	1.61	4.37
5400	10,000 S.F.	2.52	1.57	4.09
5450	Extra hazard, one floor, 2000 S.F.	11.10	3.33	14.43
5500	10,000 S.F.	6.15	2.22	8.37
5550	Each additional floor, 2000 S.F.	3.37	2.18	5.55
5600	10,000 S.F.	3.29	1.82	5.11

D40 Fire Protection

D4010 Sprinklers

Firecycle is a fixed fire protection sprinkler system utilizing water as its extinguishing agent. It is a time delayed, recycling, preaction type which automatically shuts the water off when heat is reduced below the detector operating temperature and turns the water back on when that temperature is exceeded.

The system senses a fire condition through a closed circuit electrical detector system which controls water flow to the fire automatically. Batteries supply up to 90 hours emergency power supply for system operation. The piping system is dry (until water is required) and is monitored with pressurized air. Should any leak in the system piping occur, an alarm will sound, but water will not enter the system until heat is sensed by a Firecycle detector.

D4010 390	Firecycle Sprinkler Systems	COST PER S.F.		
		MAT.	INST.	TOTAL
0520	Firecycle sprinkler systems, steel black sch. 40 pipe			
0530	Light hazard, one floor, 500 S.F.	43.50	9.30	52.80
0560	1000 S.F.	22.50	5.85	28.35
0580	2000 S.F.	13.30	4.29	17.59
0600	5000 S.F.	5.95	2.69	8.64
0620	10,000 S.F.	3.59	2.13	5.72
0640	50,000 S.F.	1.82	1.78	3.60
0660	Each additional floor of 500 S.F.	2.41	2.60	5.01
0680	1000 S.F.	1.84	2.40	4.24
0700	2000 S.F.	1.51	2.20	3.71
0720	5000 S.F.	1.34	1.89	3.23
0740	10,000 S.F.	1.32	1.74	3.06
0760	50,000 S.F.	1.28	1.60	2.88
1000	Ordinary hazard, one floor, 500 S.F.	44	9.70	53.70
1020	1000 S.F.	22.50	5.80	28.30
1040	2000 S.F.	13.45	4.44	17.89
1060	5000 S.F.	6.20	2.84	9.04
1080	10,000 S.F.	3.99	2.75	6.74
1100	50,000 S.F.	2.62	2.75	5.37
1140	Each additional floor, 500 S.F.	2.67	3	5.67
1160	1000 S.F.	1.81	2.42	4.23
1180	2000 S.F.	1.84	2.21	4.05
1200	5000 S.F.	1.69	2.07	3.76
1220	10,000 S.F.	1.55	2.04	3.59
1240	50,000 S.F.	1.47	1.82	3.29
1500	Extra hazard, one floor, 500 S.F.	47	11.15	58.15
1520	1000 S.F.	24	7.20	31.20
1540	2000 S.F.	13.85	5.45	19.30
1560	5000 S.F.	6.55	3.85	10.40
1580	10,000 S.F.	5.05	3.90	8.95

D4010 Sprinklers

D4010 390	Firecycle Sprinkler Systems	COST PER S.F.		
		MAT.	INST.	TOTAL
1600	50,000 S.F.	4.26	4.34	8.60
1660	Each additional floor, 500 S.F.	3.20	3.69	6.89
1680	1000 S.F.	2.61	3.48	6.09
1700	2000 S.F.	2.48	3.48	5.96
1720	5000 S.F.	2.09	3.08	5.17
1740	10,000 S.F.	2.29	2.82	5.11
1760	50,000 S.F.	2.31	2.74	5.05
2020	Grooved steel, black, sch. 40 pipe, light hazard, one floor			
2030	2000 S.F.	13.20	3.79	16.99
2060	10,000 S.F.	3.71	2.60	6.31
2100	Each additional floor, 2000 S.F.	1.69	1.77	3.46
2150	10,000 S.F.	1.16	1.49	2.65
2200	Ordinary hazard, one floor, 2000 S.F.	13.30	3.99	17.29
2250	10,000 S.F.	3.96	2.47	6.43
2300	Each additional floor, 2000 S.F.	1.77	1.97	3.74
2350	10,000 S.F.	1.46	1.95	3.41
2400	Extra hazard, one floor, 2000 S.F.	13.60	4.77	18.37
2450	10,000 S.F.	4.25	2.94	7.19
2500	Each additional floor, 2000 S.F.	2.24	2.84	5.08
2550	10,000 S.F.	1.87	2.42	4.29
3050	Grooved steel, black, sch. 10 pipe light hazard, one floor,			
3060	2000 S.F.	13.10	3.76	16.86
3100	10,000 S.F.	3.36	1.85	5.21
3150	Each additional floor, 2000 S.F.	1.57	1.74	3.31
3200	10,000 S.F.	1.09	1.46	2.55
3250	Ordinary hazard, one floor, 2000 S.F.	13.20	3.97	17.17
3300	10,000 S.F.	3.60	2.30	5.90
3350	Each additional floor, 2000 S.F.	1.67	1.95	3.62
3400	10,000 S.F.	1.34	1.91	3.25
3450	Extra hazard, one floor, 2000 S.F.	13.50	4.75	18.25
3500	10,000 S.F.	4.04	2.89	6.93
3550	Each additional floor, 2000 S.F.	2.16	2.82	4.98
3600	10,000 S.F.	1.75	2.39	4.14
4060	Copper tubing, type M, light hazard, one floor, 2000 S.F.	14.65	3.76	18.41
4100	10,000 S.F.	4.66	1.87	6.53
4150	Each additional floor, 2000 S.F.	3.13	1.78	4.91
4200	10,000 S.F.	2.39	1.49	3.88
4250	Ordinary hazard, one floor, 2000 S.F.	14.90	4.12	19.02
4300	10,000 S.F.	5.15	2.19	7.34
4350	Each additional floor, 2000 S.F.	3.15	1.84	4.99
4400	10,000 S.F.	2.48	1.61	4.09
4450	Extra hazard, one floor, 2000 S.F.	15.80	4.82	20.62
4500	10,000 S.F.	9.20	3.23	12.43
4550	Each additional floor, 2000 S.F.	4.47	2.89	7.36
4600	10,000 S.F.	5.60	2.64	8.24
5060	Copper tubing, type M, T-drill system, light hazard, one floor 2000 S.F.	14.50	3.57	18.07
5100	10,000 S.F.	4.29	1.58	5.87
5150	Each additional floor, 2000 S.F.	3.18	1.68	4.86
5200	10,000 S.F.	2.02	1.20	3.22
5250	Ordinary hazard, one floor, 2000 S.F.	14.30	3.64	17.94
5300	10,000 S.F.	4.85	1.96	6.81
5350	Each additional floor, 2000 S.F.	2.80	1.62	4.42
5400	10,000 S.F.	2.59	1.57	4.16
5450	Extra hazard, one floor, 2000 S.F.	14.75	4.12	18.87
5500	10,000 S.F.	6.90	2.36	9.26
5550	Each additional floor, 2000 S.F.	3.41	2.19	5.60
5600	10,000 S.F.	3.36	1.82	5.18

D4010 Sprinklers

Wet Pipe System. A system employing automatic sprinklers attached to a piping system containing water and connected to a water supply so that water discharges immediately from sprinklers opened by heat from a fire.

All areas are assumed to be open.

D4010 410	Wet Pipe Sprinkler Systems	COST PER S.F.		
		MAT.	INST.	TOTAL
0520	Wet pipe sprinkler systems, steel, black, sch. 40 pipe			
0530	Light hazard, one floor, 500 S.F.	2.83	3.06	5.89
0560	1000 S.F.	5.65	3.17	8.82
0580	2000 S.F.	5.05	3.18	8.23
0600	5000 S.F.	2.43	2.23	4.66
0620	10,000 S.F.	1.62	1.89	3.51
0640	50,000 S.F.	1.15	1.71	2.86
0660	Each additional floor, 500 S.F.	1.37	2.61	3.98
0680	1000 S.F.	1.33	2.44	3.77
0700	2000 S.F.	1.25	2.18	3.43
0720	5000 S.F.	.92	1.87	2.79
0740	10,000 S.F.	.90	1.72	2.62
0760	50,000 S.F.	.75	1.35	2.10
1000	Ordinary hazard, one floor, 500 S.F.	3.02	3.28	6.30
1020	1000 S.F.	5.60	3.11	8.71
1040	2000 S.F.	5.20	3.33	8.53
1060	5000 S.F.	2.71	2.38	5.09
1080	10,000 S.F.	2.02	2.51	4.53
1100	50,000 S.F.	1.65	2.36	4.01
1140	Each additional floor, 500 S.F.	1.61	2.94	4.55
1160	1000 S.F.	1.28	2.39	3.67
1180	2000 S.F.	1.41	2.39	3.80
1200	5000 S.F.	1.43	2.26	3.69
1220	10,000 S.F.	1.31	2.34	3.65
1240	50,000 S.F.	1.23	2.08	3.31
1500	Extra hazard, one floor, 500 S.F.	10.75	5.05	15.80
1520	1000 S.F.	6.70	4.38	11.08
1540	2000 S.F.	5.60	4.49	10.09
1560	5000 S.F.	3.57	3.92	7.49
1580	10,000 S.F.	3.07	3.73	6.80
1600	50,000 S.F.	3.38	3.60	6.98
1660	Each additional floor, 500 S.F.	2.14	3.63	5.77
1680	1000 S.F.	2.08	3.45	5.53
1700	2000 S.F.	1.95	3.45	5.40
1720	5000 S.F.	1.67	3.06	4.73
1740	10,000 S.F.	1.87	2.80	4.67
1760	50,000 S.F.	1.88	2.69	4.57
2020	Grooved steel, black sch. 40 pipe, light hazard, one floor, 2000 S.F.	4.94	2.68	7.62

D4010 Sprinklers

D4010 410	Wet Pipe Sprinkler Systems	COST PER S.F.		
		MAT.	INST.	TOTAL
2060	10,000 S.F.	1.89	1.70	3.59
2100	Each additional floor, 2000 S.F.	1.16	1.74	2.90
2150	10,000 S.F.	.74	1.47	2.21
2200	Ordinary hazard, one floor, 2000 S.F.	5	2.88	7.88
2250	10,000 S.F.	1.75	2.10	3.85
2300	Each additional floor, 2000 S.F.	1.24	1.94	3.18
2350	10,000 S.F.	1.04	1.93	2.97
2400	Extra hazard, one floor, 2000 S.F.	5.30	3.66	8.96
2450	10,000 S.F.	2.31	2.72	5.03
2500	Each additional floor, 2000 S.F.	1.71	2.81	4.52
2550	10,000 S.F.	1.45	2.40	3.85
3050	Grooved steel, black sch. 10 pipe, light hazard, one floor, 2000 S.F.	4.82	2.65	7.47
3100	10,000 S.F.	1.39	1.61	3
3150	Each additional floor, 2000 S.F.	1.04	1.71	2.75
3200	10,000 S.F.	.67	1.44	2.11
3250	Ordinary hazard, one floor, 2000 S.F.	4.92	2.86	7.78
3300	10,000 S.F.	1.63	2.06	3.69
3350	Each additional floor, 2000 S.F.	1.14	1.92	3.06
3400	10,000 S.F.	.92	1.89	2.81
3450	Extra hazard, one floor, 2000 S.F.	5.25	3.64	8.89
3500	10,000 S.F.	2.10	2.67	4.77
3550	Each additional floor, 2000 S.F.	1.63	2.79	4.42
3600	10,000 S.F.	1.33	2.37	3.70
4050	Copper tubing, type M, light hazard, one floor, 2000 S.F.	6.35	2.65	9
4100	10,000 S.F.	2.69	1.63	4.32
4150	Each additional floor, 2000 S.F.	2.60	1.75	4.35
4200	10,000 S.F.	1.97	1.47	3.44
4250	Ordinary hazard, one floor, 2000 S.F.	6.65	3.01	9.66
4300	10,000 S.F.	3.20	1.95	5.15
4350	Each additional floor, 2000 S.F.	2.96	1.96	4.92
4400	10,000 S.F.	2.35	1.75	4.10
4450	Extra hazard, one floor, 2000 S.F.	7.55	3.71	11.26
4500	10,000 S.F.	7.20	2.96	10.16
4550	Each additional floor, 2000 S.F.	3.94	2.86	6.80
4600	10,000 S.F.	5.15	2.62	7.77
5050	Copper tubing, type M, T-drill system, light hazard, one floor			
5060	2000 S.F.	6.25	2.46	8.71
5100	10,000 S.F.	2.32	1.34	3.66
5150	Each additional floor, 2000 S.F.	2.48	1.56	4.04
5200	10,000 S.F.	1.60	1.18	2.78
5250	Ordinary hazard, one floor, 2000 S.F.	6.05	2.53	8.58
5300	10,000 S.F.	2.88	1.72	4.60
5350	Each additional floor, 2000 S.F.	2.27	1.59	3.86
5400	10,000 S.F.	2.17	1.55	3.72
5450	Extra hazard, one floor, 2000 S.F.	6.50	3.01	9.51
5500	10,000 S.F.	4.98	2.14	7.12
5550	Each additional floor, 2000 S.F.	2.97	2.21	5.18
5600	10,000 S.F.	2.94	1.80	4.74

D4020 Standpipes

Roof

Roof connections with hose gate valves (for combustible roof)

Hose connections on each floor (size based on class of service)

Check Valve

Siamese inlet connections (for fire department use)

D4020 310	Wet Standpipe Risers, Class I	COST PER FLOOR		
		MAT.	INST.	TOTAL
0550	Wet standpipe risers, Class I, steel, black sch. 40, 10' height			
0560	4" diameter pipe, one floor	4,575	3,325	7,900
0580	Additional floors	1,075	1,025	2,100
0600	6" diameter pipe, one floor	7,600	5,850	13,450
0620	Additional floors	1,850	1,650	3,500
0640	8" diameter pipe, one floor	11,900	7,025	18,925
0660	Additional floors	2,700	1,975	4,675

D4020 310	Wet Standpipe Risers, Class II	COST PER FLOOR		
		MAT.	INST.	TOTAL
1030	Wet standpipe risers, Class II, steel, black sch. 40, 10' height			
1040	2" diameter pipe, one floor	1,825	1,200	3,025
1060	Additional floors	545	465	1,010
1080	2-1/2" diameter pipe, one floor	2,675	1,750	4,425
1100	Additional floors	635	540	1,175

D4020 310	Wet Standpipe Risers, Class III	COST PER FLOOR		
		MAT.	INST.	TOTAL
1530	Wet standpipe risers, Class III, steel, black sch. 40, 10' height			
1540	4" diameter pipe, one floor	4,675	3,325	8,000
1560	Additional floors	920	865	1,785
1580	6" diameter pipe, one floor	7,725	5,850	13,575
1600	Additional floors	1,900	1,650	3,550
1620	8" diameter pipe, one floor	12,000	7,025	19,025
1640	Additional floors	2,750	1,975	4,725

D40 Fire Protection

D4020 Standpipes

Roof — Roof connections with hose gate valves (for combustible roof)

Hose connections on each floor (size based on class of service)

Check valve

Siamese inlet connections (for fire department use)

D4020 330	Dry Standpipe Risers, Class I	COST PER FLOOR		
		MAT.	INST.	TOTAL
0530	Dry standpipe riser, Class I, steel, black sch. 40, 10' height			
0540	4" diameter pipe, one floor	2,775	2,700	5,475
0560	Additional floors	990	980	1,970
0580	6" diameter pipe, one floor	5,450	4,625	10,075
0600	Additional floors	1,775	1,575	3,350
0620	8" diameter pipe, one floor	8,750	5,600	14,350
0640	Additional floors	2,625	1,925	4,550

D4020 330	Dry Standpipe Risers, Class II	COST PER FLOOR		
		MAT.	INST.	TOTAL
1030	Dry standpipe risers, Class II, steel, black sch. 40, 10' height			
1040	2" diameter pipe, one floor	1,500	1,250	2,750
1060	Additional floor	470	410	880
1080	2-1/2" diameter pipe, one floor	2,125	1,450	3,575
1100	Additional floors	560	485	1,045

D4020 330	Dry Standpipe Risers, Class III	COST PER FLOOR		
		MAT.	INST.	TOTAL
1530	Dry standpipe risers, Class III, steel, black sch. 40, 10' height			
1540	4" diameter pipe, one floor	2,825	2,650	5,475
1560	Additional floors	855	885	1,740
1580	6" diameter pipe, one floor	5,500	4,625	10,125
1600	Additional floors	1,825	1,575	3,400
1620	8" diameter pipe, one floor	8,800	5,600	14,400
1640	Additional floor	2,675	1,925	4,600

D40 Fire Protection

D4020 Standpipes

D4020 410	Fire Hose Equipment	COST EACH		
		MAT.	INST.	TOTAL
0100	Adapters, reducing, 1 piece, FxM, hexagon, cast brass, 2-1/2" x 1-1/2"	55.50		55.50
0200	Pin lug, 1-1/2" x 1"	47.50		47.50
0250	3" x 2-1/2"	125		125
0300	For polished chrome, add 75% mat.			
0400	Cabinets, D.S. glass in door, recessed, steel box, not equipped			
0500	Single extinguisher, steel door & frame	110	148	258
0550	Stainless steel door & frame	205	148	353
0600	Valve, 2-1/2" angle, steel door & frame	135	99	234
0650	Aluminum door & frame	164	99	263
0700	Stainless steel door & frame	220	99	319
0750	Hose rack assy, 2-1/2" x 1-1/2" valve & 100' hose, steel door & frame	252	198	450
0800	Aluminum door & frame	375	198	573
0850	Stainless steel door & frame	495	198	693
0900	Hose rack assy & extinguisher,2-1/2"x1-1/2" valve & hose,steel door & frame	315	237	552
0950	Aluminum	475	237	712
1000	Stainless steel	575	237	812
1550	Compressor, air, dry pipe system, automatic, 200 gal., 3/4 H.P.	810	505	1,315
1600	520 gal., 1 H.P.	835	505	1,340
1650	Alarm, electric pressure switch (circuit closer)	86	25.50	111.50
2500	Couplings, hose, rocker lug, cast brass, 1-1/2"	53		53
2550	2-1/2"	69.50		69.50
3000	Escutcheon plate, for angle valves, polished brass, 1-1/2"	16.70		16.70
3050	2-1/2"	27		27
3500	Fire pump, electric, w/controller, fittings, relief valve			
3550	4" pump, 30 H.P., 500 G.P.M.	16,600	3,700	20,300
3600	5" pump, 40 H.P., 1000 G.P.M.	18,200	4,175	22,375
3650	5" pump, 100 H.P., 1000 G.P.M.	25,100	4,650	29,750
3700	For jockey pump system, add	2,675	595	3,270
5000	Hose, per linear foot, synthetic jacket, lined,			
5100	300 lb. test, 1-1/2" diameter	3.23	.46	3.69
5150	2-1/2" diameter	5.60	.54	6.14
5200	500 lb. test, 1-1/2" diameter	3.33	.46	3.79
5250	2-1/2" diameter	5.90	.54	6.44
5500	Nozzle, plain stream, polished brass, 1-1/2" x 10"	47.50		47.50
5550	2-1/2" x 15" x 13/16" or 1-1/2"	86.50		86.50
5600	Heavy duty combination adjustable fog and straight stream w/handle 1-1/2"	465		465
5650	2-1/2" direct connection	575		575
6000	Rack, for 1-1/2" diameter hose 100 ft. long, steel	60.50	59.50	120
6050	Brass	104	59.50	163.50
6500	Reel, steel, for 50 ft. long 1-1/2" diameter hose	131	84.50	215.50
6550	For 75 ft. long 2-1/2" diameter hose	212	84.50	296.50
7050	Siamese, w/plugs & chains, polished brass, sidewalk, 4" x 2-1/2" x 2-1/2"	630	475	1,105
7100	6" x 2-1/2" x 2-1/2"	810	595	1,405
7200	Wall type, flush, 4" x 2-1/2" x 2-1/2"	515	237	752
7250	6" x 2-1/2" x 2-1/2"	720	258	978
7300	Projecting, 4" x 2-1/2" x 2-1/2"	475	237	712
7350	6" x 2-1/2" x 2-1/2"	800	258	1,058
7400	For chrome plate, add 15% mat.			
8000	Valves, angle, wheel handle, 300 Lb., rough brass, 1-1/2"	94.50	55	149.50
8050	2-1/2"	174	94	268
8100	Combination pressure restricting, 1-1/2"	85	55	140
8150	2-1/2"	174	94	268
8200	Pressure restricting, adjustable, satin brass, 1-1/2"	125	55	180
8250	2-1/2"	176	94	270
8300	Hydrolator, vent and drain, rough brass, 1-1/2"	74.50	55	129.50
8350	2-1/2"	74.50	55	129.50
8400	Cabinet assy, incls. adapter, rack, hose, and nozzle	780	360	1,140

D4090 Other Fire Protection Systems

General: Automatic fire protection (suppression) systems other than water sprinklers may be desired for special environments, high risk areas, isolated locations or unusual hazards. Some typical applications would include:

1. Paint dip tanks
2. Securities vaults
3. Electronic data processing
4. Tape and data storage
5. Transformer rooms
6. Spray booths
7. Petroleum storage
8. High rack storage

Piping and wiring costs are highly variable and are not included.

D4090 910	Fire Suppression Unit Components	COST EACH		
		MAT.	INST.	TOTAL
0020	Detectors with brackets			
0040	Fixed temperature heat detector	55	78.50	133.50
0060	Rate of temperature rise detector	55	78.50	133.50
0080	Ion detector (smoke) detector	119	101	220
0200	Extinguisher agent			
0240	200 lb FM200, container	7,900	305	8,205
0280	75 lb carbon dioxide cylinder	1,300	202	1,502
0320	Dispersion nozzle			
0340	FM200 1-1/2" dispersion nozzle	68	48	116
0380	Carbon dioxide 3" x 5" dispersion nozzle	68	37.50	105.50
0420	Control station			
0440	Single zone control station with batteries	1,750	625	2,375
0470	Multizone (4) control station with batteries	3,350	1,250	4,600
0500	Electric mechanical release	170	325	495
0550	Manual pull station	61.50	112	173.50
0640	Battery standby power 10" x 10" x 17"	445	157	602
0740	Bell signalling device	85	78.50	163.50
0900	Standard low-rise sprinkler accessory package, 3 story	2,550	1,750	4,300
1000	Standard mid-rise sprinkler accessory package, 8 story	5,475	3,350	8,825
1100	Standard high-rise sprinkler accessory package, 16 story	11,700	6,575	18,275

D4090 920	FM200 Systems	COST PER C.F.		
		MAT.	INST.	TOTAL
0820	Average FM200 system, minimum			1.79
0840	Maximum			3.56

D5010 Electrical Service/Distribution

D5010 120	Overhead Electric Service, 3 Phase - 4 Wire	COST EACH		
		MAT.	INST.	TOTAL
0200	Service installation, includes breakers, metering, 20' conduit & wire			
0220	3 phase, 4 wire, 120/208 volts, 60 A	660	1,025	1,685
0240	3 phase, 4 wire, 120/208 volts, 100 A	920	1,150	2,070
0245	100 A w/circuit breaker	1,650	1,425	3,075
0280	200 A	1,850	1,600	3,450
0285	200 A w/circuit breaker	3,400	2,000	5,400
0320	400 A	3,725	3,175	6,900
0325	400 A w/circuit breaker	6,625	3,950	10,575
0360	600 A	6,200	4,700	10,900
0365	600 A, w/switchboard	10,400	5,950	16,350
0400	800 A	8,925	5,625	14,550
0405	800 A, w/switchboard	13,100	7,050	20,150
0440	1000 A	10,800	6,925	17,725
0445	1000 A, w/switchboard	15,900	8,500	24,400
0480	1200 A	14,300	7,900	22,200
0485	1200 A, w/groundfault switchboard	36,100	9,800	45,900
0520	1600 A	17,900	10,300	28,200
0525	1600 A, w/groundfault switchboard	41,600	12,400	54,000
0560	2000 A	23,700	12,500	36,200
0565	2000 A, w/groundfault switchboard	51,500	15,500	67,000
0610	1 phase, 3 wire, 120/240 volts, 100 A (no safety switch)	204	565	769
0615	100 A w/load center	550	1,100	1,650
0620	200 A	465	775	1,240
0625	200 A w/load center	1,225	1,575	2,800

Underground service conductor

Meter and
service
equipment

Safety
Switch

Ground rod

D5010 130	Underground Electric Service	COST EACH		
		MAT.	INST.	TOTAL
0950	Underground electric service including excavation, backfill, and compaction			
1000	3 phase, 4 wire, 277/480 volts, 2000 A	43,300	19,900	63,200
1050	2000 A w/groundfault switch	69,000	22,300	91,300
1100	1600 A	34,400	16,100	50,500
1150	1600 A w/groundfault switchboard	58,000	18,200	76,200
1200	1200 A	26,200	12,800	39,000
1250	1200 A w/groundfault switchboard	48,000	14,700	62,700
1400	800 A	18,800	10,700	29,500
1450	800 A w/ switchboard	23,000	12,100	35,100
1500	600 A	11,900	9,800	21,700
1550	600 A w/ switchboard	16,100	11,000	27,100
1600	1 phase, 3 wire, 120/240 volts, 200 A	3,375	3,400	6,775
1650	200 A w/load center	4,125	4,200	8,325
1700	100 A	2,250	2,775	5,025
1750	100 A w/load center	3,025	3,575	6,600

D5010 Electrical Service/Distribution

D5010 230	Feeder Installation	COST PER L.F.		
		MAT.	INST.	TOTAL
0200	Feeder installation 600 V, including RGS conduit and XHHW wire, 60 A	6.30	11.55	17.85
0240	100 A	11.75	15	26.75
0280	200 A	28	22.50	50.50
0320	400 A	56	45.50	101.50
0360	600 A	103	74	177
0400	800 A	135	87.50	222.50
0440	1000 A	168	115	283
0480	1200 A	207	148	355
0520	1600 A	271	175	446
0560	2000 A	335	230	565
1200	Branch installation 600 V, including EMT conduit and THW wire, 15 A	1.10	5.65	6.75
1240	20 A	1.32	5.95	7.27
1280	30 A	1.67	6.20	7.87
1320	50 A	3.50	8.60	12.10
1360	65 A	4.36	9.15	13.51
1400	85 A	6.80	10.60	17.40
1440	100 A	7.60	10.80	18.40
1480	130 A	11.30	12.65	23.95
1520	150 A	14.05	14.50	28.55
1560	200 A	19.15	16.55	35.70

D5010 Electrical Service/Distribution

D5010 240	Switchgear	COST EACH		
		MAT.	INST.	TOTAL
0190	Switchgear installation, including switchboard, panels, & circuit breaker			
0200	120/208 V, 1 phase, 400 A	8,050	2,675	10,725
0240	600 A	10,500	2,975	13,475
0280	800 A	13,500	3,425	16,925
0300	1000 A	16,500	3,825	20,325
0320	1200 A	17,500	4,325	21,825
0360	1600 A	26,600	4,975	31,575
0400	2000 A	31,800	5,425	37,225
0500	277/480 V, 3 phase, 400 A	11,400	4,975	16,375
0520	600 A	16,000	5,675	21,675
0540	800 A	18,900	6,275	25,175
0560	1000 A	23,500	6,975	30,475
0580	1200 A	25,700	7,825	33,525
0600	1600 A	34,800	8,575	43,375
0620	2000 A	42,500	9,475	51,975

D50 Electrical

D5010 Electrical Service/Distribution

- Panelboard
- Conduit
- Safety switch

D5010 250	Panelboard	COST EACH		
		MAT.	INST.	TOTAL
0900	Panelboards, NQOD, 4 wire, 120/208 volts w/conductor & conduit			
1000	100 A, 0 stories, 0' horizontal	1,275	1,325	2,600
1020	1 stories, 25' horizontal	1,700	1,900	3,600
1040	5 stories, 50' horizontal	2,525	3,025	5,550
1060	10 stories, 75' horizontal	3,475	4,350	7,825
1080	225A, 0 stories, 0' horizontal	2,375	1,750	4,125
2000	1 stories, 25' horizontal	3,650	2,775	6,425
2020	5 stories, 50' horizontal	6,175	4,800	10,975
2040	10 stories, 75' horizontal	9,100	7,175	16,275
2060	400A, 0 stories, 0' horizontal	3,500	2,625	6,125
2080	1 stories, 25' horizontal	5,175	4,300	9,475
3000	5 stories, 50' horizontal	8,500	7,600	16,100
3020	10 stories, 75' horizontal	17,800	14,900	32,700
3040	600 A, 0 stories, 0' horizontal	5,200	3,125	8,325
3060	1 stories, 25' horizontal	8,875	5,850	14,725
3080	5 stories, 50' horizontal	16,200	11,200	27,400
4000	10 stories, 75' horizontal	36,200	22,100	58,300
4010	Panelboards, NEHB, 4 wire, 277/480 volts w/conductor, conduit, & safety switch			
4020	100 A, 0 stories, 0' horizontal, includes safety switch	3,050	1,825	4,875
4040	1 stories, 25' horizontal	3,475	2,400	5,875
4060	5 stories, 50' horizontal	4,300	3,525	7,825
4080	10 stories, 75' horizontal	5,275	4,850	10,125
5000	225 A, 0 stories, 0' horizontal	4,600	2,225	6,825
5020	1 stories, 25' horizontal	5,875	3,250	9,125
5040	5 stories, 50' horizontal	8,400	5,300	13,700
5060	10 stories, 75' horizontal	11,300	7,650	18,950
5080	400 A, 0 stories, 0' horizontal	7,675	3,425	11,100
6000	1 stories, 25' horizontal	9,350	5,100	14,450
6020	5 stories, 50' horizontal	12,700	8,400	21,100
6040	10 stories, 75' horizontal	22,000	15,700	37,700
6060	600 A, 0 stories, 0' horizontal	11,300	4,350	15,650
6080	1 stories, 25' horizontal	15,000	7,075	22,075
7000	5 stories, 50' horizontal	22,300	12,500	34,800
7020	10 stories, 75' horizontal	42,300	23,400	65,700

D5020 Lighting and Branch Wiring

Duplex Wall Receptacle

Undercarpet Receptacle System

D5020 110	Receptacle (by Wattage)	COST PER S.F.		
		MAT.	INST.	TOTAL
0190	Receptacles include plate, box, conduit, wire & transformer when required			
0200	2.5 per 1000 S.F., .3 watts per S.F.	.40	1.39	1.79
0240	With transformer	.47	1.46	1.93
0280	4 per 1000 S.F., .5 watts per S.F.	.45	1.63	2.08
0320	With transformer	.56	1.74	2.30
0360	5 per 1000 S.F., .6 watts per S.F.	.53	1.93	2.46
0400	With transformer	.67	2.07	2.74
0440	8 per 1000 S.F., .9 watts per S.F.	.56	2.13	2.69
0480	With transformer	.78	2.35	3.13
0520	10 per 1000 S.F., 1.2 watts per S.F.	.59	2.33	2.92
0560	With transformer	.87	2.61	3.48
0600	16.5 per 1000 S.F., 2.0 watts per S.F.	.69	2.89	3.58
0640	With transformer	1.14	3.35	4.49
0680	20 per 1000 S.F., 2.4 watts per S.F.	.73	3.15	3.88
0720	With transformer	1.28	3.71	4.99

D5020 115	Receptacles, Floor	COST PER S.F.		
		MAT.	INST.	TOTAL
0200	Receptacle systems, underfloor duct, 5' on center, low density	7	3.07	10.07
0240	High density	7.55	3.96	11.51
0280	7' on center, low density	5.55	2.64	8.19
0320	High density	6.10	3.53	9.63
0400	Poke thru fittings, low density	1.20	1.50	2.70
0440	High density	2.37	3	5.37
0520	Telepoles, using Romex, low density	1.19	.93	2.12
0560	High density	2.38	1.86	4.24
0600	Using EMT, low density	1.27	1.24	2.51
0640	High density	2.51	2.46	4.97
0720	Conduit system with floor boxes, low density	1.29	1.04	2.33
0760	High density	2.58	2.10	4.68
0840	Undercarpet power system, 3 conductor with 5 conductor feeder, low density	1.49	.38	1.87
0880	High density	2.95	.79	3.74

D5020 Lighting and Branch Wiring

Description: Table D5020 130 includes the cost for switch, plate, box, conduit in slab or EMT exposed and copper wire. Add 20% for exposed conduit.

No power required for switches.

Federal energy guidelines recommend the maximum lighting area controlled per switch shall not exceed 1000 S.F. and that areas over 500 S.F. shall be so controlled that total illumination can be reduced by at least 50%.

D5020 130	Wall Switch by Sq. Ft.	COST PER S.F.		
		MAT.	INST.	TOTAL
0200	Wall switches, 1.0 per 1000 S.F.	.06	.23	.29
0240	1.2 per 1000 S.F.	.07	.25	.32
0280	2.0 per 1000 S.F.	.10	.36	.46
0320	2.5 per 1000 S.F.	.13	.45	.58
0360	5.0 per 1000 S.F.	.26	.98	1.24
0400	10.0 per 1000 S.F.	.54	1.98	2.52

D5020 135	Miscellaneous Power	COST PER S.F.		
		MAT.	INST.	TOTAL
0200	Miscellaneous power, to .5 watts	.03	.11	.14
0240	.8 watts	.05	.16	.21
0280	1 watt	.07	.21	.28
0320	1.2 watts	.08	.25	.33
0360	1.5 watts	.09	.29	.38
0400	1.8 watts	.10	.34	.44
0440	2 watts	.12	.40	.52
0480	2.5 watts	.15	.50	.65
0520	3 watts	.18	.58	.76

D5020 140	Central A. C. Power (by Wattage)	COST PER S.F.		
		MAT.	INST.	TOTAL
0200	Central air conditioning power, 1 watt	.07	.25	.32
0220	2 watts	.08	.29	.37
0240	3 watts	.12	.32	.44
0280	4 watts	.18	.43	.61
0320	6 watts	.33	.56	.89
0360	8 watts	.43	.61	1.04
0400	10 watts	.61	.71	1.32

D5020 Lighting and Branch Wiring

Motor Installation

D5020 145	Motor Installation	COST EACH		
		MAT.	INST.	TOTAL
0200	Motor installation, single phase, 115V, 1/3 HP motor size	615	955	1,570
0240	1 HP motor size	635	955	1,590
0280	2 HP motor size	685	1,000	1,685
0320	3 HP motor size	765	1,025	1,790
0360	230V, 1 HP motor size	615	965	1,580
0400	2 HP motor size	650	965	1,615
0440	3 HP motor size	730	1,050	1,780
0520	Three phase, 200V, 1-1/2 HP motor size	730	1,050	1,780
0560	3 HP motor size	785	1,150	1,935
0600	5 HP motor size	850	1,275	2,125
0640	7-1/2 HP motor size	885	1,300	2,185
0680	10 HP motor size	1,400	1,650	3,050
0720	15 HP motor size	1,800	1,825	3,625
0760	20 HP motor size	2,325	2,100	4,425
0800	25 HP motor size	2,375	2,125	4,500
0840	30 HP motor size	3,675	2,500	6,175
0880	40 HP motor size	4,700	2,950	7,650
0920	50 HP motor size	7,875	3,450	11,325
0960	60 HP motor size	8,250	3,625	11,875
1000	75 HP motor size	10,800	4,150	14,950
1040	100 HP motor size	28,800	4,900	33,700
1080	125 HP motor size	29,700	5,375	35,075
1120	150 HP motor size	34,600	6,325	40,925
1160	200 HP motor size	35,300	7,500	42,800
1240	230V, 1-1/2 HP motor size	690	1,050	1,740
1280	3 HP motor size	745	1,150	1,895
1320	5 HP motor size	810	1,275	2,085
1360	7-1/2 HP motor size	810	1,275	2,085
1400	10 HP motor size	1,250	1,550	2,800
1440	15 HP motor size	1,450	1,700	3,150
1480	20 HP motor size	2,175	2,050	4,225
1520	25 HP motor size	2,325	2,100	4,425
1560	30 HP motor size	2,375	2,125	4,500
1600	40 HP motor size	4,400	2,875	7,275
1640	50 HP motor size	4,700	3,050	7,750
1680	60 HP motor size	7,875	3,450	11,325
1720	75 HP motor size	9,750	3,925	13,675
1760	100 HP motor size	11,200	4,375	15,575
1800	125 HP motor size	29,300	5,050	34,350
1840	150 HP motor size	31,400	5,750	37,150
1880	200 HP motor size	33,400	6,400	39,800
1960	460V, 2 HP motor size	855	1,050	1,905
2000	5 HP motor size	910	1,150	2,060
2040	10 HP motor size	940	1,275	2,215

D5020 Lighting and Branch Wiring

D5020 145	Motor Installation	COST EACH		
		MAT.	INST.	TOTAL
2080	15 HP motor size	1,225	1,450	2,675
2120	20 HP motor size	1,300	1,575	2,875
2160	25 HP motor size	1,425	1,650	3,075
2200	30 HP motor size	1,775	1,775	3,550
2240	40 HP motor size	2,300	1,900	4,200
2280	50 HP motor size	2,600	2,100	4,700
2320	60 HP motor size	3,775	2,475	6,250
2360	75 HP motor size	4,450	2,750	7,200
2400	100 HP motor size	4,975	3,050	8,025
2440	125 HP motor size	8,175	3,475	11,650
2480	150 HP motor size	10,100	3,875	13,975
2520	200 HP motor size	12,000	4,375	16,375
2600	575V, 2 HP motor size	855	1,050	1,905
2640	5 HP motor size	910	1,150	2,060
2680	10 HP motor size	940	1,275	2,215
2720	20 HP motor size	1,225	1,450	2,675
2760	25 HP motor size	1,300	1,575	2,875
2800	30 HP motor size	1,775	1,775	3,550
2840	50 HP motor size	1,925	1,825	3,750
2880	60 HP motor size	3,725	2,475	6,200
2920	75 HP motor size	3,775	2,475	6,250
2960	100 HP motor size	4,450	2,750	7,200
3000	125 HP motor size	7,700	3,400	11,100
3040	150 HP motor size	8,175	3,475	11,650
3080	200 HP motor size	10,500	3,925	14,425

D5020 155	Motor Feeder	COST PER L.F.		
		MAT.	INST.	TOTAL
0200	Motor feeder systems, single phase, feed up to 115V 1HP or 230V 2 HP	2.27	7.40	9.67
0240	115V 2HP, 230V 3HP	2.43	7.50	9.93
0280	115V 3HP	2.81	7.80	10.61
0360	Three phase, feed to 200V 3HP, 230V 5HP, 460V 10HP, 575V 10HP	2.41	7.95	10.36
0440	200V 5HP, 230V 7.5HP, 460V 15HP, 575V 20HP	2.65	8.15	10.80
0520	200V 10HP, 230V 10HP, 460V 30HP, 575V 30HP	3.22	8.60	11.82
0600	200V 15HP, 230V 15HP, 460V 40HP, 575V 50HP	4.48	9.85	14.33
0680	200V 20HP, 230V 25HP, 460V 50HP, 575V 60HP	6.85	12.50	19.35
0760	200V 25HP, 230V 30HP, 460V 60HP, 575V 75HP	7.65	12.70	20.35
0840	200V 30HP	8.70	13.10	21.80
0920	230V 40HP, 460V 75HP, 575V 100HP	11.30	14.35	25.65
1000	200V 40HP	12.75	15.35	28.10
1080	230V 50HP, 460V 100HP, 575V 125HP	16	16.95	32.95
1160	200V 50HP, 230V 60HP, 460V 125HP, 575V 150HP	18.70	18	36.70
1240	200V 60HP, 460V 150HP	23	21	44
1320	230V 75HP, 575V 200HP	26.50	22	48.50
1400	200V 75HP	30	22.50	52.50
1480	230V 100HP, 460V 200HP	40.50	26	66.50
1560	200V 100HP	55	32.50	87.50
1640	230V 125HP	55	32.50	87.50
1720	200V 125HP, 230V 150HP	64	41	105
1800	200V 150HP	78	44	122
1880	200V 200HP	98	46.50	144.50
1960	230V 200HP	100	48.50	148.50

431

D5020 Lighting and Branch Wiring

Fluorescent Fixture

Incandescent Fixture

D5020 210	Fluorescent Fixtures (by Wattage)	COST PER S.F.		
		MAT.	INST.	TOTAL
0190	Fluorescent fixtures recess mounted in ceiling			
0195	T12, standard 40 watt lamps			
0200	1 watt per S.F., 20 FC, 5 fixtures @40 watts per 1000 S.F.	.77	1.84	2.61
0240	2 watt per S.F., 40 FC, 10 fixtures @40 watt per 1000 S.F.	1.52	3.62	5.14
0280	3 watt per S.F., 60 FC, 15 fixtures @40 watt per 1000 S.F	2.30	5.45	7.75
0320	4 watt per S.F., 80 FC, 20 fixtures @40 watt per 1000 S.F.	3.04	7.20	10.24
0400	5 watt per S.F., 100 FC, 25 fixtures @40 watt per 1000 S.F.	3.82	9.05	12.87
0450	T8, energy saver 32 watt lamps			
0500	0.8 watt per S.F., 20 FC, 5 fixtures @32 watt per 1000 S.F.	.82	1.84	2.66
0520	1.6 watt per S.F., 40 FC, 10 fixtures @32 watt per 1000 S.F.	1.63	3.62	5.25
0540	2.4 watt per S.F., 60 FC, 15 fixtures @ 32 watt per 1000 S.F	2.45	5.45	7.90
0560	3.2 watt per S.F., 80 FC, 20 fixtures @32 watt per 1000 S.F.	3.25	7.20	10.45
0580	4 watt per S.F., 100 FC, 25 fixtures @32 watt per 1000 S.F.	4.08	9.05	13.13

D5020 216	Incandescent Fixture (by Wattage)	COST PER S.F.		
		MAT.	INST.	TOTAL
0190	Incandescent fixture recess mounted, type A			
0200	1 watt per S.F., 8 FC, 6 fixtures per 1000 S.F.	1.02	1.48	2.50
0240	2 watt per S.F., 16 FC, 12 fixtures per 1000 S.F.	2.04	2.96	5
0280	3 watt per S.F., 24 FC, 18 fixtures, per 1000 S.F.	3.07	4.39	7.46
0320	4 watt per S.F., 32 FC, 24 fixtures per 1000 S.F.	4.10	5.85	9.95
0400	5 watt per S.F., 40 FC, 30 fixtures per 1000 S.F.	5.10	7.35	12.45

D5020 Lighting and Branch Wiring

High Bay Fixture

Low Bay Fixture

D5020 226	H.I.D. Fixture, High Bay, 16′ (by Wattage)	COST PER S.F.		
		MAT.	INST.	TOTAL
0190	High intensity discharge fixture, 16′ above work plane			
0240	1 watt/S.F., type E, 42 FC, 1 fixture/1000 S.F.	1.05	1.55	2.60
0280	Type G, 52 FC, 1 fixture/1000 S.F.	1.05	1.55	2.60
0320	Type C, 54 FC, 2 fixture/1000 S.F.	1.23	1.73	2.96
0440	2 watt/S.F., type E, 84 FC, 2 fixture/1000 S.F.	2.13	3.15	5.28
0480	Type G, 105 FC, 2 fixture/1000 S.F.	2.13	3.15	5.28
0520	Type C, 108 FC, 4 fixture/1000 S.F.	2.47	3.46	5.93
0640	3 watt/S.F., type E, 126 FC, 3 fixture/1000 S.F.	3.20	4.69	7.89
0680	Type G, 157 FC, 3 fixture/1000 S.F.	3.20	4.69	7.89
0720	Type C, 162 FC, 6 fixture/1000 S.F.	3.71	5.20	8.91
0840	4 watt/S.F., type E, 168 FC, 4 fixture/1000 S.F.	4.26	6.30	10.56
0880	Type G, 210 FC, 4 fixture/1000 S.F.	4.26	6.30	10.56
0920	Type C, 243 FC, 9 fixture/1000 S.F.	5.40	7.20	12.60
1040	5 watt/S.F., type E, 210 FC, 5 fixture/1000 S.F.	5.30	7.85	13.15
1080	Type G, 262 FC, 5 fixture/1000 S.F.	5.30	7.85	13.15
1120	Type C, 297 FC, 11 fixture/1000 S.F.	6.65	8.95	15.60

D5020 234	H.I.D. Fixture, High Bay, 30′ (by Wattage)	COST PER S.F.		
		MAT.	INST.	TOTAL
0190	High intensity discharge fixture, 30′ above work plane			
0240	1 watt/S.F., type E, 37 FC, 1 fixture/1000 S.F.	1.19	1.97	3.16
0280	Type G, 45 FC., 1 fixture/1000 S.F.	1.19	1.97	3.16
0320	Type F, 50 FC, 1 fixture/1000 S.F.	1.01	1.54	2.55
0440	2 watt/S.F., type E, 74 FC, 2 fixtures/1000 S.F.	2.38	3.91	6.29
0480	Type G, 92 FC, 2 fixtures/1000 S.F.	2.38	3.91	6.29
0520	Type F, 100 FC, 2 fixtures/1000 S.F.	2.02	3.12	5.14
0640	3 watt/S.F., type E, 110 FC, 3 fixtures/1000 S.F.	3.61	5.95	9.56
0680	Type G, 138 FC, 3 fixtures/1000 S.F.	3.61	5.95	9.56
0720	Type F, 150 FC, 3 fixtures/1000 S.F.	3.04	4.65	7.69
0840	4 watt/S.F., type E, 148 FC, 4 fixtures/1000 S.F.	4.79	7.90	12.69
0880	Type G, 185 FC, 4 fixtures/1000 S.F.	4.79	7.90	12.69
0920	Type F, 200 FC, 4 fixtures/1000 S.F.	4.05	6.25	10.30
1040	5 watt/S.F., type E, 185 FC, 5 fixtures/1000 S.F.	6	9.85	15.85
1080	Type G, 230 FC, 5 fixtures/1000 S.F.	6	9.85	15.85
1120	Type F, 250 FC, 5 fixtures/1000 S.F.	5.05	7.80	12.85

D5020 238	H.I.D. Fixture, Low Bay, 8′-10′ (by Wattage)	COST PER S.F.		
		MAT.	INST.	TOTAL
0190	High intensity discharge fixture, 8′-10′ above work plane			
0240	1 watt/S.F., type J, 30 FC, 4 fixtures/1000 S.F.	2.41	3.19	5.60
0280	Type K, 29 FC, 5 fixtures/1000 S.F.	2.48	2.87	5.35
0400	2 watt/S.F., type J, 52 FC, 7 fixtures/1000 S.F.	4.31	5.85	10.16

D5020 Lighting and Branch Wiring

D5020 238	H.I.D. Fixture, Low Bay, 8'-10' (by Wattage)	COST PER S.F.		
		MAT.	INST.	TOTAL
0440	Type K, 63 FC, 11 fixtures/1000 S.F.	5.40	6.05	11.45
0560	3 watt/S.F., type J, 81 FC, 11 fixtures/1000 S.F.	6.65	8.85	15.50
0600	Type K, 92 FC, 16 fixtures/1000 S.F.	7.85	8.90	16.75
0720	4 watt/S.F., type J, 103 FC, 14 fixtures/1000 S.F.	8.60	11.65	20.25
0760	Type K, 127 FC, 22 fixtures/1000 S.F.	10.75	12.05	22.80
0880	5 watt/S.F., type J, 133 FC, 18 fixtures/1000 S.F.	11	14.70	25.70
0920	Type K, 155 FC, 27 fixtures/1000 S.F.	13.20	14.90	28.10

D5020 242	H.I.D. Fixture, Low Bay, 16' (by Wattage)	COST PER S.F.		
		MAT.	INST.	TOTAL
0190	High intensity discharge fixture, mounted 16' above work plane			
0240	1 watt/S.F., type J, 28 FC, 4 fixt./1000 S.F.	2.56	3.62	6.18
0280	Type K, 27 FC, 5 fixt./1000 S.F.	2.87	4.06	6.93
0400	2 watt/S.F., type J, 48 FC, 7 fixt/1000 S.F.	4.69	7	11.69
0440	Type K, 58 FC, 11 fixt/1000 S.F.	6.10	8.30	14.40
0560	3 watt/S.F., type J, 75 FC, 11 fixt/1000 S.F.	7.25	10.65	17.90
0600	Type K, 85 FC, 16 fixt/1000 S.F.	8.95	12.35	21.30
0720	4 watt/S.F., type J, 95 FC, 14 fixt/1000 S.F.	9.35	14	23.35
0760	Type K, 117 FC, 22 fixt/1000 S.F.	12.20	16.55	28.75
0880	5 watt/S.F., type J, 122 FC, 18 fixt/1000 S.F.	11.95	17.60	29.55
0920	Type K, 143 FC, 27 fixt/1000 S.F.	15.05	20.50	35.55

D50 Electrical

D5030 Communications and Security

D5030 910	Communication & Alarm Systems	COST EACH		
		MAT.	INST.	TOTAL
0200	Communication & alarm systems, includes outlets, boxes, conduit & wire			
0210	Sound system, 6 outlets	5,750	7,925	13,675
0220	12 outlets	8,275	12,700	20,975
0240	30 outlets	14,800	23,900	38,700
0280	100 outlets	45,200	80,000	125,200
0320	Fire detection systems, non-addressable, 12 detectors	3,300	6,575	9,875
0360	25 detectors	5,775	11,100	16,875
0400	50 detectors	11,400	22,000	33,400
0440	100 detectors	21,200	40,000	61,200
0450	Addressable type, 12 detectors	4,950	6,625	11,575
0452	25 detectors	8,975	11,200	20,175
0454	50 detectors	17,300	22,000	39,300
0456	100 detectors	33,600	40,500	74,100
0458	Fire alarm control panel, 8 zone, excluding wire and conduit	845	1,250	2,095
0459	12 zone	2,575	1,875	4,450
0460	Fire alarm command center, addressable without voice, excl. wire & conduit	4,825	1,100	5,925
0462	Addressable with voice	10,200	1,725	11,925
0480	Intercom systems, 6 stations	3,950	5,400	9,350
0560	25 stations	12,700	20,700	33,400
0640	100 stations	48,700	76,000	124,700
0680	Master clock systems, 6 rooms	5,125	8,850	13,975
0720	12 rooms	8,050	15,100	23,150
0760	20 rooms	11,200	21,300	32,500
0800	30 rooms	18,300	39,300	57,600
0840	50 rooms	29,700	66,500	96,200
0920	Master TV antenna systems, 6 outlets	2,625	5,600	8,225
0960	12 outlets	4,900	10,400	15,300
1000	30 outlets	11,400	24,100	35,500
1040	100 outlets	37,900	78,500	116,400

D5090   Other Electrical Systems

Generator System

A: Engine
B: Battery
C: Charger
D: Transfer Switch
E: Muffler

D5090 210	Generators (by kW)	COST PER kW		
		MAT.	INST.	TOTAL
0190	Generator sets, include battery, charger, muffler & transfer switch			
0200	Gas/gasoline operated, 3 phase, 4 wire, 277/480 volt, 7.5 kW	1,250	275	1,525
0240	11.5 kW	1,150	209	1,359
0280	20 kW	785	136	921
0320	35 kW	535	88.50	623.50
0360	80 kW	385	53.50	438.50
0400	100 kW	335	52	387
0440	125 kW	550	49	599
0480	185 kW	490	37	527
0560	Diesel engine with fuel tank, 30 kW	705	103	808
0600	50 kW	505	81	586
0640	75 kW	435	65.50	500.50
0680	100 kW	365	55.50	420.50
0760	150 kW	293	44	337
0840	200 kW	247	35.50	282.50
0880	250 kW	232	29.50	261.50
0960	350 kW	203	24	227
1040	500 kW	221	19	240

E1010 Commercial Equipment

E1010 110	Security/Vault, EACH	COST EACH		
		MAT.	INST.	TOTAL
0100	Bank equipment, drive up window, drawer & mike, no glazing, economy	7,225	1,425	8,650
0110	Deluxe	9,775	2,850	12,625
0120	Night depository, economy	7,975	1,425	9,400
0130	Deluxe	11,300	2,850	14,150
0140	Pneumatic tube systems, 2 station, standard	27,100	4,775	31,875
0150	Teller, automated, 24 hour, single unit	47,500	4,775	52,275
0160	Teller window, bullet proof glazing, 44" x 60"	4,050	875	4,925
0170	Pass through, painted steel, 72" x 40"	4,425	1,775	6,200
0300	Safe, office type, 1 hr. rating, 34" x 20" x 20"	2,500		2,500
0310	4 hr. rating, 62" x 33" x 20"	9,475		9,475
0320	Data storage, 4 hr. rating, 63" x 44" x 16"	14,700		14,700
0330	Jewelers, 63" x 44" x 16"	15,700		15,700
0340	Money, "B" label, 9" x 14" x 14"	580		580
0350	Tool and torch resistive, 24" x 24" x 20"	8,375	218	8,593
0500	Security gates-scissors type, painted steel, single, 6' high, 5-1/2' wide	233	355	588
0510	Double gate, 7-1/2' high, 14' wide	710	715	1,425

E1010 510	Mercantile Equipment, EACH	COST EACH		
		MAT.	INST.	TOTAL
0015	Barber equipment, chair, hydraulic, economy	535	23	558
0020	Deluxe	3,425	34.50	3,459.50
0100	Checkout counter, single belt	3,275	87.50	3,362.50
0110	Double belt, power take-away	4,725	97	4,822
0200	Display cases, freestanding, glass and aluminum, 3'-6" x 3' x 1'-0" deep	1,325	138	1,463
0220	Wall mounted, glass and aluminum, 3' x 4' x 1'-4" deep	2,350	221	2,571
0320	Frozen food, chest type, 12 ft. long	8,775	370	9,145

E1010 610	Laundry/Dry Cleaning, EACH	COST EACH		
		MAT.	INST.	TOTAL
0100	Laundry equipment, dryers, gas fired, residential, 16 lb. capacity	720	224	944
0110	Commercial, 30 lb. capacity, single	3,525	224	3,749
0120	Dry cleaners, electric, 20 lb. capacity	36,600	6,500	43,100
0130	30 lb. capacity	55,000	8,675	63,675
0140	Ironers, commercial, 120" with canopy, 8 roll	192,000	18,600	210,600
0150	Institutional, 110", single roll	34,300	3,125	37,425
0160	Washers, residential, 4 cycle	945	224	1,169
0170	Commercial, coin operated, deluxe	3,650	224	3,874

E10 Equipment

E1020 Institutional Equipment

E1020 110	Ecclesiastical Equipment, EACH	COST EACH		
		MAT.	INST.	TOTAL
0090	Church equipment, altar, wood, custom, plain	2,525	395	2,920
0100	Granite, custom, deluxe	41,300	5,025	46,325
0110	Baptistry, fiberglass, economy	5,650	1,450	7,100
0120	Bells & carillons, keyboard operation	21,200	8,500	29,700
0130	Confessional, wood, single, economy	3,450	920	4,370
0140	Double, deluxe	20,600	2,775	23,375
0150	Steeples, translucent fiberglas, 30" square, 15' high	8,275	1,750	10,025

E1020 130	Ecclesiastical Equipment, L.F.	COST PER L.F.		
		MAT.	INST.	TOTAL
0100	Arch. equip., church equip. pews, bench type, hardwood, economy	98	27.50	125.50
0110	Deluxe	194	37	231

E1020 210	Library Equipment, EACH	COST EACH		
		MAT.	INST.	TOTAL
0110	Library equipment, carrels, metal, economy	280	111	391
0120	Hardwood, deluxe	1,475	138	1,613

E1020 230	Library Equipment, L.F.	COST PER L.F.		
		MAT.	INST.	TOTAL
0100	Library equipment, book shelf, metal, single face, 90" high x 10" shelf	127	46	173
0110	Double face, 90" high x 10" shelf	320	100	420
0120	Charging desk, built-in, with counter, plastic laminate	330	79	409

E1020 310	Theater and Stage Equipment, EACH	COST EACH		
		MAT.	INST.	TOTAL
0200	Movie equipment, changeover, economy	510		510
0210	Film transport, incl. platters and autowind, economy	5,525		5,525
0220	Lamphouses, incl. rectifiers, xenon, 1000W	7,300	315	7,615
0230	4000W	10,400	420	10,820
0240	Projector mechanisms, 35 mm, economy	12,200		12,200
0250	Deluxe	16,700		16,700
0260	Sound systems, incl. amplifier, single, economy	3,625	695	4,320
0270	Dual, Dolby/super sound	18,200	1,575	19,775
0280	Projection screens, wall hung, manual operation, 50 S.F., economy	325	111	436
0290	Electric operation, 100 S.F., deluxe	2,175	555	2,730

E1020 320	Theater and Stage Equipment, S.F.	COST PER S.F.		
		MAT.	INST.	TOTAL
0090	Movie equipment, projection screens, rigid in wall, acrylic, 1/4" thick	46	5.40	51.40
0100	1/2" thick	53.50	8.05	61.55
0110	Stage equipment, curtains, velour, medium weight	8.70	1.84	10.54
0120	Silica based yarn, fireproof	16.70	22	38.70
0130	Stages, portable with steps, folding legs, 8" high	35.50		35.50
0140	Telescoping platforms, aluminum, deluxe	52	28.50	80.50

E1020 330	Theater and Stage Equipment, L.F.	COST PER L.F.		
		MAT.	INST.	TOTAL
0100	Stage equipment, curtain track, heavy duty	67.50	61.50	129
0110	Lights, border, quartz, colored	190	31.50	221.50

E1020 610	Detention Equipment, EACH	COST EACH		
		MAT.	INST.	TOTAL
0110	Detention equipment, cell front rolling door, 7/8" bars, 5' x 7' high	5,800	1,525	7,325
0120	Cells, prefab., including front, 5' x 7' x 7' deep	10,600	2,025	12,625

E10 Equipment

E1020 Institutional Equipment

E1020 610	Detention Equipment, EACH	COST EACH		
		MAT.	INST.	TOTAL
0130	Doors and frames, 3' x 7', single plate	4,975	760	5,735
0140	Double plate	6,125	760	6,885
0150	Toilet apparatus, incl wash basin	3,700	960	4,660
0160	Visitor cubicle, vision panel, no intercom	3,475	1,525	5,000

E1020 710	Laboratory Equipment, EACH	COST EACH		
		MAT.	INST.	TOTAL
0110	Laboratory equipment, glassware washer, distilled water, economy	6,800	720	7,520
0120	Deluxe	14,000	1,300	15,300
0140	Radio isotope	18,600		18,600

E1020 720	Laboratory Equipment, S.F.	COST PER S.F.		
		MAT.	INST.	TOTAL
0100	Arch. equip., lab equip., counter tops, acid proof, economy	42	13.50	55.50
0110	Stainless steel	151	13.50	164.50

E1020 730	Laboratory Equipment, L.F.	COST PER L.F.		
		MAT.	INST.	TOTAL
0110	Laboratory equipment, cabinets, wall, open	172	55.50	227.50
0120	Base, drawer units	540	61.50	601.50
0130	Fume hoods, not incl. HVAC, economy	540	205	745
0140	Deluxe incl. fixtures	1,700	460	2,160

E1020 810	Medical Equipment, EACH	COST EACH		
		MAT.	INST.	TOTAL
0100	Dental equipment, central suction system, economy	1,700	560	2,260
0110	Compressor-air, deluxe	8,525	1,150	9,675
0120	Chair, hydraulic, economy	2,400	1,150	3,550
0130	Deluxe	9,050	2,275	11,325
0140	Drill console with accessories, economy	2,125	355	2,480
0150	Deluxe	5,625	355	5,980
0160	X-ray unit, portable	2,375	143	2,518
0170	Panoramic unit	17,100	955	18,055
0300	Medical equipment, autopsy table, standard	10,100	670	10,770
0310	Deluxe	17,100	1,125	18,225
0320	Incubators, economy	3,250		3,250
0330	Deluxe	15,800		15,800
0700	Station, scrub-surgical, single, economy	4,800	224	5,024
0710	Dietary, medium, with ice	18,200		18,200
0720	Sterilizers, general purpose, single door, 20" x 20" x 28"	13,400		13,400
0730	Floor loading, double door, 28" x 67" x 52"	224,000		224,000
0740	Surgery tables, standard	13,200	1,025	14,225
0750	Deluxe	25,300	1,425	26,725
0770	Tables, standard, with base cabinets, economy	1,075	370	1,445
0780	Deluxe	6,100	555	6,655
0790	X-ray, mobile, economy	18,300		18,300
0800	Stationary, deluxe	244,000		244,000

E10 Equipment

E1030 Vehicular Equipment

E1030 110	Vehicular Service Equipment, EACH	COST EACH		
		MAT.	INST.	TOTAL
0110	Automotive equipment, compressors, electric, 1-1/2 H.P., std. controls	485	1,050	1,535
0120	5 H.P., dual controls	2,950	1,550	4,500
0130	Hoists, single post, 4 ton capacity, swivel arms	6,050	3,900	9,950
0140	Dual post, 12 ton capacity, adjustable frame	10,400	820	11,220
0150	Lube equipment, 3 reel type, with pumps	9,550	3,125	12,675
0160	Product dispenser, 6 nozzles, w/vapor recovery, not incl. piping, installed	26,900		26,900
0800	Scales, dial type, built in floor, 5 ton capacity, 8' x 6' platform	8,250	3,325	11,575
0810	10 ton capacity, 9' x 7' platform	11,800	4,750	16,550
0820	Truck (including weigh bridge), 20 ton capacity, 24' x 10'	12,600	5,525	18,125

E1030 210	Parking Control Equipment, EACH	COST EACH		
		MAT.	INST.	TOTAL
0110	Parking equipment, automatic gates, 8 ft. arm, one way	4,625	1,150	5,775
0120	Traffic detectors, single treadle	1,975	525	2,500
0130	Booth for attendant, economy	8,175		8,175
0140	Deluxe	30,300		30,300
0150	Ticket printer/dispenser, rate computing	9,750	895	10,645
0160	Key station on pedestal	875	305	1,180

E1030 310	Loading Dock Equipment, EACH	COST EACH		
		MAT.	INST.	TOTAL
0110	Dock bumpers, rubber blocks, 4-1/2" thick, 10" high, 14" long	64	21.50	85.50
0120	6" thick, 20" high, 11" long	161	42.50	203.50
0130	Dock boards, H.D., 5' x 5', aluminum, 5000 lb. capacity	1,400		1,400
0140	16,000 lb. capacity	1,775		1,775
0150	Dock levelers, hydraulic, 7' x 8', 10 ton capacity	10,200	1,500	11,700
0160	Dock lifters, platform, 6' x 6', portable, 3000 lb. capacity	10,100		10,100
0170	Dock shelters, truck, scissor arms, economy	1,825	555	2,380
0180	Deluxe	2,625	1,100	3,725

E10 Equipment

E1090 Other Equipment

E1090 210	Solid Waste Handling Equipment, EACH	COST EACH		
		MAT.	INST.	TOTAL
0110	Waste handling, compactors, single bag, 250 lbs./hr., hand fed	16,400	650	17,050
0120	Heavy duty industrial, 5 C.Y. capacity	35,200	3,125	38,325
0130	Incinerator, electric, 100 lbs./hr., economy	70,500	3,125	73,625
0140	Gas, 2000 lbs./hr., deluxe	440,000	25,600	465,600
0150	Shredder, no baling, 35 tons/hr.	331,000		331,000
0160	Incl. baling, 50 tons/day	661,000		661,000

E1090 350	Food Service Equipment, EACH	COST EACH		
		MAT.	INST.	TOTAL
0110	Kitchen equipment, bake oven, single deck	5,550	151	5,701
0120	Broiler, without oven	4,025	151	4,176
0130	Commercial dish washer, semiautomatic, 50 racks/hr.	8,450	930	9,380
0140	Automatic, 275 racks/hr.	37,900	3,950	41,850
0150	Cooler, beverage, reach-in, 6 ft. long	3,800	202	4,002
0160	Food warmer, counter, 1.65 kw	560		560
0170	Fryers, with submerger, single	1,325	173	1,498
0180	Double	2,600	242	2,842
0185	Ice maker, 1000 lb. per day, with bin	5,375	1,200	6,575
0190	Kettles, steam jacketed, 20 gallons	8,300	265	8,565
0200	Range, restaurant type, burners, 2 ovens and 24" griddle	4,700	202	4,902
0210	Range hood, incl. carbon dioxide system, economy	5,750	405	6,155
0220	Deluxe	42,000	1,200	43,200

E1090 360	Food Service Equipment, S.F.	COST PER S.F.		
		MAT.	INST.	TOTAL
0110	Refrigerators, prefab, walk-in, 7'-6" high, 6' x 6'	170	20	190
0120	12' x 20'	107	10.10	117.10

E1090 410	Residential Equipment, EACH	COST EACH		
		MAT.	INST.	TOTAL
0110	Arch. equip., appliances, range, cook top, 4 burner, economy	305	105	410
0120	Built in, single oven 30" wide, economy	815	105	920
0130	Standing, single oven-21" wide, economy	420	87.50	507.50
0140	Double oven-30" wide, deluxe	2,775	87.50	2,862.50
0150	Compactor, residential, economy	705	111	816
0160	Deluxe	1,025	184	1,209
0170	Dish washer, built-in, 2 cycles, economy	249	325	574
0180	4 or more cycles, deluxe	1,175	650	1,825
0190	Garbage disposer, sink type, economy	94.50	130	224.50
0200	Deluxe	239	130	369
0210	Refrigerator, no frost, 10 to 12 C.F., economy	430	87.50	517.50
0220	21 to 29 C.F., deluxe	3,400	291	3,691
0300	Washing machine, automatic	1,600	670	2,270

E1090 610	School Equipment, EACH	COST EACH		
		MAT.	INST.	TOTAL
0110	School equipment, basketball backstops, wall mounted, wood, fixed	1,425	970	2,395
0120	Suspended type, electrically operated	6,475	1,875	8,350
0130	Bleachers-telescoping, manual operation, 15 tier, economy (per seat)	99.50	34.50	134
0140	Power operation, 30 tier, deluxe (per seat)	380	61	441
0150	Weight lifting gym, universal, economy	550	875	1,425
0160	Deluxe	16,200	1,750	17,950
0170	Scoreboards, basketball, 1 side, economy	2,625	825	3,450
0180	4 sides, deluxe	6,250	11,400	17,650
0800	Vocational shop equipment, benches, metal	570	221	791
0810	Wood	630	221	851

E10 Equipment

E1090 Other Equipment

E1090 610	School Equipment, EACH	COST EACH		
		MAT.	INST.	TOTAL
0820	Dust collector, not incl. ductwork, 6' diam.	4,650	590	5,240
0830	Planer, 13" x 6"	1,100	277	1,377

E1090 620	School Equipment, S.F.	COST PER S.F.		
		MAT.	INST.	TOTAL
0110	School equipment, gym mats, naugahyde cover, 2" thick	3.89		3.89
0120	Wrestling, 1" thick, heavy duty	5.90		5.90

E1090 810	Athletic, Recreational, and Therapeutic Equipment, EACH	COST EACH		
		MAT.	INST.	TOTAL
0050	Bowling alley with gutters	49,900	11,600	61,500
0060	Combo. table and ball rack	1,250		1,250
0070	Bowling alley automatic scorer	10,900		10,900
0110	Sauna, prefabricated, incl. heater and controls, 7' high, 6' x 4'	5,375	845	6,220
0120	10' x 12'	13,200	1,850	15,050
0130	Heaters, wall mounted, to 200 C.F.	690		690
0140	Floor standing, to 1000 C.F., 12500 W	3,625	209	3,834
0610	Shooting range incl. bullet traps, controls, separators, ceilings, economy	38,500	7,325	45,825
0620	Deluxe	57,000	12,300	69,300
0650	Sport court, squash, regulation, in existing building, economy			40,400
0660	Deluxe			45,000
0670	Racketball, regulation, in existing building, economy	43,800	10,500	54,300
0680	Deluxe	47,400	21,000	68,400
0700	Swimming pool equipment, diving stand, stainless steel, 1 meter	7,475	410	7,885
0710	3 meter	12,800	2,775	15,575
0720	Diving boards, 16 ft. long, aluminum	4,300	410	4,710
0730	Fiberglass	3,125	410	3,535
0740	Filter system, sand, incl. pump, 6000 gal./hr.	2,075	745	2,820
0750	Lights, underwater, 12 volt with transformer, 300W	320	625	945
0760	Slides, fiberglass with aluminum handrails & ladder, 6' high, straight	3,700	690	4,390
0780	12' high, straight with platform	15,000	920	15,920

E1090 820	Athletic, Recreational, and Therapeutic Equipment, S.F.	COST PER S.F.		
		MAT.	INST.	TOTAL
0110	Swimming pools, residential, vinyl liner, metal sides	17.80	7.75	25.55
0120	Concrete sides	21.50	14.05	35.55
0130	Gunite shell, plaster finish, 350 S.F.	39.50	29	68.50
0140	800 S.F.	32	16.80	48.80
0150	Motel, gunite shell, plaster finish	48.50	36.50	85
0160	Municipal, gunite shell, tile finish, formed gutters	187	63.50	250.50

E2010 Fixed Furnishings

E2010 310	Window Treatment, EACH	COST EACH		
		MAT.	INST.	TOTAL
0110	Furnishings, blinds, exterior, aluminum, louvered, 1'-4" wide x 3'-0" long	75	55.50	130.50
0120	1'-4" wide x 6'-8" long	150	61.50	211.50
0130	Hemlock, solid raised, 1-'4" wide x 3'-0" long	79	55.50	134.50
0140	1'-4" wide x 6'-9" long	177	61.50	238.50
0150	Polystyrene, louvered, 1'-3" wide x 3'-3" long	71.50	55.50	127
0160	1'-3" wide x 6'-8" long	139	61.50	200.50
0200	Interior, wood folding panels, louvered, 7" x 20" (per pair)	86	32.50	118.50
0210	18" x 40" (per pair)	145	32.50	177.50

E2010 320	Window Treatment, S.F.	COST PER S.F.		
		MAT.	INST.	TOTAL
0110	Furnishings, blinds-interior, venetian-aluminum, stock, 2" slats, economy	5.20	.94	6.14
0120	Custom, 1" slats, deluxe	7.20	1.26	8.46
0130	Vertical, PVC or cloth, T&B track, economy	9.05	1.20	10.25
0140	Deluxe	28	1.38	29.38
0150	Draperies, unlined, economy	3.38		3.38
0160	Lightproof, deluxe	9.70		9.70
0510	Shades, mylar, wood roller, single layer, non-reflective	7.55	.81	8.36
0520	Metal roller, triple layer, heat reflective	10.80	.81	11.61
0530	Vinyl, light weight, 4 ga.	.65	.81	1.46
0540	Heavyweight, 6 ga.	1.97	.81	2.78
0550	Vinyl coated cotton, lightproof decorator shades	3.23	.81	4.04
0560	Woven aluminum, 3/8" thick, light and fireproof	6.35	1.58	7.93

E2010 420	Fixed Floor Grilles and Mats, S.F.	COST PER S.F.		
		MAT.	INST.	TOTAL
0110	Floor mats, recessed, inlaid black rubber, 3/8" thick, solid	23.50	2.82	26.32
0120	Colors, 1/2" thick, perforated	28.50	2.82	31.32
0130	Link-including nosings, steel-galvanized, 3/8" thick	22	2.82	24.82
0140	Vinyl, in colors	25.50	2.82	28.32

E2010 510	Fixed Multiple Seating, EACH	COST EACH		
		MAT.	INST.	TOTAL
0110	Seating, painted steel, upholstered, economy	145	31.50	176.50
0120	Deluxe	465	39.50	504.50
0400	Seating, lecture hall, pedestal type, economy	206	50.50	256.50
0410	Deluxe	675	76.50	751.50
0500	Auditorium chair, veneer construction	214	50.50	264.50
0510	Fully upholstered, spring seat	268	50.50	318.50

E2020 Moveable Furnishings

E2020 210	Furnishings/EACH	COST EACH		
		MAT.	INST.	TOTAL
0200	Hospital furniture, beds, manual, economy	865		865
0210	Deluxe	2,750		2,750
0220	All electric, economy	1,775		1,775
0230	Deluxe	4,850		4,850
0240	Patient wall systems, no utilities, economy, per room	1,550		1,550
0250	Deluxe, per room	2,100		2,100
0300	Hotel furnishings, standard room set, economy, per room	2,750		2,750
0310	Deluxe, per room	9,750		9,750
0500	Office furniture, standard employee set, economy, per person	635		635
0510	Deluxe, per person	2,525		2,525
0550	Posts, portable, pedestrian traffic control, economy	142		142
0560	Deluxe	460		460
0700	Restaurant furniture, booth, molded plastic, stub wall and 2 seats, economy	400	277	677
0710	Deluxe	1,525	370	1,895
0720	Upholstered seats, foursome, single-economy	815	111	926
0730	Foursome, double-deluxe	1,875	184	2,059

E2020 220	Furniture and Accessories, L.F.	COST PER L.F.		
		MAT.	INST.	TOTAL
0210	Dormitory furniture, desk top (built-in),laminated plastc, 24" deep, economy	47.50	22	69.50
0220	30" deep, deluxe	267	27.50	294.50
0230	Dressing unit, built-in, economy	183	92	275
0240	Deluxe	550	138	688
0310	Furnishings, cabinets, hospital, base, laminated plastic	288	111	399
0320	Stainless steel	530	111	641
0330	Countertop, laminated plastic, no backsplash	49.50	27.50	77
0340	Stainless steel	161	27.50	188.50
0350	Nurses station, door type, laminated plastic	335	111	446
0360	Stainless steel	635	111	746
0710	Restaurant furniture, bars, built-in, back bar	216	111	327
0720	Front bar	297	111	408
0910	Wardrobes & coatracks, standing, steel, single pedestal, 30" x 18" x 63"	116		116
0920	Double face rack, 39" x 26" x 70"	125		125
0930	Wall mounted rack, steel frame & shelves, 12" x 15" x 26"	77.50	7.85	85.35
0940	12" x 15" x 50"	47.50	4.12	51.62

F10 Special Construction

F1010 Special Structures

F1010 120	Air-Supported Structures, S.F.	COST PER S.F.		
		MAT.	INST.	TOTAL
0110	Air supported struc., polyester vinyl fabric, 24oz., warehouse, 5000 S.F.	27.50	.35	27.85
0120	50,000 S.F.	12.90	.28	13.18
0130	Tennis, 7,200 S.F.	26	.29	26.29
0140	24,000 S.F.	18.05	.29	18.34
0150	Woven polyethylene, 6 oz., shelter, 3,000 S.F.	7.25	.58	7.83
0160	24,000 S.F.	6.15	.29	6.44
0170	Teflon coated fiberglass, stadium cover, economy	63	.15	63.15
0180	Deluxe	75	.21	75.21
0190	Air supported storage tank covers, reinf. vinyl fabric, 12 oz., 400 S.F.	25.50	.49	25.99
0200	18,000 S.F.	8.10	.44	8.54

F1010 210	Pre-Engineered Structures, EACH	COST EACH		
		MAT.	INST.	TOTAL
0600	Radio towers, guyed, 40 lb. section, 50' high, 70 MPH basic wind speed	3,225	1,425	4,650
0610	90 lb. section, 400' high, wind load 70 MPH basic wind speed	40,000	16,100	56,100
0620	Self supporting, 60' high, 70 MPH basic wind speed	5,000	2,825	7,825
0630	190' high, wind load 90 MPH basic wind speed	32,800	11,300	44,100
0700	Shelters, aluminum frame, acrylic glazing, 8' high, 3' x 9'	4,175	1,250	5,425
0710	9' x 12'	6,575	1,950	8,525

F1010 320	Other Special Structures, S.F.	COST PER S.F.		
		MAT.	INST.	TOTAL
0110	Swimming pool enclosure, translucent, freestanding, economy	16.25	5.55	21.80
0120	Deluxe	98.50	15.80	114.30
0510	Tension structures, steel frame, polyester vinyl fabric, 12,000 S.F.	14.90	2.55	17.45
0520	20,800 S.F.	14.60	2.30	16.90

F1010 330	Special Structures, EACH	COST EACH		
		MAT.	INST.	TOTAL
0110	Kiosks, round, 5' diam., 1/4" fiberglass wall, 8' high	25,100		25,100
0120	Rectangular, 5' x 9', 1" insulated dbl. wall fiberglass, 7'-6" high	25,600		25,600
0220	Silos, steel prefab, 30,000 gal., painted, economy	24,300	5,700	30,000
0230	Epoxy-lined, deluxe	50,500	11,400	61,900

F1010 340	Special Structures, S.F.	COST PER S.F.		
		MAT.	INST.	TOTAL
0110	Comfort stations, prefab, mobile on steel frame, economy	207		207
0120	Permanent on concrete slab, deluxe	211	46.50	257.50
0210	Domes, bulk storage, wood framing, wood decking, 50' diam.	79.50	1.76	81.26
0220	116' diam.	38.50	2.03	40.53
0230	Steel framing, metal decking, 150' diam.	30.50	11.60	42.10
0240	400' diam.	25	8.90	33.90
0250	Geodesic, wood framing, wood panels, 30' diam.	32.50	2	34.50
0260	60' diam.	22.50	1.24	23.74
0270	Aluminum framing, acrylic panels, 40' diam.			103
0280	Aluminum panels, 400' diam.			23.50
0310	Garden house, prefab, wood, shell only, 48 S.F.	55.50	5.55	61.05
0320	200 S.F.	35.50	23	58.50
0410	Greenhouse, shell-stock, residential, lean-to, 8'-6" long x 3'-10" wide	37.50	32.50	70
0420	Freestanding, 8'-6" long x 13'-6" wide	27	10.25	37.25
0430	Commercial-truss frame, under 2000 S.F., deluxe			15.90
0440	Over 5,000 S.F., economy			16.05
0450	Institutional-rigid frame, under 500 S.F., deluxe			39
0460	Over 2,000 S.F., economy			12.90
0510	Hangar, prefab, galv. roof and walls, bottom rolling doors, economy	11.35	4.60	15.95
0520	Electric bifolding doors, deluxe	11.75	6	17.75

F10 Special Construction

F1020 Integrated Construction

F1020 110 — Integrated Construction, EACH

		COST EACH		
		MAT.	**INST.**	**TOTAL**
0110	Integrated ceilings, radiant electric, 2' x 4' panel, manila finish	99	25	124
0120	ABS plastic finish	112	48.50	160.50

F1020 120 — Integrated Construction, S.F.

		COST PER S.F.		
		MAT.	**INST.**	**TOTAL**
0110	Integrated ceilings, Luminaire, suspended, 5' x 5' modules, 50% lighted	4.32	13.25	17.57
0120	100% lighted	6.45	24	30.45
0130	Dimensionaire, 2' x 4' module tile system, no air bar	2.61	2.21	4.82
0140	With air bar, deluxe	3.83	2.21	6.04
0220	Pedestal access floors, stl pnls, no stringers, vinyl covering, >6000 S.F.	13.40	1.73	15.13
0230	With stringers, high pressure laminate covering, under 6000 S.F.	12.35	1.84	14.19
0240	Carpet covering, under 6000 S.F.	13.70	1.84	15.54
0250	Aluminum panels, no stringers, no covering	35.50	2.21	37.71

F1020 250 — Special Purpose Room, EACH

		COST EACH		
		MAT.	**INST.**	**TOTAL**
0110	Portable booth, acoustical, 27 db 1000 hz., 15 S.F. floor	3,925		3,925
0120	55 S.F. flr.	8,050		8,050

F1020 260 — Special Purpose Room, S.F.

		COST PER S.F.		
		MAT.	**INST.**	**TOTAL**
0110	Anechoic chambers, 7' high, 100 cps cutoff, 25 S.F.			2,600
0120	200 cps cutoff, 100 S.F.			1,350
0130	Audiometric rooms, under 500 S.F.	57.50	22.50	80
0140	Over 500 S.F.	55	18.45	73.45
0300	Darkrooms, shell, not including door, 240 S.F., 8' high	31	9.20	40.20
0310	64 S.F., 12' high	79	17.30	96.30
0510	Music practice room, modular, perforated steel, under 500 S.F.	35.50	15.80	51.30
0520	Over 500 S.F.	30	13.85	43.85

F1020 330 — Special Construction, L.F.

		COST PER L.F.		
		MAT.	**INST.**	**TOTAL**
0110	Spec. const., air curtains, shipping & receiving, 8' high x 5' wide, economy	740	130	870
0120	20' high x 8' wide, heated, deluxe	3,150	325	3,475
0130	Customer entrance, 10' high x 5' wide, economy	1,100	130	1,230
0140	12' high x 4' wide, heated, deluxe	2,000	163	2,163

F1030 Special Construction Systems

F1030 120	Sound, Vibration, and Seismic Construction, S.F.	COST PER S.F.		
		MAT.	INST.	TOTAL
0020	Special construction, acoustical, enclosure, 4" thick, 8 psf panels	36	23	59
0030	Reverb chamber, 4" thick, parallel walls	51	27.50	78.50
0110	Sound absorbing panels, 2'-6" x 8', painted metal	13.55	7.70	21.25
0120	Vinyl faced	10.60	6.90	17.50
0130	Flexible transparent curtain, clear	8.20	9.10	17.30
0140	With absorbing foam, 75% coverage	11.40	9.10	20.50
0150	Strip entrance, 2/3 overlap	8.35	14.45	22.80
0160	Full overlap	10.50	16.95	27.45
0200	Audio masking system, plenum mounted, over 10,000 S.F.	.80	.29	1.09
0210	Ceiling mounted, under 5,000 S.F.	1.39	.52	1.91

F1030 210	Radiation Protection, EACH	COST EACH		
		MAT.	INST.	TOTAL
0110	Shielding, lead x-ray protection, radiography room, 1/16" lead, economy	10,100	3,725	13,825
0120	Deluxe	12,200	6,225	18,425
0210	Deep therapy x-ray, 1/4" lead, economy	28,300	11,700	40,000
0220	Deluxe	34,900	15,500	50,400

F1030 220	Radiation Protection, S.F.	COST PER S.F.		
		MAT.	INST.	TOTAL
0110	Shielding, lead, gypsum board, 5/8" thick, 1/16" lead	10.20	6.55	16.75
0120	1/8" lead	24	7.50	31.50
0130	Lath, 1/16" thick	11.05	6.90	17.95
0140	1/8" thick	22	7.75	29.75
0150	Radio frequency, galvanized steel, prefab type, economy	4.91	2.95	7.85
0160	Radio frequency, door, copper/wood laminate, 4" x 7'	6,350	740	7,100

F1030 910	Other Special Construction Systems, EACH	COST EACH		
		MAT.	INST.	TOTAL
0110	Disappearing stairways, folding, pine, 8'-6" ceiling	166	138	304
0220	Automatic electric, wood, 8' to 9' ceiling	9,275	1,100	10,375
0300	Fireplace prefabricated, freestanding or wall hung, economy	1,325	425	1,750
0310	Deluxe	3,000	615	3,615
0320	Woodburning stoves, cast iron, economy	1,050	850	1,900
0330	Deluxe	3,025	1,375	4,400

F10 Special Construction

F1040 Special Facilities

F1040 210	Ice Rinks, EACH	COAT EACH		
		MAT.	INST.	TOTAL
0100	Ice skating rink, 85' x 200', 55° system, 5 mos., 100 ton			632,500
0110	90° system, 12 mos., 135 ton			715,000
0120	Dash boards, acrylic screens, polyethylene coated plywood	148,500	37,300	185,800
0130	Fiberglass and aluminum construction	170,500	37,300	207,800

F1040 510	Liquid & Gas Storage Tanks, EACH	COST EACH		
		MAT.	INST.	TOTAL
0100	Tanks, steel, ground level, 100,000 gal.			226,500
0110	10,000,000 gal.			5,151,500
0120	Elevated water, 50,000 gal.		185,000	445,500
0130	1,000,000 gal.		862,500	1,973,500
0150	Cypress wood, ground level, 3,000 gal.	10,200	11,000	21,200
0160	Redwood, ground level, 45,000 gal.	77,000	29,900	106,900

F10 Special Construction

F1040 Special Facilities

F1040 910	Special Construction, EACH	COST EACH		
		MAT.	INST.	TOTAL
0110	Special construction, bowling alley incl. pinsetter, scorer etc., economy	45,400	11,100	56,500
0120	Deluxe	57,500	12,300	69,800
0130	For automatic scorer, economy, add	9,050		9,050
0140	Deluxe, add	10,900		10,900
0300	Control tower, modular, 12' x 10', incl. instrumentation, economy			774,000
0310	Deluxe			1,080,000
0400	Garage costs, residential, prefab, wood, single car economy	5,625	1,100	6,725
0410	Two car deluxe	11,800	2,225	14,025
0500	Hangars, prefab, galv. steel, bottom rolling doors, economy (per plane)	15,000	5,100	20,100
0510	Electrical bi-folding doors, deluxe (per plane)	14,000	7,025	21,025

G1030 Site Earthwork

G1030 805	Trenching Common Earth	COST PER L.F.		
		MAT.	INST.	TOTAL
1310	Trenching, common earth, no slope, 2' wide, 2' deep, 3/8 C.Y. bucket		2.52	2.52
1330	4' deep, 3/8 C.Y. bucket		4.97	4.97
1360	10' deep, 1 C.Y. bucket		11.09	11.09
1400	4' wide, 2' deep, 3/8 C.Y. bucket		5.20	5.20
1420	4' deep, 1/2 C.Y. bucket		8.67	8.67
1450	10' deep, 1 C.Y. bucket		23.85	23.85
1480	18' deep, 2-1/2 C.Y. bucket		37.90	37.90
1520	6' wide, 6' deep, 5/8 C.Y. bucket w/trench box		23.45	23.45
1540	10' deep, 1 C.Y. bucket		31.45	31.45
1570	20' deep, 3-1/2 C.Y. bucket		53.50	53.50
1640	8' wide, 12' deep, 1-1/2 C.Y. bucket w/trench box		44.50	44.50
1680	24' deep, 3-1/2 C.Y. bucket		85	85
1730	10' wide, 20' deep, 3-1/2 C.Y. bucket w/trench box		75.50	75.50
1740	24' deep, 3-1/2 C.Y. bucket		99	99
3500	1 to 1 slope, 2' wide, 2' deep, 3/8 C.Y. bucket		4.97	4.97
3540	4' deep, 3/8 C.Y. bucket		14.73	14.73
3600	10' deep, 1 C.Y. bucket		66.50	66.50
3800	4' wide, 2' deep, 3/8 C.Y. bucket		7.66	7.66
3840	4' deep, 1/2 C.Y. bucket		17.12	17.12
3900	10' deep, 1 C.Y. bucket		84	84
4030	6' wide, 6' deep, 5/8 C.Y. bucket w/trench box		46.35	46.35
4050	10' deep, 1 C.Y. bucket		75.50	75.50
4080	20' deep, 3-1/2 C.Y. bucket		227.50	227.50
4500	8' wide, 12' deep, 1-1/2 C.Y. bucket w/trench box		113	113
4650	24' deep, 3-1/2 C.Y. bucket		345	345
4800	10' wide, 20' deep, 3-1/2 C.Y. bucket w/trench box		232.50	232.50
4850	24' deep, 3-1/2 C.Y. bucket		366	366

G1030 806	Trenching Loam & Sandy Clay	COST PER L.F.		
		MAT.	INST.	TOTAL
1310	Trenching, loam & sandy clay, no slope, 2' wide, 2' deep, 3/8 C.Y. bucket		2.48	2.48

G1030 807	Trenching Sand & Gravel	COST PER L.F.		
		MAT.	INST.	TOTAL
1310	Trenching, sand & gravel, no slope, 2' wide, 2' deep, 3/8 C.Y. bucket		2.33	2.33
1320	3' deep, 3/8 C.Y. bucket		3.85	3.85
1330	4' deep, 3/8 C.Y. bucket		4.59	4.59
1340	6' deep, 3/8 C.Y. bucket		5.75	5.75
1350	8' deep, 1/2 C.Y. bucket		7.60	7.60
1360	10' deep, 1 C.Y. bucket		8.15	8.15
1400	4' wide, 2' deep, 3/8 C.Y. bucket		4.78	4.78
1410	3' deep, 3/8 C.Y. bucket		7.05	7.05
1420	4' deep, 1/2 C.Y. bucket		8	8
1430	6' deep, 1/2 C.Y. bucket		12.55	12.55
1440	8' deep, 1/2 C.Y. bucket		18.05	18.05
1450	10' deep, 1 C.Y. bucket		18.05	18.05
1460	12' deep, 1 C.Y. bucket		22.50	22.50
1470	15' deep, 1-1/2 C.Y. bucket		26.50	26.50
1480	18' deep, 2-1/2 C.Y. bucket		30.50	30.50
1520	6' wide, 6' deep, 5/8 C.Y. bucket w/trench box		21.50	21.50
1530	8' deep, 3/4 C.Y. bucket		28.50	28.50
1540	10' deep, 1 C.Y. bucket		28	28
1550	12' deep, 1-1/2 C.Y. bucket		31.50	31.50
1560	16' deep, 2 C.Y. bucket		42.50	42.50
1570	20' deep, 3-1/2 C.Y. bucket		49	49
1580	24' deep, 3-1/2 C.Y. bucket		60.50	60.50
1640	8' wide, 12' deep, 1-1/2 C.Y. bucket w/trench box		41.50	41.50
1650	15' deep, 1-1/2 C.Y. bucket		52	52

G1030 Site Earthwork

G1030 807	Trenching Sand & Gravel	COST PER L.F.		
		MAT.	INST.	TOTAL
1660	18' deep, 2-1/2 C.Y. bucket		61.50	61.50
1680	24' deep, 3-1/2 C.Y. bucket		79.50	79.50
1730	10' wide, 20' deep, 3-1/2 C.Y. bucket w/trench box		79.50	79.50
1740	24' deep, 3-1/2 C.Y. bucket		99	99
1780	12' wide, 20' deep, 3-1/2 C.Y. bucket w/trench box		95	95
1790	25' deep, 3-1/2 C.Y. bucket		124	124
1800	1/2:1 slope, 2' wide, 2' deep, 3/8 C.Y. bucket		3.46	3.46
1810	3' deep, 3/8 C.Y. bucket		6	6
1820	4' deep, 3/8 C.Y. bucket		9.15	9.15
1840	6' deep, 3/8 C.Y. bucket		14.25	14.25
1860	8' deep, 1/2 C.Y. bucket		23	23
1880	10' deep, 1 C.Y. bucket		28.50	28.50
2300	4' wide, 2' deep, 3/8 C.Y. bucket		5.90	5.90
2310	3' deep, 3/8 C.Y. bucket		9.55	9.55
2320	4' deep, 1/2 C.Y. bucket		11.95	11.95
2340	6' deep, 1/2 C.Y. bucket		22	22
2360	8' deep, 1/2 C.Y. bucket		36	36
2380	10' deep, 1 C.Y. bucket		41	41
2400	12' deep, 1 C.Y. bucket		68.50	68.50
2430	15' deep, 1-1/2 C.Y. bucket		77	77
2460	18' deep, 2-1/2 C.Y. bucket		123	123
2840	6' wide, 6' deep, 5/8 C.Y. bucket w/trench box		32.50	32.50
2860	8' deep, 3/4 C.Y. bucket		47.50	47.50
2880	10' deep, 1 C.Y. bucket		52	52
2900	12' deep, 1-1/2 C.Y. bucket		63.50	63.50
2940	16' deep, 2 C.Y. bucket		99	99
2980	20' deep, 3-1/2 C.Y. bucket		132	132
3020	24' deep, 3-1/2 C.Y. bucket		184	184
3100	8' wide, 12' deep, 1-1/4 C.Y. bucket w/trench box		74	74
3120	15' deep, 1-1/2 C.Y. bucket		103	103
3140	18' deep, 2-1/2 C.Y. bucket		134	134
3180	24' deep, 3-1/2 C.Y. bucket		202	202
3270	10' wide, 20' deep, 3-1/2 C.Y. bucket w/trench box		163	163
3280	24' deep, 3-1/2 C.Y. bucket		222	222
3370	12' wide, 20' deep, 3-1/2 C.Y. bucket w/trench box		181	181
3380	25' deep, 3-1/2 C.Y. bucket		257	257
3500	1:1 slope, 2' wide, 2' deep, 3/8 C.Y. bucket		6.20	6.20
3520	3' deep, 3/8 C.Y. bucket		8.55	8.55
3540	4' deep, 3/8 C.Y. bucket		13.65	13.65
3560	6' deep, 3/8 C.Y. bucket		14.25	14.25
3580	8' deep, 1/2 C.Y. bucket		38	38
3600	10' deep, 1 C.Y. bucket		49	49
3800	4' wide, 2' deep, 3/8 C.Y. bucket		7.05	7.05
3820	3' deep, 3/8 C.Y. bucket		12.15	12.15
3840	4' deep, 1/2 C.Y. bucket		15.90	15.90
3860	6' deep, 1/2 C.Y. bucket		31.50	31.50
3880	8' deep, 1/2 C.Y. bucket		54	54
3900	10' deep, 1 C.Y. bucket		64	64
3920	12' deep, 1 C.Y. bucket		91	91
3940	15' deep, 1-1/2 C.Y. bucket		127	127
3960	18' deep, 2-1/2 C.Y. bucket		169	169
4030	6' wide, 6' deep, 5/8 C.Y. bucket w/trench box		43	43
4040	8' deep, 3/4 C.Y. bucket		66.50	66.50
4050	10' deep, 1 C.Y. bucket		75.50	75.50
4060	12' deep, 1-1/2 C.Y. bucket		96	96
4070	16' deep, 2 C.Y. bucket		156	156
4080	20' deep, 3-1/2 C.Y. bucket		215	215
4090	24' deep, 3-1/2 C.Y. bucket		305	305

G1030 Site Earthwork

G1030 807	Trenching Sand & Gravel	COST PER L.F.		
		MAT.	INST.	TOTAL
4500	8' wide, 12' deep, 1-1/2 C.Y. bucket w/trench box		107	107
4550	15' deep, 1-1/2 C.Y. bucket		153	153
4600	18' deep, 2-1/2 C.Y. bucket		205	205
4650	24' deep, 3-1/2 C.Y. bucket		325	325
4800	10' wide, 20' deep, 3-1/2 C.Y. bucket w/trench box		245	245
4850	24' deep, 3-1/2 C.Y. bucket		345	345
4950	12' wide, 20' deep, 3-1/2 C.Y. bucket w/trench box		261	261
4980	25' deep, 3-1/2 C.Y. bucket		390	390

G1030 815	Pipe Bedding	COST PER L.F.		
		MAT.	INST.	TOTAL
1440	Pipe bedding, side slope 0 to 1, 1' wide, pipe size 6" diameter	1.69	.88	2.57
1460	2' wide, pipe size 8" diameter	3.66	1.88	5.54
1500	Pipe size 12" diameter	3.82	1.96	5.78
1600	4' wide, pipe size 20" diameter	9.15	4.70	13.85
1660	Pipe size 30" diameter	9.60	4.94	14.54
1680	6' wide, pipe size 32" diameter	16.45	8.45	24.90
1740	8' wide, pipe size 60" diameter	32.50	16.70	49.20
1780	12' wide, pipe size 84" diameter	60	30.50	90.50

G20 Site Improvements

G2010 Roadways

G2010 210	Roadway Pavement	COST PER L.F.		
		MAT.	INST.	TOTAL
1500	Roadways, bituminous conc. paving, 2-1/2″ thick, 20′ wide	69.50	62	131.50
1580	30′ wide	88	69.50	157.50
1800	3″ thick, 20′ wide	74	60	134
1880	30′ wide	95	70.50	165.50
2100	4″ thick, 20′ wide	81.50	60	141.50
2180	30′ wide	106	70.50	176.50
2400	5″ thick, 20′ wide	98.50	61	159.50
2480	30′ wide	132	72	204
3000	6″ thick, 20′ wide	95	62.50	157.50
3080	30′ wide	127	74	201
3300	8″ thick 20′ wide	114	63	177
3380	30′ wide	155	75	230
3600	12″ thick 20′ wide	144	67	211
3700	32′ wide	211	84	295

G2010 Roadways

The Bituminous Roadway Systems are listed for pavement thicknesses between 3-1/2" and 7" and crushed stone bases from 3" to 22" in depth. Systems costs are expressed per linear foot for varying widths of two and multi-lane roads. Earth moving is not included. Granite curbs and line painting are added as required system components.

G2010 232	Bituminous Roadways Crushed Stone	COST PER L.F.		
		MAT.	INST.	TOTAL
1050	Bitum. roadway, two lanes, 3-1/2" th. pvmt., 3" th. crushed stone, 24' wide	78	42	120
1100	28' wide	85.50	43	128.50
1150	32' wide	93.50	47	140.50
1300	9" thick crushed stone, 24' wide	96	43.50	139.50
1350	28' wide	107	48	155
1400	32' wide	117	53	170
1550	4" thick pavement., 4" thick crushed stone, 24' wide	86.50	40.50	127
1600	28' wide	95.50	44.50	140
1650	32' wide	104	49	153
1800	10" thick crushed stone , 24' wide	104	44.50	148.50
1850	28' wide	116	50	166
1900	32' wide	129	54.50	183.50

G2020 Parking Lots

G2020 210	Parking Lots Gravel Base	COST PER CAR		
		MAT.	INST.	TOTAL
1500	Parking lot, 90° angle parking, 3" bituminous paving, 6" gravel base	820	420	1,240
1540	10" gravel base	950	485	1,435
1560	4" bituminous paving, 6" gravel base	1,025	445	1,470
1600	10" gravel base	1,175	515	1,690
1620	6" bituminous paving, 6" gravel base	1,375	475	1,850
1660	10" gravel base	1,500	540	2,040
1800	60° angle parking, 3" bituminous paving, 6" gravel base	820	420	1,240
1840	10" gravel base	950	485	1,435
1860	4" bituminous paving, 6" gravel base	1,025	445	1,470
1900	10" gravel base	1,175	515	1,690
1920	6" bituminous paving, 6" gravel base	1,375	475	1,850
1960	10" gravel base	1,500	540	2,040
2200	45° angle parking, 3" bituminous paving, 6" gravel base	840	425	1,265
2240	10" gravel base	975	500	1,475
2260	4" bituminous paving, 6" gravel base	1,050	455	1,505
2300	10" gravel base	1,200	525	1,725
2320	6" bituminous paving, 6" gravel base	1,400	490	1,890
2360	10" gravel base	1,550	555	2,105

G20 Site Improvements

G2040 Site Development

G2040 810	Flagpoles	COST EACH		
		MAT.	INST.	TOTAL
0110	Flagpoles, on grade, aluminum, tapered, 20' high	1,075	675	1,750
0120	70' high	8,875	1,700	10,575
0130	Fiberglass, tapered, 23' high	650	675	1,325
0140	59' high	5,700	1,500	7,200
0150	Concrete, internal halyard, 20' high	1,300	540	1,840
0160	100' high	17,100	1,350	18,450

G2040 950	Other Site Development, EACH	COST EACH		
		MAT.	INST.	TOTAL
0110	Grandstands, permanent, closed deck, steel, economy (per seat)			60.50
0120	Deluxe (per seat)			121
0130	Composite design, economy (per seat)			143
0140	Deluxe (per seat)			286

G30 Site Mechanical Utilities

G3030 Storm Sewer

G3030 210	Manholes & Catch Basins	COST PER EACH		
		MAT.	INST.	TOTAL
1920	Manhole/catch basin, brick, 4' I.D. riser, 4' deep	1,125	1,725	2,850
1980	10' deep	2,600	4,050	6,650
3200	Block, 4' I.D. riser, 4' deep	1,100	1,400	2,500
3260	10' deep	2,425	3,350	5,775
4620	Concrete, cast-in-place, 4' I.D. riser, 4' deep	1,250	2,375	3,625
4680	10' deep	3,000	5,700	8,700
5820	Concrete, precast, 4' I.D. riser, 4' deep	1,500	1,275	2,775
5880	10' deep	3,075	2,925	6,000
6200	6' I.D. riser, 4' deep	3,300	1,900	5,200
6260	10' deep	6,600	4,450	11,050

G4020 Site Lighting

Light Pole

G4020 210	Light Pole (Installed)	COST EACH		
		MAT.	INST.	TOTAL
0200	Light pole, aluminum, 20' high, 1 arm bracket	1,225	1,125	2,350
0240	2 arm brackets	1,350	1,125	2,475
0280	3 arm brackets	1,475	1,175	2,650
0320	4 arm brackets	1,600	1,175	2,775
0360	30' high, 1 arm bracket	2,125	1,425	3,550
0400	2 arm brackets	2,275	1,425	3,700
0440	3 arm brackets	2,400	1,475	3,875
0480	4 arm brackets	2,525	1,475	4,000
0680	40' high, 1 arm bracket	2,625	1,900	4,525
0720	2 arm brackets	2,750	1,900	4,650
0760	3 arm brackets	2,900	1,950	4,850
0800	4 arm brackets	3,025	1,950	4,975
0840	Steel, 20' high, 1 arm bracket	1,450	1,200	2,650
0880	2 arm brackets	1,550	1,200	2,750
0920	3 arm brackets	1,575	1,225	2,800
0960	4 arm brackets	1,725	1,225	2,950
1000	30' high, 1 arm bracket	1,700	1,525	3,225
1040	2 arm brackets	1,800	1,525	3,325
1080	3 arm brackets	1,825	1,575	3,400
1120	4 arm brackets	1,950	1,575	3,525
1320	40' high, 1 arm bracket	2,225	2,075	4,300
1360	2 arm brackets	2,325	2,075	4,400
1400	3 arm brackets	2,350	2,100	4,450
1440	4 arm brackets	2,475	2,100	4,575

Reference Section

All the reference information is in one section making it easy to find what you need to know ... and easy to use the book on a daily basis. This section is visually identified by a vertical gray bar on the edge of pages.

In this Reference Section, we've included General Conditions; Location Factors for adjusting costs to the region you are in; Historical Cost Indexes for cost comparisons over time; a Glossary; an explanation of all Abbreviations used in the book; and sample Estimating Forms.

Table of Contents

General Conditions, Overhead & Profit

The total building costs in the Commercial/Industrial/Institutional section include a 10% allowance for general conditions and a 15% allowance for the general contractor's overhead and profit and contingencies.

General contractor overhead includes indirect costs such as permits, Workers' Compensation, insurances, supervision and bonding fees. Overhead will vary with the size of project, the contractor's operating procedures and location. Profits will vary with economic activity and local conditions.

Contingencies provide for unforeseen construction difficulties which include material shortages and weather. In all situations, the appraiser should give consideration to possible adjustment of the 25% factor used in developing the Commercial/Industrial/Institutional models.

Architectural Fees

Tabulated below are typical percentage fees by project size, for good professional architectural service. Fees may vary from those listed depending upon degree of design difficulty and economic conditions in any particular area.

Rates can be interpolated horizontally and vertically. Various portions of the same project requiring different rates should be adjusted proportionately. For alterations, add 50% to the fee for the first $500,000 of project cost and add 25% to the fee for project cost over $500,000.

Architectural fees tabulated below include Engineering fees.

Insurance Exclusions

Many insurance companies exclude from coverage such items as architect's fees, excavation, foundations below grade, underground piping and site preparation. Since exclusions vary among insurance companies, it is recommended that for greatest accuracy each exclusion be priced separately using the unit-in-place section.

As a rule of thumb, exclusions can be calculated at 9% of total building cost plus the appropriate allowance for architect's fees.

Building Types	Total Project Size in Thousands of Dollars						
	100	250	500	1,000	5,000	10,000	50,000
Factories, garages, warehouses, repetitive housing	9.0%	8.0%	7.0%	6.2%	5.3%	4.9%	4.5%
Apartments, banks, schools, libraries, offices, municipal buildings	12.2	12.3	9.2	8.0	7.0	6.6	6.2
Churches, hospitals, homes, laboratories, museums, research	15.0	13.6	12.7	11.9	9.5	8.8	8.0
Memorials, monumental work, decorative furnishings	—	16.0	14.5	13.1	10.0	9.0	8.3

Location Factors

Costs shown in *RSMeans Square Foot Costs* are based on national averages for materials and installation. To adjust these costs to a specific location, simply multiply the base cost by the factor for that city.

The data is arranged alphabetically by state and postal zip code numbers. For a city not listed, use the factor for a nearby city with similar economic characteristics.

STATE/ZIP	CITY	Residential	Commercial
ALABAMA			
350-352	Birmingham	.86	.87
354	Tuscaloosa	.77	.80
355	Jasper	.72	.79
356	Decatur	.77	.80
357-358	Huntsville	.83	.84
359	Gadsden	.73	.79
360-361	Montgomery	.75	.79
362	Anniston	.76	.82
363	Dothan	.75	.77
364	Evergreen	.73	.77
365-366	Mobile	.81	.84
367	Selma	.72	.76
368	Phenix City	.73	.79
369	Butler	.73	.77
ALASKA			
995-996	Anchorage	1.22	1.20
997	Fairbanks	1.24	1.19
998	Juneau	1.21	1.17
999	Ketchikan	1.27	1.25
ARIZONA			
850,853	Phoenix	.86	.88
851,852	Mesa/Tempe	.84	.86
855	Globe	.76	.82
856-857	Tucson	.83	.86
859	Show Low	.78	.82
860	Flagstaff	.85	.88
863	Prescott	.76	.82
864	Kingman	.83	.86
865	Chambers	.76	.82
ARKANSAS			
716	Pine Bluff	.78	.82
717	Camden	.69	.75
718	Texarkana	.73	.76
719	Hot Springs	.70	.75
720-722	Little Rock	.82	.83
723	West Memphis	.78	.81
724	Jonesboro	.76	.81
725	Batesville	.74	.77
726	Harrison	.74	.76
727	Fayetteville	.70	.76
728	Russellville	.76	.77
729	Fort Smith	.81	.82
CALIFORNIA			
900-902	Los Angeles	1.08	1.07
903-905	Inglewood	1.05	1.03
906-908	Long Beach	1.04	1.03
910-912	Pasadena	1.04	1.03
913-916	Van Nuys	1.06	1.04
917-918	Alhambra	1.07	1.04
919-921	San Diego	1.03	1.03
922	Palm Springs	1.03	1.02
923-924	San Bernardino	1.04	1.01
925	Riverside	1.06	1.05
926-927	Santa Ana	1.05	1.02
928	Anaheim	1.06	1.05
930	Oxnard	1.07	1.06
931	Santa Barbara	1.07	1.06
932-933	Bakersfield	1.05	1.04
934	San Luis Obispo	1.06	1.03
935	Mojave	1.03	1.01
936-938	Fresno	1.11	1.07
939	Salinas	1.14	1.10
940-941	San Francisco	1.25	1.23
942,956-958	Sacramento	1.13	1.10
943	Palo Alto	1.16	1.13
944	San Mateo	1.22	1.15
945	Vallejo	1.17	1.13
946	Oakland	1.22	1.17
947	Berkeley	1.24	1.17
948	Richmond	1.23	1.14
949	San Rafael	1.21	1.15
950	Santa Cruz	1.16	1.12
951	San Jose	1.22	1.17
952	Stockton	1.11	1.08
953	Modesto	1.10	1.07

STATE/ZIP	CITY	Residential	Commercial
CALIFORNIA (CONT'D)			
954	Santa Rosa	1.19	1.15
955	Eureka	1.11	1.06
959	Marysville	1.10	1.06
960	Redding	1.11	1.07
961	Susanville	1.10	1.06
COLORADO			
800-802	Denver	.90	.94
803	Boulder	.92	.92
804	Golden	.88	.92
805	Fort Collins	.88	.93
806	Greeley	.86	.90
807	Fort Morgan	.90	.91
808-809	Colorado Springs	.87	.92
810	Pueblo	.89	.92
811	Alamosa	.85	.90
812	Salida	.88	.91
813	Durango	.88	.92
814	Montrose	.86	.91
815	Grand Junction	.92	.93
816	Glenwood Springs	.89	.93
CONNECTICUT			
060	New Britain	1.10	1.09
061	Hartford	1.10	1.10
062	Willimantic	1.10	1.10
063	New London	1.10	1.07
064	Meriden	1.10	1.08
065	New Haven	1.10	1.10
066	Bridgeport	1.11	1.10
067	Waterbury	1.10	1.09
068	Norwalk	1.12	1.13
069	Stamford	1.12	1.13
D.C.			
200-205	Washington	.94	.97
DELAWARE			
197	Newark	1.03	1.03
198	Wilmington	1.03	1.03
199	Dover	1.02	1.04
FLORIDA			
320,322	Jacksonville	.82	.85
321	Daytona Beach	.87	.88
323	Tallahassee	.79	.82
324	Panama City	.78	.82
325	Pensacola	.82	.85
326,344	Gainesville	.81	.86
327-328,347	Orlando	.88	.89
329	Melbourne	.88	.91
330-332,340	Miami	.87	.89
333	Fort Lauderdale	.87	.88
334,349	West Palm Beach	.87	.86
335-336,346	Tampa	.91	.91
337	St. Petersburg	.81	.86
338	Lakeland	.88	.91
339,341	Fort Myers	.87	.88
342	Sarasota	.90	.90
GEORGIA			
300-303,399	Atlanta	.87	.88
304	Statesboro	.70	.78
305	Gainesville	.77	.81
306	Athens	.75	.81
307	Dalton	.75	.80
308-309	Augusta	.81	.82
310-312	Macon	.79	.82
313-314	Savannah	.81	.82
315	Waycross	.79	.81
316	Valdosta	.75	.81
317,398	Albany	.77	.81
318-319	Columbus	.80	.83
HAWAII			
967	Hilo	1.18	1.15
968	Honolulu	1.20	1.17

STATE/ZIP	CITY	Residential	Commercial
STATES & POSS.			
969	Guam	.92	.99
IDAHO			
832	Pocatello	.84	.88
833	Twin Falls	.70	.80
834	Idaho Falls	.72	.81
835	Lewiston	.92	.96
836-837	Boise	.85	.88
838	Coeur d'Alene	.90	.94
ILLINOIS			
600-603	North Suburban	1.17	1.13
604	Joliet	1.20	1.15
605	South Suburban	1.17	1.13
606-608	Chicago	1.22	1.17
609	Kankakee	1.11	1.08
610-611	Rockford	1.11	1.09
612	Rock Island	.96	.97
613	La Salle	1.09	1.07
614	Galesburg	1.01	1.00
615-616	Peoria	1.05	1.03
617	Bloomington	1.03	1.01
618-619	Champaign	1.05	1.03
620-622	East St. Louis	1.01	1.00
623	Quincy	1.01	.97
624	Effingham	1.02	.98
625	Decatur	1.04	1.01
626-627	Springfield	1.04	1.02
628	Centralia	1.02	.99
629	Carbondale	.98	.96
INDIANA			
460	Anderson	.91	.90
461-462	Indianapolis	.94	.93
463-464	Gary	1.04	1.02
465-466	South Bend	.91	.90
467-468	Fort Wayne	.90	.88
469	Kokomo	.91	.88
470	Lawrenceburg	.87	.87
471	New Albany	.87	.86
472	Columbus	.92	.89
473	Muncie	.92	.90
474	Bloomington	.94	.91
475	Washington	.91	.89
476-477	Evansville	.91	.92
478	Terre Haute	.91	.92
479	Lafayette	.93	.90
IOWA			
500-503,509	Des Moines	.91	.92
504	Mason City	.78	.84
505	Fort Dodge	.76	.81
506-507	Waterloo	.83	.87
508	Creston	.85	.89
510-511	Sioux City	.85	.87
512	Sibley	.74	.80
513	Spencer	.75	.81
514	Carroll	.80	.86
515	Council Bluffs	.86	.92
516	Shenandoah	.81	.86
520	Dubuque	.85	.90
521	Decorah	.78	.83
522-524	Cedar Rapids	.92	.93
525	Ottumwa	.83	.86
526	Burlington	.88	.88
527-528	Davenport	.94	.95
KANSAS			
660-662	Kansas City	.97	.97
664-666	Topeka	.77	.84
667	Fort Scott	.88	.87
668	Emporia	.82	.85
669	Belleville	.80	.84
670-672	Wichita	.77	.83
673	Independence	.87	.86
674	Salina	.79	.85
675	Hutchinson	.79	.82
676	Hays	.81	.84
677	Colby	.82	.84
678	Dodge City	.81	.86
679	Liberal	.80	.84
KENTUCKY			
400-402	Louisville	.91	.91
403-405	Lexington	.90	.90

STATE/ZIP	CITY	Residential	Commercial
KENTUCKY (CONT'D)			
406	Frankfort	.89	.91
407-409	Corbin	.82	.84
410	Covington	.95	.95
411-412	Ashland	.94	.96
413-414	Campton	.89	.88
415-416	Pikeville	.92	.93
417-418	Hazard	.88	.88
420	Paducah	.88	.89
421-422	Bowling Green	.89	.90
423	Owensboro	.90	.92
424	Henderson	.91	.90
425-426	Somerset	.84	.87
427	Elizabethtown	.87	.88
LOUISIANA			
700-701	New Orleans	.87	.88
703	Thibodaux	.82	.84
704	Hammond	.76	.79
705	Lafayette	.82	.83
706	Lake Charles	.83	.84
707-708	Baton Rouge	.83	.84
710-711	Shreveport	.76	.79
712	Monroe	.72	.78
713-714	Alexandria	.72	.78
MAINE			
039	Kittery	.91	.90
040-041	Portland	.96	.96
042	Lewiston	.96	.95
043	Augusta	.93	.91
044	Bangor	.95	.95
045	Bath	.92	.90
046	Machias	.96	.92
047	Houlton	.96	.92
048	Rockland	.96	.91
049	Waterville	.91	.89
MARYLAND			
206	Waldorf	.89	.91
207-208	College Park	.87	.92
209	Silver Spring	.88	.90
210-212	Baltimore	.91	.93
214	Annapolis	.90	.93
215	Cumberland	.89	.90
216	Easton	.84	.85
217	Hagerstown	.88	.91
218	Salisbury	.83	.83
219	Elkton	.90	.90
MASSACHUSETTS			
010-011	Springfield	1.05	1.04
012	Pittsfield	1.04	1.03
013	Greenfield	1.04	1.03
014	Fitchburg	1.13	1.08
015-016	Worcester	1.14	1.10
017	Framingham	1.16	1.11
018	Lowell	1.16	1.13
019	Lawrence	1.18	1.14
020-022, 024	Boston	1.22	1.18
023	Brockton	1.16	1.12
025	Buzzards Bay	1.14	1.09
026	Hyannis	1.14	1.11
027	New Bedford	1.16	1.12
MICHIGAN			
480,483	Royal Oak	1.02	.99
481	Ann Arbor	1.03	1.01
482	Detroit	1.06	1.03
484-485	Flint	.97	.96
486	Saginaw	.93	.94
487	Bay City	.94	.95
488-489	Lansing	.95	.96
490	Battle Creek	.93	.93
491	Kalamazoo	.92	.92
492	Jackson	.93	.95
493,495	Grand Rapids	.89	.91
494	Muskegon	.90	.90
496	Traverse City	.83	.87
497	Gaylord	.81	.85
498-499	Iron mountain	.90	.92
MINNESOTA			
550-551	Saint Paul	1.11	1.09
553-555	Minneapolis	1.14	1.11
556-558	Duluth	1.07	1.04

Location Factors

STATE/ZIP	CITY	Residential	Commercial
MINNESOTA (CONT'D)			
559	Rochester	1.05	1.03
560	Mankato	1.00	1.00
561	Windom	.93	.95
562	Willmar	.98	.99
563	St. Cloud	1.07	1.07
564	Brainerd	.97	1.00
565	Detroit Lakes	.96	.99
566	Bemidji	.95	.99
567	Thief River Falls	.95	.96
MISSISSIPPI			
386	Clarksdale	.75	.78
387	Greenville	.82	.84
388	Tupelo	.77	.80
389	Greenwood	.78	.78
390-392	Jackson	.82	.83
393	Meridian	.80	.83
394	Laurel	.77	.80
395	Biloxi	.80	.81
396	Mccomb	.75	.78
397	Columbus	.76	.79
MISSOURI			
630-631	St. Louis	1.03	1.03
633	Bowling Green	.96	.97
634	Hannibal	.93	.95
635	Kirksville	.89	.96
636	Flat River	.96	.97
637	Cape Girardeau	.89	.95
638	Sikeston	.88	.94
639	Poplar Bluff	.87	.94
640-641	Kansas City	1.05	1.04
644-645	St. Joseph	.97	.97
646	Chillicothe	.96	.96
647	Harrisonville	.99	.99
648	Joplin	.90	.91
650-651	Jefferson City	.93	.94
652	Columbia	.92	.96
653	Sedalia	.92	.95
654-655	Rolla	.97	.95
656-658	Springfield	.89	.92
MONTANA			
590-591	Billings	.88	.92
592	Wolf Point	.86	.91
593	Miles City	.87	.89
594	Great Falls	.89	.92
595	Havre	.84	.90
596	Helena	.87	.91
597	Butte	.88	.92
598	Missoula	.88	.90
599	Kalispell	.86	.89
NEBRASKA			
680-681	Omaha	.91	.91
683-685	Lincoln	.84	.87
686	Columbus	.84	.85
687	Norfolk	.89	.88
688	Grand Island	.88	.90
689	Hastings	.90	.89
690	Mccook	.82	.85
691	North Platte	.88	.89
692	Valentine	.82	.85
693	Alliance	.82	.85
NEVADA			
889-891	Las Vegas	1.03	1.04
893	Ely	1.00	.99
894-895	Reno	.93	.96
897	Carson City	.93	.96
898	Elko	.92	.92
NEW HAMPSHIRE			
030	Nashua	.97	.97
031	Manchester	.97	.98
032-033	Concord	.96	.96
034	Keene	.87	.87
035	Littleton	.91	.88
036	Charleston	.85	.85
037	Claremont	.86	.84
038	Portsmouth	.96	.96

STATE/ZIP	CITY	Residential	Commercial
NEW JERSEY			
070-071	Newark	1.13	1.12
072	Elizabeth	1.15	1.11
073	Jersey City	1.12	1.10
074-075	Paterson	1.13	1.12
076	Hackensack	1.12	1.10
077	Long Branch	1.11	1.09
078	Dover	1.13	1.11
079	Summit	1.13	1.11
080,083	Vineland	1.10	1.09
081	Camden	1.11	1.10
082,084	Atlantic City	1.14	1.09
085-086	Trenton	1.11	1.10
087	Point Pleasant	1.10	1.09
088-089	New Brunswick	1.13	1.10
NEW MEXICO			
870-872	Albuquerque	.82	.88
873	Gallup	.82	.88
874	Farmington	.82	.88
875	Santa Fe	.83	.89
877	Las Vegas	.82	.87
878	Socorro	.82	.87
879	Truth/Consequences	.81	.85
880	Las Cruces	.81	.84
881	Clovis	.82	.87
882	Roswell	.82	.88
883	Carrizozo	.82	.88
884	Tucumcari	.83	.87
NEW YORK			
100-102	New York	1.35	1.32
103	Staten Island	1.30	1.29
104	Bronx	1.31	1.28
105	Mount Vernon	1.12	1.13
106	White Plains	1.15	1.14
107	Yonkers	1.16	1.16
108	New Rochelle	1.16	1.14
109	Suffern	1.11	1.08
110	Queens	1.31	1.29
111	Long Island City	1.34	1.30
112	Brooklyn	1.35	1.30
113	Flushing	1.33	1.30
114	Jamaica	1.33	1.29
115,117,118	Hicksville	1.20	1.21
116	Far Rockaway	1.32	1.30
119	Riverhead	1.21	1.21
120-122	Albany	.96	.98
123	Schenectady	.98	.98
124	Kingston	1.05	1.08
125-126	Poughkeepsie	1.17	1.12
127	Monticello	1.05	1.07
128	Glens Falls	.91	.93
129	Plattsburgh	.95	.93
130-132	Syracuse	.97	.98
133-135	Utica	.94	.95
136	Watertown	.92	.96
137-139	Binghamton	.97	.99
140-142	Buffalo	1.05	1.02
143	Niagara Falls	1.03	1.01
144-146	Rochester	.97	.98
147	Jamestown	.90	.93
148-149	Elmira	.91	.95
NORTH CAROLINA			
270,272-274	Greensboro	.85	.80
271	Winston-Salem	.87	.81
275-276	Raleigh	.86	.80
277	Durham	.85	.80
278	Rocky Mount	.82	.78
279	Elizabeth City	.80	.77
280	Gastonia	.85	.80
281-282	Charlotte	.85	.80
283	Fayetteville	.90	.82
284	Wilmington	.86	.79
285	Kinston	.81	.77
286	Hickory	.80	.77
287-288	Asheville	.82	.79
289	Murphy	.83	.76
NORTH DAKOTA			
580-581	Fargo	.77	.86
582	Grand Forks	.72	.81
583	Devils Lake	.76	.82
584	Jamestown	.70	.77
585	Bismarck	.76	.85

Location Factors

STATE/ZIP	CITY	Residential	Commercial
NORTH DAKOTA (CONT'D)			
586	Dickinson	.74	.83
587	Minot	.82	.89
588	Williston	.74	.82
OHIO			
430-432	Columbus	.95	.95
433	Marion	.90	.90
434-436	Toledo	.99	.98
437-438	Zanesville	.90	.91
439	Steubenville	.93	.95
440	Lorain	.95	.95
441	Cleveland	1.00	.99
442-443	Akron	.97	.97
444-445	Youngstown	.94	.94
446-447	Canton	.92	.93
448-449	Mansfield	.91	.91
450	Hamilton	.93	.92
451-452	Cincinnati	.93	.92
453-454	Dayton	.93	.92
455	Springfield	.94	.92
456	Chillicothe	.97	.94
457	Athens	.91	.92
458	Lima	.92	.93
OKLAHOMA			
730-731	Oklahoma City	.77	.81
734	Ardmore	.76	.79
735	Lawton	.79	.81
736	Clinton	.75	.80
737	Enid	.75	.80
738	Woodward	.75	.79
739	Guymon	.66	.67
740-741	Tulsa	.75	.78
743	Miami	.85	.82
744	Muskogee	.69	.71
745	Mcalester	.72	.74
746	Ponca City	.75	.78
747	Durant	.75	.78
748	Shawnee	.74	.78
749	Poteau	.76	.79
OREGON			
970-972	Portland	.99	1.00
973	Salem	.98	1.00
974	Eugene	.99	.99
975	Medford	.97	.99
976	Klamath Falls	.98	.99
977	Bend	1.00	.99
978	Pendleton	.98	.98
979	Vale	.96	.92
PENNSYLVANIA			
150-152	Pittsburgh	1.01	1.02
153	Washington	.96	.99
154	Uniontown	.94	.99
155	Bedford	.91	.96
156	Greensburg	.96	.99
157	Indiana	.93	.98
158	Dubois	.92	.98
159	Johnstown	.92	.97
160	Butler	.93	.96
161	New Castle	.93	.95
162	Kittanning	.94	.97
163	Oil City	.90	.94
164-165	Erie	.94	.95
166	Altoona	.88	.94
167	Bradford	.90	.96
168	State College	.91	.94
169	Wellsboro	.92	.95
170-171	Harrisburg	.95	.97
172	Chambersburg	.89	.93
173-174	York	.93	.96
175-176	Lancaster	.92	.92
177	Williamsport	.84	.87
178	Sunbury	.93	.95
179	Pottsville	.93	.96
180	Lehigh Valley	1.02	1.03
181	Allentown	1.04	1.03
182	Hazleton	.93	.97
183	Stroudsburg	.92	.98
184-185	Scranton	.97	.99
186-187	Wilkes-Barre	.94	.96
188	Montrose	.91	.96
189	Doylestown	1.06	1.06

STATE/ZIP	CITY	Residential	Commercial
PENNSYLVANIA (CONT'D)			
190-191	Philadelphia	1.16	1.14
193	Westchester	1.11	1.08
194	Norristown	1.10	1.10
195-196	Reading	.98	.99
PUERTO RICO			
009	San Juan	.72	.78
RHODE ISLAND			
028	Newport	1.09	1.07
029	Providence	1.09	1.07
SOUTH CAROLINA			
290-292	Columbia	.81	.76
293	Spartanburg	.80	.75
294	Charleston	.83	.80
295	Florence	.76	.75
296	Greenville	.79	.75
297	Rock Hill	.78	.74
298	Aiken	.99	.85
299	Beaufort	.78	.73
SOUTH DAKOTA			
570-571	Sioux Falls	.76	.82
572	Watertown	.73	.78
573	Mitchell	.74	.77
574	Aberdeen	.75	.79
575	Pierre	.74	.80
576	Mobridge	.73	.78
577	Rapid City	.76	.81
TENNESSEE			
370-372	Nashville	.84	.87
373-374	Chattanooga	.82	.84
375,380-381	Memphis	.82	.85
376	Johnson City	.71	.80
377-379	Knoxville	.75	.80
382	Mckenzie	.74	.81
383	Jackson	.77	.83
384	Columbia	.77	.83
385	Cookeville	.71	.80
TEXAS			
750	Mckinney	.72	.77
751	Waxahackie	.78	.80
752-753	Dallas	.84	.85
754	Greenville	.69	.74
755	Texarkana	.72	.77
756	Longview	.67	.73
757	Tyler	.73	.79
758	Palestine	.75	.78
759	Lufkin	.78	.79
760-761	Fort Worth	.83	.82
762	Denton	.74	.75
763	Wichita Falls	.77	.79
764	Eastland	.74	.76
765	Temple	.74	.75
766-767	Waco	.77	.79
768	Brownwood	.72	.76
769	San Angelo	.72	.76
770-772	Houston	.84	.87
773	Huntsville	.77	.79
774	Wharton	.72	.79
775	Galveston	.82	.84
776-777	Beaumont	.81	.81
778	Bryan	.73	.80
779	Victoria	.71	.74
780	Laredo	.74	.78
781-782	San Antonio	.81	.83
783-784	Corpus Christi	.78	.79
785	Mc Allen	.75	.76
786-787	Austin	.78	.80
788	Del Rio	.72	.76
789	Giddings	.78	.78
790-791	Amarillo	.79	.82
792	Childress	.79	.79
793-794	Lubbock	.79	.81
795-796	Abilene	.75	.79
797	Midland	.76	.78
798-799,885	El Paso	.74	.77
UTAH			
840-841	Salt Lake City	.80	.87
842,844	Ogden	.79	.85
843	Logan	.79	.86

Location Factors

STATE/ZIP	CITY	Residential	Commercial
UTAH (CONT'd)			
845	Price	.77	.85
846-847	Provo	.79	.86
VERMONT			
050	White River Jct.	.84	.84
051	Bellows Falls	.90	.90
052	Bennington	.90	.89
053	Brattleboro	.92	.92
054	Burlington	.89	.91
056	Montpelier	.90	.89
057	Rutland	.89	.90
058	St. Johnsbury	.85	.84
059	Guildhall	.85	.83
VIRGINIA			
220-221	Fairfax	1.03	.93
222	Arlington	1.05	.94
223	Alexandria	1.07	.94
224-225	Fredericksburg	1.04	.92
226	Winchester	1.03	.92
227	Culpeper	1.03	.91
228	Harrisonburg	.88	.86
229	Charlottesville	.92	.86
230-232	Richmond	.97	.87
233-235	Norfolk	.99	.87
236	Newport News	.98	.86
237	Portsmouth	.92	.85
238	Petersburg	.96	.86
239	Farmville	.86	.79
240-241	Roanoke	.97	.85
242	Bristol	.85	.80
243	Pulaski	.83	.80
244	Staunton	.88	.83
245	Lynchburg	.95	.85
246	Grundy	.82	.79
WASHINGTON			
980-981,987	Seattle	1.02	1.04
982	Everett	1.03	1.01
983-984	Tacoma	1.00	1.02
985	Olympia	.99	1.00
986	Vancouver	.97	.99
988	Wenatchee	.93	.96
989	Yakima	.97	.99
990-992	Spokane	.97	.93
993	Richland	.97	.96
994	Clarkston	.92	.92
WEST VIRGINIA			
247-248	Bluefield	.90	.91
249	Lewisburg	.90	.92
250-253	Charleston	.96	.95
254	Martinsburg	.87	.90
255-257	Huntington	.97	.97
258-259	Beckley	.90	.93
260	Wheeling	.92	.96
261	Parkersburg	.92	.95
262	Buckhannon	.92	.95
263-264	Clarksburg	.91	.95
265	Morgantown	.92	.95
266	Gassaway	.92	.95
267	Romney	.90	.93
268	Petersburg	.91	.93
WISCONSIN			
530,532	Milwaukee	1.08	1.04
531	Kenosha	1.02	1.00
534	Racine	1.00	1.00
535	Beloit	.97	.98
537	Madison	.97	.98
538	Lancaster	.95	.94
539	Portage	.94	.95
540	New Richmond	.98	.95
541-543	Green Bay	1.01	.98
544	Wausau	.95	.94
545	Rhinelander	.94	.95
546	La Crosse	.95	.96
547	Eau Claire	.98	.97
548	Superior	.96	.96
549	Oshkosh	.90	.90
WYOMING			
820	Cheyenne	.80	.85
821	Yellowstone Nat. Pk.	.80	.83
822	Wheatland	.77	.84

STATE/ZIP	CITY	Residential	Commercial
WYOMING (CONT'd)			
823	Rawlins	.81	.86
824	Worland	.80	.83
825	Riverton	.80	.84
826	Casper	.74	.83
827	Newcastle	.80	.84
828	Sheridan	.80	.85
829-831	Rock Springs	.84	.86
CANADIAN FACTORS (reflect Canadian currency)			
ALBERTA			
	Calgary	1.17	1.15
	Edmonton	1.15	1.15
	Fort McMurray	1.15	1.12
	Lethbridge	1.13	1.09
	Lloydminster	1.08	1.05
	Medicine Hat	1.08	1.05
	Red Deer	1.08	1.05
BRITISH COLUMBIA			
	Kamloops	1.07	1.06
	Prince George	1.06	1.07
	Vancouver	1.11	1.11
	Victoria	1.06	1.05
MANITOBA			
	Brandon	1.03	.99
	Portage la Prairie	1.03	.98
	Winnipeg	1.03	1.03
NEW BRUNSWICK			
	Bathurst	.95	.94
	Dalhousie	.96	.94
	Fredericton	1.02	.99
	Moncton	.96	.96
	Newcastle	.96	.94
	St. John	1.03	1.00
NEWFOUNDLAND			
	Corner Brook	.96	.97
	St Johns	1.05	1.04
NORTHWEST TERRITORIES			
	Yellowknife	1.09	1.08
NOVA SCOTIA			
	Bridgewater	.98	.98
	Dartmouth	.98	.99
	Halifax	1.02	1.03
	New Glasgow	.98	.98
	Sydney	.97	.96
	Truro	.98	.97
	Yarmouth	.98	.98
ONTARIO			
	Barrie	1.14	1.08
	Brantford	1.15	1.09
	Cornwall	1.14	1.07
	Hamilton	1.15	1.10
	Kingston	1.14	1.08
	Kitchener	1.09	1.04
	London	1.13	1.09
	North Bay	1.12	1.06
	Oshawa	1.13	1.07
	Ottawa	1.14	1.09
	Owen Sound	1.13	1.07
	Peterborough	1.13	1.07
	Sarnia	1.15	1.09
	Sault Ste Marie	1.08	1.04
	St. Catharines	1.10	1.04
	Sudbury	1.07	1.03
	Thunder Bay	1.12	1.04
	Timmins	1.12	1.06
	Toronto	1.17	1.12
	Windsor	1.11	1.03
PRINCE EDWARD ISLAND			
	Charlottetown	.92	.95
	Summerside	.92	.93
QUEBEC			
	Cap-de-la-Madeleine	1.14	1.03
	Charlesbourg	1.14	1.03
	Chicoutimi	1.16	1.05
	Gatineau	1.13	1.03

Location Factors

STATE/ZIP	CITY	Residential	Commercial
QUEBEC (CONT'D)			
	Granby	1.13	
	Hull	1.13	1.03
	Joliette	1.14	1.03
	Laval	1.13	1.03
	Montreal	1.18	1.10
	Quebec	1.17	1.09
	Rimouski	1.16	1.05
	Rouyn-Noranda	1.13	1.03
	Saint Hyacinthe	1.13	1.03
	Sherbrooke	1.13	1.03
	Sorel	1.14	1.03
	St Jerome	1.13	1.03
	Trois Rivieres	1.14	1.03
SASKATCHEWAN			
	Moose Jaw	.95	.94
	Prince Albert	.94	.93
	Regina	1.07	1.06
	Saskatoon	1.06	1.02
YUKON			
	Whitehorse	.96	.97

Historical Cost Indexes

The following tables are the estimated Historical Cost Indexes based on a 30-city national average with a base of 100 on January 1, 1993.

The indexes may be used to:

1. Estimate and compare construction costs for different years in the same city.
2. Estimate and compare construction costs in different cities for the same year.
3. Estimate and compare construction costs in different cities for different years.
4. Compare construction trends in any city with the national average.

EXAMPLES

1. Estimate and compare construction costs for different years in the same city.

   A. To estimate the construction cost of a building in Lexington, KY in 1970, knowing that it cost $915,000 in 2013.

   Index Lexington, KY in 1970 = 26.9
   Index Lexington, KY in 2013 = 177.8

   $$\frac{\text{Index 1970}}{\text{Index 2013}} \quad x \quad \text{Cost 2013} \quad = \quad \text{Cost 1970}$$

   $$\frac{26.9}{177.8} \quad x \quad \$915,000 \quad = \quad \$138,500$$

   Construction Cost in Lexington, KY in 1970 = $138,500

   B. To estimate the current construction cost of a building in Boston, MA that was built in 1980 for $900,000.

   Index Boston, MA in 1980 = 64.0
   Index Boston, MA in 2013 = 223.4

   $$\frac{\text{Index 2013}}{\text{Index 1980}} \quad x \quad \text{Cost 1980} \quad = \quad \text{Cost 2013}$$

   $$\frac{223.4}{64.0} \quad x \quad \$900,000 \quad = \quad \$3,141,500$$

   Construction Cost in Boston in 2013 = $3,141,500

2. Estimate and compare construction costs in different cities for the same year.

   To compare the construction cost of a building in Topeka, KS in 2013 with the known cost of $800,000 in Baltimore, MD in 2013

   Index Topeka, KS in 2013   = 165.2
   Index Baltimore, MD in 2013 = 183.8

   $$\frac{\text{Index Topeka}}{\text{Index Baltimore}} \quad x \quad \text{Cost Baltimore} \quad = \quad \text{Cost Topeka}$$

   $$\frac{165.2}{183.8} \quad x \quad \$800,000 \quad = \quad \$719,000$$

   Construction Cost in Topeka in 2013 = $719,000

3. Estimate and compare construction costs in different cities for different years.

   To compare the construction cost of a building in Detroit, MI in 2013 with the known construction cost of $5,000,000 for the same building in San Francisco, CA in 1980.

   Index Detroit, MI in 2013   = 203.8
   Index San Francisco, CA in 1980 = 75.2

   $$\frac{\text{Index Detroit 2013}}{\text{Index San Francisco 1980}} \quad x \quad \text{Cost San Francisco 1980} = \text{Cost Detroit 2013}$$

   $$\frac{203.8}{75.2} \quad x \quad \$5,000,000 \quad = \quad \$13,550,500$$

   Construction Cost in Detroit in 2013   = $13,550,500

4. Compare construction trends in any city with the national average.

   To compare the construction cost in Las Vegas, NV from 1975 to 2013 with the increase in the National Average during the same time period.

   Index Las Vegas, NV for 1975   =   42.8   For 2013 =   205.6
   Index 30 City Average for 1975 =   43.7   For 2013 =   197.6

   A. National Average escalation = $\dfrac{\text{Index — 30 City 2013}}{\text{Index — 30 City 1975}}$
      From 1975 to 2013

   $$= \quad \frac{197.6}{43.7}$$

   National Average escalation
   From 1975 to 2013   =   4.52 or increased by 352%

   B. Escalation for Las Vegas, NV = $\dfrac{\text{Index Las Vegas, NV 2013}}{\text{Index Las Vegas, NV 1975}}$
      From 1975 to 2013

   $$= \quad \frac{205.6}{42.8}$$

   Las Vegas escalation
   From 1975 to 2013   =   4.80 or increased by 380%

   Conclusion: Construction costs in Las Vegas are higher than National average costs and increased at a greater rate from 1975 to 2013 than the National Average.

Historical Cost Indexes

Year	National 30 City Average	Birmingham	Huntsville	Mobile	Montgomery	Tuscaloosa	Anchorage	Phoenix	Tuscon	Fort Smith	Little Rock	Anaheim	Bakersfield	Fresno	Los Angeles	Oxnard
		Alabama					**Alaska**	**Arizona**		**Arkansas**		**California**				
Jan 2013	197.6E	172.0E	165.4E	165.9E	156.6E	157.3E	236.7E	174.7E	168.8E	161.4E	163.7E	208.1E	206.2E	211.0E	211.8E	208.7E
2012	194.0	169.1	162.7	163.2	153.8	154.6	232.5	172.2	166.4	158.7	161.0	204.1	202.9	207.3	207.2	204.7
2011	185.7	163.0	156.3	156.9	147.6	148.3	225.1	163.8	159.5	152.0	154.1	196.1	194.6	199.8	199.2	197.0
2010	181.6	159.6	152.9	153.1	144.4	144.9	218.3	160.7	157.3	150.3	152.5	192.8	190.8	195.0	194.9	192.8
2009	182.5	162.7	157.2	155.6	148.8	149.4	222.9	161.3	156.8	149.2	156.3	194.5	192.3	195.0	196.6	194.5
2008	171.0	150.3	146.9	143.7	138.4	138.8	210.8	152.2	148.4	138.6	145.4	182.7	179.8	183.2	184.7	182.6
2007	165.0	146.9	143.3	140.3	134.8	135.4	206.8	147.7	143.1	134.8	141.7	175.1	173.7	176.7	177.5	176.2
2006	156.2	135.7	133.6	126.7	124.0	122.2	196.4	137.9	134.8	123.2	127.2	166.8	164.9	169.8	167.3	167.4
2005	146.7	127.9	125.4	119.3	116.6	114.6	185.6	128.5	124.1	115.4	119.4	156.5	153.0	157.9	157.1	156.4
2004	132.8	115.9	112.9	107.0	104.9	102.9	167.0	116.5	113.6	103.8	108.6	142.1	139.5	143.8	142.0	140.9
2003	129.7	113.1	110.7	104.8	102.6	100.8	163.5	113.9	110.6	101.8	105.8	139.4	137.2	142.1	139.6	138.7
2002	126.7	110.0	103.4	103.6	101.7	99.7	159.5	113.3	110.2	100.4	102.1	136.5	133.4	136.0	136.4	136.3
2001	122.2	106.0	100.5	100.6	98.3	95.9	152.6	109.0	106.4	97.3	98.7	132.5	129.3	132.8	132.4	132.4
2000	118.9	104.1	98.9	99.1	94.2	94.6	148.3	106.9	104.9	94.4	95.7	129.4	125.2	129.4	129.9	129.7
1999	116.6	101.2	97.4	97.7	92.5	92.8	145.9	105.5	103.3	93.1	94.4	127.9	123.6	126.9	128.7	128.2
1998	113.6	96.2	94.0	94.8	90.3	89.8	143.8	102.1	101.0	90.6	91.4	125.2	120.3	123.9	125.8	124.7
1997	111.5	94.6	92.4	93.3	88.8	88.3	142.0	101.8	100.7	89.3	90.1	124.0	119.1	122.3	124.6	123.5
1996	108.9	90.9	91.4	92.3	87.5	86.5	140.4	98.3	97.4	86.5	86.7	121.7	117.2	120.2	122.4	121.5
1995	105.6	87.8	88.0	88.8	84.5	83.5	138.0	96.1	95.5	85.0	85.6	120.1	115.7	117.5	120.9	120.0
1994	103.0	85.9	86.1	86.8	82.6	81.7	134.0	93.7	93.1	82.4	83.9	118.1	113.6	114.5	119.0	117.5
1993	100.0	82.8	82.7	86.1	82.1	78.7	132.0	90.9	90.9	80.9	82.4	115.0	111.3	112.9	115.6	115.4
1990	93.2	79.4	77.6	82.7	78.4	75.2	125.8	86.4	87.0	77.1	77.9	107.7	102.7	103.3	107.5	107.4
1985	81.8	71.1	70.5	72.7	70.9	67.9	116.0	78.1	77.7	69.6	71.0	95.4	92.2	92.6	94.6	96.6
1980	60.7	55.2	54.1	56.8	56.8	54.0	91.4	63.7	62.4	53.1	55.8	68.7	69.4	68.7	67.4	69.9
1975	43.7	40.0	40.9	41.8	40.1	37.8	57.3	44.5	45.2	39.0	38.7	47.6	46.6	47.7	48.3	47.0
1970	27.8	24.1	25.1	25.8	25.4	24.2	43.0	27.2	28.5	24.3	22.3	31.0	30.8	31.1	29.0	31.0
1965	21.5	19.6	19.3	19.6	19.5	18.7	34.9	21.8	22.0	18.7	18.5	23.9	23.7	24.0	22.7	23.9
1960	19.5	17.9	17.5	17.8	17.7	16.9	31.7	19.9	20.0	17.0	16.8	21.7	21.5	21.8	20.6	21.7
1955	16.3	14.8	14.7	14.9	14.9	14.2	26.6	16.7	16.7	14.3	14.5	18.2	18.1	18.3	17.3	18.2
1950	13.5	12.2	12.1	12.3	12.3	11.7	21.9	13.8	13.8	11.8	11.6	15.1	15.0	15.1	14.3	15.1
1945	8.6	7.8	7.7	7.8	7.8	7.5	14.0	8.8	8.8	7.5	7.4	9.6	9.5	9.6	9.1	9.6
1940	6.6	6.0	6.0	6.1	6.0	5.8	10.8	6.8	6.8	5.8	5.7	7.4	7.4	7.4	7.0	7.4

Year	National 30 City Average	Riverside	Sacramento	San Diego	San Francisco	Santa Barbara	Stockton	Vallejo	Colorado Springs	Denver	Pueblo	Bridge-Port	Bristol	Hartford	New Britain	New Haven
		California							**Colorado**			**Connecticut**				
Jan 2013	197.6E	208.1E	216.3E	203.0E	242.1E	208.8E	212.6E	223.0E	182.6E	185.4E	180.8E	217.0E	216.9E	217.3E	215.5E	217.4E
2012	194.0	204.1	211.9	198.9	237.6	204.8	208.8	218.9	179.4	182.2	177.9	213.9	213.3	214.4	212.9	214.7
2011	185.7	195.8	203.1	192.3	227.9	196.2	201.3	210.3	171.7	174.1	170.7	205.2	204.2	205.1	203.9	205.8
2010	181.6	192.7	196.7	188.2	223.0	192.9	195.9	204.6	170.5	172.6	168.1	198.8	199.1	199.8	198.8	200.7
2009	182.5	193.1	198.1	191.6	224.9	193.6	194.5	202.7	168.4	172.0	166.7	199.0	197.7	198.7	197.4	199.3
2008	171.0	181.3	186.0	179.9	210.6	181.5	183.2	191.6	158.8	161.9	157.9	185.8	184.8	185.7	184.5	186.1
2007	165.0	174.7	179.1	173.6	201.1	175.2	178.2	185.9	152.6	155.9	152.3	179.6	178.3	179.6	178.0	179.5
2006	156.2	166.0	172.2	164.0	191.2	166.4	171.0	177.9	146.1	149.2	145.0	169.5	168.0	169.5	167.7	169.8
2005	146.7	155.4	161.1	153.8	179.7	155.7	160.0	167.2	138.3	141.1	136.4	160.8	159.4	159.6	159.1	160.8
2004	132.8	141.0	147.2	139.0	163.6	141.4	144.2	148.6	125.6	127.2	123.5	143.6	142.9	142.4	142.7	144.3
2003	129.7	138.7	144.9	136.6	162.1	139.2	141.8	146.4	123.1	123.7	120.7	141.3	140.2	140.3	140.0	141.6
2002	126.7	135.3	138.4	134.2	157.9	135.9	136.9	143.8	118.7	121.3	116.7	133.8	133.5	133.1	133.3	134.8
2001	122.2	131.4	135.5	129.7	151.8	131.5	134.1	140.2	113.3	117.2	113.1	128.6	128.5	128.5	128.3	128.6
2000	118.9	128.0	131.5	127.1	146.9	128.7	130.7	137.1	109.8	111.8	108.8	122.7	122.6	122.9	122.4	122.7
1999	116.6	126.5	129.4	124.9	145.1	127.2	127.9	135.3	107.0	109.1	107.1	121.1	121.3	121.2	121.1	121.3
1998	113.6	123.7	125.9	121.3	141.9	123.7	124.7	132.5	103.3	106.5	103.7	119.1	119.4	120.0	119.7	120.0
1997	111.5	122.6	124.7	120.3	139.2	122.4	123.3	130.6	101.1	104.4	102.0	119.2	119.5	119.9	119.7	120.0
1996	108.9	120.6	122.4	118.4	136.8	120.5	121.4	128.1	98.5	101.4	99.7	117.4	117.6	117.9	117.8	118.1
1995	105.6	119.2	119.5	115.4	133.8	119.0	119.0	122.5	96.1	98.9	96.8	116.0	116.5	116.9	116.3	116.5
1994	103.0	116.6	114.8	113.4	131.4	116.4	115.5	119.7	94.2	95.9	94.6	114.3	115.0	115.3	114.7	114.8
1993	100.0	114.3	112.5	111.3	129.6	114.3	115.2	117.5	92.1	93.8	92.6	108.7	107.8	108.4	106.3	106.1
1990	93.2	107.0	104.9	105.6	121.8	106.4	105.4	111.4	86.9	88.8	88.8	96.3	95.9	96.6	95.9	96.5
1985	81.8	94.8	92.2	94.2	106.2	93.2	93.7	97.0	79.6	81.0	80.4	86.1	86.2	87.2	86.1	86.3
1980	60.7	68.7	71.3	68.1	75.2	71.1	71.2	71.9	60.7	60.9	59.5	61.5	60.7	61.9	60.7	61.4
1975	43.7	47.3	49.1	47.7	49.8	46.1	47.6	46.5	43.1	42.7	42.5	45.2	45.5	46.0	45.2	46.0
1970	27.8	31.0	32.1	30.7	31.6	31.3	31.8	31.8	27.6	26.1	27.3	29.2	28.2	29.6	28.2	29.3
1965	21.5	23.9	24.8	24.0	23.7	24.1	24.5	24.5	21.3	20.9	21.0	22.4	21.7	22.6	21.7	23.2
1960	19.5	21.7	22.5	21.7	21.5	21.9	22.3	22.2	19.3	19.0	19.1	20.0	19.8	20.0	19.8	20.0
1955	16.3	18.2	18.9	18.2	18.0	18.4	18.7	18.6	16.2	15.9	16.0	16.8	16.6	16.8	16.6	16.8
1950	13.5	15.0	15.6	15.0	14.9	15.2	15.4	15.4	13.4	13.2	13.2	13.9	13.7	13.8	13.7	13.9
1945	8.6	9.6	10.0	9.6	9.5	9.7	9.9	9.8	8.5	8.4	8.4	8.8	8.7	8.8	8.7	8.9
1940	6.6	7.4	7.7	7.4	7.3	7.5	7.6	7.6	6.6	6.5	6.5	6.8	6.8	6.8	6.8	6.8

Historical Cost Indexes

Year	National 30 City Average	Connecticut Norwalk	Stamford	Waterbury	Delaware Wilming-ton	D.C. Washing-ton	Florida Fort Lau-derdale	Jackson-ville	Miami	Orlando	Talla-hassee	Tampa	Georgia Albany	Atlanta	Colum-bus	Macon
Jan 2013	197.6E	222.5E	222.8E	216.1E	203.8E	192.2E	173.6E	168.4E	176.4E	175.5E	160.9E	180.6E	160.8E	173.9E	163.3E	162.0E
2012	194.0	219.8	220.0	213.4	201.0	189.6	170.9	165.8	173.7	172.8	158.5	177.8	158.3	170.7	160.6	159.3
2011	185.7	211.0	211.2	204.6	193.3	182.3	164.4	159.4	167.0	166.1	151.7	171.4	151.8	164.0	154.3	152.7
2010	181.6	198.3	203.8	199.4	187.7	179.0	160.3	155.4	163.3	162.5	148.1	167.5	148.3	160.1	150.9	149.4
2009	182.5	198.8	204.3	198.4	188.9	181.4	160.7	152.0	165.2	163.9	145.1	165.1	152.0	164.5	155.6	153.5
2008	171.0	185.2	189.2	185.3	177.5	170.4	149.6	142.5	152.9	153.0	135.3	155.6	140.4	153.2	143.9	142.5
2007	165.0	178.6	183.3	179.1	173.4	163.1	145.6	139.0	148.9	148.1	131.7	152.0	135.8	148.0	140.1	138.1
2006	156.2	169.7	173.7	169.1	158.3	153.0	136.0	126.5	136.7	134.3	118.4	136.6	123.6	139.9	126.3	124.3
2005	146.7	160.9	164.1	160.5	149.6	142.7	126.0	119.2	127.1	126.0	110.7	127.7	115.7	131.7	118.9	116.1
2004	132.8	143.6	146.5	143.2	136.1	126.8	114.8	107.8	115.6	113.2	100.6	116.6	102.8	119.6	102.0	105.0
2003	129.7	140.6	145.1	140.8	133.0	124.6	109.0	105.6	110.5	107.8	98.0	103.5	100.5	115.8	99.1	102.7
2002	126.7	133.5	136.2	133.9	129.4	120.3	107.1	104.6	107.4	106.7	97.3	102.8	99.7	113.0	98.3	101.9
2001	122.2	128.2	131.5	128.8	124.8	115.9	104.3	100.8	104.6	103.8	94.8	100.3	96.4	109.1	95.5	99.0
2000	118.9	121.5	126.4	123.2	117.4	113.8	102.4	99.0	101.8	101.2	93.6	98.9	94.9	106.1	94.1	97.0
1999	116.6	120.0	122.0	121.2	116.6	111.5	101.4	98.0	100.8	100.1	92.5	97.9	93.5	102.9	92.8	95.8
1998	113.6	118.8	120.8	120.1	112.3	109.6	99.4	96.2	99.0	98.4	90.9	96.3	91.4	100.8	90.4	93.3
1997	111.5	118.9	120.9	120.2	110.7	106.4	98.6	95.4	98.2	97.2	89.9	95.5	89.9	98.5	88.7	91.6
1996	108.9	117.0	119.0	118.4	108.6	105.4	95.9	92.6	95.9	95.1	88.0	93.6	86.6	94.3	83.9	88.3
1995	105.6	115.5	117.9	117.3	106.1	102.3	94.0	90.9	93.7	93.5	86.5	92.2	85.0	92.0	82.6	86.8
1994	103.0	113.9	116.4	115.8	105.0	99.6	92.2	88.9	91.8	91.5	84.5	90.2	82.3	89.6	80.6	83.7
1993	100.0	108.8	110.6	104.8	101.5	96.3	87.4	86.1	87.1	88.5	82.1	87.7	79.5	85.7	77.8	80.9
1990	93.2	96.3	98.9	95.1	92.5	90.4	83.9	81.1	84.0	82.9	78.4	85.0	75.0	80.4	74.0	76.9
1985	81.8	85.3	86.8	85.6	81.1	78.8	76.7	73.4	78.3	73.9	70.6	77.3	66.9	70.3	66.1	68.1
1980	60.7	60.7	60.9	62.3	58.5	59.6	55.3	55.8	56.5	56.7	53.5	57.2	51.7	54.0	51.1	51.3
1975	43.7	44.7	45.0	46.3	42.9	43.7	42.1	40.3	43.2	41.5	38.1	41.3	37.5	38.4	36.2	36.5
1970	27.8	28.1	28.2	28.9	27.0	26.3	25.7	22.8	27.0	26.2	24.5	24.2	23.8	25.2	22.8	23.4
1965	21.5	21.6	21.7	22.2	20.9	21.8	19.8	17.4	19.3	20.2	18.9	18.6	18.3	19.8	17.6	18.0
1960	19.5	19.7	19.7	20.2	18.9	19.4	18.0	15.8	17.6	18.3	17.2	16.9	16.7	17.1	16.0	16.4
1955	16.3	16.5	16.5	17.0	15.9	16.3	15.1	13.2	14.7	15.4	14.4	14.1	14.0	14.4	13.4	13.7
1950	13.5	13.6	13.7	14.0	13.1	13.4	12.5	11.0	12.2	12.7	11.9	11.7	11.5	11.9	11.0	11.3
1945	8.6	8.7	8.7	8.9	8.4	8.6	7.9	7.0	7.8	8.1	7.6	7.5	7.4	7.6	7.0	7.2
1940	6.6	6.7	6.7	6.9	6.4	6.6	6.1	5.4	6.0	6.2	5.8	5.7	5.7	5.8	5.5	5.6

Year	National 30 City Average	Georgia Savan-nah	Hawaii Hono-lulu	Idaho Boise	Poca-tello	Illinois Chicago	Decatur	Joliet	Peoria	Rock-ford	Spring-field	Indiana Ander-son	Evans-ville	Fort Wayne	Gary	Indian-apolis
Jan 2013	197.6E	161.7E	232.0E	173.5E	174.4E	230.1E	199.2E	226.3E	203.1E	215.4E	200.7E	177.5E	180.8E	174.1E	200.5E	183.0E
2012	194.0	158.9	227.9	170.4	171.5	226.2	195.7	222.7	199.7	211.8	197.2	175.2	177.8	171.8	197.9	180.6
2011	185.7	151.7	219.0	162.8	163.6	217.6	187.9	216.8	193.4	205.8	189.4	168.4	169.7	165.0	191.0	173.0
2010	181.6	147.9	214.6	161.7	162.4	210.6	181.2	208.3	186.2	197.6	182.2	162.8	165.9	160.0	183.7	168.1
2009	182.5	152.8	218.6	162.6	163.4	209.0	182.2	205.9	186.2	196.5	184.0	164.9	166.9	163.0	185.2	170.3
2008	171.0	141.1	205.5	153.2	152.8	195.7	168.7	185.5	172.2	180.2	168.5	152.4	155.8	151.0	169.0	159.2
2007	165.0	136.9	200.4	148.0	148.1	186.5	160.5	177.7	166.0	175.0	159.2	149.0	152.6	147.7	165.4	154.1
2006	156.2	124.0	191.9	141.6	141.2	177.8	153.8	168.7	155.2	163.3	152.8	139.8	144.4	139.1	154.5	144.6
2005	146.7	116.9	181.2	133.8	132.5	164.4	145.0	158.3	146.5	151.2	145.4	131.4	136.2	131.0	146.4	137.4
2004	132.8	105.6	161.2	122.0	120.5	149.3	129.3	145.1	133.9	138.1	130.3	120.9	123.5	120.1	132.2	125.0
2003	129.7	103.0	159.4	120.2	118.7	146.2	127.4	143.5	132.4	137.3	128.4	119.7	121.4	118.2	131.4	122.5
2002	126.7	102.0	157.2	118.3	117.1	141.2	123.8	138.5	129.6	132.9	125.3	117.5	119.0	116.4	129.1	120.5
2001	122.2	99.0	150.0	114.3	113.4	135.8	120.1	133.7	124.3	127.8	119.8	113.4	115.6	112.1	123.4	116.4
2000	118.9	97.5	144.8	112.9	112.1	131.2	115.1	124.6	119.0	122.2	116.2	109.8	111.5	108.4	117.8	113.2
1999	116.6	96.0	143.0	110.2	109.6	129.6	113.0	122.6	116.4	120.7	113.8	107.1	109.3	106.8	112.8	110.6
1998	113.6	93.7	140.4	107.4	107.0	125.2	110.1	119.8	113.8	115.5	111.1	105.0	107.2	104.5	111.1	108.0
1997	111.5	92.0	139.8	104.6	104.7	121.3	107.8	117.5	111.5	113.1	108.9	101.9	104.4	101.8	110.3	105.2
1996	108.9	88.6	134.5	102.2	102.1	118.8	106.6	116.2	109.3	111.5	106.5	100.0	102.1	99.9	107.5	102.7
1995	105.6	87.4	130.3	99.5	98.2	114.2	98.5	110.5	102.3	103.6	98.1	96.4	97.2	95.0	100.7	100.1
1994	103.0	85.3	124.0	94.8	95.0	111.3	97.3	108.9	100.9	102.2	97.0	93.6	95.8	93.6	99.1	97.1
1993	100.0	82.0	122.0	92.2	92.1	107.6	95.6	106.8	98.9	99.6	95.2	91.2	94.3	91.5	96.7	93.9
1990	93.2	77.9	104.7	88.2	88.1	98.4	90.9	98.4	93.7	94.0	90.1	84.6	89.3	83.4	88.4	87.1
1985	81.8	68.9	94.7	78.0	78.0	82.4	81.9	83.4	83.7	83.0	81.5	75.2	79.8	75.0	77.8	77.1
1980	60.7	52.2	68.9	60.3	59.5	62.8	62.3	63.4	64.5	61.6	61.1	56.5	59.0	56.7	59.8	57.9
1975	43.7	36.9	44.6	40.8	40.5	45.7	43.1	44.5	44.7	42.7	42.4	39.5	41.7	39.9	41.9	40.6
1970	27.8	21.0	30.4	26.7	26.6	29.1	28.0	28.6	29.0	27.5	27.6	25.4	26.4	25.5	27.1	26.2
1965	21.5	16.4	21.8	20.6	20.5	22.7	21.5	22.1	22.4	21.2	21.3	19.5	20.4	19.7	20.8	20.7
1960	19.5	14.9	19.8	18.7	18.6	20.2	19.6	20.0	20.3	19.2	19.3	17.7	18.7	17.9	18.9	18.4
1955	16.3	12.5	16.6	15.7	15.6	16.9	16.4	16.8	17.0	16.1	16.2	14.9	15.7	15.0	15.9	15.5
1950	13.5	10.3	13.7	13.0	12.9	14.0	13.6	13.9	14.0	13.3	13.4	12.3	12.9	12.4	13.1	12.8
1945	8.6	6.6	8.8	8.3	8.2	8.9	8.6	8.9	9.0	8.5	8.6	7.8	8.3	7.9	8.4	8.1
1940	6.6	5.1	6.8	6.4	6.3	6.9	6.7	6.8	6.9	6.5	6.6	6.0	6.4	6.1	6.5	6.3

473

Historical Cost Indexes

Year	National 30 City Average	Indiana Muncie	Indiana South Bend	Indiana Terre Haute	Iowa Cedar Rapids	Iowa Daven-port	Iowa Des Moines	Iowa Sioux City	Iowa Water-loo	Kansas Topeka	Kansas Wichita	Kentucky Lexing-ton	Kentucky Louis-ville	Louisiana Baton Rouge	Louisiana Lake Charles	Louisiana New Orleans
Jan 2013	197.6E	178.7E	176.8E	182.0E	183.0E	187.9E	182.4E	171.3E	172.0E	165.2E	164.2E	177.8E	180.5E	167.0E	1664. E	174.6E
2012	194.0	175.8	174.3	178.9	180.3	184.6	179.2	168.8	169.2	162.3	161.4	174.8	177.5	164.4	163.8	171.7
2011	185.7	169.2	168.1	171.3	173.8	178.1	172.7	162.3	162.8	156.3	155.3	167.8	170.8	156.9	156.5	162.1
2010	181.6	163.1	163.5	167.8	170.4	175.6	170.2	159.0	160.2	152.8	151.3	158.7	166.9	153.7	153.4	160.3
2009	182.5	164.5	167.1	168.0	165.9	172.0	162.1	156.3	147.6	154.5	152.4	160.7	167.5	156.9	154.4	161.9
2008	171.0	153.0	153.1	155.9	157.3	163.4	152.4	147.8	139.2	144.9	143.1	152.0	156.8	144.0	141.5	149.3
2007	165.0	149.7	149.5	152.7	152.4	157.6	148.9	144.0	135.0	141.0	138.7	147.6	150.9	139.4	138.0	145.2
2006	156.2	141.2	141.3	144.0	145.8	151.1	142.9	137.3	128.4	134.7	132.9	129.0	143.0	129.4	129.8	135.8
2005	146.7	131.0	133.8	135.2	135.7	142.1	133.5	129.4	120.0	125.9	125.6	121.5	135.1	120.8	117.7	126.2
2004	132.8	120.0	120.6	123.0	122.7	128.6	121.6	116.2	108.3	112.6	113.4	110.1	120.3	105.6	109.0	115.3
2003	129.7	118.7	119.2	121.7	120.9	126.4	120.2	114.6	105.9	109.4	111.3	107.8	118.7	103.3	106.1	112.4
2002	126.7	116.7	117.1	119.9	116.0	122.1	116.7	111.4	103.7	107.8	109.6	106.2	116.4	102.5	105.0	110.6
2001	122.2	112.8	111.6	115.5	112.5	117.5	113.2	107.7	100.6	104.3	105.2	103.3	112.7	99.4	101.3	104.6
2000	118.9	109.1	106.9	110.8	108.8	112.6	108.9	99.0	98.1	101.0	101.1	101.4	109.3	97.8	99.7	102.1
1999	116.6	105.7	103.7	107.9	104.1	109.2	107.6	96.7	96.4	98.7	99.3	99.7	106.6	96.1	97.6	99.5
1998	113.6	103.6	102.3	106.0	102.5	106.8	103.8	95.1	94.8	97.4	97.4	97.4	101.5	94.7	96.1	97.7
1997	111.5	101.3	101.2	104.6	101.2	105.6	102.5	93.8	93.5	96.3	96.2	96.1	99.9	93.3	96.8	96.2
1996	108.9	99.4	99.1	102.1	99.1	102.0	99.1	90.9	90.7	93.6	93.8	94.4	98.3	91.5	94.9	94.3
1995	105.6	95.6	94.6	97.2	95.4	96.6	94.4	88.2	88.9	91.1	91.0	91.6	94.6	89.6	92.7	91.6
1994	103.0	93.2	93.0	95.8	93.0	92.3	92.4	85.9	86.8	89.4	88.1	89.6	92.2	87.7	89.9	89.5
1993	100.0	91.0	90.7	94.4	91.1	90.5	90.7	84.1	85.2	87.4	86.2	87.3	89.4	86.4	88.5	87.8
1990	93.2	83.9	85.1	89.1	87.1	86.5	86.9	80.4	80.3	83.2	82.6	82.8	82.6	82.0	84.1	84.5
1985	81.8	75.2	76.0	79.2	75.9	77.6	77.1	72.1	72.3	75.0	74.7	75.5	74.5	74.9	78.5	78.2
1980	60.7	56.1	58.3	59.7	62.7	59.6	61.8	59.3	57.7	58.9	58.0	59.3	59.8	59.1	60.0	57.2
1975	43.7	39.2	40.3	41.9	43.1	41.0	43.1	41.0	40.0	42.0	43.1	42.7	42.5	40.4	40.6	41.5
1970	27.8	25.3	26.2	27.0	28.2	26.9	27.6	26.6	25.9	27.0	25.5	26.9	25.9	25.0	26.7	27.2
1965	21.5	19.5	20.2	20.9	21.7	20.8	21.7	20.5	20.0	20.8	19.6	20.7	20.3	19.4	20.6	20.4
1960	19.5	17.8	18.3	18.9	19.7	18.8	19.5	18.6	18.2	18.9	17.8	18.8	18.4	17.6	18.7	18.5
1955	16.3	14.9	15.4	15.9	16.6	15.8	16.3	15.6	15.2	15.8	15.0	15.8	15.4	14.8	15.7	15.6
1950	13.5	12.3	12.7	13.1	13.7	13.1	13.5	12.9	12.6	13.1	12.4	13.0	12.8	12.2	13.0	12.8
1945	8.6	7.8	8.1	8.4	8.7	8.3	8.6	8.2	8.0	8.3	7.9	8.3	8.1	7.8	8.3	8.2
1940	6.6	6.0	6.3	6.5	6.7	6.4	6.6	6.3	6.2	6.4	6.1	6.4	6.3	6.0	6.4	6.3

Year	National 30 City Average	Louisiana Shreve-port	Maine Lewis-ton	Maine Portland	Maryland Balti-more	Mass. Boston	Mass. Brockton	Mass. Fall River	Mass. Law-rence	Mass. Lowell	Mass. New Bedford	Mass. Pitts-field	Mass. Spring-field	Mass. Wor-cester	Michigan Ann Arbor	Michigan Dear-born
Jan 2013	197.6E	155.8E	186.6E	188.4E	183.8E	233.4E	221.3E	221.8E	225.7E	223.4E	221.0E	202.9E	205.4E	218.2E	199.4E	202.6E
2012	194.0	153.2	180.7	182.4	180.8	229.3	217.2	217.8	221.6	219.3	217.0	200.2	201.9	215.2	196.1	199.2
2011	185.7	146.7	167.4	169.1	173.6	219.3	208.2	206.0	211.6	210.1	205.2	191.2	194.5	206.6	186.7	189.4
2010	181.6	145.1	162.1	163.7	168.3	214.3	202.7	200.7	204.9	204.7	200.0	185.9	189.4	201.1	183.3	186.8
2009	182.5	147.6	159.4	161.8	169.1	211.8	199.2	197.6	201.9	201.9	196.9	185.1	187.1	198.2	179.2	186.9
2008	171.0	135.9	149.1	150.9	157.7	198.6	186.4	185.1	188.7	189.1	184.5	171.3	173.7	184.5	169.6	176.4
2007	165.0	132.4	146.4	148.2	152.5	191.8	180.4	179.7	182.0	181.7	179.1	166.8	169.1	180.4	166.8	172.5
2006	156.2	125.2	140.2	139.8	144.9	180.4	171.6	170.4	172.3	173.5	170.4	157.3	159.7	171.6	161.1	165.5
2005	146.7	117.4	131.7	131.4	135.8	169.6	161.3	159.6	162.2	162.3	159.6	147.6	150.7	158.8	147.5	155.8
2004	132.8	105.7	119.8	119.4	121.4	154.1	144.4	143.6	146.4	146.7	143.6	131.9	136.5	143.2	135.9	142.1
2003	129.7	103.9	117.7	117.3	118.1	150.2	139.6	140.6	143.3	143.6	140.5	128.9	133.8	139.0	134.1	139.0
2002	126.7	102.1	117.5	117.1	115.6	145.6	136.0	135.0	136.9	137.3	134.9	124.7	128.3	134.9	131.4	134.3
2001	122.2	98.3	114.6	114.3	111.7	140.9	132.1	131.4	133.1	132.9	131.3	120.3	124.5	130.1	126.8	129.8
2000	118.9	96.2	105.7	105.4	107.7	138.9	129.4	128.4	129.7	130.3	128.3	116.7	121.0	127.3	124.3	125.8
1999	116.6	94.8	105.0	104.7	106.4	136.2	126.7	126.3	127.3	127.4	126.2	115.0	118.8	123.8	117.5	122.7
1998	113.6	92.0	102.8	102.5	104.1	132.8	125.0	124.7	125.0	125.3	124.6	114.2	117.1	122.6	116.0	119.7
1997	111.5	90.1	101.7	101.4	102.2	132.1	124.4	124.7	125.1	125.6	124.7	114.6	117.4	122.9	113.5	117.9
1996	108.9	88.5	99.5	99.2	99.6	128.7	121.6	122.2	121.7	121.1	122.2	112.7	115.0	120.2	112.9	116.2
1995	105.6	86.7	96.8	96.5	96.1	128.6	119.6	117.7	120.5	119.6	117.2	110.7	112.7	114.8	106.1	110.5
1994	103.0	85.0	95.2	94.9	94.4	124.9	114.2	113.7	116.8	116.9	111.6	109.2	111.1	112.9	105.0	109.0
1993	100.0	83.6	93.0	93.0	93.1	121.1	111.7	111.5	114.9	114.3	109.4	107.1	108.9	110.3	102.8	105.8
1990	93.2	80.3	88.5	88.6	85.6	110.9	103.7	102.9	105.2	102.8	102.6	98.7	101.4	103.2	93.2	96.5
1985	81.8	73.6	76.7	77.0	72.7	92.8	88.8	88.7	89.4	88.2	88.6	85.0	85.6	86.7	80.1	82.7
1980	60.7	58.7	57.3	58.5	53.6	64.0	63.7	64.1	63.4	62.7	63.1	61.8	62.0	62.3	62.9	64.0
1975	43.7	40.5	41.9	42.1	39.8	46.6	45.8	45.7	46.2	45.7	46.1	45.5	45.8	46.0	44.1	44.9
1970	27.8	26.4	26.5	25.8	25.1	29.2	29.1	29.2	29.1	28.9	29.0	28.6	28.5	28.7	28.5	28.9
1965	21.5	20.3	20.4	19.4	20.2	23.0	22.5	22.5	22.5	22.2	22.4	22.0	22.4	22.1	22.0	22.3
1960	19.5	18.5	18.6	17.6	17.5	20.5	20.4	20.4	20.4	20.2	20.3	20.0	20.1	20.1	20.0	20.2
1955	16.3	15.5	15.6	14.7	14.7	17.2	17.1	17.1	17.1	16.9	17.1	16.8	16.9	16.8	16.7	16.9
1950	13.5	12.8	12.9	12.2	12.1	14.2	14.1	14.1	14.1	14.0	14.1	13.8	14.0	13.9	13.8	14.0
1945	8.6	8.1	8.2	7.8	7.7	9.1	9.0	9.0	9.0	8.9	9.0	8.9	9.0	8.9	8.8	8.9
1940	6.6	6.3	6.3	6.0	6.0	7.0	6.9	7.0	7.0	6.9	6.9	6.8	6.9	6.8	6.8	6.9

Historical Cost Indexes

Year	National 30 City Average	Michigan						Minnesota			Mississippi		Missouri			
		Detroit	Flint	Grand Rapids	Kala-mazoo	Lansing	Sagi-naw	Duluth	Minne-apolis	Roches-ter	Biloxi	Jackson	Kansas City	St. Joseph	St. Louis	Spring-field
Jan 2013	197.6E	203.8E	185.4E	179.4E	181.9E	189.3E	186.2E	206.2E	218.3E	203.1E	160.1E	163.1E	204.5E	192.5E	201.7E	181.1E
2012	194.0	200.2	186.3	176.4	178.9	185.9	182.8	203.1	214.7	199.7	157.6	160.5	200.8	189.0	198.0	177.7
2011	185.7	190.3	178.5	169.9	171.9	178.0	174.6	195.7	208.1	193.8	151.4	154.3	191.1	178.8	190.6	168.3
2010	181.6	187.5	176.2	160.6	167.3	175.8	171.9	193.1	203.8	188.9	148.0	151.0	186.1	174.2	185.9	164.3
2009	182.5	187.8	176.0	156.7	166.3	173.7	168.6	191.8	203.1	188.0	152.8	156.6	185.6	172.3	187.2	163.7
2008	171.0	177.0	164.4	140.2	155.7	161.6	158.7	177.4	190.6	175.0	141.8	147.1	175.5	164.2	176.2	152.9
2007	165.0	172.9	161.7	137.2	152.6	158.7	156.1	174.2	184.5	171.4	137.7	131.2	169.0	157.8	170.6	145.3
2006	156.2	165.8	153.8	131.1	146.0	152.5	150.6	166.2	173.9	161.9	123.3	117.1	162.0	152.4	159.3	139.4
2005	146.7	156.2	145.1	123.1	134.9	142.0	140.6	157.3	164.6	152.8	116.0	109.6	151.3	143.1	149.4	131.7
2004	132.8	141.9	129.6	112.8	122.5	129.7	127.6	139.1	150.1	137.0	105.3	99.3	135.1	126.9	135.5	116.3
2003	129.7	138.7	127.9	110.9	120.8	127.7	126.0	136.7	146.5	134.5	101.4	96.7	131.9	124.8	133.3	113.9
2002	126.7	134.2	126.9	107.4	119.9	125.3	124.4	131.9	139.5	130.0	101.0	96.3	128.4	122.7	129.6	112.5
2001	122.2	129.4	122.4	104.3	115.2	120.1	119.7	131.4	136.1	124.9	98.7	93.9	121.8	114.7	125.5	106.7
2000	118.9	125.3	119.8	102.7	111.6	117.6	115.9	124.1	131.1	120.1	97.2	92.9	118.2	111.9	122.4	104.3
1999	116.6	122.6	113.7	100.9	106.1	111.5	110.1	120.3	126.5	117.1	95.7	91.8	114.9	106.1	119.8	101.1
1998	113.6	119.5	112.3	99.7	104.9	110.1	108.7	117.7	124.6	115.5	92.1	89.5	108.2	104.2	115.9	98.9
1997	111.5	117.6	110.8	98.9	104.2	108.7	107.5	115.9	121.9	114.1	90.9	88.3	106.4	102.4	113.2	97.3
1996	108.9	116.0	109.7	93.9	103.5	107.8	106.9	115.0	120.4	113.5	87.1	85.4	103.2	99.9	110.1	94.6
1995	105.6	110.1	104.1	91.1	96.4	98.3	102.3	100.3	111.9	102.5	84.2	83.5	99.7	96.1	106.3	89.5
1994	103.0	108.5	102.9	89.7	95.0	97.1	101.1	99.1	109.3	101.1	82.4	81.7	97.3	92.8	103.2	88.2
1993	100.0	105.4	100.7	87.7	92.7	95.3	98.8	99.8	106.7	99.9	80.0	79.4	94.5	90.9	99.9	86.2
1990	93.2	96.0	91.7	84.0	87.4	90.2	88.9	94.3	100.5	95.6	76.4	75.6	89.5	86.8	94.0	82.6
1985	81.8	81.6	80.7	75.8	79.4	80.0	80.5	85.2	87.9	86.5	69.4	68.5	78.2	77.3	80.8	73.2
1980	60.7	62.5	63.1	59.8	61.7	60.3	63.3	63.4	64.1	64.8	54.3	54.3	59.1	60.4	59.7	56.7
1975	43.7	45.8	44.9	41.8	43.7	44.1	44.8	45.2	46.0	45.3	37.9	37.7	42.7	45.4	44.8	41.1
1970	27.8	29.7	28.5	26.7	28.1	27.8	28.7	28.9	29.5	28.9	24.4	21.3	25.3	28.3	28.4	26.0
1965	21.5	22.1	21.9	20.6	21.7	21.5	22.2	22.3	23.4	22.3	18.8	16.4	20.6	21.8	21.8	20.0
1960	19.5	20.1	20.0	18.7	19.7	19.5	20.1	20.3	20.5	20.2	17.1	14.9	19.0	19.8	19.5	18.2
1955	16.3	16.8	16.7	15.7	16.5	16.3	16.9	17.0	17.2	17.0	14.3	12.5	16.0	16.6	16.3	15.2
1950	13.5	13.9	13.8	13.0	13.6	13.5	13.9	14.0	14.2	14.0	11.8	10.3	13.2	13.7	13.5	12.6
1945	8.6	8.9	8.8	8.3	8.7	8.6	8.9	8.9	9.0	8.9	7.6	6.6	8.4	8.8	8.6	8.0
1940	6.6	6.8	6.8	6.4	6.7	6.6	6.9	6.9	7.0	6.9	5.8	5.1	6.5	6.7	6.7	6.2

Year	National 30 City Average	Montana		Nebraska		Nevada		New Hampshire		New Jersey					NM	NY
		Billings	Great Falls	Lincoln	Omaha	Las Vegas	Reno	Man-chester	Nashua	Camden	Jersey City	Newark	Pater-son	Trenton	Albu-querque	Albany
Jan 2013	197.6E	181.6E	182.5E	172.5E	180.1E	205.6E	189.6E	192.6E	191.7E	216.3E	218.0E	222.1E	220.3E	216.7E	173.6E	194.1E
2012	194.0	178.6	179.5	169.3	177.1	202.6	186.8	189.2	188.3	213.6	215.2	219.2	217.5	213.7	171.0	191.2
2011	185.7	168.8	170.8	162.8	169.6	195.7	179.8	176.8	176.1	205.5	207.0	210.3	209.2	206.4	163.3	180.9
2010	181.6	166.5	168.6	159.3	165.8	193.7	175.6	172.5	172.0	201.2	203.3	206.3	205.3	200.8	162.5	177.8
2009	182.5	165.3	166.0	161.7	165.0	191.5	176.5	174.0	173.5	196.8	200.5	203.6	202.0	198.6	163.0	178.1
2008	171.0	153.1	154.6	152.0	154.8	176.2	167.0	162.5	161.9	184.0	187.3	189.6	188.7	185.6	152.6	166.0
2007	165.0	147.7	147.8	147.7	150.3	166.7	161.4	159.3	158.7	179.3	183.3	185.2	184.6	181.7	146.7	159.4
2006	156.2	140.5	140.8	132.4	140.7	160.0	154.5	146.7	146.4	167.7	171.0	173.6	172.2	169.9	140.1	151.6
2005	146.7	131.9	132.4	124.1	132.8	149.5	145.4	136.8	136.5	159.0	162.4	163.9	163.0	161.5	130.4	142.2
2004	132.8	118.2	117.9	112.1	119.4	137.1	130.5	124.2	123.9	143.4	145.4	147.2	146.6	146.0	118.3	128.6
2003	129.7	115.8	115.8	110.2	117.3	133.8	128.3	122.4	122.2	142.0	144.3	145.7	145.0	143.0	116.4	126.8
2002	126.7	114.6	114.5	108.2	115.0	131.9	126.4	119.9	119.6	135.9	138.5	140.2	140.1	137.6	114.6	122.6
2001	122.2	117.5	117.7	101.3	111.6	127.8	122.5	116.2	116.2	133.4	136.7	136.9	136.8	136.0	111.4	119.2
2000	118.9	113.7	113.9	98.8	107.0	125.8	118.2	111.9	111.9	128.4	130.5	132.6	132.4	130.6	109.0	116.5
1999	116.6	112.1	112.7	96.6	104.8	121.9	114.4	109.6	109.6	125.3	128.8	131.6	129.5	129.6	106.7	114.6
1998	113.6	109.7	109.1	94.9	101.2	118.1	111.4	110.1	110.0	124.3	127.7	128.9	128.5	127.9	103.8	113.0
1997	111.5	107.4	107.6	93.3	99.5	114.6	109.8	108.6	108.6	121.9	125.1	126.4	126.4	125.2	100.8	110.0
1996	108.9	108.0	107.5	91.4	97.4	111.8	108.9	106.5	106.4	119.1	122.5	122.5	123.8	121.8	98.6	108.3
1995	105.6	104.7	104.9	85.6	93.4	108.5	105.0	100.9	100.8	107.0	112.2	111.9	112.1	111.2	96.3	103.6
1994	103.0	100.2	100.7	84.3	91.0	105.3	102.2	97.8	97.7	105.4	110.6	110.1	110.6	108.2	93.4	102.4
1993	100.0	97.9	97.8	82.3	88.7	102.8	99.9	95.4	95.4	103.7	109.0	108.2	109.0	106.4	90.1	99.8
1990	93.2	92.9	93.8	78.8	85.0	96.3	94.5	86.3	86.3	93.0	93.5	93.8	97.2	94.5	84.9	93.2
1985	81.8	83.9	84.3	71.5	77.5	87.6	85.0	78.1	78.1	81.3	83.3	83.5	84.5	82.8	76.7	79.5
1980	60.7	63.9	64.6	58.5	63.5	64.6	63.0	56.6	56.0	58.6	60.6	60.1	60.0	58.9	59.0	59.5
1975	43.7	43.1	43.8	40.9	43.2	42.8	41.9	41.3	40.8	42.3	43.4	44.2	43.9	43.8	40.3	43.9
1970	27.8	28.5	28.9	26.4	26.8	29.4	28.0	26.2	25.6	27.2	27.8	29.0	27.8	27.4	26.4	28.3
1965	21.5	22.0	22.3	20.3	20.6	22.4	21.6	20.6	19.7	20.9	21.4	23.8	21.4	21.3	20.6	22.3
1960	19.5	20.0	20.3	18.5	18.7	20.2	19.6	18.0	17.9	19.0	19.4	19.4	19.4	19.2	18.5	19.3
1955	16.3	16.7	17.0	15.5	15.7	16.9	16.4	15.1	15.0	16.0	16.3	16.3	16.3	16.1	15.6	16.2
1950	13.5	13.9	14.0	12.8	13.0	14.0	13.6	12.5	12.4	13.2	13.5	13.5	13.5	13.3	12.9	13.4
1945	8.6	8.8	9.0	8.1	8.3	8.9	8.7	8.0	7.9	8.4	8.6	8.6	8.6	8.5	8.2	8.5
1940	6.6	6.8	6.9	6.3	6.4	6.9	6.7	6.2	6.1	6.5	6.6	6.6	6.6	6.6	6.3	6.6

Historical Cost Indexes

Year	National 30 City Average	New York Binghamton	Buffalo	New York	Rochester	Schenectady	Syracuse	Utica	Yonkers	Charlotte	Durham	Greensboro	Raleigh	Winston-Salem	N. Dakota Fargo	Ohio Akron
Jan 2013	197.6E	195.6E	201.1E	260.6E	194.2E	194.1E	192.8E	188.0E	229.7E	158.3E	157.8E	158.4E	157.4E	159.3E	170.0E	191.0E
2012	194.0	192.9	198.1	256.3	191.2	191.4	189.6	185.2	227.1	156.4	155.6	156.3	155.4	157.2	167.3	187.7
2011	185.7	176.1	188.1	245.6	181.9	181.5	180.8	173.5	218.9	142.6	143.8	142.2	142.8	140.9	160.3	180.8
2010	181.6	173.8	184.0	241.4	179.6	179.0	176.7	170.5	217.1	140.5	141.6	140.0	140.6	138.8	156.2	174.5
2009	182.5	172.9	184.4	239.9	179.6	178.8	176.9	171.3	217.5	145.1	145.8	144.0	145.1	135.3	154.3	175.7
2008	171.0	160.7	173.9	226.8	168.0	167.2	165.9	161.6	202.1	135.8	136.6	134.6	135.3	133.8	145.0	165.4
2007	165.0	155.7	168.6	215.2	163.5	159.8	159.8	155.0	196.5	132.8	133.7	131.7	132.1	131.0	139.8	159.0
2006	156.2	147.5	159.2	204.5	155.2	151.1	150.9	146.4	186.0	125.1	124.6	123.8	124.2	123.0	133.2	152.8
2005	146.7	137.5	149.4	194.0	147.3	142.2	141.9	136.9	176.1	110.5	112.1	112.1	112.1	111.1	125.3	144.4
2004	132.8	123.5	136.1	177.7	132.0	128.4	127.3	124.4	161.8	98.9	100.0	100.1	100.3	99.4	113.0	131.9
2003	129.7	121.8	132.8	173.4	130.3	126.7	124.7	121.7	160.0	96.2	97.5	97.5	97.8	96.8	110.6	129.6
2002	126.7	119.0	128.5	170.1	127.1	123.0	121.8	118.4	154.8	94.8	96.1	96.1	96.3	95.5	106.3	127.2
2001	122.2	116.0	125.2	164.4	123.1	120.1	118.6	115.3	151.4	91.5	92.9	92.9	93.3	92.3	103.1	123.5
2000	118.9	112.4	122.3	159.2	120.0	117.5	115.1	112.4	144.8	90.3	91.6	91.6	91.9	91.0	97.9	117.8
1999	116.6	108.6	120.2	155.9	116.8	114.7	113.7	108.5	140.6	89.3	90.2	90.3	90.5	89.0	95.2	113.1
1998	113.6	109.0	119.1	154.4	117.2	114.2	113.6	108.6	141.4	88.2	89.1	89.2	89.4	89.0	95.2	113.1
1997	111.5	107.2	115.7	150.3	115.2	111.2	110.6	107.0	138.8	86.8	87.7	87.8	87.9	87.6	93.5	110.6
1996	108.9	105.5	114.0	148.0	113.3	109.5	108.2	105.2	137.2	85.0	85.9	86.0	86.1	85.8	91.7	107.9
1995	105.6	99.3	110.1	140.7	106.6	104.6	104.0	97.7	129.4	81.8	82.6	82.6	82.7	82.6	88.4	103.4
1994	103.0	98.1	107.2	137.1	105.2	103.5	102.3	96.5	128.1	80.3	81.1	81.0	81.2	81.1	87.0	102.2
1993	100.0	95.7	102.2	133.3	102.0	100.8	99.6	94.1	126.3	78.2	78.9	78.9	78.9	78.8	85.8	100.3
1990	93.2	87.0	94.1	118.1	94.6	93.9	91.0	85.6	111.4	74.8	75.4	75.5	75.5	75.3	81.3	94.6
1985	81.8	77.5	83.2	94.9	81.6	80.1	81.0	77.0	92.8	66.9	67.6	67.7	67.6	67.5	73.4	86.8
1980	60.7	58.0	60.6	66.0	60.5	60.3	61.6	58.5	65.9	51.1	52.2	52.5	51.7	50.8	57.4	62.3
1975	43.7	42.5	45.1	49.5	44.8	43.7	44.8	42.7	47.1	36.1	37.0	37.0	37.3	36.1	39.3	44.7
1970	27.8	27.0	28.9	32.8	29.2	27.8	28.5	26.9	30.0	20.9	23.7	23.6	23.4	23.0	25.8	28.3
1965	21.5	20.8	22.2	25.5	22.8	21.4	21.9	20.7	23.1	16.0	18.2	18.2	18.0	17.7	19.9	21.8
1960	19.5	18.9	19.9	21.5	19.7	19.5	19.9	18.8	21.0	14.4	16.6	16.6	16.4	16.1	18.1	19.9
1955	16.3	15.8	16.7	18.1	16.5	16.3	16.7	15.8	17.6	12.1	13.9	13.9	13.7	13.5	15.2	16.6
1950	13.5	13.1	13.8	14.9	13.6	13.5	13.8	13.0	14.5	10.0	11.5	11.5	11.4	11.2	12.5	13.7
1945	8.6	8.3	8.8	9.5	8.7	8.6	8.8	8.3	9.3	6.4	7.3	7.3	7.2	7.1	8.0	8.8
1940	6.6	6.4	6.8	7.4	6.7	6.7	6.8	6.4	7.2	4.9	5.6	5.6	5.6	5.5	6.2	6.8

Year	National 30 City Average	Ohio Canton	Cincinnati	Cleveland	Columbus	Dayton	Lorain	Springfield	Toledo	Youngstown	Oklahoma Lawton	Oklahoma City	Tulsa	Oregon Eugene	Portland	PA Allentown
Jan 2013	197.6E	182.8E	181.8E	194.8E	186.8E	180.8E	187.4E	181.2E	194.0E	186.3E	160.7E	160.9E	153.8E	193.6E	195.3E	203.0E
2012	194.0	180.2	178.6	191.4	183.5	177.6	184.1	178.0	190.4	183.3	158.1	158.0	151.3	190.1	191.8	199.3
2011	185.7	172.1	171.3	184.0	175.7	168.4	176.8	168.4	183.1	176.5	151.7	151.6	144.8	184.9	186.6	192.2
2010	181.6	167.5	166.5	180.2	170.3	163.4	173.0	164.6	176.5	171.7	149.5	149.4	143.1	181.5	182.9	187.9
2009	182.5	168.4	168.1	181.6	171.2	165.3	174.0	166.0	178.5	172.8	152.5	153.3	146.8	181.3	183.6	188.3
2008	171.0	158.4	159.0	170.6	160.2	157.5	163.9	157.5	168.2	161.5	141.3	140.6	136.2	172.3	174.7	175.7
2007	165.0	153.2	152.4	164.9	155.1	149.4	158.1	150.3	161.6	156.7	136.2	135.6	132.5	168.5	169.5	169.4
2006	156.2	147.0	144.9	156.9	146.9	142.9	151.8	143.4	154.5	150.3	129.5	129.5	125.7	160.5	161.7	160.1
2005	146.7	137.7	136.8	147.9	138.0	134.5	143.3	134.8	145.5	141.1	121.6	121.6	117.8	150.9	152.2	150.2
2004	132.8	125.6	124.3	135.6	126.2	121.1	131.5	122.0	132.8	129.3	109.3	109.4	107.3	136.8	137.8	133.9
2003	129.7	123.7	121.3	132.7	123.7	119.4	126.1	120.0	130.4	126.2	107.5	107.8	105.0	134.2	135.9	130.3
2002	126.7	120.9	119.2	130.0	120.9	117.5	124.3	118.0	128.4	123.6	106.3	106.0	104.1	132.2	133.9	128.1
2001	122.2	117.7	115.9	125.9	117.1	114.0	120.4	114.7	123.8	119.9	102.0	102.1	99.7	129.8	131.2	123.5
2000	118.9	112.7	110.1	121.3	112.5	109.0	115.0	109.2	115.7	114.1	98.7	98.9	97.5	126.3	127.4	119.9
1999	116.6	110.6	107.9	118.7	109.5	107.1	112.0	106.4	113.6	111.9	97.5	97.4	96.2	120.9	124.3	117.8
1998	113.6	108.5	105.0	114.8	106.8	104.5	109.4	104.1	111.2	109.0	94.3	94.5	94.5	120.0	122.2	115.7
1997	111.5	106.4	102.9	112.9	104.7	102.6	107.4	102.2	108.5	107.2	92.9	93.2	93.0	118.1	119.6	113.9
1996	108.9	103.2	100.2	110.1	101.1	98.8	103.8	98.4	105.1	104.1	90.9	91.3	91.2	114.4	116.1	112.0
1995	105.6	98.8	97.1	106.4	99.1	94.6	97.1	92.0	100.6	100.1	85.3	88.0	89.0	112.2	114.3	108.3
1994	103.0	97.7	95.0	104.8	95.4	93.0	96.0	90.1	99.3	98.9	84.0	86.5	87.6	107.6	109.2	106.4
1993	100.0	95.9	92.3	101.9	93.9	90.6	93.9	87.7	98.3	97.2	81.3	83.9	84.8	107.2	108.8	103.6
1990	93.2	92.1	86.7	95.1	88.1	85.9	89.4	83.4	92.9	91.8	77.1	79.9	80.0	98.0	99.4	95.0
1985	81.8	83.7	78.2	86.2	77.7	76.3	80.7	74.3	84.7	82.7	71.2	73.9	73.8	88.7	90.1	81.9
1980	60.7	60.6	59.8	61.0	58.6	56.6	60.8	56.2	64.0	61.5	52.2	55.5	57.2	68.9	68.4	58.7
1975	43.7	44.0	43.7	44.6	42.2	40.5	42.7	40.0	44.9	44.5	38.6	38.6	39.2	44.6	44.9	42.3
1970	27.8	27.8	28.2	29.6	26.7	27.0	27.5	25.4	29.1	29.6	24.1	23.4	24.1	30.4	30.0	27.3
1965	21.5	21.5	20.7	21.1	20.2	20.0	21.1	19.6	21.3	21.0	18.5	17.8	20.6	23.4	22.6	21.0
1960	19.5	19.5	19.3	19.6	18.7	18.1	19.2	17.8	19.4	19.1	16.9	16.1	18.1	21.2	21.1	19.1
1955	16.3	16.3	16.2	16.5	15.7	15.2	16.1	14.9	16.2	16.0	14.1	13.5	15.1	17.8	17.7	16.0
1950	13.5	13.5	13.3	13.6	13.0	12.5	13.3	12.3	13.4	13.2	11.7	11.2	12.5	14.7	14.6	13.2
1945	8.6	8.6	8.5	8.7	8.3	8.0	8.5	7.8	8.6	8.4	7.4	7.1	8.0	9.4	9.3	8.4
1940	6.6	6.7	6.5	6.7	6.4	6.2	6.5	6.1	6.6	6.5	5.7	5.5	6.1	7.2	7.2	6.5

Historical Cost Indexes

Year	National 30 City Average	Pennsylvania						RI	South Carolina		South Dakota		Tennessee			
		Erie	Harrisburg	Philadelphia	Pittsburgh	Reading	Scranton	Providence	Charleston	Columbia	Rapid City	Sioux Falls	Chattanooga	Knoxville	Memphis	Nashville
Jan 2013	197.6E	185.4E	191.4E	224.8E	200.4E	194.3E	195.0E	212.2E	157.4E	150.0E	159.3E	161.2E	166.3E	158.4E	168.6E	170.6E
2012	194.0	182.0	187.9	221.8	196.8	190.8	191.5	209.3	155.3	148.1	156.7	158.8	163.8	155.8	166.3	167.5
2011	185.7	176.0	180.9	212.8	187.7	184.3	185.3	196.8	149.4	142.3	149.5	151.3	157.3	149.3	158.4	160.7
2010	181.6	171.0	175.2	209.3	182.3	179.8	180.8	192.8	147.4	140.2	147.3	149.0	152.4	144.7	154.2	156.7
2009	182.5	171.9	176.5	209.5	181.7	180.7	181.3	192.9	150.8	144.3	148.2	151.1	155.7	148.0	157.2	159.8
2008	171.0	161.0	165.7	196.2	169.2	169.3	167.9	178.2	142.1	134.8	138.4	141.3	136.7	133.4	146.0	147.5
2007	165.0	156.0	159.4	188.4	163.2	164.9	163.1	174.4	139.3	131.7	129.5	133.0	133.7	130.6	143.1	144.4
2006	156.2	148.7	150.7	177.8	155.5	155.0	153.9	163.6	122.4	118.7	122.8	126.6	127.0	123.9	136.5	135.5
2005	146.7	140.6	141.5	166.4	146.6	145.5	142.9	155.6	110.2	109.6	115.2	118.5	116.0	115.1	128.6	127.9
2004	132.8	127.0	126.2	148.4	133.1	128.7	129.0	138.8	98.9	98.4	103.9	107.3	105.0	104.5	115.6	115.8
2003	129.7	124.5	123.8	145.5	130.5	126.3	126.0	135.9	96.1	95.7	101.3	104.0	102.8	102.3	111.1	110.9
2002	126.7	122.1	121.1	142.0	127.8	123.4	124.2	131.1	94.8	93.9	100.2	103.2	102.2	101.2	107.6	109.2
2001	122.2	118.9	118.4	136.9	124.1	120.4	121.1	127.8	92.1	91.3	96.8	99.7	98.2	96.5	102.6	104.3
2000	118.9	114.6	114.0	132.1	120.9	115.9	117.7	122.8	89.9	89.1	94.2	98.0	96.9	95.0	100.8	100.8
1999	116.6	113.8	112.8	129.8	119.6	114.8	116.3	121.6	89.0	88.1	92.4	95.8	95.9	93.4	99.7	98.6
1998	113.6	109.8	110.5	126.6	117.1	112.5	113.7	120.3	88.0	87.2	90.7	93.6	94.9	92.4	97.2	96.8
1997	111.5	108.5	108.8	123.3	113.9	110.8	112.3	118.9	86.5	85.6	89.3	92.1	93.6	90.8	96.1	94.9
1996	108.9	106.8	105.7	120.3	110.8	108.7	109.9	117.2	84.9	84.1	87.2	90.0	91.6	88.8	94.0	91.9
1995	105.6	99.8	100.6	117.1	106.3	103.6	103.8	111.1	82.7	82.2	84.0	84.7	89.2	86.0	91.2	87.6
1994	103.0	98.2	99.4	115.2	103.7	102.2	102.6	109.6	81.2	80.6	82.7	83.1	87.4	84.1	89.0	84.8
1993	100.0	94.9	96.6	107.4	99.0	98.6	99.8	108.2	78.4	77.9	81.2	81.9	85.1	82.3	86.8	81.9
1990	93.2	88.5	89.5	98.5	91.2	90.8	91.3	94.1	73.7	74.5	77.2	78.4	79.9	75.1	81.3	77.1
1985	81.8	79.6	77.2	82.2	81.5	77.7	80.0	83.0	65.9	66.9	69.7	71.2	72.5	67.7	74.3	66.7
1980	60.7	59.4	56.9	58.7	61.3	57.7	58.4	59.2	50.6	51.1	54.8	57.1	55.0	51.6	55.9	53.1
1975	43.7	43.4	42.2	44.5	44.5	43.6	41.9	42.7	35.0	36.4	37.7	39.6	39.1	37.3	40.7	37.0
1970	27.8	27.5	25.8	27.5	28.7	27.1	27.0	27.3	22.8	22.9	24.8	25.8	24.9	22.0	23.0	22.8
1965	21.5	21.1	20.1	21.7	22.4	20.9	20.8	21.9	17.6	17.6	19.1	19.9	19.2	17.1	18.3	17.4
1960	19.5	19.1	18.6	19.4	19.7	19.0	18.9	19.1	16.0	16.0	17.3	18.1	17.4	15.5	16.6	15.8
1955	16.3	16.1	15.6	16.3	16.5	15.9	15.9	16.0	13.4	13.4	14.5	15.1	14.6	13.0	13.9	13.3
1950	13.5	13.3	12.9	13.5	13.6	13.2	13.1	13.2	11.1	11.1	12.0	12.5	12.1	10.7	11.5	10.9
1945	8.6	8.5	8.2	8.6	8.7	8.4	8.3	8.4	7.1	7.1	7.7	8.0	7.7	6.8	7.3	7.0
1940	6.6	6.5	6.3	6.6	6.7	6.4	6.5	6.5	5.4	5.5	5.9	6.2	6.0	5.3	5.6	5.4

Year	National 30 City Average	Texas														Utah
		Abilene	Amarillo	Austin	Beaumont	Corpus Christi	Dallas	El Paso	Fort Worth	Houston	Lubbock	Odessa	San Antonio	Waco	Wichita Falls	Ogden
Jan 2013	197.6E	155.0E	161.4E	157.1E	160.9E	154.6E	167.9E	151.5E	162.1E	171.0E	159.2E	151.9E	163.7E	156.1E	155.2E	168.0E
2012	194.0	152.5	158.6	154.3	158.2	151.8	165.2	149.1	159.4	167.8	156.6	149.1	160.7	153.8	152.9	165.6
2011	185.7	145.9	151.9	147.6	152.4	145.0	157.9	142.5	152.8	160.8	150.0	142.5	152.6	147.3	146.6	158.7
2010	181.6	144.2	150.1	144.7	150.5	142.1	155.1	140.8	149.6	157.3	148.1	140.7	147.9	145.8	145.0	157.5
2009	182.5	143.5	147.3	146.4	149.8	141.7	155.5	141.1	150.1	160.9	144.6	139.1	150.8	146.7	146.1	155.1
2008	171.0	132.4	137.0	137.5	140.3	133.2	144.1	130.7	138.3	149.1	134.7	128.9	141.0	136.1	136.1	144.8
2007	165.0	128.5	132.2	131.6	137.1	128.8	138.6	126.2	134.8	146.0	130.2	125.0	136.4	132.1	132.1	140.0
2006	156.2	121.6	125.7	125.5	130.1	122.0	131.1	120.3	127.6	138.2	123.2	118.3	129.8	125.2	125.3	133.4
2005	146.7	113.8	117.3	117.9	121.3	114.4	123.7	112.5	119.4	129.0	115.5	110.5	121.3	116.1	117.3	126.0
2004	132.8	102.9	106.3	105.7	108.5	102.9	112.0	101.5	108.4	115.9	104.7	99.9	108.4	104.8	105.1	115.2
2003	129.7	100.5	104.4	103.9	106.1	100.4	109.3	99.5	105.5	113.4	102.3	97.6	104.8	102.7	103.0	112.7
2002	126.7	99.9	101.9	102.4	104.5	98.9	107.9	98.7	104.6	111.5	100.5	95.9	103.7	100.5	101.4	110.8
2001	122.2	93.4	98.4	99.8	102.3	96.6	103.8	95.5	100.9	107.8	97.6	93.4	100.5	97.8	97.3	107.7
2000	118.9	93.4	98.1	99.1	101.6	96.9	102.7	92.4	99.9	106.0	97.6	93.4	99.4	97.3	96.9	104.6
1999	116.6	91.8	94.5	96.0	99.7	94.0	101.0	90.7	97.6	104.6	96.0	92.1	98.0	94.8	95.5	103.3
1998	113.6	89.8	92.7	94.2	97.9	91.8	97.9	88.4	94.5	101.3	93.3	90.1	94.8	92.7	92.9	98.5
1997	111.5	88.4	91.3	92.8	96.8	90.3	96.1	87.0	93.3	100.1	91.9	88.8	93.4	91.4	91.5	96.1
1996	108.9	86.8	89.6	90.9	95.3	88.5	94.1	86.7	91.5	97.9	90.2	87.1	92.3	89.7	89.9	94.1
1995	105.6	85.2	87.4	89.3	93.7	87.4	91.4	85.2	89.5	95.9	88.4	85.6	88.9	86.4	86.8	92.2
1994	103.0	83.3	85.5	87.0	91.8	84.6	89.6	82.2	87.4	93.4	87.0	83.9	87.0	84.9	85.4	89.4
1993	100.0	81.4	83.4	84.9	90.0	82.8	87.8	80.2	85.5	91.1	84.8	81.9	85.0	83.0	83.5	87.1
1990	93.2	78.0	80.1	81.3	86.5	79.3	84.5	76.7	82.1	85.4	81.5	78.6	80.7	79.6	80.3	83.4
1985	81.8	71.1	72.5	74.5	79.3	72.3	77.6	69.4	75.1	79.6	74.0	71.2	73.9	71.7	73.3	75.2
1980	60.7	53.4	55.2	54.5	57.6	54.5	57.9	53.1	57.0	59.4	55.6	57.2	55.0	54.9	55.4	62.2
1975	43.7	37.6	39.0	39.0	39.6	38.1	40.7	38.0	40.4	41.2	38.9	37.9	39.0	38.6	38.0	40.0
1970	27.8	24.5	24.9	24.9	25.7	24.5	25.5	23.7	25.9	25.4	25.1	24.6	23.3	24.8	24.5	26.8
1965	21.5	18.9	19.2	19.2	19.9	18.9	19.9	19.0	19.9	20.0	19.4	19.0	18.5	19.2	18.9	20.6
1960	19.5	17.1	17.4	17.4	18.1	17.1	18.2	17.0	18.1	18.2	17.6	17.3	16.8	17.4	17.2	18.8
1955	16.3	14.4	14.6	14.6	15.1	14.4	15.3	14.3	15.2	15.2	14.8	14.5	14.1	14.6	14.4	15.7
1950	13.5	11.9	12.1	12.1	12.5	11.9	12.6	11.8	12.5	12.6	12.2	12.0	11.6	12.1	11.9	13.0
1945	8.6	7.6	7.7	7.7	8.0	7.6	8.0	7.5	8.0	8.0	7.8	7.6	7.4	7.7	7.6	8.3
1940	6.6	5.9	5.9	5.9	6.1	5.8	6.2	5.8	6.2	6.2	6.0	5.9	5.7	5.9	5.8	6.4

Historical Cost Indexes

Year	National 30 City Average	Utah Salt Lake City	Vermont Burlington	Vermont Rutland	Virginia Alexandria	Virginia Newport News	Virginia Norfolk	Virginia Richmond	Virginia Roanoke	Washington Seattle	Washington Spokane	Washington Tacoma	West Virginia Charleston	West Virginia Huntington	Wisconsin Green Bay	Wisconsin Kenosha
Jan 2013	197.6E	171.4E	177.0E	175.0E	185.4E	170.6E	171.9E	171.6E	168.7E	205.2E	184.6E	198.4E	186.8E	189.8E	192.8E	198.3E
2012	194.0	168.8	172.4	170.5	182.4	168.3	169.7	169.4	166.4	201.9	181.6	194.3	183.4	186.4	190.0	195.5
2011	185.7	161.8	158.8	157.1	174.5	159.8	161.1	159.0	154.2	194.1	175.5	187.6	177.5	180.3	181.4	187.6
2010	181.6	160.9	156.3	154.6	170.9	156.7	158.0	156.6	151.8	191.8	171.5	186.0	171.6	176.1	179.0	185.1
2009	182.5	160.4	158.3	156.7	172.5	160.0	161.9	160.7	155.6	188.5	173.0	186.3	174.6	177.4	175.5	182.5
2008	171.0	149.7	147.2	145.8	161.8	150.8	150.8	150.9	145.9	176.9	162.4	174.9	162.8	165.7	165.6	172.2
2007	165.0	144.6	144.4	143.2	155.0	146.2	146.1	147.3	142.9	171.4	156.7	168.7	158.7	160.1	159.4	164.8
2006	156.2	137.7	131.7	130.7	146.2	133.7	134.6	134.8	129.7	162.9	150.1	160.7	149.4	151.4	152.7	157.1
2005	146.7	129.4	124.6	123.8	136.5	123.5	124.4	125.4	112.2	153.9	141.9	151.5	140.8	141.0	144.4	148.3
2004	132.8	117.8	113.0	112.3	121.5	108.9	110.2	110.9	99.7	138.0	127.6	134.5	124.8	125.6	128.9	132.4
2003	129.7	116.0	110.7	110.1	119.5	104.8	106.2	108.6	97.0	134.9	125.9	133.0	123.3	123.6	127.4	130.9
2002	126.7	113.7	109.0	108.4	115.1	102.9	104.1	106.6	95.2	132.7	123.9	131.4	121.2	120.9	123.0	127.3
2001	122.2	109.1	105.7	105.2	110.8	99.8	100.3	102.9	92.1	127.9	120.3	125.7	114.6	117.5	119.1	123.6
2000	118.9	106.5	98.9	98.3	108.1	96.5	97.6	100.2	90.7	124.6	118.3	122.9	111.5	114.4	114.6	119.1
1999	116.6	104.5	98.2	97.7	106.1	95.6	96.5	98.8	89.8	123.3	116.7	121.6	110.6	113.4	112.1	115.8
1998	113.6	99.5	97.8	97.3	104.1	93.7	93.9	97.0	88.3	119.4	114.3	118.3	106.7	109.0	109.5	112.9
1997	111.5	97.2	96.6	96.3	101.2	91.6	91.7	92.9	86.9	118.1	111.7	117.2	105.3	107.7	105.6	109.1
1996	108.9	94.9	95.1	94.8	99.7	90.2	90.4	91.6	85.5	115.2	109.2	114.3	103.1	104.8	103.8	106.4
1995	105.6	93.1	91.1	90.8	96.3	86.0	86.4	87.8	82.8	113.7	107.4	112.8	95.8	97.2	97.6	97.9
1994	103.0	90.2	89.5	89.3	93.9	84.6	84.8	86.3	81.4	109.9	104.0	108.3	94.3	95.3	96.3	96.2
1993	100.0	87.9	87.6	87.6	91.6	82.9	83.0	84.3	79.5	107.3	103.9	106.7	92.6	93.5	94.0	94.3
1990	93.2	84.3	83.0	82.9	86.1	76.3	76.7	77.6	76.1	100.1	98.5	100.5	86.1	86.8	86.7	87.8
1985	81.8	75.9	74.8	74.9	75.1	68.7	68.8	69.5	67.2	88.3	89.0	91.2	77.7	77.7	76.7	77.4
1980	60.7	57.0	55.3	58.3	57.3	52.5	52.4	54.3	51.3	67.9	66.3	66.7	57.7	58.3	58.6	58.3
1975	43.7	40.1	41.8	43.9	41.7	37.2	36.9	37.1	37.1	44.9	44.4	44.5	41.0	40.0	40.9	40.5
1970	27.8	26.1	25.4	26.8	26.2	23.9	21.5	22.0	23.7	28.8	29.3	29.6	26.1	25.8	26.4	26.5
1965	21.5	20.0	19.8	20.6	20.2	18.4	17.1	17.2	18.3	22.4	22.5	22.8	20.1	19.9	20.3	20.4
1960	19.5	18.4	18.0	18.8	18.4	16.7	15.4	15.6	16.6	20.4	20.8	20.8	18.3	18.1	18.4	18.6
1955	16.3	15.4	15.1	15.7	15.4	14.0	12.9	13.1	13.9	17.1	17.4	17.4	15.4	15.2	15.5	15.6
1950	13.5	12.7	12.4	13.0	12.7	11.6	10.7	10.8	11.5	14.1	14.4	14.4	12.7	12.5	12.8	12.9
1945	8.6	8.1	7.9	8.3	8.1	7.4	6.8	6.9	7.3	9.0	9.2	9.2	8.1	8.0	8.1	8.2
1940	6.6	6.3	6.1	6.4	6.2	5.7	5.3	5.3	5.7	7.0	7.1	7.1	6.2	6.2	6.3	6.3

Year	National 30 City Average	Washington Madison	Washington Milwaukee	Washington Racine	Wyoming Cheyenne	Canada Calgary	Canada Edmonton	Canada Hamilton	Canada London	Canada Montreal	Canada Ottawa	Canada Quebec	Canada Toronto	Canada Vancouver	Canada Winnipeg
Jan 2013	197.6E	194.0E	206.0E	197.9E	167.8E	224.1E	224.8E	215.4E	212.9E	215.2E	213.7E	213.1E	220.7E	218.2E	204.3E
2012	194.0	191.2	202.2	194.9	165.0	219.1	219.6	211.1	208.2	209.6	209.7	207.5	216.6	214.0	200.7
2011	185.7	183.4	191.9	187.0	155.3	212.6	212.4	204.2	200.1	202.1	202.0	200.6	209.6	207.1	192.6
2010	181.6	181.3	187.1	184.4	154.4	200.6	200.7	197.7	193.7	193.9	195.6	192.7	200.7	192.7	182.9
2009	182.5	180.0	187.3	182.8	155.2	205.4	206.6	203.0	198.9	198.8	200.1	197.4	205.6	199.6	188.5
2008	171.0	168.4	176.3	173.0	146.5	190.2	191.0	194.3	188.6	186.7	188.0	186.8	194.6	184.9	174.4
2007	165.0	160.7	168.9	164.7	141.3	183.1	184.4	186.8	182.4	181.7	182.4	182.0	187.7	180.6	170.1
2006	156.2	152.8	158.8	157.2	128.0	163.6	164.8	169.0	164.9	158.6	163.7	159.0	170.6	169.3	155.9
2005	146.7	145.2	148.4	147.9	118.9	154.4	155.7	160.0	156.2	149.2	155.1	150.1	162.6	159.4	146.9
2004	132.8	129.7	134.2	132.7	104.7	138.8	139.4	142.6	140.4	134.5	141.0	135.7	146.0	141.7	129.7
2003	129.7	128.4	131.1	131.5	102.6	133.6	134.2	139.1	137.0	130.2	137.4	131.4	142.3	137.0	124.4
2002	126.7	124.5	128.2	126.5	101.7	122.5	122.2	136.3	134.1	127.2	134.9	128.4	139.8	134.6	121.4
2001	122.2	120.6	123.9	123.3	99.0	117.5	117.4	131.5	129.1	124.4	130.2	125.5	134.7	130.2	117.2
2000	118.9	116.6	120.5	118.7	98.1	115.9	115.8	130.0	127.5	122.8	128.8	124.1	133.1	128.4	115.6
1999	116.6	115.9	117.4	115.5	96.9	115.3	115.2	128.1	125.6	120.8	126.8	121.9	131.2	127.1	115.2
1998	113.6	110.8	113.4	112.7	95.4	112.5	112.4	126.3	123.9	119.0	124.7	119.6	128.5	123.8	113.7
1997	111.5	106.1	110.1	109.3	93.2	110.7	110.6	124.4	121.9	114.6	122.8	115.4	125.7	121.9	111.4
1996	108.9	104.4	107.1	106.5	91.1	109.1	109.0	122.6	120.2	112.9	121.0	113.6	123.9	119.0	109.7
1995	105.6	96.5	103.9	97.8	87.6	107.4	107.4	119.9	117.5	110.8	118.2	111.5	121.6	116.2	107.6
1994	103.0	94.5	100.6	96.1	85.4	106.7	106.6	116.5	114.2	109.5	115.0	110.2	117.8	115.1	105.5
1993	100.0	91.3	96.7	93.8	82.9	104.7	104.5	113.7	111.9	106.9	112.0	107.0	114.9	109.2	102.8
1990	93.2	84.3	88.9	87.3	79.1	98.0	97.1	103.8	101.4	99.0	102.7	96.8	104.6	103.2	95.1
1985	81.8	74.3	77.4	77.0	72.3	90.2	89.1	85.8	84.9	82.4	83.9	79.9	86.5	89.8	83.4
1980	60.7	56.8	58.8	58.1	56.9	64.9	63.3	63.7	61.9	59.2	60.9	59.3	60.9	65.0	61.7
1975	43.7	40.7	43.3	40.7	40.6	42.2	41.6	42.9	41.6	39.7	41.5	39.0	42.2	42.4	39.2
1970	27.8	26.5	29.4	26.5	26.0	28.9	28.6	28.5	27.8	25.6	27.6	26.0	25.6	26.0	23.1
1965	21.5	20.6	21.8	20.4	20.0	22.3	22.0	22.0	21.4	18.7	21.2	20.1	19.4	20.5	17.5
1960	19.5	18.1	19.0	18.6	18.2	20.2	20.0	20.0	19.5	17.0	19.3	18.2	17.6	18.6	15.8
1955	16.3	15.2	15.9	15.6	15.2	17.0	16.8	16.7	16.3	14.3	16.2	15.3	14.8	15.5	13.3
1950	13.5	12.5	13.2	12.9	12.6	14.0	13.9	13.8	13.5	11.8	13.4	12.6	12.2	12.8	10.9
1945	8.6	8.0	8.4	8.2	8.0	9.0	8.8	8.8	8.6	7.5	8.5	8.0	7.8	8.2	7.0
1940	6.6	6.2	6.5	6.3	6.2	6.9	6.8	6.8	6.6	5.8	6.6	6.2	6.0	6.3	5.4
1940	6.6	6.2	6.5	6.3	6.2	6.9	6.8	6.8	6.6	5.8	6.6	6.2	6.0	6.3	5.4

Glossary

Accent lighting—Fixtures or directional beams of light arranged to bring attention to an object or area.

Acoustical material—A material fabricated for the sole purpose of absorbing sound.

Addition—An expansion to an existing structure generally in the form of a room, floor(s) or wing(s). An increase in the floor area or volume of a structure.

Aggregate—Materials such as sand, gravel, stone, vermiculite, perlite and fly ash (slag) which are essential components in the production of concrete, mortar or plaster.

Air conditioning system—An air treatment to control the temperature, humidity and cleanliness of air and to provide for its distribution throughout the structure.

Air curtain—A stream of air directed downward to prevent the loss of hot or cool air from the structure and inhibit the entrance of dust and insects. Generally installed at loading platforms.

Anodized aluminum—Aluminum treated by an electrolytic process to produce an oxide film that is corrosion resistant.

Arcade—A covered passageway between buildings often with shops and offices on one or both sides.

Ashlar—A square-cut building stone or a wall constructed of cut stone.

Asphalt paper—A sheet paper material either coated or saturated with asphalt to increase its strength and resistance to water.

Asphalt shingles—Shingles manufactured from saturated roofing felts, coated with asphalt and topped with mineral granules to prevent weathering.

Assessment ratio—The ratio between the market value and assessed valuation of a property.

Back-up—That portion of a masonry wall behind the exterior facing-usually load bearing.

Balcony—A platform projecting from a building either supported from below or cantilevered. Usually protected by railing.

Bay—Structural component consisting of beams and columns occurring consistently throughout the structure.

Beam—A horizontal structural framing member that transfers loads to vertical members (columns) or bearing walls.

Bid—An offer to perform work described in a contract at a specified price.

Board foot—A unit or measure equal in volume to a board one foot long, one foot wide and one inch thick.

Booster pump—A supplemental pump installed to increase or maintain adequate pressure in the system.

Bowstring roof—A roof supported by trusses fabricated in the shape of a bow and tied together by a straight member.

Brick veneer—A facing of brick laid against, but not bonded to, a wall.

Bridging—A method of bracing joists for stiffness, stability and load distribution.

Broom finish—A method of finishing concrete by lightly dragging a broom over freshly placed concrete.

Built-up roofing—Installed on flat or extremely low-pitched roofs, a roof covering composed of plies or laminations of saturated roofing felts alternated with layers of coal tar pitch or asphalt and surfaced with a layer of gravel or slag in a thick coat of asphalt and finished with a capping sheet.

Caisson—A watertight chamber to expedite work on foundations or structure below water level.

Cantilever—A structural member supported only at one end.

Carport—An automobile shelter having one or more sides open to the weather.

Casement window—A vertical opening window having one side fixed.

Caulking—A resilient compound generally having a silicone or rubber base used to prevent infiltration of water or outside air.

Cavity wall—An exterior wall usually of masonry having an inner and outer wall separated by a continuous air space for thermal insulation.

Cellular concrete—A lightweight concrete consisting of cement mixed with gas-producing materials which, in turn, forms bubbles resulting in good insulating qualities.

Chattel mortgage—A secured interest in a property as collateral for payment of a note.

Clapboard—A wood siding used as exterior covering in frame construction. It is applied horizontally with grain running lengthwise. The thickest section of the board is on the bottom.

Clean room—An environmentally controlled room usually found in medical facilities or precision manufacturing spaces where it is essential to eliminate dust, lint or pathogens.

Close studding—A method of construction whereby studs are spaced close together and the intervening spaces are plastered.

Cluster housing—A closely grouped series of houses resulting in a high density land use.

Cofferdam—A watertight enclosure used for foundation construction in waterfront areas. Water is pumped out of the cofferdam allowing free access to the work area.

Collar joint—Joint between the collar beam and roof rafters.

Column—A vertical structural member supporting horizontal members (beams) along the direction of its longitudinal axis.

Combination door (windows)—Door (windows) having interchangeable screens and glass for seasonal use.

Common area—Spaces either inside or outside the building designated for use by the occupant of the building but not the general public.

Common wall—A wall used jointly by two dwelling units.

Compound wall—A wall constructed of more than one material.

Concrete—A composite material consisting of sand, coarse aggregate (gravel, stone or slag), cement and water that when mixed and allowed to harden forms a hard stone-like material.

Concrete floor hardener—A mixture of chemicals applied to the surface of concrete to produce a dense, wear-resistant bearing surface.

Conduit—A tube or pipe used to protect electric wiring.

Coping—The top cover or capping on a wall.

Craneway—Steel or concrete column and beam supports and rails on which a crane travels.

Curtain wall—A non-bearing exterior wall not supported by beams or girders of the steel frame.

Dampproofing—Coating of a surface to prevent the passage of water or moisture.

Dead load—Total weight of all structural components plus permanently attached fixtures and equipment.

Decibel—Unit of acoustical measurement.

Decorative block—A concrete masonry unit having a special treatment on its face.

Deed restriction—A restriction on the use of a property as set forth in the deed.

Depreciation—A loss in property value caused by physical aging, functional or economic obsolescence.

Direct heating—Heating of spaces by means of exposed heated surfaces (stove, fire, radiators, etc.).

Distributed load—A load that acts evenly over a structural member.

Dock bumper—A resilient material attached to a loading dock to absorb the impact of trucks backing in.

Dome—A curved roof shape spanning an area.

Dormer—A structure projecting from a sloping roof.

Double-hung window—A window having two vertical sliding sashes-one covering the upper section and one covering the lower.

Downspout—A vertical pipe for carrying rain water from the roof to the ground.

Drain tile–A hollow tile used to drain water-soaked soil.

Dressed size–In lumber about 3/8″ - 1/2″ less than nominal size after planing and sawing.

Drop panel–The depressed surface on the bottom side of a flat concrete slab which surrounds a column.

Dry pipe sprinkler system–A sprinkler system that is activated with water only in case of fire. Used in areas susceptible to freezing or to avoid the hazards of leaking pipes.

Drywall–An interior wall constructed of gypsum board, plywood or wood paneling. No water is required for application.

Duct–Pipe for transmitting warm or cold air. A pipe containing electrical cable or wires.

Dwelling–A structure designed as living quarters for one or more families.

Easement–A right of way or free access over land owned by another.

Eave–The lower edge of a sloping roof that overhangs the side wall of the structure.

Economic rent–That rent on a property sufficient to pay all operating costs exclusive of services and utilities.

Economic life–The term during which a structure is expected to be profitable. Generally shorter than the physical life of the structure.

Economic obsolescence–Loss in value due to unfavorable economic influences occurring from outside the structure itself.

Effective age–The age a structure appears to be based on observed physical condition determined by degree of maintenance and repair.

Elevator–A mechanism for vertical transport of personnel or freight equipped with car or platform.

Ell–A secondary wing or addition to a structure at right angles to the main structure.

Eminent domain–The right of the state to take private property for public use.

Envelope–The shape of a building indicating volume.

Equity–Value of an owner's interest calculated by subtracting outstanding mortgages and expenses from the value of the property.

Escheat–The assumption of ownership by the state, of property whose owner cannot be determined.

Facade–The exterior face of a building sometimes decorated with elaborate detail.

Face brick–Brick manufactured to present an attractive appearance.

Facing block–A concrete masonry unit having a decorative exterior finish.

Feasibility study–A detailed evaluation of a proposed project to determine its financial potential.

Felt paper–Paper sheathing used on walls as an insulator and to prevent infiltration of moisture.

Fenestration–Pertaining to the density and arrangement of windows in a structure.

Fiberboard–A building material composed of wood fiber compressed with a binding agent, produced in sheet form.

Fiberglass–Fine spun filaments of glass processed into various densities to produce thermal and acoustical insulation.

Field house–A long structure used for indoor athletic events.

Finish floor–The top or wearing surface of the floor system.

Fireproofing–The use of fire-resistant materials for the protection of structural members to ensure structural integrity in the event of fire.

Fire stop–A material or member used to seal an opening to prevent the spread of fire.

Flashing–A thin, impervious material such as copper or sheet metal used to prevent air or water penetration.

Float finish–A concrete finish accomplished by using a flat tool with a handle on the back.

Floor drain–An opening installed in a floor for removing excess water into a plumbing system.

Floor load–The live load the floor system has been designed for and which may be applied safely.

Flue–A heat-resistant enclosed passage in a chimney to remove gaseous products of combustion from a fireplace or boiler.

Footing–The lowest portion of the foundation wall that transmits the load directly to the soil.

Foundation–Below grade wall system that supports the structure.

Foyer–A portion of the structure that serves as a transitional space between the interior and exterior.

Frontage–That portion of a lot that is placed facing a street, public way or body of water.

Frost action–The effects of freezing and thawing on materials and the resultant structural damage.

Functional obsolescence–An inadequacy caused by outmoded design, dated construction materials or over or undersized areas, all of which cause excessive operating costs.

Furring strips–Wood or metal channels used for attaching gypsum or metal lath to masonry walls as a finish or for leveling purposes.

Gambrel roof (mansard)–A roof system having two pitches on each side.

Garden apartment–Two or three story apartment building with common outside areas.

General contractor–The prime contractor responsible for work on the construction site.

Girder–A principal horizontal supporting member.

Girt–A horizontal framing member for adding rigid support to columns which also acts as a support for sheathing or siding.

Grade beam–That portion of the foundation system that directly supports the exterior walls of the building.

Gross area–The total enclosed floor area of a building.

Ground area–The area computed by the exterior dimensions of the structure.

Ground floor–The floor of a building in closest proximity to the ground.

Grout–A mortar containing a high water content used to fill joints and cavities in masonry work.

Gutter–A wood or sheet metal conduit set along the building eaves used to channel rainwater to leaders or downspouts.

Gunite–A concrete mix placed pneumatically.

Gypsum–Hydrated calcium sulphate used as both a retarder in Portland cement and as a major ingredient in plaster of Paris. Used in sheets as a substitute for plaster.

Hall–A passageway providing access to various parts of a building–a large room for entertainment or assembly.

Hangar–A structure for the storage or repair of aircraft.

Head room–The distance between the top of the finished floor to the bottom of the finished ceiling.

Hearth–The floor of a fireplace and the adjacent area of fireproof material.

Heat pump–A mechanical device for providing either heating or air conditioning.

Hip roof–A roof whose four sides meet at a common point with no gabled ends.

I-beam–A structural member having a cross-section resembling the letter "I".

Insulation–Material used to reduce the effects of heat, cold or sound.

Jack rafter–An unusually short rafter generally found in hip roofs.

Jalousie–Adjustable glass louvers which pivot simultaneously in a common frame.

Jamb–The vertical member on either side of a door frame or window frame.

Joist–Parallel beams of timber, concrete or steel used to support floor and ceiling.

Junction box–A box that protects splices or joints in electrical wiring.

Kalamein door–Solid core wood doors clad with galvanized sheet metal.

Kip–A unit of weight equal to 1,000 pounds.

Lally column–A concrete-filled steel pipe used as a vertical support.

Laminated beam–A beam built up by gluing together several pieces of timber.

Landing–A platform between flights of stairs.

Lath–Strips of wood or metal used as a base for plaster.

Lead-lined door, sheetrock–Doors or sheetrock internally lined with sheet lead to provide protection from X-ray radiation.

Lean-to–A small shed or building addition with a single pitched roof attached to the exterior wall of the main building.

Lintel–A horizontal framing member used to carry a load over a wall opening.

Live load–The moving or movable load on a structure composed of furnishings, equipment or personnel weight.

Load bearing partition–A partition that supports a load in addition to its own weight.

Loading dock leveler–An adjustable platform for handling off-on loading of trucks.

Loft building–A commercial/industrial type building containing large, open unpartitioned floor areas.

Louver window–A window composed of a series of sloping, overlapping blades or slats that may be adjusted to admit varying degrees of air or light.

Main beam–The principal load bearing beam used to transmit loads directly to columns.

Mansard roof–A roof having a double pitch on all four sides, the lower level having the steeper pitch.

Mansion–An extremely large and imposing residence.

Masonry–The utilization of brick, stone, concrete or block for walls and other building components.

Mastic–A sealant or adhesive compound generally used as either a binding agent or floor finish.

Membrane fireproofing–A coating of metal lath and plaster to provide resistance to fire and heat.

Mesh reinforcement–An arrangement of wire tied or welded at their intersection to provide strength and resistance to cracking.

Metal lath–A diamond-shaped metallic base for plaster.

Mezzanine–A low story situated between two main floors.

Mill construction–A heavy timber construction that achieves fire resistance by using large wood structural members, noncombustible bearing and non-bearing walls and omitting concealed spaces under floors and roof.

Mixed occupancy–Two or more classes of occupancy in a single structure.

Molded brick–A specially shaped brick used for decorative purposes.

Monolithic concrete–Concrete poured in a continuous process so there are no joints.

Movable partition–A non-load bearing demountable partition that can be relocated and can be either ceiling height or partial height.

Moving walk–A continually moving horizontal passenger carrying device.

Multi-zone system–An air conditioning system that is capable of handling several individual zones simultaneously.

Net floor area–The usable, occupied area of a structure excluding stairwells, elevator shafts and wall thicknesses.

Non-combustible construction–Construction in which the walls, partitions and the structural members are of non-combustible materials.

Nurses call system–An electrically operated system for use by patients or personnel for summoning a nurse.

Occupancy rate–The number of persons per room, per dwelling unit, etc.

One-way joist connection–A framing system for floors and roofs in a concrete building consisting of a series of parallel joists supported by girders between columns.

Open web joist–A lightweight prefabricated metal chord truss.

Parapet–A low guarding wall at the point of a drop. That portion of the exterior wall that extends above the roof line.

Parging–A thin coat of mortar applied to a masonry wall used primarily for waterproofing purposes.

Parquet–Inlaid wood flooring set in a simple design.

Peaked roof–A roof of two or more slopes that rises to a peak.

Penthouse–A structure occupying usually half the roof area and used to house elevator, HVAC equipment, or other mechanical or electrical systems.

Pier–A column designed to support a concentrated load.

Pilaster–A column usually formed of the same material and integral with, but projecting from, a wall.

Pile–A concrete, wood or steel column usually less than 2′ in diameter which is driven or otherwise introduced into the soil to carry a vertical load or provide vertical support.

Pitched roof–A roof having one or more surfaces with a pitch greater than 10°.

Plenum–The space between the suspended ceiling and the floor above.

Plywood–Structural wood made of three or more layers of veneer and bonded together with glue.

Porch–A structure attached to a building to provide shelter for an entranceway or to serve as a semi-enclosed space.

Post and beam framing–A structural framing system where beams rest on posts rather than bearing walls.

Post-tensioned concrete–Concrete that has the reinforcing tendons tensioned after the concrete has set.

Pre-cast concrete–Concrete structural components fabricated at a location other than in-place.

Pre-stressed concrete–Concrete that has the reinforcing tendons tensioned prior to the concrete setting.

Purlins–A horizontal structural member supporting the roof deck and resting on the trusses, girders, beams or rafters.

Rafters–Structural members supporting the roof deck and covering.

Resilient flooring–A manufactured interior floor covering material in either sheet or tile form, that is resilient.

Ridge–The horizontal line at the junction of two sloping roof edges.

Rigid frame–A structural framing system in which all columns and beams are rigidly connected; there are no hinged joints.

Roof drain–A drain designed to accept rainwater and discharge it into a leader or downspout.

Rough floor–A layer of boards or plywood nailed to the floor joists that serves as a base for the finished floor.

Sanitary sewer–A sewer line designed to carry only liquid or waterborne waste from the structure to a central treatment plant.

Sawtooth roof–Roof shape found primarily in industrial roofs that creates the appearance of the teeth of a saw.

Semi-rigid frame–A structural system wherein the columns and beams are so attached that there is some flexibility at the joints.

Septic tank–A covered tank in which waste matter is decomposed by natural bacterial action.

Service elevator–A combination passenger and freight elevator.

Sheathing–The first covering of exterior studs by boards, plywood or particle board.

Shoring–Temporary bracing for structural components during construction.

Span–The horizontal distance between supports.

Specification–A detailed list of materials and requirements for construction of a building.

Storm sewer–A system of pipes used to carry rainwater or surface waters.

Stucco–A cement plaster used to cover exterior wall surfaces usually applied over a wood or metal lath base.

Stud–A vertical wooden structural component.

Subfloor–A system of boards, plywood or particle board laid over the floor joists to form a base.

Superstructure–That portion of the structure above the foundation or ground level.

Terrazzo–A durable floor finish made of small chips of colored stone or marble, embedded in concrete, then polished to a high sheen.

Tile–A thin piece of fired clay, stone or concrete used for floor, roof or wall finishes.

Unit cost–The cost per unit of measurement.

Vault–A room especially designed for storage.

Veneer–A thin surface layer covering a base of common material.

Vent–An opening serving as an outlet for air.

Wainscot–The lower portion of an interior wall whose surface differs from that of the upper wall.

Wall bearing construction–A structural system where the weight of the floors and roof are carried directly by the masonry walls rather than the structural framing system.

Waterproofing–Any of a number of materials applied to various surfaces to prevent the infiltration of water.

Weatherstrip–A thin strip of metal, wood or felt used to cover the joint between the door or window sash and the jamb, casing or sill to keep out air, dust, rain, etc.

Wing–A building section or addition projecting from the main structure.

X-ray protection–Lead encased in sheetrock or plaster to prevent the escape of radiation.

A	Area Square Feet; Ampere	Brk., brk	Brick	Csc	Cosecant
AAFES	Army and Air Force Exchange Service	brkt	Bracket	C.S.F.	Hundred Square Feet
		Brng.	Bearing	CSI	Construction Specifications Institute
ABS	Acrylonitrile Butadiene Stryrene; Asbestos Bonded Steel	Brs.	Brass		
		Brz.	Bronze	CT	Current Transformer
A.C., AC	Alternating Current; Air-Conditioning; Asbestos Cement; Plywood Grade A & C	Bsn.	Basin	CTS	Copper Tube Size
		Btr.	Better	Cu	Copper, Cubic
		Btu	British Thermal Unit	Cu. Ft.	Cubic Foot
		BTUH	BTU per Hour	cw	Continuous Wave
ACI	American Concrete Institute	Bu.	bushels	C.W.	Cool White; Cold Water
ACR	Air Conditioning Refrigeration	BUR	Built-up Roofing	Cwt.	100 Pounds
ADA	Americans with Disabilities Act	BX	Interlocked Armored Cable	C.W.X.	Cool White Deluxe
AD	Plywood, Grade A & D	°C	degree centegrade	C.Y.	Cubic Yard (27 cubic feet)
Addit.	Additional	c	Conductivity, Copper Sweat	C.Y./Hr.	Cubic Yard per Hour
Adj.	Adjustable	C	Hundred; Centigrade	Cyl.	Cylinder
af	Audio-frequency	C/C	Center to Center, Cedar on Cedar	d	Penny (nail size)
AFUE	Annual Fuel Utilization Efficiency	C-C	Center to Center	D	Deep; Depth; Discharge
AGA	American Gas Association	Cab	Cabinet	Dis., Disch.	Discharge
Agg.	Aggregate	Cair.	Air Tool Laborer	Db	Decibel
A.H., Ah	Ampere Hours	Cal.	caliper	Dbl.	Double
A hr.	Ampere-hour	Calc	Calculated	DC	Direct Current
A.H.U., AHU	Air Handling Unit	Cap.	Capacity	DDC	Direct Digital Control
A.I.A.	American Institute of Architects	Carp.	Carpenter	Demob.	Demobilization
AIC	Ampere Interrupting Capacity	C.B.	Circuit Breaker	d.f.t.	Dry Film Thickness
Allow.	Allowance	C.C.A.	Chromate Copper Arsenate	d.f.u.	Drainage Fixture Units
alt., alt	Alternate	C.C.F.	Hundred Cubic Feet	D.H.	Double Hung
Alum.	Aluminum	cd	Candela	DHW	Domestic Hot Water
a.m.	Ante Meridiem	cd/sf	Candela per Square Foot	DI	Ductile Iron
Amp.	Ampere	CD	Grade of Plywood Face & Back	Diag.	Diagonal
Anod.	Anodized	CDX	Plywood, Grade C & D, exterior glue	Diam., Dia	Diameter
ANSI	American National Standards Institute			Distrib.	Distribution
		Cefi.	Cement Finisher	Div.	Division
APA	American Plywood Association	Cem.	Cement	Dk.	Deck
Approx.	Approximate	CF	Hundred Feet	D.L.	Dead Load; Diesel
Apt.	Apartment	C.F.	Cubic Feet	DLH	Deep Long Span Bar Joist
Asb.	Asbestos	CFM	Cubic Feet per Minute	dlx	Deluxe
A.S.B.C.	American Standard Building Code	c.g.	Center of Gravity	Do.	Ditto
Asbe.	Asbestos Worker	CHW	Chilled Water; Commercial Hot Water	DOP	Dioctyl Phthalate Penetration Test (Air Filters)
ASCE.	American Society of Civil Engineers				
A.S.H.R.A.E.	American Society of Heating, Refrig. & AC Engineers	C.I., CI	Cast Iron	Dp., dp	Depth
		C.I.P., CIP	Cast in Place	D.P.S.T.	Double Pole, Single Throw
ASME	American Society of Mechanical Engineers	Circ.	Circuit	Dr.	Drive
		C.L.	Carload Lot	DR	Dimension Ratio
ASTM	American Society for Testing and Materials	CL	Chain Link	Drink.	Drinking
		Clab.	Common Laborer	D.S.	Double Strength
Attchmt.	Attachment	Clam	Common maintenance laborer	D.S.A.	Double Strength A Grade
Avg., Ave.	Average	C.L.F.	Hundred Linear Feet	D.S.B.	Double Strength B Grade
AWG	American Wire Gauge	CLF	Current Limiting Fuse	Dty.	Duty
AWWA	American Water Works Assoc.	CLP	Cross Linked Polyethylene	DWV	Drain Waste Vent
Bbl.	Barrel	cm	Centimeter	DX	Deluxe White, Direct Expansion
B&B, BB	Grade B and Better; Balled & Burlapped	CMP	Corr. Metal Pipe	dyn	Dyne
		CMU	Concrete Masonry Unit	e	Eccentricity
B&S	Bell and Spigot	CN	Change Notice	E	Equipment Only; East; emissivity
B.&W.	Black and White	Col.	Column	Ea.	Each
b.c.c.	Body-centered Cubic	CO₂	Carbon Dioxide	EB	Encased Burial
B.C.Y.	Bank Cubic Yards	Comb.	Combination	Econ.	Economy
BE	Bevel End	comm.	Commercial, Communication	E.C.Y	Embankment Cubic Yards
B.F.	Board Feet	Compr.	Compressor	EDP	Electronic Data Processing
Bg. cem.	Bag of Cement	Conc.	Concrete	EIFS	Exterior Insulation Finish System
BHP	Boiler Horsepower; Brake Horsepower	Cont., cont	Continuous; Continued, Container	E.D.R.	Equiv. Direct Radiation
		Corr.	Corrugated	Eq.	Equation
B.I.	Black Iron	Cos	Cosine	EL	elevation
bidir.	bidirectional	Cot	Cotangent	Elec.	Electrician; Electrical
Bit., Bitum.	Bituminous	Cov.	Cover	Elev.	Elevator; Elevating
Bit., Conc.	Bituminous Concrete	C/P	Cedar on Paneling	EMT	Electrical Metallic Conduit; Thin Wall Conduit
Bk.	Backed	CPA	Control Point Adjustment		
Bkrs.	Breakers	Cplg.	Coupling	Eng.	Engine, Engineered
Bldg., bldg	Building	CPM	Critical Path Method	EPDM	Ethylene Propylene Diene Monomer
Blk.	Block	CPVC	Chlorinated Polyvinyl Chloride		
Bm.	Beam	C.Pr.	Hundred Pair	EPS	Expanded Polystyrene
Boil.	Boilermaker	CRC	Cold Rolled Channel	Eqhv.	Equip. Oper., Heavy
bpm	Blows per Minute	Creos.	Creosote	Eqlt.	Equip. Oper., Light
BR	Bedroom	Crpt.	Carpet & Linoleum Layer	Eqmd.	Equip. Oper., Medium
Brg.	Bearing	CRT	Cathode-ray Tube	Eqmm.	Equip. Oper., Master Mechanic
Brhe.	Bricklayer Helper	CS	Carbon Steel, Constant Shear Bar Joist	Eqol.	Equip. Oper., Oilers
Bric.	Bricklayer			Equip.	Equipment
				ERW	Electric Resistance Welded

E.S.	Energy Saver	H	High Henry	Lath.	Lather
Est.	Estimated	HC	High Capacity	Lav.	Lavatory
esu	Electrostatic Units	H.D., HD	Heavy Duty; High Density	lb.; #	Pound
E.W.	Each Way	H.D.O.	High Density Overlaid	L.B., LB	Load Bearing; L Conduit Body
EWT	Entering Water Temperature	HDPE	High density polyethelene plastic	L. & E.	Labor & Equipment
Excav.	Excavation	Hdr.	Header	lb./hr.	Pounds per Hour
excl	Excluding	Hdwe.	Hardware	lb./L.F.	Pounds per Linear Foot
Exp., exp	Expansion, Exposure	H.I.D., HID	High Intensity Discharge	lbf/sq.in.	Pound-force per Square Inch
Ext., ext	Exterior; Extension	Help.	Helper Average	L.C.L.	Less than Carload Lot
Extru.	Extrusion	HEPA	High Efficiency Particulate Air Filter	L.C.Y.	Loose Cubic Yard
f.	Fiber stress			Ld.	Load
F	Fahrenheit; Female; Fill	Hg	Mercury	LE	Lead Equivalent
Fab., fab	Fabricated; fabric	HIC	High Interrupting Capacity	LED	Light Emitting Diode
FBGS	Fiberglass	HM	Hollow Metal	L.F.	Linear Foot
F.C.	Footcandles	HMWPE	high molecular weight polyethylene	L.F. Nose	Linear Foot of Stair Nosing
f.c.c.	Face-centered Cubic			L.F. Rsr	Linear Foot of Stair Riser
f'c.	Compressive Stress in Concrete; Extreme Compressive Stress	HO	High Output	Lg.	Long; Length; Large
		Horiz.	Horizontal	L & H	Light and Heat
F.E.	Front End	H.P., HP	Horsepower; High Pressure	LH	Long Span Bar Joist
FEP	Fluorinated Ethylene Propylene (Teflon)	H.P.F.	High Power Factor	L.H.	Labor Hours
		Hr.	Hour	L.L., LL	Live Load
F.G.	Flat Grain	Hrs./Day	Hours per Day	L.L.D.	Lamp Lumen Depreciation
F.H.A.	Federal Housing Administration	HSC	High Short Circuit	lm	Lumen
Fig.	Figure	Ht.	Height	lm/sf	Lumen per Square Foot
Fin.	Finished	Htg.	Heating	lm/W	Lumen per Watt
FIPS	Female Iron Pipe Size	Htrs.	Heaters	LOA	Length Over All
Fixt.	Fixture	HVAC	Heating, Ventilation & Air-Conditioning	log	Logarithm
FJP	Finger jointed and primed			L-O-L	Lateralolet
Fl. Oz.	Fluid Ounces	Hvy.	Heavy	long.	longitude
Flr.	Floor	HW	Hot Water	L.P., LP	Liquefied Petroleum; Low Pressure
FM	Frequency Modulation; Factory Mutual	Hyd.; Hydr.	Hydraulic	L.P.F.	Low Power Factor
		Hz	Hertz (cycles)	LR	Long Radius
Fmg.	Framing	I.	Moment of Inertia	L.S.	Lump Sum
FM/UL	Factory Mutual/Underwriters Labs	IBC	International Building Code	Lt.	Light
Fdn.	Foundation	I.C.	Interrupting Capacity	Lt. Ga.	Light Gauge
FNPT	Female National Pipe Thread	ID	Inside Diameter	L.T.L.	Less than Truckload Lot
Fori.	Foreman, Inside	I.D.	Inside Dimension; Identification	Lt. Wt.	Lightweight
Foro.	Foreman, Outside	I.F.	Inside Frosted	L.V.	Low Voltage
Fount.	Fountain	I.M.C.	Intermediate Metal Conduit	M	Thousand; Material; Male; Light Wall Copper Tubing
fpm	Feet per Minute	In.	Inch		
FPT	Female Pipe Thread	Incan.	Incandescent	M²CA	Meters Squared Contact Area
Fr	Frame	Incl.	Included; Including	m/hr.; M.H.	Man-hour
F.R.	Fire Rating	Int.	Interior	mA	Milliampere
FRK	Foil Reinforced Kraft	Inst.	Installation	Mach.	Machine
FSK	Foil/scrim/kraft	Insul., insul	Insulation/Insulated	Mag. Str.	Magnetic Starter
FRP	Fiberglass Reinforced Plastic	I.P.	Iron Pipe	Maint.	Maintenance
FS	Forged Steel	I.P.S., IPS	Iron Pipe Size	Marb.	Marble Setter
FSC	Cast Body; Cast Switch Box	IPT	Iron Pipe Threaded	Mat; Mat'l.	Material
Ft., ft	Foot; Feet	I.W.	Indirect Waste	Max.	Maximum
Ftng.	Fitting	J	Joule	MBF	Thousand Board Feet
Ftg.	Footing	J.I.C.	Joint Industrial Council	MBH	Thousand BTU's per hr.
Ft lb.	Foot Pound	K	Thousand; Thousand Pounds; Heavy Wall Copper Tubing, Kelvin	MC	Metal Clad Cable
Furn.	Furniture			MCC	Motor Control Center
FVNR	Full Voltage Non-Reversing	K.A.H.	Thousand Amp. Hours	M.C.F.	Thousand Cubic Feet
FVR	Full Voltage Reversing	kcmil	Thousand Circular Mils	MCFM	Thousand Cubic Feet per Minute
FXM	Female by Male	KD	Knock Down	M.C.M.	Thousand Circular Mils
Fy.	Minimum Yield Stress of Steel	K.D.A.T.	Kiln Dried After Treatment	MCP	Motor Circuit Protector
g	Gram	kg	Kilogram	MD	Medium Duty
G	Gauss	kG	Kilogauss	MDF	Medium-density fibreboard
Ga.	Gauge	kgf	Kilogram Force	M.D.O.	Medium Density Overlaid
Gal., gal.	Gallon	kHz	Kilohertz	Med.	Medium
gpm, GPM	Gallon per Minute	Kip	1000 Pounds	MF	Thousand Feet
Galv., galv	Galvanized	KJ	Kiljoule	M.F.B.M.	Thousand Feet Board Measure
GC/MS	Gas Chromatograph/Mass Spectrometer	K.L.	Effective Length Factor	Mfg.	Manufacturing
		K.L.F.	Kips per Linear Foot	Mfrs.	Manufacturers
Gen.	General	Km	Kilometer	mg	Milligram
GFI	Ground Fault Interrupter	KO	Knock Out	MGD	Million Gallons per Day
GFRC	Glass Fiber Reinforced Concrete	K.S.F.	Kips per Square Foot	MGPH	Thousand Gallons per Hour
Glaz.	Glazier	K.S.I.	Kips per Square Inch	MH, M.H.	Manhole; Metal Halide; Man-Hour
GPD	Gallons per Day	kV	Kilovolt		
gpf	Gallon per flush	kVA	Kilovolt Ampere	MHz	Megahertz
GPH	Gallons per Hour	kVAR	Kilovar (Reactance)	Mi.	Mile
GPM	Gallons per Minute	KW	Kilowatt	MI	Malleable Iron; Mineral Insulated
GR	Grade	KWh	Kilowatt-hour	MIPS	Male Iron Pipe Size
Gran.	Granular	L	Labor Only; Length; Long; Medium Wall Copper Tubing	mj	Mechanical Joint
Grnd.	Ground			m	Meter
GVW	Gross Vehicle Weight	Lab.	Labor	mm	Millimeter
GWB	Gypsum wall board	lat	Latitude	Mill.	Millwright
				Min., min.	Minimum, minute

Misc.	Miscellaneous	PDCA	Painting and Decorating Contractors of America	SC	Screw Cover
ml	Milliliter, Mainline			SCFM	Standard Cubic Feet per Minute
M.L.F.	Thousand Linear Feet	P.E., PE	Professional Engineer;	Scaf.	Scaffold
Mo.	Month		Porcelain Enamel;	Sch., Sched.	Schedule
Mobil.	Mobilization		Polyethylene; Plain End	S.C.R.	Modular Brick
Mog.	Mogul Base	Perf.	Perforated	S.D.	Sound Deadening
MPH	Miles per Hour	PEX	Cross linked polyethylene	SDR	Standard Dimension Ratio
MPT	Male Pipe Thread	Ph.	Phase	S.E.	Surfaced Edge
MRGWB	Moisture Resistant Gypsum	P.I.	Pressure Injected	Sel.	Select
	Wallboard	Pile.	Pile Driver	SER, SEU	Service Entrance Cable
MRT	Mile Round Trip	Pkg.	Package	S.F.	Square Foot
ms	Millisecond	Pl.	Plate	S.F.C.A.	Square Foot Contact Area
M.S.F.	Thousand Square Feet	Plah.	Plasterer Helper	S.F. Flr.	Square Foot of Floor
Mstz.	Mosaic & Terrazzo Worker	Plas.	Plasterer	S.F.G.	Square Foot of Ground
M.S.Y.	Thousand Square Yards	plf	Pounds Per Linear Foot	S.F. Hor.	Square Foot Horizontal
Mtd., mtd., mtd	Mounted	Pluh.	Plumbers Helper	SFR	Square Feet of Radiation
Mthe.	Mosaic & Terrazzo Helper	Plum.	Plumber	S.F. Shlf.	Square Foot of Shelf
Mtng.	Mounting	Ply.	Plywood	S4S	Surface 4 Sides
Mult.	Multi; Multiply	p.m.	Post Meridiem	Shee.	Sheet Metal Worker
M.V.A.	Million Volt Amperes	Pntd.	Painted	Sin.	Sine
M.V.A.R.	Million Volt Amperes Reactance	Pord.	Painter, Ordinary	Skwk.	Skilled Worker
MV	Megavolt	pp	Pages	SL	Saran Lined
MW	Megawatt	PP, PPL	Polypropylene	S.L.	Slimline
MXM	Male by Male	P.P.M.	Parts per Million	Sldr.	Solder
MYD	Thousand Yards	Pr.	Pair	SLH	Super Long Span Bar Joist
N	Natural; North	P.E.S.B.	Pre-engineered Steel Building	S.N.	Solid Neutral
nA	Nanoampere	Prefab.	Prefabricated	SO	Stranded with oil resistant inside
NA	Not Available; Not Applicable	Prefin.	Prefinished		insulation
N.B.C.	National Building Code	Prop.	Propelled	S-O-L	Socketolet
NC	Normally Closed	PSF, psf	Pounds per Square Foot	sp	Standpipe
NEMA	National Electrical Manufacturers	PSI, psi	Pounds per Square Inch	S.P.	Static Pressure; Single Pole; Self-
	Assoc.	PSIG	Pounds per Square Inch Gauge		Propelled
NEHB	Bolted Circuit Breaker to 600V.	PSP	Plastic Sewer Pipe	Spri.	Sprinkler Installer
NFPA	National Fire Protection Association	Pspr.	Painter, Spray	spwg	Static Pressure Water Gauge
NLB	Non-Load-Bearing	Psst.	Painter, Structural Steel	S.P.D.T.	Single Pole, Double Throw
NM	Non-Metallic Cable	P.T.	Potential Transformer	SPF	Spruce Pine Fir
nm	Nanometer	P. & T.	Pressure & Temperature	S.P.S.T.	Single Pole, Single Throw
No.	Number	Ptd.	Painted	SPT	Standard Pipe Thread
NO	Normally Open	Ptns.	Partitions	Sq.	Square; 100 Square Feet
N.O.C.	Not Otherwise Classified	Pu	Ultimate Load	Sq. Hd.	Square Head
Nose.	Nosing	PVC	Polyvinyl Chloride	Sq. In.	Square Inch
NPT	National Pipe Thread	Pvmt.	Pavement	S.S.	Single Strength; Stainless Steel
NQOD	Combination Plug-on/Bolt on	PRV	Pressure Relief Valve	S.S.B.	Single Strength B Grade
	Circuit Breaker to 240V.	Pwr.	Power	sst, ss	Stainless Steel
N.R.C., NRC	Noise Reduction Coefficient/	Q	Quantity Heat Flow	Sswk.	Structural Steel Worker
	Nuclear Regulator Commission	Qt.	Quart	Sswl.	Structural Steel Welder
N.R.S.	Non Rising Stem	Quan., Qty.	Quantity	St.; Stl.	Steel
ns	Nanosecond	Q.C.	Quick Coupling	STC	Sound Transmission Coefficient
nW	Nanowatt	r	Radius of Gyration	Std.	Standard
OB	Opposing Blade	R	Resistance	Stg.	Staging
OC	On Center	R.C.P.	Reinforced Concrete Pipe	STK	Select Tight Knot
OD	Outside Diameter	Rect.	Rectangle	STP	Standard Temperature & Pressure
O.D.	Outside Dimension	recpt.	receptacle	Stpi.	Steamfitter, Pipefitter
ODS	Overhead Distribution System	Reg.	Regular	Str.	Strength; Starter; Straight
O.G.	Ogee	Reinf.	Reinforced	Strd.	Stranded
O.H.	Overhead	Req'd.	Required	Struct.	Structural
O&P	Overhead and Profit	Res.	Resistant	Sty.	Story
Oper.	Operator	Resi.	Residential	Subj.	Subject
Opng.	Opening	RF	Radio Frequency	Subs.	Subcontractors
Orna.	Ornamental	RFID	Radio-frequency identification	Surf.	Surface
OSB	Oriented Strand Board	Rgh.	Rough	Sw.	Switch
OS&Y	Outside Screw and Yoke	RGS	Rigid Galvanized Steel	Swbd.	Switchboard
OSHA	Occupational Safety and Health	RHW	Rubber, Heat & Water Resistant;	S.Y.	Square Yard
	Act		Residential Hot Water	Syn.	Synthetic
Ovhd.	Overhead	rms	Root Mean Square	S.Y.P.	Southern Yellow Pine
OWG	Oil, Water or Gas	Rnd.	Round	Sys.	System
Oz.	Ounce	Rodm.	Rodman	t.	Thickness
P.	Pole; Applied Load; Projection	Rofc.	Roofer, Composition	T	Temperature; Ton
p.	Page	Rofp.	Roofer, Precast	Tan	Tangent
Pape.	Paperhanger	Rohe.	Roofer Helpers (Composition)	T.C.	Terra Cotta
P.A.P.R.	Powered Air Purifying Respirator	Rots.	Roofer, Tile & Slate	T & C	Threaded and Coupled
PAR	Parabolic Reflector	R.O.W.	Right of Way	T.D.	Temperature Difference
P.B., PB	Push Button	RPM	Revolutions per Minute	Tdd	Telecommunications Device for
Pc., Pcs.	Piece, Pieces	R.S.	Rapid Start		the Deaf
P.C.	Portland Cement; Power Connector	Rsr	Riser	T.E.M.	Transmission Electron Microscopy
P.C.F.	Pounds per Cubic Foot	RT	Round Trip	temp	Temperature, Tempered, Temporary
PCM	Phase Contrast Microscopy	S.	Suction; Single Entrance; South	TFFN	Nylon Jacketed Wire

TFE	Tetrafluoroethylene (Teflon)	U.L., UL	Underwriters Laboratory	w/	With
T. & G.	Tongue & Groove;	Uld.	unloading	W.C., WC	Water Column; Water Closet
	Tar & Gravel	Unfin.	Unfinished	W.F.	Wide Flange
Th., Thk.	Thick	UPS	Uninterruptible Power Supply	W.G.	Water Gauge
Thn.	Thin	URD	Underground Residential	Wldg.	Welding
Thrded	Threaded		Distribution	W. Mile	Wire Mile
Tilf.	Tile Layer, Floor	US	United States	W-O-L	Weldolet
Tilh.	Tile Layer, Helper	USGBC	U.S. Green Building Council	W.R.	Water Resistant
THHN	Nylon Jacketed Wire	USP	United States Primed	Wrck.	Wrecker
THW.	Insulated Strand Wire	UTMCD	Uniform Traffic Manual For Control	W.S.P.	Water, Steam, Petroleum
THWN	Nylon Jacketed Wire		Devices	WT., Wt.	Weight
T.L., TL	Truckload	UTP	Unshielded Twisted Pair	WWF	Welded Wire Fabric
T.M.	Track Mounted	V	Volt	XFER	Transfer
Tot.	Total	VA	Volt Amperes	XFMR	Transformer
T-O-L	Threadolet	V.C.T.	Vinyl Composition Tile	XHD	Extra Heavy Duty
tmpd	Tempered	VAV	Variable Air Volume	XHHW,	Cross-Linked Polyethylene Wire
T.S.	Trigger Start	VC	Veneer Core	XLPE	Insulation
Tr.	Trade	VDC	Volts Direct Current	XLP	Cross-linked Polyethylene
Transf.	Transformer	Vent.	Ventilation	Xport	Transport
Trhv.	Truck Driver, Heavy	Vert.	Vertical	Y	Wye
Trlr	Trailer	V.F.	Vinyl Faced	yd	Yard
Trlt.	Truck Driver, Light	V.G.	Vertical Grain	yr	Year
TTY	Teletypewriter	VHF	Very High Frequency	Δ	Delta
TV	Television	VHO	Very High Output	%	Percent
T.W.	Thermoplastic Water Resistant	Vib.	Vibrating	~	Approximately
	Wire	VLF	Vertical Linear Foot	∅	Phase; diameter
UCI	Uniform Construction Index	VOC	Volitile Organic Compound	@	At
UF	Underground Feeder	Vol.	Volume	#	Pound; Number
UGND	Underground Feeder	VRP	Vinyl Reinforced Polyester	<	Less Than
UHF	Ultra High Frequency	W	Wire; Watt; Wide; West	>	Greater Than
U.I.	United Inch			Z	zone

RESIDENTIAL
COST ESTIMATE

OWNER'S NAME: _____

APPRAISER: _____

RESIDENCE ADDRESS: _____

PROJECT: _____

CITY, STATE, ZIP CODE: _____

DATE: _____

CLASS OF CONSTRUCTION
- ☐ ECONOMY
- ☐ AVERAGE
- ☐ CUSTOM
- ☐ LUXURY

RESIDENCE TYPE
- ☐ 1 STORY
- ☐ 1-1/2 STORY
- ☐ 2 STORY
- ☐ 2-1/2 STORY
- ☐ 3 STORY
- ☐ BI-LEVEL
- ☐ TRI-LEVEL

CONFIGURATION
- ☐ DETACHED
- ☐ TOWN/ROW HOUSE
- ☐ SEMI-DETACHED

OCCUPANCY
- ☐ ONE STORY
- ☐ TWO FAMILY
- ☐ THREE FAMILY
- ☐ OTHER _____

EXTERIOR WALL SYSTEM
- ☐ WOOD SIDING—WOOD FRAME
- ☐ BRICK VENEER—WOOD FRAME
- ☐ STUCCO ON WOOD FRAME
- ☐ PAINTED CONCRETE BLOCK
- ☐ SOLID MASONRY (AVERAGE & CUSTOM)
- ☐ STONE VENEER—WOOD FRAME
- ☐ SOLID BRICK (LUXURY)
- ☐ SOLID STONE (LUXURY)

*LIVING AREA (Main Building)

First Level	_____ S.F.
Second Level	_____ S.F.
Third Level	_____ S.F.
Total	_____ S.F.

*LIVING AREA (Wing or Ell) ( )

First Level	_____ S.F.
Second Level	_____ S.F.
Third Level	_____ S.F.
Total	_____ S.F.

*LIVING AREA (Wing or Ell) ( )

First Level	_____ S.F.
Second Level	_____ S.F.
Third Level	_____ S.F.
Total	_____ S.F.

*Basement Area is not part of living area.

MAIN BUILDING	COSTS PER S.F. LIVING AREA
Cost per Square Foot of Living Area, from Page _____	$
Basement Addition: _____ % Finished, _____ % Unfinished	+
Roof Cover Adjustment: _____ Type, Page _____ (Add or Deduct)	( )
Central Air Conditioning: ☐ Separate Ducts ☐ Heating Ducts, Page _____	+
Heating System Adjustment: _____ Type, Page _____ (Add or Deduct)	( )
Main Building: Adjusted Cost per S.F. of Living Area	$

MAIN BUILDING TOTAL COST

$ _____ /S.F. (Cost per S.F. Living Area) × _____ S.F. (Living Area) × _____ (Town/Row House Multiplier, Use 1 for Detached) = $ _____ (TOTAL COST)

WING OR ELL ( ) _____ STORY	COSTS PER S.F. LIVING AREA
Cost per Square Foot of Living Area, from Page _____	$
Basement Addition: _____ % Finished, _____ % Unfinished	+
Roof Cover Adjustment: _____ Type, Page _____ (Add or Deduct)	( )
Central Air Conditioning: ☐ Separate Ducts ☐ Heating Ducts, Page _____	+
Heating System Adjustment: _____ Type, Page _____ (Add or Deduct)	( )
Wing or Ell ( ): Adjusted Cost per S.F. of Living Area	$

WING OR ELL ( ) TOTAL COST

$ _____ /S.F. (Cost per S.F. Living Area) × _____ S.F. (Living Area) = $ _____ (TOTAL COST)

WING OR ELL ( ) _____ STORY	COSTS PER S.F. LIVING AREA
Cost per Square Foot of Living Area, from Page _____	$
Basement Addition: _____ % Finished, _____ % Unfinished	+
Roof Cover Adjustment: _____ Type, Page _____ (Add or Deduct)	( )
Central Air Conditioning: ☐ Separate Ducts ☐ Heating Ducts, Page _____	+
Heating System Adjustment: _____ Type, Page _____ (Add or Deduct)	( )
Wing or Ell ( ): Adjusted Cost per S.F. of Living Area	$

WING OR ELL ( ) TOTAL COST

$ _____ /S.F. (Cost per S.F. Living Area) × _____ S.F. (Living Area) = $ _____ (TOTAL COST)

TOTAL THIS PAGE _____

Page 1 of 2

RESIDENTIAL
COST ESTIMATE

Total Page 1				$
		QUANTITY	UNIT COST	
Additional Bathrooms: _____ Full, _____ Half				
Finished Attic: _____ Ft. x _____ Ft.		S.F.		+
Breezeway: ☐ Open ☐ Enclosed _____ Ft. x _____ Ft.		S.F.		+
Covered Porch: ☐ Open ☐ Enclosed _____ Ft. x _____ Ft.		S.F.		+
Fireplace: ☐ Interior Chimney ☐ Exterior Chimney ☐ No. of Flues ☐ Additional Fireplaces				+
Appliances:				+
Kitchen Cabinets Adjustment:	(+/−)			
☐ Garage ☐ Carport: _____ Car(s) Description _____ (+/−)				
Miscellaneous:				+

ADJUSTED TOTAL BUILDING COST $ [   ]

REPLACEMENT COST	
ADJUSTED TOTAL BUILDING COST	$ _____
Site Improvements	
(A) Paving & Sidewalks	$ _____
(B) Landscaping	$ _____
(C) Fences	$ _____
(D) Swimming Pool	$ _____
(E) Miscellaneous	$ _____
TOTAL	$ _____
Location Factor	x _____
Location Replacement Cost	$ _____
Depreciation	−$ _____
LOCAL DEPRECIATED COST	$ _____

INSURANCE COST	
ADJUSTED TOTAL BUILDING COST	$ _____
Insurance Exclusions	
(A) Footings, Site Work, Underground Piping	−$ _____
(B) Architects' Fees	−$ _____
Total Building Cost Less Exclusion	$ _____
Location Factor	x _____
LOCAL INSURABLE REPLACEMENT COST	$ _____

SKETCH AND ADDITIONAL CALCULATIONS

CII APPRAISAL

1. SUBJECT PROPERTY: _____
2. BUILDING: _____
3. ADDRESS: _____
4. BUILDING USE: _____

5. DATE: _____
6. APPRAISER: _____

7. YEAR BUILT: _____

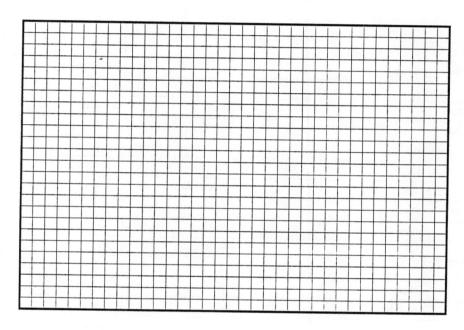

8. EXTERIOR WALL CONSTRUCTION: _____
9. FRAME: _____
10. GROUND FLOOR AREA: _____ S.F.   11. GROSS FLOOR AREA (EXCL. BASEMENT): _____ S.F.
12. NUMBER OF STORIES: _____   13. STORY HEIGHT: _____
14. PERIMETER: _____ L.F.   15. BASEMENT AREA: _____ S.F.
16. GENERAL COMMENTS: _____

_____
_____
_____
_____
_____
_____
_____

	NO.	DESCRIPTION		UNIT	UNIT COST	NEW SF COST	MODEL SF COST	+/- CHANGE
A SUBSTRUCTURE								
A	1010	Standard Foundations	Bay size:	S.F. Gnd.				
A	1030	Slab on Grade	Material: Thickness:	S.F. Slab				
A	2010	Basement Excavation	Depth: Area:	S.F. Gnd.				
A	2020	Basement Walls		L.F. Walls				
B SHELL								
		B10 Superstructure						
B	1010	Floor Construction	Elevated floors:	S.F. Floor				
				S.F. Floor				
B	1020	Roof Construction		S.F. Roof				
		B20 Exterior Enclosure						
B	2010	Exterior walls Material:	Thickness: % of wall	S.F. Walls				
		Material:	Thickness: % of wall	S.F. Walls				
B	2020	Exterior Windows Type:	% of wall	S.F. Wind.				
		Type:	% of wall Each	S.F. Wind.				
B	2030	Exterior Doors Type:	Number:	Each				
		Type:	Number:	Each				
		B30 Roofing						
B	3010	Roof Coverings Material:		S.F. Roof				
		Material:		S.F. Roof				
B	3020	Roof Openings		S.F. Opng.				
C INTERIORS								
C	1010	Partitions: Material:	Density:	S.F. Part.				
		Material:	Density:					
C	1020	Interior Doors Type:	Number:	Each				
C	1030	Fittings		Each				
C	2010	Stair Construction		Flight				
C	3010	Wall Finishes Material:	% of Wall	S.F. Walls				
		Material:		S.F. Walls				
C	3020	Floor Finishes Material:		S.F. Floor				
		Material:		S.F. Floor				
		Material:		S.F. Floor				
C	3030	Ceiling Finishes Material:		S.F. Ceil.				

	NO.	SYSTEM/COMPONENT	DESCRIPTION	UNIT	UNIT COST	NEW S.F. COST	MODEL S.F. COST	+/- CHANGE
D	**SERVICES**							
	D10 Conveying							
D	1010	Elevators & Lifts	Type: Capacity: Stops:	Each				
D	1020	Escalators & Moving Walks	Type:	Each				
	D20 Plumbing							
D	2010	Plumbing		Each				
D	2020	Domestic Water Distribution		S.F. Floor				
D	2040	Rain Water Drainage		S.F. Roof				
	D30 HVAC							
D	3010	Energy Supply		Each				
D	3020	Heat Generating Systems	Type:	S.F. Floor				
D	3030	Cooling Generating Systems	Type:	S.F. Floor				
D	3090	Other HVAC Sys. & Equipment		Each				
	D40 Fire Protection							
D	4010	Sprinklers		S.F. Floor				
D	4020	Standpipes		S.F. Floor				
	D50 Electrical							
D	5010	Electrical Service/Distribution		S.F. Floor				
D	5020	Lighting & Branch Wiring		S.F. Floor				
D	5030	Communications & Security		S.F. Floor				
D	5090	Other Electrical Systems		S.F. Floor				
E	**EQUIPMENT & FURNISHINGS**							
E	1010	Commercial Equipment		Each				
E	1020	Institutional Equipment		Each				
E	1030	Vehicular Equipment		Each				
E	1090	Other Equipment		Each				
F	**SPECIAL CONSTRUCTION**							
F	1020	Integrated Construction		S.F.				
F	1040	Special Facilities		S.F.				
G	**BUILDING SITEWORK**							

ITEM		Total Change
17	Model 020 total ____ $ _____	
18	Adjusted S.F. cost ____ $ _____ item 17 +/- changes	
19	Building area - from item 11 _____ S.F. x adjusted S.F. cost...$...	
20	Basement area - from item 15 _____ S.F. x S.F. cost $ ....$..	
21	Base building sub-total - item 20 + item 19 ....$.....	
22	Miscellaneous addition (quality, etc.) ....$......	
23	Sub-total - item 22 + 21 ....$.....	
24	General conditions -25 % of item 23 ....$.....	
25	Sub-total - item 24 + item 23..$.....	
26	Architects fees _____ % of item 25 ....$	
27	Sub-total - item 26 + item 27 ....$.....	
28	Location modifier ....x...	
29	Local replacement cost - item 28 x item 27 ....$.....	
30	Depreciation _____ % of item 29 ....$	
31	Depreciated local replacement cost - item 29 less item 30 ....$.....	
32	Exclusions ....$.......	
33	Net depreciated replacement cost - item 31 less item 32 ....$....	

Base Cost per square foot floor area: *(from square foot table)*

Specify source:

Page: _____ , Model # _____ , Area _____ S.F.,
Exterior wall _____ , Frame _____

Adjustments for exterior wall variation:

Size Adjustment:
(Interpolate)

Height Adjustment: _____ + _____ = _____

Adjusted Base Cost per square foot:

Building Cost $ _____ x _____ = _____
Adjusted Base Cost   *Floor Area*
per square foot

Basement Cost $ _____ x _____ = _____
Basement Cost   *Basement Area*

Lump Sum Additions

TOTAL BUILDING COST *(Sum of above costs)* _____

Modifications: *(complexity, workmanship, size)* +/- _____ % _____

Location Modifier: City _____ Date _____ x _____

Local cost of replacement _____

Less depreciation: _____ - _____

Local cost of replacement less depreciation $ _____

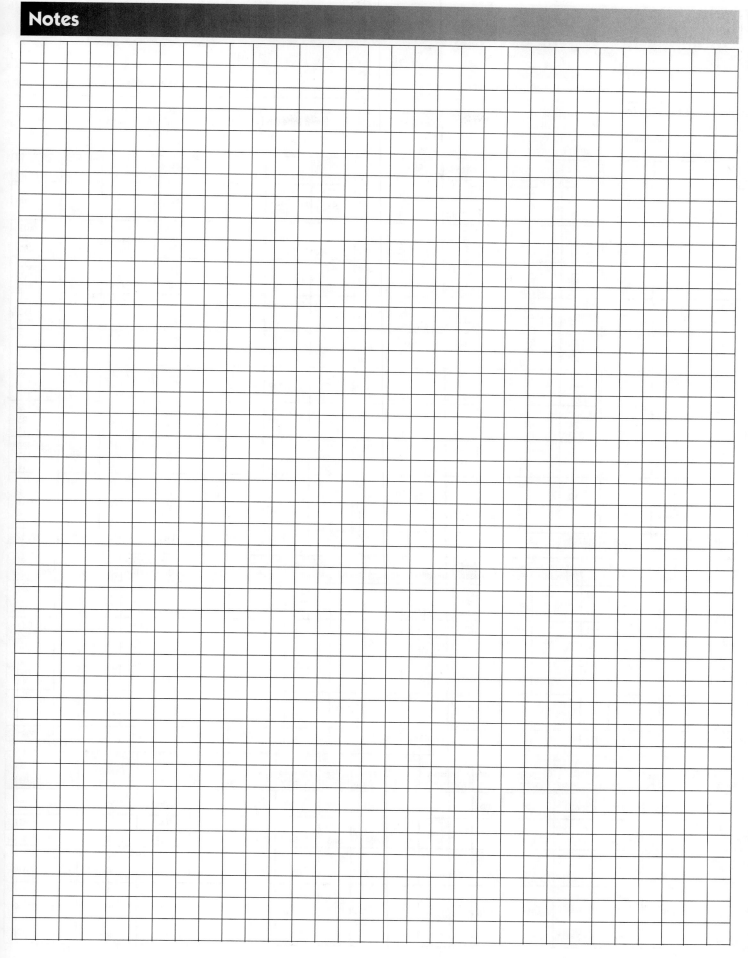

For more information visit the RSMeans website at **www.rsmeans.com**

Unit prices according to the latest MasterFormat!

Book Selection Guide The following table provides definitive information on the content of each cost data publication. The number of lines of data provided in each unit price or assemblies division, as well as the number of reference tables and crews, is listed for each book. The presence of other elements such as equipment rental costs, historical cost indexes, city cost indexes, square foot models, or cross-referenced indexes is also indicated. You can use the table to help select the RSMeans book that has the quantity and type of information you most need in your work.

Unit Cost Divisions	Building Construction	Mechanical	Electrical	Commercial Renovation	Square Foot	Site Work Landsc.	Green Building	Interior	Concrete Masonry	Open Shop	Heavy Construction	Light Commercial	Facilities Construction	Plumbing	Residential
1	573	395	410	499		517	124	319	463	572	507	233	1038	411	176
2	755	281	87	706		1005	180	366	223	751	744	447	1197	291	258
3	1678	346	223	1053		1474	1004	323	2037	1678	1680	469	1774	315	376
4	900	22	0	698		701	183	573	1135	876	591	490	1147	0	405
5	1844	159	156	1040		809	1777	1047	712	1835	1031	924	1845	204	720
6	2294	67	69	1881		411	534	1421	300	2273	383	1968	1908	22	2480
7	1467	215	125	1501		563	776	572	497	1464	26	1178	1529	224	942
8	2122	98	48	2351		299	1157	1768	657	2066	0	1790	2581	0	1410
9	1927	72	26	1720		313	449	2012	389	1868	15	1639	2171	54	1416
10	987	17	10	610		215	27	807	156	987	29	497	1073	230	221
11	1022	208	165	486		126	54	878	28	1007	0	231	1042	169	110
12	563	0	2	325		248	150	1698	12	552	19	371	1748	23	324
13	727	117	112	246		346	128	251	66	726	253	82	754	71	80
14	277	36	0	225		28	0	259	0	276	0	12	296	18	6
21	78	0	16	30		0	0	248	0	78	0	60	429	435	224
22	1140	7318	154	1105		1536	1100	605	20	1102	1659	830	7152	9029	659
23	1211	7086	599	846		158	940	695	38	1132	111	752	5176	1860	416
26	1344	499	10087	1019		816	644	1132	55	1329	562	1218	10,011	444	618
27	72	0	266	34		13	0	60	0	72	39	52	268	0	4
28	96	58	124	71		0	21	78	0	99	0	40	139	44	25
31	1482	735	610	803		3207	290	7	1208	1426	3284	600	1545	660	611
32	770	54	0	836		4341	341	362	274	741	1796	381	1647	166	428
33	509	1014	461	207		2027	41	0	231	508	1937	119	1557	1205	117
34	109	0	20	4		192	0	0	31	64	170	0	129	0	0
35	18	0	0	0		327	0	0	0	18	442	0	83	0	0
41	52	0	0	24		7	0	22	0	52	30	0	59	14	0
44	75	83	0	0		0	0	0	0	0	0	0	75	75	0
46	23	16	0	0		274	181	0	0	23	264	0	33	33	0
48	13	0	28	0		4	28	0		13	4	28	13		26
Totals	24128	18896	13798	18320		19957	10,129	15503	8532	23588	15576	14411	48419	15997	12052

Assem Div	Building Construction	Mechanical	Electrical	Commercial Renovation	Square Foot	Site Work Landscape	Assemblies	Green Building	Interior	Concrete Masonry	Heavy Construction	Light Commercial	Facilities Construction	Plumbing	Asm Div	Residential
A		15	0	188	150	577	598	0	0	536	571	154	24	0	1	368
B		0	0	848	2483	0	5643	56	329	1966	368	2074	174	0	2	211
C		0	0	647	863	0	1233	0	1558	146	0	763	245	0	3	588
D		1067	886	712	1864	72	2464	265	826	0	0	1348	1084	1083	4	839
E		0	0	86	260	0	298	0	5	0	0	257	5	0	5	392
F		0	0	0	115	0	115	0	0	0	0	115	3	0	6	357
G		527	447	318	202	3456	668	0	0	535	1431	200	293	707	7	296
															8	760
															9	80
															10	0
															11	0
															12	0
Totals		1609	1333	2799	5937	4105	11019	321	2718	3183	2370	4911	1828	1790		3891

Reference Section	Building Construction Costs	Mechanical	Electrical	Commercial Renovation	Square Foot	Site Work Landscape	Assem.	Green Building	Interior	Concrete Masonry	Open Shop	Heavy Construction	Light Commercial	Facilities Construction	Plumbing	Resi.
Reference Tables	yes	yes	yes	yes	no	yes	yes	yes	yes	yes	yes	yes	yes	yes	yes	yes
Models					109								48			28
Crews	558	558	558	536		558		558	558	558	534	558	534	536	558	534
Equipment Rental Costs	yes	yes	yes	yes		yes		yes	yes	yes	yes	yes	yes	yes	yes	
Historical Cost Indexes	yes	yes	yes	yes	yes	yes	yes	yes	yes	yes	yes	yes	yes	yes	yes	no
City Cost Indexes	yes	yes	yes	yes	yes	yes	yes	yes	yes	yes	yes	yes	yes	yes	yes	yes

RSMeans Building Construction Cost Data 2013

Offers you unchallenged unit price reliability in an easy-to-use format. Whether used for verifying complete, finished estimates or for periodic checks, it supplies more cost facts better and faster than any comparable source. Over 23,000 unit prices have been updated for 2013. The City Cost Indexes and Location Factors cover over 930 areas, for indexing to any project location in North America. Order and get *RSMeans Quarterly Update Service* FREE. You'll have year-long access to the RSMeans Estimating **HOTLINE** FREE with your subscription. Expert assistance when using RSMeans data is just a phone call away.

$186.95 per copy | Available Sept. 2012 | Catalog no. 60013

RSMeans Mechanical Cost Data 2013

Total unit and systems price guidance for mechanical construction. . . materials, parts, fittings, and complete labor cost information. Includes prices for piping, heating, air conditioning, ventilation, and all related construction.

Plus new 2013 unit costs for:

- Thousands of installed HVAC/controls assemblies components

- "On-site" Location Factors for over 930 cities and towns in the U.S. and Canada

- Crews, labor, and equipment

$183.95 per copy | Available Oct. 2012 | Catalog no. 60023

RSMeans Square Foot Costs 2013

Accurate and Easy To Use

- **Updated price information** based on nationwide figures from suppliers, estimators, labor experts, and contractors

- "How-to-Use" sections, with **clear examples** of commercial, residential, industrial, and institutional structures

- Realistic graphics, offering true-to-life illustrations of building projects

- Extensive information on using square foot cost data, including sample estimates and alternate pricing methods

$197.95 per copy | Available Oct. 2012 | Catalog no. 60053

RSMeans Green Building Cost Data 2013

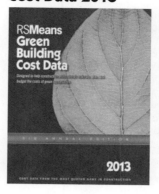

Estimate, plan, and budget the costs of green building for both new commercial construction and renovation work with this first edition of RSMeans Green Building Cost Data. More than 9,000 unit costs for a wide array of green building products plus assemblies costs. Easily identified cross references to LEED and Green Globes building rating systems criteria.

$155.95 per copy | Available Nov. 2012 | Catalog no. 60553

RSMeans Facilities Construction Cost Data 2013

For the maintenance and construction of commercial, industrial, municipal, and institutional properties. Costs are shown for new and remodeling construction and are broken down into materials, labor, equipment, and overhead and profit. Special emphasis is given to sections on mechanical, electrical, furnishings, site work, building maintenance, finish work, and demolition.

More than 47,000 unit costs, plus assemblies costs and a comprehensive Reference Section are included.

$476.95 per copy | Available Nov. 2012 | Catalog no. 60203

RSMeans Commercial Renovation Cost Data 2013

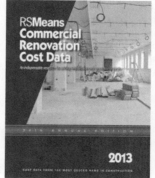

Commercial/Multifamily Residential

Use this valuable tool to estimate commercial and multifamily residential renovation and remodeling.

Includes: New costs for hundreds of unique methods, materials, and conditions that only come up in repair and remodeling, PLUS:

- Unit costs for more than 17,000 construction components
- Installed costs for more than 2,000 assemblies
- More than 930 "on-site" localization factors for the U.S. and Canada

$152.95 per copy | Available Oct. 2012 | Catalog no. 60043

RSMeans Electrical Cost Data 2013

Pricing information for every part of electrical cost planning. More than 13,000 unit and systems costs with design tables; clear specifications and drawings; engineering guides; illustrated estimating procedures; complete labor-hour and materials costs for better scheduling and procurement; and the latest electrical products and construction methods.

- A variety of special electrical systems including cathodic protection
- Costs for maintenance, demolition, HVAC/mechanical, specialties, equipment, and more

$188.95 per copy | Available Oct. 2012 | Catalog no. 60033

RSMeans Electrical Change Order Cost Data 2013

RSMeans Electrical Change Order Cost Data provides you with electrical unit prices exclusively for pricing change orders—based on the recent, direct experience of contractors and suppliers. Analyze and check your own change order estimates against the experience others have had doing the same work. It also covers productivity analysis and change order cost justifications. With useful information for calculating the effects of change orders and dealing with their administration.

$183.95 per copy | Available Dec. 2012 | Catalog no. 60233

RSMeans Assemblies Cost Data 2013

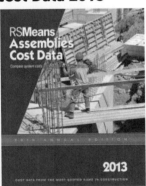

RSMeans Assemblies Cost Data takes the guesswork out of preliminary or conceptual estimates. Now you don't have to try to calculate the assembled cost by working up individual component costs. We've done all the work for you.

Presents detailed illustrations, descriptions, specifications, and costs for every conceivable building assembly—over 350 types in all—arranged in the easy-to-use UNIFORMAT II system. Each illustrated "assembled" cost includes a complete grouping of materials and associated installation costs, including the installing contractor's overhead and profit.

$305.95 per copy | Available Sept. 2012 | Catalog no. 60063

RSMeans Open Shop Building Construction Cost Data 2013

The latest costs for accurate budgeting and estimating of new commercial and residential construction. . . renovation work. . . change orders. . . cost engineering.

RSMeans Open Shop "BCCD" will assist you to:

- Develop benchmark prices for change orders
- Plug gaps in preliminary estimates and budgets
- Estimate complex projects
- Substantiate invoices on contracts
- Price ADA-related renovations

$160.95 per copy | Available Dec. 2012 | Catalog no. 60153

RSMeans Residential Cost Data 2013

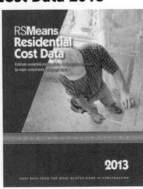

Contains square foot costs for 28 basic home models with the look of today, plus hundreds of custom additions and modifications you can quote right off the page. Includes costs for more than 700 residential systems. Complete with blank estimating forms, sample estimates, and step-by-step instructions.

Now contains line items for cultured stone and brick, PVC trim lumber, and TPO roofing.

$134.95 per copy | Available Oct. 2012 | Catalog no. 60173

RSMeans Site Work & Landscape Cost Data 2013

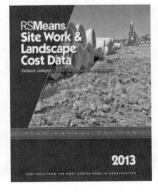

Includes unit and assemblies costs for earthwork, sewerage, piped utilities, site improvements, drainage, paving, trees and shrubs, street openings/repairs, underground tanks, and more. Contains 78 types of assemblies costs for accurate conceptual estimates.

Includes:

- Estimating for infrastructure improvements
- Environmentally-oriented construction
- ADA-mandated handicapped access
- Hazardous waste line items

$181.95 per copy | Available Dec. 2012 | Catalog no. 60283

RSMeans Facilities Maintenance & Repair Cost Data 2013

RSMeans Facilities Maintenance & Repair Cost Data gives you a complete system to manage and plan your facility repair and maintenance costs and budget efficiently. Guidelines for auditing a facility and developing an annual maintenance plan. Budgeting is included, along with reference tables on cost and management, and information on frequency and productivity of maintenance operations.

The only nationally recognized source of maintenance and repair costs. Developed in cooperation with the Civil Engineering Research Laboratory (CERL) of the Army Corps of Engineers.

$417.95 per copy | Available Nov. 2012 | Catalog no. 60303

RSMeans Concrete & Masonry Cost Data 2013

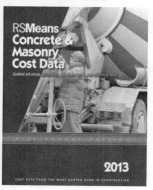

Provides you with cost facts for virtually all concrete/masonry estimating needs, from complicated formwork to various sizes and face finishes of brick and block—all in great detail. The comprehensive Unit Price Section contains more than 8,000 selected entries. Also contains an Assemblies [Cost] Section, and a detailed Reference Section that supplements the cost data.

$171.95 per copy | Available Dec. 2012 | Catalog no. 60113

RSMeans Construction Cost Indexes 2013

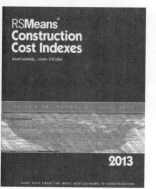

What materials and labor costs will change unexpectedly this year? By how much?

- Breakdowns for 318 major cities
- National averages for 30 key cities
- Expanded five major city indexes
- Historical construction cost indexes

$362.00 per year (subscription) | Catalog no. 50143
$90.50 individual quarters | Catalog no. 60143 A,B,C,D

RSMeans Light Commercial Cost Data 2013

Specifically addresses the light commercial market, which is a specialized niche in the construction industry. Aids you, the owner/designer/contractor, in preparing all types of estimates—from budgets to detailed bids. Includes new advances in methods and materials.

Assemblies Section allows you to evaluate alternatives in the early stages of design/planning.

Over 13,000 unit costs ensure that you have the prices you need. . . when you need them.

$139.95 per copy | Available Nov. 2012 | Catalog no. 60183

RSMeans Labor Rates for the Construction Industry 2013

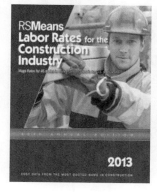

Complete information for estimating labor costs, making comparisons, and negotiating wage rates by trade for more than 300 U.S. and Canadian cities. With 46 construction trades listed by local union number in each city, and historical wage rates included for comparison. Each city chart lists the county and is alphabetically arranged with handy visual flip tabs for quick reference.

$415.95 per copy | Available Dec. 2012 | Catalog no. 60123

RSMeans Interior Cost Data 2013

Provides you with prices and guidance needed to make accurate interior work estimates. Contains costs on materials, equipment, hardware, custom installations, furnishings, and labor costs. . . for new and remodel commercial and industrial interior construction, including updated information on office furnishings, and reference information.

$192.95 per copy | Available Nov. 2012 | Catalog no. 60093

RSMeans Heavy Construction Cost Data 2013

A comprehensive guide to heavy construction costs. Includes costs for highly specialized projects such as tunnels, dams, highways, airports, and waterways. Information on labor rates, equipment, and materials costs is included. Features unit price costs, systems costs, and numerous reference tables for costs and design.

$186.95 per copy | Available Dec. 2012 | Catalog no. 60163

RSMeans Plumbing Cost Data 2013

Comprehensive unit prices and assemblies for plumbing, irrigation systems, commercial and residential fire protection, point-of-use water heaters, and the latest approved materials. This publication and its companion, *RSMeans Mechanical Cost Data*, provide full-range cost estimating coverage for all the mechanical trades.

Now contains updated costs for potable water and radiant heat systems and high efficiency fixtures.

$188.95 per copy | Available Oct. 2012 | Catalog no. 60213

2013 RSMeans Seminar Schedule

Note: call for exact dates and details.

Location	Dates	Location	Dates
Las Vegas, NV	March	Las Vegas, NV	October
Washington, DC	April	Orlando, FL	November
San Francisco, CA	June	San Diego, CA	December
Washington, DC	September		
Dallas, TX	September		

1-800-334-3509, Press 1

2013 Regional Seminar Schedule

Note: call for exact dates and details.

Location	Dates	Location	Dates
Seattle, WA	January and August	Columbus, GA	June
Dallas/Ft. Worth, TX	January	El Segundo, CA	August
Austin, TX	February	Jacksonville FL	September
Oklahoma City, OK	February	St. Louis, MO	October
Anchorage, AK	March and September	Houston, TX	October
Phoenix, AZ	March	Vancouver, BC	November
San Bernardino, CA	April	Baltimore, MD	November
Kansas City, MO	April	San Antonio, TX	December
Toronto	May	Raleigh, NC	December
Denver, CO	May		
Bethesda, MD	June		

1-800-334-3509, Press 1

Registration Information

Register early... Save up to $100! Register 30 days before the start date of a seminar and save $100 off your total fee. *Note: This discount can be applied only once per order. It cannot be applied to team discount registrations or any other special offer.*

How to register Register by phone today! The RSMeans toll-free number for making reservations is **1-800-334-3509, Press 1.**

Individual seminar registration fee - $935. *Means CostWorks®* **training registration fee - $375.** To register by mail, complete the registration form and return, with your full fee, to: RSMeans Seminars, 700 Longwater Drive, Norwell, MA 02061.

Government pricing All federal government employees save off the regular seminar price. Other promotional discounts cannot be combined with the government discount.

Team discount program for two to four seminar registrations. Call for pricing: 1-800-334-3509, Press 1

Multiple course discounts When signing up for two or more courses, call for pricing.

Refund policy Cancellations will be accepted up to ten business days prior to the seminar start. There are no refunds for cancellations received later than ten working days prior to the first day of the seminar. A $150 processing fee will be applied for all cancellations. Written notice of cancellation is required. Substitutions can be made at any time before the session starts. **No-shows are subject to the full seminar fee.**

AACE approved courses Many seminars described and offered here have been approved for 14 hours (1.4 recertification credits) of credit by the AACE

International Certification Board toward meeting the continuing education requirements for recertification as a Certified Cost Engineer/Certified Cost Consultant.

AIA Continuing Education We are registered with the AIA Continuing Education System (AIA/CES) and are committed to developing quality learning activities in accordance with the CES criteria. Many seminars meet the AIA/CES criteria for Quality Level 2. AIA members may receive (14) learning units (LUs) for each two-day RSMeans course.

Daily course schedule The first day of each seminar session begins at 8:30 a.m. and ends at 4:30 p.m. The second day begins at 8:00 a.m. and ends at 4:00 p.m. Participants are urged to bring a hand-held calculator, since many actual problems will be worked out in each session.

Continental breakfast Your registration includes the cost of a continental breakfast, and a morning and afternoon refreshment break. These informal segments allow you to discuss topics of mutual interest with other seminar attendees. (You are free to make your own lunch and dinner arrangements.)

Hotel/transportation arrangements RSMeans arranges to hold a block of rooms at most host hotels. To take advantage of special group rates when making your reservation, be sure to mention that you are attending the RSMeans seminar. You are, of course, free to stay at the lodging place of your choice. (**Hotel reservations and transportation arrangements should be made directly by seminar attendees.**)

Important Class sizes are limited, so please register as soon as possible.

Note: Pricing subject to change.

Registration Form

ADDS-1000

Call 1-800-334-3509, Press 1 to register or FAX this form 1-800-632-6732. Visit our website: www.rsmeans.com

Please register the following people for the RSMeans construction seminars as shown here. We understand that we must make our own hotel reservations if overnight stays are necessary.

☐ Full payment of $_____enclosed.

☐ Bill me.

Please print name of registrant(s).

(To appear on certificate of completion)

_____

_____

_____

P.O. #: _____
GOVERNMENT AGENCIES MUST SUPPLY PURCHASE ORDER NUMBER OR TRAINING FORM.

Firm name_____

Address_____

City/State/Zip_____

Telephone no._____ Fax no._____

E-mail address_____

Charge registration(s) to: ☐ MasterCard ☐ VISA ☐ American Express

Account no._____Exp. date_____

Cardholder's signature_____

Seminar name_____

Seminar City _____

Please mail check to: RSMeans Seminars, 700 Longwater Drive, Norwell, MA 02061 USA

Professional Development

RSMeans Online™ Training

Construction estimating is vital to the decision-making process at each state of every project. Means CostWorks.com works the way you do. It's systematic, flexible and intuitive. In this one day class you will see how you can estimate any phase of any project faster and better.

Some of what you'll learn:
- Customizing Means CostWorks.com
- Making the most of RSMeans "Circle Reference" numbers
- How to integrate your cost data
- Generate reports, exporting estimates to MS Excel, sharing, collaborating and more

Also available: RSMeans Online™ training webinar

Facilities Construction Estimating

In this *two-day* course, professionals working in facilities management can get help with their daily challenges to establish budgets for all phases of a project.

Some of what you'll learn:
- Determining the full scope of a project
- Identifying the scope of risks and opportunities
- Creative solutions to estimating issues
- Organizing estimates for presentation and discussion
- Special techniques for repair/remodel and maintenance projects
- Negotiating project change orders

Who should attend: facility managers, engineers, contractors, facility tradespeople, planners, and project managers.

Scheduling with Microsoft Project

Two days of hands-on training gives you a look at the basics to putting it all together with MS Project to create a complete project schedule. Learn to work better and faster, manage changes and updates, enhance tracking, generate actionable reports, and boost control of your time and budget.

Some of what you'll learn:
- Defining task/activity relationships: link tasks, establish predecessor/successor relationships and create logs
- Baseline scheduling: the essential what/when
- Using MS Project to determine the critical path
- Integrate RSMeans data with MS Project
- The dollar-loaded schedule: enhance the reliability of this "must-know" technique for scheduling public projects.

Who should attend: contractors' project-management teams, architects and engineers, project owners and their representatives, and those interested in improving their project planning and scheduling and management skills.

Maintenance & Repair Estimating for Facilities

This *two-day* course teaches attendees how to plan, budget, and estimate the cost of ongoing and preventive maintenance and repair for existing buildings and grounds.

Some of what you'll learn:
- The most financially favorable maintenance, repair, and replacement scheduling and estimating
- Auditing and value engineering facilities
- Preventive planning and facilities upgrading
- Determining both in-house and contract-out service costs
- Annual, asset-protecting M&R plan

Who should attend: facility managers, maintenance supervisors, buildings and grounds superintendents, plant managers, planners, estimators, and others involved in facilities planning and budgeting.

Practical Project Management for Construction Professionals

In this *two-day* course, acquire the essential knowledge and develop the skills to effectively and efficiently execute the day-to-day responsibilities of the construction project manager.

Covers:
- General conditions of the construction contract
- Contract modifications: change orders and construction change directives
- Negotiations with subcontractors and vendors
- Effective writing: notification and communications
- Dispute resolution: claims and liens

Who should attend: architects, engineers, owner's representatives, project managers.

Mechanical & Electrical Estimating

This *two-day* course teaches attendees how to prepare more accurate and complete mechanical/electrical estimates, avoiding the pitfalls of omission and double-counting, while understanding the composition and rationale within the RSMeans mechanical/electrical database.

Some of what you'll learn:
- The unique way mechanical and electrical systems are interrelated
- M&E estimates—conceptual, planning, budgeting, and bidding stages
- Order of magnitude, square foot, assemblies, and unit price estimating
- Comparative cost analysis of equipment and design alternatives

Who should attend: architects, engineers, facilities managers, mechanical and electrical contractors, and others who need a highly reliable method for developing, understanding, and evaluating mechanical and electrical contracts.

Professional Development

Unit Price Estimating

This interactive *two-day* seminar teaches attendees how to interpret project information and process it into final, detailed estimates with the greatest accuracy level.

The most important credential an estimator can take to the job is the ability to visualize construction and estimate accurately.

Some of what you'll learn:

- Interpreting the design in terms of cost
- The most detailed, time-tested methodology for accurate pricing
- Key cost drivers—material, labor, equipment, staging, and subcontracts
- Understanding direct and indirect costs for accurate job cost accounting and change order management

Who should attend: corporate and government estimators and purchasers, architects, engineers, and others who need to produce accurate project estimates.

RSMeans CostWorks® CD Training

This one-day course helps users become more familiar with the functionality of *RSMeans CostWorks* program. Each menu, icon, screen, and function found in the program is explained in depth. Time is devoted to hands-on estimating exercises.

Some of what you'll learn:

- Searching the database using all navigation methods
- Exporting RSMeans data to your preferred spreadsheet format
- Viewing crews, assembly components, and much more
- Automatically regionalizing the database

This training session requires you to bring a laptop computer to class.

When you register for this course you will receive an outline for your laptop requirements.

Also offering web training for CostWorks CD!

Facilities Est. Using RSMeans CostWorks® CD

Combines hands-on skill building with best estimating practices and real-life problems. Brings you up-to-date with key concepts, and provides tips, pointers, and guidelines to save time and avoid cost oversights and errors.

Some of what you'll learn:

- Estimating process concepts
- Customizing and adapting RSMeans cost data
- Establishing scope of work to account for all known variables
- Budget estimating: when, why, and how
- Site visits: what to look for—what you can't afford to overlook
- How to estimate repair and remodeling variables

This training session requires you to bring a laptop computer to class.

Who should attend: facility managers, architects, engineers, contractors, facility tradespeople, planners, project managers and anyone involved with JOC, SABRE, or IDIQ.

Conceptual Estimating Using RSMeans CostWorks® CD

This *two day* class uses the leading industry data and a powerful software package to develop highly accurate conceptual estimates for your construction projects. All attendees must bring a laptop computer loaded with the current year *Square Foot Model Costs* and the *Assemblies Cost Data* CostWorks titles.

Some of what you'll learn:

- Introduction to conceptual estimating
- Types of conceptual estimates
- Helpful hints
- Order of magnitude estimating
- Square foot estimating
- Assemblies estimating

Who should attend: architects, engineers, contractors, construction estimators, owner's representatives, and anyone looking for an electronic method for performing square foot estimating.

Assessing Scope of Work for Facility Construction Estimating

This *two-day* practical training program addresses the vital importance of understanding the SCOPE of projects in order to produce accurate cost estimates for facility repair and remodeling.

Some of what you'll learn:

- Discussions of site visits, plans/specs, record drawings of facilities, and site-specific lists
- Review of CSI divisions, including means, methods, materials, and the challenges of scoping each topic
- Exercises in SCOPE identification and SCOPE writing for accurate estimating of projects
- Hands-on exercises that require SCOPE, take-off, and pricing

Who should attend: corporate and government estimators, planners, facility managers, and others who need to produce accurate project estimates.

Unit Price Est. Using RSMeans CostWorks® CD

Step-by-step instruction and practice problems to identify and track key cost drivers—material, labor, equipment, staging, and subcontractors—for each specific task. Learn the most detailed, time-tested methodology for accurately "pricing" these variables, their impact on each other and on total cost.

Some of what you'll learn:

- Unit price cost estimating
- Order of magnitude, square foot, and assemblies estimating
- Quantity takeoff
- Direct and indirect construction costs
- Development of contractor's bill rates
- How to use *RSMeans Building Construction Cost Data*

This training session requires you to bring a laptop computer to class.

Who should attend: architects, engineers, corporate and government estimators, facility managers, and government procurement staff.

Complete
Book of Framing, 2nd Edition

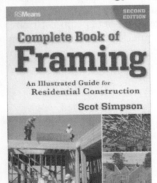

by Scot Simpson

This updated, easy-to-learn guide to rough carpentry and framing is written by an expert with more than thirty years of framing experience. Starting with the basics, this book begins with types of lumber, nails, and what tools are needed, followed by detailed, fully illustrated steps for framing each building element. Framer-Friendly Tips throughout the book show how to get a task done right—and more easily.

$29.95 per copy | 352 pages, softcover | Catalog no. 67353A

Estimating Building Costs, 2nd Edition

For the Residential & Light Commercial Construction Professional
by Wayne J. DelPico

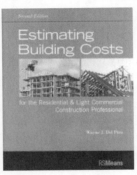

This book guides readers through the entire estimating process, explaining in detail how to put together a reliable estimate that can be used not only for budgeting, but also for developing a schedule, managing a project, dealing with contingencies, and ultimately making a profit.

Completely revised and updated to reflect the new CSI MasterFormat 2010 system, this practical guide describes estimating techniques for each building system and how to apply them according to the latest industry standards.

$65.00 per copy | 398 pages, softcover | Catalog no. 67343A

Risk Management
for Design and Construction

This book introduces risk as a central pillar of project management and shows how a project manager can be prepared for dealing with uncertainty. Written by experts in the field, *Risk Management for Design and Construction* uses clear, straightforward terminology to demystify the concepts of project uncertainty and risk.

Highlights include:

- Integrated cost and schedule risk analysis

- An introduction to a ready-to-use system of analyzing a project's risks and tools to proactively manage risks

- A methodology that was developed and used by the Washington State Department of Transportation

- Case studies and examples on the proper application of principles

- Information about combining value analysis with risk analysis

$125.00 per copy | Over 250 pages, softcover | Catalog no. 67359

How to Estimate with Means Data
& CostWorks, 4th Edition

by RSMeans and
Saleh A. Mubarak, Ph.D.

This step-by-step guide takes you through all the major construction items with extensive coverage of site work, concrete and masonry, wood and metal framing, doors and windows, and other divisions. The only construction cost estimating handbook that uses the most popular source of construction data, RSMeans, this indispensible guide features access to the instructional version of CostWorks in electronic form, enabling you to practice techniques to solve real-world estimating problems.

$70.00 per copy | 292 pages, softcover | Includes CostWorks CD
Catalog no. 67324C

How Your House Works, 2nd Edition

by Charlie Wing

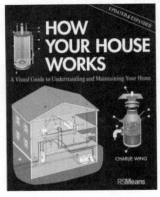

Knowledge of your home's systems helps you control repair and construction costs and makes sure the correct elements are being installed or replaced. This book uncovers the mysteries behind just about every major appliance and building element in your house. See-through, cross-section drawings in full color show you exactly how these things should be put together and how they function, including what to check if they don't work. It just might save you having to call in a professional.

$22.95 per copy | 192 pages, softcover | Catalog no. 67351A

RSMeans Cost Data,
Student Edition

This book provides a thorough introduction to cost estimating in a self-contained print and online package. With clear explanations and a hands-on, example-driven approach, it is the ideal reference for students and new professionals.

Features include:

- Commercial and residential construction cost data in print and online formats

- Complete how-to guidance on the essentials of cost estimating

- A supplemental website with plans, problem sets, and a full sample estimate

$99.00 per copy | 512 pages, softcover | Catalog no. 67363

Value Engineering: Practical Applications

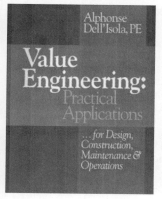

For Design, Construction, Maintenance & Operations
by Alphonse Dell'Isola, PE

A tool for immediate application—for engineers, architects, facility managers, owners, and contractors. Includes making the case for VE—the management briefing; integrating VE into planning, budgeting, and design; conducting life cycle costing; using VE methodology in design review and consultant selection; case studies; VE workbook; and a life cycle costing program on disk.

$79.95 per copy | Over 450 pages, illustrated, softcover | Catalog no. 67319A

The Building Professional's Guide to Contract Documents

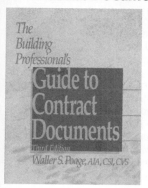

3rd Edition
by Waller S. Poage, AIA, CSI, CVS

A comprehensive reference for owners, design professionals, contractors, and students

- Structure your documents for maximum efficiency.
- Effectively communicate construction requirements.
- Understand the roles and responsibilities of construction professionals.
- Improve methods of project delivery.

$70.00 per copy | 400 pages Diagrams and construction forms, hardcover
Catalog no. 67261A

Cost Planning & Estimating for Facilities Maintenance

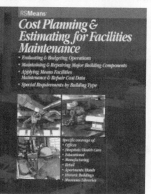

In this unique book, a team of facilities management authorities shares their expertise on:

- Evaluating and budgeting maintenance operations
- Maintaining and repairing key building components
- Applying *RSMeans Facilities Maintenance & Repair Cost Data* to your estimating

Covers special maintenance requirements of the ten major building types

$89.95 per copy | Over 475 pages, hardcover | Catalog no. 67314

Facilities Operations & Engineering Reference

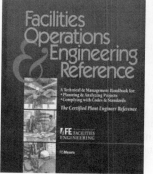

by the Association for Facilities Engineering and RSMeans

An all-in-one technical reference for planning and managing facility projects and solving day-to-day operations problems. Selected as the official certified plant engineer reference, this handbook covers financial analysis, maintenance, HVAC and energy efficiency, and more.

$109.95 per copy | Over 700 pages, illustrated, hardcover | Catalog no. 67318

Green Home Improvement

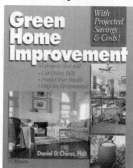

by Daniel D. Chiras, PhD

With energy costs rising and environmental awareness increasing, people are looking to make their homes greener. This book, with 65 projects and actual costs and projected savings help homeowners prioritize their green improvements.

Projects range from simple water savers that cost only a few dollars, to bigger-ticket items such as HVAC systems. With color photos and cost estimates, each project compares options and describes the work involved, the benefits, and the savings.

$34.95 | 320 pages, illustrated, softcover | Catalog no. 67355

Life Cycle Costing for Facilities

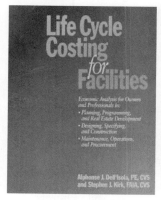

by Alphonse Dell'Isola and Dr. Steven Kirk

Guidance for achieving higher quality design and construction projects at lower costs! Cost-cutting efforts often sacrifice quality to yield the cheapest product. Life cycle costing enables building designers and owners to achieve both. The authors of this book show how LCC can work for a variety of projects — from roads to HVAC upgrades to different types of buildings.

$99.95 per copy | 396 pages, hardcover | Catalog no. 67341

Interior Home Improvement Costs

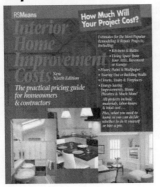

9th Edition

Updated estimates for the most popular remodeling and repair projects—from small, do-it-yourself jobs to major renovations and new construction. Includes: kitchens & baths; new living space from your attic, basement, or garage; new floors, paint, and wallpaper; tearing out or building new walls; closets, stairs, and fireplaces; new energy-saving improvements, home theaters, and more!

$24.95 per copy | 250 pages, illustrated, softcover | Catalog no. 67308E

Exterior Home Improvement Costs

9th Edition

Updated estimates for the most popular remodeling and repair projects—from small, do-it-yourself jobs, to major renovations and new construction. Includes: curb appeal projects—landscaping, patios, porches, driveways, and walkways; new windows and doors; decks, greenhouses, and sunrooms; room additions and garages; roofing, siding, and painting; "green" improvements to save energy & water.

$24.95 per copy | Over 275 pages, illustrated, softcover | Catalog no. 67309E

Builder's Essentials: Plan Reading & Material Takeoff

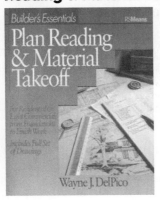

For Residential and Light Commercial Construction
by Wayne J. DelPico

A valuable tool for understanding plans and specs, and accurately calculating material quantities. Step-by-step instructions and takeoff procedures based on a full set of working drawings.

$35.95 per copy | Over 420 pages, softcover | Catalog no. 67307

Means Unit Price Estimating Methods

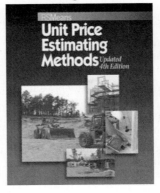

New 4th Edition

This new edition includes up-to-date cost data and estimating examples, updated to reflect changes to the CSI numbering system and new features of RSMeans cost data. It describes the most productive, universally accepted ways to estimate, and uses checklists and forms to illustrate shortcuts and timesavers. A model estimate demonstrates procedures. A new chapter explores computer estimating alternatives.

$65.00 per copy | Over 350 pages, illustrated, softcover | Catalog no. 67303B

Concrete Repair and Maintenance Illustrated

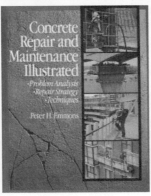

by Peter Emmons

Hundreds of illustrations show users how to analyze, repair, clean, and maintain concrete structures for optimal performance and cost effectiveness. From parking garages to roads and bridges to structural concrete, this comprehensive book describes the causes, effects, and remedies for concrete wear and failure. Invaluable for planning jobs, selecting materials, and training employees, this book is a must-have for concrete specialists, general contractors, facility managers, civil and structural engineers, and architects.

$69.95 per copy | 300 pages, illustrated, softcover | Catalog no. 67146

Total Productive Facilities Management

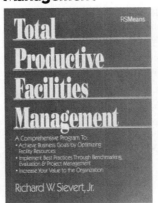

by Richard W. Sievert, Jr.

Today, facilities are viewed as strategic resources. . . elevating the facility manager to the role of asset manager supporting the organization's overall business goals. Now, Richard Sievert Jr., in this well-articulated guidebook, sets forth a new operational standard for the facility manager's emerging role. . . a comprehensive program for managing facilities as a true profit center.

$79.95 per copy | 275 pages, softcover | Catalog no. 67321

Reference Books

Means Illustrated Construction Dictionary

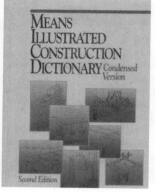

Condensed, 2nd Edition

The best portable dictionary for office or field use—an essential tool for contractors, architects, insurance and real estate personnel, facility managers, homeowners, and anyone who needs quick, clear definitions for construction terms. The second edition has been further enhanced with updates and hundreds of new terms and illustrations . . . in keeping with the most recent developments in the construction industry.

Now with a quick-reference Spanish section. Includes tools and equipment, materials, tasks, and more!

$59.95 per copy | Over 500 pages, softcover | Catalog no. 67282A

Facilities Planning & Relocation

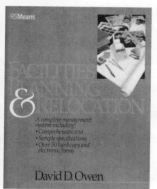

by David D. Owen

A complete system for planning space needs and managing relocations. Includes step-by-step manual, over 50 forms, and extensive reference section on materials and furnishings.

New lower price and user-friendly format.

$89.95 per copy | Over 450 pages, softcover | Catalog no. 67301

Electrical Estimating Methods

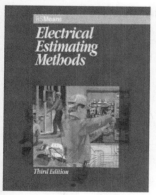

3rd Edition

Expanded edition includes sample estimates and cost information in keeping with the latest version of the CSI MasterFormat and UNIFORMAT II. Complete coverage of fiber optic and uninterruptible power supply electrical systems, broken down by components, and explained in detail. Includes a new chapter on computerized estimating methods. A practical companion to *RSMeans Electrical Cost Data.*

$64.95 per copy | Over 325 pages, hardcover | Catalog no. 67230B

Square Foot & UNIFORMAT Assemblies Estimating Methods

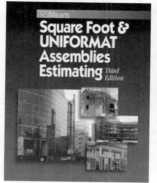

3rd Edition

Develop realistic square foot and assemblies costs for budgeting and construction funding. The new edition features updated guidance on square foot and assemblies estimating using UNIFORMAT II. An essential reference for anyone who performs conceptual estimates.

$69.95 per copy | Over 300 pages, illustrated, softcover | Catalog no. 67145B

Mechanical Estimating Methods

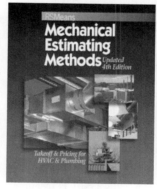

4th Edition

Completely updated, this guide assists you in making a review of plans, specs, and bid packages, with suggestions for takeoff procedures, listings, substitutions, and pre-bid scheduling for all components of HVAC. Includes suggestions for budgeting labor and equipment usage. Compares materials and construction methods to allow you to select the best option.

$64.95 per copy | Over 350 pages, illustrated, softcover | Catalog no. 67294B

Residential & Light Commercial Construction Standards

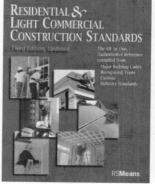

3rd Edition, Updated by RSMeans and contributing authors

This book provides authoritative requirements and recommendations compiled from leading professional associations, industry publications, and building code organizations. This all-in-one reference helps establish a standard for workmanship, quickly resolve disputes, and avoid defect claims. Updated third edition includes new coverage of green building, seismic, hurricane, and mold-resistant construction.

$59.95 | Over 550 pages, illustrated, softcover | Catalog no. 67322B

Construction Business Management

by Nick Ganaway

Only 43% of construction firms stay in business after four years. Make sure your company thrives with valuable guidance from a pro with 25 years of success as a commercial contractor. Find out what it takes to build all aspects of a business that is profitable, enjoyable, and enduring. With a bonus chapter on retail construction.

$49.95 per copy | 200 pages, softcover | Catalog no. 67352

Project Scheduling & Management for Construction

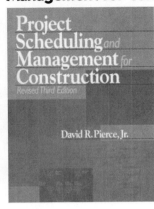

3rd Edition
by David R. Pierce, Jr.

A comprehensive yet easy-to-follow guide to construction project scheduling and control—from vital project management principles through the latest scheduling, tracking, and controlling techniques. The author is a leading authority on scheduling, with years of field and teaching experience at leading academic institutions. Spend a few hours with this book and come away with a solid understanding of this essential management topic.

$64.95 per copy | Over 300 pages, illustrated, hardcover | Catalog no. 67247B

The Practice of Cost Segregation Analysis

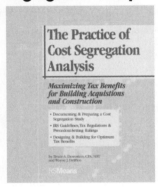

by Bruce A. Desrosiers and Wayne J. DelPico

This expert guide walks you through the practice of cost segregation analysis, which enables property owners to defer taxes and benefit from "accelerated cost recovery" through depreciation deductions on assets that are properly identified and classified.

With a glossary of terms, sample cost segregation estimates for various building types, key information resources, and updates via a dedicated website, this book is a critical resource for anyone involved in cost segregation analysis.

$99.95 per copy | Over 225 pages | Catalog no. 67345

Preventive Maintenance for Multi-Family Housing

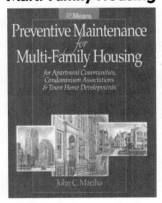

by John C. Maciha

Prepared by one of the nation's leading experts on multi-family housing.

This complete PM system for apartment and condominium communities features expert guidance, checklists for buildings and grounds maintenance tasks and their frequencies, a reusable wall chart to track maintenance, and a dedicated website featuring customizable electronic forms. A must-have for anyone involved with multi-family housing maintenance and upkeep.

$89.95 per copy | 225 pages | Catalog no. 67346

Green Building: Project Planning & Cost Estimating

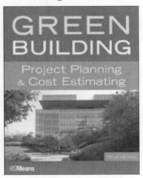

3rd Edition

Since the widely read first edition of this book, green building has gone from a growing trend to a major force in design and construction.

This new edition has been updated with the latest in green building technologies, design concepts, standards, and costs. Full-color with all new case studies—plus a new chapter on commercial real estate.

$99.95 per copy | Over 450 pages, softcover | Catalog no. 67338B

Job Order Contracting Expediting Construction Project Delivery

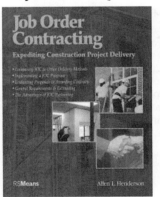

by Allen Henderson

Expert guidance to help you implement JOC—fast becoming the preferred project delivery method for repair and renovation, minor new construction, and maintenance projects in the public sector and in many states and municipalities. The author, a leading JOC expert and practitioner, shows how to:

- Establish a JOC program
- Evaluate proposals and award contracts
- Handle general requirements and estimating
- Partner for maximum benefits

$89.95 per copy | 192 pages, illustrated, hardcover | Catalog no. 67348

Construction Supervision

This brand new title inspires supervisory excellence with proven tactics and techniques applied by thousands of construction supervisors over the past decade. Recognizing the unique and critical role the supervisor plays in project success, the book's leadership guidelines carve out a practical blueprint for motivating work performance and increasing productivity through effective communication.

Features:

- A unique focus on field supervision and crew management

- Coverage of supervision from the foreman to the superintendent level

- An overview of technical skills whose mastery will build confidence and success for the supervisor

- A detailed view of "soft" management and communication skills

$90.00 per copy | Over 450 pages, softcover | Catalog no. 67358

Building & Renovating Schools

This all-inclusive guide covers every step of the school construction process—from initial planning, needs assessment, and design, right through moving into the new facility. A must-have resource for anyone concerned with new school construction or renovation. With square foot cost models for elementary, middle, and high school facilities, and real-life case studies of recently completed school projects.

The contributors to this book—architects, construction project managers, contractors, and estimators who specialize in school construction—provide start-to-finish, expert guidance on the process.

$99.95 per copy | Over 425 pages, hardcover | Catalog no. 67342

The Homeowner's Guide to Mold

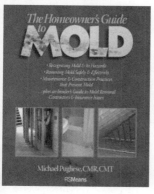

By Michael Pugliese

Expert guidance to protect your health and your home.

Mold, whether caused by leaks, humidity or flooding, is a real health and financial issue—for homeowners and contractors.

This full-color book explains:

- Construction and maintenance practices to prevent mold

- How to inspect for and remove mold

- Mold remediation procedures and costs

- What to do after a flood

- How to deal with insurance companies

$21.95 per copy | 144 pages, softcover | Catalog no. 67344

Universal Design Ideas for Style, Comfort & Safety

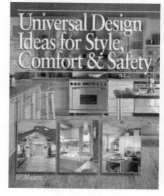

by RSMeans and Lexicon Consulting, Inc.

Incorporating universal design when building or remodeling helps people of any age and physical ability more fully and safely enjoy their living spaces. This book shows how universal design can be artfully blended into the most attractive homes. It discusses specialized products like adjustable countertops and chair lifts, as well as simple ways to enhance a home's safety and comfort. With color photos and expert guidance, every area of the home is covered. Includes budget estimates that give an idea how much projects will cost.

$21.95 | 160 pages, illustrated, softcover | Catalog no. 67354

Landscape Estimating Methods

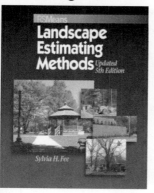

5th Edition

Answers questions about preparing competitive landscape construction estimates, with up-to-date cost estimates and the new MasterFormat classification system. Expanded and revised to address the latest materials and methods, including new coverage on approaches to green building. Includes:

- Step-by-step explanation of the estimating process

- Sample forms and worksheets that save time and prevent errors

$69.95 per copy | Over 350 pages, softcover | Catalog no. 67295C

Plumbing Estimating Methods

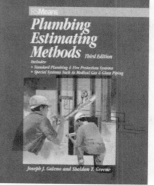

3rd Edition
by Joseph Galeno and Sheldon Greene

Updated and revised! This practical guide walks you through a plumbing estimate, from basic materials and installation methods through change order analysis. *Plumbing Estimating Methods* covers residential, commercial, industrial, and medical systems, and features sample takeoff and estimate forms and detailed illustrations of systems and components.

$59.95 per copy | 330+ pages, softcover | Catalog no. 67283B

Reference Books

For more information visit the RSMeans website at **www.rsmeans.com**

Understanding & Negotiating Construction Contracts

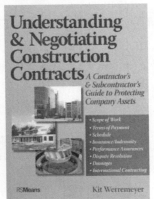

by Kit Werremeyer

Take advantage of the author's 30 years' experience in small-to-large (including international) construction projects. Learn how to identify, understand, and evaluate high risk terms and conditions typically found in all construction contracts—then negotiate to lower or eliminate the risk, improve terms of payment, and reduce exposure to claims and disputes. The author avoids "legalese" and gives real-life examples from actual projects.

$74.95 per copy | 300 pages, softcover | Catalog no. 67350

Means Illustrated Construction Dictionary

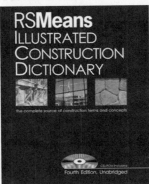

Unabridged 4th Edition, with CD-ROM

Long regarded as the industry's finest, *Means Illustrated Construction Dictionary* is now even better. With nearly 20,000 terms and more than 1,400 illustrations and photos, it is the clear choice for the most comprehensive and current information. The companion CD-ROM that comes with this new edition adds extra features such as: larger graphics and expanded definitions.

$99.95 per copy | Over 790 pages, illust., hardcover | Catalog no. 67292B

RSMeans Estimating Handbook

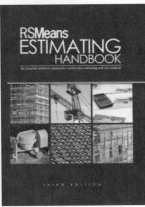

3rd Edition

Widely used in the industry for tasks ranging from routine estimates to special cost analysis projects, this handbook has been completely updated and reorganized with new and expanded technical information.

RSMeans Estimating Handbook will help construction professionals:

- Evaluate architectural plans and specifications
- Prepare accurate quantity takeoffs
- Compare design alternatives and costs
- Perform value engineering
- Double-check estimates and quotes
- Estimate change orders

$99.95 per copy | Over 900 pages, hardcover | Catalog No. 67276B

Spanish/English Construction Dictionary

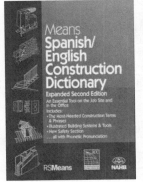

2nd Edition
by RSMeans and the International Code Council

This expanded edition features thousands of the most common words and useful phrases in the construction industry with easy-to-follow pronunciations in both Spanish and English. Over 800 new terms, phrases, and illustrations have been added. It also features a new stand-alone "Safety & Emergencies" section, with colored pages for quick access.

Unique to this dictionary are the systems illustrations showing the relationship of components in the most common building systems for all major trades.

$23.95 per copy | Over 400 pages | Catalog no. 67327A

The Gypsum Construction Handbook

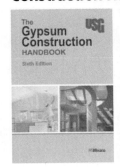

by USG

An invaluable reference of construction procedures for gypsum drywall, cement board, veneer plaster, and conventional plaster. This new edition includes the newest product developments, installation methods, fire- and sound-rated construction information, illustrated framing-to-finish application instructions, estimating and planning information, and more. Great for architects and engineers; contractors, builders, and dealers; apprentices and training programs; building inspectors and code officials; and anyone interested in gypsum construction. Features information on tools and safety practices, a glossary of construction terms, and a list of important agencies and associations. Also available in Spanish.

$29.95 | Over 575 pages, illustrated, softcover | Catalog no. 67357A

	COST ESTIMATING BOOKS	Unit Price	Total
563	Assemblies Cost Data 2013	$305.95	
013	Building Construction Cost Data 2013	186.95	
0113	Concrete & Masonry Cost Data 2013	171.95	
50143	Construction Cost Indexes 2013 (subscription)	362.00	
60143A	Construction Cost Index–January 2013	90.50	
60143B	Construction Cost Index–April 2013	90.50	
60143C	Construction Cost Index–July 2013	90.50	
60143D	Construction Cost Index–October 2013	90.50	
60343	Contr. Pricing Guide: Resid. R & R Costs 2013	39.95	
60233	Electrical Change Order Cost Data 2013	183.95	
60033	Electrical Cost Data 2013	188.95	
60203	Facilities Construction Cost Data 2013	476.95	
60303	Facilities Maintenance & Repair Cost Data 2013	417.95	
60553	Green Building Cost Data 2013	155.95	
60163	Heavy Construction Cost Data 2013	186.95	
60093	Interior Cost Data 2013	192.95	
60123	Labor Rates for the Const. Industry 2013	415.95	
60183	Light Commercial Cost Data 2013	139.95	
60023	Mechanical Cost Data 2013	183.95	
60153	Open Shop Building Const. Cost Data 2013	160.95	
60213	Plumbing Cost Data 2013	188.95	
60043	Commercial Renovation Cost Data 2013	152.95	
60173	Residential Cost Data 2013	134.95	
60283	Site Work & Landscape Cost Data 2013	181.95	
60053	Square Foot Costs 2013	197.95	
62012	Yardsticks for Costing (2012)	176.95	
62013	Yardsticks for Costing (2013)	185.95	
	REFERENCE BOOKS		
67329	Bldrs Essentials: Best Bus. Practices for Bldrs	29.95	
67298A	Bldrs Essentials: Framing/Carpentry 2nd Ed.	24.95	
67298AS	Bldrs Essentials: Framing/Carpentry Spanish	24.95	
67307	Bldrs Essentials: Plan Reading & Takeoff	35.95	
67342	Building & Renovating Schools	99.95	
67261A	The Building Prof. Guide to Contract Documents	70.00	
67339	Building Security: Strategies & Costs	89.95	
67353A	Complete Book of Framing, 2nd Ed.	29.95	
67146	Concrete Repair & Maintenance Illustrated	69.95	
67352	Construction Business Management	49.95	
67358	Construction Supervision	90.00	
67363	Cost Data Student Edition	99.00	
67314	Cost Planning & Est. for Facil. Maint.	89.95	
67328A	Designing & Building with the IBC, 2nd Ed.	99.95	
67230B	Electrical Estimating Methods, 3rd Ed.	64.95	
67343A	Est. Bldg. Costs for Resi. & Lt. Comm., 2nd Ed.	65.00	
67276B	Estimating Handbook, 3rd Ed.	99.95	
67318	Facilities Operations & Engineering Reference	109.95	
67301	Facilities Planning & Relocation	89.95	
67338B	Green Building: Proj. Planning & Cost Est., 3rd Ed.	99.95	
67355	Green Home Improvement	34.95	
67357A	The Gypsum Construction Handbook	29.95	
67357S	The Gypsum Construction Handbook (Spanish)	29.95	

Qty.	Book no.	REFERENCE BOOKS (Cont.)	Unit Price	Total
	67148	Heavy Construction Handbook	99.95	
	67308E	Home Improvement Costs–Int. Projects, 9th Ed.	24.95	
	67309E	Home Improvement Costs–Ext. Projects, 9th Ed.	24.95	
	67344	Homeowner's Guide to Mold	21.95	
	67324C	How to Est.w/Means Data & CostWorks, 4th Ed.	70.00	
	67351A	How Your House Works, 2nd Ed.	22.95	
	67282A	Illustrated Const. Dictionary, Condensed, 2nd Ed.	59.95	
	67292B	Illustrated Const. Dictionary, w/CD-ROM, 4th Ed.	99.95	
	67348	Job Order Contracting	89.95	
	67295C	Landscape Estimating Methods, 5th Ed.	69.95	
	67341	Life Cycle Costing for Facilities	99.95	
	67294B	Mechanical Estimating Methods, 4th Ed.	64.95	
	67283B	Plumbing Estimating Methods, 3rd Ed.	59.95	
	67345	Practice of Cost Segregation Analysis	99.95	
	67346	Preventive Maint. for Multi-Family Housing	89.95	
	67326	Preventive Maint. Guidelines for School Facil.	149.95	
	67247B	Project Scheduling & Manag. for Constr., 3rd Ed.	64.95	
	67322B	Resi. & Light Commercial Const. Stds., 3rd Ed.	59.95	
	67359	Risk Management for Design & Construction	125.00	
	67327A	Spanish/English Construction Dictionary, 2nd Ed.	23.95	
	67145B	Sq. Ft. & Assem. Estimating Methods, 3rd Ed.	69.95	
	67321	Total Productive Facilities Management	79.95	
	67350	Understanding and Negotiating Const. Contracts	74.95	
	67303B	Unit Price Estimating Methods, 4th Ed.	65.00	
	67354	Universal Design	21.95	
	67319A	Value Engineering: Practical Applications	79.95	

MA residents add 6.25% state sales tax	
Shipping & Handling**	
Total (U.S. Funds)*	

Prices are subject to change and are for U.S. delivery only. *Canadian customers may call for current prices. **Shipping & handling charges: Add 7% of total order for check and credit card payments. Add 9% of total order for invoiced orders.

Send order to: ADDV-1000

Name (please print) _____

Company _____

☐ Company

☐ Home Address _____

_____

City/State/Zip _____

Phone # _____ P.O. # _____

(Must accompany all orders being billed)

RSMeans Project Cost Report

By filling out this report, your project data will contribute to the database that supports the RSMeans® Project Cost Square Foot D
When you fill out this form, RSMeans will provide a $30 discount off one of the RSMeans products advertised in the preceding pa
Please complete the form including all items where you have cost data, and all the items marked (☑).

$30.00 Discount per product for each report you submit

Project Description (NEW construction only, please)

☑ Building Use (Office School...)_____

☑ Address (City, State)_____

_____

☑ Total Building Area (SF) _____

☑ Ground Floor (SF) _____

☑ Frame (Wood, Steel...) _____

☑ Exterior Wall (Brick, Tilt-up...) _____

☑ Basement: (checkone) ☐ Full ☐ Partial ☐ None

☑ Number of Stories _____

☑ Floor-to-Floor Height_____

☑ Volume (C.F.) _____

% Air Conditioned _____ Tons_____

Total Project Cost _____ $ _____

Owner _____

Architect_____

General Contractor _____

☑ Bid Date _____

☑ Typical Bay Size _____

☑ Occupant Capacity_____

☑ Labor Force: _____ % Union _____ % Non-Union

☑ Project Description (Circle one number in each line.)

1. Economy 2. Average 3. Custom 4. Luxury

1. Square 2. Rectangular 3. Irregular 4. Very Irregular

Comments _____

_____

A	☑	General Conditions	$
B	☑	Site Work	$
C	☑	Concrete	$
D	☑	Masonry	$
E	☑	Metals	$
F	☑	Wood & Plastics	$
G	☑	Thermal &Moisture Protection	$
GR		Roofing & Flashing	$
H	☑	Doors and Windows	$
J	☑	Finishes	$
JP		Painting & Wall Covering	$

K	☑	Specialties	$
L	☑	Equipment	$
M	☑	Furnishings	$
N	☑	SpecialConstruction	$
P	☑	ConveyingSystems	$
Q	☑	Mechanical	$
QP		Plumbing	$
QB		HVAC	$
R	☑	Electrical	$
S	☑	Mech./Elec. Combined	$

Please specify the RSMeans product you wish to receive.
Complete the address information.

Product Name _____

Product Number _____

Your Name _____

Title _____

Company_____
  ☐ Company
  ☐ Home Street Address_____

City, State, Zip _____

Email Address _____

Return by mail or fax 888-492-6770.

Method of Payment:

Credit Card # _____

Expiration Date _____

Check_____

Purchase Order _____

Phone Number _____
(if you would prefer a sales call)

RSMeans
a division of Reed Construction Data
Square Foot Costs Department
700 Longwater Drive
Norwell, MA 02061